Fourth Edition

Fundamentals of
ENVIRONMENTAL AND TOXICOLOGICAL CHEMISTRY

Sustainable Science

Fourth Edition

Fundamentals of
ENVIRONMENTAL AND TOXICOLOGICAL CHEMISTRY

Sustainable Science

Stanley E. Manahan

CRC Press
Taylor & Francis Group
Boca Raton London New York

CRC Press is an imprint of the
Taylor & Francis Group, an **informa** business

CRC Press
Taylor & Francis Group
6000 Broken Sound Parkway NW, Suite 300
Boca Raton, FL 33487-2742

Library of Congress Cataloging-in-Publication Data

Manahan, Stanley E.
    Fundamentals of environmental and toxicological chemistry : sustainable science / Stanley E. Manahan.
-- Fourth edition.
    pages cm
    Includes bibliographical references and index.
    ISBN 978-1-4665-5316-3 (hardback)
    1. Environmental chemistry. I. Title.

TD193.M36 2013
540--dc23   2012034014

Visit the Taylor & Francis Web site at
http://www.taylorandfrancis.com

and the CRC Press Web site at
http://www.crcpress.com

Contents

Preface

This book covers environmental chemistry, including toxicological chemistry, at the university level. Readers with a basic knowledge of general chemistry and organic chemistry can readily understand the material in the text. Furthermore, for readers who may not have this background, basic chapters are included at the end of the text that will enable them to acquire the fundamentals of general and organic chemistry required to master the material in the text. The main features of this book are as follows:

- Integration of toxicological chemistry along with environmental chemistry
- Organization based on the five spheres of Earth's environment
- Discussion of each sphere of the environment based on the nature, pollution, and sustainability of the sphere
- Emphasis on sustainability
- Relation of environmental/toxicological chemistry to the practice of industrial hygiene
- Importance of abundant, nonpolluting sustainable energy sources
- Basic chapters on general chemistry and organic chemistry for readers needing a background in these topics
- Availability of PowerPoint presentations for each chapter of the book
- Availability of an online course covering the book

This book views the environment as consisting of five strongly interacting spheres: (1) the hydrosphere (water), (2) the atmosphere (air), (3) the geosphere (solid Earth), (4) the biosphere (life), and (5) the anthrosphere (the part of the environment made and used by humans). A prime concern with the environment is the toxic effects of pollutants, so aspects of toxicological chemistry are included along with environmental chemistry. The environmental/toxicological chemistry of each of the spheres of the environment is discussed in clusters. The first chapter in each cluster defines and explains a particular environmental sphere within the context of its basic environmental and toxicological chemistry. Pollution and threats of human activities to each sphere are covered, followed by a discussion of ways in which human activities may be directed toward sustaining the sphere, preventing its deterioration, and enhancing its quality for the future.

Chapter 1 begins with the definition of environmental chemistry and then defines and outlines each of the five major environmental spheres and the interactions between them. Such interactions occur largely through biogeochemical cycles, which are defined in this chapter with the carbon cycle as a specific example.

An important feature of this book is the integration of toxicological chemistry throughout. To enable the integration of toxicological chemistry with the material in this book, Chapter 2 explains the basics of toxicological chemistry and how it relates to environmental chemistry.

The next three chapters involve the hydrosphere. Chapter 3 explains the nature of the hydrosphere and the major aspects of its environmental chemistry. Chapter 4 deals specifically with water pollution and includes some aspects of the toxicological chemistry of the hydrosphere. Chapter 5 addresses the sustainability of the hydrosphere and water as nature's most renewable resource.

Chapters 6 through 8 address the atmosphere. Chapter 6 explains the atmosphere as one of the five spheres of the environment and discusses the composition of air, the structure of the atmosphere, and the importance of the atmosphere for protecting life on Earth. Chapter 7 addresses air pollutants and their environmental and toxicological chemistry. Chapter 8 outlines how atmospheric quality can be sustained and enhanced.

Chapters 9 through 11 address the geosphere. Chapter 9 explains the nature of the geosphere, including its physical configuration and chemical composition. Soil in which food is grown is crucial to life on Earth; it is discussed in Chapter 10. Sustainability of the geosphere is described in Chapter 11. The biosphere is discussed as a distinct environmental sphere in Chapter 12. Sustaining the biosphere is discussed in Chapter 13.

Chapter 14 explains the anthrosphere, which is the part of the environment made and operated by humans. This chapter explains how the anthrosphere has become such an important influence on Earth's environment that a new epoch, the Anthropocene, is developing in which human influences are determining the status of Earth's environment, including climate. The anthrosphere as a source and receptor of pollutants is covered in Chapter 15. Chapter 16 covers the means of sustaining the anthrosphere, including the practice of industrial ecology and green chemistry. Chapter 17 discusses renewable, abundant, and nonpolluting energy, a crucial aspect of sustaining the anthrosphere. Environmental chemical analysis is discussed in Chapter 18. This chapter also briefly introduces workplace monitoring in the practice of industrial hygiene.

The last two chapters of this book are made available for readers who may need some more background in basic chemistry. General chemistry is covered in Chapter 19. Basic principles of organic chemistry are presented in Chapter 20.

PowerPoint presentations for each chapter are available to the reader. The author may be contacted at manahans@missouri.edu.

Author

Stanley E. Manahan is a professor emeritus of chemistry at the University of Missouri-Columbia, where he has been on the faculty since 1965. He earned his AB in chemistry from Emporia State University in Kansas in 1960 and his PhD in analytical chemistry from the University of Kansas in 1965. Since 1968, his primary research and professional activities have been in environmental chemistry, with recent emphasis on hazardous waste treatment. His latest research involves the gasification of wastes and sewage sludge and crop by-product biomass for energy production.

Dr. Manahan has taught courses on environmental chemistry, hazardous wastes, toxicological chemistry, and analytical chemistry and has lectured on these topics throughout the United States as an American Chemical Society Local Sections tour speaker and in a number of countries, including France, Italy, Austria, Japan, Mexico, and Venezuela. He has written books on environmental chemistry (*Environmental Chemistry*, 9th ed., 2010, Taylor & Francis/CRC Press, and *Fundamentals of Environmental Chemistry*, 3rd ed., 2009, Taylor & Francis/CRC Press); green chemistry (*Green Chemistry and the Ten Commandments of Sustainability*, 3rd ed., 2010, ChemChar Research); water chemistry (*Water Chemistry: Green Science and Technology of Nature's Most Renewable Resource*, 2011, Taylor & Francis/CRC Press); energy (*Energy: Environmental Toxicological Chemistry for a Sustainable Energy Future*, 2012, Amazon Kindle); general chemistry (*Fundamentals of Sustainable Chemical Science*, 2009, Taylor & Francis/CRC Press); environmental geology (*Environmental Geology and Geochemistry*, 2011, Amazon Kindle and Barnes & Noble Nook Books); the Anthropocene (*Environmental Chemistry of the Anthropocene: A World Made by Humans*, 2011, Amazon Kindle and Barnes & Noble Nook Books); climate change (*Environmental Chemistry of Global Climate Change*, 2011, Amazon Kindle and Barnes & Noble Nook Books), environmental science (*Environmental Science: Sustainability in the Anthropocene*, 2011, Amazon Kindle and Barnes & Noble Nook Books); hazardous wastes and industrial ecology (*Industrial Ecology: Environmental Chemistry and Hazardous Waste*, 1999, Lewis Publishers/CRC Press); toxicological chemistry (*Toxicological Chemistry and Biochemistry*, 3rd ed., 2002, Lewis Publishers/CRC Press); applied chemistry; and quantitative chemical analysis. Dr. Manahan is the author or coauthor of approximately 90 research articles.

1 Environmental Chemistry and the Five Spheres of the Environment

1.1 WHAT IS ENVIRONMENTAL CHEMISTRY?

Environmental chemistry is the discipline that describes the origin, transport, reactions, effects, and fate of chemical species in the hydrosphere, atmosphere, geosphere, biosphere, and anthrosphere.[1] This definition is illustrated for a typical pollutant species in Figure 1.1, which shows the following: (1) Coal, which contains sulfur in the form of organically bound sulfur and pyrite, FeS_2, is mined from the geosphere. (2) The coal is burned in a power plant that is part of the anthrosphere and the sulfur is converted to sulfur dioxide, SO_2, by atmospheric chemical processes. (3) The sulfur dioxide and its reaction products are moved by wind and air currents in the atmosphere. (4) Atmospheric chemical processes convert SO_2 to sulfuric acid, H_2SO_4. (5) The sulfuric acid falls from the atmosphere as acidic acid rain. (6) The sulfur dioxide in the atmosphere may adversely affect biospheric organisms, including asthmatic humans who inhale it, and the sulfuric acid in the acid rain may be toxic to plants and fish in the hydrosphere and may have a corrosive effect on structures and electrical equipment in the anthrosphere. (7) The sulfuric acid ends up in a sink, either soil in the geosphere or a body of water in the hydrosphere. In these locations, the H_2SO_4 may continue having toxic effects, including leaching phytotoxic (toxic to plants) aluminum ion from soil and rock in the geosphere and poisoning fish fingerlings in the hydrosphere.

1.2 ENVIRONMENTAL RELATIONSHIPS IN ENVIRONMENTAL CHEMISTRY

To understand environmental chemistry, it is important to understand the five environmental spheres within and among which environmental chemical processes occur (Figure 1.2). These are outlined here and each is discussed in more detail throughout the remainder of this book as well as in a book on green chemistry.[2]

As discussed in more detail in Chapters 3 through 5, the **hydrosphere** contains Earth's water (chemical formula, H_2O). By far, the largest portion of the hydrosphere is in the oceans. Water circulates within Earth's environment through the solar-powered **hydrologic cycle** beginning with water vaporized into the atmosphere by energy from the sun. The water vapor and cloud droplets of water are carried through the atmosphere from which they fall back to Earth as rain or some form of frozen water. This precipitation produces rivers, is held temporarily in lakes or reservoirs, infiltrates Earth's solid surface to accumulate in underground **aquifers**, and is stored for centuries in ice pack in glaciers, such as those in the Antarctic and Greenland ice caps, and in mountain glaciers, such as those in the Himalayan Mountains in Asia. A small portion of Earth's water is contained in organisms and another small fraction is held in the atmosphere. Most of the hydrosphere either rests on or is located beneath the surface of the geosphere, and the characteristics of water, especially water in underground aquifers, are very much influenced by contact with minerals in the geosphere.

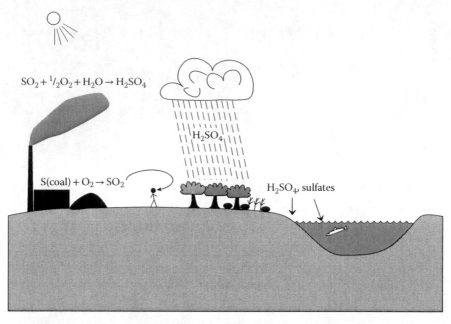

FIGURE 1.1 Illustration of the definition of environmental chemistry exemplified by the life cycle of a typical pollutant, sulfur dioxide. Sulfur present in fuel, almost always coal, is oxidized to gaseous sulfur dioxide, which is emitted to the atmosphere with stack gas. Sulfur dioxide is an air pollutant that may affect human respiration and may be phytotoxic (toxic to plants). Of greater importance is the oxidation of sulfur dioxide in the atmosphere to sulfuric acid, the main ingredient of acid rain. Acidic precipitation may adversely affect plants, materials, and water, where excessive acidity may kill fish. Eventually, the sulfuric acid or sulfate salts end up in water or in soil.

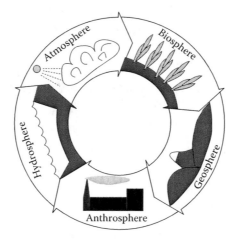

FIGURE 1.2 The environment may be considered as consisting of five spheres representing water, air, earth, life, and technology (the anthrosphere is made and operated by humans). Materials and energy are exchanged among these spheres, largely through biogeochemical cycles.

Chapters 6 through 8 discuss the environmental chemistry of Earth's atmosphere. There is really no definite altitude as to where the atmosphere ends, but most of it is within just a few kilometers of Earth's surface. In fact, if Earth were the size of a geography classroom globe, virtually all of the mass of the air in the atmosphere would be within a layer the thickness of the varnish on the surface of the globe. Exclusive of water vapor, the atmosphere is composed of mostly elemental nitrogen, N_2, with about one-fourth amount of elemental oxygen, O_2. Slightly less than 1% of the atmosphere by volume is elemental argon and only about 0.04% of the atmosphere by volume is carbon dioxide, CO_2, a very significant constituent because of its ability to retain Earth's heat. The content of water vapor in the atmosphere varies, usually within a range of 1–3% by volume. The atmosphere has a very important relationship with the hydrosphere as a conduit for water moving through the hydrologic cycle. The atmosphere is crucial to the biosphere as a source of elemental oxygen for organisms requiring this element for their metabolism, as a reservoir of carbon dioxide, as a carbon source for plants performing photosynthesis, and as a source of nitrogen for organisms that fix this element as a key constituent of proteins. The atmosphere provides the anthrosphere with oxygen for combustion, argon as a non-reactive noble gas, elemental nitrogen for extremely cold liquid nitrogen and as a raw material for chemical synthesis of ammonia (NH_3). The atmosphere also serves the anthrosphere as a sink for waste products, especially carbon dioxide from fossil fuel combustion. The geosphere acts as a sink for atmospheric contaminants, especially particles, and emits gases to the atmosphere, especially sulfur dioxide (SO_2) and hydrogen sulfide (H_2S) from volcanoes.

Chapters 9 through 11 discuss the environmental chemistry of solid earth, including the rocks and minerals of which the geosphere is largely composed. Actually, Earth is not so solid because a few kilometers in depth below its surface, it becomes plastic and at greater depths liquid rock. Although humans have been able to penetrate only a few kilometers below Earth's surface, evidence of the high temperatures and molten nature below a few kilometers is provided by the emissions of molten rock (lava) from volcanoes and the shifting of continental plates floating on the plastic rock layer manifested by earthquakes. The intimate connection of the geosphere with the biosphere is especially evident with respect to soil on Earth's surface upon which grow the plants that provide most of the food consumed by organisms (see Chapter 10). The geosphere surface provides watersheds that collect water for the hydrosphere. The geosphere is a source of metals, other critical minerals, and fossil fuels required by the anthrosphere.

The biosphere is discussed in Chapters 12 and 13, and an important specific aspect of the biosphere, toxicological chemistry, is the topic of Chapter 2. The biosphere is strongly influenced by the other four environmental spheres and, in turn, strongly affects these spheres. For example, productivity of biomass by plants in the biosphere is strongly influenced by the nature and fertility of geospheric soil. Organisms are very much involved with weathering of geospheric rock, the process by which soil is produced. The oxygen that makes up about 20% of the atmosphere, which was originally released by the photosynthesis of microscopic photosynthetic bacteria. Organisms in the biosphere can be exposed to potentially toxic substances through the water they drink, the hydrosphere in which fish live, air from the atmosphere that animals must breathe, exposure of plant leaf surfaces to phytotoxic substances (those toxic to plants) carried by the atmosphere, uptake of toxic substances by plants growing in soil on the geosphere, and emissions released from the anthrosphere.

Chapters 14 through 16 deal with the anthrosphere, the part of the environment constructed and operated by humans. Meeting the material and energy needs of the anthrosphere and handling its waste products safely and sustainably is a major challenge. A majority of substances of concern for their toxicities are made in, processed by, or released from the anthrosphere. A particularly important aspect of the anthrosphere is the sustainable production of energy discussed in Chapter 17.

1.3 ENVIRONMENTAL SPHERES AND BIOGEOCHEMICAL CYCLES

One of the most important ways of relating the environmental chemistry of the five environmental spheres is through **biogeochemical cycles**. These are commonly expressed in terms of key elements, including essential nutrient elements. Often, as is the case with the nitrogen cycle, they contain an atmospheric component, though in some cases, such as the phosphorous cycle, the atmospheric component is not significant. Before the appearance of humans on Earth, the anthrospheric compartment did not exist, but now, as is the case of carbon dioxide emitted to the atmosphere by fossil fuel combustion, the anthrosphere is a very significant component.

Figure 1.3 illustrates an important example of a biogeochemical cycle, the carbon cycle. As shown in the figure, a small, but very significant fraction of Earth's carbon is held in the atmosphere as CO_2 gas. This gas is transferred to the biosphere through the leaf surfaces of plants that photosynthetically convert it to biomass using solar energy. It also enters the geosphere by dissolving in surface water; Earth's oceans constitute a large sink for atmospheric carbon dioxide. Carbon dioxide enters the atmosphere from the biosphere as organisms produce it as a product of their respiratory biochemical oxidation of organic matter and from the anthrosphere by the combustion of fossil fuels. Volcanoes and geothermal vents (such as those in Yellowstone National Park) emit carbon dioxide from the geosphere to the atmosphere. Sudden emissions of large quantities of geospheric carbon dioxide underlying volcanic lakes have killed many people in Africa. Carbon dioxide dissolved in water as HCO_3^- ion is converted to CO_3^{2-} ion, which in the presence of dissolved Ca^{2+} ion precipitates $CaCO_3$ (limestone) that ends up as solid rock in the geosphere. Carbon goes back into the hydrosphere as acidic CO_2 from the atmosphere or from the biodegradation of organic matter, which reacts with solid $CaCO_3$ to produce dissolved HCO_3^-.

Other important cycles of matter are linked to the carbon cycle. The oxygen cycle describes movement of oxygen in various chemical forms through the five environmental spheres. At 21% elemental oxygen by volume, the atmosphere is a vast reservoir of this element. This oxygen becomes chemically bound as carbon dioxide by respiration processes of organisms and by combustion.

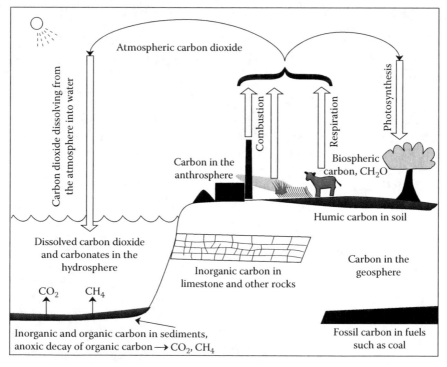

FIGURE 1.3 Carbon in various chemical forms circulates throughout the environment by way of the carbon cycle, an important and typical biogeochemical cycle. Carbon is present in the atmosphere as carbon dioxide, which is incorporated into biomass in the biosphere by plant photosynthesis. Carbon occurs in the geosphere as organic matter, in fossil fuels, and in inorganic rocks, particularly calcium and magnesium carbonates. Carbon is present in water as dissolved carbon dioxide and bicarbonate ion. Carbon-containing fuels are burned in the anthrosphere, a process that releases carbon dioxide to the atmosphere. Living organisms including humans "burn" carbon-containing foods and release carbon dioxide to the atmosphere as well.

Photosynthesis adds oxygen to the atmosphere. Oxygen is a component of biomass in the biosphere and most rocks in Earth's crust are composed of oxygen-containing compounds. With its chemical formula of H_2O, water in the hydrosphere is predominantly oxygen. In addition to the carbon and oxygen cycles described above, three other important life-element cycles are those of nitrogen, sulfur, and phosphorus:

1. **Nitrogen cycle**: Biochemically bound nitrogen is essential for life molecules, including proteins and nucleic acids. Although the atmosphere is about 80% elemental N_2 by volume, this molecule is so stable that it is difficult to split it apart so that N can combine with other elements. This process is performed in the anthrosphere by the synthesis of NH_3, from N_2 and H_2 over a catalyst at high temperatures and very high pressures. Furthermore, air pollutant NO and NO_2 produced by the reaction of N_2 and O_2 under the extreme conditions in internal combustion engines. In contrast, some bacteria, including *Rhizobium* bacteria growing on the roots of legume plants, convert atmospheric nitrogen to nitrogen compounds under very mild conditions (compared to those by which nitrogen is fixed by anthrospheric processes) just below the soil surface. Plants convert nitrogen in NH_4^+ and NO_3^- to biochemically bound N. As part of the nitrogen cycle, biochemically bound nitrogen is released as NH_4^+ by the biodegradation of organic compounds. The nitrogen cycle is completed by microorganisms that use NO_3^- as a substitute for O_2 in energy-yielding metabolic processes and release molecular N_2 gas to the atmosphere. Other than nitrogen fixation in the anthrosphere and formation of nitrogen oxides in the atmosphere from lightning discharges, most transitions in the nitrogen cycle are carried out by organisms, especially microorganisms.

2. **Sulfur cycle**: The sulfur cycle includes both chemical and biochemical processes and involves all spheres of the environment. Chemically combined sulfur enters the atmosphere as pollutant H_2S and SO_2 gases, which are also emitted by natural sources including volcanoes. Large quantities of H_2S are produced by anoxic microorganisms degrading organic sulfur compounds and using sulfate, SO_4^{2-}, as an oxidizing agent and discharged to the atmosphere. Accumulation of this highly toxic gas in confined areas can result in fatal exposures to humans. Globally, a major flux of sulfur to the atmosphere is in the form of volatile dimethyl sulfide, $(CH_3)_2S$, produced by marine microorganisms. The major atmospheric pollutant sulfur compound is SO_2 released in the combustion of sulfur-containing fuels, especially coal. In the atmosphere, gaseous sulfur compounds are oxidized to sulfate, largely in the forms of H_2SO_4 (pollutant acid rain) and corrosive ammonium salts (NH_4HSO_4), which settle from the atmosphere or are washed out with precipitation. The geosphere is a vast reservoir of sulfur minerals, including sulfate salts ($CaSO_4$), sulfide salts (FeS), and even elemental sulfur. Sulfur is a relatively minor, though essential constituent of biomolecules, occurring in two essential amino acids, but various sulfur compounds are processed by oxidation-reduction biochemical reactions of microorganisms.

3. **Phosphorus cycle**: Unlike all the exogenous cycles with an atmospheric component discussed above, the phosphorus cycle is endogenous with no significant participation in the atmosphere. It is an essential life element and ingredient of deoxyribonucleic acid (DNA) as well as adenosine triphosphate (ATP) and adenosine diphosphate (ADP) through which energy is transferred in organisms. Dissolved phosphate in the hydrosphere is required as a nutrient for aquatic organisms, although excessive phosphate may result in too much algal growth causing an unhealthy condition called eutrophication. Phosphorus is abundant in the geosphere, especially as the mineral hydroxyapatite, $Ca_5OH(PO_4)_3$. Significant deposits of phosphorus-rich material have been formed from the feces of birds and bats (guano).

1.4 EARTH'S NATURAL CAPITAL

The very small group of humans who have been privileged to view Earth from outer space have been struck with a sense of awe at the sight. Photographs of Earth taken at altitudes high enough to capture its entirety reveal a marvelous sphere, largely blue in color, white where covered by clouds, and with desert regions showing up in shades of brown, yellow, and red. But Earth is far more than a beautiful globe that inspires artists and poets. In a very practical sense, it is a source of the life support systems that sustain humans and all other known forms of life. Earth obviously provides the substances required for life, including water, atmospheric oxygen, and carbon dioxide, from which billions of tons of biomass are made each year by photosynthesis, and ranging all the way down to the trace levels of micronutrients such as iodine and chromium that organisms require for their metabolic processes. But more than materials are involved. Earth provides temperature conditions conducive to life and a shield against incoming ultraviolet radiation, its potentially deadly photons absorbed by molecules in the atmosphere and their energy dissipated as heat. Earth also has a good capacity to deal with waste products that are discharged into the atmosphere, into water, or into the geosphere.

The capacity of Earth to provide materials, protection, and conditions conducive to life is known as its **natural capital**, which can be regarded as the sum of two major components: **natural resources** and **ecosystem services**. Early hunter-gatherer and agricultural human societies made few demands upon Earth's natural capital. As shown in Figure 1.4, as the industrial revolution developed from around 1800, natural resources were abundant and production of material goods was limited largely by labor and the capacity of machines to process materials. But now, population is in excess, computerized machines have an enormous capacity to process materials, the economies of once impoverished countries including India and China are becoming highly industrialized, and the availability of natural capital is the limiting factor in production, including availability of natural resources, the vital life support ability of ecological systems, and the capacity of the natural environment to absorb the by-products of industrial production, most notably greenhouse gas carbon dioxide.

To sustain Earth and its natural capital for future generations, economic systems must evolve in the future such that they provide adequate and satisfying standards of living while increasing well-being, productivity, wealth, and capital and at the same time reducing waste, consumption of

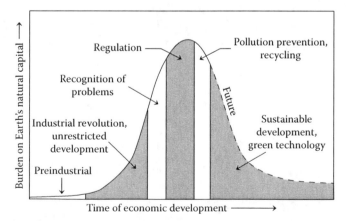

FIGURE 1.4 Stages of economic development with respect to utilization of Earth's natural capital. The preindustrial impact of human activities was very low. As the industrial revolution gathered force from about 1800, unrestricted development put a rapidly increasing burden on natural capital, which continued during an era in which there was recognition of the problem. This eventually led to regulations that began to reduce the impact on natural capital. To an extent, the regulatory approach was supplemented by pollution prevention and recycling. In an optimistic view of the future, sustainable development and green technology will further reduce the burden on natural capital even with increased economic development.

resources, and adverse environmental effects. The traditional capitalist economic system has proven powerful in delivering consumer goods and services using the leverage of individual and corporate incentives. Future systems must evolve in a manner that preserves these economic drivers while incorporating sustainable practices such as recycling wastes back into the raw material stream and emphasizing the provision of services rather than just material goods. In so doing, they can emulate nature's systems through the application of the principles of green chemistry and the practice of industrial ecology (see Chapter 16).

1.5 ENVIRONMENTAL CHEMISTRY AND GREEN CHEMISTRY

In the earlier years during which environmental chemistry was recognized as a distinct discipline, emphasis was placed on finding and quantifying pollutants, capturing or destroying potential pollutants after they were made (so-called end-of-pipe pollution control), and remediating polluted areas, such as by adding lime to a lake made too acidic by acid rainfall. Following regulations that were put forth as the result of pollution control legislation, this approach was called a **command and control** means of pollution control.

The limitations of a command and control system for environmental protection have become more obvious even as the system has become more successful. In industrialized societies with good, well-enforced regulations, the easy and inexpensive measures that can be taken to reduce environmental pollution and exposure to harmful chemicals have been implemented. Therefore, small increases in environmental protection now require relatively large investments in money and effort. Is there a better way? There is, indeed. The better way is through the practice of green chemistry.

Green chemistry can be defined as the practice of chemical science and manufacturing in a manner that is sustainable, safe, and nonpolluting and that consumes minimum amounts of materials and energy while producing little or no waste material. This definition of green chemistry is illustrated in Figure 1.5. The practice of green chemistry begins with recognition that the production, processing, use, and eventual disposal of chemical products may cause harm when performed incorrectly. In accomplishing its objectives, green chemistry and green chemical engineering may modify or totally redesign chemical products and processes with the objective of minimizing wastes and the use or generation of particularly dangerous materials. Those who practice green chemistry recognize that they are responsible for any effects on the world that their chemicals or

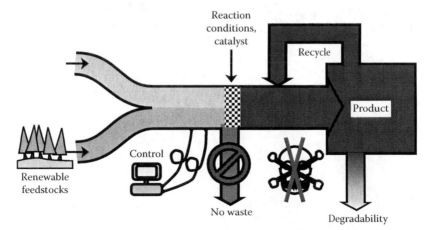

FIGURE 1.5 Green chemistry emphasizes renewable feedstocks, exacting control to maximize efficiency, mild reaction conditions, maximum recycling of materials, minimal wastes, and degradability of products that might enter the environment. To the extent possible, green chemistry avoids use and production of otherwise dangerous materials. Green chemical processes are expedited by catalysts that enable specific reactions to occur, make possible milder reaction conditions, and minimize energy consumption.

chemical processes may have. Far from being economically regressive and a drag on profits, green chemistry is about increasing profits and promoting innovation while protecting human health and the environment.

To a degree, we are still finding out what green chemistry is. That is because it is a rapidly evolving and developing subdiscipline in the field of chemistry. And it is a very exciting time for those who are practitioners of this developing science. Basically, green chemistry harnesses a vast body of chemical knowledge and applies it to the production, use, and ultimate disposal of chemicals in a way that minimizes consumption of materials; exposure of living organisms, including humans, to toxic substances; and damage to the environment. And it does so in a manner that is economically feasible and cost-effective. In one sense, green chemistry is the most efficient possible practice of chemistry and the least costly when all of the costs of doing chemistry, including hazards and potential environmental damage, are taken into account.

Green chemistry is sustainable chemistry in several important respects. Often, the practice of green chemistry is less costly in strictly **economic** terms than the conventional practice of chemistry and invariably less when the costs to the environment are factored in. By efficiently using materials, maximum recycling, and minimum use of virgin raw materials, green chemistry is sustainable with respect to **materials**. By reducing insofar as possible, or even totally eliminating their production, green chemistry is sustainable with respect to **wastes**.

Green chemistry is obviously strongly related to environmental chemistry and toxicological chemistry. It is a key discipline in pollution prevention and sustainability. Reference is made to the practice of green chemistry in later parts of this book and it is discussed in more detail in Chapter 16.

1.6 AS WE ENTER INTO THE ANTHROPOCENE

Environmental chemistry has a strong role to play in preserving our planet in these challenging times in which the Earth is undergoing significant, perhaps drastic, change, especially with respect to climate. Earth is entering the new age of the **Anthropocene**, an evolving epoch in Earth's lifetime. The existence of the Anthropocene was first suggested in 2000 by Paul Crutzen (who shared the 1995 Nobel Prize for his work on stratospheric ozone depletion caused by chlorofluorocarbons) and his colleague Eugene Stoermer.[3] The argument was made convincingly that the relatively hospitable Holocene epoch in which modern humans have been living since the end of the last ice age about 10,000 years ago is ending and that Earth is entering a new epoch, the Anthropocene, in which conditions are determined largely by what humans do with their growing capacity to change global conditions, a change that poses an enormous challenge for humankind.

The **Earth System** is a term used to describe the interacting processes that determine the state and dynamics of Planet Earth including its transition into the Anthropocene. These processes have physical, chemical, and biological components strongly tied with biogeochemical cycles. **Earth System Science** is the study of the interactions among various parts of the five spheres of the environment (hydrosphere, atmosphere, geosphere, biosphere, and anthrosphere) to enable understanding and prediction of global environmental changes.[4] What is known about Earth System Science and what can be expected in the future are based on both paleoenvironmental studies (geological strata, fossils, and ice core data) and increasingly sophisticated computer models that can forecast future trends.

The Earth System is complex, integrated, and self-regulated. As shown in Figure 1.6, the Earth System is essentially closed with respect to materials, but it exhibits a strong external energy flux with incoming radiant energy from the sun and outgoing radiant energy primarily at longer wavelengths. It is the balance between these two energy flows that largely determines conditions on Earth, especially its climate and suitability to maintain life including human life.

Of particular importance in the Earth System are surface water and air, the "two great fluids" in Earth's environment. The great fluids can move and transport materials and energy. Air heated in equatorial regions expands and flows away from the equator carrying heat energy as

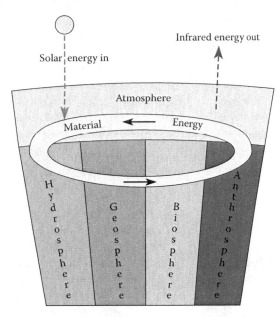

FIGURE 1.6 The Earth System is essentially a closed system with respect to matter, which circulates within the five spheres of the environment. Energy enters the Earth System in the form of sunlight and leaves primarily as infrared radiation. The balance between these two flows largely determines Earth's climate and conditions leading to the Anthropocene.

sensible heat in the air molecules and latent heat in water vapor toward polar regions. A plume of water called the Gulf Stream heated in the Caribbean region flows northward near the surface of the Atlantic along the east coast of North America and releases heat off the coast of Europe before sinking and flowing back at greater depths (the thermohaline circulation of the North Atlantic). This phenomenon is responsible in part for the relatively warm temperatures of Ireland, England, and Western Europe despite their more northern latitudes, and its possible demise is of great concern with respect to global climate change. In addition to large quantities of water, flowing rivers carry sediments and are very much involved in the transport of waterborne pollutants.

Within the Earth System, materials and energy are cycled and transformed in complex and dynamic ways through processes involving various forces and feedbacks. The kinds and predominance of organisms that operate within the system are strongly influenced by external factors including temperature, sunlight available for photosynthesis, and water. The various components of the system itself, including the organisms in it, exert massive effects upon the system; are active participants in it through physical, chemical, biological, and ecological processes; and do not just respond passively. The greatest change to the Earth System caused by organisms was the production of the atmosphere's oxygen by photosynthetic cyanobacteria billions of years ago. In the modern era, humans are the organisms that are having the greatest effects on the Earth System leading to the transition to the Anthropocene epoch.

External environmental factors, the influence of organisms, and the activities of humans in determining the Earth System are strongly tied together and mutually interacting, often in ways that are hard to determine. Such interactions may be illustrated by the growth of crops to satisfy human needs. A modern example is the intensive cultivation of corn in the United States to provide carbohydrate for the biosynthesis by yeast fermentation of ethanol fuel, an application that is now taking about one-third of the U.S. corn crop. The surface areas of the geosphere where fields of corn are located are cultivated and altered to provide the platform upon which the corn is grown. Solar energy and atmospheric carbon dioxide are utilized for photosynthesis to produce the corn biomass. Increased demand for fertilizer in the form of chemically combined nitrogen means that more ammonia is synthesized using atmospheric nitrogen and impacting the nitrogen cycle. The corn production cycle is strongly dependent on the availability of water from rainfall, a source that may be supplemented by withdrawing water from underground aquifers. Growing so much corn means that a relatively larger fraction of the biosphere consists of a single kind of plant that reduces biodiversity.

The time scale employed is important in studying and understanding the Earth System. In some cases, the time scale is very large, for example, hundreds of thousands of years in deducing past climate fluctuations from ice core data and millions of years from fossil records. It is important (and in a sense a bit frightening) to note that some major changes in the Earth System in the past have occurred very abruptly, within a decade or so.

QUESTIONS AND PROBLEMS

In answering all questions, it is assumed that the reader has access to the Internet from which general information, statistics, constants, and mathematical formulas required to solve problems may be obtained. These questions are designed to promote inquiry and thought rather than just finding material in the text. So in some cases, there may be several "right" answers. Therefore, if your answer reflects intellectual effort and a search for information from available sources, it may be considered to be "right."

1. Much of what is known about Earth's past history is based on paleoenvironmental studies. Doing some research on the Internet, suggest what is meant by these studies. How can past climatic conditions, temperature, and atmospheric carbon dioxide levels be inferred going back hundreds of thousands of years based on ice cores and even millions of years based on fossils?

2. The idea of climate change caused by human activities appears to be relatively recent. However, it was proposed quite some time ago in a paper entitled "On the Influence of Carbonic Acid in the Air upon the Temperature of the Ground." When was this paper published and who was the author? What were his credentials and credibility?

3. The definition of environmental chemistry shown in Figure 1.1 could very well be illustrated with nitrogen oxides, NO and NO_2, emitted to the atmosphere. What would be the sources of these gaseous nitrogen oxides? Which secondary air pollutant would they form interacting with volatile hydrocarbons in the sunlight? Could acid rain result from these oxides and, if so, what would be the formula of the acid?

4. A number of reputable scientists now believe that the Holocene is ending and a new era is beginning. What is the Holocene? What is the new era that may well be replacing it and how does it relate to the material in this chapter? What are some of the environmental implications of this change?

5. In the late 1800s, there was concern that within the nitrogen biogeochemical cycle, not enough of the atmosphere's inexhaustible store of nitrogen was being "fixed" to chemical forms that could be utilized by plants and that food shortages would result from a shortage of fixed nitrogen. What happened to change this situation? In what respect did this development save many lives and how did it also make possible the loss of millions of people in warfare after about 1900?

6. In what respect is the term "solid earth" a misnomer? What are some specific events in 2011 that cast some doubt on "solid earth?" How did one of these events specifically impact the anthrosphere and perhaps change the course of future energy developments?

7. In what important, fundamental respect does the phosphorus cycle differ from the carbon, oxygen, and nitrogen cycles?

8. Most people are aware that atmospheric carbon dioxide contributes to global warming and climate change. In what respect, however, is the atmosphere's carbon dioxide part of Earth's natural capital, that is, where would we be without it? What crucial natural

phenomenon causes a slight, but perceptible change in atmospheric carbon dioxide levels over the course of a year?

9. Figure 1.1 illustrates the definition of environmental chemistry in terms of a common pollutant. What command and control regulations have been implemented in limiting this source of pollution? What "end-of-pipe" measures have been used? Suggest how the practices of green chemistry might serve as alternatives to these measures.

10. As it applies to environmental processes, the term "sink" is mentioned several times in this chapter. In what sense is Earth's ability to act as a sink part of its natural capital? Explain.

11. In dealing with pollution and the potential for pollution, three approaches are pollution prevention, end-of-pipe measures, and remediation. What do these terms mean in terms of pollution control? Which is the most desirable, and which is the least? Explain.

12. With respect to increased production of corn to provide fuel ethanol, it is stated in this chapter that "Increased demand for fertilizer in the form of chemically combined nitrogen means that more ammonia is synthesized using atmospheric nitrogen and impacting the nitrogen cycle." With respect to which resource of Earth's capital is the synthetic production of nitrogen fertilizer a problem and in respect to which resource is it *not* a problem? Explain.

LITERATURE CITED

1. Manahan, Stanley E., *Environmental Chemistry*, 9th ed., Taylor & Francis/CRC Press, Boca Raton, FL, 2009.
2. Manahan, Stanley E., *Green Chemistry and the Ten Commandments of Sustainability*, 3rd ed., ChemChar Research, Columbia, MO, 2011.
3. Manahan, Stanley E., *Anthropocene: Environmental Chemistry of the World Made by Humans*, ChemChar Research, Columbia, MO, 2011.
4. Oldfield, Frank, and Will Steffen, The Earth System, in *Global Change and the Earth System: A Planet Under Pressure*, Will Steffen, Ed., Springer-Verlag, New York, 2004.

SUPPLEMENTARY REFERENCES

Allenby, Braden, *Reconstructing Earth: Technology and Environment in the Age of Humans,* Island Press, Washington, 2005.

Baird, Colin, and Michael Cann, *Environmental Chemistry*, 5th ed., W. H. Freeman, New York, 2012.

Concepción, Jiménez-González, and David J. C. Constable, *Green Chemistry and Engineering: A Practical Design Approach*, Wiley, Hoboken, NJ, 2011.

Ehlers, Eckart, Thomas Krafft, and C. Moss, *Earth System Science in the Anthropocene: Emerging Issues and Problems*, Springer, New York, 2010.

Florinsky, Igor V., Ed., *Man and the Geosphere*, Nova Science Publishers, Hauppauge, NY, 2009.

Girard, James E., *Principles of Environmental Chemistry*, 2nd ed., Jones and Bartlett Publishers, Sudbury, MA, 2010.

Hanrahan, Grady, *Key Concepts in Environmental Chemistry*, Academic Press, Waltham, MA, 2011.

Hites, Ronald A., and Jonathan D. Raff, *Elements of Environmental Chemistry*, 2nd ed., Wiley, Hoboken, NJ, 2012.

Manahan, Stanley E., *Environmental Chemistry*, 9th ed., Taylor & Francis/CRC Press, Boca Raton, FL, 2010.

Manahan, Stanley E., *Environmental Science and Technology*, 2nd ed., Taylor & Francis/CRC Press, Boca Raton, FL, 2006.

Manahan, Stanley E., *Fundamentals of Environmental Chemistry*, 3rd ed., Taylor & Francis/CRC Press, Boca Raton, FL, 2008.

Silivanch, Annalise, *Rebuilding America's Infrastructure*, Rosen Publications, New York, 2011.

VanLoon, Gary W., *Environmental Chemistry: A Global Perspective*, Oxford University Press, Oxford, UK, 2010.

2 Fundamentals of Biochemistry and Toxicological Chemistry

2.1 LIFE CHEMICAL PROCESSES

Biochemistry is the science of chemical processes that occur in living organisms.[1] There are two major reasons to introduce biochemistry at this point. The first of these is that by its nature, biochemistry is a sustainable chemical and biological science. This is because over eons of evolution, organisms that carry out biochemical processes sustainably have evolved. Because the enzymes that carry out biochemical processes can function only under mild conditions, temperature in particular, biochemical processes take place under safe conditions, avoiding the high temperatures, high pressures, and corrosive and reactive chemicals that often characterize synthetic chemical operations. Therefore, it is appropriate to refer to **green biochemistry**, an important area of sustainable chemical science. Biochemical processes not only are profoundly influenced by chemical species in the environment, but they also largely determine the nature of these species, their degradation, and even their syntheses, particularly in the aquatic and soil environments. The study of such phenomena forms the basis of **environmental biochemistry**. **Biochemicals** are molecules that are made by living organisms through biological processes. The major types of biochemicals are discussed in Sections 2.3–2.6 of this chapter.

The second important reason to introduce biochemistry here is that in the practice of environmental chemistry, it is essential to know the potential toxic effects of various materials, a subject addressed by **toxicological chemistry**.[2] Aspects of toxicological chemistry are discussed throughout this book and the topic is introduced and outlined in this chapter.

2.2 BIOCHEMISTRY AND THE CELL

For the most part, biochemical processes occur within cells, the very small units that living organisms are composed of.[3] Cells are discussed in more detail as basic units of life in Chapter 12; in this chapter, they are regarded as what chemical engineers would call "unit operations" for carrying out biochemical processes. The ability of organisms to carry out chemical processes is truly amazing, even more so when one considers that many of them occur in single-celled organisms. Photosynthetic cyanobacteria consisting of individual cells less than a micrometer (μm) in size can make all the complex biochemicals they need to exist and reproduce using sunlight for energy and simple inorganic substances such as CO_2, K^+ ion, NO_3^- ion, and HPO_4^{2-} ion for raw materials (Figure 2.1). Soon after conditions on Earth became hospitable to life, these photosynthetic bacteria produced the oxygen that now composes about 20% of Earth's atmosphere. Fossilized stromatolites (bodies of sedimentary materials bound together by films produced by microorganisms) produced by cyanobacteria have been demonstrated dating back 2.8 billion years, and the remarkable cyanobacteria that convert atmospheric carbon dioxide to biomass and atmospheric N_2 to chemically fixed N might have been on Earth as long as 3.5 billion years ago.

Many organisms consist of single cells or individual cells growing together in colonies. Bacteria, yeasts, protozoa, and some algae consist of single cells. Other than these microorganisms, organisms are composed of many cells that have different functions. Liver cells, muscle cells, brain cells, and skin cells in the human body are quite different from each other and do different things. Two major kinds of cells are eukaryotic cells, which have a nucleus, and prokaryotic cells, which do

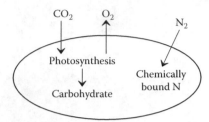

FIGURE 2.1 Cyanobacteria are remarkable organisms that within a single "simple" prokaryotic cell carry out all the biochemical processes needed to convert atmospheric carbon dioxide to carbohydrate and biomass and that can split the chemically very stable atmospheric nitrogen molecule and convert the nitrogen to chemically and biochemically bound nitrogen.

not. Prokaryotic cells are found predominately in single-celled bacteria. Eukaryotic cells occur in multicellular plants and animals—higher life forms.

Cell structure has an important influence on determining the nature of biomaterials. Muscle cells consist largely of strong structural proteins capable of contracting and movement. Bone cells secrete a protein mixture that mineralizes with calcium and phosphate to produce solid bone. The walls of cells in plants are largely composed of strong cellulose, which makes up the sturdy structure of wood.

2.3   CARBOHYDRATES

Carbohydrates are biochemicals consisting of carbon, hydrogen, and oxygen with the approximate simple formula CH_2O. One of the most common carbohydrates is the simple sugar glucose shown in Figure 2.2. Units of glucose and other simple sugars called monosaccharides join together in chains with the loss of a water molecule for each linkage to produce macromolecular polysaccharides. These include starch and cellulose in plants and starch-like glycogen in animals.

Glucose, a carbohydrate and simple sugar, is the biological material generated from water and carbon dioxide when solar energy in sunlight is utilized in photosynthesis. The overall reaction is

$$6CO_2 + 6H_2O \rightarrow C_6H_{12}O_6 + 6O_2 \qquad (2.1)$$

This is obviously an extremely important reaction because it is the one by which inorganic molecules are used to synthesize high-energy carbohydrate molecules that are in turn converted to the vast number of biomolecules that comprise living systems. There are other simple sugars, including fructose, mannose, and galactose, that have the same simple formula as glucose, $C_6H_{12}O_6$, but which must be converted to glucose before being utilized by organisms for energy. Consisting of a molecule of glucose and fructose linked together (with the loss of a water molecule), common table sugar, $C_{12}H_{22}O_{11}$, is a disaccharide.

Starch molecules, which may consist of several hundred glucose units joined together, are readily broken down by organisms to produce simple sugars used for energy and to produce biomass. For example, humans readily digest starch in potatoes or bread to produce glucose used for energy (or to make fat tissue).

The chemical formula of starch is $(C_6H_{10}O_5)_n$, where n may represent a number as large as several hundred. What this means is that the very large starch molecule consists of as many as several hundred units of $C_6H_{10}O_5$ from glucose joined together. For example, if n is 100, there are 6 times 100 carbon atoms, 10 times 100 hydrogen atoms, and 5 times 100 oxygen atoms in the molecule. Its chemical formula is $C_{600}H_{1000}O_{500}$. The atoms in a starch molecule are actually present as linked rings represented by the structural formula shown in Figure 2.2. Starch occurs in many foods such as bread, potatoes, and cereals. It is readily digested by animals, including humans.

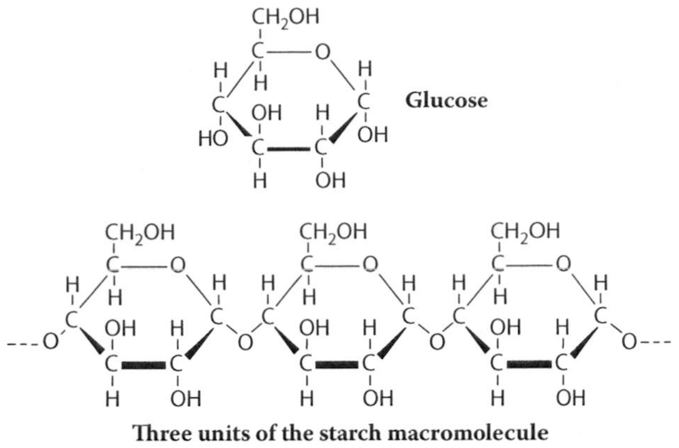

Three units of the starch macromolecule

FIGURE 2.2 Glucose, a monosaccharide, or simple sugar, and a segment of the starch molecule, which is formed when glucose molecules polymerize with the elimination of one H_2O molecule per glucose monomer.

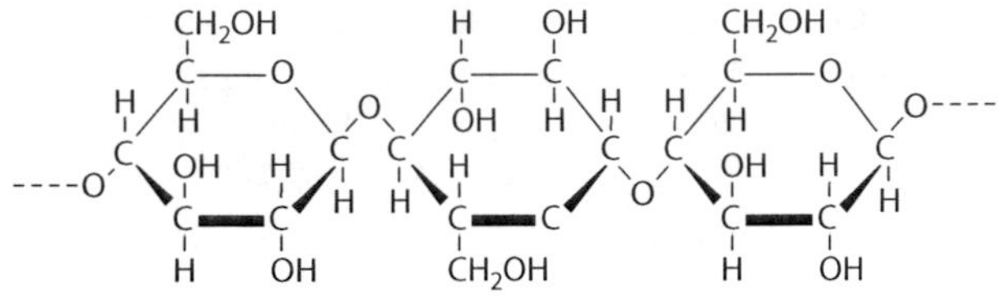

FIGURE 2.3 A segment of a cellulose molecule. These molecules are biosynthesized from glucose with the loss of one H_2O for each linkage formed.

Cellulose is a polysaccharide that is also made up of $C_6H_{10}O_5$. Molecules of cellulose are huge, with molecular masses of around 400,00 atomic mass units, u. The cellulose structure (Figure 2.3) is similar to that of starch. Cellulose is produced by plants and forms the structural material of plant cell walls. Wood is about 60% cellulose, and cotton contains over 90% of cellulose. Fibers of cellulose are extracted from wood and pressed together to make paper.

Humans and most other animals cannot digest cellulose because they lack the enzyme needed to hydrolyze the oxygen linkages between the glucose molecules. Ruminant animals (cattle, sheep, goats, and moose) have bacteria in their stomachs that break down cellulose into products that can be used by the animals. Fungi and termites existing synergistically with cellulose-degrading bacteria biodegrade huge quantities of cellulose.

Carbohydrates are potentially very important in green chemistry and sustainable chemical science. Firstly, they are a concentrated form of organic energy synthesized and stored by plants as part of the process by which plants capture solar energy through photosynthesis. Carbohydrates can be utilized directly for energy or fermented to produce ethanol, C_2H_6O, a combustible alcohol that is added to gasoline or can even be used in place of gasoline. Secondly, carbohydrates are a source of organic raw material that can be converted to other organic molecules to make plastics and other useful materials.

2.4 PROTEINS

Proteins are macromolecules that are composed of nitrogen, carbon, hydrogen, and oxygen along with smaller quantities of sulfur. The small molecules of which proteins are made are composed of 20 naturally occurring amino acids. The simplest of these, glycine, is shown in the first structure in Figure 2.4, along with two other amino acids. As shown in Figure 2.4, amino acids join together with the loss of a molecule of H_2O for each linkage formed. The three amino acids in Figure 2.4 are shown linked together as they would be in a protein in the bottom structure in the figure. Many hundreds of amino acid units may be present in a protein molecule.

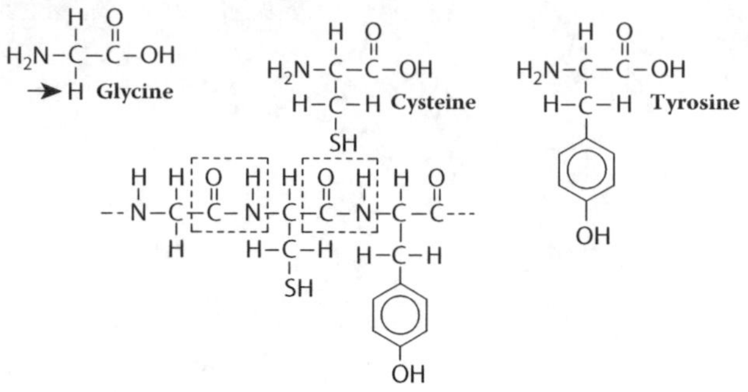

FIGURE 2.4 Three amino acids. Glycine is the simplest amino acid. All others have the basic glycine structure except that different groups are substituted for the H designated in glycine by an arrow. The lower structure shows these three amino acids are linked together in a macromolecule chain composing a protein. For each linkage, one molecule of H_2O is lost. The peptide linkage holding amino acids together in proteins is outlined by a dashed rectangle.

The three-dimensional structures of protein molecules are of utmost importance and largely determine what the proteins do in living systems and how they are recognized by other biomolecules. Enzymes, special proteins that act as catalysts to enable biochemical reactions to occur, recognize the substrates upon which they act by the complementary shapes of the enzyme molecules and substrate molecules. There are several levels of protein structure. The first of these is determined by the order of amino acids in the protein macromolecule. Folding of protein molecules and pairing of two different protein molecules further determine structure. The loss of protein structure, called denaturation, can be very damaging to the proteins and the organism in which they are contained.

Two major kinds of proteins are tough fibrous proteins, which compose hair, tendons, muscles, feathers, and silk, and spherical or oblong-shaped globular proteins, such as hemoglobin in blood or the proteins, which comprise enzymes. Proteins serve many functions. These include nutrient proteins, such as casein in milk; structural proteins, such as collagen in tendons; contractile proteins, such as those in muscles; and regulatory proteins, such as insulin, which regulate biochemical processes.

Some proteins are very valuable biomaterials for pharmaceutical, nutritional, and other applications, and their synthesis is an important aspect of green chemistry. The production of specific proteins has been greatly facilitated in recent years by the application of genetic engineering to transfer to bacteria the genes that direct the synthesis of specific proteins. The best example is insulin, a protein injected into diabetics to control blood sugar. Insulin injected for blood glucose control used to be isolated from the pancreas of slaughtered cattle and hogs. Although this enabled many diabetics to live normal lives, the process of getting the insulin was cumbersome, supply was limited, and the insulin from this source was not exactly the same as that made in the human body, which often caused the body to have an allergic response to it as a foreign protein. The transfer through recombinant DNA technology of the human gene for insulin production into prolific *Escherichia coli* bacteria has enabled large-scale production of human insulin by the bacteria.

2.5 LIPIDS: FATS, OILS, AND HORMONES

Lipids differ from most other kinds of biomolecules in that they are repelled by water. Lipids can be extracted from biological matter by organic solvents such as diethyl ether or toluene. Recall that proteins and carbohydrates are distinguished largely by chemically similar characteristics and structures. However, lipids have a variety of chemical structures that share the common physical characteristic of solubility in organic solvents. Many of the commonly encountered lipid fats and oils are esters of glycerol alcohol, $CH_2(OH)CH(OH)CH_2(OH)$, and long-chain carboxylic acids (fatty acids) such as stearic acid, $CH_3(CH_2)_{16}CO_2H$. The glycerol molecule has three $-OH$ groups to each of which a fatty acid molecule may be joined through the carboxylic acid group with the loss of a water molecule for each linkage that is formed. Figure 2.5 shows a fat molecule formed from three stearic acid molecules and a glycerol molecule. Such a molecule is one of many possible triglycerides. Also shown in this figure is cetyl palmitate, the major ingredient of spermaceti wax extracted from the sperm of whale blubber and used in some cosmetics and pharmaceutical preparations. Cholesterol shown in Figure 2.5 is one of several important lipid steroids, which share the ring structure composed of rings of 5 and 6 carbon atoms shown in the figure for cholesterol.

Although the structures shown in Figure 2.5 are diverse, they all share a common characteristic. This similarity is the preponderance of hydrocarbon chains and rings so that lipid molecules largely resemble hydrocarbons. Their hydrocarbon-like molecules make lipids soluble in organic solvents.

Some of the steroid lipids are particularly important because they act as hormones, chemical messengers that convey information from one part of an organism to another. Major examples of steroid hormones are cholesterol, testosterone (male sex hormone), and estrogens (female sex hormones). Steroid lipids readily penetrate the membranes that enclose cells, which are especially permeable to more hydrophobic lipid materials. Hormones start and stop a number of body functions and regulate the expression of many genes. In addition to steroid lipids, many hormones including insulin and human growth hormone are proteins. Hormones are given off by ductless glands in the body called endocrine glands.

Lipids are important in green chemistry and toxicological chemistry for several reasons. Lipids are very much involved with toxic substances, the generation and use of which are always important in green chemistry. Poorly biodegradable substances, particularly organochlorine compounds, that are always an essential consideration in green chemistry, tend to accumulate in lipids in living organisms, a process called bioaccumulation. Lipids can be valuable raw materials and fuels.

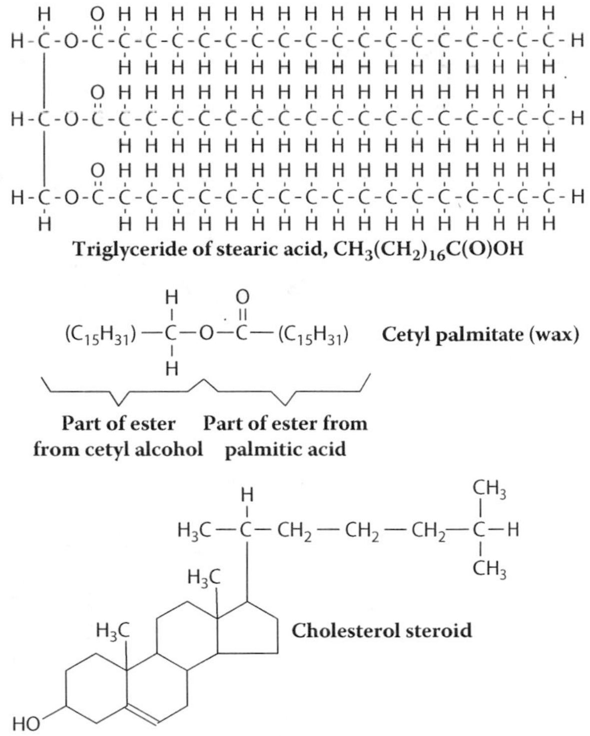

Triglyceride of stearic acid, $CH_3(CH_2)_{16}C(O)OH$

Cetyl palmitate (wax)

Part of ester   Part of ester from
from cetyl alcohol   palmitic acid

Cholesterol steroid

FIGURE 2.5   Three examples of lipids formed in biological systems. Note that a line structure is used to show the ring structure of cholesterol. The hydrocarbon-like nature of these compounds, which makes them soluble in organic compounds, is obvious. For interpretation of the line structure of cholesterol, see Chapter 20, Section 20.3.

A major kind of renewable fuel is made by hydrolyzing the long-chain fatty acids from triglycerides and attaching methyl groups to produce esters. This liquid product, commonly called biodiesel fuel, serves as a substitute for petroleum-derived liquids in diesel engines. The development and cultivation of plants that produce oils and other lipids is a major possible route to the production of renewable resources.

2.6 NUCLEIC ACIDS

Nucleic acids (Figure 2.6) are biological macromolecules that store and pass on the genetic information that organisms need to reproduce and synthesize proteins. The two major kinds of nucleic acids are DNA, which basically stays in place in the cell nucleus of an organism, and ribonucleic acid (RNA), which is spun off from DNA and functions throughout a cell. Molecules of nucleic acids contain three basic kinds of materials. The first of these is a simple sugar, 2-deoxy-β-D-ribofuranose (deoxyribose) contained in DNA and β-D-ribofuranose (ribose) contained in RNA. The second major kind of ingredient consists of nitrogen-containing bases: cytosine, adenine, and guanine, which occur in both DNA and RNA; thymine, which occurs only in DNA; and uracil, which occurs only in RNA. The third constituent of both DNA and RNA is inorganic phosphate, PO_4^{3-}. These three kinds of substances occur as repeating units called nucleotides joined together in astoundingly long chains in the nucleic acid polymer as shown in Figure 2.6.

The remarkable way in which DNA operates to pass on genetic information and perform other functions essential for life is the result of the structure of the DNA molecule. In 1953, James D. Watson and Francis Crick deduced that DNA consisted of two strands of material counterwound around each other in a structure known as an α-helix (Figure 2.7), a remarkable bit of insight that earned Watson and Crick the Nobel Prize in 1962. These strands are held together by hydrogen bonds between complementary nitrogenous bases. Taken apart, the two strands resynthesize complementary strands, a process that occurs during reproduction of cells in living organisms.

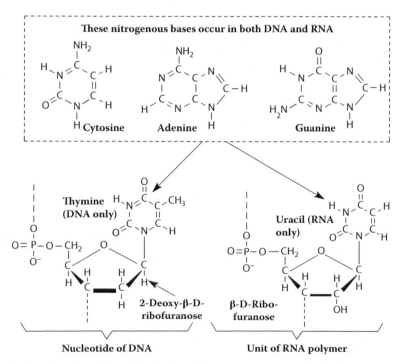

FIGURE 2.6 Basic units of nucleic acid polymers. These units act as a code in directing reproduction and other activities of organisms. Dashed lines show bonds to next nucleotide unit.

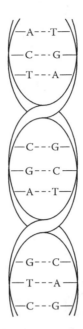

FIGURE 2.7 Representation of the double-helix structure of DNA showing the allowed base pairs held together by hydrogen bonding between the phosphate/sugar polymer "backbones" of the two strands of DNA. The letters stand for adenine (A), cytosine (C), guanine (G), and thymine (T). The dashed lines, --- , represent hydrogen bonds.

In directing protein synthesis, DNA becomes partially unraveled and generates a complementary strand of material in the form of RNA, which in turn directs protein synthesis in the cell.

Consideration of nucleic acids and their function is very important in the development of green chemistry and the practice of sustainable chemical science. One aspect of this relationship is that the toxicity hazards of many chemical substances result from potential effects of these substances on DNA. Of most concern is the ability of some substances to alter DNA and cause the uncontrolled cell replication that is cancer. Also of concern is the ability of some chemical substances called mutagens to alter DNA such that undesirable characteristics are passed on to offspring.

Another important consideration with DNA as it relates to green chemistry is the ability that humans now have to transfer DNA between organisms, popularly called genetic engineering. An important example is the development of bacteria that have the DNA transferred from humans to make human insulin. This technology of recombinant DNA is discussed in more detail in Chapter 13.

2.7 ENZYMES

Catalysts are substances that speed up a chemical reaction without being consumed in the reaction. Catalysis is one of the most important aspects of green chemistry because the ability to make reactions go faster as well as more efficiently, safely, and specifically means that less energy and few raw materials are used and less waste is produced. Biochemical catalysts called **enzymes** include some of the most sophisticated catalysts. Enzymes speed up biochemical reactions by as much as ten- to a hundred-million-fold. They often enable reactions to take place that otherwise would not occur, that is, they tend to be very selective in the reactions they promote. One of the greatest advantages of enzymes as catalysts is that they have evolved to function under the benign conditions under which organisms exist. This optimum temperature range is generally from about the freezing point of water (0°C) to slightly above body temperature (up to about 40°C). Chemical reactions go faster at higher temperatures, so there is considerable interest in enzymes isolated from microorganisms that thrive at temperatures near the boiling point of water (100°C) in hot water pools heated by underground thermal activity such as are found in Yellowstone National Park in the United States.

Enzymes are proteinaceous substances. Their structure is highly specific so that they bind with whatever they act upon (a substrate). The basic mechanism of enzyme action is shown in Figure 2.8. As indicated in the figure, an enzyme recognizes a substrate by its shape, bonds with the substrate to produce an enzyme–substrate complex, causes a change such as splitting a substrate into two products with addition of water (hydrolysis), and then emerges unchanged to do the same thing again. The basic process can be represented as follows:

$$\text{enzyme} + \text{substrate} \leftrightarrow \text{enzyme} - \text{substrate complex} \leftrightarrow \text{enzyme} + \text{product} \qquad (2.2)$$

Note that the arrows in the formula for enzyme reaction point both ways. This means that the reaction is **reversible**. An enzyme–substrate complex can simply go back to the enzyme and the substrate. The products of an enzymatic reaction can react with the enzyme to form the enzyme–substrate complex again. It, in turn, may again form the enzyme and the substrate. Therefore, the same enzyme may act to cause a reaction to go either way.

For some enzymes to work, they must first be attached to coenzymes. Coenzymes normally are not protein materials. Some of the vitamins are important coenzymes.

The names of enzymes are based on what they do and where they occur. For example, gastric protease, commonly called pepsin, is an enzyme released by the stomach (gastric), which splits protein molecules as part of the digestion process (protease). Similarly, the enzyme produced by the pancreas that breaks down fats (lipids) is called pancreatic lipase. Its common name is steapsin. In general, lipase enzymes cause lipid triglycerides to dissociate and form glycerol and fatty acids.

Lipase and protease enzymes are hydrolyzing enzymes, which enable the splitting of molecules of high-molecular-mass biological compounds with the addition of water, one of the most important types of the reactions involved in digestion of food carbohydrates, proteins, and fats. Recall that the higher carbohydrates humans eat are largely disaccharides (sucrose, or table sugar) and polysaccharides (starch). These are formed by the joining of units of simple sugars, $C_6H_{12}O_6$, with the elimination of an H_2O molecule at the linkage where they join. Proteins are formed by the condensation

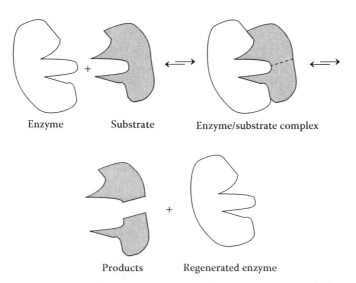

FIGURE 2.8 Representation of the "lock-and-key" mode of enzyme action, which enables the very high specificity of enzyme-catalyzed reactions.

of amino acids, again with the elimination of a water molecule at each linkage. Fats are esters that are produced when glycerol and fatty acids link together. A water molecule is lost for each of these linkages when a protein, fat, or carbohydrate is synthesized. For these substances to be used as a food source, the reverse process must be catalyzed by hydrolyzing enzymes to break down large, complicated molecules of protein, fat, or carbohydrate to simple, soluble substances that can penetrate a cell membrane and take part in chemical processes in the cell.

An important biochemical process is the shortening of carbon atom chains, such as those in fatty acids, commonly by the elimination of CO_2 from carboxylic acids. For example, pyruvate decarboxylase enzyme removes CO_2 from pyruvic acid

$$H-\overset{\overset{\displaystyle H}{|}}{\underset{\underset{\displaystyle H}{|}}{C}}-\overset{\overset{\displaystyle O}{||}}{C}-\overset{\overset{\displaystyle O}{||}}{C}-OH \xrightarrow[\text{decarboxylase}]{\text{Pyruvate}} H-\overset{\overset{\displaystyle H}{|}}{\underset{\underset{\displaystyle H}{|}}{C}}-\overset{\overset{\displaystyle O}{||}}{C}-H + CO_2$$

Pyruvic acid   **Acetaldehyde**

to produce a compound with one less carbon. It is by such carbon-by-carbon breakdown reactions that long-chain compounds are eventually degraded to CO_2 in the body. Another important consequence of this kind of reaction is the biodegradation of long-chain hydrocarbons by the action of microorganisms in the water and soil environments.

Energy is exchanged in living systems largely by oxidation and reduction mediated by oxidoreductase enzymes. **Cellular respiration** is an oxidation reaction in which a carbohydrate, $C_6H_{12}O_6$, is broken down to carbon dioxide and water with the release of energy:

$$C_6H_{12}O_6 + 6O_2 \rightarrow 6CO_2 + 6H_2O + \text{energy} \qquad (2.3)$$

Actually, such an overall reaction occurs in living systems by a complicated series of individual steps including oxidation. The enzymes that bring about oxidation in the presence of free O_2 are called oxidases.

In addition to the major types of enzymes discussed above, there are numerous other enzymes that perform various functions. Isomerases form isomers of particular compounds. For example, isomerases convert several simple sugars with the formula $C_6H_{12}O_6$ to glucose, the only sugar that can be used directly for cell processes. Transferase enzymes move chemical groups from one molecule to another, lyase enzymes remove chemical groups without hydrolysis and participate in the formation of C=C bonds or addition of species to such bonds, and ligase enzymes work in conjunction with ATP (a high-energy molecule that plays a crucial role in energy-yielding, glucose-oxidizing metabolic processes) to link molecules together with the formation of bonds such as carbon–carbon or carbon–sulfur bonds.

Enzymes are affected by the conditions and media in which they operate. Among these is the hydrogen ion concentration (pH). An interesting example is gastric protease, which requires the acid environment of the stomach to work well, but stops working when it passes into the much more alkaline medium of the small intestine. This prevents damage to the intestine walls, which would occur if the enzyme tried to digest them. Part of the damage to the esophagus from reflux esophagitis (acid reflux) is due to the action of gastric protease enzyme that flows back into the esophagus from the stomach with the acidic stomach juices. As noted in Section 2.1, temperature is critical for enzyme function. Not surprisingly, the enzymes in the human body work best at around 37°C (98.6°F), which is the normal body temperature. Heating these enzymes to around 60°C permanently destroys them. Some bacteria that thrive in hot springs have enzymes that work best at temperatures as high as that of boiling water. Other "cold-seeking" bacteria have enzymes adapted to temperatures near the freezing point of water.

2.7.1 Effects of Toxic Substances on Enzymes

Toxic substances may destroy enzymes or alter them so that they function improperly or not at all. Among the many toxic substances that act adversely with enzymes are heavy metals, cyanide, and various organic compounds such as insecticidal parathion. Many enzyme-active sites through which an enzyme recognizes and bonds with a substrate contain –SH groups. Toxic heavy metal ions such as Pb^{2+} or Hg^{2+} are "sulfur seekers" that bind to the sulfur in the enzyme-active site causing the enzyme not to function. A particularly potent class of toxic substances consists of the organophosphate "nerve gases" such as sarin, which inhibit the acetylcholinesterase enzyme required to stop nerve impulses. Very small doses of sarin stop respiration by binding with acetylcholinesterase and causing it not to work.

2.8 BIOCHEMICAL PROCESSES IN METABOLISM

So far, this chapter has discussed the cells in which biochemical processes occur, the major categories of biochemicals, and the enzymes that catalyze biochemical reactions. Biochemical processes involve the alteration of biomolecules, their synthesis, and their breakdown to provide the raw materials for new biomolecules, processes that fall under the category of **metabolism**. Metabolic processes may be divided into two major categories: anabolism (synthesis) and catabolism (degradation of substances). An organism may use metabolic processes to yield energy or to modify the constituents of biomolecules. Metabolism is discussed in this chapter as it affects biochemicals and in Chapter 12 as it applies to the function of organisms in the biosphere.

Metabolism is a very important consideration in green chemistry and sustainability. Toxic substances that impair metabolism pose a danger to humans and other organisms and attempts are made to avoid such substances in the practice of green chemistry. Exposures to environmental pollutants that impair metabolism endanger humans and other organisms; the control of such pollutants is an important aspect of environmental chemistry. Metabolic processes are used to make renewable raw materials and to modify substances to give desired materials. The complex metabolic process of photosynthesis provides the food that forms the base of essentially all food webs and is increasingly being called upon to provide renewable raw materials for manufacturing and fuels (see biofuels, Chapter 17, Section 17.17).

2.8.1 Energy-Yielding and Processing Processes

The processing of energy is obviously one of the most important metabolic functions of organisms. The metabolic processes by which organisms acquire and utilize energy are complex, generally involving numerous steps and various enzymes. Organisms can process and utilize energy by one of the following three major processes:

1. Respiration, in which organic compounds undergo catabolism
2. Fermentation, which differs from respiration in not having an electron transport chain
3. Photosynthesis, in which light energy captured by plant and algal chloroplasts is used to synthesize sugars from carbon dioxide and water

There are two major pathways in respiration. **Oxic respiration** (called aerobic respiration in the older literature) requires molecular oxygen, whereas **anoxic respiration** (anaerobic respiration) occurs in the absence of molecular oxygen. Oxic respiration uses the **Krebs cycle** to obtain energy from the reaction given above for cellular respiration:

$$C_6H_{12}O_6 + 6O_2 \rightarrow 6CO_2 + 6H_2O + \text{energy} \qquad (2.3)$$

About half of the energy released is converted to short-term stored chemical energy, particularly through the synthesis of ATP shown in Figure 2.9.

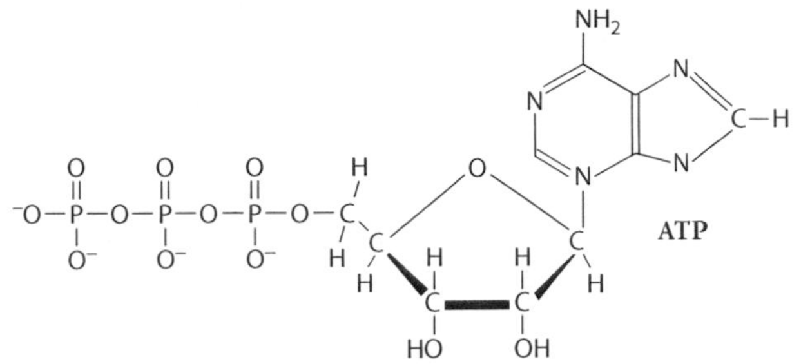

FIGURE 2.9 Adenosine triphosphate, a molecule strongly involved with energy transfer in living organisms.

The highly energized ATP molecule is sometimes described as the "molecular unit of currency" for the transfer of energy within cells during metabolism. It releases its energy when it loses a phosphate group and reverts to ADP and other precursors. ATP is used by enzymes and proteins for cell processes, including biosynthetic reactions (anabolism), cell division, and motility (e.g., occurs in moving protozoa cells). In so doing, ATP is continually being produced and reconverted back to its precursor species in an organism. Some studies have suggested that the human body processes its own mass in ATP during a single day! Whereas ATP is used for very short-term energy storage and processing, for longer-term energy storage, glycogen or starch polysaccharides are synthesized, and for still longer-term energy storage, lipids (fats) are generated and retained by the organism.

As noted above in this section, fermentation differs from respiration in not having an electron transport chain, and organic compounds are the final electron acceptors rather than O_2 in the energy-yielding process. Many biochemical processes including some used to make commercial products are fermentations. A common example of fermentation is the production of ethanol from sugars by yeasts growing in the absence of molecular oxygen:

$$C_6H_{12}O_6 \rightarrow 2CO_2 + 2C_2H_5OH \tag{2.4}$$

Photosynthesis is an energy-capture process in which light energy ($h\nu$) captured by plant and algal chloroplasts is used to synthesize sugars from carbon dioxide and water:

$$6CO_2 + 6H_2O + h\nu \rightarrow C_6H_{12}O_6 + 6O_2 \tag{2.5}$$

When it is dark, plants cannot get the energy that they need from sunlight but still must carry on basic metabolic processes using stored food. Plant cells, like animal cells, contain mitochondria in which stored food is converted to energy by cellular respiration.

Nonphotosynthetic organisms depend on organic matter produced by plants for their food and are said to be heterotrophic. They act as "middlemen" in the chemical reaction between oxygen and food material, using the energy from the reaction to carry out their life processes. Plant cells, which use sunlight as a source of energy and CO_2 as a source of carbon, are classified as **autotrophic**. In contrast, animal cells must depend on organic material manufactured by plants for their food. These are called **heterotrophic** cells.

Biochemical conversions involving energy are very important in the practice of green chemistry and sustainability. The most obvious connection is the capture of solar energy as chemical energy by photosynthesis. As discussed in Chapter 17, photosynthetically produced biomass can serve as a source of chemically fixed carbon for the synthesis of chemical fuels, including synthetic natural gas, gasoline, diesel fuel, and ethanol. A tantalizing possibility is to use recombinant DNA techniques to increase by several-fold the very low efficiency of photosynthesis by most plants. Fermentation has a strong role to play in sustainable energy development. As shown in Reaction 2.4, fermentation of glucose produces ethanol, which can be used as fuel. Anoxic fermentation of biomass (abbreviated {CH_2O})

from sources such as sewage sludge or food wastes yields methane (natural gas), the cleanest burning of all hydrocarbon fuels:

$$2\{CH_2O\} \rightarrow CH_4 + CO_2 \tag{2.6}$$

2.9 TOXIC SUBSTANCES, TOXICOLOGY, AND TOXICOLOGICAL CHEMISTRY

Toxicology is the science of **poisons** or **toxicants**, substances that damage or destroy living tissue or that cause biochemical processes to malfunction. The remainder of this chapter discusses the chemical aspects of toxicology, that is, **toxicological chemistry**.

There are numerous ways in which toxicants may detrimentally affect living organisms, such as the examples shown later in the chapter (Figure 2.17). One such effect is inhibition of enzymes, the substances in organisms that act as catalysts to enable biochemical processes to occur. Another example consists of alterations of the production and function of hormones, biological molecules that are produced and distributed in organisms to regulate biochemical processes.

Although many toxic substances are foreign to living systems and are called **xenobiotic** substances, others are produced by organisms. Many toxicants, especially xenobiotics, have an affinity for lipids, meaning that they undergo bioaccumulation in the fat tissue of animals and cross the lipid cell membranes readily. Probably, most substances classified as toxic require activation by biochemical processes to have any toxic effects and are properly called **protoxicants**.

2.9.1 EXPOSURE TO TOXIC SUBSTANCES

The modes and routes of exposure to toxic substances (Figure 2.10) are very important in determining toxic effects. Exposure to toxicants may be either acute or chronic or either local or systemic. **Acute exposure** occurs over a short period of time and normally requires a relatively high level of

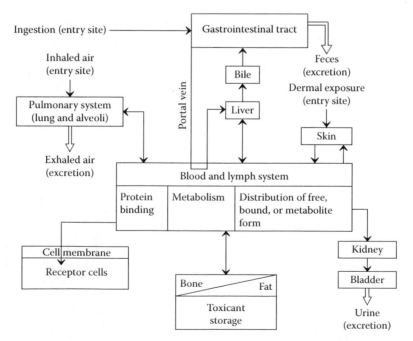

FIGURE 2.10 Major sites of exposure, metabolism, and storage and routes of distribution and elimination of toxic substances in the body.

toxicants. **Chronic exposure** is exposure to relatively lower levels of substances and takes place over long periods of time, for example, exposure of lungs to cigarette smoke through many years of smoking. **Local exposure** takes place at a specific location such as an acid burn from nitric acid spilled on a person's hand. **Systemic exposure** is the term applied to poison that is distributed throughout an organism. An example would be inhalation of carbon monoxide that does not directly harm the lungs where it enters the body, but converts blood hemoglobin to methemoglobin, which is useless for carrying oxygen through the bloodstream and deprives the brain of oxygen that it needs to function.

There are several routes by which organisms are exposed to toxicants, as shown for human exposure in Figure 2.10. Inhalation is the most direct route for human exposure because the thickness of only one cell in lung alveoli (the innermost air cavities in the lungs) separates air inhaled into the lungs from the blood that circulates through the lungs. This pulmonary route of exposure is particularly significant for environmental air pollutants. Humans may absorb toxic substances through the skin (percutaneous or dermal route) and ingest them by mouth (oral route), both of which are of particular concern for exposure of children to environmental pollutants, such as lead from smelter dust that gets into the soil. Exposures can occur, though rarely, by the rectal or vaginal routes. Intentional exposure to toxic substances, such as in testing substances for toxicity or drugs for their effects, is often by intravenous or intramuscular injection.

2.9.2 Distribution of Toxic Substances

The sites of entry of toxicants into the body and their subsequent distribution have a major influence on their toxic effects. It was noted in Section 2.9.1 that the pulmonary route affords direct access of toxicants to the bloodstream and the effects of a direct-acting toxicant may be manifested very rapidly. Absorption through the skin affords similar direct access of toxicants to the blood and lymph systems. (Advantage is taken of this route through the use of skin patches worn to continuously deliver low doses of pharmaceuticals to the bloodstream, for example, a combination of norelgestromin and ethinyl estradiol hormones that function as contraceptives.)

Toxicants that are ingested generally are absorbed through the small intestine walls and are transported to the liver. The liver is the main site of toxicant metabolism and is where some poisonous substances are converted to less toxic forms more readily eliminated from the body whereas other substances are converted to toxic species. Toxic species are distributed around the body by the blood and lymph system, which can lead to systemic poisoning at sites remote from the entry of the substance into the body. Bone and adipose tissue (fat) are major sites of storage of toxicants. Bone accumulates heavy metals including lead and some radioactive materials, especially strontium-90, which biochemically behaves like calcium. Radioactive iodine accumulates in the thyroid and can cause thyroid cancer. Lipophilic toxicants, such as polychlorinated biphenyls (PCBs), that are poorly soluble in water tend to accumulate in adipose tissue.

2.9.3 Dose–Response Relationship

The **dose–response** plot (Figure 2.11) is one of the most important relationships in toxicology. Such a plot can be prepared by dosing a uniform population of test subjects with increasing levels of toxicant and observing response, usually death. For such a curve, the dose corresponding statistically to the death of 50% of the test subjects is denoted as the LD_{50}. Most commonly, toxicities of substances are expressed as LD_{50} values where the test subjects are male rats.

2.9.4 Toxicities

As shown by the examples in Figure 2.12, toxicities of different substances vary appreciably. Note that in the figure, the estimated LD_{50} values are on a log scale and that toxicities of substances vary over several orders of magnitude. A striking illustration of this variability is shown on the right of

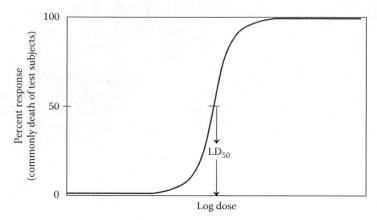

FIGURE 2.11   A dose–response plot of percentage of a uniform population of subjects responding in a speci-fied way (most commonly death) versus log dose. The dose at which statistically half of the subjects die is designated as the LD_{50}.

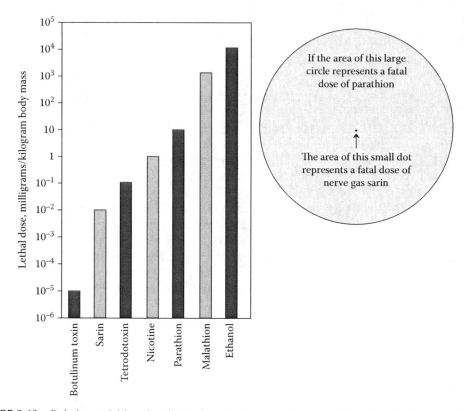

FIGURE 2.12   Relative toxicities of various substances commonly expressed in units of milligrams dose per kilogram of body mass required to kill 50% of test subjects (LD_{50}). Also shown is a comparison of the toxicity of insecticidal parathion to that of another organophosphate compound, nerve gas sarin, in which the toxic dose of parathion is represented by the area of the large circle and that of sarin by the area of the small dot.

Figure 2.12 comparing the toxicities of insecticidal parathion, which is now no longer used because of its toxicity to mammals, and nerve gas sarin, a military poison. In this figure, the area of the large cir-cle represents a lethal dose of parathion and the barely visible area of the smaller circle is proportional to the lethal dose of nerve gas sarin, showing the much higher toxicity of sarin compared to parathion.

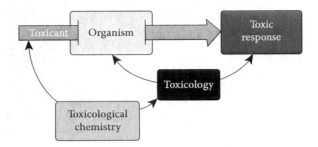

FIGURE 2.13 Toxicology is the science dealing with various aspects of the effects of poisonous substances on organisms. Toxicological chemistry relates the chemical nature of toxicants and protoxicants to their toxic effects on organisms.

2.10 TOXICOLOGICAL CHEMISTRY

As noted in Section 2.1, the relationship between the chemical nature of substances and their toxic effects is addressed by the topic of toxicological chemistry (Figure 2.13). In addition to their chemical properties, toxicological chemistry addresses the sources and uses of toxic substances and the chemical aspects of their exposure, fate, and disposal. The theory of toxicological chemistry is treated by the science of **quantitative structure–activity relationships** (**QSAR**), which relates the chemical nature of substances to their reactions in biological systems.[4]

2.10.1 REACTIONS OF TOXICANTS AND PROTOXICANTS IN LIVING SYSTEMS

An important aspect of toxicological chemistry is that of the reactions that toxicants and protoxicants undergo in a living system before they even have any toxic effects. These are divided into Phase I and Phase II reactions.

Typically, lipid-soluble toxicants and protoxicants are converted by **Phase I reactions** to species that are more polar and water-soluble and more easily eliminated from the body through urine (Figure 2.14) compared to the species from which they are made. This usually occurs through attachment of an −OH group. A Phase I reaction is generally catalyzed by the cytochrome P-450 enzyme system associated with cellular **endoplasmic reticulum**, which occurs most abundantly in the liver of vertebrates. A typical Phase I reaction is the production of phenol from benzene:

$$\text{(benzene)} + \{O\} \longrightarrow \text{(phenol)}-OH \qquad (2.7)$$

Phase II reactions use enzymes to attach an **endogenous conjugating agent** (one that occurs naturally in an organism) to a toxicant, which may be (though not necessarily) a Phase I reaction product. The resulting **conjugation product** is usually less toxic than the xenobiotic reactant, though in some cases it is more toxic. Also, conjugation tends to make a less lipid-soluble product that is more soluble in water and therefore more readily eliminated from an organism. There are several conjugating agents, the most common of which is glucuronide attached by the action of the glucuronyltransferase enzyme (Figure 2.15). Other common conjugating agents include glutathione (attached by glutathionetransferase enzyme), sulfate (sulfotransferase enzyme), and acetyl (acetylation by acetyltransferase enzymes).

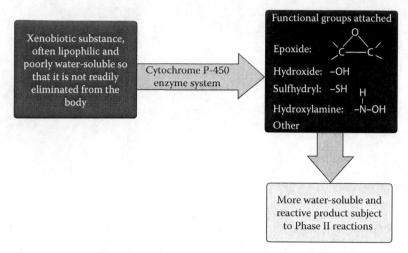

FIGURE 2.14 Illustration of Phase I reactions in which the cytochrome P-450 enzyme system attaches a functional group, typically –OH. The Phase I reaction product is generally more water-soluble and is amenable to Phase II reactions in which a conjugating agent is attached.

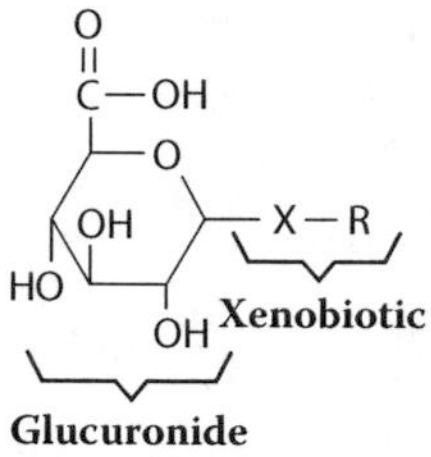

FIGURE 2.15 Glucuronide conjugate formed from a xenobiotic, HX-R. For example, if the xenobiotic compound conjugated is phenol, HXR is HOC_6H_5, X is the O atom, and R represents the phenyl group, C_6H_5. For interpretation of the line formulas in this figure, see Chapter 20.

2.11 KINETIC PHASE OF XENOBIOTIC METABOLISM

In an organism, toxicants and protoxicants are ingested, undergo metabolic processes (Phase I and Phase II reactions), bind with biochemical species such as blood hemoglobin or DNA, and are excreted. Normally, through binding with endogenous biomolecules, toxicants exert some sort of toxic effect. These various processes are conveniently divided between the kinetic phase and the dynamic phase.

Figure 2.16 illustrates the **kinetic phase** during which a toxicant or its metabolite may be absorbed, metabolized, stored temporarily (such as in adipose tissue), distributed (typically through the bloodstream), and excreted. An **active parent compound** may pass through the kinetic phase unchanged and be excreted. It may also be converted by Phase I and Phase II reactions to a detoxified metabolite or changed by enzymatic action to an active metabolite capable of having some sort of toxic effect. One of the more significant kinetic phase processes is conversion of a nontoxic substance to a toxic active metabolite.

2.12 DYNAMIC PHASE OF TOXICANT ACTION

A toxicant or toxic metabolite manifests a toxic effect in the **dynamic phase** (Figure 2.17). There are three major steps in the dynamic phase: (1) a **primary reaction** in which the toxic substance binds with a receptor, generally a biomolecule in a **target organ** or tissue; (2) a **biochemical response** resulting from the binding of the toxicant; and (3) an **observable effect** or symptom of poisoning.

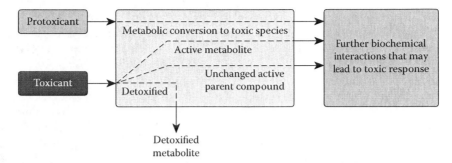

FIGURE 2.16 Illustration of the kinetic phase of metabolism for toxicants and protoxicants. In this phase, a protoxicant may be metabolically converted to a toxic species. A toxicant may be detoxified and excreted without doing harm, remain unchanged as an active parent compound that may have a toxic effect, or converted to another active metabolite that is potentially toxic.

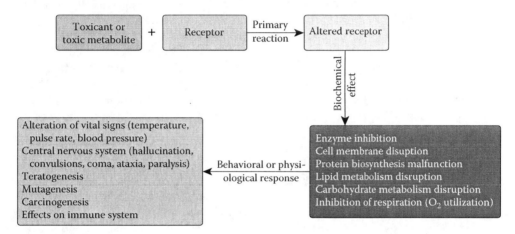

FIGURE 2.17 The dynamic phase of toxicant action.

A typical primary reaction is the binding of carbon monoxide, CO, with blood hemoglobin designated Hb:

$$HbO_2 + CO \leftrightarrow HbCO + O_2 \qquad (2.8)$$

Hemoglobin has a much stronger affinity for CO than it has for O_2, which it normally carries through the bloodstream, so it prevents hemoglobin from transporting O_2 through the bloodstream from the lungs to the tissues. This chemical reaction is written with a double arrow denoting that it is **reversible**, whereas other primary reactions such as an acid burn to skin by concentrated sulfuric acid are irreversible. Treatment of carbon monoxide poisoning with pure or even pressurized oxygen can reverse the binding of carbon monoxide to hemoglobin and lead to recovery.

A biochemical response to carbon monoxide exposure is deprivation of tissues of O_2 needed to carry out metabolic processes. In this case, brain cells are deprived of oxygen, the subject may lose consciousness, and permanent brain damage or death may result. An observable effect of carbon monoxide poisoning may range from lethargy through unconsciousness and even death.

Among the numerous biochemical effects that may result from the binding of a toxicant to a receptor, one of the most common is impairment of enzyme function. This may occur as the result of direct binding of the toxicant to the active site on the enzyme or binding to enzyme substrates or to coenzymes that are required for enzymes to function. Alteration of cell membranes or carriers in cell membranes by toxicants may be harmful. Interference with the metabolism of carbohydrates, proteins, and lipids may occur, in the case of lipids resulting in excess lipid accumulation, a condition commonly called "fatty liver." The production of essential proteins can be impaired by binding of toxicants with the DNA in cells that direct protein biosynthesis. Respiration, the overall process by which electrons are transferred to molecular oxygen in the biochemical oxidation of energy-yielding substrates, may be impaired or even stopped by binding of the enzymes and other species involved by toxicants. Regulatory processes directed by hormones and enzymes that are essential to an organism's proper function may be impaired by binding with toxicants. Interference with nerve function is a common mode of toxicant action. An example is peripheral neuropathy that results when organic solvents destroy the sheaths around peripheral nerves, including those in the hands and feet.

There are many manifestations of poisoning, and these are useful in determining the kind of toxicant. Often, the most immediate of these are alterations in the **vital signs** of **temperature**, **pulse rate**, **respiratory rate**, and **blood pressure**. The central nervous system is commonly affected by toxicants as manifested by **convulsions**, **paralysis**, **hallucinations**, and **ataxia** (lack of coordination of voluntary movements of the body). Behavioral abnormalities may result from central nervous system damage, and the victim may become agitated, hyperactive, or disoriented as well as suffer from delirium. The subject may go into a coma or even die from damage to the central nervous system. Other manifestations of poisoning include abnormal skin color, excessively dry or moist skin, abnormal appearance or behavior of eyes, strange odors, and gastrointestinal tract effects among which are pain, vomiting, or paralytic ileus (stoppage of the normal peristalsis movement of the intestines).

The symptoms of poisoning in the preceding discussion are generally manifested shortly after exposure. Other often more important effects are chronic effects that are the result of exposure to environmental toxicants. These include mutations, birth defects, cancer, effects on the immune system (including suppression and hyperactivity effects), gastrointestinal illness, cardiovascular disease, hepatic (liver) disease, central and peripheral nervous system effects, and skin abnormalities including rash and dermatitis. Long-term effects of environmental exposure to toxicants are often subclinical in nature and may require sophisticated testing for diagnosis. These include some kinds of damage to the immune system, chromosomal abnormalities, modification of functions of liver enzymes, and slowing of conduction of nerve impulses. Table 2.1 lists some of the major target

TABLE 2.1
Examples of Target Systems, Toxicants, and Effects of Toxic Substances

System Affected	Toxicant	Effect
Respiratory system	Cigarette smoke and asbestos	Emphysema and cancer
Skin	Coal tar constituents	Skin cancer
Liver	Vinyl chloride	Hemangiosarcoma (cancer)
Blood	Aniline and nitrobenzene	Methemoglobinemia
Immune system	Allergens such as beryllium	Hypersensitivity
Endocrine system	Bisphenol-A (plasticizer)	Disruption of system function
Nervous system	Organophosphates (sarin)	Acetylcholinesterase inhibition
Kidney	Ethylene glycol	Calcium oxalate deposits in tubules
Bladder	Aromatic amines from coal tar	Bladder cancer

systems of toxic substances, important examples of toxicants that affect them, and the effects resulting from their exposure to specific toxicants.

2.13 MUTAGENESIS AND CARCINOGENESIS

Because of the generally similar mechanisms by which they occur, **mutagenesis**, the process by which toxicants cause inheritable mutations, and **carcinogenesis**, the process by which toxicants cause cancer, are considered together. Although other mechanisms can cause these conditions, both generally result from alterations in cellular DNA.

2.13.1 MUTATIONS FROM CHEMICAL EXPOSURE

Mutagens are chemical species that alter DNA to produce traits that can be inherited. Mutation is a natural process that occurs randomly, but from an environmental perspective, the mutations of concern are those caused by xenobiotic substances in the environment. Because mutagens also often cause cancer and birth defects (see Section 2.14), they are of major toxicological concern.

The huge molecules of DNA contain the four nitrogenous bases adenine, guanine, cytosine, and thymine, the order of which in the DNA macromolecule directs the synthesis of proteins and enzymes through the intermediate formation of a substance known as RNA. Vital life processes may be affected by exchange, addition, or deletion of any of the bases in DNA. A mutation occurs when these changes are passed on to progeny. A xenobiotic compound that causes such changes is called a mutagen.

One way in which DNA may be altered to produce mutations is replacement of the $-NH_2$ group present in adenine, guanine, and cytosine by the $-OH$ group. Nitrous acid, HNO_2, commonly used to induce mutations in experimental studies, acts in this manner. As a result of this alteration, DNA does not function in the intended manner and a mutation may result. A second kind of mutation occurs through **alkylation**, usually the attachment of a methyl group, $-CH_3$, to an N atom on one of the nitrogenous bases on DNA. Figure 2.18 shows some of the xenobiotic compounds that are mutagens due to their ability to act as alkylating agents.

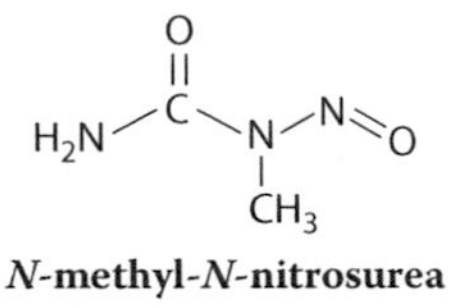

N-methyl-*N*-nitrosurea

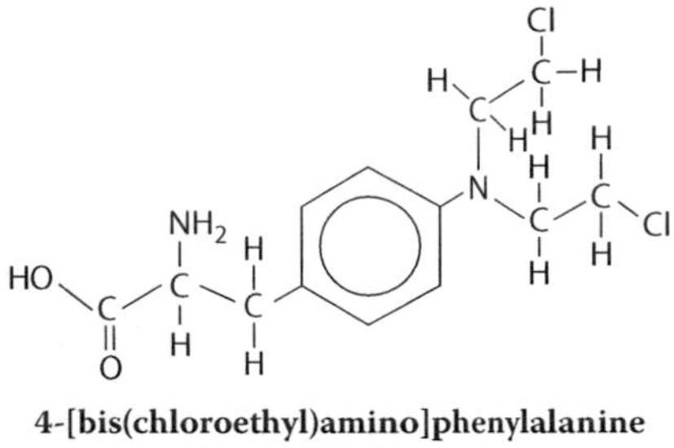

4-[bis(chloroethyl)amino]phenylalanine
Melphalan

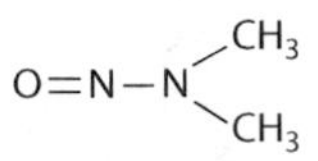

Dimethylnitrosamine

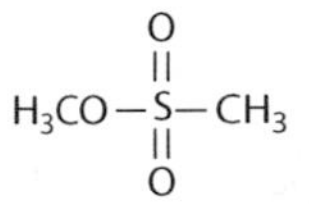

Methyl methanesulfonate

FIGURE 2.18 Examples of simple alkylating agents capable of causing mutations, including *N*-methyl-*N*-nitrosurea, commonly used as a model compound for alkylation studies; Melphalan, a chemotherapeutic agent used to treat cancer; dimethylnitrosamine; and methyl methanesulfonate.

2.13.2 CARCINOGENESIS

Cancer occurs when the body's own (somatic) cells undergo uncontrolled replication and growth. As with mutations, cancer often results from alterations in cellular DNA. Many cancers occur as apparently random events. Study of the human genome has shown genetic characteristics indicative of a propensity to certain kinds of cancers. For experimental purposes, animals have been bred that have inherited tendencies toward certain kinds of cancers. Some biological agents including hepadnaviruses or retroviruses are known to cause cancer.

Insofar as environmental and toxicological chemistry are concerned, physical and chemical carcinogens are of most concern. The major physical factor is ionizing radiation including x-rays and gamma radiation. Before the dangers were realized, people working with radium were poisoned by ingestion of this radioactive element, an emitter of alpha particles, a deadly form of ionizing radiation that caused bone cancer when the element accumulated in bones. Another radioactive element of toxicological concern is radon, a noble gas and alpha particle emitter that can infiltrate dwellings from underground formations and cause lung cancer when inhaled. Some chemical agents including nitrosamines, polycyclic aromatic hydrocarbons, and vinyl chloride are known to cause cancer in humans, and many more are suspected carcinogens largely based on evidence from animal studies.

The study of chemical agents in causing cancer is called **chemical carcinogenesis** and is the branch of environmental toxicology that receives the most attention. The history of chemical carcinogenesis goes back to a classical study published in 1775 by Sir Percival Pott, surgeon general to King George III of England, showing that chimney sweeps in England developed cancer of the scrotum from exposure to carcinogens released in the burning of coal in stoves and fireplaces. Around 1900, Ludwig Rehn, a German surgeon, reported elevated levels of bladder cancer in German dye workers exposed to 2-naphthylamine.

2-Naphthylamine

This and related compounds were extracted from coal tar and used to synthesize dyes. Tobacco juice was reported to be carcinogenic in 1915, the carcinogenic properties of tobacco smoke were reported in 1939, and those of asbestos in 1960. After about 1920, fatal bone cancers developed in young women who ingested radioactive radium from painting watch and instrument dials with radium-containing luminescent paint. In the early 1970s, in what is arguably the most clear-cut evidence of cancer caused by a chemical agent, vinyl chloride, C_2H_3Cl, widely used to manufacture polyvinylchloride plastic polymer, was shown to cause hemangiosarcoma (liver angiosarcoma) in workers exposed to the vapors of this volatile compound.

Probably, most chemicals regarded as carcinogens are actually **procarcinogens** that require biochemical activation to produce **ultimate carcinogens**, the agents that actually initiate cancer. Some agents are **primary** or **direct-acting carcinogens** that do not require bioactivation. Some examples of each of them are shown in Figure 2.19.

Figure 2.20 shows the relatively complex process by which a chemical carcinogen or procarcinogen causes metastatic cancer. There are two major stages, an **initiation stage** followed by a **promotional stage**. **Genotoxic carcinogens**, which are also mutagens, are **DNA-reactive species** that initiate cancer by acting directly on DNA. Most genotoxic carcinogen species are either inherently electrophilic (attracted to electron-rich regions of the DNA molecule) or, more commonly, are biochemically activated to produce reactive electrophilic species that form adducts with DNA, especially the potent alkylating agent $^+CH_3$. These adducts cause gene mutations. **Epigenetic carcinogens** initiate cancer by mechanisms that do not involve reaction with DNA. Most epigenetic carcinogens are actually **promoters** that cause cancer to develop after it is initiated.

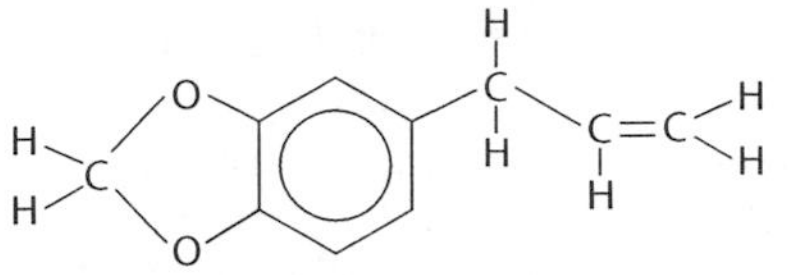

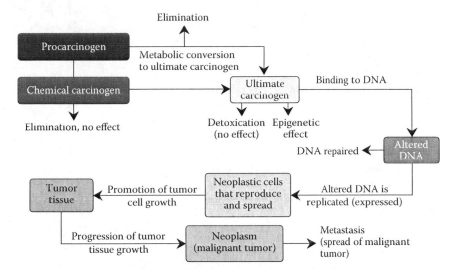

Safrole (from sassafras) a natural product that requires bioactivation

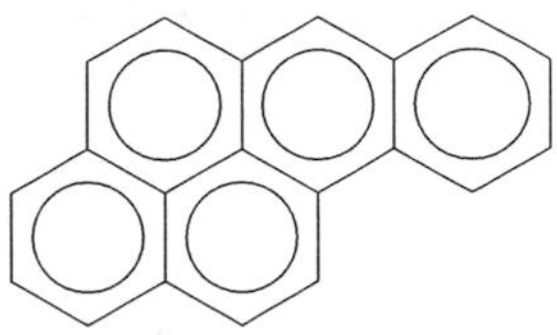

Benzo(a)pyrene, a synthetic chemical that requires bioactivation

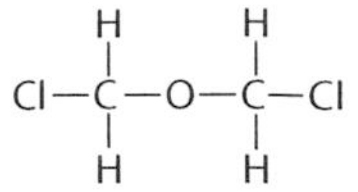

Bis(chloromethyl)ether, a synthetic compound that does not require bioactivation

FIGURE 2.19 Carcinogens may come from natural and synthetic sources. Most require biochemical activation to cause cancer, although some are direct-acting carcinogens. Safrole is a natural product that requires bioactivation; benzo(a)pyrene is a polycyclic aromatic compound made by both natural and synthetic processes, which is converted to a carcinogenic form by metabolic processes; and bis(chloromethyl)ether is a synthetic compound that is direct-acting as a carcinogen.

FIGURE 2.20 Outline of the process by which a carcinogen or procarcinogen may cause cancer. There are several steps in which the process of causing cancer can be stopped. A procarcinogen can be metabolically converted to a carcinogen, but it or a carcinogen may be eliminated without harm. An ultimate carcinogen, which may be formed metabolically from a procarcinogen, may be detoxified without ill effect. The ultimate carcinogen may exert an epigenetic effect that does not involve binding with DNA or it may alter DNA, which can be repaired by repair enzymes. Altered DNA that is not repaired may form neoplastic cancer cells that reproduce and spread. These can lead to tumor cell growth and development of tumor tissue. Progression of tumor tissue growth can occur, leading to a neoplasm (malignant tumor). Ultimately, metastasis, the spread of malignant tumor tissue, may occur.

A crucial aspect of carcinogenesis is the determination of which materials are carcinogenic, especially to humans. It should be emphasized that most cancers are not caused by carcinogenic agents, but rather by natural processes in which there is a strong genetic component for many types of cancer. Relatively few substances (e.g., vinyl chloride mentioned earlier) are known human carcinogens as established by a cause and effect relationship from documented human exposures to relatively high levels of the carcinogen. Studies on animals in which cancer develops from very high exposures over relatively short time periods are used to infer carcinogenicity to humans from much lower exposures over long periods of time, a practice with significant uncertainties.

The most commonly used alternative to animal tests for carcinogenicity is the **Bruce Ames test**, a test for mutagenicity based on the assumption that mutagens are likely to be carcinogenic. The numerous variations of this test make use of a strain of *Salmonella* bacteria that have been selected to require histidine in the media in which they are grown rather than making this amino acid from other amino acids as the wild strain may do. Inoculated onto histidine-deficient media in the presence of a potential mutagen, often along with a homogenate of raw liver that contains enzymes capable of converting substances to carcinogenic forms, some of the bacteria revert to the natural strain and form colonies, the numbers of which are proportional to the mutagenicity of the test compound. Although it is useful, the Bruce Ames test has been shown to give both false-positive and false-negative results for carcinogenicity.

2.14   DEVELOPMENTAL EFFECTS AND TERATOGENESIS

The organisms most vulnerable to toxicants are the very young ranging from the fertilized eggs to adulthood. In general, the earlier the stage of development, the more vulnerable the organism. Damage to embryonic and fetal cells and to egg or sperm germ cells may result in birth defects. Chemical substances that cause such defects are called **teratogens**.[5]

A variety of biochemical effects caused by xenobiotics may be responsible for teratogenesis. These include enzyme inhibition, interference with energy supply, deprivation of the fetus of vitamins and other essential substrates, and altered permeability of the placental membrane.

Fetuses exposed to toxic substances in utero are especially vulnerable to the effects of toxicants resulting in retarded fetal growth, birth defects, and maladies such as diabetes and coronary artery disease after birth. Exposure occurs when a toxic substance passes through the placental barrier and enters the fetal bloodstream. Fetuses are vulnerable to toxicants because they have relatively ineffective detoxification enzyme systems and their developing organs are more subject to damage than are mature organs.

Children are generally more vulnerable to toxic xenobiotics than are adults. Arguably, the toxic substance that has received more attention than any other in this respect is lead. Children may be exposed to lead in drinking water (which is why lead is now banned in solder used to connect water pipe), from inhalation of polluted air (more common before the 1970s after which lead was banned as an octane booster in gasoline), from ingestion of lead-contaminated soil (once common around lead smelters), and from ingestion of old paint, which once contained lead-based pigments. Thus, there exist vectors from the other environmental spheres by which children may be exposed to lead. In addition to detrimental effects on blood and kidneys, lead is suspected of causing mental retardation in children.

2.15   TOXIC EFFECTS ON THE IMMUNE SYSTEM

The **immune system** of the body serves a number of valuable functions, mostly related to defense against foreign agents. Included among immune system functions are the following: (1) defense against viral, bacterial, and protozoal infectious agents; (2) destruction and neutralization of neoplastic (cancerous) cells; and (3) resistance to xenobiotic toxicants.

Adverse effects on the immune system are among the more important aspects of toxicological chemistry.[6] Such effects may be divided into two major categories. One of these is **immunosuppression**, which impairs the immune system's ability to resist the effects of toxicants, to fight disease-causing agents, and to impede the development of cancerous cells, such as those responsible for leukemia or lymphoma.

As those who suffer from allergies well know, overstimulation of the immune system can cause significant ill effects. Known as **allergy** or **hypersensitivity**, the self-destructive overstimulation of the immune system can result from xenobiotic substance exposure. Some metals including

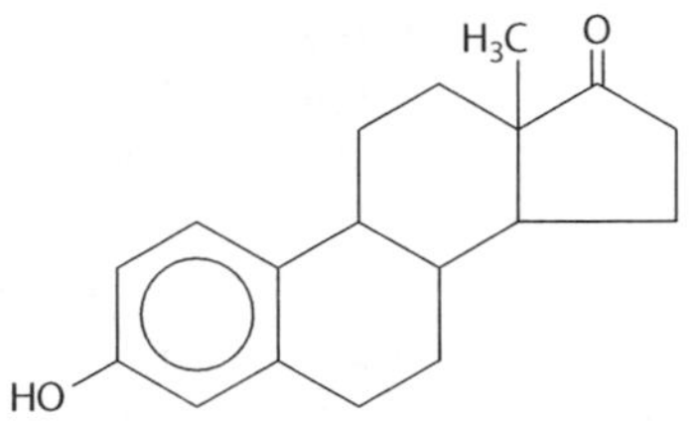

Estrone (a natural estrogen) 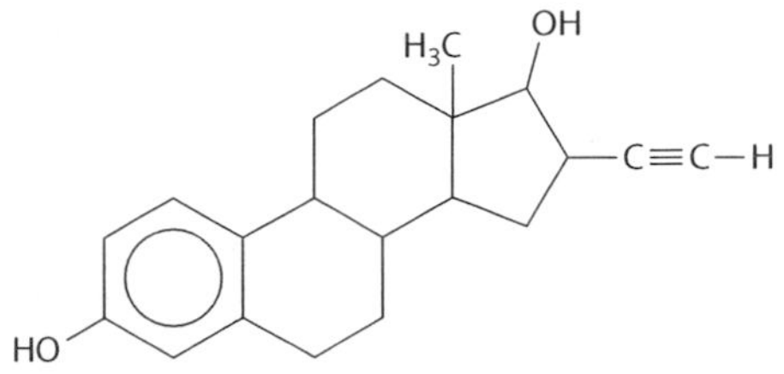   **17α-Ethynylestradiol (oral contraceptives)**

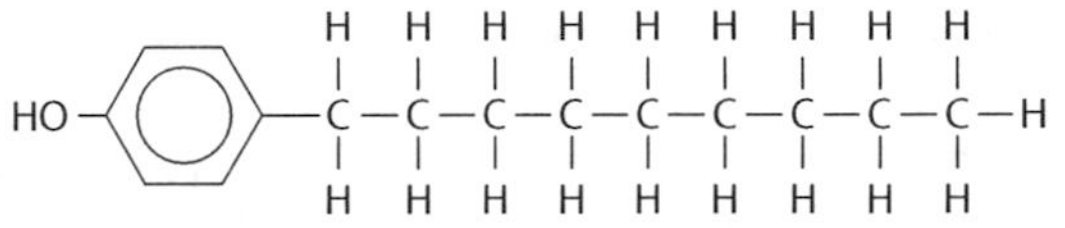

p-Nonylphenol (from surfactants) 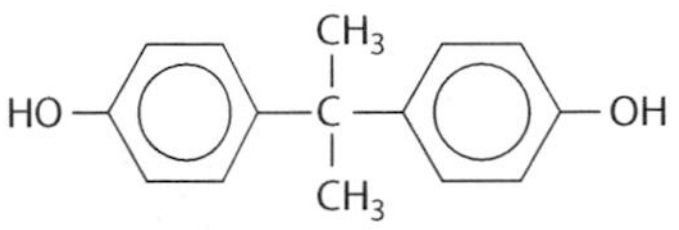   **Bisphenol A (epoxy and polycarbonate resins)**

FIGURE 2.21 Examples of estrogenic agents found in water. Estrone is a natural estrogen, 17α-ethynylestradiol is an ingredient of oral contraceptives, p-nonylphenol comes from surfactants used in cleaning agents, and bisphenol A is an ingredient in some epoxy and polycarbonate resins.

beryllium, chromium, and nickel can cause hypersensitivity. Organic xenobiotics that cause hypersensitivity include some pesticides, plasticizers, and resins.

2.16   DAMAGE TO THE ENDOCRINE SYSTEM

The endocrine system regulates an organism's metabolism and reproduction. Some toxic substances and pollutants disrupt endocrine system function, commonly manifested by abnormal behavior of the reproductive system including general dysfunction, alterations in secondary sexual characteristics, and abnormal blood serum steroid levels.[7] Of particular significance are **hormonally active agents** that exhibit hormonelike behavior. The most significant of these are **estrogens** that act in a manner similar to the female sex hormone estrogen, examples of which are shown in Figure 2.21. Among these agents are estrogen, itself, 17α-ethynylestradiol that is an ingredient of oral contraceptives, and various industrial and consumer product chemicals, of which bisphenol A, a plasticizer widely used to improve the properties of plastics, is one of the most prominent examples. Estrogenic agents from nonnatural sources are called **xenoestrogens**.

Estrogenic agents are present in wastewaters in relatively low concentrations from sources such as metabolites excreted in urine and from cleaning agents. As a result, water-dwelling creatures including fish, frogs, and alligators get exposed to estrogenic agents. The effects of these substances on fish have been studied extensively leading to observations of reproductive dysfunction, alterations in secondary sex characteristics, and abnormal serum steroid levels.[8]

2.17   HEALTH HAZARDS OF TOXIC SUBSTANCES

Much of what is known about toxicology is based on relatively acute maladies resulting from brief, intense exposures to toxicants from readily recognized sources and that have developed over short periods of time. Of greater importance, especially with respect to exposure to environmental toxicants, are chronic and often relatively less severe maladies that become manifested long after exposure. These are difficult to assess because they result from long-term exposure to often uncertain sources of toxicants with low occurrence above background levels of disease and with generally long latency periods. But their overall impact on the affected populations is generally greater than that of the more acute cases of disease caused by toxicant exposure.

The key step in assessing the effects of toxic substances is to establish a relationship between toxicant-caused disease and exposure to the toxic agent. In some cases, this is done by determining levels of the toxicant or its metabolites in organisms. The kinds of toxicants for which this is possible include heavy metals, radioactive elements, mineral asbestos, and silica (SiO_2 dust) responsible for lung silicosis. An example of a toxicant metabolite that can be measured is *trans-trans* muconic acid indicative of exposure to benzene. Although relatively unspecific with regard to cause and effect, symptoms such as skin rashes or subclinical effects such as chromosomal damage can be used to narrow the possibilities of toxicant exposure. Another indicator of exposure is production of substances that do not contain the toxicant or its metabolites, but are relatively specific for particular kinds of toxicants. For example, the presence of methemoglobin, a derivative of hemoglobin in which the iron is in the +3 rather than the +2 oxidation state, is typical of poisoning from nitrite, aniline, or nitrobenzene.

In some cases, **epidemiological studies** of particular kinds of maladies are indicative of probable exposures of populations to environmental pollutants. In cases where a particular agent is suspected, evidence can be sought of elevated levels of diseases known to be caused by the agent. Another approach that can be employed when a particular agent is not suspected is to look for abnormally large occurrences of particular maladies—spontaneous abortions, birth defects, and particular types of cancer—in a limited geographic area. After establishing these **clusters** of disease, the source of an environmental toxicant may be sought. Although often cited as evidence of environmental exposure, epidemiological studies have some notable shortcomings. One of the main ones of these is the uncertainty of the correlation between an ill effect and a specific toxicant; for example, skin rash may be caused by a large variety of toxicants. Often, the long latency periods between exposure and the appearance of symptoms mean that the source is gone before symptoms are even noticed. Another complication is the occurrence of particular maladies even in the absence of toxicants that may cause them.

2.17.1 HEALTH RISK ASSESSMENT

An important aspect of toxicological chemistry is the estimation of health risks due to exposure to toxicants. Normally, what is needed is an estimate of the occurrence of a toxic effect in a small percentage of people after long-term exposure to low levels of a toxicant after a long latency period. However, the risks are usually estimated based on exposure to animals, usually rats, to relatively high doses of toxicant for relatively short periods of time, necessitated by the short lifetimes of test animals. Based on the results of these studies, risks to human populations are made using linear or curvilinear projections. Obviously, there is a high degree of uncertainty in this approach.

2.18 STRUCTURE–ACTIVITY RELATIONSHIPS IN TOXICOLOGICAL CHEMISTRY

An important aspect of toxicological chemistry that relates chemical structure to toxic effects is the **QSAR**.[9] Increasingly, sophisticated computerized calculations of QSAR are leading to predictions of the effects of toxic substances and can play an important role in determining the avoidance of exposure to toxic substances in the practice of green chemistry. The following are chemical features of substances that may be indicative of toxic effects and that may be taken into consideration in doing QSAR calculations:

- **Corrosive substances** that exhibit extremes of acidity, basicity, dehydrating ability, or oxidizing power such that they will tend to damage tissue.
- **Reactive substances** that contain functional groups likely to undergo adverse biochemical processes. An interesting comparison is that between two 3-carbon alcohols, propyl alcohol ($CH_3CH_2CH_2OH$) and allyl alcohol ($H_2C=CHCH_2OH$). Propyl alcohol is only about 1/100 as toxic as allyl alcohol. The toxicity of allyl alcohol is due to the reactive double bond between carbon atoms.
- **Heavy metals** such as lead or mercury may be quite toxic because of their ability to bind with proteins and enzymes, especially through the −SH group.

- **Binding species** bond to biomolecules causing adverse effects. A simple example is the strong attraction of carbon monoxide to hemoglobin so that the hemoglobin in blood is prevented from transferring oxygen to tissues. Reversible binding of methyl groups to DNA can result in mutations and cancer.
- **Lipid-soluble compounds**, such as PCBs, may readily traverse cell membranes and undergo bioaccumulation in the tissue.
- **Immunotoxicants** are toxic because of their chemical structures, which cause adverse reactions in an organism's immune system.

2.19   TOXICOLOGICAL CHEMISTRY AND ECOTOXICOLOGY

As discussed further in Chapter 12, in nature, organisms interact with each other and their surroundings in a generally steady-state and sustainable manner. These interactions are described by the science of **ecology**, and the organisms and their environment constitute **ecosystems**. Toxic substances tend to perturb ecosystems so that it is important to consider the combination of ecology and toxicology, the discipline of **ecotoxicology**. Ecotoxicology may be considered at various levels ranging all the way from biochemical processes at the molecular level to ecosystems as a whole.

An important aspect of ecotoxicology involves the pathways of toxicants into ecosystems and is very much tied with environmental chemistry. Toxic substances are transferred among the five environmental spheres as shown in Figure 2.22. The greatest concern is with the transfer of toxic substances into the biosphere, which can occur from water, air, the geosphere, and the anthrosphere. An important phenomenon is bioaccumulation in which toxic substances accumulate in tissue. The

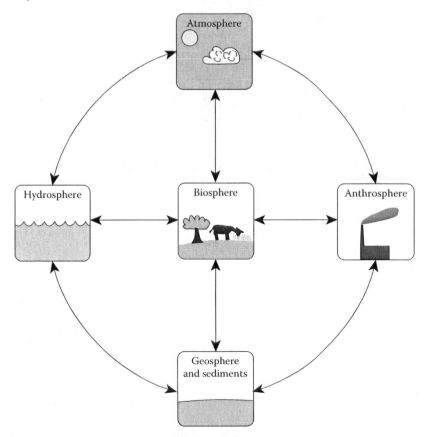

FIGURE 2.22   Transfers of substances among the various environmental spheres, especially those involving the biosphere, are very important in determining their ecotoxicological effects.

relatively high solubility of some organic toxicants in lipid (fat) fish tissue can lead to bioconcentration of such substances in organisms and biomagnification as substances are transferred up the food chain. These concepts are especially applicable to xenobiotic substances transferring to organisms from water as discussed in Chapter 3, Section 3.13.

2.19.1 EFFECTS OF TOXICANTS ON ECOSYSTEMS

Exposure to toxic substances may adversely affect individual organisms altering their populations and ecosystems, leading to disruption of communities of organisms and effects on whole ecosystems. This occurred, for example, where the exposure of birds of prey to dichlorodiphenyltrichloroethane (DDT) resulted in eggs with very weak shells and severely diminished populations of hawks and eagles in ecosystems. With reduced populations of predators, rodent populations increased, adversely affecting some terrestrial ecosystems.

The Arctic ecosystems impacted by toxic acidic precipitation have been studied extensively. From about 1940 to 1970, the normal pH values of 6.5–7.5 in some Scandinavian lakes fell to within a range of 5.5–6.0, primarily due to sulfuric acid produced from sulfur dioxide released to the atmosphere during coal combustion. The result was greatly reduced populations of valuable salmon and trout. Lakes in which the water had pH values below 5.0 became essentially devoid of fish populations. There were also adverse effects on phytoplankton and zooplankton populations with a loss of population diversity such that only a few species of algae predominated.

2.19.2 BIOMARKERS OF EXPOSURE TO TOXIC SUBSTANCES

It is important to have means to assess whether or not organisms have been exposed to toxic materials. This can often be done through **biomarkers of exposure** exhibited by organisms.[10] A biomarker of exposure may consist of the presence of a toxic substance itself in an organism, a metabolic product of the toxic substance, or an effect that can be directly attributed to a toxicant. Heavy metal poisoning is commonly assessed by analyzing for the toxic metal in tissue. The occurrence of *para*-aminophenol in blood or urine may be indicative of aniline poisoning. Changes in sexual characteristics of aquatic organisms, such as feminization of juvenile males, may indicate exposure to endocrine disruptors.

2.20 TOXIC AGENTS THAT MAY BE USED IN TERRORIST ATTACKS

Many substances pose hazards with respect to the potential for attacks on human beings including terrorist attacks. These include the following toxic agents:

- Cyanide such as gaseous HCN or salts including KCN. In 1982, 12 people in the Chicago area died from cyanide that was placed in Tylenol, a case that remains unsolved.
- Nerve gas agents that interfere with the action of acetylcholinesterase enzyme essential for nerve function. These are primarily organophosphate compounds, including sarin, VX, and Russian VX. A total of 13 people died and about 6000 sought medical help from a 1995 attack with sarin on the Tokyo subway system.
- Toxic substances including sulfur mustard, an extreme irritant and blister agent used as a military poison; ricin, a protein poison isolated from the castor bean; rodenticidal tetramethylenedisulfotetramine, now banned from commerce; and rodenticide brodifacoum.

An important aspect of dealing with chemical threats is the ability to trace their sources, which is addressed by the discipline of **chemical forensics**.[11] With increasingly sophisticated instrumentation, chemical forensics uses information such as the presence of impurities, ratios of carbon and oxygen isotopes, and ratios of stereoisomers (see the discussion of chirality in Chapter 14, Section 14.11, to trace sources of toxicants such as those that may be used in terrorist attacks).

QUESTIONS AND PROBLEMS

1. Glucose and fructose are both simple sugars with a formula of $C_6H_{12}O_6$, yet they are distinct compounds. What distinguishes them? What is a particularly important place in the human body in which glucose is found and why are the levels of glucose important? Do humans consume fructose and, if so, what may be the source of this carbohydrate?

2. Although starch in sources such as flour and potatoes is an important nutrient, the starch molecule is too large and insoluble to enter the bloodstream? Considering its chemical formula and the monomers of which starch is composed, suggest what happens to starch before it can be utilized as a nutrient.

3. What is a peptide linkage? In what kind of biomolecule are peptide linkages found and what function do they serve in such a molecule?

4. Although the triglyceride of stearic acid and the cholesterol steroid shown in Figure 2.5 are not very similar structurally, both are classified as lipids. Explain.

5. DNA is described in the text as consisting of "astoundingly long" chains. See if you can find any "gee whiz" information on the Internet that describes how "astoundingly long" molecules of DNA may be.

6. What is recombinant DNA technology? Is it relatively new? For what purposes is it being used?

7. Suggest how an isomerase enzyme may be involved in making fructose available for human nutrition.

8. What molecule in living organisms is responsible for intermediate energy storage? What substance is it converted to when it releases energy?

9. In a sense, photosynthesis is the reverse of oxic respiration. Explain.

10. Exposure to vinyl chloride is known to cause a specific kind of liver cancer in humans. After doing some Internet research on the subject, explain the statement that vinyl chloride is a *xenobiotic* compound and a *protoxicant*.

11. Classify each of the following toxic effects regarding whether or not exposure to the toxic agent is acute or chronic and whether they are local or systemic:
    a. Death from inhaling carbon monoxide from an automobile left running in a garage.
    b. Cancer of the mouth from long-term use of chewing tobacco.
    c. Peripheral neuropathy from working in an atmosphere contaminated with hydrocarbon solvents over many decades.
    d. A lesion caused by spilling concentrated nitric acid on the skin.

12. Suggest why the bladder is the organ that is likely to develop cancer from exposure to aromatic amines.

13. Look up the structure of lung alveoli and suggest why they are particularly significant sites of exposure to toxicants.

14. Suggest explanations for (a) accumulation of PCBs in fat tissue, (b) occurrence of bone cancer from exposure to radioactive radium, and (c) which organ is most likely to develop cancer from exposure to radioactive iodine.

15. Though highly toxic, botulinum toxin has some important pharmaceutical and cosmetic applications. After a search on the Internet, explain what some of these are.

16. What are endogenous conjugating agents? What is their function with respect to toxic substances?

17. What was the "thalidomide tragedy"? To what class of toxic substances does thalidomide belong?

18. After an Internet search regarding the nature of radium and radon, suggest why areas in which there is radon infiltration into dwellings may also have problems with radium in water supplies.

19. The most common effect of carbon monoxide poisoning is damage to the brain, which may be fatal. Does carbon monoxide attack brain tissue directly? How does carbon monoxide result in brain damage?
20. What is the rationale of using a test for mutagenicity to indicate carcinogenicity? What is this test called?
21. What are xenoestrogens? Look up on the Internet the status of research involving at least one common plasticizer suspected of being a xenoestrogen.
22. What is the evidence from studies of organisms that live in water, including fish, frogs, and alligators, that wastewaters may be contaminated by xenoestrogens?
23. What is biomagnification? How is lipid tissue involved in biomagnification processes?

LITERATURE CITED

1. Voet, Donald, Judith G. Voet, and Charlotte W. Pratt, *Fundamentals of Biochemistry: Life at the Molecular Level*, 4th ed., Wiley, Hoboken, NJ, 2012.
2. Manahan, Stanley E., *Toxicological Chemistry and Biochemistry*, 3rd ed., Taylor & Francis/CRC Press, Boca Raton, FL, 2002.
3. Rogers, Kara, Ed., *The Cell*, Rosen Educational Services, New York, 2011.
4. Walker, John D., Ed., *Quantitative Structure-Activity Relationships for Pollution Prevention, Toxicity Screening, Risk Assessment, and Web Applications*, SETAC Press, Pensacola, FL, 2003.
5. Holmes, Lewis B., Human Teratogens: Update 2010, *Birth Defects Research, Part A: Clinical and Molecular Teratology* **91**, 1–7 (2011).
6. Winans, Bethany, Michael C. Humble, and Lawrence B. Paige, Environmental Toxicants and the Developing Immune System: A Missing Link in the Global Battle against Infectious Disease?, *Reproductive Toxicology* **31**, 327–336 (2011).
7. Nakamura, Kazuo, and Hiroko Kariyazono, Influence of Endocrine-Disrupting Chemicals on the Immune System, *Journal of Health Science* **56**, 361–373 (2010).
8. Norris, David O., and James A. Carr, *Endocrine Disruption: Biological Basis for Health Effects in Wildlife and Humans*, Oxford University Press, New York, 2006.
9. Peyret, T., and K. Krishnan, QSARs for PBPK Modelling of Environmental Contaminants, *Environmental Research* **22**, 129–169 (2011).
10. Paustenbach, Dennis, and David Galbraith, Biomonitoring and Biomarkers: Exposure Assessment Will Never be the Same, *Environmental Health Perspectives* **114**, 1143–1149 (2006).
11. Halford, Bethany, Tracing a Threat, *Chemical and Engineering News* **90**, 10–15 (February 12, 2012).

SUPPLEMENTARY REFERENCES

Boelsterli, Urs A., *Mechanistic Toxicology: The Molecular Basis of How Chemicals Disrupt Biological Targets*, 2nd ed., Taylor & Francis/CRC Press, Boca Raton, FL, 2007.

Elliott, William H., and Daphne C. Elliott, *Biochemistry and Molecular Biology*, 4th ed., Oxford University Press, New York, 2009.

Garrett, Reginald H., and Charles M. Grisham, *Biochemistry*, 4th ed., Brooks/Cole, Belmont, CA, 2009.

Hodgson, Ernest, *A Textbook of Modern Toxicology*, 4th ed., Wiley, Hoboken, NJ, 2010.

Klaassen, Curtis, *Casarett & Doull's Toxicology: The Basic Science of Poisons*, 7th ed., McGraw-Hill Professional, New York, 2007.

Landis, Wayne G., Ruth M. Sofield, and Ming-Ho Yu, *Introduction to Environmental Toxicology: Molecular Substructures to Ecological Landscapes*, 4th ed., Taylor & Francis/CRC Press, Boca Raton, FL, 2010.

Lu, Frank C., and Sam Kacew, *Lu's Basic Toxicology: Fundamentals, Target Organs, and Risk Assessment*, 5th ed., Informa Healthcare, London, 2009.

Markandey, D. K., and N. Rajvaida, *Environmental Biochemistry*, APH Publishing, New Delhi, 2005.

Murray, Lindsay, *Toxicology Handbook*, 2nd ed., Churchill Livingstone, Philadelphia, 2011

Newman, Michael C., *Fundamentals of Ecotoxicology*, 3rd ed., Taylor & Francis/CRC Press, Boca Raton, FL, 2009.

Pratt, Charlotte W., and Kathleen Cornely, *Essential Biochemistry*, Wiley, Hoboken, NJ, 2011.

Raven, Peter H., Linda R. Berg, and David M. Hassenzahl, *Environment*, 6th ed., Wiley, Hoboken, NJ, 2008.

Richards, Ira S., *Principles and Practice of Toxicology in Public Health*, Jones & Bartlett, Sudbury, MA, 2007.

Rose, Vernon E., and Barbara Cohrssen, *Patty's Industrial Hygiene* (four volume set), Wiley, Hoboken, NJ, 2010.

Smart, Robert C., and Ernest Hodgson, Eds., *Molecular and Biochemical Toxicology*, 4th ed., Wiley, Hoboken, NJ, 2008.

Timbrell, John A., *Introduction to Toxicology*, 3rd ed., Taylor & Francis, London, 2001.

Timbrell, John A., *Principles of Biochemical Toxicology*, 4th ed., Taylor & Francis/CRC Press, Boca Raton, FL, 2008.

Walker, C. H., Steve P. Hopkin, R. M. Sibly, and D. B. Peakall, *Principles of Ecotoxicology*, 3rd ed., Taylor & Francis/CRC Press, Boca Raton, FL, 2005.

Watkins, John B., and Curtis Klaassen, *Casarett and Doulls Essentials of Toxicology*, 2nd ed., McGraw Hill Professional, New York, 2010.

Wiley-VCH, *Ullmann's Industrial Toxicology*, Wiley, Hoboken, NJ, 2005.

3 Environmental and Toxicological Chemistry of the Hydrosphere

3.1 H₂O: SIMPLE FORMULA, REMARKABLE MOLECULE

The chemical formula of water, H_2O, is probably the best known of all compounds. This simple formula represents a substance that is unique and complex in its behavior.[1] These special properties are due to the molecular structure of the H_2O molecule, which is represented in Figure 3.1. There are four pairs of electrons in the outer electron shell of the O atom in the H_2O molecule, two of which compose the bonds between the H and O atoms and two of which are lone pairs. These pairs are located as far apart as possible around an imaginary sphere representing the outer electron shell of the O atoms, which results in the two H–O bonds being located at an angle rather than in a straight line. The side of the molecule with the two H atoms has a partial positive charge and the side with the two nonbonding pairs has a partial negative charge, so the molecule is **polar**. This polarity and the ability of the H atoms on one molecule to form hydrogen bonds in which a hydrogen atom acts as a bridge between two O atoms on separate H_2O molecules determine the remarkable chemical and physical diversity of water.

Especially because of their hydrogen bonding capability, water molecules are strongly attracted to each other. This means that a large amount of heat energy must be put into a mass of water to enable the molecules to move more rapidly with the rise in temperature. This gives water a very high **heat capacity**. A very large amount of energy must be put into a mass of ice to break the hydrogen bonds holding the molecules in place in the solid as it melts, and an equally large amount of heat energy is released when liquid water freezes. Thus, water has a very high **latent heat of fusion**. Even more energy per unit mass is required to convert liquid water to vapor (steam), and an equal amount of energy is released when water vapor condenses to liquid. This means that water has a very high **heat of vaporization**.

The ability of water to absorb, release, and store heat is crucial to its role in the environment and its practical uses. Water's high heat capacity stabilizes temperatures of organisms and geographical areas. Steam produced in a boiler can be transferred through insulated pipes to remote locations and condensed to release heat. The heat released when atmospheric water condenses warms the surrounding air and is the driving force behind tropical storms. Europe owes its relatively mild weather despite its northern latitudes to the heat carried by water across the North Atlantic Ocean from the Gulf of Mexico. As the water releases the heat it carries and cools along the European coasts, its density increases and it flows at lower ocean depths back to the Gulf of Mexico to repeat the cycle. Water's high latent heat of fusion stabilizes temperatures of bodies of water at water's freezing point (0°C).

Water's unique properties make it essential to life and determine its behavior in the hydrosphere and in interactions with all other environmental spheres, including its many uses in the anthrosphere. As noted earlier in this section, these properties are due primarily to water's molecular structure, including its polar character and ability to form hydrogen bonds. The more important characteristics of water pertinent to its environmental behavior, uses, and interactions with all the environmental spheres are summarized in Table 3.1.

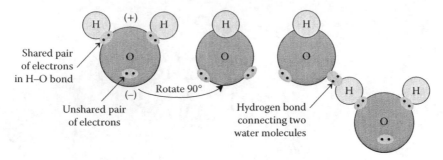

FIGURE 3.1 Because of the arrangement of the two bonding pairs and the two nonbonding pairs of electrons as far as possible from each other around the sphere of the oxygen atom in the water molecule, the molecule is polar. The nonbonding pairs of electrons can form hydrogen bonds with hydrogen atoms in other water molecules. This hydrogen bonding capability and the polar nature of the water molecule are responsible for the unique solvent behavior, heat/temperature behavior, and other properties of water.

TABLE 3.1
Important Properties of Water

Property	Effects and Significance
Excellent solvent	Transport of nutrients and waste products, making biological processes possible in an aqueous medium
Highest dielectric constant of any common liquid	High solubility of ionic substances and their ionization in solution
Higher surface tension than any other liquid	Controlling factor in physiology; governs drop and surface phenomena
Transparent to visible and longer-wavelength fraction of ultraviolet light	Colorless, allowing light required for photosynthesis to reach considerable depths in bodies of water
Maximum density as a liquid at 4°C	Ice floats; vertical circulation restricted in stratified bodies of water
Higher heat of evaporation than any other material	Determines transfer of heat and water molecules between the atmosphere and bodies of water
Higher latent heat of fusion than any other liquid except ammonia	Temperature stabilized at the freezing point of water
Higher heat capacity than any other liquid except ammonia	Stabilization of temperatures of organisms and geographical regions

3.2 HYDROSPHERE

The hydrosphere is composed of water, chemical formula H_2O. Water participates in one of the great natural cycles of matter, the **hydrologic cycle**, which is illustrated in Figure 3.2.[2] Basically, the hydrologic cycle is powered by solar energy that evaporates water as atmospheric water vapor from the oceans and bodies of freshwater from where it may be carried by wind currents through the atmosphere to fall as rain, snow, or some other forms of precipitation in areas far from the source. In addition to carrying water, the hydrologic cycle conveys the energy absorbed as latent heat when water is evaporated by solar energy and the energy released as heat when the water condenses to form precipitation.

There is a strong connection between the hydrosphere, where water is found, and the **geosphere**, or land: human activities affect both. For example, the disturbance of land by the conversion of grasslands or forests to agricultural land or the intensification of agricultural production may reduce vegetation cover, decreasing **transpiration** (loss of water vapor by plants) and affecting the microclimate. The result is increased rain runoff, erosion, and accumulation of silt in bodies of water.

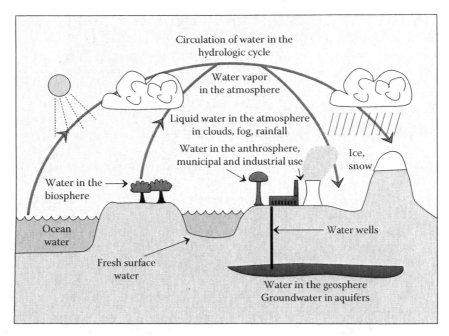

FIGURE 3.2 The hydrosphere overlaps strongly with all the other environmental spheres. This illustration shows water in bodies of water: underground as groundwater, in a snowpack, in plants in the biosphere, as droplets and vapor in the atmosphere, and in water distribution systems and cooling towers in the anthrosphere. Water's cycle in the environment is the hydrologic cycle shown here, in which solar energy evaporates water from the oceans, surface bodies of water, soil, and plants (transpiration); and the water in the atmosphere is carried for a distance (sometimes thousands of kilometers), falls back to Earth as precipitation, infiltrates in part into groundwater, moves on Earth's surface as rivers, flows back in part to the oceans, and then evaporates to renew the cycle.

The nutrient cycles may be accelerated, leading to nutrient enrichment of surface waters. This, in turn, can profoundly affect the chemical and biological characteristics of bodies of water.

Earth could have lost most of its water by now except for one very fortunate atmospheric feature: the very cold tropopause boundary at the upper part of the atmospheric troposphere. At a temperature well below the freezing point of water, this region converts water vapor to ice that remains in the troposphere and participates in the hydrologic cycle. If this were not the case, the water vapor would infiltrate the next higher atmospheric layer, the stratosphere, where highly energetic solar ultraviolet radiation would split H atoms off the H_2O molecules. These very light atoms and the H_2 molecules formed from them would have diffused into space, leaving Earth with an arid Martian landscape. Instead, there is probably a net influx of water into Earth's atmosphere from meteorites, which are largely composed of water.

Although water is not destroyed on Earth, the hydrosphere can certainly suffer damage by human activities as discussed in Chapters 4 and 5. One of the main human activities detrimental to the hydrosphere is excessive utilization of water in regions deficient in rainfall. Withdrawal of irrigation water from rivers in arid regions has reduced some once-mighty rivers to trickles by the time they reach the ocean. The water is not destroyed, but it evaporates and in some cases infiltrates the soil.

3.3　OCCURRENCE OF WATER

Water occurs throughout all the spheres of the environment. To name just a few of these locations, water is present in the atmosphere as water vapor, in the geosphere as soil moisture, in the biosphere as water in organisms, and in the anthrosphere in storage tanks and water distribution systems.

There are several major parts of the hydrosphere where water is accessible and potentially available for use. Where water is found in the hydrosphere has a lot to do with its availability, its chemistry, biological processes in water, and its pollution by harmful substances. The most important of these compartments of the hydrosphere are discussed in Sections 3.3.1 through 3.3.4.

3.3.1 STANDING BODIES OF WATER

The physical condition of a body of water strongly influences the chemical and biological processes that occur in the water. **Surface water** occurs primarily in streams, lakes, and reservoirs. Lakes may be classified as **oligotrophic**, **eutrophic**, or **dystrophic**, an order that often parallels the life cycle of lakes. Oligotrophic lakes are deep, generally clear, deficient in nutrients, and do not have much biological activity. Eutrophic lakes have more nutrients, support more life, and are more turbid. Dystrophic lakes are shallow and clogged with plant life and normally contain colored water of low pH. **Wetlands** are flooded areas in which the water is shallow enough to enable the growth of bottom-rooted plants.

Some constructed reservoirs are very similar to lakes, whereas others differ a great deal from lakes. Reservoirs with a large volume relative to their inflow and outflow are called **storage reservoirs**. Reservoirs with a large rate of flow-through compared to their volume are called **run-of-the-river reservoirs**. The physical, chemical, and biological properties of water in the two types of reservoirs may vary appreciably. Water in storage reservoirs more closely resembles lake water, whereas water in run-of-the-river reservoirs is much like river water. Impounding water may have profound effects on its quality.

Estuaries constitute another type of body of water, consisting of arms of the ocean into which streams flow. The mixing of fresh- and saltwater gives estuaries unique chemical and biological properties. Because they are the breeding grounds of much marine life, the preservation of estuaries is very important for the health of the biosphere.

Water's unique temperature-density relationship results in the formation of distinct layers within nonflowing bodies of water, as shown in Figure 3.3. During summer, a surface layer (**epilimnion**) is heated by solar radiation and, because of its lower density, floats upon the bottom layer, or **hypolimnion**. This phenomenon is called **thermal stratification**. When an appreciable temperature difference exists between the two layers, they do not mix but behave independently and have very different chemical and biological properties. The epilimnion, which is exposed to light, may have a heavy growth of algae. As a result of its exposure to the atmosphere and (during daylight hours) because of the photosynthetic activity of algae, the epilimnion contains relatively higher levels of dissolved oxygen, and it is said to be **oxic** or, in older terminology, **aerobic**. In the hypolimnion, bacterial action on biodegradable organic material consumes oxygen and may cause the water to

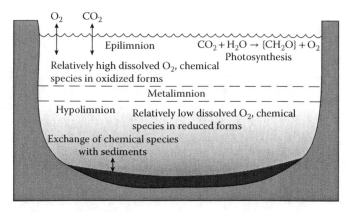

FIGURE 3.3   Stratification of a lake.

become **anoxic (anaerobic)**, that is, essentially free of oxygen. As a consequence, chemical species in a relatively reduced form tend to predominate in the hypolimnion.

The shear plane, or layer between epilimnion and hypolimnion, is called the **metalimnion**. During autumn, when the epilimnion cools, a point is reached at which the temperatures of epilimnion and hypolimnion are equal. This disappearance of thermal stratification causes the entire body of water to behave as a hydrological unit, and the resultant mixing is known as **overturn**. An overturn also generally occurs in the spring. During an overturn, the chemical and physical characteristics of the body of water become much more uniform, and a number of chemical, physical, and biological changes may result. Biological activity may increase from the mixing of nutrients. Changes in water composition during overturn may cause disruption in water treatment processes.

The chemistry and biology of Earth's vast oceans are unique because of their high salt content, their great depth, and other factors. Oceanographic chemistry is a discipline in its own right. The environmental problems of the oceans have increased greatly in recent years because of the release of pollutants to oceans, oil spills, and increased utilization of natural resources from oceans.

3.3.2 FLOWING WATER

Surface water that flows in streams and rivers originates from precipitation that initially falls on areas of land called the **watershed**. Watershed protection has become one of the most important aspects of water conservation and management. To a large extent, the quantity and quality of available water from a watershed depend on the nature of the watershed. An important characteristic of a good watershed is the ability to retain water for a significant length of time. This reduces flooding, allows for a steady flow of runoff water, and maximizes the recharge of water into groundwater reservoirs (aquifer recharge). Runoff is slowed and stabilized by several means: one method is to minimize cultivation and forest cutting on steeply sloping portions of watersheds, and another is to use terraces and grass-planted waterways on cultivated land. The preservation of wetlands maximizes aquifer recharge, stabilizes runoff, and reduces the turbidity of runoff water. Small impoundments in the feeder streams of watersheds have similar beneficial effects.

Rivers in their natural state are free flowing. Unfortunately, the free-flowing characteristics of many of the world's finest rivers have been lost to development for power generation, water supply, and other purposes. Many beautiful river valleys have been flooded by reservoirs, and many rivers have been largely spoiled by straightening channels and other measures designed to improve navigation. A major adverse effect from dam construction consists of the loss of highly productive farmland in river floodplains. Esthetically, an unfortunate case was the flooding in the early 1900s of the Hetch-Hetchy Valley in Yosemite National Park in the Sierra Nevada mountains of California by a dam designed to produce hydroelectric power and water for San Francisco, California. More recently, proposals have been made to drain the valley in an attempt to restore it to some of its original beauty.

3.3.3 SEDIMENTATION BY FLOWING WATER

The action of flowing water in streams cuts away stream banks and carries sedimentary materials over great distances. Sedimentary materials may be carried by flowing water in streams as the following:

- **Dissolved load** from sediment-forming minerals in solution
- **Suspended load** from solid sedimentary materials carried along in suspension
- **Bed load** dragged along the bottom of the stream channel

The transport of calcium carbonate as dissolved calcium bicarbonate provides a straightforward example of dissolved load transport. Water with a high dissolved carbon dioxide concentration (usually present as the result of bacterial action) in contact with calcium carbonate formations contains Ca^{2+} and HCO_3^- ions. Flowing water containing calcium in this form may become more basic by loss of CO_2 to the atmosphere, consumption of CO_2 by algal growth, or contact with dissolved bases, resulting in the deposition of solid $CaCO_3$:

$$Ca^{2+} + 2HCO_3^- \rightarrow CaCO_3(s) + CO_2(g) + H_2O \qquad (3.1)$$

Most flowing water that contains dissolved load originates underground, where it dissolves minerals from the rock strata that it flows through.

Most sediments are transported by streams as suspended load, which is obvious from the appearance of mud in the flowing water of rivers draining agricultural areas or finely divided rock in Alpine streams fed by melting glaciers. Under normal conditions, finely divided silt, clay, or sand make up most of the suspended load, although larger particles are transported in rapidly flowing water. The degree and rate of movement of suspended sedimentary material in streams are functions of the velocity of water flow and the settling velocity of the particles in suspension.

Bed load is moved along the bottom of a stream by the action of water "pushing" the particles along. Particles carried as bed load do not move continuously. The grinding action of such particles is an important factor in stream erosion.

Typically, about two-thirds of the sediments carried by a stream are transported in suspension, about one-fourth in solution, and the remaining relatively small fraction as bed load. The ability of a stream to carry water increases with both the overall rate of flow of the water (mass per unit time) and the velocity of the water. Both the velocities are higher under flood conditions, so floods are particularly important in the transport of sediments.

Streams mobilize sedimentary materials through **erosion, transport** materials along with stream flow, and release them in a solid form during **deposition**. Deposits of stream-borne sediments are called **alluvium**. As conditions such as lowered stream velocity begin to favor deposition, larger, more settleable particles are released first. This results in **sorting** such that particles of similar size and type tend to occur together in alluvium deposits. Much sediment is deposited in floodplains where streams overflow their banks.

3.3.4 Groundwater

Most **groundwater** originates as **meteoritic** water from precipitation in the form of rain or snow and enters underground aquifers through **infiltration** (Figure 3.4). The rock and soil layer in which all pores are filled with liquid water is called the **zone of saturation**, the top of which is defined as the **water table**. Water infiltrates into aquifers in areas called **recharge zones**. Groundwater may dissolve minerals from the formations through which it passes. Most microorganisms originally present in groundwater are filtered out as it seeps through mineral formations. Occasionally, the content of undesirable salts becomes excessively high in groundwater, although it is generally superior to surface water as a domestic water source. Groundwater is a vital resource in its own right; it plays a crucial role in geochemical processes, such as the formation of secondary minerals. The nature, quality, and mobility of groundwater are all strongly dependent on the rock formations in which the water is held. Physically, an important characteristic of such formations is their **porosity**, which determines the percentage of rock volume available to contain water. A second important physical characteristic is **permeability**, which describes the ease of flow of the water through the rock. High permeability is usually associated with high porosity. However, clays, which are common secondary mineral constituents of soil, tend to have low permeability even when a large percentage of their volume is filled with water.

Groundwater that is used for water supply is usually taken from a **water well**. Poor design and mismanagement of water wells can result in problems of water pollution, land subsidence where the water is pumped out, and severely decreased production. As an example, when soluble iron and

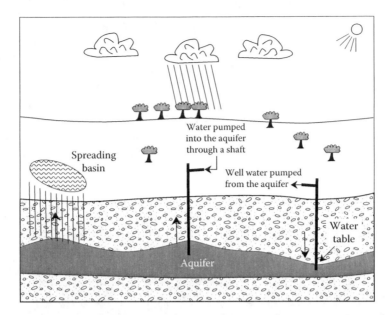

FIGURE 3.4 Aspects of groundwater in an aquifer: groundwater is contained in underground aquifers beneath Earth's surface. The level that water reaches in a well drilled into the aquifer defines the water table. The water table can be lowered by pumping water from an aquifer and raised by pumping water into an aquifer or from natural infiltration of water in a spreading basin on the surface.

manganese are present in groundwater, exposure to air at the well wall can result in the formation of solid deposits of insoluble iron and manganese oxides produced by bacterially catalyzed oxidation processes that result from their contact with oxygen in the air. The deposits fill the spaces that water must traverse to enter the well. As a result, they can seriously impede the flow of water into the well from the water-bearing aquifer. This creates major water source problems for municipalities using groundwater for water supply. As a result of this problem, chemical or mechanical cleaning of wells, drilling of new wells, or even acquisition of new water sources may be required.

3.4 WATER SUPPLY AND AVAILABILITY

It is interesting to note that, despite what appears to be enormous amounts of water in major rainfall and flooding events, 97.5% of Earth's water remains in the oceans and only 2.5% is freshwater. Furthermore, the greater portion of freshwater, 1.7% of Earth's total water, is held immobilized in ice caps in polar regions and in Greenland. This leaves only 0.77% of Earth's water, commonly designated as freshwater, potentially accessible for human use.

As shown in Figure 3.2 and discussed in Section 3.3, freshwater occurs in several places that can serve as water sources. Surface water is found on land in natural lakes, in rivers, in impoundments or reservoirs produced by damming rivers, and under the ground as groundwater. Groundwater is an especially important resource and is susceptible to contamination by human activities.

As discussed in more detail in Chapter 5, water is a key resource in the maintenance of sustainability. Shortages of water from climate-induced droughts have been responsible for the declines of major civilizations. In some parts of the world, water is in such short supply that the daily per capita consumption is less than the amount of water used by a person in an industrialized country during a single tooth brushing with the water left running in the bathroom sink. Variations in water supply cause severe problems for humans and other life forms throughout the world. Devastating floods displace and even kill large numbers of people throughout the world and destroy homes and other structures. At the other extreme of the precipitation scale, severe droughts curtail plant productivity resulting in food shortages for humans and other animals, often necessitating the slaughter of farm animals. It is feared that the severity of both droughts and occasional floods will become much worse as the result of global warming brought on by rising carbon dioxide levels in the atmosphere (see Chapter 8).

Both water quality and water quantity are important factors in sustaining healthy and prosperous human populations. Water is an important vector for disease. In the past, waterborne cholera and typhoid have killed millions of people, and these and other waterborne diseases, especially dysentery, are still problems in less developed areas lacking proper sanitation. The prevention of water pollution has been a major objective of the environmental movement, and avoiding the discharge of harmful water-polluting chemicals is one of the main objectives of the practice of green technology. Water supplies are a concern with respect to terrorism because of their potential for deliberate contamination with biological or chemical agents.

The water that humans use is primarily fresh surface water and groundwater, the sources of which may differ from each other significantly. In arid regions, a small fraction of the water supply comes from the ocean, making use of huge water desalination plants, which is a source that is likely to become more important as the world's supply of freshwater dwindles relative to demand. Saline or brackish groundwater may also be utilized in some areas.

In the continental United States, an average of approximately 1.48×10^{13} L of water fall as precipitation each day, which translates to 76 cm per year. Of that amount, approximately 1.02×10^{13} L per day, or 53 cm per year, are lost by evaporation and transpiration. Thus, the water theoretically available for use is approximately 4.6×10^{12} L per day, or only 23 cm per year. At present, the United States uses 1.6×10^{12} L per day, or 8 cm of the average annual precipitation, an almost tenfold increase from the usage of 1.66×10^{11} L per day in 1900. Even more striking is the per capita increase from about 40 L per day in 1900 to around 600 L per day by the end of the last century. Much of this increase is due to high agricultural and industrial use, each of which accounts for approximately 46% of the total consumption. Municipal use consumes the remaining 8%.

For centuries, humans have endeavored to manage water by measures such as building reservoirs to store water for future use and dikes and dams to control flooding. The results of these measures have been mixed. Typically, construction of reservoirs to provide water for arid regions has been successful and has enabled development in these areas, which may occur for many years without significant problems. The adverse effects of severe, prolonged droughts are made worse by the fact that control of water supplies has enabled excessive growth in water-deficient areas. The Las Vegas metropolitan region of the United States and Mexico City in Mexico are examples of metropolitan regions that have outgrown the natural water capital available to them. Also, the construction of river dikes has enabled agricultural and other development in flood-prone areas. But when a record flooding event occurs, such as the 500-year flood that occurred along the Missouri River in 1993, failure of the protective systems causes much greater devastation than would otherwise be the case. Following this particular event, sensible actions were taken in some areas where farm property along the Missouri River was purchased by government agencies and allowed to revert to a wildlife habitat in its natural state, which included periodic flooding.

Problems with water supply are illustrated in Figure 3.5, which shows rainfall patterns in the continental United States. It is seen that the eastern United States has generally adequate rainfall. However, the western United States is water deficient. Furthermore, some of the fastest growing areas of the United States are among the most water-deficient areas in the country, including southern California, Arizona, Nevada, and Colorado. Even much more severe water supply problems exist in other parts of the world, such as sections of Africa and the area of Palestine and Israel.

The world abounds with examples of groundwater depletion, which is one of the most obvious manifestations of water overuse. Water pumped from below the ground in Mexico City, which is built on an old lakebed, has caused much of the city to sink, damaging many of its structures. In the United States, groundwater has been wastefully depleted from the High Plains Aquifer commonly called the Ogallala Aquifer, most of which underlies regions of Nebraska, Kansas, Oklahoma, and Texas. Groundwater depletion related to sustainability is discussed in more detail in Chapter 5.

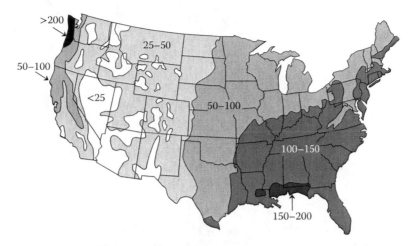

FIGURE 3.5 Water distribution in the continental United States showing water deficiencies in the rapidly growing southwestern states: precipitation amounts are in centimeters per year.

3.5 LIFE AND ITS INFLUENCE ON ENVIRONMENTAL CHEMISTRY IN THE HYDROSPHERE

Aquatic life is very important because all life forms require water, much of the biosphere exists in the hydrosphere, aquatic organisms are a key source of food for humans, and much of the chemistry that takes place in natural water (water in the environment) occurs through the biochemical processes of organisms. One of the main concerns with respect to sustainability is the reduction of seafood production due to overfishing, the effects of climate change, and pollution. As discussed further in Chapter 17, aquatic algae, which are prolific producers of biomass, are likely to be called upon to produce increasing amounts of renewable fuel and biomass material in the future. The role that aquatic organisms play in determining the environmental chemistry of the hydrosphere is introduced here. The toxicological chemistry and other aspects of organisms in the hydrosphere, including **biodegradation** and **bioaccumulation**, are discussed in Sections 3.12 through 3.14.

Figure 3.6 shows the major aspects of aquatic life. Before discussing the living organisms (**biota**) in water, it is useful to define some terms that apply to them. **Plankton** are small plants, small animals, and single-celled organisms that float, drift, or propel themselves weakly in water; **phytoplankton** perform photosynthesis, and animal plankton are called **zooplankton**. The hydrosphere supports a large variety of invertebrate organisms, such as crustaceans, and vertebrate organisms including fish. Bottom-rooted water plants grow exposed to sunlight near the surface. **Autotrophic** biota powered by sunlight or chemical energy produce biomass from simple inorganic molecules including dissolved CO_2. They are called **producers**. The most important of these organisms are photosynthetic algae, which produce biomass (abbreviated as $\{CH_2O\}$) by the following biochemical reaction in which $h\nu$ stands for solar energy:

$$CO_2 + H_2O + h\nu \rightarrow \{CH_2O\} + O_2(g) \tag{3.2}$$

Heterotrophic organisms metabolize organic matter and are important in completing elemental cycles by breaking down complex biomass back to simple inorganic species, including CO_2, NH_4^+, NO_3^-, SO_4^{2-}, and $H_2PO_4^-/HPO_4^{2-}$, which are sources of C, N, S, and P for autotrophic organisms. Aquatic microorganisms are very important in the biodegradation of xenobiotic toxicants in the hydrosphere (see Section 3.12).

Although relatively high productivity of biomass by photosynthetic organisms is needed to produce the food that forms the basis of aquatic food chains, excessive productivity can result in too

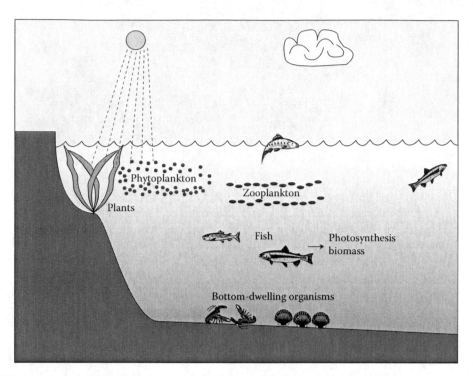

FIGURE 3.6 A variety of life forms exist in a healthy body of water. Free-floating algae (phytoplankton) produce biomass that forms the basis of the food chain in the aquatic ecosystem. Very small animals including single-celled protozoa (zooplankton) are suspended in the water or move through it. A variety of organisms, especially shellfish, dwell largely in the bottom regions of the water. Fishes are higher on the food chain, and land animals that feed on aquatic life are even higher.

much biomass, which can choke a body of water, decay, and use up the oxygen required by fish. Such a condition is called **eutrophication**. A common water pollution problem, eutrophication is the result of excessive nutrients, especially inorganic phosphorus and nitrogen, in the water.

Dissolved oxygen (DO) is a very important constituent of water required by fish and some other forms of aquatic life and largely determines the extent and kinds of life in a body of water. An important water quality parameter related to oxygen is **biochemical oxygen demand (BOD)**, which is the amount of oxygen utilized when the organic matter in a given volume of water is degraded biologically. Excessive BOD may make water unfit to support fish and other organisms that require water.

3.5.1 Aquatic Organisms and Chemical Transitions in the Hydrosphere

Aquatic organisms, especially single-celled microorganisms, are very much involved in chemical transitions in the hydrosphere. Most of them are **oxidation-reduction** reactions (see Section 3.7).

Two very important microbial transitions involve carbon. The first of these is photosynthesis, which is powered by solar energy from the sun (represented by light photon energy, $h\nu$, in which h is Planck's constant and ν is the frequency of light radiation), in which carbon dioxide, largely from the atmosphere, is converted to biomass, represented by the general formula $\{CH_2O\}$:

$$CO_2 + H_2O + h\nu \rightarrow \{CH_2O\} + O_2(g) \tag{3.3}$$

A second important microbially mediated reaction is the opposite of photosynthesis in which biomass is **oxidized** biochemically by elemental oxygen, converting the carbon in biomass to carbon dioxide:

$$\{CH_2O\} + O_2(g) \rightarrow CO_2 + H_2O \tag{3.4}$$

The same process acts in the biodegradation of pollutant chemicals in the hydrosphere.

Several important microbially mediated elemental transitions in the hydrosphere involve nitrogen. **Nitrogen fixation** is the process in which elemental nitrogen from the atmosphere is converted to organic or ammoniacal nitrogen:

$$3\{CH_2O\} + 2N_2 + 3H_2O + 4H^+ \rightarrow 3CO_2 + 4NH_4^+ \tag{3.5}$$

It is carried out by *Azotobacter*, several species of *Clostridium*, and *Cyanobacteria*. The chemical conversion of elemental nitrogen to a chemically combined form in a chemical manufacturing operation requires catalysts, extremely high pressures, and high temperatures; in comparison, the same conversion by bacteria under very mild conditions is a remarkably green, environmentally friendly process.

Nitrification, the bacterial conversion of NH_4^+ to NO_3^-, is an important process in nature because it provides nitrogen in the nitrate form that algae and other plants can utilize:

$$2O_2 + NH_4^+ \rightarrow NO_3^- + 2H^+ + H_2O \tag{3.6}$$

Bacterially mediated **denitrification**

$$4NO_3^- + 5\{CH_2O\} + 4H^+ \rightarrow 2N_2 + 5CO_2 + 7H_2O \tag{3.7}$$

is the process by which nitrate is reduced by bacteria to gaseous nitrogen and returned to the atmosphere. In addition to completing the nitrogen cycle, denitrification is important in wastewater treatment because it removes fixed nitrogen from wastewater effluent, which, when released to the aquatic environment, would cause excessive algal growth and eutrophication.

The microbial transition of inorganic sulfur species is an important process in the hydrosphere. An example is the reduction of sulfate to H_2S with biomass acting as the reducing agent:

$$SO_4^{2-} + 2\{CH_2O\} + 2H^+ \rightarrow H_2S + 2CO_2 + 2H_2O \tag{3.8}$$

This is a process carried out by *Desulfovibrio* acting with other bacteria. Because of the presence of sulfate in seawater, this process for the formation of hydrogen sulfide is a significant source of atmospheric sulfur and a source of the pollutant H_2S in coastal areas. In areas where this occurs, the sediment is often black in color due to the formation of FeS. In the presence of elemental oxygen, *Thiobacillus thiooxidans* and other bacteria may oxidize hydrogen sulfide to sulfate ion:

$$H_2S + 2O_2 \rightarrow 2H^+ + SO_4^{2-} \tag{3.9}$$

This is a reaction that produces strong sulfuric acid. *Thiobacillus thiooxidans* is remarkably acid tolerant and can live in solutions containing up to 1 mole of H^+ per liter.

Some metals, especially iron, are acted upon by bacteria in water. *Ferrobacillus*, *Gallionella*, and some forms of *Sphaerotilus* bacteria obtain energy for their metabolic needs by oxidizing iron(II) to iron(III) with molecular oxygen:

$$4FeCO_3(s) + O_2 + 6H_2O \rightarrow 4Fe(OH)_3(s) + 4CO_2 \tag{3.10}$$

Acid mine drainage (see Chapter 4) is an environmentally important pollutant that results from bacterial action on iron compounds from coal mines. Acid mine water, due to the presence of sulfuric acid and acidic hydrated Fe^{3+} ion, forms when bacteria act upon pyrite, FeS_2, a sulfur-containing mineral associated with coal. The first step is the oxidation of pyrite:

$$2FeS_2\,(s) + 2H_2O + 7O_2 \rightarrow 2Fe^{2+} + 4H^+ + 4SO_4^{2-} \qquad (3.11)$$

The iron(II) ion product of this reaction is oxidized to iron(III) ion:

$$4Fe^{2+} + O_2 + 4H^+ \rightarrow 4Fe^{3+} + 2H_2O \qquad (3.12)$$

This is also catalyzed by bacteria. The Fe^{3+} product of this reaction acts chemically on pyrite:

$$FeS_2\,(s) + 14Fe^{3+} + 8H_2O \rightarrow 15Fe^{2+} + 2SO_4^{2-} + 16H^+ \qquad (3.13)$$

This produces more sulfuric acid, puts additional Fe^{2+} into solution, and completes the cycle for the dissolution of pyrite. The Fe(III) ion exists in solution as the hydrated species $Fe(H_2O)_6^{3+}$ and forms a precipitate of iron(III) hydroxide:

$$Fe(H_2O)_6^{3+} \rightarrow Fe(OH)_3\,(s) + 3H^+ + 3H_2O \qquad (3.14)$$

This releases H^+ ion and contributes to the acidity of acid mine water. The $Fe(OH)_3$ precipitate is a semigelatinous orange material that remains in and along stream beds as an unsightly deposit. The sulfuric acid from acid mine water is toxic to water animal and plant life and tends to leach Al^{3+} ion from the geosphere. This species is phytotoxic (toxic to plants).

3.5.2 MICROBIAL ACTION ON ORGANIC MATTER IN THE HYDROSPHERE

An environmentally important function of microorganisms in the hydrosphere is detoxication (also called **detoxification**) and degradation of water contaminants, discussed further in Sections 3.12 through 3.15. The complete biodegradation of water pollutants is often the result of several kinds of microorganisms acting in sequence. For example, *Micrococcus*, *Pseudmonas*, *Mycobacterium*, and *Nocardia* may all be involved in the biodegradation of petroleum.

The main mechanism for the biodegradation of water pollutants is oxidation, which is shown in a general sense in Reaction 3.4. For example, bacteria act on octane, an ingredient of gasoline:

$$2C_8H_{18} + 25O_2 \xrightarrow{\text{Bacterial oxidation}} 16CO_2 + 18H_2O \qquad (3.15)$$

This is an overall reaction that results in the conversion of the hydrocarbon completely to the simple inorganic compounds carbon dioxide and water, a process of **mineralization**. The partial oxidation of organic pollutants to intermediate organic compounds also commonly occurs.

Microbially mediated processes other than oxidation may also occur with water pollutants. In anoxic regions of water where O_2 is absent, **reduction** may take place. One of the most common biochemical processes mediated by microorganisms and operating on pollutants in water is **hydrolysis** in which a molecule is split in two with the addition of a water molecule. The products of hydrolysis are often more amenable to further biodegradation, such as oxidation, than is the parent compound.

3.6 ENVIRONMENTAL CHEMISTRY OF THE HYDROSPHERE

The environmental chemistry of the hydrosphere is rich and complex. It is strongly influenced by the unique chemical properties of water, especially the polar nature of the water molecule and hydrogen bonding discussed in Section 3.1. It involves a variety of chemical processes, including

acid-base, precipitation, oxidation-reduction, and metal chelation. To a large extent, these phenomena are described by **chemical equilibrium** calculations, such as those involving the ionization of weak acids or the dissolution of slightly soluble salts.

As discussed with some examples in Section 3.5, the chemistry of water is strongly influenced by biochemical processes largely involving aquatic microorganisms, especially with respect to oxidation-reduction phenomena. The chemistry of water depends on where the water is located, such as in a turbulent stream exposed to atmospheric oxygen, the bottom regions of a quiescent body of water isolated from O_2, groundwater in intimate contact with underground formations, or oceanic saltwater with its high concentration of ions.

An overview of the environmental chemistry of water may be had by the examination of Figure 3.7, which shows a body of water stratified by the temperature/density relationship that develops in water, especially in the summer months (see also Figure 3.3). Exposed to atmospheric oxygen and light, the top layer, the epilimnion, normally has a significant concentration of dissolved oxygen and is oxic. Photosynthetic algae thrive in the epilimnion and during daylight produce O_2 by photosynthesis. Isolated from atmospheric oxygen, the hypolimnion located in the bottom regions of the body of water becomes anoxic as O_2 is consumed by bacteria. Chemically oxidized species including CO_2, NO_3^-, and SO_4^{2-} predominate in the epilimnion, and chemically reduced species such as CH_4, NH_4^+, and H_2S are found in the hypolimnion.

A major factor in the chemistry of the aquatic system shown in Figure 3.7 is the biochemical photosynthetic production of organic matter represented as $\{CH_2O\}$. Organic matter is a biochemical reducing agent, and when it sinks into the hypolimnion it is oxidized by microorganism-mediated processes that, for example, reduce NO_3^- and SO_4^{2-} to NH_4^+ and H_2S, respectively. Two important microbially mediated oxidation-reduction reactions of $\{CH_2O\}$ are reaction with dissolved O_2

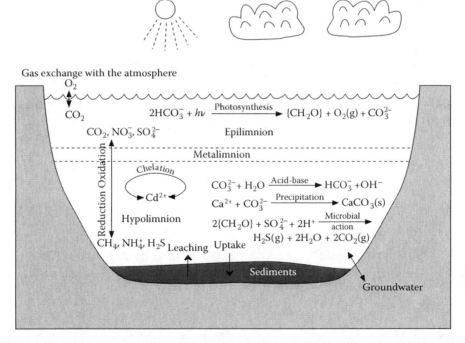

FIGURE 3.7 Water chemistry as affected by the stratification of a body of water in which a warmer, less dense layer of water in the epilimnion resides above a cooler, denser hypolimnion: this stratification prevents oxygen from penetrating to lower depths where chemically reduced species (CH_4, NH_4^+, H_2S) predominate. In the oxic epilimnion, chemically oxidized species (CO_2, NO_3^-, SO_4^{2-}) are present.

$$\{CH_2O\} + O_2 \rightarrow CO_2 + H_2O \tag{3.16}$$

which depletes dissolved oxygen in water, making the hypolimnion anoxic, and methane fermentation

$$2\{CH_2O\} \rightarrow CH_4 + CO_2 \tag{3.17}$$

which produces combustible methane gas. As noted in Section 3.5, the ability of $\{CH_2O\}$ to react with dissolved O_2 is a measure of the potential of water to become depleted in the oxygen needed by fish and other aquatic organisms and is expressed as BOD.

The photosynthetic biochemical production of biomass results in some important chemical reactions. As illustrated in Figure 3.7, algae performing photosynthesis use dissolved HCO_3^- ion as a carbon source and in so doing produce carbonate ion, CO_3^{2-}. Two additional reactions of carbonate ion are shown in Figure 3.7. One is its hydrolysis reaction with H_2O molecules back to HCO_3^- with the production of OH^- ion. This makes the water basic, an important aquatic acid-base reaction. The second reaction of carbonate is its reaction with dissolved Ca^{2+} to produce solid $CaCO_3$, an important precipitation reaction in water that has been responsible for the formation of large deposits of limestone.

An important property of degradable biomass in water is its ability to act as a reducing agent in oxidation-reduction reactions mediated by microorganisms, such as the aforementioned reaction with dissolved O_2. As shown in Figure 3.7, utilization of SO_4^{2-} ion as an alternative to O_2 as a source of oxygen produces odorous, toxic hydrogen sulfide, H_2S.

3.7   ACID-BASE PHENOMENA IN THE HYDROSPHERE

Acid-base phenomena in water involve the exchange of H^+ ions. An important water quality parameter is the presence of species capable of accepting H^+ ions, which is called **alkalinity**; it is a characteristic that is important in the biology, chemistry, and chemical treatment of water. Normally due to the presence of bicarbonate ion, HCO_3^-, alkalinity serves as a pH buffer and reservoir for inorganic carbon required for algal photosynthesis. Most natural waters are somewhat alkaline. The following three reactions show water alkalinity in the form of bicarbonate, carbonate, and hydroxide ions accepting H^+ ions:

$$HCO_3^- + H^+ \rightarrow CO_2 + H_2O \tag{3.18}$$

$$CO_3^{2-} + H^+ \rightarrow HCO_3^- \tag{3.19}$$

$$OH^- + H^+ \rightarrow H_2O \tag{3.20}$$

A characteristic that is analogous to alkalinity is **acidity**, the capability of species in water to produce H^+ ion or to neutralize OH^- ion. The most common acidic component of water is dissolved CO_2, which produces weakly acidic water. Relatively more acidic water with a low pH generally indicates pollution. Acidity may be due to dissolved hydrated metal ions, such as $Fe(H_2O)_6^{3+}$, which readily release H^+ ion. Strong acids including H_2SO_4 and HCl in water are called **free mineral acids**. Water containing a substantial concentration of free mineral acid may be quite toxic to most forms of aquatic life. The pollutant acid mine water contains an appreciable concentration of free mineral acid, and free mineral acid may enter bodies of water from the atmosphere as acid rain.

3.7.1 CARBON DIOXIDE IN WATER

An important solute in water that is associated with both acidity and alkalinity is dissolved carbon dioxide, CO_2, which is almost always present in natural water from contact with atmospheric air or as a product of microbial biodegradation of organic matter. Present in the atmosphere at a level of about 390 ppm of dry air (and increasing due to release from the anthrosphere at a rate of almost 2 ppm per year), atmospheric CO_2 gas makes rainwater from even a totally unpolluted atmosphere slightly acidic. At 25°C, in water in equilibrium with unpolluted air containing 390 ppm carbon dioxide, the concentration of dissolved $CO_2(aq)$ is 1.276×10^{-5} mol/L (M), a value that is used for subsequent calculations in this chapter.

Dissolved carbon dioxide at high levels can be toxic to aquatic life. Three volcanic lakes in Africa, Lake Nyos and Lake Monoun in Cameroon and Lake Kivu in Rwanda, are saturated with carbon dioxide gas from underground sources. Occasionally, people and livestock are killed by exposure to the gas along the lakeshores. In the worst incident by far, 1700 people and several thousand livestock were asphyxiated in 1986 by an abrupt release of carbon dioxide from the super-saturated bottom layer of Lake Nyos.

Most dissolved CO_2 in water is present as neutral $CO_2(aq)$ and not as H_2CO_3 as it is often shown. The following reactions and equations describe the CO_2–HCO_3^-–CO_3^{2-} system in water. Dissolved CO_2 in water acts as an acid reacting with water as follows:

$$CO_2 + H_2O \leftrightarrow H^+ + HCO_3^- \tag{3.21}$$

This is a chemical equilibrium reaction and the double arrows denote that it is reversible, and it lies primarily to the left as shown by the relatively low value of its equilibrium constant at 25°C:

$$K_{a1} = \frac{[H^+][HCO_3^-]}{[CO_2]} = 4.45 \times 10^{-7} \quad pK_{a1} = 6.35 \tag{3.22}$$

The bicarbonate ion, HCO_3^-, can also ionize to produce H^+:

$$HCO_3^- \longleftrightarrow H^+ + CO_3^{2-} \tag{3.23}$$

But it is an extremely weak acid as shown by the very low value of its equilibrium constant expression:

$$K_{a2} = \frac{[H^+][HCO_3^{2-}]}{[CO_3^-]} = 4.69 \times 10^{-11} \quad pK_{a2} = 10.33 \tag{3.24}$$

The predominance of the three major species in the CO_2–HCO_3^-–CO_3^{2-} system is a function of pH as shown by the **distribution of species** diagram in Figure 3.8. At low pH, dissolved CO_2 predominates as is the case with carbonated beverages or any mineral water "with gas." At relatively basic pH values, CO_3^{2-} may predominate. Throughout the normal pH range of natural waters and drinking water, HCO_3^- is the predominant species and is the species responsible for most of the alkalinity in the water. The pH ranges in which various species predominate are shown by a distribution of species diagram with pH as a master variable, as illustrated in Figure 3.8. The plots in this diagram may be calculated from the aforementioned K_a expressions.

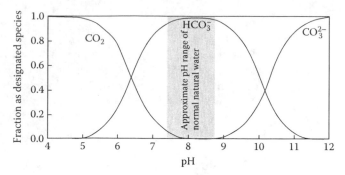

FIGURE 3.8 Distribution of species diagram for the CO_2–HCO_3^-–CO_3^{2-} system as a function of pH: at pH values below pK_{a1}, CO_2 predominates; when pH = pK_{a1}, the fraction of CO_2 equals the fraction of HCO_3^-; at pH = $1/2(pK_{a1} + pK_{a2})$, the fraction of HCO_3^- is at its maximum value; when pH = pK_{a2}, the fraction of HCO_3^- equals the fraction of CO_3^{2-}; and as pH exceeds pK_{a2}, the fraction of CO_3^{2-} approaches 1.

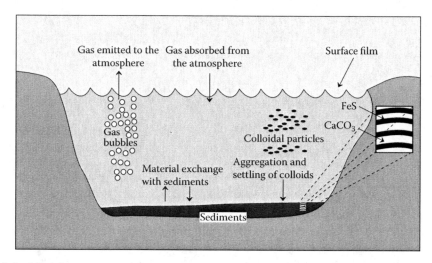

FIGURE 3.9 Most important environmental chemical processes in the hydrosphere involve interactions between water and another phase. Aquatic biochemical processes take place inside the cells of organisms suspended in water, materials are exchanged between water and sediments, gases are emitted to and absorbed from the atmosphere, very small colloidal particles are suspended in water and aggregate to form solids that settle into sediments, and water-immiscible liquids such as hydrocarbons may be present as films on the water surface. The inset shows a phenomenon observed in some sediments in which layers of white $CaCO_3$ precipitated as the result of photosynthesis during the summer alternate with black layers of FeS produced by the reaction of Fe^{2+} and H_2S formed by anoxic bacterial processes during the winter.

3.8 SOLUBILITY AND PHASE INTERACTIONS

Rather than occurring homogeneously in solution, most significant chemical and biochemical phenomena in water involve interactions between species in water and another phase (Figure 3.9). Typically, solid biomass produced photosynthetically in the hydrosphere is generated within a suspended algal cell and the process involves the exchange of dissolved solids and gases between the surrounding water and the cell. The biodegradation of organic matter in water (often in the form of small particles suspended in water) occurs inside or on the surface of bacterial cells and involves the exchange of matter with the aqueous phase. Solids and gases are produced by chemical reactions in water. To a large extent, iron and other important trace-level substances are transported through the hydrosphere in the form of small (**colloidal**) particles. Some organic pollutants such as petroleum or water-immiscible herbicides are often present as a water-immiscible film on the water surface. Sediments are washed from the geosphere surface into bodies of water and may be formed from chemical and biochemical processes in water. A body of water typically consists of the water itself, very small particles of colloidal matter suspended in the water, and sediments. Colloidal matter may be in the form of solids, gases, or water-immiscible liquids and often is very reactive because of its high surface area to volume ratio. An important aspect of phase interactions involves solubilities of gases and solids in water.

3.8.1 GAS SOLUBILITIES

Dissolved gases—O_2 for fish and CO_2 for photosynthetic algae—are crucial to the welfare of living species in water. Fish require dissolved oxygen, dissolved carbon dioxide supports photosynthesis, and supersaturated nitrogen transferred from water to fish blood can cause bubbles that can kill the fish. The biodegradation of organic matter in water consumes dissolved oxygen and may reduce its levels below those required by fish, resulting in fish kills. Therefore, biodegradable organic matter, such as improperly treated sewage, although not particularly toxic in itself, may result in fish kills due to oxygen deprivation.

Gas solubilities are calculated with equilibrium calculations using **Henry's law**, which states that *the solubility of a gas in a liquid is proportional to the partial pressure of that gas in contact with the liquid*. Mathematically, Henry's law applies to the process of a gas from the atmosphere or some other gas phase going into solution without any additional chemical reaction

$$X(g) \longleftrightarrow X(aq) \tag{3.25}$$

and is expressed mathematically as

$$[X(aq)] = KP_X \tag{3.26}$$

where [X(aq)] is the aqueous concentration of the gas in moles per liter, P_X is the partial pressure of the gas in atmospheres, and K is the Henry's law constant that applies to a specific gas at a given temperature in units of moles per liter per atmosphere. (Other units of concentration and pressure are commonly used as well.)

As an example of a gas solubility calculation, consider the concentration of O_2 in water in equilibrium with air at 1 atm total pressure and a temperature of 25°C at which temperature the Henry's law constant for O_2 is 1.28×10^{-3} mol $\cdot$ L^{-1} $\cdot$ atm^{-1}. It is first necessary to correct for the partial pressure of water at 25°C, which is 0.0313 atm. Making this correction and applying the knowledge that dry air is 20.95% by volume of O_2 yields the following equation:

$$P_{O_2} = (1.0000 \text{ atm} - 0.0313 \text{ atm}) \times 0.2095 = 0.2029 \text{ atm} \tag{3.27}$$

$$[O_2(aq)] = K \times P_{O_2} = 1.28 \times 10^{-3} \text{ mol} \times L^{-1} \text{ atm}^{-1} \times 0.2029 \text{ atm} \tag{3.28}$$

$$[O_2(aq)] = 2.60 \times 10^{-4} \text{ mol} \cdot L^{-1} \tag{3.29}$$

Converting to commonly used units of milligrams per liter for concentration shows that at 25°C and atmospheric pressure the concentration of dissolved O_2 in water is only 8.32 mg/L. The solubility of O_2 decreases with increasing water temperature. Thus, water in equilibrium with air cannot contain a high level of dissolved oxygen compared to many other solute species. Oxygen-consuming processes in the water may cause the dissolved oxygen level to rapidly approach zero unless some efficient mechanism for the reaeration of water is operative, such as turbulent flow in a shallow stream or air pumped into the aeration tank of an activated sludge secondary-sewage treatment facility. At higher temperatures, the decreased solubility of O_2, combined with the increased respiration rate of aquatic organisms, often results in severe oxygen depletion.

3.8.2 Carbon Dioxide and Carbonate Species in Water

Figure 3.10 illustrates the important relationships among dissolved CO_2 in water, dissolved carbonate species (HCO_3^-, CO_3^{2-}), and solid carbonate minerals, particularly limestone ($CaCO_3$) and dolomite ($CaCO_3 \cdot MgCO_3$). These species are very important in the chemistry of the hydrosphere. As examples, many minerals are deposited as salts of CO_3^{2-} ion and algae in water utilize dissolved CO_2 and HCO_3^- in the synthesis of biomass. The equilibrium relationships among CO_2 gas in the atmosphere, dissolved CO_2 and carbonate species in water, and solid carbonate minerals largely determine the pH, alkalinity, and hardness of water.

Since air is only about 0.039% CO_2 by volume, the concentration of this gas that dissolves in water from the atmosphere is relatively low. Much of the dissolved CO_2 in water comes from the microbial decay of organic matter in water and in soil. If the concentration of free carbon dioxide in water from these or subterranean (volcanic) sources is too high, the respiration and gas exchange of aquatic organisms may be adversely affected and fish and other organisms may even die as a result. The formation of HCO_3^- and CO_3^{2-} significantly increases the solubility of carbon dioxide in water.

The reaction between dissolved carbon dioxide in water and carbonate minerals such as limestone, $CaCO_3$,

$$CaCO_3(s) + CO_2(aq) + H_2O \leftrightarrow Ca^{2+} + 2HCO_3^- \tag{3.30}$$

determines the alkalinity, pH, and dissolved calcium concentration for water in contact with $CaCO_3$ mineral. From a Henry's law calculation, it may be shown that the concentration of CO_2 in water in equilibrium with air, that is, 0.039% CO_2, is 1.276×10^{-5} M. Dissolved carbon dioxide reacts as a weak acid

$$CO_2 + H_2O \leftrightarrow H^+ + HCO_3^- \tag{3.31}$$

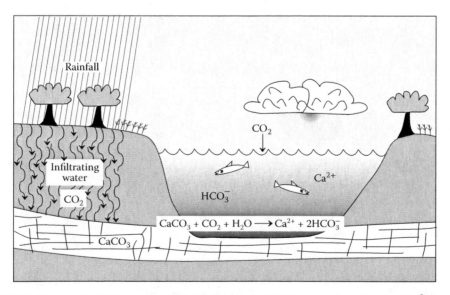

FIGURE 3.10 Carbon dioxide in water reacts with limestone, $CaCO_3$, to produce dissolved Ca^{2+} (water hardness) and dissolved HCO_3^- (water alkalinity), a reaction that buffers the water pH at a slightly basic level. The carbon dioxide may come from the atmosphere or from decaying organic matter in water or soil.

with the following acid dissociation constant:

$$K_{a1} = \frac{[H^+][HCO_3^-]}{[CO_2]} = 4.45 \times 10^{-7} \quad pK_{a1} = 6.35 \tag{3.32}$$

The reaction of HCO_3^- to produce H^+ and the equilibrium constant expression for that reaction are as follows:

$$HCO_3^- \longleftrightarrow H^+ + CO_3^{2-} \tag{3.33}$$

$$K_{a2} = \frac{[H^+][CO_3^{2-}]}{[HCO_3^-]} = 4.69 \times 10^{-11} \quad pK_{a2} = 10.33 \tag{3.34}$$

The solubility product of calcium carbonate is

$$K_{sp} = [Ca^{2+}][CO_3^{2-}] = 4.47 \times 10^{-9} \tag{3.35}$$

From these reactions and equilibrium constant expressions, it may be shown that the reaction between calcium carbonate and dissolved CO_2

$$CaCO_3(s) + CO_2(aq) + H_2O \longleftrightarrow Ca^{2+} + 2HCO_3^- \tag{3.36}$$

has the following equilibrium constant expression:

$$K' = \frac{[Ca^{2+}][HCO_3^-]^2}{[CO_2]} = \frac{K_{sp}K_{a1}}{K_{a2}} = 4.24 \times 10^{-5} \tag{3.37}$$

The stoichiometry of Reaction 3.30 gives $[HCO_3^-] = 2[Ca^{2+}]$, a bicarbonate ion concentration that is twice that of calcium. Substitution of the value of CO_2 concentration of 1.276×10^{-5} M into the expression for K' yields a value of 5.14×10^{-4} M for $[Ca^{2+}]$ and a value of 1.03×10^{-3} M for $[HCO_3^-]$. Substitution into the expression for K_{sp} yields a value of 8.70×0^{-6} M for $[CO_3^{2-}]$. Further substitution into the product $K_{a1}K_{a2}$

$$K_{a1}K_{a2} = \frac{[H^+]^2[CO_3^{2-}]}{[CO_2]} = 2.09 \times 10^{-17} \tag{3.38}$$

yields $[H^+] = 5.54 \times 10^{-9}$ M (pH 8.26). The value of $[HCO_3^-]$ is essentially equal to the alkalinity and a much higher contribution than either CO_3^{2-} or OH^-. These calculations show that for water in equilibrium with CO_2 from the atmosphere and with solid $CaCO_3$, the pH should be slightly basic and both hardness and alkalinity should be around 1×10^{-3} M, which are values close to those observed in many natural waters.

3.8.3 SEDIMENTS

Sediments are the layers of solid and semisolid material in the bottom of bodies of water. Varying in composition from pure mineral matter to pure organic matter, sediments are typically mixtures of clay, silt, sand, organic matter, and various minerals. The most straightforward process for sediment formation is erosion of geospheric material into bodies of water. Other physical, chemical, and biological processes may result in sediment formation. Typically, as the pH rises in water as the result of photosynthetic activity, solid calcium carbonate is precipitated:

$$Ca^{2+} + 2HCO_3^- + h\nu \longleftrightarrow \{CH_2O\} + CaCO_3(s) + O_2(g) \tag{3.39}$$

Another biochemical process that produces sedimentary material occurs when anoxic bacteria in sediments degrade biomass using sulfate ion, SO_4^{2-}, as an oxygen source to produce H_2S (a biochemically mediated reaction that is shown in Figure 3.7) and other bacteria reduce insoluble iron(III) oxides to soluble Fe^{2+}. The following reaction then occurs to produce black, solid FeS in sediments:

$$Fe^{2+} + H_2S \rightarrow FeS(s) + 2H^+ \tag{3.40}$$

In some lakes, this has resulted in the formation of alternate layers of white $CaCO_3$ formed as the result of photosynthesis in the summer and black FeS produced by anoxic processes during the winter, as shown in the inset in Figure 3.9.

A beneficial effect of the formation of H_2S and FeS produced by anoxic processes in sediments is the reaction of both these species with heavy metal ion pollutants to remove heavy metals from solution as precipitates such as PbS and CdS.

Sediments can be sources of toxicants and are an important consideration in toxicological chemistry. Although heavy metal sulfides such as PbS and CdS are removed from water into sediments, when the sediments are stirred up the sulfides can be oxidized to toxic soluble forms. The dense, toxic pollutants polychlorinated biphenyls (PCBs) have accumulated in Hudson River sediments as discussed in Chapter 4, Section 4.12. As noted in Section 3.11, anoxic bacterial processes in sediments may convert insoluble inorganic mercury to mobile methylmercury compounds that contaminate fish tissue. Bottom-feeding organisms may bioaccumulate metal and organic pollutants that have accumulated in sediments.

3.8.4 COLLOIDS IN WATER

Very small suspended colloidal particles are important constituents of natural waters and sediments. Colloidal particles are composed of a variety of materials including minerals (especially clays), algae cells, bacteria cells, proteinaceous materials, and water-immiscible pollutants. Colloidal particles suspended in water (1) range in diameter from about 0.001 μm to about 1 μm; (2) have characteristics intermediate between those of suspended matter and true solutions; (3) tend to be very reactive because of their high surface-to-volume ratios; (4) have a high interfacial energy; (5) exhibit a high surface/charge density ratio; and (6) scatter white light as a light blue hue observed at right angles to the incident light, the **Tyndall effect**, due to the fact that colloidal particles have about the same dimensions as the wavelength of visible light.

Colloidal particles are of three general types, as illustrated in Figure 3.11: **hydrophilic colloids** tend to remain in suspension because of their strong interactions with water; they are generally macromolecules such as proteins and some synthetic polymers. Exemplified by clay particles and petroleum droplets, **hydrophobic colloids** are generally repelled by water and remain in suspension because of their electrical charges, which tend to keep the particles away from each other. Special aggregates of ions and molecules called **micelles** constitute the third kind of colloidal particles. Soap ions, such as sodium stearate,

$$CH_3CH_2CH_2CH_2CH_2CH_2CH_2CH_2CH_2CH_2CH_2CH_2CH_2CH_2CH_2CH_2CH_2CO_2^-Na^+$$

form micelles. An examination of the formula of sodium stearate shows that it has a "split personality" with a long hydrocarbon chain "tail" that tends to impart an oil-like water-repelling character

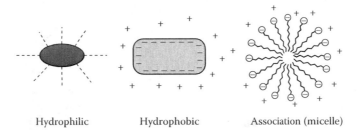

Hydrophilic   Hydrophobic   Association (micelle)

FIGURE 3.11   Representations of hydrophilic, hydrophobic, and association colloidal particles: the hydrophilic colloids are bound to water as indicated by the dashed lines. Positively charged counter ions surround negatively charged hydrophobic and association colloidal particles. The micelle of the association colloid may be visualized as a cluster of organophilic hydrocarbon chains inside a ball with anionic heads attracted to water on the surface of the ball.

to the compound and a charged ionic "head" that is attracted to water. As a result, the ions form spherical micelles in which the hydrocarbon chains cluster together inside the particle and the $-CO_2^-$ ionic groups are on the surface of the sphere, their charges neutralized by random distributions of positively charged counter ions (Na^+).

Colloids and their stabilities are very much involved in the environmental and toxicological chemistry of the hydrosphere. Many water pollutants are associated with colloids and transported by them. Pollutants such as spilled crude oil may be suspended as a colloidal material. The stability of colloids is a very important aspect of their behavior. For example, colloidal mineral and organic matter carried by river water contacts saltwater when the river is discharged to the sea, neutralizing the charge that keeps the particles in suspension and forming sediments. Bacterial cells are largely suspended as colloids in water. The bacteria involved in biodegrading sewage are held in suspension in aeration tanks where the sewage is treated, but the suspended bacterial cells agglomerate to produce sewage sludge before the treated water is discharged.

3.9   OXIDATION REDUCTION

Oxidation-reduction (**redox**) reactions are those in which exchanges of electrons (e^-) occur along with changes of oxidation states of reactants (see Chapter 19, Section 19.3). Oxidation originated as a term to describe the acquisition of electrons by elemental oxygen atoms from other atoms as shown by the following reaction in which each oxygen atom gains two electrons from a calcium atom to produce ions:

$$2Ca + O_2 \rightarrow 2\left\{Ca^{2+}O^{2-}\right\} \qquad (3.41)$$

In this process, each calcium atom is **oxidized** with a loss of two electrons to produce a Ca^{2+} **cation** and each oxygen atom is **reduced** with a gain of two electrons to produce an *oxide anion*, O^{2-}. In the calcium oxide product, the **oxidation state** of calcium is +2 and that of oxygen is –2. Although it is easiest to view oxidation-reduction as an exchange of electrons to produce ions, it is not necessary for ions to be formed for oxidation-reduction to occur. For example, when elemental carbon burns in the presence of O_2

$$C + O_2 \rightarrow CO_2 \qquad (3.42)$$

the elemental C is oxidized and the elemental oxygen is reduced. Since the oxidation state of each of the two O atoms in CO_2 is –2, for neutrality the oxidation state of the C atom in CO_2 is +4. The reaction of elemental H_2 with a substance is reduction as shown by the reaction

$$C + 2H_2 \rightarrow CH_4 \tag{3.43}$$

in which each H atom changes from its 0 oxidation state in H_2 (the oxidation state of all atoms in the elemental state is 0) to +1 in CH_4 and the oxidation state of C changes from 0 in its elemental state to −4 in CH_4.

Oxidation-reduction phenomena, usually mediated by microorganisms, are highly significant in the microbiology, environmental chemistry, and toxicological chemistry of the hydrosphere. One of the most common processes is the microorganism-mediated oxidation of organic matter, {CH_2O}:

$$\{CH_2O\}(\text{becomes oxidized}) + O_2(\text{becomes reduced}) \rightarrow CO_2 + H_2O \tag{3.44}$$

Occurring in a body of water, this reaction depletes dissolved oxygen, resulting in fish kills. It is an important reaction for the biodegradation of wastes in sewage treatment. Another oxidation-reduction process that is important in the hydrosphere is

$$Fe(OH)_3(s) + 3H^+ + e^- \rightarrow Fe^{2+}(aq) + 3H_2O \tag{3.45}$$

which occurs in anoxic bottom regions in a body of water, such as a reservoir used for municipal water supply, and places soluble iron(II) in solution. The soluble iron(II) must be removed prior to water use; otherwise, the reverse of the aforementioned reaction can take place in a bathroom fixture or washing machine and leave an intractable residue of unsightly brown iron(III) oxide (rust).

Oxidation-reduction processes are very important in the nitrogen cycle. The initial product of the biodegradation of nitrogen-containing biomass (protein) and the form normally applied to farm lands as fertilizer is ammonium ion, NH_4^+. To be assimilated as a nutrient by algae or in field crops, it normally requires microbial oxidation to NO_3^-:

$$NH_4^+ + 2O_2 \rightarrow NO_3^- + 2H^+ + H_2O \tag{3.46}$$

Many other examples can be cited of the ways in which the types, rates, and equilibria of redox reactions largely determine the nature of important solute species in water.

The degree to which a water solution is oxidizing or reducing depends on the solute species present and reflects the activity of the electron, e^-. If the electron activity is high the solution is reducing, and if it is low the solution is oxidizing. Like H^+ ion concentration, the activity of the electron in solution may vary over many orders of magnitude. As with pH, which is $-\log[H^+]$, for which each unit change represents a 10-fold change in activity, it is convenient to represent the relative reducing and oxidizing tendencies in solution by pE, the negative log of electron activity. A high, relatively positive pE value, such as the one encountered in water in equilibrium with atmospheric O_2, indicates an oxidizing solution, and a low, relatively negative pE value, such as the one in sediments where H_2S is produced by anoxic bacteria, indicates a reducing solution.

The kinds of species that are stable at equilibrium in water are a function of both pE and pH as displayed by a pE-pH diagram. Figure 3.12 shows a simplified pE-pH diagram for iron species in water where the possible species are Fe^{2+} ion in solution, Fe^{3+} ion in solution, solid $Fe(OH)_3$, and solid $Fe(OH)_2$. At pE values higher than those along the upper dashed line water is thermodynamically unstable toward oxidation, and below the lower dashed line water is thermodynamically unstable toward reduction. So, above the upper line water decomposes to yield O_2 and below the lower line water tends to decompose to yield H_2. The Fe^{3+} ion exists in a stable state only in a very oxidizing, acidic medium, of which acid mine water in equilibrium with atmospheric oxygen is a common example, whereas the region of stability for Fe^{2+} ion is relatively large as reflected by the common occurrence of soluble iron(II), which is a potentially troublesome water contaminant, in

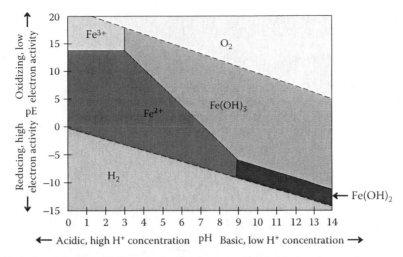

FIGURE 3.12 A simplified pE-pH diagram for iron in water: in this system, the maximum total soluble iron concentration is 1.0×10^{-5} mol/L. In acidic, oxidizing water (low pH, high pE), there is a small region where Fe^{3+} is a stable species in solution. At lower pE values, soluble Fe^{2+} has a relatively large area of stability. Solid $Fe(OH)_3$ is the predominant species over a large area. Solid $Fe(OH)_2$ exhibits a small region of stability; in most anoxic water systems, solid FeS or $FeCO_3$ tends to predominate in this area. At high pE values above the upper dashed line the H_2O molecule decomposes giving off O_2, and at low pE values the H_2O molecule decomposes giving off H_2.

oxygen-deficient groundwaters. Highly insoluble $Fe(OH)_3$ is the predominant iron species over a very wide pE-pH range, and as a result soluble iron species are usually not a contamination problem in oxic waters.

The pE-pH diagram shown in Figure 3.12 can be used to illustrate some aspects of the behavior of iron in the hydrosphere. For example, anoxic groundwater at around pH 7 with a relatively high concentration of Fe^{2+} brought to the surface from a well and exposed to atmospheric oxygen enters an environment in which solid $Fe(OH)_3$ is the stable iron species, resulting in the precipitation of this compound.

3.9.1 pE and Toxicological Chemistry

The degree to which water, sediments, and soil are in an oxidizing environment or a reducing environment can have important effects on toxicants. In general, low pE (reducing) media are more likely to produce toxic substances or prevent the biodegradation of toxicants. Some of the more important influences of pE on potentially toxic substances are the following:

- At low pE, highly toxic hydrogen sulfide gas, H_2S, may be produced when anoxic bacteria use sulfate ion, SO_4^{2-}, as an oxidant in the absence of molecular O_2.
- At low pE, methane gas, CH_4, is produced by the anoxic biodegradation of organic matter (see Reaction 3.17) in the absence of molecular O_2. Although methane gas is not particularly toxic, it can act as an asphyxiant when it displaces air. Workers digging wells have been killed from subterranean methane.
- Under low pE conditions, some bacteria produce toxic substances. Although not generally an environmental problem, *Clostridium botulinum* bacteria growing anoxically can produce deadly botulism toxin. Cases of human poisoning from this source have generally been from improperly canned food.
- At low pE, biodegradation of toxic substances is generally slower.

3.10 METAL IONS IN WATER

Although metal ions in aqueous solution are usually written as M^{n+}, they are always bound with four to six water molecules as represented by the formula for the hydrated metal cation $M(H_2O)_x^{n+}$. Metal ions are stabilized in water by binding with water molecules and tend to reach a state of higher stability by binding with stronger bases (electron-donor partners) that might be present in the solution. This can occur with reactions such as acid-base

$$Fe(H_2O)_6^{3+} \longleftrightarrow FeOH(H_2O)_5^{2+} + H^+ \qquad (3.47)$$

precipitation

$$Fe(H_2O)_6^{3+} \longleftrightarrow Fe(OH)_3(s) + 3H_2O + 3H^+ \qquad (3.48)$$

and oxidation-reduction reactions:

$$Fe(H_2O)_6^{2+} \longleftrightarrow Fe(OH)_3(s) + 3H_2O + 3H^+ + e^- \qquad (3.49)$$

3.10.1 CALCIUM AND HARDNESS IN WATER

The Ca^{2+} ion is usually the most abundant cation in freshwater systems. Of the cations found in most freshwater systems, calcium generally has the highest concentration. The Ca^{2+} ion is a key species in geochemical cycles of matter. Calcium gets into water by its contact with a number of minerals such as $CaSO_4 \cdot 2H_2O$; anhydrite, $CaSO_4$; dolomite, $CaCO_3 \cdot MgCO_3$; and calcite and aragonite, which are different mineral forms of $CaCO_3$. The equilibrium between dissolved carbon dioxide and calcium carbonate minerals is important in determining several natural water chemistry parameters such as alkalinity, pH, and dissolved calcium concentration (Figure 3.10). As illustrated in Figure 3.10, CO_2 in water dissolves calcium from its carbonate minerals:

$$CaCO_3(s) + CO_2(aq) + H_2O \longleftrightarrow Ca^{2+} + 2HCO_3^- \qquad (3.50)$$

Water with dissolved calcium in which the anion is HCO_3^- is said to contain **temporary hardness**; in general, **water hardness** is manifested by the presence of Ca^{2+} and Mg^{2+} ions, which form precipitates with soap. When CO_2 is lost from water in which the anion is HCO_3^-, solid $CaCO_3$ is precipitated by the reverse of the aforementioned reaction. This occurs, for example, when such water is boiled, resulting in the formation of a cloudy suspension of solid $CaCO_3$.

3.11 COMPLEXATION AND SPECIATION OF METALS

Metal ions in water gain stability by forming **metal complexes** in which species such as Cl^- bind reversibly to metal ions and especially **metal chelates** in which metal ions are bound in two or more places by organic substances. The solubilities, transport properties, and biological effects of chelated metal species and of **organometallic** compounds are often vastly different from those of the metal ions themselves. The **chelating agents** that bind with metals may come from pollutant sources, such as the nitrilotriacetate anion that is shown to be bound with Zn^{2+} ion in Figure 3.13.

The most common naturally occurring chelating agents are humic substances. These complex materials are produced by the partial biodegradation of biomass, especially wood. They are complex large organic molecules with numerous aromatic ring structures containing oxygen in functional

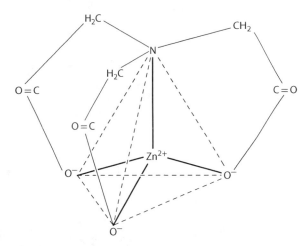

FIGURE 3.13 The nitrilotriacetate anion binding in four places to the Zn^{2+} metal ion: sodium nitrilotriacetate was once strongly advocated as a substitute for phosphates in household detergent builders, but was not used in the United States because of concern over its tendency to bind with and transport toxic heavy metal ions. (From Manahan, S.E., *Environmental Chemistry*, Taylor & Francis/CRC Press, Boca Raton, FL, 2009. With permission.)

groups including carboxyl ($-CO_2H$) and phenolic hydroxyl ($-OH$). These groups can lose H^+ to produce negatively charged groups capable of bonding with metal ions as shown here for the chelation of Fe^{2+} ion:

$$\text{(3.51)}$$

The most important humic substances in water are the lower-molecular-mass fulvic acids. These species tend to chelate Fe^{2+} ion, producing a yellow material called *gelbstoffe* (German for yellow stuff), which is responsible for much of the undesirable color found in some water. Metal ions bound with fulvic acid are hard to remove from water and, since iron is a very undesirable water impurity, drastic measures such as destruction of the fulvic acid with chlorine may be required to remove the chelated iron. Figure 3.14 illustrates a typical fulvic acid molecule.

Chelation is an example of **speciation** of metals in water, which plays a crucial role in their environmental chemistry and toxicological chemistry. Another example of metal species in water consists of organometallic compounds containing carbon-to-metal bonds. The most notorious examples of organometallic species in water are methylated mercury species, the water-soluble monomethylmercury ion, $HgCH_3^+$, and soluble and volatile dimethylmercury, $Hg(CH_3)_2$. These two species were responsible for the surprisingly high levels of mercury found in water and especially fish lipid tissue around 1970. Investigators were puzzled by these high levels because under conditions in natural waters, inorganic mercury exists as insoluble species including HgS and mercury oxides. It was subsequently shown that anoxic bacteria in sediments were methylating mercury, leading to soluble species that could get into water and fish tissue. In addition to methylated organometallic compounds of mercury, tin, selenium, and arsenic produced by the action of anoxic bacteria, organometallic compounds may enter the environment directly as industrial pollutants alright. Until its use was banned starting in the 1970s, one of the most common chemicals of this type was highly toxic tetraethyllead, $Pb(C_2H_5)_4$, which was added to gasoline as an octane booster. More

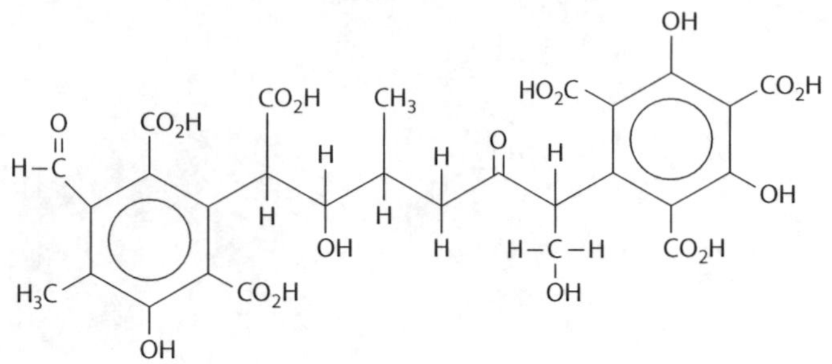

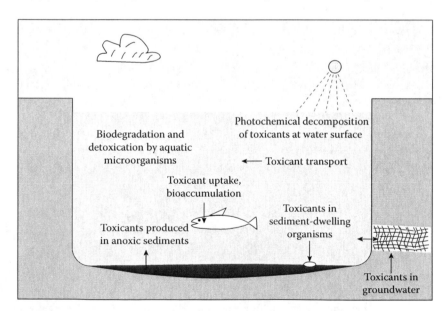

FIGURE 3.14   A hypothetical structural formula of a molecule of fulvic acid: fulvic acid forms soluble complex ions and chelates with metals in water. The molecule contains some aromatic rings, that is, carboxylic acid ($-CO_2H$) groups and phenolic ($-OH$) groups.

recently, organotin compounds used in biocidal paints to prevent the growth of organisms on boat and ship hulls were found to be common water pollutants are caused toxicity problems in sediment-dwelling organisms.

3.12   TOXICOLOGICAL CHEMISTRY IN THE HYDROSPHERE

Figure 3.15 shows various aspects of toxicological chemistry in the hydrosphere. Various parts of the hydrosphere are involved in the production, movement, storage, and degradation of toxic substances. As discussed in Chapter 4, pollution of the hydrosphere by toxicants that may harm fish that live in water or that may make the water hazardous for humans and other organisms to drink is a major consideration in environmental protection. Some important toxic substances are produced by biochemical processes in the hydrosphere. An important example shown in Figure 3.7 is toxic hydrogen sulfide, H_2S, which is generated by bacteria in anoxic bottom regions of bodies of water or in sediments by bacteria growing in isolation from atmospheric air and using sulfate ion, SO_4^{2-}, as a

FIGURE 3.15   Some aspects of the toxicological chemistry of the hydrosphere: the aspects included are toxicant transport through the hydrosphere; toxicant uptake and bioaccumulation by water-dwelling organisms (fish) and sediment-dwelling organisms (some shellfish); production of toxicants, such as those by anoxic bacteria; biodegradation and detoxification by microorganisms; photochemical decomposition of toxicants exposed to sunlight on the surface; and toxicant exchange between water and geospheric mineral formations.

source of oxygen. Water can leach toxic heavy metal ions, such as Cd^{2+}, from geological formations and transport them to where they may come in contact with organisms. The prevention of water quality degradation resulting from contamination by toxic substances has been a major objective of the environmental movement, and avoiding the discharge of toxic water-polluting chemicals is one of the main objectives of the practice of green technology.

The hydrosphere is a very important conduit for the transport of agents that are toxic or that may cause illness. The potential of water in this respect is illustrated by waterborne diseases that have killed millions in the past and still sicken and kill large numbers of people, particularly in developing regions. In the past, cholera and typhoid carried by water have been especially deadly waterborne diseases. Now the greatest problem is dysentery; waterborne cases of this malady are a leading cause of infant death around the world. Water supplies are a concern with respect to terrorism because of their potential for deliberate contamination with biological or chemical agents.

Inadequately regulated industrial development has resulted in pollution of drinking water sources by toxic industrial chemicals in the poorer sections of the world. Water supplies are a concern with respect to terrorism because of their potential for deliberate contamination with biological or chemical agents.

3.13   CHEMICAL INTERACTIONS WITH ORGANISMS IN THE HYDROSPHERE

One of the most important aspects of xenobiotic compounds in the hydrosphere is their interactions with water-dwelling organisms including fish and sediment-dwelling shellfish. Bioaccumulation is the term given to the uptake and concentration of xenobiotic materials by living organisms. Bioaccumulation can lead to **biomagnification** in which xenobiotic substances become successively more concentrated in the tissues of organisms higher in the food chain. This usually occurs with poorly degradable, lipid-soluble organic compounds. Suppose, for example, that such a compound comes in contact with lake water; accumulates in solid detritus in the water; sinks to the sediment; and is eaten by small burrowing creatures in the sediment, which are eaten by small fish. The small fish may be eaten by larger fish, which in turn are consumed as food by birds. At each step, the xenobiotic substance may become more concentrated in the organism and may reach harmful concentrations in the birds at the top of the food chain. This is basically what happened with dichlorodiphenyltrichloroethane (DDT), the indiscriminate use of which almost caused the extinction of eagles and hawks.

The most straightforward case of bioaccumulation is **bioconcentration**, which occurs when a substance dissolved in water enters the body of a fish or some other aquatic organism by passive processes (basically, it just dissolves in the organism) and is carried to bodies of lipid in the organism in the blood flow (see Figure 3.16). The model of bioconcentration assumes that the organism taking up the compound does not metabolize the compound, which is a good assumption for refractory organic compounds such as DDT or PCBs. It also assumes that uptake is by nondietary routes, including diffusion through the skin and especially through the gills of fish. The model of bioconcentration applies especially to substances that have low water solubilities (though they are high enough to make the compound available for uptake) and high lipid solubilities. This model of bioconcentration assumes a dynamic equilibrium between the xenobiotic substance dissolved in water and the same substance dissolved in lipid tissue. It is called the **hydrophobicity model** because of the hydrophobic (water-hating) nature of the substance being taken up.

Bioconcentration may be expressed by **bioconcentration factors**; it is defined as follows:

$$\text{Binconcentration factor} = \frac{\text{Concentration of xenobiotic in lipid}}{\text{Concentration of xenobiotic in water}} \qquad (3.52)$$

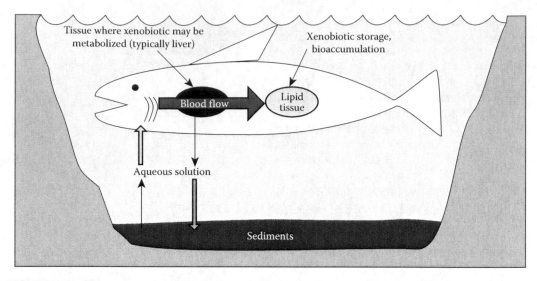

FIGURE 3.16   Illustration of the distribution of a hydrophobic xenobiotic between sediment, solution in water, and lipid tissue in a water-dwelling organism: the weights of the vertical arrows reflect the relative solubilities of a hydrophobic xenobiotic in sediment, aqueous solution, and lipid tissue, showing preference for sediment and lipid tissue over aqueous solution.

Typical bioconcentration factors for highly lipid-soluble PCBs and hexachlorobenzene in sunfish, trout, and minnows range from somewhat more than 1,000 to around 50,000.

3.14   BIODEGRADATION IN THE HYDROSPHERE

Biodegradation is mentioned as a process mediated by microorganisms in Section 3.5. Biodegradation by the action of enzymes in bacteria, fungi, and protozoa is the process by which biomass from deceased organisms is broken down, ultimately to simple inorganic constituents, thus completing the cycle that begins when atmospheric carbon dioxide is used to produce biomass by photosynthesis. Much biodegradation occurs in the hydrosphere and in sediments in bodies of water.

The first of the two general categories of biodegradation is the utilization by microorganisms of organic matter that can be metabolized for energy and as material to synthesize additional biomass. This is the route taken by microorganisms in the degradation of biomass from other organisms and, to a lesser extent, in the biodegradation of some xenobiotic materials. The second way in which microorganisms metabolize environmental chemicals is through **cometabolism**, a process in which an organism's enzymes act upon the substances as a "sideline" of their normal metabolic processes. The substances that are cometabolized are called secondary substrates because they are not the main compounds for which the enzymatic processes are designed. As an example of cometabolism, *Phanerochaete chrysosporium* fungi (white rot fungi) biodegrade organochlorine compounds, including PCBs and dioxins, using an enzyme system that normally functions to break down degradation-resistant lignin in wood.

Biodegradation may involve relatively minor changes such as addition, deletion, or modification of a functional group. Complete biodegradation or **mineralization** converts organic molecules to simple inorganic species including CO_2 for carbon, NH_4^+ or NO_3^- for nitrogen, HPO_4^{2-} for phosphorus, and SO_4^{2-} for sulfur. Mineralization completes elemental cycles in the environment.

Detoxification (or detoxication) is of particular importance with respect to environmental toxicological chemistry. Reaction 3.53 shows the conversion of insecticidal paraoxon, a potent neurotoxin (see insecticidal organophosphates in Chapter 4, Section 4.11), to *p*-nitrophenol, which is only about $0.005 \times$ as toxic. In some cases, microorganisms actually produce more toxic materials. This occurs in sediments in the hydrosphere when inorganic mercury compounds that are relatively harmless by virtue of their low water solubilities are converted by anoxic bacteria to highly toxic, mobile methylmercury species, $Hg(CH_3)_2$ and $HgCH_3^+$ (see the discussion on methylmercury water pollutants in Chapter 4, Section 4.3).

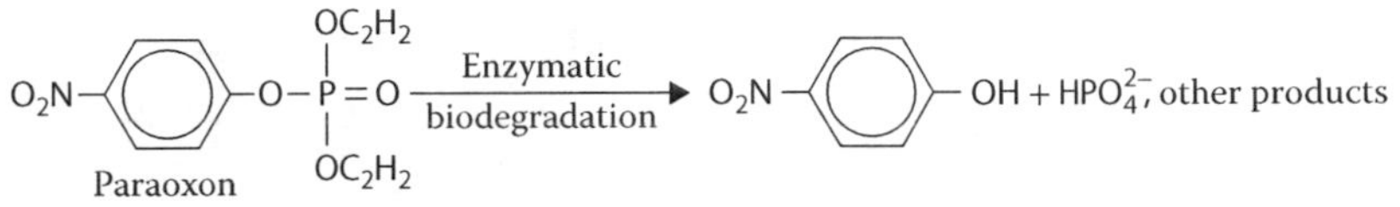

$$O_2N - \underset{\text{Paraoxon}}{\underset{|}{\overset{OC_2H_2}{\overset{|}{\bigcirc - O - P = O}}}} \xrightarrow[\text{biodegradation}]{\text{Enzymatic}} O_2N - \bigcirc - OH + HPO_4^{2-}, \text{other products} \qquad (3.53)$$

When considering the environmental and toxicological chemistry of toxicants in the hydrosphere, the effectiveness and rate of biodegradation are quite important. Both physical properties (e.g., water solubility) and chemical properties (e.g., the presence of functional groups amenable to attack) of the substance being degraded are important. The compound in question has to be biodegradable. As shown by the following examples, compounds with straight hydrocarbon chains are much more amenable to biodegradation than are those with branched chains:

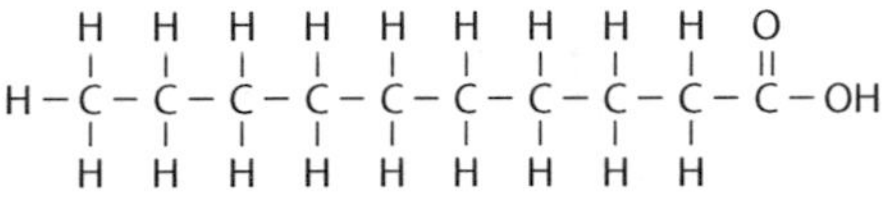

Highly biodegradable compound with a straight hydrocarbon chain and a functional group amenable to biological attack

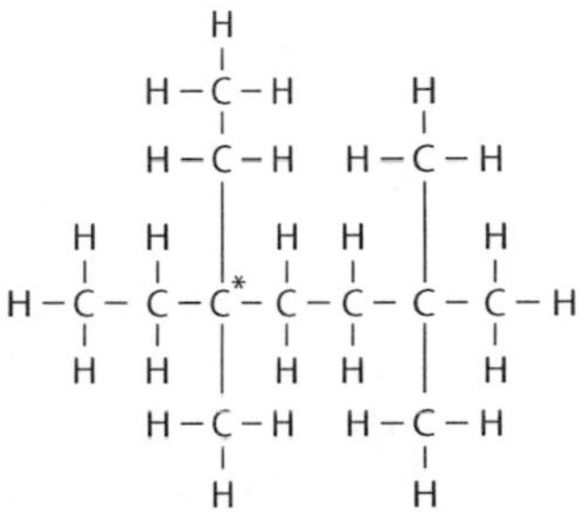

Highly branched, poorly biodegradable compound. The presence of a quaternary carbon atom bound with four other carbons, as denoted by the asterisk above, makes a compound especially resistant to biodegradation

Phenol, a biocidal compound and the first commonly used disinfectant, is an example of a normally nonbiodegradable substance that can be degraded under appropriate conditions. Acclimated bacteria will break down phenol in dilute water solution. When bacteria are used in waste treatment processes to break down substances such as phenol that are normally biorefractory, acclimated microorganisms are often gathered from sites where the substances in question have been present for some time. For example, sites of crude oil spills serve as sources of bacteria that can biodegrade petroleum wastes.

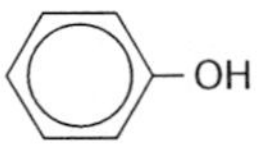

$\bigcirc$—OH **Phenol**

An important consideration in green chemistry is the biodegradability of substances commonly denoted as "consumables" that are likely to be discarded to the environment, especially the hydrosphere. A long-standing example is provided by the surfactants that came into widespread use in detergents in the mid-1900s. These substances were very popular replacements for soap, which works poorly in hard (high Ca^{2+} content) water because of the formation of precipitates of calcium salts of soap anions. Biodegradability of compounds is an important consideration in green chemistry. This is especially true of consumable materials that are dissipated to the environment. One of

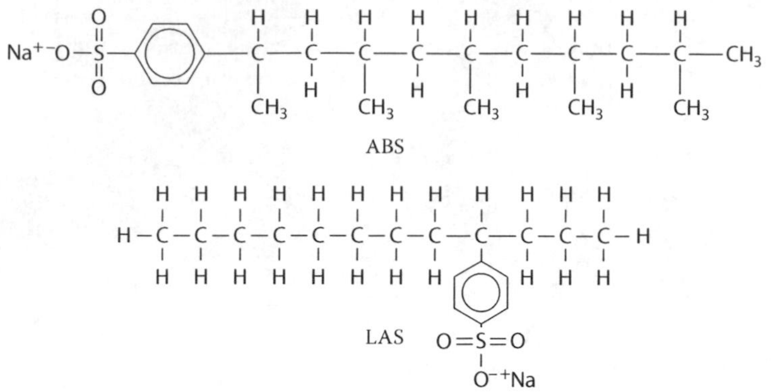

FIGURE 3.17 Surfactants that have been used in household detergent formulations: the alkyl benzene sulfonate surfactant (ABS) originally used did not undergo biodegradation very well and caused problems due to its persistence, which is the result of its branched-chain structure. The ABS was replaced by linear alkyl sulfonate (LAS), which is biodegradable because of its straight-chain structure.

the most common examples of the use of a biodegradable material as a consumable material is the substitution in household detergents of biodegradable linear alkyl sulfonate (LAS) surfactant, which has a readily biodegradable straight hydrocarbon chain as part of its molecular structure, in place of alkyl benzene sulfonate (ABS) surfactant, which has a poorly biodegradable branched chain. As shown in Figure 3.17, the first surfactants used were ABS compounds. These performed very well to "make water wetter"; but within a few years of their introduction, problems of foaming appeared in wastewater treatment plants and in receiving waters and there was some evidence of harm to aquatic life traced to the persistence of ABS compounds, which were poorly biodegradable because of their branched-chain structures. In an exercise in green chemistry several decades before the term was even coined, ABS surfactants were replaced by LAS compounds, which are readily biodegradable because of their straight-chain structures.

QUESTIONS AND PROBLEMS

Access to and use of the Internet is assumed in answering all questions, including general information, statistics, constants, and mathematical formulas required to solve problems. These questions are designed to promote inquiry and thought rather than just finding material in the chapter. So in some cases, there may be several "right" answers. Therefore, if your answer reflects intellectual effort and a search for information from available sources, it may be considered to be right.

1. Look up proposals to restore the Hetch-Hetchy Valley in Yosemite National Park to its former state. How might this affect water supply to parts of California? What might be some of the benefits of the restoration of this valley to its former state?
2. Look up and explain the significance of the name Mulholland in relationship to water. How did Mulholland affect history?
3. Paradoxically, pollution by a strong acid, such as HCl, of groundwater in contact with limestone ($CaCO_3$) can lead to an increase in the alkalinity of the water. Using chemical reactions, explain how this may occur.
4. Tests can be performed on water that show the presence of BOD, and other tests that chemically oxidize organic matter to produce CO_2 can show total organic carbon (TOC). When applied to a particular sample of water, these two sets of tests showed relatively high TOC and relatively low BOD. What does this say about the nature of the organic pollutants in the water?

5. Agricultural fertilizers normally add nitrogen, phosphorus, and potassium to soil. Explain how fertilizer runoff into a body of water can eventually lead to increased BOD pollution.

6. The development of flameless atomic absorption analysis for mercury enabled very sensitive tests for this element around 1970, and they showed surprisingly high levels of this toxic element in some fish samples. The inorganic chemistry of mercury suggests that mercury compounds should precipitate and settle into sediments where they are unavailable to fish. What then was the explanation for the high mercury levels found in some fish around 1970?

7. Phosphate in the form of $H_2PO_4^-$ and HPO_4^{2-} ions is the substance that is usually removed from secondary sewage effluent to prevent excessive algal growth and eutrophication in receiving waters. Of several possible algal nutrients, why is phosphate chosen for this purpose? Show with a chemical reaction the most common means of phosphorous removal.

8. Membrane filtration processes can be very effective in removing residual BOD from secondary wastewater effluent. What does this suggest regarding the nature of contaminants responsible for the BOD?

9. Using the Internet, gather information regarding the use of wastewater for irrigation. Is this a practice that is used, and if so where does it usually take place? What are some of its benefits? What are some of its risks?

10. Water is used for both its special solvent properties and its ability to absorb, transfer, and release heat energy. Explain on the basis of the characteristics of the water molecule how these two uses are related.

11. How far back into history does the use of waterpower go? Which civilizations were the first to use it? Explain why during the mid-1800s waterpower development slowed, only to start growing rapidly around the late 1800s and early 1900s.

12. What is plaster of Paris? Show with a chemical reaction how water is employed as a chemical reagent in making objects from plaster of Paris.

LITERATURE CITED

1. Manahan, Stanley E., *Water Chemistry: Green Science and Technology of Nature's Most Renewable Resource*, Taylor & Francis/CRC Press, Boca Raton, FL, 2010.
2. Manahan, Stanley E., *Environmental Science and Technology: A Sustainable Approach to Green Science and Technology*, 2nd ed., Taylor & Francis/CRC Press, Boca Raton, FL, 2006.
3. Manahan, Stanley E., *Environmental Chemistry*, 9th ed., Taylor & Francis/CRC Press, Boca Raton, FL, 2009.

SUPPLEMENTARY REFERENCES

Amjad, Sahid, *The Science and Technology of Industrial Water Treatment*, Taylor & Francis/CRC Press, Boca Raton, FL, 2010.

Benjamin, Mark M., *Water Chemistry*, Waveland Press, Inc., Long Grove, IL, 2002.

Berk, Zeki, *Water Science for Food, Health, Agriculture and Environment*, Technomic Publishing, Lancaster, PA, 2001.

Brezonik, Patrick L., and William A. Arnold, *Water Chemistry: An Introduction to the Chemistry of Natural and Engineered Aquatic System*, Oxford University Press, Oxford, UK, 2011.

Bundschuh, Jochen, and Jan Hoinkis, Eds., *Renewable Energy Applications for Freshwater Production*, Taylor & Francis/CRC Press, Boca Raton, FL, 2012.

Crittenden, John, *Water Treatment: Design and Principles*, Wiley, New York, 2005.

Dodds, Walter K., and Matt R. Whiles, *Freshwater Ecology: Concepts and Environmental Applications of Limnology*, 2nd ed., Academic Press, Boston, 2010.

Dodson, Stanley, I., *Introduction to Limnology*, McGraw-Hill, New York, 2005.

Essington, Michael E., *Soil and Water Chemistry: An Integrative Approach*, Taylor & Francis/CRC Press, Boca Raton, FL, 2004.

Hammer, Mark, *Water and Wastewater Technology*, 6th ed., Prentice Hall, Upper Saddle River, NJ, 2007.

Kalff, Jacob, *Limnology: Inland Water Ecosystems*, Prentice Hall, Upper Saddle River, NJ, 2002.

Karamouz, Mohammed, Azadeh Ahmadi, and Masih Akhbari, *Groundwater Hydrology*, Taylor & Francis/CRC Press, Boca Raton, FL, 2011.

Kumar, Arvind, Ed., *Aquatic Ecosystems*, A.P.H. Publishing Corp., New Delhi, 2003.

Namiesnik, Jacek, and Piotr Szefer, Eds., *Analytical Measurements in Aquatic Environments*, Taylor & Francis/CRC Press, London, 2010.

Nollet, Leo M. L., *Handbook of Water Analysis*, 2nd ed., Taylor & Francis/CRC Press, London or Boca Raton, FL, 2008.

Salina, Irena, Ed., Written in Water: Messages of Hope for Earth's Most Precious Resource, *National Geographic*, Washington, DC 2010.

Spellman, Frank R., and Joanne E. Drinan, *The Drinking Water Handbook*, Taylor & Francis/CRC Press, Boca Raton, FL, 2012.

Spellman, Frank R., *The Science of Water: Concepts and Applications*, 2nd ed., Taylor & Francis/CRC Press, Boca Raton, FL, 2007.

Sullivan, Patrick, Franklin J. Agardy, and James J. J. Clark, *The Environmental Science of Drinking Water*, Butterworth-Heineman, Burlington, MA, 2005.

Treidel, Holger, Jose Luis Martin-Bordes, and Jason J. Gurdak, Eds., *Climate Change Effects on Groundwater Resources*, Taylor & Francis/CRC Press, Boca Raton, FL, 2011.

Trimble, Stanley W., *Encyclopedia of Water Science*, 2nd ed., Taylor & Francis/CRC Press, London or Boca Raton, FL, 2008.

Tundisi, Jose Galizia, and Takako Matsumura Tundisi, *Limnology*, Taylor & Francis/CRC Press, Boca Raton, FL, 2012.

Water Environment Federation, *Industrial Wastewater Management, Treatment, and Disposal*, 3rd ed., McGraw-Hill Professional, New York, 2008.

Weiner, Eugene R., *Applications of Environmental Aquatic Chemistry: A Practical Guide*, 2nd ed., Taylor & Francis/CRC Press, Boca Raton, FL, 2012.

Welch, E. B., and Jean Jacoby, *Pollutant Effects in Fresh Waters: Applied Limnology*, Taylor & Francis, London, 2007.

4 Pollution of the Hydrosphere

4.1 NATURE AND TYPES OF WATER POLLUTANTS

Throughout history, the quality of drinking water has been a factor in determining human welfare. Fecal pollution of drinking water has frequently caused waterborne diseases that have decimated the populations of whole cities. Unwholesome water polluted by sewage has caused great hardship for people forced to drink it or use it for irrigation. Although waterborne diseases are now well controlled in technologically advanced countries, the shortage of safe drinking water is still a major problem in regions afflicted by strife and poverty, and dysentery spread by pathogens in drinking water takes an especially high death toll of children in impoverished parts of the world.

An ongoing concern with water safety now is the potential presence of chemical pollutants. These may include organic chemicals, inorganics, and heavy metals from industrial, urban runoff, and agricultural sources. Water pollutants can be divided among some general categories, as summarized in Table 4.1. Most of these categories of pollutants, and several subcategories, are discussed in this chapter.

4.1.1 MARKERS OF WATER POLLUTION

Markers of water pollution show the presence of pollution sources. These include herbicides indicative of agricultural runoff, fecal coliform bacteria that are characteristic of pollution from sewage, and pharmaceuticals, pharmaceutical metabolites, and even caffeine that show contamination by domestic wastewater.

Biomarkers of water pollution are organisms that live in or are closely associated with bodies of water and that provide evidence of pollution either from accumulation in the organism of water pollutants or their metabolites or from effects on the organism owing to pollutant exposure. Fish are the most common bioindicators of water pollution, and fish lipid (fat) tissue is commonly analyzed for persistent organic water pollutants, which tend to become concentrated in lipid tissue. The osprey, a large, widely distributed fish-eating bird at the top of the food chain, has proven to be a good water pollution indicator species through chemical and biochemical analysis of its feathers, eggs, blood, and organs along with observations of behavior, nesting habits, and populations.[1]

4.2 ELEMENTAL POLLUTANTS

Table 4.2 lists the most important **trace elements** (those encountered at levels of a few parts per million or less) found in natural waters and wastewaters. Typical of the behavior for many substances in the aquatic environment, some of these elements are essential plant and animal nutrients at lower levels but toxic at higher levels. Several of these, such as lead or mercury, have such toxicological and environmental significance that they are discussed in detail in separate sections.

A few of the **heavy metals** are among the most harmful of the elemental pollutants and are of particular concern because of their toxicities to humans. These elements are, in general, the transition metals and some of the representative elements, such as lead and tin, in the lower right-hand corner of the periodic table. Heavy metals include essential elements like iron and toxic metals like cadmium and mercury. Most of them have a tremendous affinity for sulfur and disrupt enzyme function by forming bonds with sulfur groups in enzymes. Protein carboxylic acid ($-CO_2H$) and

TABLE 4.1
General Types of Water Pollutants

Class of Pollutant	Significance
Trace elements	Health, aquatic biota, toxicity
Heavy metals	Health, aquatic biota, toxicity
Organically bound metals	Metal transport
Radionuclides	Toxicity
Inorganic pollutants	Toxicity, aquatic biota
Asbestos	Human health
Algal nutrients	Eutrophication
Acidity, alkalinity, salinity (in excess)	Water quality, aquatic life
Trace organic and refractory pollutants	Toxicity
Polychlorinated biphenyls	Possible biological effects
Pesticides	Toxicity, aquatic biota, wildlife
Petroleum wastes	Effect on wildlife, esthetics
Sewage, human and animal wastes	Water quality, oxygen levels
Biochemical oxygen demand	Water quality, oxygen levels
Pathogens	Health effects
Detergents	Eutrophication, wildlife, esthetics
Chemical carcinogens	Incidence of cancer
Sediments	Water quality, aquatic biota, wildlife
Taste, odor, and color	Esthetics

TABLE 4.2
Important Trace Elements in Natural Waters and Wastewaters

Element	Sources	Effects and Significance
Arsenic	Mining by-product, chemical waste	Toxic, possibly carcinogenic
Beryllium	Coal, industrial wastes	Toxic
Boron	Coal, detergents, wastes	Toxic
Chromium	Metal plating	Essential as Cr(III), toxic as Cr(VI)
Copper	Metal plating, mining, industrial waste	Essential trace element, toxic to plants and algae at higher levels
Fluorine (F^-)	Natural geological sources, wastes, water additive	Prevents tooth decay at around 1 mg/L, toxic at higher levels
Iodine (I^-)	Industrial wastes, natural brines, seawater intrusion	Prevents goiter
Iron	Industrial wastes, corrosion, acid mine water, microbial action	Essential nutrient, damages fixtures by staining
Lead	Industrial waste, mining, fuels	Toxic, harmful to wildlife
Manganese	Industrial wastes, acid mine water, microbial action	Toxic to plants, damages fixtures by staining
Mercury	Industrial waste, mining, coal	Toxic, mobilized as methyl mercury compounds by anoxic bacteria
Molybdenum	Industrial wastes, natural sources	Essential to plants, toxic to animals
Selenium	Natural sources, coal	Essential at lower levels, toxic at higher levels
Zinc	Industrial waste, metal plating, plumbing	Essential element, toxic to plants at higher levels

amino ($-NH_2$) groups are also chemically bound by heavy metals. Cadmium, copper, lead, and mercury ions bind to cell membranes, hindering transport processes through the cell wall. Heavy metals may also precipitate phosphate biocompounds or catalyze their decomposition.

Some of the **metalloids**, elements on the borderline between metals and nonmetals, are significant water pollutants. Arsenic, selenium, and antimony are of particular interest.

Inorganic chemicals' manufacture has the potential to contaminate water with trace elements. Among the industries regulated for potential trace element pollution of water are those producing chlor-alkali, hydrofluoric acid, sodium dichromate (sulfate process and chloride ilmenite process), aluminum fluoride, chrome pigments, copper sulfate, nickel sulfate, sodium bisulfate, sodium hydrosulfate, sodium bisulfite, titanium dioxide, and hydrogen cyanide.

4.3 HEAVY METALS

4.3.1 CADMIUM

Pollutant **cadmium** in water may arise from mining wastes and industrial discharges, especially from metal plating. Chemically, cadmium is very similar to zinc, and these two metals frequently undergo geochemical processes together. Both metals are found in water in the +2 oxidation state. Cadmium and zinc are common water and sediment pollutants in harbors surrounded by industrial installations. Concentrations of more than 100 ppm dry mass of sediment have been found in harbor sediments.

Cadmium has a long half-life of more than 10 years in the human body, with about 50% of the body burden of this heavy metal residing in the kidney. Cadmium causes tubule dysfunction and injury in the kidney. Other effects of acute cadmium poisoning in humans include high blood pressure, kidney damage, damage to testicular tissue, and destruction of red blood cells. Much of the physiological action of cadmium is due to its chemical similarity to zinc. Cadmium may replace zinc in some enzymes, thereby altering the stereostructure of the enzyme and impairing its catalytic activity, causing disease symptoms.

Cadmium is alleged to be the cause of itai-itai (in Japanese, literally "ouch-ouch") disease that was first recognized in the 1940s in the Jinzu river basin of Japan. Affecting primarily post-menopausal women, itai-itai disease is manifested by skeletal, blood, and kidney disorders. The bones of victims become so fragile that even the act of sneezing can cause painful rib fracture, hence the name of the malady. Water in the Jinzu basin became contaminated with cadmium from cadmium mining activities that began in the 1930s, which resulted in cadmium-contaminated drinking water. Use of the contaminated water to irrigate rice fields resulted in cadmium contamination of the rice cereal, a staple food in the area, and led to cadmium poisoning of those who consumed the cereal.

4.3.2 LEAD

Inorganic **lead** arising from a number of industrial and mining sources and formerly from leaded gasoline occurs in water in the +2 oxidation state. In addition to pollutant sources, lead-bearing limestone and galena (PbS) contribute lead to natural waters in some locations. Evidence from hair samples and other sources indicates that body burdens of this toxic metal have decreased during recent decades, largely the result of less lead used in plumbing and other products that come in contact with food or drink.

Acute lead poisoning in humans causes severe dysfunction in the kidneys, reproductive system, liver, and the brain and central nervous system, leading to sickness or even death. Lead poisoning from environmental exposure is thought to have caused mental retardation in many children. Lead poisoning causes multiple hematological (blood) effects including anemia. Peripheral neuropathy including symptoms of foot drop and wrist drop has been observed in workers exposed to lead in an industrial setting. The victim of lead poisoning may have headaches and sore muscles and feel generally fatigued and irritable.

Except in isolated cases, lead is probably not a major problem in drinking water, although the potential exists in cases where old lead pipe is still in use. Lead used to be a constituent of solder and some pipe-joint formulations, and so in some cases, household water does have some contact with lead. Water that has stood in household plumbing for some time may have elevated levels of lead (along with zinc, cadmium, and copper) and should be drained for a while before use.

4.3.3 MERCURY

Because of its toxicity, mobilization as methylated (organometallic) forms by anoxic bacteria, and other pollution factors, **mercury** generates a great deal of concern as a heavy-metal pollutant. Mercury is found as a trace component of many minerals, with continental rocks containing an average of around 80 parts per billion, or slightly less, of this element. Fossil fuel coal and lignite contain mercury, often at levels of 100 parts per billion or even higher, and emissions from the combustion of these fuels are a major source of environmental mercury. Mercury that enters the atmosphere from the combustion of coal and from waste incinerators may end up as a water pollutant, often as methylated organometallic mercury (see Reaction 4.1).

Metallic mercury once was commonly used as an electrode in the electrolytic generation of chlorine gas, in laboratory vacuum apparatus, and in other applications, and significant quantities of inorganic mercury(I) and mercury(II) compounds were used annually. Organic mercury compounds used to be widely applied as pesticides, particularly fungicides, but these uses have been essentially eliminated because of health concerns.

Globally, one of the greatest sources of mercury pollution has been mercury used to extract gold from gold-bearing ores. It is estimated that each year 15 million miners in approximately 40 developing countries use 650–1000 metric tons of mercury to extract gold. This use subjects many local areas to mercury contamination and contributes significantly to the global environmental burden of mercury and exposes the miners, many of whom are children, to toxic mercury. The problem has been made worse in recent years as industrialized countries have reduced mercury use, with the result that surplus mercury has been made available on the world market for gold extraction. In recognition of this problem, in 2008, the European Union adopted a mercury export ban to take effect in 2011 and the United States adopted a similar ban to be effective in 2013.

The toxicity of mercury was tragically illustrated in the Minamata Bay area of Japan during the period 1953–1960. A total of 111 cases of mercury poisoning and 43 deaths were reported among people who had consumed seafood from the bay that had been contaminated with mercury waste that drained into Minamata Bay from a chemical plant. Congenital defects were observed in 19 babies whose mothers had consumed seafood contaminated with mercury. The level of metal in the contaminated seafood was 5–20 parts per million.

Among the toxicological effects of mercury are neurological damage, including irritability, paralysis, blindness, or insanity; chromosome breakage; and birth defects. The milder symptoms of mercury poisoning such as depression and irritability have a psychopathological character. Because of the resemblance of these symptoms to common human behavioral problems, mild mercury poisoning may escape detection. Some forms of mercury are relatively nontoxic, and they were formerly used as medicines, for example, in the treatment of syphilis. Other forms of mercury, particularly organic compounds, are highly toxic.

Because there are few major natural sources of mercury and since most inorganic compounds of this element are relatively insoluble, it was assumed for some time that mercury was not a serious water pollutant. However, in 1970, alarming mercury levels were discovered in fish in Lake Saint Clair located between Michigan and Ontario, Canada. A subsequent survey by the U.S. Federal Water Quality Administration revealed a number of other waters contaminated with mercury. It was found that several chemical plants, particularly caustic chemical manufacturing operations for the production of chlorine and sodium hydroxide, were each releasing up to 14 or more kilograms of mercury in wastewaters each day.

The unexpectedly high concentrations of mercury found in water and in fish tissues result from the formation of soluble monomethylmercury ion, CH_3Hg^+, and volatile dimethylmercury, $(CH_3)_2Hg$, by anoxic bacteria in sediments. Mercury from these compounds becomes concentrated in fish lipid (fat) tissue, and the concentration factor from water to fish may exceed 10^3. The methylating agent by which inorganic mercury is converted to methylmercury compounds is methylcobalamin, a vitamin B_{12} analog:

$$HgCl_2 \xrightarrow{\text{Methylcobalamin}} CH_3HgCl + Cl^- \tag{4.1}$$

It is believed that the bacteria that synthesize methane produce methylcobalamin as an intermediate in the synthesis. Thus, waters and sediments in which anoxic decay leading to methane formation is occurring provide the conditions under which methylmercury production occurs. In neutral or alkaline waters, volatile dimethylmercury, $(CH_3)_2Hg$, may be formed.

4.4 METALLOIDS

The most significant water pollutant metalloid element is arsenic, a toxic element that has been the chemical villain of more than a few murder plots. Acute arsenic poisoning can result from the ingestion of more than about 100 mg of the element. Chronic poisoning occurs with the ingestion of small amounts of arsenic over a long period of time. There is some evidence that this element is also carcinogenic.

Arsenic occurs in the Earth's crust at an average level of 2–5 ppm. The combustion of fossil fuels, particularly coal, introduces significant quantities of arsenic into the environment, much of which reaches natural waters. Arsenic occurs with phosphate minerals and enters into the environment along with some phosphorus compounds. Some formerly used pesticides, particularly those from before World War II, contain highly toxic arsenic compounds. The most common of these are lead arsenate, $Pb_3(AsO_4)_2$; sodium arsenite, Na_3AsO_3; and Paris Green, $Cu_3(AsO_3)_2$. Another major source of arsenic is mine tailings. Arsenic produced as a by-product of copper, gold, and lead refining exceeds the commercial demand for arsenic, and it accumulates as waste material.

Like mercury, arsenic may be converted to more mobile and toxic methyl derivatives by bacteria, first by reduction of H_3AsO_4 to H_3AsO_3, then by methylation to produce $CH_3AsO(OH)_2$ (methylarsinic acid), $(CH_3)_2AsO(OH)$ (dimethyarsinic acid), and $(CH_3)_2AsH$ (dimethylarsine).

In what may have been the largest mass poisoning of a human population in history, between 35 million and 77 million people of the approximately 160 million inhabitants of Bangladesh were exposed to potentially toxic levels of arsenic in drinking water. This catastrophic public health problem has resulted from well-intentioned programs funded initially by the United Nations Children's Fund to install shallow tube wells that provided a source of drinking water free from disease-causing pathogens. By 1987, numerous cases of arsenic-induced skin lesions characterized by pigmentation changes, predominantly on the upper chest, arms, and legs, and keratoses of the palms of the hands and soles of the feet were being observed, characteristics of arsenic poisoning, that lead to the discovery that arsenic-contaminated drinking water from the tube wells was responsible. Since the initial discovery of arsenic poisoning, huge numbers of new cases have been revealed, and it is possible that tens of thousands of people in Bangladesh will die prematurely because of exposure to arsenic in drinking water. Other countries with acute problems from arsenic in drinking water include Vietnam (where large numbers of tube wells were drilled more recently than those in Bangladesh) Argentina, Chile, China, Mexico, Taiwan, and Thailand.

The geochemical conditions that result in arsenic contamination of water are often associated with the presence of iron, sulfur, and organic matter in (alluvial) deposits produced by water. Iron released from rocks eroded by river water forms iron oxide deposits on rock particle surfaces.

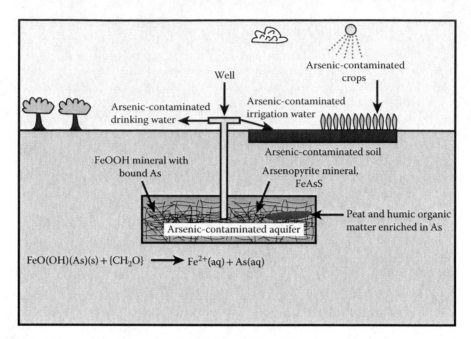

FIGURE 4.1 Tube wells sunk into arsenic-containing aquifer formations may yield water contaminated by toxic soluble arsenic. Use of the water for irrigation can contaminate soil with arsenic and provide a pathway for arsenic to get into food crops such as rice. The problem has been especially acute in Bangladesh, but has occurred in other parts of the world as well.

The iron oxide accumulates arsenic and concentrates it from the river water. These particles then become buried along with degradable organic matter in sediments, and the insoluble iron(III) in the iron oxides is converted to soluble iron(II) by the anoxic reducing conditions under which the organic matter biodegrades. This releases bound arsenic that can get into well water. The oxidation of arsenopyrite, FeAsS, can release arsenic as can the decomposition of arsenic-contaminated organic matter (peat) in underground deposits. The pathway for arsenic from underground sources to get into drinking water, irrigation water, and food crops is shown in Figure 4.1.

4.5   ORGANICALLY BOUND METALS

The binding of metals including toxic heavy metals by organic entities is discussed in Section 3.11. Metal chelates form by the reversible binding of metal ions with organic chelating agents to produce metal chelates such as the nitrilotriacetate chelate of zinc shown in Figure 3.13. Another form of organically bound metals and metalloids consists of organometallic compounds, such as the methylmercury compounds discussed in Section 4.3. Organometallic compounds have a carbon–metal bond, although compounds with similar properties may involve linking of metal atoms with organic groups through other kinds of atoms, especially oxygen.

The interaction of trace metals with organic compounds in natural waters and wastewaters is an important aspect of water pollution. It may be noted that metal–organic interactions may involve organic species of both pollutant (such as EDTA) and natural (such as fulvic acids, Figure 3.14) origin. These interactions are influenced by, and sometimes play a role in, redox equilibria, formation and dissolution of precipitates, colloid formation and stability, acid-base reactions, and microorganism-mediated reactions in water. Metal–organic interactions may increase or decrease the toxicity of metals in aquatic ecosystems, and they have a strong influence on the growth of algae in water.

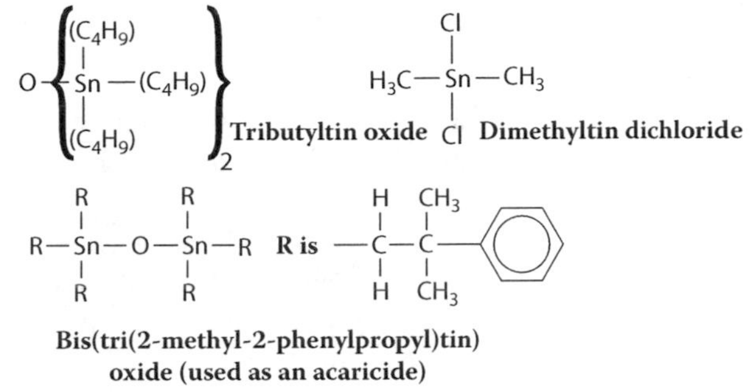

FIGURE 4.2 Examples of organotin compounds.

4.5.1 ORGANOTIN COMPOUNDS

Of all the metals, tin has had the greatest number of organometallic compounds in commercial use, with global production that reached 40,000 metric tons per year before applications were restricted because of concerns over water pollution. In addition to synthetic organotin compounds, methylated tin species can be produced biologically in the environment. Figure 4.2 gives some examples of the many known organotin compounds.

Major industrial uses of organotin compounds in the past included applications of tin compounds in fungicides, acaricides, disinfectants, antifouling paints, stabilizers to lessen the effects of heat and light in PVC plastics, catalysts, and precursors for the formation of films of SnO_2 on glass. Tributyl tin (TBT) chloride and related TBT compounds have bactericidal, fungicidal, and insecticidal properties and formerly were of particular environmental significance because of their use as industrial biocides. In addition to TBT chloride, other TBT compounds used as biocides included the hydroxide, oxide, naphthenate, bis(tributyltin) oxide, and tris(tributylstannyl) phosphate. TBT was once widely used in boat and ship hull coatings to prevent the growth of fouling organisms. Other applications have included preservation of wood, leather, paper, and textiles. Antifungal TBT compounds have been used as slimicides in cooling tower water.

The many applications of organotin compounds for a variety of uses have posed a significant potential for environmental pollution. The greatest environmental and toxicological concern with organotin compounds has been their uses as antifouling agents, especially TBT compounds used as ingredients in coatings on ship hulls and other structures in contact with seawater. **Biofouling**, in which marine organisms produce fouling communities on surfaces in contact with water, is a tremendous economic problem, especially in that it adds significantly to the consumption of fuel for ship propulsion and slows the movement of ships through water. Such biofouling may be divided into the following two categories: (1) **microfouling**, in which microorganism biofilms form on surfaces, and (2) **macrofouling**, which involves the attachment of larger organisms including barnacles, bryozoans, mussels, polychaete worms, and seaweeds.[2] Organotin compounds used in antifouling agents have been linked to endocrine disruption in shellfish, oysters, and snails. Because of such concerns, several countries, including the United States, England, and France, prohibited TBT application on vessels smaller than 25 meters in length during the 1980s, and efforts continue to ban organotin compounds in antifouling coatings applied to ship hulls.

4.6 INORGANIC SPECIES AS WATER POLLUTANTS

Some important inorganic water pollutants are mentioned earlier as part of the discussion of pollutant trace elements. Inorganic pollutants that contribute acidity, alkalinity, or salinity to water are considered separately in this chapter. Still another class is that of algal nutrients. This leaves unclassified, however, some important inorganic pollutant species, of which cyanide ion, CN^-, is probably the most important. Others include ammonia, carbon dioxide, hydrogen sulfide, nitrite, and sulfite.

4.6.1 CYANIDE

Cyanide, a deadly poisonous substance, exists in water as HCN, a weak acid, with a K_a of 6×10^{-10}. The cyanide ion has a strong affinity for many metal ions, forming relatively less-toxic ferrocyanide, $Fe(CN)_6^{4-}$, with iron(II), for example. Volatile HCN is very toxic and has been used in gas chamber executions in the United States.

Cyanide has some significant industrial uses, especially for metal cleaning and electroplating. It is also one of the main gas and coke scrubber effluent pollutants from gas works and coke ovens. Cyanide is also used in some mineral-processing operations, especially in extracting gold from low-grade ores. In this application, called "heap leaching," the gold ore is placed in a pile and a solution of sodium cyanide is sprayed over the ore for a prolonged period of time. In the presence of atmospheric oxygen, the following reaction occurs to extract the gold:

$$2Au(s) + 8CN^-(aq) + O_2(g) + 2H_2O \rightarrow 2Au(CN)_4^{2-}(aq) + 4OH^-(aq) \qquad (4.2)$$

Spraying large piles of gold ore with a toxic cyanide solution in the open is a somewhat dicey operation, despite which very few workers have died from accidental cyanide exposure in the process. Some significant fish kills have resulted from discharge of cyanide from mineral processing operations into waterways.

4.6.2 AMMONIA AND OTHER INORGANIC WATER POLLUTANTS

Excessive levels of ammoniacal nitrogen cause water-quality problems. **Ammonia** is the initial product of the decay of nitrogenous organic wastes, and its presence frequently indicates the presence of such wastes. It is a normal constituent of low-pE groundwaters and is sometimes added to drinking water as an aid to disinfection, where it reacts with chlorine to provide residual chlorine (see Section 5.11). Since the pK_a of ammonium ion, NH_4^+, is 9.26, most ammonia in water is present as NH_4^+ rather than as NH_3.

Hydrogen sulfide, H_2S, is a product of the anoxic decay of organic matter containing sulfur. It is also produced in the anoxic reduction of sulfate by microorganisms (see Chapter 3, Reaction 3.8) and is evolved as a gaseous pollutant from geothermal waters. Wastes from chemical plants, paper mills, textile mills, and tanneries may also contain H_2S. Its presence is easily detected by its characteristic rotten-egg odor. In water, H_2S is a weak diprotic acid with pK_{a1} of 6.99 and pK_{a2} of 12.92; S^{2-} is not present in normal natural waters. The sulfide ion has tremendous affinity for many heavy metals, and heavy metals present in water containing H_2S will precipitate as metal sulfides.

Hydrogen sulfide is a very toxic gas, killing more people on average than hydrogen cyanide, usually from releases in natural gas production and in industrial settings. If the geothermally active Yellowstone National Park in the United States was an industrial installation, it is likely that authorities would consider placing off-limits some areas in which the geothermal waters are laden with hydrogen sulfide because of emissions of the lethal gas, its odor readily detected by visitors.

Free **carbon dioxide**, CO_2, is frequently present in water at high levels as the result of decay of organic matter. It is also added to softened water during water treatment as part of a recarbonation process (see Chapter 5). Excessive carbon dioxide levels may make water more corrosive and may be harmful to aquatic life.

As noted in Section 3.7, in 1986, carbon dioxide of volcanic origin released suddenly from bottom regions of Lake Nyos in the central African nation of Cameroon asphyxiated about 1700 people and thousands of livestock; 37 people had been killed 2 years before by a similar release from nearby Lake Monoun. The physical conditions leading to these releases are interesting, in that under high pressure, lake water can dissolve about five times its volume of carbon dioxide gas. Because of the high molecular mass of CO_2, the presence of so much of the dissolved gas in the bottom regions of a lake produces a dense hypolimnion layer (see lake stratification in Figure 3.3) that may be stable for years. However, formation and nucleation of carbon dioxide gas bubbles in this layer produces a much less dense mixture of water and gaseous CO_2. The result can be a cascading positive feedback effect, in which the bottom layer begins to rise abruptly as more gas is released, and a total overturn of the lake water can occur very quickly, which is what happened with the two lakes mentioned earlier. Efforts have now been made to immerse pipes into the bottom regions of the lakes, initiating an upward flow of water buoyed by the presence of carbon dioxide bubbles, thus resulting in a controlled release of dissolved carbon dioxide and preventing a catastrophic release of the gas at a later time.

Nitrite ion, NO_2^-, occurs in water as an intermediate oxidation state of nitrogen over a relatively narrow pE range. Nitrite is added to some industrial process water as a corrosion inhibitor. However, it rarely occurs in drinking water at levels over 0.1 mg/L.

Sulfite ion, SO_3^{2-}, is found in some industrial wastewaters. Sodium sulfite is commonly added to boiler feedwaters as an oxygen scavenger:

$$2SO_3^{2-} + O_2 \rightarrow 2SO_4^{2-} \tag{4.3}$$

Since pK_{a1} of sulfurous acid is 1.76 and pK_{a2} is 7.20, sulfite exists as either HSO_3^- or SO_3^{2-} in natural waters, depending upon the pH. It may be noted that hydrazine, N_2H_4, also functions as an oxygen scavenger:

$$N_2H_4 + O_2 \rightarrow 2H_2O + N_2(g) \tag{4.4}$$

Perchlorate ion, ClO_4^-, emerged as a water pollution problem in some areas in the 1990s when advances in ion chromatography enabled its detection in the low parts per billion range of concentrations. Ammonium perchlorate, NH_4ClO_4, has been widely manufactured as an oxidizer for use in solid rocket propellants, and contamination from ammonium perchlorate manufacturing facilities has been regarded as the major source of contamination. Perchlorate in water is very unreactive, and all common perchlorate salts other than $KClO_4$ are soluble, so it is difficult to remove them. Physiologically, it competes with iodide ion and diminishes essential iodide uptake by the thyroid. The U.S. Environmental Protection Agency has recommended a drinking water standard for perchlorate of 1 part per billion (μg/L).

4.6.3 ASBESTOS IN WATER

The toxicity of inhaled asbestos is well established. The fibers scar lung tissue and cancer eventually develops, often 20 or 30 years after exposure. It is not known for sure whether asbestos is toxic in drinking water. This has been a matter of considerable concern because of the dumping of taconite (iron ore tailings) containing asbestos-like fibers into Lake Superior. The fibers have been found in drinking waters of cities around the lake. After having dumped the tailings into Lake Superior since 1952, the Reserve Mining Company at Silver Bay on Lake Superior solved the problem in 1980 by constructing a 6-square-mile containment basin inland from the lake. This $370-million facility keeps the taconite tailings covered with a 3-meter layer of water to prevent escape of fiber dust.

4.7 ALGAL NUTRIENTS AND EUTROPHICATION

The term **eutrophication**, derived from the Greek word meaning "well-nourished," describes a condition of lakes or reservoirs involving excess algal growth. Although some algal productivity is necessary to support the food chain in an aquatic ecosystem, excess growth under eutrophic conditions may eventually lead to severe deterioration of the body of water. The first step in eutrophication of a body of water is an input of plant nutrients (Table 4.3) from watershed runoff or sewage. The nutrient-rich body of water then produces a great deal of plant biomass by photosynthesis, along with a smaller

TABLE 4.3

Essential Plant Nutrients: Sources and Functions

Nutrient	Source	Function
Macronutrients		
Carbon (CO_2)	Atmosphere, decay	Biomass constituent
Hydrogen	Water	Biomass constituent
Oxygen	Water	Biomass constituent
Nitrogen (NO_3^-)	Decay, pollutants, atmosphere	Protein constituent (from nitrogen-fixing organisms)
Phosphorus	Decay, minerals, (phosphate)	DNA/RNA constituent pollutants
Potassium	Minerals, pollutants	Metabolic function
Sulfur (sulfate)	Minerals	Proteins, enzymes
Magnesium	Minerals	Metabolic function
Calcium	Minerals	Metabolic function
Micronutrients		
B, Cl, Co, Cu, Fe, Mo, Mn, Na, Si, V, Zn	Minerals, pollutants	Metabolic function and constituent of enzymes

amount of animal biomass. Dead biomass accumulates in the bottom of the lake, where it partially decays, recycling nutrient carbon dioxide, phosphorus, nitrogen, and potassium. In more shallow regions of the lake, bottom-rooted plants begin to grow, accelerating the accumulation of solid material in the basin. Eventually a marsh is formed, which finally fills in to produce a meadow or forest.

Eutrophication is often a natural phenomenon; for instance, it was responsible for much of the plant growth that resulted in the formation of huge deposits of coal and peat millions of years ago. However, human activity can greatly accelerate the process. To understand why this is so, consider that most of the nutrients required for plant and algae growth shown in Table 4.3 are available in adequate amounts from natural sources. The nutrients most likely to be limiting are the "fertilizer" elements: nitrogen, phosphorus, and potassium. These are all present in sewage and are, of course, found in runoff from heavily fertilized fields. They are also constituents of various kinds of industrial wastes. Each of these elements can also come from natural sources—phosphorus and potassium from mineral formations and nitrogen fixed by bacteria, photosynthetic cyanobacteria, or discharge of lightning in the atmosphere.

In some cases, nitrogen or even carbon may be limiting nutrients, the presence of which determines the rate of algal growth. This is particularly true of nitrogen in seawater. In seawater, micronutrients, particularly iron, may be limiting.

In most cases in freshwaters, the single plant nutrient most likely to be limiting is phosphorus, and it is generally named as the culprit in excessive eutrophication. Household detergents were once a common source of phosphate in wastewater, and eutrophication control has concentrated upon eliminating phosphates from detergents, removing phosphate at the sewage treatment plant, and preventing phosphate-laden sewage effluents from entering bodies of water, which would enable the excessive growth of algae that can cause eutrophication.

4.8   ACIDITY, ALKALINITY, AND SALINITY

Aquatic biota are sensitive to extremes of pH. Largely because of osmotic effects, they cannot live in a medium having a salinity to which they are not adapted. Thus, a freshwater fish soon succumbs in the ocean, and sea fish normally cannot live in freshwater, although some fish, most notably salmon, can function in both media. Excess salinity soon kills plants not adapted to it. There are,

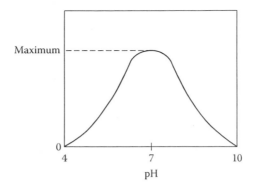

FIGURE 4.3   A generalized plot of the growth of an aquatic organism as a function of pH.

of course, ranges in salinity and pH in which organisms live and at the extremes of which they may exist without really thriving (see Figure 4.3).

The most common source of **pollutant acid** in water is acid mine drainage. The sulfuric acid in such drainage arises from the microbial oxidation of pyrite or other sulfide minerals as described in Section 3.5. The values of pH encountered in acid-polluted water may fall below 3, a condition deadly to most forms of aquatic life except the culprit bacteria mediating the pyrite and iron(II) oxidation, which thrive under very low pH conditions. Industrial wastes frequently have the potential to contribute strong acid to water. Sulfuric acid produced by the air oxidation of pollutant sulfur dioxide (see Chapter 7) enters natural waters as acidic rainfall. In cases where the water does not have contact with a basic mineral, such as limestone, the water pH may become dangerously low. This condition occurs in some Canadian and Scandinavian lakes, for example.

Excess **alkalinity** and frequently accompanying high pH generally are not introduced directly into water from anthrospheric sources. However, in many geographic areas, the soil and mineral strata are alkaline and impart a high alkalinity to water. Human activity can aggravate the situation, for example, by exposure of alkaline overburden from strip mining to surface water or groundwater. Excess alkalinity in water is manifested by a characteristic fringe of white salts at the edges of a body of water or on the banks of a stream.

Water **salinity** may be increased by a number of human activities. Water passing through a municipal water system picks up salt from the process of recharging water softeners with sodium chloride. Salts can leach from spoil piles. One of the major environmental constraints on the production of shale oil, for example, is the high percentage of leachable sodium sulfate in the piles of spent shale. Careful control of these wastes is necessary to prevent further saline pollution of water in potential shale oil producing areas where salinity is already a problem. Irrigation adds a great deal of salt to water, a phenomenon responsible for the Salton Sea in California, and is a source of conflict between the United States and Mexico over saline contamination of the Rio Grande and Colorado rivers. Irrigation and intensive agricultural production have caused saline seeps in some of the western states. These occur when water seeps into a slight depression in tilled, sometimes irrigated, fertilized land, carrying salts (particularly sodium, magnesium, and calcium sulfates) along with it. The water evaporates in the dry summer heat, leaving behind a salt-laden area, which no longer supports much plant growth. With time, these areas spread, destroying the productivity of cropland.

4.9   OXYGEN, OXIDANTS, AND REDUCTANTS

Oxygen is a vitally important species in water (see Chapter 3). In water, oxygen is consumed rapidly by the oxidation of organic matter, $\{CH_2O\}$:

$$\{CH_2O\} + O_2 \xrightarrow{\text{Microorganisms}} CO_2 + H_2O \qquad (4.5)$$

Unless the water is reaerated efficiently, as by turbulent flow in a shallow stream, it rapidly loses oxygen and will not support higher forms of aquatic life.

In addition to the microorganism-mediated oxidation of organic matter, oxygen in water may be consumed by the biooxidation of nitrogenous material

$$NH_4^+ + 2O_2 \rightarrow 2H^+ + NO_3^- + H_2O \qquad (4.6)$$

and by the chemical or biochemical oxidation of chemical reducing agents:

$$4Fe^{2+} + O_2 + 10H_2O \rightarrow 4Fe(OH)_3(s) + 8H^+ \qquad (4.7)$$

$$2SO_3^{2-} + O_2 \rightarrow 2SO_4^{2-} \qquad (4.8)$$

All these processes contribute to the deoxygenation of water.

As noted in Section 3.5, the degree of oxygen consumption by microbially mediated oxidation of contaminants in water is called the **biochemical oxygen demand** (or biological oxygen demand), **BOD**. This parameter is commonly measured by determining the quantity of oxygen utilized by suitable aquatic microorganisms during a five-day period. The consumption of O_2 by microorganisms metabolizing matter responsible for BOD tends to make an aquatic medium more reducing (lower pE, see Section 3.9) and has a large influence on the chemistry of natural water and wastewater.

The addition of oxidizable pollutants to streams produces a typical oxygen sag curve as shown in Figure 4.4. Initially, a well-aerated, unpolluted stream is relatively free of oxidizable material; the oxygen level is high; and the bacterial population is relatively low. With the addition of oxidizable pollutant (BOD), the oxygen level drops because reaeration cannot keep up with oxygen consumption. In the decomposition zone, the bacterial population rises. The septic zone is characterized by a high bacterial population and very low oxygen levels. The septic zone terminates when the oxidizable pollutant is exhausted and then the recovery zone begins. In the recovery zone, the bacterial population decreases and the dissolved oxygen level increases until the water regains its original condition.

Although BOD is a reasonably realistic measure of water quality insofar as oxygen is concerned, the test for determining it is time-consuming and cumbersome to perform. Total organic carbon (TOC) is frequently measured by catalytically oxidizing carbon in the water and measuring the CO_2 that is evolved. It has become a popular test in place of BOD because TOC is readily determined instrumentally.

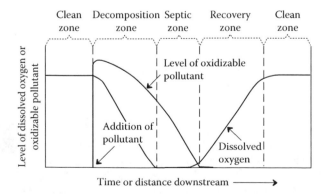

FIGURE 4.4 Oxygen sag curve resulting from the addition of oxidizable pollutant material to a stream.

4.10 ORGANIC POLLUTANTS

4.10.1 SEWAGE

As shown in Table 4.4, sewage from domestic, commercial, food-processing, and industrial sources contains a wide variety of pollutants, including organic pollutants. Some of these pollutants, particularly oxygen-demanding substances—oil, grease, and solids—are removed by primary and secondary sewage treatment processes discussed in Chapter 5. Others, such as salts and refractory (degradation-resistant) organics, are not efficiently removed.

Disposal of inadequately treated sewage can cause severe problems. For example, offshore disposal of sewage, once commonly practiced by coastal cities, results in the formation of beds of sewage residues. Municipal sewage typically contains about 0.1% solids, even after treatment, and these settle out in the ocean in a typical pattern, illustrated in Figure 4.5. The warm sewage water rises in the cold hypolimnion and is carried in one direction or another by tides or currents. It does not rise above the thermocline (metalimnion); instead, it spreads out as a cloud from which the solids rain down on the ocean floor. Aggregation of sewage colloids is aided by dissolved salts in seawater (see Section 3.8), thus, promoting the formation of sludge-containing sediment.

Another major disposal problem with sewage is the sludge produced as a product of the sewage treatment process (see Chapter 5). This sludge contains organic material, which continues to degrade slowly, refractory organics, and heavy metals. The amounts of sludge produced are truly staggering. For example, the city of Chicago produces about 3 million tons of sludge each year. A major consideration in the safe disposal of such amounts of sludge is the presence of potentially dangerous components such as heavy metals.

Careful control of sewage sources is needed to minimize sewage pollution problems. Particularly, heavy metals and refractory organic compounds need to be controlled at the source to enable use of sewage, or treated sewage effluents, for irrigation, recycling to the water system, or groundwater recharge.

TABLE 4.4
Some of the Primary Constituents of Sewage from a City Sewage System

Constituent	Potential Sources	Effects in Water
Oxygen-demanding substances	Mostly organic materials, particularly human feces and urine	Consume dissolved oxygen
Refractory organics	Industrial wastes, household products	Toxic to aquatic life
Viruses	Human wastes	Cause disease (possibly cancer); major deterrent to sewage recycle through water systems
Detergents	Household detergents	Esthetics, prevent grease and oil removal, toxic to aquatic life
Phosphates	Detergents	Algal nutrients
Grease and oil	Cooking, food processing, industrial wastes	Esthetics, harmful to some aquatic life
Salts	Human wastes, water softeners, industrial wastes	Increase water salinity
Heavy metals	Industrial wastes, chemical laboratories	Toxicity
Chelating agents	Some detergents, industrial wastes	Heavy metal ion solubilization and transport
Solids	All sources	Esthetics, harmful to aquatic life

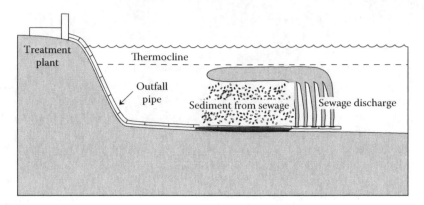

FIGURE 4.5   Settling of solids from an ocean-floor sewage effluent discharge.

Soaps, detergents, and associated chemicals are potential sources of organic pollutants. These pollutants are discussed briefly here.

4.10.2 SOAPS AND DETERGENTS

Soaps are salts of higher fatty acids, such as sodium stearate, $C_{17}H_{35}COO^-Na^+$. The cleaning action of soap results largely from its emulsifying power and its ability to lower the surface tension of water. This concept may be understood by considering the dual nature of the soap anion. An examination of its structure shows that the stearate ion consists of an ionic carboxyl "head" and a long hydrocarbon "tail":

$$\text{CO}_2^-\text{Na}^+$$

In the presence of oils, fats, and other water-insoluble organic materials, the tendency is for the "tail" of the anion to dissolve in the organic matter, whereas the "head" remains in aquatic solution. Thus, the soap emulsifies, or suspends, organic material in water. In the process, the anions form micelles, a form of very small colloidal particles discussed in Section 3.8 and shown in Figure 3.11.

The primary disadvantage of soap as a cleaning agent comes from its reaction with divalent cations to form insoluble salts of fatty acids:

$$2C_{17}H_{35}COO^-Na^+\left(aq\right)+Ca^{2+}\left(aq\right)\rightarrow Ca\left(C_{17}H_{35}CO_2\right)_2\left(s\right)+2Na^+\left(aq\right) \qquad (4.9)$$

These insoluble solids, usually salts of magnesium or calcium, are not at all effective as cleaning agents. In addition, the insoluble "curds" form unsightly deposits on clothing and in washing machines. If sufficient soap is used, all of the divalent cations may be removed by their reaction with soap, and the water containing excess soap will have good cleaning qualities. This is the approach commonly used when soap is used with unsoftened water in the bathtub or washbasin, where the insoluble calcium and magnesium salts can be tolerated. However, in applications such as washing clothing, the water must be softened by the removal of calcium and magnesium or their complexation by substances such as polyphosphates.

The triphosphate anion, an example of the polyphosphates that act as chelating agents for water hardness ions of calcium and magnesium

Although the formation of insoluble calcium and magnesium salts has essentially resulted in the elimination of soap as a cleaning agent for clothing, dishes, and most other materials, it has distinct advantages from the environmental standpoint. As soon as soap gets into sewage or an aquatic system, it generally precipitates as calcium and magnesium salts. Hence, any effects that soap might have in solution are eliminated. With eventual biodegradation, the soap is completely eliminated from the environment. Therefore, aside from the occasional formation of unsightly scum, soap does not cause any substantial pollution problems.

Synthetic **detergents** have good cleaning properties and do not form insoluble salts with "hardness ions" such as calcium and magnesium. Such synthetic detergents have the additional advantage of being the salts of relatively strong acids, and therefore, they do not precipitate out of acidic waters as insoluble acids, an undesirable characteristic of soaps. The potential of detergents to contaminate water is high because of their heavy use throughout the consumer, institutional, and industrial markets. Hundreds of millions of kilograms of synthetic detergents are used annually throughout the world, most of which is discarded with wastewater.

The key ingredient of detergents is the **surfactant** or surface-active agent, which acts in effect to make water "wetter" and a better cleaning agent. Surfactants concentrate at interfaces of water with gases (air), solids (dirt), and immiscible liquids (oil). They do so because of their **amphiphilic structure**, meaning that one part of the molecule is a polar or ionic group (head) with a strong affinity for water and the other part is a hydrocarbon group (tail) with an aversion to water. This kind of structure is illustrated below for the structure of alkyl benzene sulfonate (ABS) surfactant:

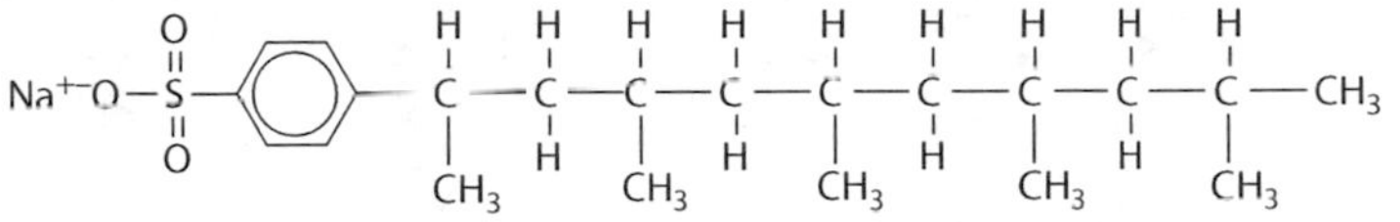

Until the early 1960s, ABS was the most common surfactant used in detergent formulations. However, it suffered the distinct disadvantage of being only very slowly biodegradable because of its branched-chain structure, which microorganisms biodegrade poorly. An objectionable manifestation of the nonbiodegradable detergents was the "head" of foam that began to appear in glasses of drinking water in areas where sewage was recycled through the domestic water supply. Spectacular beds of foam appeared near sewage outflows and in sewage treatment plants, and at least one fatality was reported when an individual fell in a bed of foam at a sewage treatment plant and was asphyxiated by the gases in the foam. Occasionally, the entire aeration tank of an activated sludge plant would be smothered by a blanket of foam. Among the other undesirable effects of persistent detergents upon waste treatment processes were lowered surface tension of water, deflocculation of colloids, flotation of solids, emulsification of grease and oil, and destruction of useful bacteria. Consequently, ABS was replaced by a biodegradable surfactant, linear alkyl sulfonate (LAS).

LAS has the general structural formula illustrated as follows, where the benzene ring may be attached at any point on the alkyl chain except at the ends:

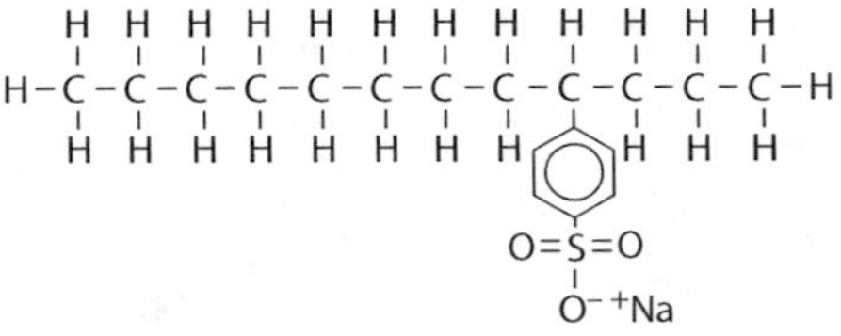

LAS is more biodegradable than ABS because the alkyl portion of LAS is not branched and does not contain the tertiary carbon, which is so detrimental to biodegradability. Since LAS has replaced ABS in detergents, the problems arising from the surface-active agent in the detergents (such as toxicity to fish fingerlings) have greatly diminished and the levels of surface-active agents found in water have decreased markedly. Some detergent surfactants are nonionic. One type that has proven troublesome consists of the alkylphenol polyethoxylates:

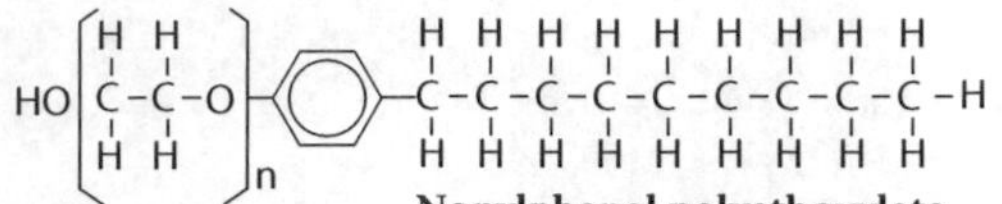

Nonylphenol polyethoxylate

The alkylphenol polyethoxylates are very useful as detergents, dispersing agents, emulsifiers, solubilizers, and wetting agents, leading to annual use of millions of kilograms in the United States. These substances and their degradation products in which the polyethoxylate chains have been shortened by microbially mediated hydrolysis tend to survive biological sewage treatment and to be discharged with the sewage effluent. They also accumulate in sewage sludge, much of which is disposed on agricultural lands. These products are thought to be xenoestrogens (estrogen hormone mimickers, see estrogenic substances in Section 4.13), and their potential entry into the food chain from sludge-treated soil is of concern. Because of these concerns, the uses of these compounds have been severely restricted in some European countries.

Detergent **builders** added to detergents to bind to hardness ions, making the detergent solution alkaline and greatly improving the action of the detergent surfactant, can cause environmental problems. A commercial solid detergent contains only 10%–30% surfactant. Detergent formulations also contain complexing agents added to complex calcium and to function as builders. Other ingredients include ion exchangers, alkalies (sodium carbonate), anticorrosive sodium silicates, amide foam stabilizers, soil-suspending carboxymethylcellulose, bleaches, fabric softeners, enzymes, optical brighteners, fragrances, dyes, and diluent sodium sulfate. The polyphosphates formerly used in builders have caused the most concern as environmental pollutants, although these problems have largely been resolved as polyphosphates have been phased out.

Increasing demands on the performance of detergents have led to the use of enzymes in detergent formulations destined for both domestic and commercial applications. To a degree, enzymes can take the place of chlorine bleach and phosphates, both of which can have detrimental environmental consequences. Lipases and cellulases are the most useful enzymes for detergent applications.

4.10.3 NATURALLY OCCURRING CHLORINATED AND BROMINATED COMPOUNDS

Although halogenated organic compounds in water, such as those discussed as pesticides in Section 4.11, are normally considered to be from anthropogenic sources, more than 2000 such compounds have been identified from natural sources.[3] These are produced largely by marine species, especially some kinds of red algae, probably as chemical defense agents. Some marine microorganisms, worms, sponges, and tunicates are also known to produce organochlorine and organobromine compounds. Various types of these compounds have been detected in samples in Arctic regions, including air, fish, seabird eggs, marine mammals, and Eskimo women's milk. An example of such a compound commonly encountered in the marine environment is 1,2'-bi-1H-pyrrole,2,3,3',4,4',5,5'-heptachloro-1'-methyl ($C_9H_3N_2Cl_7$) and its analogous bromine compound:

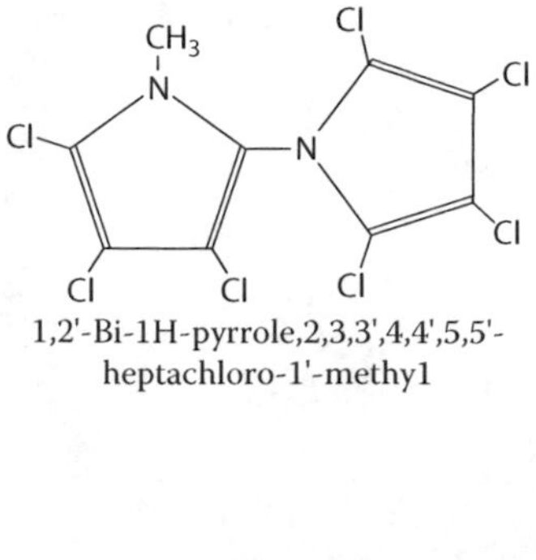

1,2'-Bi-1H-pyrrole,2,3,3',4,4',5,5'-
heptachloro-1'-methyl

4.10.4 MICROBIAL TOXINS

Bacteria and protozoa in water can produce toxins that can cause illness or even death. Toxins produced in rivers, lakes, and reservoirs by cyanobacteria including *Anabaena*, *Microcystis*, and *Nodularia* have caused adverse health effects in Australia, Brazil, England, and elsewhere in the world. There are about 40 species of cyanobacteria that produce toxins from six chemical groups. *Cylindrospermopsin* toxin (below) produced by cyanobacteria has poisoned people who have consumed water contaminated by the toxin. Surface scums of cyanobacteria are likely to have especially high levels of cyanobacteria toxins.

Cylindrospermopsin

Most of the protozoans that produce toxins belong to the order **dinoflagellata**, which are predominantly marine species. The cells of these organisms are enclosed in cellulose envelopes, which often have beautiful patterns on them. Among the effects caused by toxins from these organisms are gastrointestinal, respiratory, and skin disorders in humans; mass kills of various marine animals; and paralytic conditions caused by eating infested shellfish that can afflict humans, sometimes with fatal consequences.

The marine growth of dinoflagellates is characterized by occasional incidents in which they multiply at such an explosive rate that they color the water yellow, olive-green, or red by their vast numbers. In 1946, some sections of the Florida coast became so afflicted by "red tide" that the water became viscous, and for many miles, the beaches were littered with the remains of dead fish, shellfish, turtles, and other marine organisms. The sea spray in these areas became so irritating that coastal schools and resorts were closed.

4.11 PESTICIDES IN WATER

The introduction of DDT during World War II marked the beginning of a period of very rapid growth in pesticide use. Pesticides are used for many different purposes. Chemicals used in the control of invertebrates include **insecticides**, **molluscicides** for the control of snails and slugs, and **nematicides** for the control of microscopic roundworms. Vertebrates are controlled by **rodenticides**, which kill rodents; **avicides** used to repel birds; and **piscicides** used in fish control. **Herbicides** are used to kill plants. Plant **growth regulators**, **defoliants**, and **plant desiccants** are used for various purposes in the cultivation of plants. **Fungicides** are used against fungi, **bactericides** against bacteria, **slimicides** against slime-causing organisms in water, and **algicides** against algae. The quantities of pesticides used around the world each year amount to hundreds of millions of kilograms. Large quantities of pesticides are used in agricultural applications and for nonagricultural applications including forestry, landscaping, gardening, food distribution, and home pest control. Insecticide production has remained about level during the last three or four decades. However, insecticides and fungicides are the most important pesticides with respect to human exposure in food because they are applied shortly before or even after harvesting. Herbicide production has increased as chemicals have increasingly replaced cultivation of land in the control of weeds, and herbicides now account for the majority of agricultural pesticides. The potential exists for large quantities of pesticides to enter water either directly, in applications such as mosquito control, or indirectly, primarily from drainage of agricultural lands.

Several classes of pesticides and other chemicals are of particular concern as water pollutants because of their potential effects. These are (1) highly biodegradation-resistant compounds, (2) known or probable carcinogens, (3) toxicants with adverse reproductive or developmental effects, (4) neurotoxins including cholinesterase inhibitors, (5) substances with high acute toxicities, (6) known groundwater contaminants. Table 4.5 lists some of the most commonly used pesticides that may be of concern as water pollutants.

TABLE 4.5

Pesticides That May Be Found as Water Pollutants[1]

Pesticide	Use	Compound Type
2,4-D	Herbicide	Chlorophenoxyacetic acid
Acephate	Insecticide	Organophosphate
Acetochlor	Herbicide	Chloroacetanilide
Alachlor	Herbicide	Chloroacetanilide
Aldicarb	Insecticide	Carbamate
Allethrin	Insecticide	Pyrethroid
Atrazine	Herbicide	Triazine
Azadirachtin	Insecticide, nematicide	Complex botanical compound
Azoxystrobin	Fungicide	Strobin
Captan	Fungicide	Thiophthalimide
Carbaryl	Insecticide, plant growth regulator, nematicide	Carbamate
Carbofuran	Insecticide, nematicide	Carbamate
Chlorothalonil	Fungicide	Substituted benzene
Chloropicrin	Fumigant	Chloronitromethane
Chlorpyrifos	Insecticide, nematicide	Organophosphate
Copper hydroxide	Fumigant	Inorganic compound
Copper sulfate	Fumigant	Inorganic compound
Cypermethrin	Insecticide	Pyrethroid
Deltamethrin	Insecticide	Pyrethroid
Diazinon	Insecticide	Organophosphorus
Dichloropropene	Fumigant	Organochlorine compound
Diclofopmethyl	Herbicide	Chlorophenoxy acid/ester
Dimethenamid	Herbicide	Organo S, N, O, Cl
Dicamba	Herbicide	Organochlorine
Dichlorvos	Insecticide	Organophosphate
Diquat	Herbicide	Organonitrogen
Diuron	Herbicide	Urea
EPTC	Herbicide	Thiocarbamate
Ethephon	Plant growth regulator	Organophosphate
Fenitrothion	Insecticide	Organophosphate
Fenvalerate	Insecticide	Pyrethroid
Glyphosate	Herbicide	Phosphonoglycine
Imidacloprid	Insecticide	Nicotine-based neonicotinoid
Iprodione	Fungicide	Dicarboximide
Kresoxim-methyl	Fungicide	Strobin
Linuron	Herbicide	Urea
Malathion	Insecticide	Organophosphate
Mancozeb	Fumigant	Dithiocarbamate
MCPA and MCPP	Herbicide	Chlorophenoxyacetic acid
Mecoprop	Herbicide	Chlorphenoxy compound

TABLE 4.5 (*Continued*)
Pesticides That May Be Found as Water Pollutants[1]

Pesticide	Use	Compound Type
Metam-sodium, Metam-potassium	Fumigant, herbicide, microbiocide, algacide	Dithiocarbamate
Methiocarb	Insecticide, molluscicide	Carbamate
Metolachlor	Herbicide	Chloroacetanilide
Metribuzin	Herbicide	Triazinone
MSMA	Herbicide, defoliant	Organoarsenic
Paraquat	Herbicide	Organonitrogen
Pelargonic acid	Herbicide	9-C carboxylic acid
Pendimethalin	Herbicide	Organonitrogen
Phosmet	Insecticide	Organophosphate
Prometon	Herbicide	Triazine
Propanil	Herbicide	Anilide
Propiconazole	Fungicide	Azole
Simazine	Herbicide	Triazine
Spinosad	Insecticide	Compound from bacteria Macrocyclic lactone
Sulfuryl fluoride	Fumigant	SO_2F_2
Tebuconazole	Fungicide	Azole
Tebuthiuron	Herbicide	Urea
Terbacil	Herbicide	Uracil
Terbuthylazine	Algicide, herbicide, microbiocide	Triazine
Thifensulfuron-methyl	Herbicide	Sulfonylurea
Thiophanate-methyl	Fungicide	Benzimidazole
Tri-allate	Herbicide	Thiocarbamate
Triclopyr	Herbicide	Chloropyridinyl
Trifloxystrobin	Fungicide	Strobin
Trifluralin	Herbicide	2,6-Dinitroaniline

[1] For additional information and structural formulas, see Pesticide Action Network: http://www.pesticideinfo.org/.

Degradation (often hydrolysis) products of some pesticides are encountered in water at levels similar to, or even higher than, the parent pesticides. In some cases, the degradation products are more toxic than the parent pesticides. An example of a pesticide degradation product commonly found in water is aminomethyl phosphonic acid produced from glyphosate (brand name Roundup, see Figure 4.11 later in the chapter), which is the most widely produced pesticide in the world.

4.11.1 Natural Product Insecticides, Pyrethrins, and Pyrethroids

Plants provide several significant classes of insecticides including **nicotine** from tobacco, **rotenone** from certain legume roots, and **pyrethrins** (see structural formulas in Figure 4.6). Pyrethrins and their synthetic analogs represent both the oldest and some of the newer insecticides. Extracts of dried chrysanthemum or pyrethrum flowers, which contain pyrethrin I and related compounds, have been known for their insecticidal properties for a long time and may have even been used as botanical insecticides in China almost 2000 years ago. The most important commercial sources of insecticidal pyrethrins are chrysanthemum varieties grown in Kenya. Pyrethrins have several advantages as

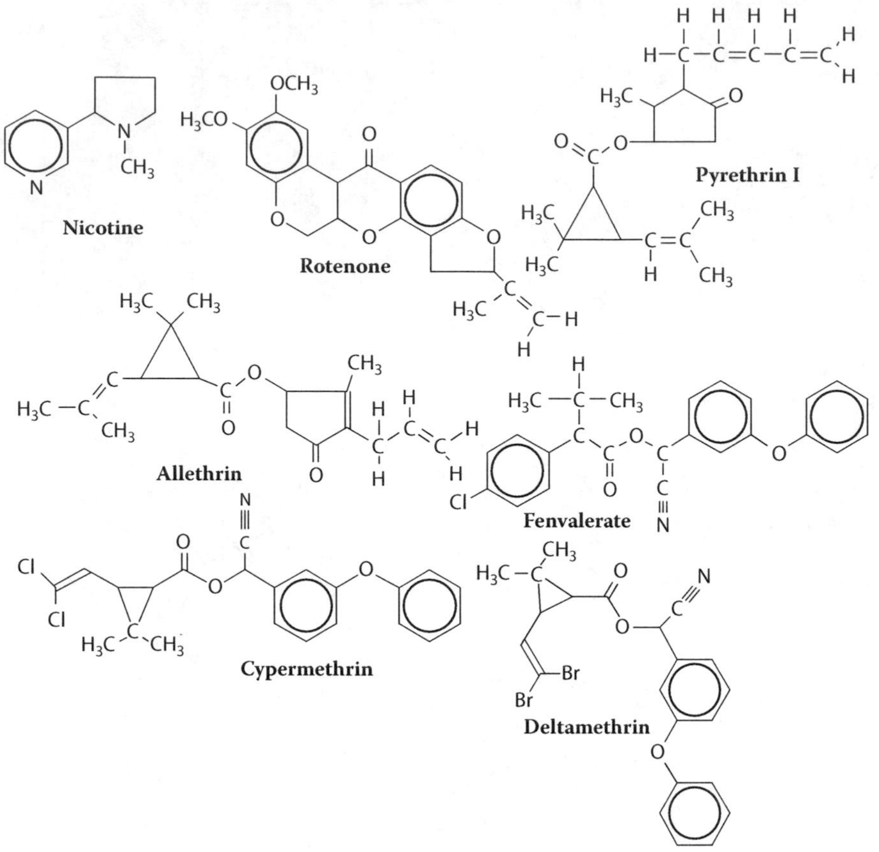

FIGURE 4.6 Common botanical insecticides and synthetic analogs of the pyrethrins.

insecticides, including facile enzymatic degradation, which makes them relatively safe for mammals; ability to rapidly paralyze ("knock down") flying insects; and good biodegradability characteristics.

Synthetic analogs of the pyrethrins, **pyrethroids**, have been widely produced as insecticides during recent years. The first of these was allethrin, and another common example is fenvalerate (see structural formulas in Figure 4.6). Other examples of insecticidal pyrethroids that have water pollution potential include cypermethrin and deltamethrin. Because of the ways that they are applied and their biodegradabilities, these substances generally are not found to be significant water pollutants.

4.11.2　DDT and Organochlorine Insecticides

Chlorinated hydrocarbon or organochlorine insecticides are hydrocarbon compounds in which various numbers of hydrogen atoms have been replaced by Cl atoms (Figure 4.7). Of the organochlorine insecticides, the most notable has been **DDT** (dichlorodiphenyltrichloroethane or 1,1,1-trichloro-2,2-bis(4-chlorophenyl)ethane), which was used in massive quantities following World War II. It has a low acute toxicity to mammals, although there is some evidence that it might be carcinogenic. It is a very persistent insecticide and accumulates in food chains. It has been banned in the United States since 1972.

Many organochlorine insecticides were widely used in decades past, but are now banned because of their toxicities and, particularly, their accumulation and persistence in food chains. Now largely of historical interest, these banned insecticides include methoxychlor (once a popular replacement for DDT), dieldrin, endrin, chlordane, aldrin, heptachlor, toxaphene, lindane, and endosulfan (one of the last to be phased out of general use).

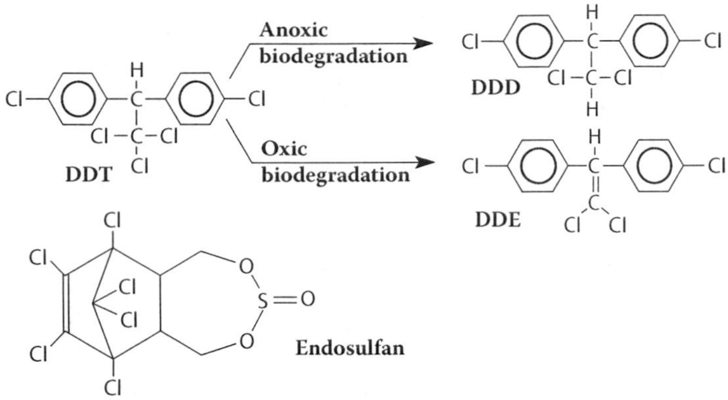

FIGURE 4.7 Two examples of organochlorine insectides. DDT was the most notable of these because of its history of environmental effects. Endosulfan has been one of the last of these to be phased out of use.

4.11.3 ORGANOPHOSPHATE INSECTICIDES

Organophosphate insecticides are insecticidal organic compounds that contain phosphorus, some of which are organic esters of orthophosphoric acid, such as paraoxon:

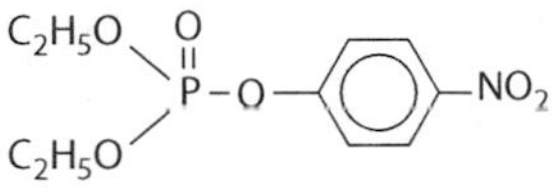

More commonly, insecticidal phosphorus compounds are phosphorothionate and phosphorodithioate compounds, such as those shown in Figure 4.8, which have an =S group rather than an =O group bonded to P.

The toxicities of organophosphate insecticides vary a great deal. Their major toxic effect is inhibition of acetylcholinesterase, an enzyme essential for nerve function. For example, as little as 120 mg of parathion has been known to kill an adult human and a dose of 2 mg has killed a child. Most accidental poisonings have occurred by absorption through the skin. Since its use began, several hundred people have been killed by parathion.

Numerous, mostly neurological symptoms of poisoning by organophosphorus insecticides can be observed. Just a few of these include excessive salivation, rhinorrhea ("runny nose"), tremor, headache, nausea, respiratory distress, and in severe cases, seizures, respiratory depression, incontinence, loss of consciousness, and even death.

In contrast to the high toxicity of methyl parathion, **malathion** shows how differences in structural formula can cause pronounced differences in the properties of organophosphate pesticides. Malathion has two carboxyester linkages, which are hydrolyzable by carboxylase enzymes to relatively nontoxic products as shown by the following reaction:

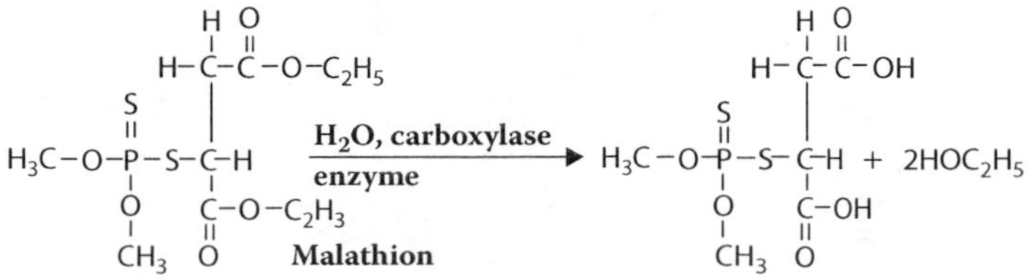

The enzymes that accomplish malathion hydrolysis are possessed by mammals, but not by insects, and so mammals can detoxify malathion and insects cannot. The result is that malathion

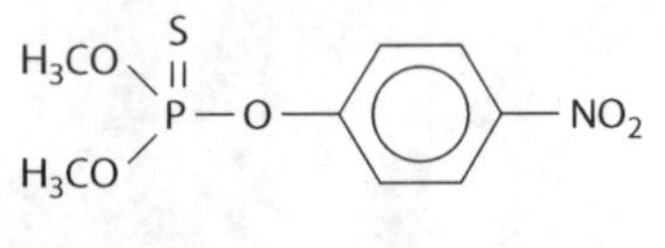

Methyl parathion

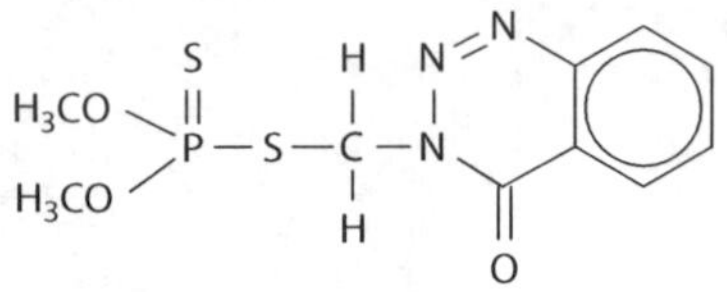

Azinphos-methyl

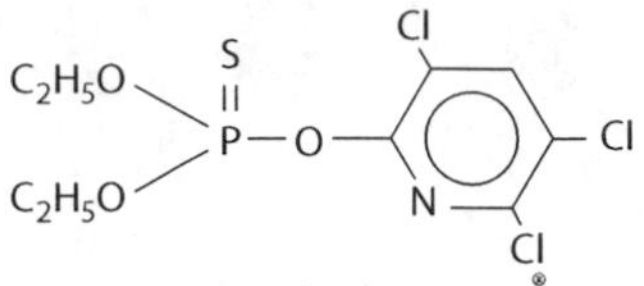

Chlorpyrifos (Dursban®)

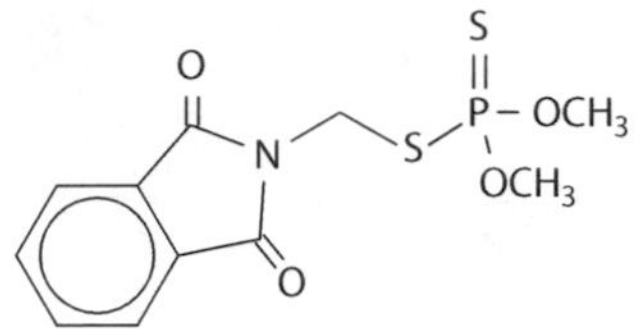

Phosmet

FIGURE 4.8 Examples of phosphorothionate (methyl parathion and chlorpyrifos) and phosphorodithioate (azinphos-methyl, phosmet) insecticides. Once widely used, methyl parathion is no longer allowed for use because of its neurotoxicity.

has selective insecticidal activity. For example, although malathion is a very effective insecticide, its LD_{50} (dose required to kill 50% of test subjects) for adult male rats is about 100 times that of parathion, reflecting the much lower mammalian toxicity of malathion compared to some of the more toxic organophosphate insecticides, such as parathion.

Unlike the organohalide compounds they largely displaced, the organophosphates readily undergo biodegradation and do not bioaccumulate. Because of their high biodegradability and restricted use, organophosphates are of comparatively little significance as water pollutants.

4.11.4 Carbamates

Pesticidal organic derivatives of carbamic acid, for which the formula is shown in Figure 4.9, are known collectively as **carbamates**. Carbamate pesticides have been widely used because some are more biodegradable than the formerly popular organochlorine insecticides and have lower dermal toxicities than most common organophosphate pesticides.

Carbaryl has been widely used as an insecticide on lawns or gardens. It has a low toxicity to mammals. **Carbofuran** has high water solubility and acts as a plant systemic insecticide. As a plant systemic insecticide, it is taken up by the roots and leaves of plants so that insects are poisoned by the plant material on which they feed.

Carbamates are toxic to animals because they inhibit acetylcholinesterase. Unlike some of the organophosphate insecticides, they do so without the need for undergoing a prior biotransformation

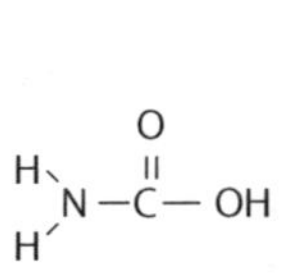

Carbamic acid

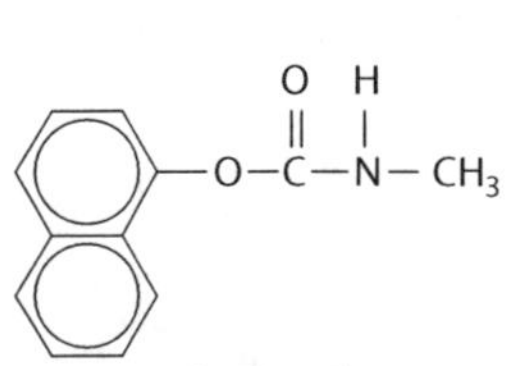

Carbaryl

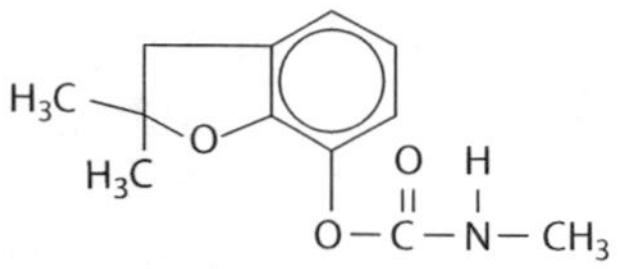

Carbofuran

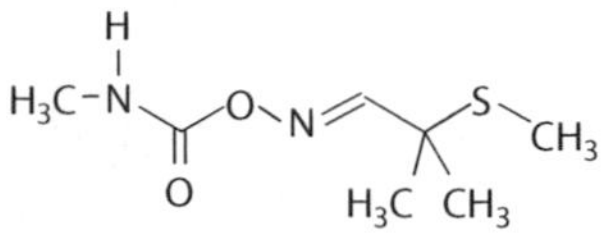

Aldicarb

FIGURE 4.9 Carbamic acid and three insecticidal carbamates.

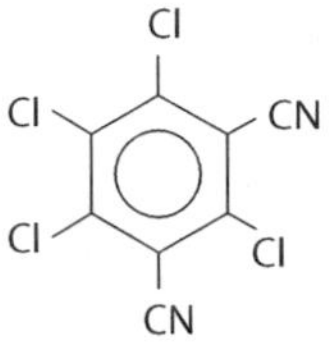

Chlorothalonil

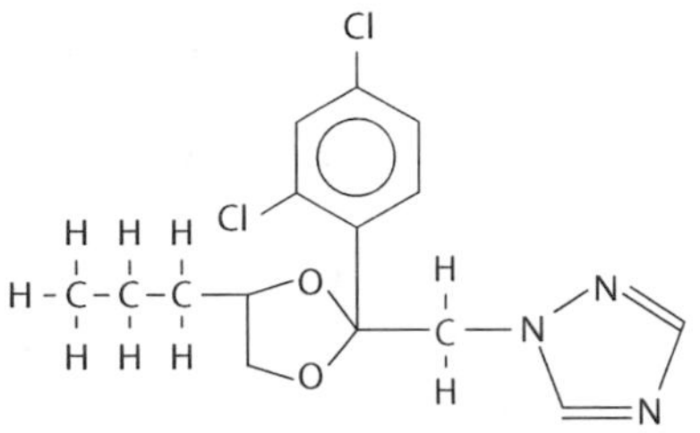

Propiconazole, a triazole fungicide

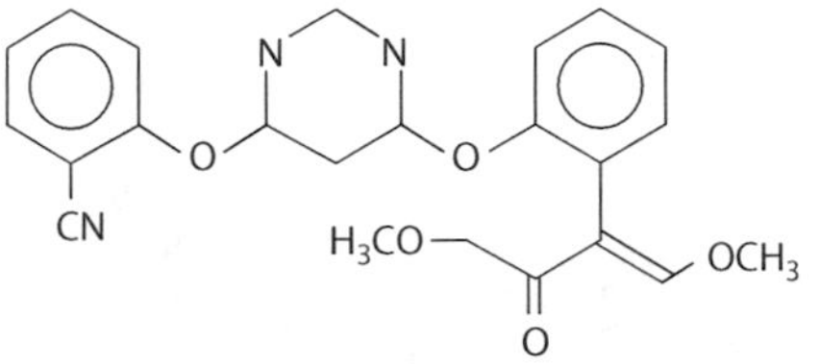

Azoxystrobin, a strobilurin fungicide

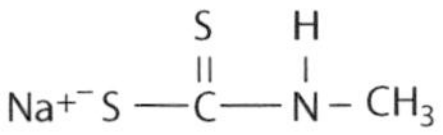

Metam-sodium

FIGURE 4.10 Four examples of widely used fungicides that may be of concern as potential water pollutants.

and are, therefore, classified as direct inhibitors. Their inhibition of acetylcholinesterase is relatively reversible. Loss of acetylcholinesterase inhibition activity may result from hydrolysis of the carbamate ester, which can occur metabolically.

4.11.5 Fungicides

Fungicides are widely applied to cereal and food crops to prevent fungal infections of these crops. Because of this, fungicides have the potential to contaminate water. Structural formulas of four commonly used fungicides are shown in Figure 4.10. Of the ones shown, chlorothalonil has been used for more than 30 years, with annual applications of more than 5 million kg in the United States. It is typically applied at a rate of 1 kg per hectare per application with 4–9 applications per year. The strobilurin fungicides exemplified by azoxystrobin and the triazole fungicides exemplified by propiconazole came into use during the 1990s and have been effective, although some problems with resistance developed by target organisms have been encountered.

4.11.6 Herbicides

Herbicides are substances that kill vegetation, most commonly weeds that interfere with crop growth. There are many chemical formulas of herbicidal compounds, representatives of which are shown in Figure 4.11. For the structural formulas of other herbicides including those mentioned in this chapter, the reader is referred to Internet sources, such as the Pesticide Action Network Database.[4]

Herbicides are applied over millions of acres of farmland worldwide and are widespread water pollutants as a result of this intensive use. Herbicides are commonly found in surface water and groundwater. Among those that have been reported in surface water and groundwater are atrazine, simazine, and cyanazine, widely used to control weeds on corn and soybeans in the "Corn Belt" states of Kansas, Nebraska, Iowa, Illinois, and Missouri as well as agricultural regions throughout the world. Other herbicides found in water include prometon, metolachlor, metribuzin, tebuthiuron, trifluran, alachlor, and the atrazine metabolites deisopropylatrazine and deethylatrazine. Although glyphosate is the most widely used herbicide to control weeds on crops genetically engineered to resist its effects, it is rarely found at levels of concern in water because of its very strong affinity for soil solids.

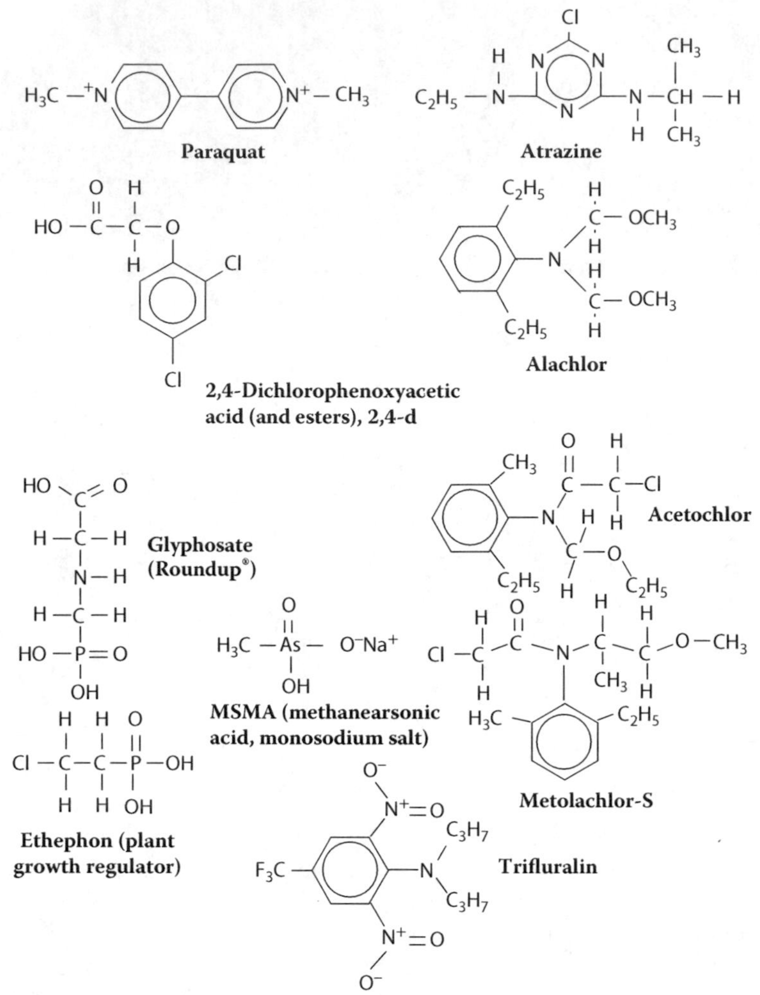

FIGURE 4.11 Structural formulas of some common herbicides. Organoarsenic compounds such as MSMA are no longer registered for herbicidal use.

Paraquat, which was registered for use in 1965, was once one of the most widely used herbicides, though now largely discontinued in the United States. It is reputed to be one of the most toxic herbicides that have ever been in widespread commercial use and may have been responsible for several hundred human deaths, primarily in suicides. Exposure to fatal or dangerous levels of paraquat can occur by all pathways, including inhalation of spray, skin contact, ingestion, and even suicidal hypodermic injections. Its toxicity is primarily due to acute systemic effects, with the lung the major target organ and secondarily the kidney.[5] Fortunately, paraquat is almost never found in drinking water sources.

A number of important herbicides contain three heterocyclic nitrogen atoms in ring structures and are, therefore, called **triazines,** of which atrazine shown in Figure 4.11 is the best known example. Triazine herbicides inhibit photosynthesis. Selectivity is gained by the inability of target plants to metabolize and detoxify the herbicide. Widely applied to kill weeds in corn, atrazine has been commonly encountered as a water pollutant in corn-growing regions. Another member of this class is simazine, which has many uses to control annual grasses and broadleaf weeds in a variety of crops and orchards and for algae and submerged weed control in bodies of water. The acute toxicities of the commonly used triazine herbicides are relatively low; there is some concern over their potential to disrupt endocrines.

The **chlorophenoxy herbicides**, including 2,4-dichlorophenoxyacetic acid (2,4-D) shown in Figure 4.11 and the related compounds 2,4,5-trichlorophenoxyacetic acid (2,4,5-T) and mecoprop, were manufactured on a large scale for weed and brush control and as military defoliants. At one time, 2,4,5-T (now banned), known as the infamous "Agent Orange," was used to defoliate vegetation in the Vietnam War and was of particular concern because of contaminant TCDD (see below) present as a manufacturing by-product.

Many herbicides that are environmentally significant do not fall into the classifications described earlier. Nitroaniline herbicides are characterized by the presence of $-NO_2$ and a substituted $-NH_2$ group on a benzene ring as shown for trifluralin in Figure 4.11. This class of herbicides is widely represented in agricultural applications and includes benefin (Balan®), oryzalin (Surflan®), and pendimethalin (Prowl®). Other types of herbicides include substituted ureas, carbamates, and thiocarbamates.

Until about 1960, arsenic trioxide and other inorganic arsenic compounds (see Section 4.4) were used to kill weeds. Because of the incredibly high rates of application of up to several hundred kilograms per acre and because arsenic is nonbiodegradable, the potential still exists for arsenic pollution of surface water and groundwater from fields formerly dosed with inorganic arsenic. In the past, organic arsenicals, such as MSMA shown in Figure 4.11 and the related compound cacodylic acid, have also been widely applied to kill weeds. Toxicity is a major concern with arsenic-containing pesticides. Acute effects include adverse effects to the gastrointestinal tract, the central nervous system, muscles, the cardiovascular system, the renal (kidney) system, liver, and blood-forming tissues. Chronic exposure to arsenic compounds also causes a wide variety of adverse health effects including muscle weakness, skin hyperpigmentation, peripheral neuropathy, liver toxicity, renal toxicity, hematologic (blood) abnormalities, and cancer.

4.11.7 By-Products of Pesticide Manufacture

A number of water pollution and health problems have been associated with the manufacture of organochlorine pesticides. The most notorious by-products of pesticide manufacture are **polychlorinated dibenzodioxins**. From 1 to 8 Cl atoms may be substituted for H atoms on dibenzo-*p*-dioxin (Figure 4.12), giving a total of 75 possible chlorinated derivatives. Commonly referred to as "dioxins," these species have a high environmental and toxicological significance. Of the dioxins, the most notable pollutant and hazardous waste compound is **2,3,7,8-tetrachlorodibenzo-*p*-dioxin** (**TCDD**), often referred to simply as "**dioxin**." This compound was produced as a low-level contaminant in the manufacture of some aryl, oxygen-containing organohalide compounds such as chlorophenoxy herbicides (mentioned previously in this section) synthesized by processes used until the 1960s.

TCDD has a very low vapor pressure, a high melting point of 305°C, and a water solubility of only 0.2 μg/L. It is chemically unreactive, stable thermally up to about 700°C, and is poorly biodegradable. It is very toxic to some animals, with an LD_{50} of only about 0.6 μg/kg body mass in male guinea pigs. (The type and degree of its toxicity to humans is largely unknown; it is known to cause a severe skin condition called chloracne). Because of its properties, TCDD

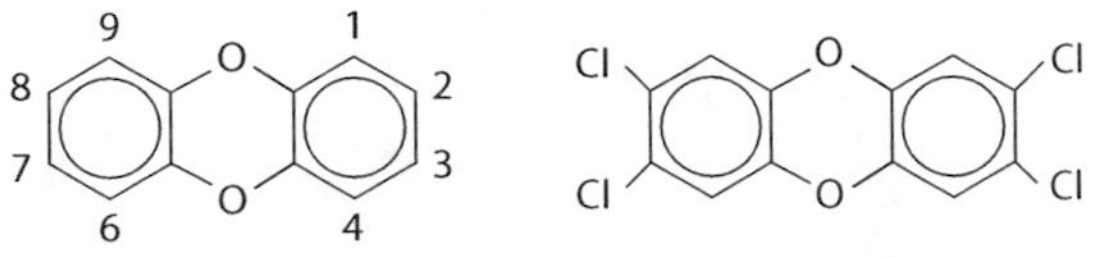

Dibenzo-*p*-dioxin   **2,3,7,8-Tetrachlorodibenzo-*p*-dioxin**

FIGURE 4.12 Dibenzo-*p*-dioxin and 2,3,7,8-tetrachlorodibenzo-*p*-dioxin (TCDD), often called simply "dioxin." In the structure of dibenzo-*p*-dioxin, each number refers to a numbered carbon atom to which an H atom is bound, and the names of derivatives are based upon the carbon atoms where another group has been substituted for the H atoms, as is seen by the structural formula and name of 2,3,7,8-tetrachlorodibenzo-*p*-dioxin.

is a stable, persistent environmental pollutant and hazardous waste constituent of considerable concern. It has been identified in some municipal incineration emissions, in which it is believed to form when chlorine from the combustion of organochlorine compounds reacts with carbon in the incinerator.

TCDD contamination has resulted from improper waste disposal, the most notable case of which resulted from the spraying of waste oil mixed with TCDD on roads and horse arenas in Missouri in the early 1970s. Contamination of the soil in Times Beach, Missouri, resulted in the whole town being bought out and its topsoil dug up and incinerated at a cost exceeding $100 million.

One of the greater environmental disasters ever to result from pesticide manufacture involved the production of Kepone, with the following structural formula:

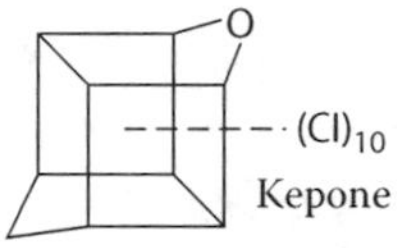

This pesticide has been used for the control of banana-root borer, tobacco wireworm, ants, and cockroaches. Kepone exhibits acute, delayed, and cumulative toxicity in birds, rodents, and humans, and it causes cancer in rodents. It was manufactured in Hopewell, Virginia, during the mid-1970s. During this time, workers were exposed to Kepone and are alleged to have suffered health problems as a result. As much as 53,000 kg of Kepone may have been dumped through the sewage system of Hopewell into the James River. The cost of dredging the river to remove this waste was estimated at a prohibitive cost of several billion dollars.

4.12   POLYCHLORINATED BIPHENYLS

First discovered as environmental pollutants in 1966, **polychlorinated biphenyls** (**PCB** compounds) have been found throughout the world in water, sediments, bird tissue, and fish tissue. These compounds constitute an important class of special wastes. They are made by substituting from 1 to 10 Cl atoms onto the biphenyl aryl structure as shown on the left in Figure 4.13. This substitution can produce 209 different compounds (congeners), of which one example is shown on the right in Figure 4.13.

Polychlorinated biphenyls have very high chemical, thermal, and biological stability; low vapor pressure; and high dielectric constants. These properties have led to the use of PCBs as coolant-insulation fluids in transformers and capacitors; for the impregnation of cotton and asbestos; as plasticizers; and as additives to some epoxy paints. The same properties that made extraordinarily stable PCBs so useful also contributed to the widespread dispersion and accumulation of these substances in the environment. By regulations issued in the United States under the authority of the Toxic Substances Control Act passed in 1976, the manufacture of PCBs was discontinued in the United States, and their uses and disposal were strictly controlled. Some degree of biodegradation of PCBs in the environment does occur.

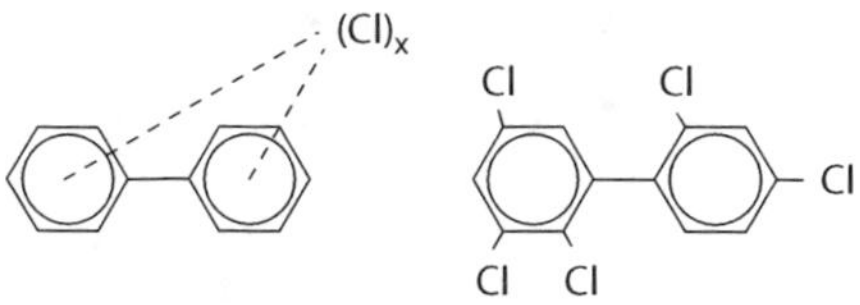

FIGURE 4.13   General formula of polychlorinated biphenyls (left, where X may range from 1 to 10) and a specific 5-chlorine congener (right).

Substitutes for PCBs for electrical applications have been developed. Disposal of PCBs from discarded electrical equipment and other sources have caused problems, particularly since PCBs can survive ordinary incineration by escaping as vapors through the smokestack. However, they can be destroyed by special incineration processes.

PCBs are especially prominent pollutants in the sediments of the Hudson River as a result of waste discharges from two capacitor manufacturing plants that operated about 60 km upstream from the southernmost dam on the river from 1950 to 1976. The river sediments downstream from the plants exhibit PCB levels of about 10 ppm, 1–2 orders of magnitude higher than levels commonly encountered in river and estuary sediments. In 2002, General Electric Co. was ordered to dredge and decontaminate sections of the Hudson River polluted with PCBs at a cost estimated at the time of around $100 million. In 2009, a trial dredging of a contaminated section of the river was tried and the data analyzed in 2010. The trial was judged a success and dredging was resumed in 2011 with an estimated completion time of 7 years and a total cost of around $1 billion.

An interesting toxicological chemistry aspect of the Hudson River PCBs has been the observation of PCB resistance in the Atlantic tomcod, a small, rapidly reproducing bottom-feeding fish that lives in the PCB-contaminated sections of the Hudson River.[6] The Hudson River Atlantic tomcods have undergone a natural selection process in which now virtually all of that population have a modified gene that produces a protein called the aryl hydrocarbon receptor2 (AHR2) that binds poorly to PCBs, thus weakening the toxic effects of the PCBs. The same gene appears in a very small percentage of the populations of the same fish species in the Shinnecock Bay on the south shore of Long Island and in Connecticut's Nantic River and has probably long been present in some of the Hudson River populations where it came to predominate because of the ability of the fish that possess it to resist PCBs. Remarkably, this evolution has taken place within only 20–50 generations of the Hudson River tomcods.

4.13　EMERGING WATER POLLUTANTS, PHARMACEUTICALS, AND HOUSEHOLD WASTES

The continued development of new products used for various purposes has led to interest in **emerging pollutants** of various kinds, which may be of concern in water.[7] A factor in finding emerging pollutants has been the development of sophisticated methods of water analysis, including especially liquid chromatography combined with high resolution mass spectrometry, which is very effective in identifying specific water contaminants at low levels. The materials found are often degradation or reaction products of parent compounds. Table 4.6 lists the major categories of emerging water contaminants of concern.

Prominent among the emerging water contaminants are **nanomaterials** consisting of very small entities in the 1–100 nm size range. Nanomaterials of various kinds have unique properties among which may be high thermal stability, low permeability, high strength, and high conductivity. These and other properties lead to rapidly increasing uses in electronics, automobiles, apparel, sunscreens, cosmetics, water purification, and other products. Little is known about the potential pollution effects and toxicities of nanomaterials, so their potential effects as water pollutants are of significant concern.

Another class of emerging pollutants consists of **siloxanes** (commonly called silicones), including octamethylcyclotetrasiloxane, decamethylcyclopentasiloxane, and dodecamethylcyclohexasiloxane. Siloxanes are thermally and chemically very stable, leading to their uses as coolants in transformers, protective encapsulating materials in semiconductors, lubricants, coatings, and sealants. Siloxanes are widely used in personal care products including deodorants, cosmetics, soaps, hair conditioners and hair dyes, and other products such as

TABLE 4.6

Classes of Emerging Water Contaminants of Concern as Pollutants

Contaminant	Significance
Pharmaceuticals	Discarded into wastewaters or from human wastes in sewage
Pharmaceutical metabolites	From human wastes in sewage
Hormones	From human wastes in sewage
Endocrine disruptors	Synthetic compounds that disrupt endocrine function
Nanomaterials	A wide variety of inorganic and organic substances in the 1–100 nm size range with many developing uses
Sucralose	Highly persistent artificial sweetener from chlorinating sucrose
Disinfection by-products	By-products from the reaction of chlorine disinfectant with a variety of contaminants
Personal care products	Cosmetics, sunscreens, ultraviolet filters
Brominated flame retardants	Refractory brominated organic compounds
Benzotriazoles	Complexing agents and anticorrosive additives
1,4-Dioxane	Widely manufactured cyclic ether
Siloxanes	Organosilicon compounds with many uses
Naphthenic acids	Commonly released in the processing of oil sands
Musks	A variety of aromatic substances produced by some animals and plants and synthetic analogs and used in perfumes
Perchlorate	Used as rocket propellant, interferes with iodide in thyroid function
Pesticide metabolites	In some cases potentially more harmful than parent pesticides
Emerging pesticides	New classes of herbicides, insecticides, and other pesticides

water repellent windshield coatings, detergent antifoaming agents, and even food additives. Siloxanes are resistant to biodegradation and as a result are encountered in water that has received wastewater.

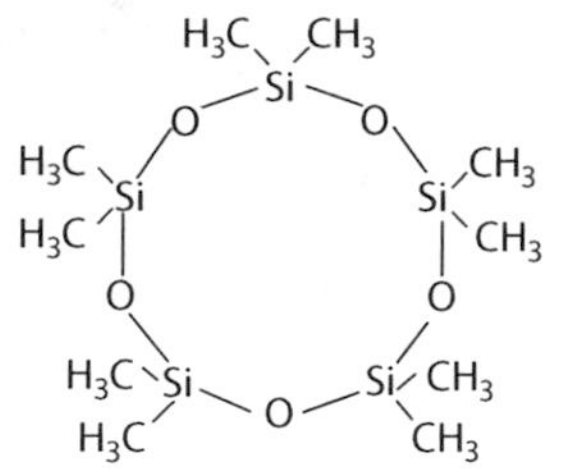

**Decamethylcyclopentasiloxane,
a cyclic siloxane**

Disinfection by-products are of some concern as water pollutants. These are compounds containing halogens and nitrogen that result from reactions of water disinfectants including chlorine, hypochlorite, and chlorine dioxide. In addition to exposure in drinking water, humans may be exposed to these substances through skin contact in bathing or swimming and as vapors emitted from water during showers. Brominated and iodated compounds are formed in chlorinated water by reactions of bromide or iodide in the water, usually present in areas of seawater or saline groundwater intrusion.

The most common disinfection by-products are the **trihalomethanes**—chloroform ($CHCl_3$), dibromochloromethane ($CHClBr_2$), bromodichloromethane ($CHCl_2Br$), and tribromomethane ($CHBr_3$). These are all by-products of water chlorination and are Group B carcinogens (shown to cause cancer in laboratory animals). By far, the most abundant of these in water systems is trichloromethane (chloroform). Dibromochloromethane is regarded as posing about 10 times the risk of cancer as chloroform, and tribromomethane is thought to pose just a slightly greater cancer risk than chloroform. As of 2011, the maximum allowable limit for total trihalomethanes in drinking water was 80 μg/L.

Various substances associated with household wastes are found in water, especially in treated sewage discharges. These materials include steroids, surfactants, flame retardants, fragrances, plasticizers, and pharmaceuticals and their metabolites. It should be noted that, although significant numbers of such compounds are found, they are generally at sub-part-per-billion levels and are detectable only by the remarkable capability of modern analytical instrumentation. Typical of "exotic" organics found in groundwater and water supplies detected are cholesterol, nicotine metabolite cotinine, β-sitosterol (a natural plant sterol), 1,7-dimethylxanthine (caffeine metabolite), bisphenol-A plasticizer, and fire retardant (2-chloroethyl) phosphate.

Pharmaceutical compounds and their partial degradation products are discharged with sewage as wastes from human ingestion and from their being deliberately discarded into wastewater.[8] The quantities of these substances in sewage in developed countries can reach of the order of 100 metric tons per year. Levels of common pharmaceuticals of around 1 μg/L have been observed in river water. Although these are relatively low values, they are of some concern because of the biological activity inherent to pharmaceutical products, which are increasingly designed for higher potency, bioavailability, and resistance to degradation. Figure 4.14 shows some examples of the most common pharmaceuticals and their degradation products that have been observed in water. One of the greater concerns with pharmaceuticals in water is the possibility that some are endocrine disruptors that may interfere with hormonal function, especially in fish and other aquatic organisms as discussed further.

A study of the occurrence of pharmaceuticals and endocrine disrupting chemicals in sources of drinking water for more than 28 million people in the United States showed that the most commonly detected chemicals were atenolol, atrazine, carbamazepine, estrone, gemfibrozil, meprobamate, naproxen, phenytoin, sulfamethoxazole, TCEP, and trimethoprim, generally occurring at levels below 10 ng/L.[9] The study suggested that potential contamination by pharmaceuticals and endocrine disrupting chemicals and the effectiveness of treatment measures for their removal may be shown by the presence and levels of atenolol, atrazine, DEET, estrone, meprobamate, and trimethoprim acting as indicator compounds.

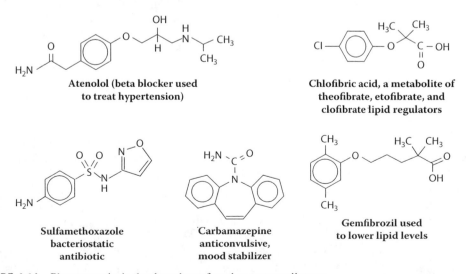

Atenolol (beta blocker used to treat hypertension)

Chlofibric acid, a metabolite of theofibrate, etofibrate, and clofibrate lipid regulators

Sulfamethoxazole bacteriostatic antibiotic

Carbamazepine anticonvulsive, mood stabilizer

Gemfibrozil used to lower lipid levels

FIGURE 4.14 Pharmaceuticals that have been found as water pollutants.

4.13.1 BACTERICIDES

Bactericides used in cleaning and consumer products may be encountered in water. One of the most common of these is triclosan,

**Triclosan (methyl triclosan has a –CH$_3$ group
substituted for the H designated H*)**

widely used in antibacterial soaps and other consumer items such as shampoo, deodorants, lotions, toothpaste, sportswear, shoes, carpets, and even refuse containers. This compound and its methyl derivative commonly occur at low levels in natural waters that receive wastewater.

4.13.2 ESTROGENIC SUBSTANCES IN WASTEWATER EFFLUENTS

Estrogenic substances compose a class of water pollutants of particular concern commonly found in sewage and even treated sewage effluent. These substances can disrupt the crucial endocrine gland activities that regulate the metabolism and reproductive functions of organisms. Aquatic organisms including fish, frogs, and reptiles such as alligators exposed to such substances may exhibit reproductive dysfunction, alterations in secondary sex characteristics, and abnormal serum steroid levels. Such substances include exogenous estrogenic substances, among which are 17α-ethynyl estradiol, diethylstilbestrol, mestranol, levonorgestrel, and norethindrone used in oral contraceptives, treatment of hormonal disorders, and cancer treatment. Some synthetic substances other than hormones also act as estrogen disruptors. Of prime concern as water pollutants are nonionic surfactant polyethoxylates, mentioned in the discussion of detergents in Section 4.10.2, and their major degradation product, persistent nonylphenol. Although these substances are orders of magnitude less potent than hormonal substances, annual usage of millions of kilograms of nonionic surfactants make them a significant factor as water pollutants.

The most obvious effect of pharmaceuticals and their metabolic products in water has been feminization of male fish observed downstream from treated sewage discharges resulting from estrogenic compounds in wastewater. First noted in England and later in the United States and Europe, these male fish have been observed to produce proteins associated with egg production by female fish and to produce early-stage eggs in their testes. These effects are largely attributed to residues of synthetic 17α-ethinylestradiol and the natural hormone 17β-estradiol, used in oral contraceptives. The glucuronide and sulfate conjugates of 17α-ethinylestradiol are excreted with urine and cleaved by bacteria in water to regenerate the original compound.

4.13.3 BIOREFRACTORY ORGANIC POLLUTANTS

Millions of tons of organic compounds are manufactured globally each year, often taking the place of natural products that were formerly used. Significant quantities of several thousand synthetic organic compounds appear as water pollutants. Most of these compounds, particularly the less biodegradable ones, are substances to which living organisms have not been exposed until recent decades. Often, their effects upon organisms are not well known, particularly for long-term exposures at very low levels. The potential exists for synthetic organics to cause genetic damage, cancer, or other ill effects. On the positive side, organic pesticides enable a level of agricultural productivity without which millions would starve. Although synthetic organic chemicals are essential to the operation of a modern society, it is very important to know about their environmental harm and potential as pollutants.

Biorefractory organics are the organic compounds of most concern in wastewater, particularly when they are found in sources of drinking water. These are poorly biodegradable substances and are sometimes referred to as **persistent organic pollutants** (POP), prominent among which are aromatic or chlorinated hydrocarbons. Biorefractory compounds that may be found in water include benzene, chloroform, methyl chloride, styrene, tetrachloroethylene, trichloroethane, and toluene. In addition to their potential toxicity, biorefractive compounds can cause taste and odor problems in water. They are not completely removed by biological treatment, and water contaminated with these compounds must be treated by physical and chemical means, including air stripping, solvent extraction, ozonation, and carbon adsorption.

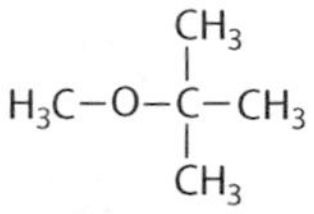

$$H_3C-O-\underset{\underset{CH_3}{|}}{\overset{\overset{CH_3}{|}}{C}}-CH_3$$

Methyl *tert*-butyl ether

Methyl *tert*-butyl ether (MTBE) was once used as a gasoline octane booster, but was phased out after it appeared as a low-level water pollutant in the United States. Levels of this chemical in recreational lakes and reservoirs were attributed largely to emissions of unburned fuel from recreational motorboats and personal watercraft having two-cycle engines that discharge their exhausts directly to the water.

Perfluorinated organic compounds constitute a unique class of POP. They occur as completely fluorinated hydrocarbon derivatives, such as CF_4, in the atmosphere, where they are regarded as atmospheric pollutants and potential greenhouse gases (see Chapter 7). Other perfluorinated compounds that are organic acids or their salts have been encountered as water pollutants. Most commonly cited are salts of perfluorooctane sulfonic acid; others include salts of perfluorinated carboxylic acids, such as perfluorohexanoic acid:

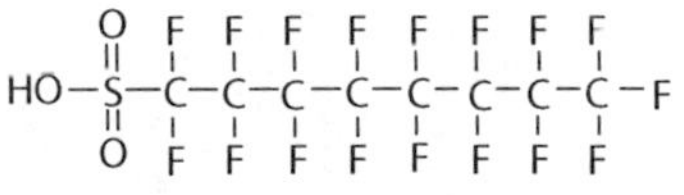

Perfluorooctane sulfonic acid 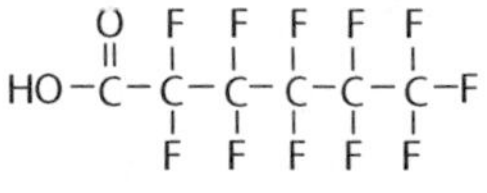   **Perfluorohexanoic acid**

Perfluorocarbons have been used commercially since the 1950s, primarily as coatings to resist soil and grease in paper products, fabrics, carpet material, and leather. Scotchgard fabric protector once manufactured by 3M Corporation contained perfluorooctane sulfonates, but this use has been discontinued. Perfluorocarbons have also been used as surfactants in oil drilling fluids and firefighting foams. Other applications have included alkaline cleaners, floor polish formulations, etching baths, and even denture cleaners. Perfluorocarbons have been detected in water, fish blood and liver, and human blood.

Brominated compounds have been recognized as significant environmental and water pollutants in recent years and have even been found in mothers' milk in some countries. These compounds have been manufactured as flame retardants, largely for use in polymers and textiles. The most common of the brominated compounds likely to be encountered as pollutants are polybrominated diphenyl ethers and tetrabromobisphenol:

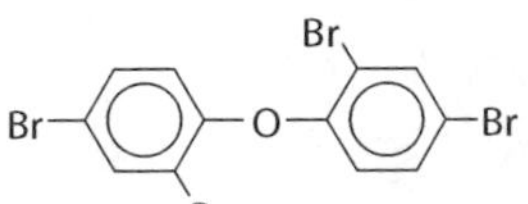

2,2',4,4' tetrabromodiphenyl ether 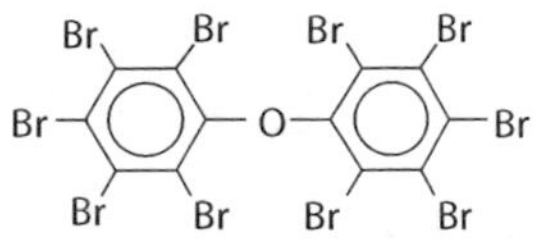   **Decabromodiphenyl ether**

Benzotriazole and tolyltriazoles (structural formulas as follows) are complexing agents for metals that are widely used as anticorrosive additives by forming a thin complexing film on metal surfaces, thereby protecting the metal from corrosion. These compounds are used in a variety of products including hydraulic fluids, cooling fluids, antifreeze formulations, and aircraft de-icer fluids and for silver protection in dishwasher detergents. Because of these uses, the triazoles are commonly encountered as "down-the-drain" chemicals that get into wastewaters. Their widespread use, high water solubility, and poor biodegradability make them among the most widely encountered chemicals in waters receiving treated wastewater.[10]

Benzotriazole   **Tolyltriazoles**

Naphthenic acid is a complex and variable mixture of carboxylic acids of molar masses in the approximate range of 180–350 and typically containing one $-CO_2H$ group and one 5- or 6-membered ring per molecule that is a by-product of petroleum refining, oil sand extraction, and coal. Naphthenic acids recovered from petroleum refining are used for solvents, lubricants, corrosion inhibitors, metal naphthenate synthesis, fuel additives, de-icing, dust control, wood preservation, and road stabilization. Water pollution from naphthenic acids is most severe in the oil sands processing area of Alberta, Canada, where caustic hot water is used to wash heavy hydrocarbon crude oil from sand, leaving huge quantities of tailings of clay, sand, and water contaminated with 80–120 mg/L naphthenic acids. The acids are toxic to aquatic organisms and are endocrine disrupting substances.

4-Methyl-1-cyclohexane carboxylic acid,
a typical low-molar-mass naphthenic acid

Two chemically similar synthetic organonitrogen compounds that have been detected in some wastewaters are the industrial compounds melamine and cyanuric acid. From their structural formulas below, it is seen that these compounds are triazine compounds that have very high nitrogen contents. Mixed with resins, melamine has fire retardant properties, because when heated to high temperatures, it releases N_2 gas, which tends to extinguish flames. Several million kilograms of melamine and cyanuric acid are produced worldwide annually as components or precursors in a variety of products including bleaches, disinfectants, and herbicides and as by-products in the production of pesticides, paints, building materials, textiles, and other products. These compounds were the agents involved in some tragic incidents involving pet food and infant formula starting in 2007, when hundreds of pets, mostly cats, died of renal failure. In 2008, in China, numerous puzzling incidents of infant illness and death were observed. These unfortunate events were found to be due to the presence of melamine and cyanuric acid in pet food and infant formula, deliberately added to raise the nitrogen content so that the foods would appear to have relatively higher protein contents. In 2009, two individuals in China were executed and several others imprisoned for their roles in producing the tainted products.

Melamine   **Cyanuric acid**

4.14   RADIONUCLIDES IN THE AQUATIC ENVIRONMENT

The massive production of **radionuclides** (radioactive isotopes) by weapons and nuclear reactors since World War II has been accompanied by increasing concern about the effects of radioactivity upon health and the environment. As illustrated in Figure 4.15 and by the specific examples shown in Table 4.7, radionuclides are produced as fission products of heavy nuclei of such elements as uranium or plutonium and are also produced by the reaction of neutrons with stable nuclei. The ultimate disposition of radionuclides formed in large quantities as waste products in nuclear power generation poses challenges with regard to the widespread use of nuclear power. Artificially produced radionuclides are also widely used in industrial and medical applications, particularly as "tracers." Radionuclides may enter aquatic systems from both artificial and natural sources, and their transport, reactions, and biological concentration in aquatic ecosystems can be a water pollution concern.

Radionuclides differ from other nuclei in that they emit **ionizing radiation**—alpha particles, beta particles, and gamma rays. The most massive of these emissions is the **alpha particle**, a helium nucleus of atomic mass 4, consisting of two neutrons and two protons. The symbol for an alpha particle is shown as the product of Reaction 4.10. An example of alpha production is found in the radioactive decay of uranium-238:

$$\ce{^{238}_{92}U} \rightarrow \ce{^{234}_{90}Th} + \ce{^{4}_{2}\alpha} \left(\text{alpha particle}\right) \tag{4.10}$$

This transformation occurs when a uranium nucleus, atomic number 92 and atomic mass 238, loses an alpha particle, atomic number 2 and atomic mass 4 (a helium atom nucleus), to yield a thorium nucleus, atomic number 90 and atomic mass 234.

Beta radiation consists of either highly energetic, negative electrons or their positively charged counterparts, called positrons:

$$\ce{^{0}_{-1}\beta} \text{ Beta particle} \qquad \ce{^{0}_{1}\beta} \text{ Positron}$$

A typical beta emitter, chlorine-38, may be produced by irradiating chlorine with neutrons. The chlorine-37 nucleus, natural abundance 24.5%, absorbs a neutron to produce chlorine-38 and gamma radiation:

$$\ce{^{37}_{17}Cl} + \ce{^{1}_{0}n} \rightarrow \ce{^{38}_{17}Cl} + \gamma \tag{4.11}$$

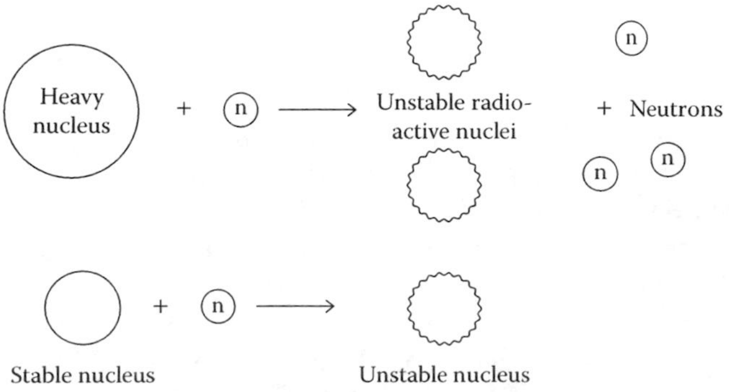

FIGURE 4.15   A heavy nucleus, such as that of ^{235}U, may absorb a neutron and break up (undergo fission), yielding lighter radioactive nuclei. A stable nucleus may absorb a neutron to produce a radioactive nucleus.

TABLE 4.7

Radionuclides That May Be Encountered in Water

Radionuclide	Half-Life	Nuclear Reaction, Description, Source
Naturally Occurring and from Cosmic Reactions		
Carbon-14	5730 y[1]	$^{14}N(n,p)^{14}C$,[2] thermal neutrons from cosmic or nuclear-weapon sources reacting with N_2
Silicon-32	~300 y	$^{40}Ar(p,x)^{32}Si$, nuclear spallation (splitting of the nucleus) of atmospheric argon by cosmic-ray protons
Potassium-40	~1.4×10^9 y	0.0119% of natural potassium including potassium in the body
Naturally Occurring from ^{238}U Series		
Radium-226	1620 y	Diffusion from sediments, atmosphere
Lead-210	21 y	$^{226}Ra \rightarrow 6$ steps $\rightarrow ^{210}Pb$
Thorium-230	75,200 y	$^{238}U \rightarrow 3$ steps $\rightarrow ^{230}Th$ produced in situ
Thorium-234	24 d	$^{238}U \rightarrow ^{234}Th$ produced in situ

From Reactor and Weapons Fission[3]

Strontium-90 (28 y) Iodine-131 (8 d) Cesium-137 (30 y)

Barium-140 (13 d) > Zirconium-95 (65 d) > Cerium-141 (33d) > Strontium-89 (51 d) > Ruthenium-103 (40 d) > Krypton-85 (10.3 y)

From Nonfission Sources		
Cobalt-60	5.25 y	From nonfission neutron reactions in reactors
Manganese-54	310 d	From nonfission neutron reactions in reactors
Iron-55	2.7 y	$^{56}Fe(n,2n)^{55}Fe$, from high-energy neutrons acting on iron in weapons hardware
Plutonium-239	24,300 y	$^{238}U(n,\gamma)^{239}Pu$, neutron capture by uranium

[1] Abbreviations: y, years; d, days

[2] This notation shows the isotope nitrogen-14 reacting with a neutron, n, giving off a proton, p, and forming the carbon-14 isotope; other nuclear reactions can be deduced from this notation, where x represents nuclear fragments from spallation.

[3] The first three fission-product radioisotopes listed below as products of reactor and weapons fission are of most significance because of their high yields and biological activity. The other fission products are listed in generally decreasing order of yield.

The chlorine-38 nucleus is radioactive and loses a negative **beta particle** to become an argon-38 nucleus:

$$^{38}_{17}Cl \rightarrow ^{38}_{18}Ar + ^{0}_{-1}\beta \qquad (4.12)$$

Since the negative beta particle has essentially no mass and a –1 charge, the stable product isotope, argon-38, has the same mass and a nuclear charge 1 greater than chlorine-38.

Gamma rays are electromagnetic radiation similar to x-rays, though more energetic. Since the energy of gamma radiation is often a well-defined property of the emitting nucleus, it may be used in some cases for the qualitative and quantitative analysis of radionuclides.

The primary effect of alpha particles, beta particles, and gamma rays upon materials is the production of ions; therefore, they are called **ionizing radiation**. Because of their large size, alpha particles do not penetrate matter deeply, but cause an enormous amount of ionization along their short path of penetration. Therefore, alpha particles present little hazard outside the body, but are very dangerous when ingested. Although beta particles are more penetrating than alpha particles, they produce much less ionization per unit path length. Gamma rays are much more penetrating than particulate radiation, but cause much less ionization. Their degree of penetration is proportional to their energy.

The **decay** of a specific radionuclide follows first-order kinetics; that is, the number of nuclei disintegrating in a short time interval is directly proportional to the number of radioactive nuclei present. The rate of decay, $-dN/dt$, is given by the equation

$$\text{Decay rate} = -\frac{dN}{dt} = \lambda N \tag{4.13}$$

where N is the number of radioactive nuclei present and λ is the rate constant, which has units of reciprocal time. Since the exact number of disintegrations per second is difficult to determine in the laboratory, radioactive decay is often described in terms of the measured **activity**, A, which is proportional to the absolute rate of decay. The first-order decay equation may be expressed in terms of A,

$$A = A_0 e^{-\lambda t} \tag{4.14}$$

where A is the activity at time t; A_0 is the activity when t is zero; and e is the natural logarithm base. The **half-life**, $t_{1/2}$, is generally used instead of λ to characterize a radionuclide:

$$t_{1/2} = \frac{0.693}{\lambda} \tag{4.15}$$

As the term implies, a half-life is the period of time during which half of a given number of atoms of a specific kind of radionuclide decay. Ten half-lives are required for the loss of 99.9% of the activity of a radionuclide.

Radiation damages living organisms by initiating harmful chemical reactions in tissues. For example, bonds are broken in the macromolecules that carry out life processes. In cases of acute radiation poisoning, bone marrow, which produces red blood cells, is destroyed and the concentration of red blood cells is diminished. Radiation-induced genetic damage, which may not become apparent until many years after exposure, is of great concern. As humans have learned more about the effects of ionizing radiation, the dosage level considered to be safe has been steadily lowered by regulatory agencies. Although it is possible that even the slightest exposure to ionizing radiation entails some damage, some radiation is unavoidably received from natural sources including the radioactive ^{40}K found in all humans. For the majority of the population, exposure to natural radiation exceeds that from artificial sources.

The study of the ecological and health effects of radionuclides involves consideration of many factors. Among these are the type and energy of radiation emitter and the half-life of the source. In addition, the degree to which the particular element is absorbed by living species and the chemical interactions and transport of the element in aquatic ecosystems are important factors. Radionuclides having very short half-lives may be hazardous when produced, but decay too rapidly to affect the environment into which they are introduced. Radionuclides with very long half-lives may be quite persistent in the environment, but of such low activity that little environmental damage is caused. Therefore, in general, radionuclides with intermediate half-lives are the most dangerous. They persist long enough to enter living systems while still retaining a high activity. Because they may be incorporated within living tissue, radionuclides of "life elements" are particularly dangerous. Much concern has been expressed over strontium-90, a common waste product of nuclear testing. This element is interchangeable with calcium in bone. Strontium-90 fallout drops onto pasture and cropland and is ingested by cattle. Eventually, it enters the bodies of infants and children by way of cow's milk.

Some radionuclides found in water, primarily radium and potassium-40, originate from natural sources, particularly leaching from minerals. Others come from pollutant sources, primarily nuclear power plants and testing of nuclear weapons. The levels of radionuclides found in water typically are measured in units of picocuries/liter, where a curie is 3.7×10^{10} disintegrations per second and a pico-curie is 1×10^{-12} that amount or 3.7×10^{-2} disintegrations per second (2.2 disintegrations per minute).

The radionuclide of most concern in drinking water is **radium**, Ra. Areas in the United States where significant radium contamination of water has been observed include the uranium-producing regions of the western United States, Iowa, Illinois, Wisconsin, Missouri, Minnesota, Florida, North Carolina, Virginia, and New England.

The U.S. Environmental Protection Agency specifies maximum contaminant levels (MCL) for total radium (^{226}Ra plus ^{228}Ra) in drinking water in units of pCi/L (picocuries per liter). In the past, perhaps as many as several hundred municipal water supplies in the United States have exceeded permissible levels, which has required finding alternative sources or additional treatment to remove radium. Fortunately, conventional water softening processes, which are designed to take out excessive levels of calcium, are relatively efficient in removing radium from water.

If nations continue to refrain from testing nuclear weapons above ground, this source will not contribute any more radioactivity to water. The possible contamination of water by fission-product radioisotopes from nuclear power production is of some concern. The worst such incident of radioactive contamination from a nuclear power source was the catastrophic 1986 meltdown and fire at the Chernobyl nuclear reactor in the former Soviet Union, which spread large quantities of radioactive materials over a wide area of Europe and contaminated many water supplies. A more recent incident was the result of the magnitude 9.0 earthquake that occurred on March 11, 2011, off the coast of Honshu, Japan, about 373 km (231 miles) northeast of Tokyo and the devastating tsunami that followed, which severely damaged the Fukushima Daiichi Nuclear Power Station. As a result, the electrical power infrastructure leading to the plant was damaged so that water could not be pumped to cool the reactor core and the spent fuel rods normally stored under water close to the core. Zirconium metal alloy cladding the hot fuel rods reacted with water to generate hydrogen gas, which exploded destroying structures that sheltered the reactors. Significant amounts of radioactivity were released, much of it contaminating water in the surrounding areas.

Transuranic elements are also of concern in the oceanic environment. These alpha emitters are long-lived and highly toxic. Included among these elements are various isotopes of neptunium, plutonium, americium, and curium. Specific isotopes, with half-lives in years given in parentheses, are Np-237 (2.14×10^6); Pu-236 (2.85); Pu-238 (87.8); Pu-239 (2.44×10^4); Pu-240 (6.54×10^3); Pu-241 (15); Pu-242 (3.87×10^5); Am-241 (433); Am-243 (7.37×10^6); Cm-242 (0.22); and Cm-244 (17.9).

4.15   TOXICOLOGICAL CHEMISTRY AND WATER POLLUTION

The toxicological chemistry aspects of various water pollutants are noted throughout this chapter. Toxicological chemistry as related to the hydrosphere was discussed in Section 3.12. As shown in Figures 3.15 and 3.16, an important aspect of the toxicological chemistry of water pollutants in the hydrosphere is the exchange of toxic substances among sediments, water, and the bodies of fish and other aquatic organisms and the metabolism, transport, and storage of toxicants within the organism. These processes are especially pertinent to lipophilic organic pollutants including hydrocarbons and organochlorine compounds, which may undergo bioaccumulation and bioconcentration in lipid tissue of fish.[11] Fish lipid tissue is frequently analyzed for such substances as a biomarker of pollution. An important example is the accumulation of methylmercury species in the lipid tissue of some fish, primarily older members of some larger species (see Section 4.3), to the extent that human consumption of some species should be limited, especially during pregnancy.

Aquatic organisms may be affected directly by water pollutants. One important example is the feminization of some organisms by exposure to estrogenic pollutants. Mussels and other sediment-dwelling organisms in the hydrosphere have been adversely affected by organotin water pollutants.

Another important aspect of the toxicological chemistry of water pollutants involves biotransformations and biodegradation of pollutants. Some of the compounds found as markers of water pollution are metabolites of parent pollutant species. Such metabolites, including those of some pharmaceuticals, are produced by humans and livestock on land, then discharged as wastes to water. Microbial metabolites of some widely used herbicides are encountered as water pollutants including the atrazine metabolites deisopropylatrazine and deethylatrazine and aminomethyl phosphonic acid produced from herbicidal glyphosate (see Section 4.11).

Microorganisms in water and sediments can act to detoxify organic water pollutants. As noted in Section 3.14, this most commonly takes place through the process of co-metabolism. In most favorable cases, the pollutants are completely mineralized to inorganic forms. As discussed in Section 3.9, an indirect pathway to detoxification of toxic heavy metal ions occurs when reducing conditions in the bottom regions and sediments in bodies of water convert sulfate ion, SO_4^{2-}, to H_2S, which precipitates heavy metals as biologically less available metal sulfides.

As discussed in Section 4.10, some aquatic microorganisms produce toxins that can harm or even kill fish and other organisms. Photosynthetic cyanobacteria produce a variety of toxins, such as the *Cylindrospermopsin* toxin that can poison humans who drink the water. "Blooms" of cyanobacteria in lakes and reservoirs serving as water supplies lead to warnings about the safety of drinking water supplies each summer. Also noted in Section 4.10 was the production of toxins by predominantly marine single-celled photosynthetic protozoans, which have led to kills of some marine organisms and gastrointestinal, respiratory, and skin disorders in humans. The greatest danger to humans from dinoflagellata toxins comes from the ingestion of shellfish, such as mussels and clams, that have accumulated the protozoa from seawater. In this form, the toxic material is called paralytic shellfish poison. As little as 4 mg of this toxin, the amount found in several severely infested mussels or clams, can be fatal to a human. The toxin depresses respiration and affects the heart, resulting in complete cardiac arrest in extreme cases.

QUESTIONS AND PROBLEMS

Access to and use of the Internet is assumed in answering all questions, including general information, statistics, constants, and mathematical formulas required to solve problems. These questions are designed to promote inquiry and thought rather than just finding material in the chapter. So in some cases, there may be several "right" answers. Therefore, if your answer reflects intellectual effort and a search for information from available sources, it may be considered to be right.

1. Which of the following statements is true regarding chromium in water: (a) chromium(III) is suspected of being carcinogenic; (b) chromium(III) is less likely to be found in a soluble form than chromium(VI); (c) the toxicity of chromium(III) in electroplating wastewaters is decreased by oxidation to chromium(VI); (d) chromium is not an essential trace element; (e) chromium is known to form methylated species analogous to methylmercury compounds.
2. What do mercury and arsenic have in common in regard to their interactions with bacteria in sediments?
3. What are some characteristics of radionuclides that make them especially hazardous to humans?
4. To what class do pesticides containing the following group belong?

$$-\overset{\overset{\displaystyle H}{|}}{N}-\overset{\overset{\displaystyle O}{\|}}{C}-$$

5. Consider the following compound:

$$Na^+ \ ^-O-\overset{\overset{\displaystyle O}{\|}}{\underset{\underset{\displaystyle O}{\|}}{S}}-\underset{}{\bigcirc}-\overset{H}{\underset{H}{C}}-\overset{H}{\underset{H}{C}}-\overset{H}{\underset{H}{C}}-\overset{H}{\underset{H}{C}}-\overset{H}{\underset{H}{C}}-\overset{H}{\underset{H}{C}}-\overset{H}{\underset{H}{C}}-\overset{H}{\underset{H}{C}}-\overset{H}{\underset{H}{C}}-\overset{H}{\underset{H}{C}}-H$$

Which of the following characteristics is not possessed by the compound: (a) one end of the molecule is hydrophilic and the other end is hydrophobic, (b) surface-active qualities, (c) the ability to lower surface tension of water, (d) good biodegradability, (e) tendency to cause foaming in sewage treatment plants.

6. A certain pesticide is fatal to fish fingerlings at a level of 0.50 parts per million in water. A leaking metal can containing 5.00 kg of the pesticide was dumped into a stream with a flow of 10.0 L/s moving at 1 km/h. The container leaks pesticide at a constant rate of 5 mg/s. For what distance (in km) downstream is the water contaminated by fatal levels of the pesticide by the time the container is empty?

7. Give a reason that Na_3PO_4 would not function well as a detergent builder, whereas $Na_3P_3O_{10}$ is satisfactory, though it is a source of pollutant phosphate.

8. Of the compounds $CH_3(CH_2)_{10}CO_2H$, $(CH_3)_3C(CH_2)_2CO_2H$, $CH_3(CH_2)_{10}CH_3$, and $\phi-(CH_2)_{10}CH_3$ (where ϕ represents a benzene ring), which is the most readily biodegradable?

9. A pesticide sprayer got stuck while trying to ford a stream flowing at a rate of 136 L/s. Pesticide leaked into the stream for exactly 1 hour and at a rate that contaminated the stream at a uniform 0.25 ppm of methoxychlor. How much pesticide was lost from the sprayer during this time?

10. A sample of water contaminated by the accidental discharge of a radionuclide used for medicinal purposes showed an activity of 12,436 counts per second at the time of sampling and 8,966 cps exactly 30 days later. What is the half-life of the radionuclide?

11. What are the two reasons that soap is environmentally less harmful than ABS surfactant used in detergents?

12. What is the exact chemical formula of the specific compound designated as PCB?

13. Match each compound designated by a letter with the description corresponding to it designated by a number.

(A) CdS (B) $(CH_3)_2AsH$ (C) $--(Cl)_{10}$ (D) Cl,

(1) Pollutant released to a U.S. stream by a poorly controlled manufacturing process.
(2) Insoluble form of a toxic trace element likely to be found in anoxic sediments.
(3) Common environmental pollutant formerly used as a transformer coolant.
(4) Chemical species thought to be produced by bacterial action.

14. A radioisotope has a nuclear half-life of 24 hours and a biological half-life of 16 hours (half of the element is eliminated from the body in 16 hours). A person accidentally swallowed sufficient quantities of this isotope to give an initial "whole body" count rate of 1000 counts per minute. What was the count rate after 16 hours?

15. What is the primary detrimental effect upon organisms of salinity in water arising from dissolved NaCl and Na_2SO_4?

16. Give a specific example of each of the following general classes of water pollutants: (a) trace elements, (b) metal–organic combinations, (c) pesticides.

17. A polluted water sample is suspected of being contaminated with one of the following: soap, ABS surfactant, or LAS surfactant. The sample has a very low BOD relative to its TOC. Which is the contaminant?

18. Of the following, the one that is **not** a cause of or associated with eutrophication is (a) eventual depletion of oxygen in the water, (b) excessive phosphate, (c) excessive algal growth, (d) excessive nutrients, (e) excessive O_2.

19. From the formulas below match the following: (a) Lowers surface tension of water, (b) a carbamate, (c) a herbicide, (d) a non-carbamate insecticide

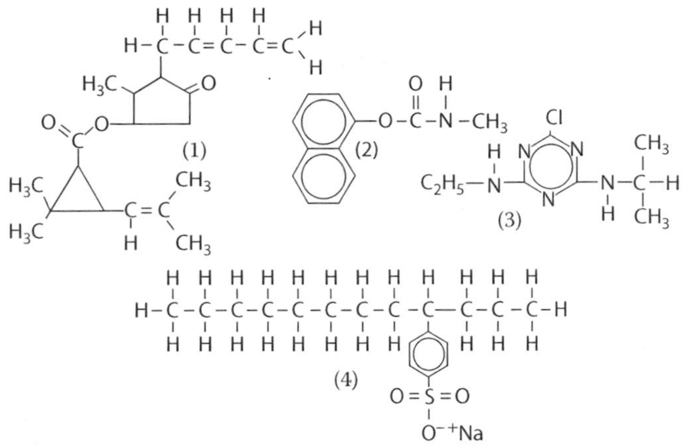

H H H H
H–C–C=C–C=C⟨H / H
H₃C
O
O=C–O
(1)
H₃C
H₃C C=C⟨CH₃ / CH₃
H CH₃

O H
–O–C–N–CH₃
(2)

Cl
H N N CH₃
C₂H₅–N N –N–C–H
N H CH₃
(3)

H H H H H H H H H H H H
H–C–C–C–C–C–C–C–C–C–C–C–C–H
H H H H H H H H H H H
(4)
O=S=O
O⁻ ⁺Na

20. Of the following heavy metals, choose the one most likely to have microorganisms involved in its mobilization in water and explain why this is so: (a) lead, (b) mercury, (c) cadmium, (d) chromium, (e) zinc.

21. Of the following, the true statement is (a) eutrophication results from the direct discharge of toxic pollutants into water, (b) treatment of a lake with phosphates is a process used to deter eutrophication, (c) alkalinity is the most frequent limiting nutrient in eutrophication, (d) eutrophication results from excessive plant or algal growth, (e) eutrophication is generally a beneficial phenomenon because it produces oxygen.

22. Of the following, the statement that is **untrue** regarding radionuclides in the aquatic environment is (a) they emit ionizing radiation, (b) they invariably come from human activities, (c) radionuclides of "life elements," such as iodine-131, are particularly dangerous, (d) normally the radionuclide of most concern in drinking water is radium, (e) they may originate from the fission of uranium nuclei.

23. Match the following pollutants on the left with effects or other significant aspects on the right:

(a) Salinity	1. Excessive productivity
(b) Alkalinity	2. Can enter water from pyrite or from the atmosphere
(c) Acidity	3. Osmotic effects on organisms
(d) Nitrate	4. From soil and mineral strata

24. After some research on the Internet, suggest three radionuclide pollutants released by the 2011 earthquake and tsunami damage to Japan's Fukushima Daiichi Nuclear Power Station and why each poses a health concern.

25. Explain why fish lipid tissue is commonly analyzed as a biomarker of water pollution. For which kinds of pollutants is it most useful? Name a class of pollutants for which it probably would not be very useful.

26. Cases have been reported in which male fish exhibit early stages of egg development and other female characteristics. For which class of water pollutants is this a biomarker?

LITERATURE CITED

1. Grove, Robert A., Charles J. Henny, and James L. Kaiser, Worldwide Sentinel Species for Assessing and Monitoring Environmental Contamination in Rivers, Lakes, Reservoirs, and Estuaries, *Journal of Toxicology and Environmental Health, Part B: Critical Reviews*, **12**, 25–44 (2008).
2. Mergel, Maria, Tributyltin, http://toxipedia.org/display/toxipedia/Tributyltin, 2011.
3. Vetter, Walter, Marine Halogenated Natural Products of Environmental Relevance, *Reviews of Environmental Contamination and Toxicology*, **188**, 1–57 (2006).
4. PAN Pesticide Database, http://www.pesticideinfo.org/, 2011.
5. Costa, Lucio C., Toxic Effects of Pesticides, in *Casarett & Doull's Essentials of Toxicology*, 2nd ed., John B. Watkins III, Ed., McGraw-Hill Professional, Kindle ed, New York, 2010, Kindle Locations 10381–10385.
6. Wirgin, Isaac, Nirmal K. Roy, Matthew Loftus, Christopher Chambers, Diana G. Franks, and Mark E. Hahn, Mechanistic Basis of Resistance to PCBs in Atlantic Tomcod from the Hudson River, *Science*, **331**, 1322–1325 (2011).
7. Richardson, Susan D., Environmental Mass Spectrometry: Emerging Contaminants and Current Issues, *Analytical Chemistry*, **82**, 4742–4774 (2010).
8. Khetan, Sushil K., and Terrence J. Collins, Human Pharmaceuticals in the Aquatic Environment: A Challenge to Green Chemistry, *Chemical Reviews*, **107**, 2319–2364 (2007).
9. Benotti, Mark J., Rebecca Trenholm, Brett Vanderford, Janie Holady, Benjamin Stanford, and Shane Snyder, Pharmaceuticals and Endocrine Disrupting Compounds in U.S. Drinking Water, *Environmental Science and Technology*, **43**, 597–603 (2009).
10. Giger, Walter, Christian Schaffner, and Hans-Petere Kohler, Benzotriazole and Tolyltriazole as Aquatic Contaminants. 1. Input and Occurrence in Rivers and Lakes, *Environmental Science and Technology*, **40**, 7186–7192 (2006).
11. Arnot, Jon A. and Frank A. P. C. Gobas, A Review of Bioconcentration Factor (BCF) and Bioaccumulation Factor (BAF) Assessments for Organic Chemicals in Aquatic Organisms, *Environmental Reviews*, **14**, 257–297 (2008).

SUPPLEMENTARY REFERENCES

Alley, E. Roberts, *Water Quality Control Handbook*, 2nd ed., McGraw-Hill, New York, 2007.

Burk, A. R., Ed., *Water Pollution: New Research*, Nova Science Publishers, New York, 2005.

Calhoun, Yael, Ed., *Water Pollution*, Chelsea House Publishers, Philadelphia, 2005.

Eckenfelder, W. Wesley, Davis L. Ford, and Andrew J. Englande, *Industrial Water Quality*, 4th ed., McGraw-Hill, New York, 2009.

Hamilton, Denis, and Stephen Crossley, Eds., *Pesticide Residues in Food and Drinking Water: Human Exposure and Risks*, Wiley, New York, 2004.

Howd, Robert A., and Anna M. Fan, *Risk Assessment for Chemicals in Drinking Water*, Wiley, Hoboken, NJ, 2007.

Gilliom, Robert, J., *Pesticides in the Nation's Streams and Ground Water, 1992–2001: The Quality of our Nation's Waters*, U.S. Geological Survey, Reston, VA, 2006.

Kaluarachchi, Jagath J., *Groundwater Contamination by Organic Pollutants: Analysis and Remediation*, American Society of Civil Engineers, Reston, VA, 2001.

Laws, Edward A., *Aquatic Pollution: An Introductory Text*, 3rd ed., John Wiley & Sons, New York, 2000.

Livingston, James V., Ed., *Focus on Water Pollution Research*, Nova Science Publishers, New York, 2006.

Mason, Christopher F., *Biology of Freshwater Pollution,* 4th ed., Prentice Hall College Division, Upper Saddle River, NJ, 2002.

Raven, Peter H., Linda R. Berg, David M. Hassenzahl, *Environment*, 6th ed., Wiley, Hoboken, NJ, 2008.

Ravenscroft, Peter, Hugh Brammer, and Keith Richards, *Arsenic Pollution: A Global Synthesis*, Blackwell, Malden, MA, 2009.

Rico, D. Prats, C.A. Brebbia, and Y. Villacampa Esteve, Eds., *Water Pollution IX*, WIT Press, Southampton, UK, 2008.

Stollenwerk, Kenneth G., and Alan H. Welch, Eds., *Arsenic in Ground Water: Geochemistry and Occurrence*, Kluwer Academic Publishers, Hingham, MA, 2003.

Sullivan, Patrick J., Franklin J. Agardy, and James J.J. Clark, *The Environmental Science of Drinking Water*, Elsevier Butterworth-Heinemann, Burlington, MA, 2005.

Viessman, Warren, Mark J. Hammer, Elizabeth M. Perez, and Paul A. Chadik *Water Supply and Pollution Control*, 8th ed., Pearson Prentice Hall, Upper Saddle River, NJ, 2008.

Wang, Lawrence K., Jiaping Paul Chen, Yung-Tse Hung, and Nazih K. Shammas, Eds., *Heavy Metals in the Environment*, Taylor & Francis/CRC Press, Boca Raton, FL, 2010.

Water Environment Research, a research publication of the Water Environment Federation, Water Environment Federation, Alexandria, VA. This journal contains many articles of interest to water science; the annual reviews are especially informative.

Wheeler, Willis B., *Pesticides in Agriculture and the Environment*, Marcel Dekker, New York, 2002.

Whitacre, David M., Ed., *Reviews of Environmental Contamination and Toxicology*, **196**, Springer-Verlag, New York, 2008 (published annually).

Xie, Yuefeng, *Disinfection Byproducts in Drinking Water: Form, Analysis, and Control*, CRC Press, Boca Raton, FL, 2002.

5 Sustaining the Hydrosphere

5.1 MORE IMPORTANT THAN OIL

As Earth's population reached the 7 billion mark at the end of October 2011, the demands that population growth place upon the planet's finite resources are matters of very serious concern, much of them involving energy, especially petroleum resources. But, arguably, a more urgent concern with respect to Earth's natural capital is water and it has become fashionable to say that water is the new petroleum. In a sense, the problems posed by shortages of water and deteriorating water quality in many areas of the world are much more challenging than those of energy. There are numerous options for supplying energy, including renewable resources such as wind energy and solar energy. But, overall, there is a finite amount of water on Earth, especially if water in the oceans is regarded as being largely off-limits as a source of usable H_2O. Although water, like petroleum, can be moved some distances, to a large extent, "you have got to use it where you find it," and the uneven distribution of water around the planet makes its utilization in an economical way rather difficult in many areas.

Water is strongly linked to both energy supply and food supply. A number of technologies that provide energy use water in areas such as cooling power plants and in hydraulic fracturing of methane-bearing tight shale deposits, and the exploitation of some sources of energy has the potential to harm water quality. Obviously, water is crucial in the production of food in the growing of crops and as a medium in which fish grow and are harvested, providing a major source of protein for human consumption.

Fortunately, in principle, water is a totally renewable resource and, unlike hydrocarbon fossil fuels, for example, water is never totally destroyed, although it may become so contaminated as to make its reuse very difficult. Water may accurately be described as "nature's most renewable resource."[1] This chapter discusses sustaining the hydrosphere and its precious content of water, that crucial part of Earth's natural capital. A critical part of this endeavor is the recycling of water, which will have to be practiced to a large extent if enough water is to be provided to meet demands for it. Much of this chapter deals with treatment processes used to bring water from diverse sources up to standards enabling its use, to restore water that has been used back to standards where it can be used again, and to provide usable freshwater from impaired sources including brackish groundwater and even seawater.

5.2 GREENING OF WATER: PURIFICATION BEFORE AND AFTER USE

For most uses, water requires treatment. The processes used to treat water are generally similar for municipal, commercial, or industrial uses although the kinds of treatment used and the degree of treatment depend on the end use of the water. This section addresses the major treatment processes applied to water supplies.

An important consideration in the treatment of water is the source of the water. Chapter 3 addresses the question of where water is found in the hydrosphere and, therefore, the potential sources of water to be treated. Generally, there are two main sources of freshwater. Surface water occurs in rivers, reservoirs, and lakes. Groundwater is contained in subsurface aquifers. The overexploitation of groundwater sources is a matter of great current concern as levels of groundwater (the water table, see Chapter 3, Figure 3.4) are being lowered at an alarming rate in many parts of the world. In many areas, the natural rate of groundwater recharge from surface infiltration is inadequate to maintain groundwater levels. A number of aquifers that are being pumped to exhaustion contain fossil water (in some cases, remaining from the last ice age) that have no means of natural recharge. In the future, the provision of adequate water supplies will require greater utilization of wastewater and saline water, which will put heavy demands on the treatment processes used.[2]

A large variety of processes are used to treat water and are discussed in this chapter. Some of these are physical processes including filtration, aeration, and adsorption on solids such as activated carbon. Chemical processes used include acid-base, precipitation, oxidation-reduction, and complexation. Treatment of drinking water almost always involves physical and chemical processes, whereas wastewater treatment usually uses all three processes with a heavy reliance on biological treatment.

Most physical and chemical processes used to treat water involve similar phenomena, regardless of their application to municipal, commercial, or industrial use. Therefore, after introductions to water treatment for municipal use, industrial use, and disposal, each major kind of treatment process is discussed as it applies to all of these applications.

5.2.1 EMERGING CONSIDERATIONS IN WATER TREATMENT

Several aspects of water treatment are becoming more important, especially as pressures increase for water reclamation and recycle. These include the emergence of increased amounts of "exotic" contaminants, including pharmaceuticals and their metabolites, personal care products, household chemicals, disinfection by-products, and nanomaterials, those in a size range of 1–100 nm used increasingly in applications such as the delivery of pharmaceuticals.

As water treatment processes become more sophisticated and water shortages necessitate use of impaired and recycled water, problems are growing with the treatment and disposal of **residuals**. These include sludge materials and highly contaminated water from the dewatering of sludges and a variety of other solid, liquid, and gaseous by-products. Residuals include turbidity-causing inorganic and organic materials removed from raw water used as a municipal source as well as materials removed in the treatment of domestic and industrial wastewater. Such residuals may include microbial biomass from algae, bacteria, protozoa, and viruses. The potential presence of pathogens needs to be considered in managing residuals. Significant amounts of residuals arise from chemicals used to treat water, such as $Al(OH)_3$ and $Fe(OH)_3$, produced when aluminum and iron salts are added to water to coagulate colloidal impurities. Membrane water treatment processes produce waste **retentates** enriched in impurities removed by membranes. As desalination of saline groundwater and ocean water becomes more common, the disposal of concentrated brine retentate from reverse osmosis treatment will grow in importance.

Another consideration of growing importance in water treatment is the development of new technologies. These include special membrane processes for water filtration, alternatives to chlorine for water disinfection, advanced oxidation of impurities, and the use of ultraviolet radiation for water disinfection and as an aid to destruction of organic contaminants by oxidants. It is important to consider the sustainability of developing techniques including costs and by-product generation.

5.3 MUNICIPAL WATER TREATMENT

The modern treatment plant that produces municipal water is often called upon to perform wonders with the water fed to it. The water may have entered the plant as a murky liquid pumped from a polluted river laden with mud and swarming with bacteria. Or, its source may have been well water, which is much too hard for domestic use because of high levels of Ca^{2+} ion or the presence of stain-producing dissolved iron and manganese. But for the most part, the product that comes from a faucet is clear, safe, and even tasteful water.

A schematic diagram of a typical municipal water treatment plant is shown in Figure 5.1. This particular facility treats water containing excessive hardness and a high level of iron. The raw water taken from wells first goes to an aerator. Contact of the water with air removes volatile solutes such

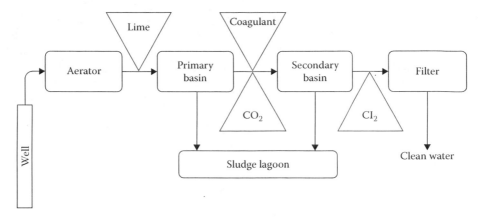

FIGURE 5.1   Schematic of a municipal water treatment plant.

as hydrogen sulfide, carbon dioxide, methane, and volatile odorous substances such as methanethiol (CH_3SH) and bacterial metabolites. Contact with oxygen also aids iron removal by oxidizing soluble iron(II) to insoluble iron(III). The addition of lime as CaO or $Ca(OH)_2$ after aeration raises the pH and results in the formation of precipitates containing the hardness ions Ca^{2+} and Mg^{2+}. These precipitates settle from the water in a primary basin. Much of the solid material remains in suspension and requires the addition of coagulants (such as iron(III) and aluminum sulfates, which form gelatinous metal hydroxides) to settle the colloidal particles. Activated silica or synthetic polyelectrolytes may also be added to stimulate coagulation or flocculation. The settling occurs in a secondary basin after the addition of carbon dioxide to lower the pH. Sludge from both the primary and secondary basins is pumped to a sludge lagoon. The water is finally chlorinated, filtered, and pumped to the city water mains.

5.3.1   Contamination in Water Distribution Systems

Although water may be pure and clean when it leaves the treatment plant, it may not remain so throughout the water distribution system, largely because of materials picked up from the piping in the system. Iron and copper piping used to distribute water is subject to corrosion processes that may result in the presence of metals in the water. Iron from iron pipe leads to "red water," and dissolved copper from water at a pH below 7 that has stood in copper pipe can produce "blue-green water." Lead-based solder was once widely used in connecting copper pipe; concerns over lead in drinking water have led to the replacement of this kind of solder with solder composed of silver and tin. The term "plumbing" is based on the Latin word for lead, reflecting the fact that in years past, much of the pipe used for water distribution was composed of lead. Although plastic water pipe does not corrode, there is a potential for plasticizers and organic additives to leach from the plastic into drinking water. Pathogens may get into water distribution systems by accident, and residual disinfectants are commonly put into water before it leaves the treatment plant to guard against waterborne disease agents.

5.4   TREATMENT OF WATER FOR INDUSTRIAL USE

Water is widely used in various process applications in industry. Other major industrial uses are boiler feedwater and cooling water. The kind and degree of treatment of water in these applications depend on the end use. As examples, although cooling water may require only minimal treatment, removal of corrosive substances and scale-forming solutes is essential for boiler feedwater, and water used in food processing must be free of pathogens and toxic substances. Inadequate treatment of water for industrial use can cause problems such as corrosion, scale formation, reduced heat transfer in heat exchangers, reduced water flow, and product contamination. These effects may be detrimental to equipment performance or cause equipment failure, increased energy costs due to inefficient heat utilization or cooling, increased costs for pumping water, and product deterioration. Obviously, the effective treatment of water at minimum cost for industrial use is a very important area of water treatment.

Numerous factors must be taken into consideration in designing and operating an industrial water treatment facility. These include the following factors:

- Water requirement
- Quantity and quality of available water sources
- Sequential use of water (successive uses for applications requiring progressively lower water quality)
- Water recycle
- Discharge standards

The various specific processes employed to treat water for industrial use are discussed in Sections 5.6 through 5.11. **External treatment**, usually applied to the plant's entire water supply, uses processes such as aeration, filtration, and clarification to remove material that might cause problems from water. Such substances include suspended or dissolved solids, hardness, and dissolved gases. Following this basic treatment, the water can be divided into different streams, some to be used without further treatment and the rest to be treated for specific applications.

Internal treatment is designed to modify the properties of water for specific applications. Examples of internal treatment are as follows:

- Reaction of dissolved oxygen with hydrazine or sulfite
- Addition of chelating agents to react with dissolved Ca^{2+} and prevent formation of calcium deposits
- Addition of precipitants, such as phosphate, used for calcium removal
- Treatment with dispersants to inhibit scale
- Addition of inhibitors to prevent corrosion
- Adjustment of pH
- Disinfection for food processing uses or to prevent bacterial growth in cooling water

5.5 WASTEWATER TREATMENT

Typical municipal sewage contains oxygen-demanding materials, sediments, grease, oil, scum, pathogenic bacteria, viruses, salts, algal nutrients, pesticides, refractory organic compounds, heavy metals, and an astonishing variety of flotsam ranging from children's socks to sponges. It is the job of the waste-treatment plant to remove as much of this material as possible.

Several characteristics are used to describe sewage. These include turbidity (international turbidity units), suspended solids (ppm), total dissolved solids (ppm), acidity (H^+ ion concentration or pH), and dissolved oxygen (in ppm O_2). Biochemical oxygen demand (BOD) is used as a measure of oxygen-demanding substances.

Current processes for the treatment of wastewater can be divided into three main categories of primary treatment, secondary treatment, and tertiary treatment. Wastewater is first subjected to **primary treatment**, which is largely a physical process including removal of grit, grease, and scum. **Secondary wastewater treatment** is concentrated on the removal of biodegradable wastes that are responsible for a BOD in the water (see Chapter 3, Section 3.5) and is usually accomplished by biological treatment with microorganisms. **Tertiary waste treatment** (sometimes called **advanced waste treatment**) is a term used to describe a variety of processes performed on the effluent from secondary waste treatment and is especially important in the treatment of water that may be recycled.

Waste from a municipal water system is normally treated in a **publicly owned treatment works** (**POTW**). In the United States, these systems are allowed to discharge only effluents that have attained a certain level of treatment, as mandated by federal law. One of the major objectives in the treatment of hazardous wastes, which usually have a high content of water, is to bring the water by-product up to a quality that can be sent to a POTW for treatment and release.

5.5.1 INDUSTRIAL WASTEWATER TREATMENT

Industrial wastewater may present special challenges in restoring it to a standard that can be discharged. Before treatment, industrial wastewater should be characterized fully and the biodegradability of wastewater constituents determined. Biological treatment such as that used to remove BOD from municipal wastewater may be employed to remove biodegradable impurities. It may be necessary to acclimate microorganisms to the degradation of constituents that are not normally biodegradable. Various pollutants may be removed from industrial wastewater by sorption onto solids, especially activated carbon. Wastewater can be treated by a variety of chemical processes, including acid-base neutralization, precipitation, and oxidation-reduction. Sometimes, these steps must precede biological treatment; for example, acidic or alkaline wastewater must be neutralized for microorganisms to thrive in it.

An important consideration in industrial wastewater treatment is the fate and disposal of residues from the treatment processes, which may contain contaminants not normally encountered in municipal wastewater. The sludges produced in biological treatment may contain hazardous constituents such as toxic heavy metal ions contained in the sludge. Spent activated carbon from industrial wastewater treatment may contain toxic organic materials and heavy metals. Another potential hazard is that from sludges formed by chemical treatment of wastewater.

5.6 REMOVAL OF SOLIDS

Sewage, industrial wastewater, and water to be used for drinking water sources commonly have various levels of suspended particulate matter, often in the form of small colloidal particles. Inorganic particles may include finely divided mineral matter, clay, or silt. Organic particles are often of biological origin, including bacteria, protozoa, viruses, finely divided leaf litter, and humus. Suspended solids can reduce water clarity as measured by water turbidity; consist of pathogenic bacteria, protozoa, viruses, or harmful organic pollutants; or serve as carriers of toxic substances. Therefore, an important process in water and wastewater treatment is the removal of suspended matter.

Relatively large solid particles are removed from water by simple **settling** and **filtration**. A special type of filtration procedure known as **microstraining** is especially effective in the removal of the very small particles. These filters are woven from extremely fine stainless steel wire. This enables preparation of filters with openings only 60–70 μm across. These openings may be reduced to 5–15 μm by partial clogging with small particles such as bacterial cells. The cost of this treatment is likely to be substantially lower than the costs of competing processes. High flow rates at low back pressures are normally achieved.

The removal of colloidal solids from water usually requires **coagulation**. Salts of aluminum and iron are the coagulants most often used in water treatment. Of these, alum or filter alum is most commonly used. This substance is a hydrated aluminum sulfate, $Al_2(SO_4)_3 \cdot 18H_2O$. When this salt is added to water, the aluminum ion hydrolyzes by reactions that consume alkalinity in the water, such as

$$Al(H_2O)_6^{3+} + 3HCO_3^- \rightarrow Al(OH)_3(s) + 3CO_2 + 6H_2O \qquad (5.1)$$

The gelatinous hydroxide thus formed carries suspended material with it as it settles. Furthermore, it is likely that positively charged hydroxyl-bridged dimers such as

$$(H_2O)_4Al \underset{O}{\overset{O}{\diamond}} Al(H_2O)_4^{4+}$$

and higher polymers are formed, which interact specifically with colloidal particles, bringing about coagulation. Sodium silicate partially neutralized by acid aids coagulation, particularly when used with alum. Metal ions in coagulants also react with virus proteins and destroy viruses in water.

Anhydrous iron(III) sulfate added to water forms iron(III) hydroxide in a reaction analogous to Reaction 5.1. An advantage of iron(III) sulfate is that it works over a wide pH range of approximately 4–11. Hydrated iron(II) sulfate, or copperas, $FeSO_4 \cdot 7H_2O$, is also commonly used as a coagulant. It forms a gelatinous precipitate of hydrated iron(III) oxide when the iron(II) is oxidized to iron(III).

Natural and synthetic polyelectrolytes are used in causing small suspended particles to flocculate to larger settleable particles. Among the natural compounds so used are starch and cellulose derivatives, proteinaceous materials, and gums composed of polysaccharides. More recently, selected synthetic polymers, including neutral polymers and both anionic and cationic polyelectrolytes that are effective flocculants, have come into use.

Coagulation-filtration is a much more effective procedure than filtration alone for the removal of suspended material from water. As the term implies, the process consists of the addition of coagulants that aggregate the particles into larger-size particles, followed by filtration. Either alum or lime, often with added polyelectrolytes, is most commonly employed for coagulation.

The filtration step of coagulation-filtration is usually performed on a medium such as sand or anthracite coal. Often, to reduce clogging, several media with progressively smaller interstitial spaces are used. One example is the **rapid sand filter**, which consists of a layer of sand supported by layers of gravel particles, the particles becoming progressively larger with increasing depth. The substance that actually filters the water is coagulated material that collects in the sand. As more material is removed, the buildup of coagulated material eventually clogs the filter and must be removed by back-flushing. Filtration of water with special membrane filters is discussed in Section 5.10.

An important class of solids that must be removed from wastewater consists of suspended solids in secondary sewage effluents that arise primarily from sludge that was not removed in the settling process (see Section 5.12). These solids account for a large part of the BOD in the effluent and may interfere with other aspects of tertiary waste treatment, such as by clogging membranes in reverse osmosis water treatment processes. The quantity of material involved may be rather high. Processes designed to remove suspended solids often will remove 10–20 mg/L of organic material from secondary sewage effluent. In addition, a small amount of the inorganic material is removed.

5.6.1 Dissolved Air Flotation

Many of the particles found in water have low densities close to or even less than those of water. Particles less dense than water have a tendency to rise to the surface from which they can be skimmed off, but this is often a slow and incomplete process. The removal of such particles can be aided by **dissolved air flotation** in which small air bubbles are formed that attach to particles causing them to float. As shown in Figure 5.2, flotation of particles with air can be accompanied by coagulation of the particles as aided by a coagulant. Water supersaturated with air under pressure is released to the bottom of a tank where bubbles are formed in a layer of milky (white) water. Bubble formation accompanied by flocculation of the particles entrains bubbles in the floc, which floats to the surface where it is skimmed off.

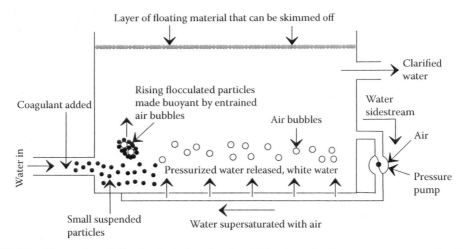

FIGURE 5.2 Illustration of dissolved air flotation in which pressurized water supersaturated with air is released into the bottom of a tank of water producing an abundance of small bubbles giving the water a white or milky appearance. Flocculated particles that entrain air bubbles are buoyant and rise to the top where the flocculated material may be skimmed off.

5.7 REMOVAL OF CALCIUM AND OTHER METALS

Calcium and magnesium salts, which generally are present in water as bicarbonates or sulfates, cause water hardness, which may require reduction or removal for municipal use or some commercial applications. One of the most common manifestations of water hardness is the insoluble "curd" formed by the reaction of soap with calcium or magnesium ions. The formation of these insoluble soap salts is discussed in Chapter 4, Section 4.10. Although ions that cause water hardness do not form insoluble products with detergents, they do adversely affect detergent performance. Therefore, calcium and magnesium must be complexed or removed from water for detergents to function properly.

Another problem caused by hard water is the formation of mineral deposits. For example, when water containing calcium and bicarbonate ions is heated, insoluble calcium carbonate is formed:

$$Ca^{2+} + 2HCO_3^- \rightarrow CaCO_3(s) + CO_2(g) + H_2O \tag{5.2}$$

This product coats the surfaces of hot water systems, clogging pipes and reducing heating efficiency. Dissolved salts such as calcium and magnesium bicarbonates and sulfates can be especially damaging in boiler feedwater. Clearly, the removal of water hardness is essential for many uses of water.

Several processes are used for softening water. On a large scale, such as in community water-softening operations, the lime–soda process is used. This process involves the treatment of water with lime, $Ca(OH)_2$, and soda ash, Na_2CO_3. Calcium is precipitated as $CaCO_3$ and magnesium as $Mg(OH)_2$. When the calcium is present primarily as "bicarbonate hardness," it can be removed by the addition of $Ca(OH)_2$ alone:

$$Ca^{2+} + 2HCO_3^- + Ca(OH)_2 \rightarrow 2CaCO_3(s) + 2H_2O \tag{5.3}$$

Some large-scale lime–soda softening plants make use of the precipitated calcium carbonate product as a source of additional lime. The calcium carbonate is first heated to at least 825°C to produce quicklime, CaO:

$$CaCO_3 + heat \rightarrow CaO + CO_2(g) \tag{5.4}$$

The quicklime is then slaked with water to produce calcium hydroxide:

$$CaO + H_2O \rightarrow Ca(OH)_2 \tag{5.5}$$

When bicarbonate ion is not present at substantial levels, a source of CO_3^{2-} must be provided at a high enough pH to prevent conversion of most of the carbonate to bicarbonate. These conditions are obtained by the addition of Na_2CO_3 (soda ash). For example, calcium present as the chloride can be removed from water by the addition of soda ash:

$$Ca^{2+} + 2Cl^- + 2Na^+ + CO_3^{2-} \rightarrow CaCO_3(s) + 2Na^+ + 2Cl^- \tag{5.6}$$

Note that the removal of bicarbonate hardness results in a net removal of soluble salts from solution, whereas removal of non-bicarbonate hardness involves the addition of at least as many equivalents of ionic material as are removed.

The precipitation of magnesium as the hydroxide requires a higher pH than the precipitation of calcium as the carbonate:

$$Mg^{2+} + 2OH^- \rightarrow Mg(OH)_2(s) \tag{5.7}$$

The high pH required can be provided by the basic carbonate ion from soda ash:

$$CO_3^{2-} + H_2O \rightarrow HCO_3^- + OH^- \tag{5.8}$$

The water softened by lime–soda softening plants usually suffers from two defects. First, because of supersaturation effects, some $CaCO_3$ and $Mg(OH)_2$ usually remain in solution. If not removed, these compounds will precipitate at a later time and cause harmful deposits or undesirable cloudiness in water. The second problem results from the use of highly basic sodium carbonate, which gives the product water an excessively high pH, up to pH 11. To overcome these problems, the water is recarbonated by bubbling CO_2 into it. The carbon dioxide converts the slightly soluble calcium carbonate and magnesium hydroxide to their soluble bicarbonate forms:

$$CaCO_3(s) + CO_2 + H_2O \rightarrow Ca^{2+} + 2HCO_3^- \tag{5.9}$$

$$Mg(OH)_2(s) + 2CO_2 \rightarrow Mg^{2+} + 2HCO_3^- \tag{5.10}$$

The CO_2 also neutralizes excess hydroxide ion:

$$OH^- + CO_2 \rightarrow HCO_3^- \tag{5.11}$$

The pH generally is brought within the range 7.5–8.5 by recarbonation, commonly using CO_2 from the combustion of carbonaceous fuel. Scrubbed stack gas from a power plant frequently is utilized. Water adjusted to a pH, alkalinity, and Ca^{2+} concentration very close to $CaCO_3$ saturation is labeled **chemically stabilized**. It neither precipitates $CaCO_3$ in water mains, which can clog the pipes, nor dissolves protective $CaCO_3$ coatings from the pipe surfaces. Water with Ca^{2+} concentration much below $CaCO_3$ saturation is called **aggressive** water.

Calcium can be removed from water very efficiently by the addition of orthophosphate:

$$5Ca^{2+} + 3PO_4^{3-} + OH^- \rightarrow Ca_5OH(PO_4)_3(s) \tag{5.12}$$

It should be pointed out that the chemical formation of a slightly soluble product for the removal of undesired solutes such as hardness ions, phosphate, iron, and manganese must be followed by sedimentation in a suitable apparatus. Frequently, coagulants must be added, and filtration employed for complete removal of these sediments.

The softening of water by ion exchange does not require the removal of all ionic solutes, just those cations responsible for water hardness. Generally, therefore, only a cation exchanger is necessary. Furthermore, the sodium rather than the hydrogen form of the cation exchanger is used, and the divalent cations are replaced by sodium ion. Sodium ion at low concentrations is harmless in water to be used for most purposes, and sodium chloride is a cheap and convenient substance with which to recharge the cation exchangers.

A number of materials have ion-exchanging properties. Among the minerals especially noted for their ion-exchange properties are the aluminum silicate minerals, or **zeolites**. An example of a zeolite that has been used commercially in water softening is glauconite, chemical formula $K_2(MgFe)_2Al_6(Si_4O_{10})_3(OH)_{12}$. Synthetic zeolites have been prepared by drying and crushing the white gel produced by mixing solutions of sodium silicate and sodium aluminate.

Structural formulas of typical synthetic ion exchangers are shown in Figures 5.3 and 5.4. The cation exchanger shown in Figure 5.3 is called a **strongly acidic cation exchanger** because the parent $-SO_3^-H^+$ group is a strong acid. When the functional group binding the cation is the $-CO_2^-$ group, the exchange resin is called a **weakly acidic cation exchanger**, because the $-CO_2H$ group is a weak acid. Figure 5.4 shows a **strongly basic anion exchanger** in which the functional group is a quaternary ammonium group, $-N^+(CH_3)_3$. In the hydroxide form, $-N^+(CH_3)_3OH^-$, the hydroxide ion is readily released, so the exchanger is classified as **strongly basic**.

The water-softening capability of a cation exchanger is shown in Figure 5.3, where sodium ion on the exchanger is exchanged for calcium ion in solution. The same reaction occurs with magnesium ion. Water softening by cation exchange is widely used, effective, and economical. However, it does cause some deterioration of wastewater quality arising from the contamination of wastewater by sodium chloride. Such contamination results from the periodic need to regenerate a water softener with sodium chloride to displace calcium and magnesium ions from the resin and replace these hardness ions with sodium ions:

$$Ca^{2+}\left\{Cat(s)\right\}_2 + 2Na^+ + 2Cl^- \rightarrow 2Na^+\left\{Cat(s)\right\} + Ca^{2+} + 2Cl^- \qquad (5.13)$$

During the regeneration process, a large excess of sodium chloride must be used—several pounds for a home water softener. Appreciable amounts of dissolved sodium chloride can be introduced into sewage by this route.

Whereas strongly acidic cation exchangers are used for the removal of water hardness as shown in Figure 5.3, acidic cation exchangers having the $-CO_2H$ group as a functional group are useful

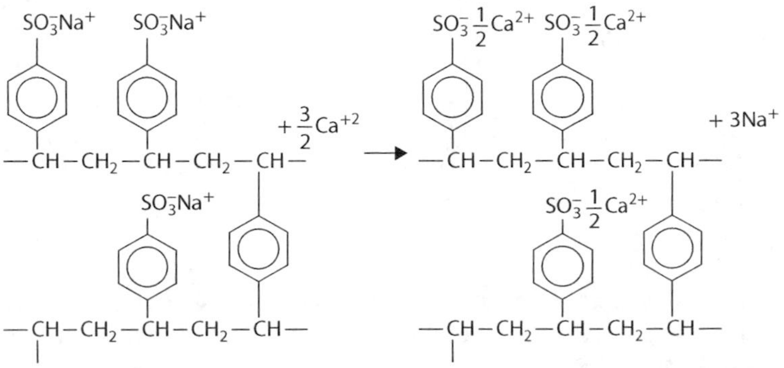

FIGURE 5.3 Strongly acidic cation exchanger. Sodium exchange for calcium in water is shown.

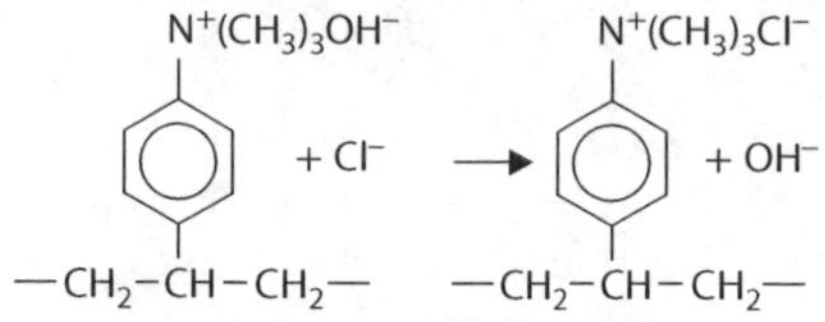

FIGURE 5.4 Strongly basic anion exchanger. Chloride exchange for hydroxide ion is shown.

for removing alkalinity. Alkalinity generally is manifested by bicarbonate ion, a species that is a sufficiently strong base to neutralize the acid of a weak acid cation exchanger:

$$2R - CO_2H + Ca^{2+} + 2HCO_3^- \rightarrow \left[R - CO_2^-\right]_2 Ca^{2+} + 2H_2O + 2CO_2 \qquad (5.14)$$

However, weak bases such as sulfate ion or chloride ion are not strong enough to remove hydrogen ion from the carboxylic acid ion exchanger. An additional advantage of these exchangers is that they can be regenerated almost stoichiometrically with dilute strong acids, thus avoiding the potential pollution problem caused by the use of excess sodium chloride to regenerate strongly acidic cation changers.

Chelation, or as it is sometimes known, **sequestration**, is an effective method of softening water without actually having to remove calcium and magnesium from the solution. A complexing agent is added, which greatly reduces the concentrations of free hydrated cations. For example, chelating hydrated calcium ion (Ca^{2+}) with excess EDTA anion (an organic species that has 6 sites that can bind to metals, represented below as Y^{4-}):

$$Ca^{2+} + Y^{4-} \rightarrow CaY^{2-} \qquad (5.15)$$

reduces the concentration of hydrated calcium ion, preventing the precipitation of calcium carbonate:

$$Ca^{2+} + CO_3^{2-} \rightarrow CaCO_3(s) \qquad (5.16)$$

Polyphosphate salts, EDTA, and nitrilotriacetic acid (NTA) (see Chapter 3, Figure 3.13) are chelating agents commonly used for water softening. Polysilicates are used to complex iron.

5.7.1 Removal of Iron and Manganese

Soluble iron and manganese are found in many groundwaters because of reducing conditions that favor the soluble +2 oxidation state of these metals. Iron is the more commonly encountered of the two metals. In groundwater, the level of iron seldom exceeds 10 mg/L, and that of manganese is rarely higher than 2 mg/L. The basic method for removing both of these metals depends on oxidation to higher insoluble oxidation states. The oxidation is generally accomplished by aeration. The rate of oxidation is pH-dependent in both cases, with a high pH favoring more rapid oxidation. The oxidation of soluble Mn(II) to insoluble MnO_2 is a complicated process. It appears to be catalyzed by solid MnO_2, which can adsorb Mn(II). This adsorbed Mn(II) is slowly oxidized on the MnO_2 surface.

Chlorine and potassium permanganate are sometimes employed as oxidizing agents for iron and manganese. There is some evidence that organic chelating agents with reducing properties hold iron(II) in a soluble form in water. In such cases, chlorine is effective because it destroys the organic compounds and enables the oxidation of iron(II).

In water with a high level of carbonate, $FeCO_3$ and $MnCO_3$ may be precipitated directly by raising the pH above about 11.5 by the addition of sodium carbonate or lime. However, this approach is less popular than oxidation.

Relatively high levels of insoluble iron(III) and manganese(IV) frequently are found in water as colloidal material, which is difficult to remove. These metals can be associated with humic colloids or "peptizing" organic material that binds to colloidal metal oxides, stabilizing the colloid.

5.7.2 REMOVAL OF HEAVY METALS

Heavy metals such as copper, cadmium, mercury, and lead are found in wastewaters from a number of industrial processes. Because of the toxicity of many heavy metals, their concentrations must be reduced to very low levels before release of the wastewater. Several approaches are used in heavy metals removal.

The form of the heavy metal has a strong effect on the efficiency of metal removal. For instance, generally soluble chromium(VI) is much more difficult to remove than chromium(III). Chelation may prevent metal removal by solubilizing metals.

Even when not specifically designed for the removal of heavy metals, some wastewater treatment processes remove appreciable quantities of the more troublesome heavy metals encountered in wastewater. Biological waste treatment effectively removes metals from water. These metals accumulate in the sludge from biological treatment, so sludge disposal must be given careful consideration.

Lime treatment, discussed earlier in this section for calcium removal, precipitates heavy metals as insoluble hydroxides, basic salts, or coprecipitated with calcium carbonate or iron(III) hydroxide. This process does not completely remove mercury, cadmium, or lead, so their removal is aided by addition of sulfide (most heavy metals are sulfide-seekers):

$$Cd^{2+} + S^{2-} \rightarrow CdS(s) \tag{5.17}$$

Heavy chlorination is frequently necessary to break down metal-solubilizing organic ligands. Lime precipitation does not normally permit recovery of metals and is sometimes undesirable from the economic viewpoint.

Electrodeposition (reduction of metal ions to metal by electrons at an electrode), **reverse osmosis** (see Section 5.10), and **ion exchange** are frequently employed for metal removal. Solvent extraction using organic-soluble chelating substances can be used to remove many metals. **Cementation**, a process by which a metal deposits by reaction of its ion with a more readily oxidized metal, can be employed:

$$Cu^{2+} + Fe(iron\ scrap) \rightarrow Fe^{2+} + Cu \tag{5.18}$$

Activated carbon adsorption effectively removes some metals from water at the part per million level. Sometimes a chelating agent is sorbed to the charcoal to increase metal removal.

Various physical–chemical treatment processes effectively remove heavy metals from wastewaters. One such treatment is lime precipitation followed by activated-carbon filtration. Activated-carbon filtration may also be preceded by treatment with iron(III) chloride to form an iron(III) hydroxide floc, which is an effective heavy metals scavenger. Similarly, alum, which forms aluminum hydroxide, may be added before activated-carbon filtration.

5.7.3 ARSENIC REMOVAL

In the United States and many other countries, the toxic metalloid arsenic is not allowed in drinking water at levels greater than 10 μg/L, a level that has been exceeded by some water supplies used in the past. Arsenic usually is present in water as arsenic(V), primarily $H_2AsO_4^-$ and $HAsO_4^{2-}$. Groundwater supplies containing the more toxic As(III) form have to be treated with chlorine, ozone, or permanganate oxidants to convert the arsenic to the +5 oxidation state before arsenic removal. Coagulation with aluminum sulfate, iron(III) salts that produce $Fe(OH)_3$, and lime effectively removes arsenic. Sorption onto granular iron(III) or aluminum oxide/hydroxides is also effective in removing arsenic. Arsenic can also be removed from water with anion exchange resins that bind with the anionic arsenic.

5.8   REMOVAL OF DISSOLVED ORGANICS

Very low levels of exotic organic compounds in drinking water are suspected of contributing to cancer and other maladies. Water disinfection processes, which by their nature involve chemically rather severe conditions, particularly of oxidation, have a tendency to produce **disinfection by-products**. Some of these are chlorinated organic compounds produced by chlorination of organics in water, especially humic substances. Removal of organics to very low levels before chlorination has been found to be effective in preventing formation of trihalomethane compounds (see Chapter 4, Section 4.13). Another major class of disinfection by-products consists of organooxygen compounds such as aldehydes, carboxylic acids, and oxoacids.

A variety of organic compounds survive, or are produced by, secondary wastewater treatment and should be considered as factors in discharge or reuse of the treated water. Almost half of these are humic substances (see Chapter 3, Section 3.11) with a molecular mass range of 1000–5000 amu. Among the remainder are found ether-extractable materials, carbohydrates, proteins, detergents, tannins, and lignins. The humic compounds, because of their high molecular mass and anionic character, influence some of the physical and chemical aspects of waste treatment. The ether-extractables contain many of the compounds that are resistant to biodegradation and are of particular concern regarding potential toxicity, carcinogenicity, and mutagenicity. In the ether extract are found many fatty acids, hydrocarbons of the *n*-alkane class, naphthalene, diphenylmethane, diphenyl, methylnaphthalene, isopropylbenzene, dodecylbenzene, phenol, phthalates, and triethylphosphate.

The standard method for the removal of dissolved organic material is adsorption on activated carbon, a product that is produced from a variety of carbonaceous materials including wood, pulp mill char, peat, and lignite. The carbon is produced by charring the raw material anoxically below 600°C, followed by an activation step consisting of partial oxidation. Carbon dioxide can be employed as an oxidizing agent at 600–700°C:

$$CO_2 + C \rightarrow 2CO \qquad (5.19)$$

Or, the carbon can be oxidized by superheated steam at 800–900°C:

$$H_2O + C \rightarrow H_2 + CO \qquad (5.20)$$

These processes develop porosity, increase the surface area, and leave the C atoms in bonding orientations that have affinities for organic compounds.

The exact mechanism by which activated carbon holds organic materials is still a subject of research. However, one reason for the effectiveness of this material as an adsorbent is its tremendous surface area. A solid cubic foot of carbon particles can have a combined pore and surface area of approximately 10 square miles!

Activated carbon comes in two general types: granulated activated carbon, consisting of particles 0.1–1 mm in diameter, and powdered activated carbon, in which most of the particles are 50–100 μm in diameter. Both forms are used to treat water.

Granular activated carbon can be employed in a fixed bed, through which water flows downward. Accumulation of particulate matter requires periodic backwashing. An expanded bed in which particles are kept slightly separated by water flowing upward can be used with less chance of clogging.

Powdered activated carbon can be added to water as needed, then removed by settling and filtration. The increased use of membrane filters that are effective in removing very small particles from water has made the use of activated carbon adsorbent more attractive.

Microorganisms tend to grow on activated carbon. This can be a problem with granular activated carbon filters that become fouled by microbial growth. However, systems have been developed in which microorganisms held on granular activated carbon are used to degrade water pollutants. Pretreatment with ozone to partially oxidize complex organic substances, such as biorefractory humic substances, can be used before treatment with microorganisms on granular activated carbon.

Regeneration of spent activated carbon can be accomplished by heating carbon to 950°C in a steam-air atmosphere. This process oxidizes adsorbed organics and regenerates the carbon surface, with an approximately 10% loss of carbon.

Removal of organics can also be accomplished by adsorbent synthetic polymers. Such polymers as Amberlite XAD-4, a copolymer of polystyrene and divinylbenzene, have hydrophobic surfaces and strongly attract relatively insoluble organic compounds such as chlorinated pesticides. The porosity of these polymers is up to 50% by volume, and the surface area may be as high as 850 m2/g. They are readily regenerated by solvents such as isopropanol and acetone. Under appropriate operating conditions, these polymers remove virtually all nonionic organic solutes; for example, phenol at 250 mg/L is reduced to less than 0.1 mg/L by appropriate treatment with Amberlite XAD-4. However, the use of adsorbent polymers is more expensive than that of activated carbon.

Oxidation of dissolved organics can be used for their removal from water. Ozone, hydrogen peroxide, molecular oxygen (with or without catalysts), chlorine and its derivatives, permanganate, or ferrate [iron(VI)] can be used as oxidants. Electrochemical oxidation may be possible in some cases. High-energy electron beams produced by high-voltage electron accelerators also have the potential to destroy organic compounds. Oxidation can be augmented by the application of ultraviolet radiation. Partial oxidation with chemical and photochemical agents can be helpful in making organic pollutants more amenable to other treatment processes, especially biodegradation.

5.8.1 REMOVAL OF HERBICIDES

Because of their widespread application and persistence, herbicides have proven to be particularly troublesome in some drinking water sources. Herbicide levels vary with season, related to times that they are applied to control weeds. The more soluble ones, such as chlorophenoxy esters, are most likely to enter drinking water sources. One of the most troublesome is atrazine, which is often manifested by its metabolite desethylatrazine. Activated carbon treatment is the best means of removing herbicides and their metabolites from drinking water sources. A problem with activated carbon is that of **preloading**, in which natural organic matter in the water loads up the carbon and hinders uptake of pollutant organics such as herbicides. Pretreatment to remove organic matter, such as flocculation and precipitation of humic substances, can significantly increase the efficacy of activated carbon for the removal of herbicides and other organics.

5.8.2 REMOVAL OF TASTE, ODOR, AND COLOR

Substances that cause taste, odor, and color in water are usually organic and must be removed to bring drinking water up to acceptable standards. Taste, odor, and color agents can come from microorganisms in the water, from microorganisms degrading organic matter in soil, and from chemical pollutant sources. Organically bound iron is a common cause of undesirable color. Very low levels of some agents can cause taste, odor, and color problems and are often hard to identify. Groundwater is often afflicted with bad odors and tastes as the consequence of anoxic processes that occur underground in the absence of air. Various tastes and odors have been described subjectively as musty, swampy, grassy, fishy, sweet, septic, medicinal, and phenolic. One of the most common causes of taste and odor in water is geosmin, a metabolite of cyanobacteria that grow in water, especially in reservoirs under hot weather conditions (see Chapter 4, Section 4.10). A variety of species of algae, flagellates, diatoms, and other organisms in water produce taste and odor agents either directly or when they biodegrade.

Various processes are used to remove agents that cause taste, odor, and color. Simple aeration can remove volatile materials such as odorous hydrogen sulfide. Oxidation that destroys organics usually removes taste, odor, and color as does adsorption of organics onto activated carbon.

5.8.3 PHOTOLYSIS

Photolysis is a process in which energetic photons of ultraviolet electromagnetic radiation are used to transfer energy to chemical species. This can result in direct destruction of contaminant compounds, or reactive intermediates produced by photolysis may react with contaminants. Photolysis usually results in oxidation of organic compounds in water and is often used in conjunction with oxidants. One of the most reactive intermediates produced by photolysis is the hydroxyl radical, $HO\cdot$, which can be generated by several photolysis reactions including the action of ultraviolet radiation on hydrogen peroxide, H_2O_2:

$$H_2O_2 + h\nu \rightarrow HO\cdot + HO \tag{5.21}$$

The hydroxyl radical reacts with organic species, usually by removing H atoms and producing unstable free radicals that undergo chain reactions that ultimately result in the destruction of organic molecules.

The ultraviolet radiation source used in water treatment by photolysis is a lamp in which mercury vapor energized by an electrical discharge or microwaves is electronically excited, giving off ultraviolet radiation at a wavelength predominantly of 254 nm. The lamp is contained in a fused silica (SiO_2) tube that is transparent to ultraviolet radiation. The ability of ultraviolet radiation in the 200–400-nm range to break chemical bonds is responsible for its uses in water treatment. Although, ultraviolet radiation can be used to destroy organic constituents, usually by augmenting oxidation, its greatest use in water treatment is in destroying microorganisms by destroying the structures of their nucleic acid (DNA, RNA) molecules so that they cannot reproduce. Photolysis is a very green process in that the photons are without mass so that no extraneous material or residues are introduced into the water. Ultraviolet irradiation is a very efficient means of delivering into molecules high energy sufficient to break chemical bonds.

5.8.4 SONOLYSIS

Sonolysis is a relatively new technique for water treatment that is somewhat analogous to photolysis. Sonolysis occurs when ultrasound waves are generated in water generating microscopic bubbles by cavitation, which then collapse, producing extremely high localized temperatures and pressures. Water contaminants are destroyed in sonolysis by pyrolysis at the high temperatures generated by the bubbles and by the formation of reactive free radicals, especially $H\cdot$ and $HO\cdot$ from splitting apart water molecules. As is the case with photolysis, these radicals can initiate chain reactions resulting in the destruction of organics. Like photolysis, sonolysis is most effective when used with oxidants such as ozone. Sonolysis is a very green technology because it is without mass and enables putting large amounts of energy into molecules without significantly heating the water that is being treated.

5.9   REMOVAL OF DISSOLVED INORGANICS

The effluent from secondary wastewater treatment generally contains 300–400 mg/L more dissolved inorganic material than does the municipal water supply. Therefore, complete water recycling requires inorganic solute removal. Even when water is not destined for immediate reuse, the removal of the inorganic nutrients phosphorus and nitrogen is highly desirable to reduce eutrophication downstream. In some cases, the removal of toxic trace metals is needed.

Methods for the removal of inorganics from water may be general, removing essentially all inorganics, or they may be specific to particular kinds of inorganic solutes. One of the most obvious methods for removing inorganics from water is distillation. However, the energy required for distillation is generally quite high, so that distillation is often not economically desirable. Distillation may require special measures to prevent volatile materials such as ammonia and odorous compounds being carried over to the distilled water product. Freezing produces very pure water, but is considered uneconomical with present technology, although there have been serious proposals to tow large icebergs to warmer regions where water would be collected as they melt (potentially a green approach). Several means of removing inorganic substances from water are discussed in this section. Membrane processes, which are generally the most cost-effective in removing inorganics from water, are discussed separately in Section 5.10.

5.9.1 ION EXCHANGE

Dissolved salts can be removed from water by ion exchange, the reversible transfer of ions between aquatic solution and a solid material capable of bonding ions, discussed as a means of softening water in Section 5.7. The removal of NaCl from solution by two ion exchange reactions is a good illustration of this process. First, the water is passed over a solid cation exchanger in the hydrogen form, represented by $H^{+-}\{Cat(s)\}$:

$$H^{+-}\left\{Cat(s)\right\} + Na^+ + Cl^- \rightarrow Na^{+-}\left\{Cat(s)\right\} + H^+ + Cl^- \qquad (5.22)$$

Next, the water is passed over an anion exchanger in the hydroxide ion form, represented by $OH^{-+}\{An(s)\}$:

$$OH^{+-}\left\{An(s)\right\} + H^+ + Cl^- \rightarrow Cl^{-+}\left\{An(s)\right\} + H_2O \qquad (5.23)$$

Thus, the cations in solution are replaced by hydrogen ion and the anions by hydroxide ion, yielding water as the product.

Demineralization by ion exchange generally produces water of a very high quality. Unfortunately, some organic compounds in wastewater foul ion exchangers, and microbial growth on the exchangers can diminish their efficiency. In addition, regeneration of the resins is expensive, and the concentrated wastes from regeneration require disposal in a manner that will not damage the environment.

5.9.2 PHOSPHORUS REMOVAL

Advanced wastewater treatment normally requires removal of phosphorus to reduce algal growth. Algae may grow at PO_4^{3-} levels as low as 0.05 mg/L. Growth inhibition requires levels well below 0.5 mg/L. Since municipal wastes typically contain approximately 25 mg/L of phosphate (as orthophosphates, polyphosphates, and insoluble phosphates), the efficiency of phosphate removal must be quite high to prevent algal growth. This removal may occur in the sewage treatment process (1) in the primary settler, (2) in the aeration chamber of the activated sludge unit, or (3) after secondary waste treatment.

Activated sludge treatment removes about 20% of the phosphorus from sewage where it ends up in the sewage sludge (biosolids). Thus, an appreciable fraction of largely biological phosphorus is removed with the sludge. Although municipal wastewater contains excess phosphorus, some wastes, such as carbohydrate wastes from sugar refineries, are so deficient in phosphorus that supplementation of the waste with inorganic phosphorus is required for proper growth of the microorganisms degrading the wastes.

Activated sludge sewage treatment plants operated under conditions of high dissolved oxygen and high pH levels in the aeration tank can remove 60–90% of the phosphorus in the sewage. At a relatively high rate of aeration in hard water, the CO_2 produced by biodegradation of wastes is swept out, the pH rises, and reactions such as the following occur:

$$5Ca^{2+} + 3HPO_4^{2-} + H_2O \rightarrow Ca_5OH(PO_4)_3(s) + 4H^+ \qquad (5.24)$$

The precipitated hydroxyapatite or other form of calcium phosphate is incorporated in the sludge floc.

Chemically, 90–95% phosphate removal is accomplished by precipitation with lime as shown by the following reaction:

$$5Ca(OH)_2 + 3HPO_4^{2-} \rightarrow Ca_5OH(PO_4)_3(s) + 3H_2O + 6OH^- \qquad (5.25)$$

Salts of Al^{3+}, Fe^{3+}, and Mg^{2+} are also capable of forming precipitates with phosphate. Phosphate can be removed from solution by adsorption on some solids, particularly activated alumina, Al_2O_3. Removals of up to 99.9% of orthophosphate have been achieved with this method.

5.9.3 Nitrogen Removal

Next to phosphorus, nitrogen is the algal nutrient most commonly removed as part of advanced wastewater treatment. Nitrogen in municipal wastewater generally is present as organic nitrogen or ammoniacal nitrogen, NH_4^+. Nitrification followed by denitrification is the most effective technique for the removal of nitrogen from wastewater. The first step is an essentially complete conversion of ammonia and organic nitrogen to nitrate under strongly anoxic conditions, achieved by more extensive than normal aeration of the sewage:

$$NH_4^+ + 2O_2\,(\text{Nitrifying bacteria}) \rightarrow NO_3^- + 2H^+ + H_2O \qquad (5.26)$$

The second step is the reduction of nitrate to nitrogen gas. This reaction is also bacterially catalyzed, in this case by denitrifying bacteria, and requires a carbon source and a reducing agent such as methanol, CH_3OH, or an organic carbon source, $\{CH_2O\}$:

$$4NO_3^- + 5\{CH_2O\} + 4H^+ \rightarrow 2N_2(g) + 5CO_2(g) + 7H_2O \qquad (5.27)$$

Typically, denitrification is carried out in an anoxic column with added methanol as a food source (microbial reducing agent). Methanol is a synthetic chemical that poses some toxicity and flammability hazards. A greener option is the use of biomass, such as carbohydrates from plants as the microbial reducing agent. One of the best organic reductants for this purpose is high-fructose corn syrup.

5.10 MEMBRANE PROCESSES AND REVERSE OSMOSIS FOR WATER PURIFICATION

Advances in materials have enabled the relatively recent development of **membrane filtration processes** for the purification of water. Increasingly applied to treatment of drinking water, specialized membrane processes are an important advance in water treatment. There are four general categories of membrane processes classified generally on the basis of the sizes and nature of substances removed (Table 5.1). Membranes are thin, typically less than 1 mm thick, and vary in their composition and the sizes of the pores through which water flows. The first three processes

TABLE 5.1

Membrane Filtration Processes for Water Purification

Name	Approximate Pore Size[a]	Materials Removed
Microfiltration	100 nm	Particles, single-cell microorganisms
Ultrafiltration	10 nm	Small colloids, viruses
Nanofiltration	1 nm	Dissolved organics, divalent ions (Ca^{2+}, Mg^{2+}, SO_4^{2-})
Reverse osmosis	—	Monovalent ions (Na^+, Cl^-)

[a] In nanometers (nm, 10^{-9} m).

listed in Table 5.1 vary according to pore size, whereas reverse osmosis, discussed in detail in Section 5.10.1, depends on the specific permeability of the membrane for water and its ability to exclude ions. Membranes are made from a variety of materials, including organic polymers, such as cellulose acetate, and inorganic ceramics composed of silicon and metal oxides fired at high temperatures.

Membrane filtration uses high pressures, which increase with decreasing pore size. Microfiltration is usually conducted at pressures below 5 atm, ultrafiltration at 2–8 atm, nanofiltration at 5–15 atm, and reverse osmosis or hyperfiltration at 15–100 atm. To accommodate high pressures, membranes are usually configured as hollow fibers. Typically, contaminated water flows through the inside of the fiber, purified water (**permeate**) flows to the outside of the fiber, and water concentrated in contaminants (**retentate**) exits from the downstream end of the fiber. This configuration also enables continuous flushing of the internal membrane surface to aid in removal of retained impurities. Some systems in which the filtration device is immersed in the wastewater operate with the water flowing from the outside to the inside of the hollow fibers.

A problem common to all membrane processes is that posed by the retentate in which impurities are concentrated. In some cases, retentate can be discharged with wastewater. Other options include evaporation of the water followed by disposal or incineration of the residue, reclamation of chemicals from industrial wastewater, and disposal in deep saline water aquifers. For the special case of desalination of seawater by reverse osmosis, the retentate is returned to the ocean, which has the potential to cause problems due to excess salinity.

5.10.1 REVERSE OSMOSIS

In its most general sense, **reverse osmosis** describes any pressure-driven process that depends on preferential diffusion of a liquid through a membrane that is selectively permeable to the liquid. Illustrated in Figures 5.5 and 5.6, reverse osmosis is a very useful and well-developed technique for the purification of water. Basically, it consists of forcing pure water through a semipermeable membrane that allows the passage of water but not of other material. This process, which is not simply sieve separation or ultrafiltration, depends on the preferential sorption of water on the surface of a porous cellulose acetate or polyamide membrane. Pure water from the sorbed layer is forced through pores in the membrane under pressure. If the thickness of the sorbed water layer is d, the pore diameter for optimum separation should be $2d$. The optimum pore diameter depends on the thickness of the sorbed pure water layer and may be several times the diameters of the solute and solvent molecules.

Fouling caused by various materials can cause problems with reverse osmosis treatment of water. Although the relatively small ions constituting the salts dissolved in wastewater readily pass through the membranes, large organic ions (e.g., proteins) and charged colloids migrate to the membrane surfaces, often fouling or plugging the membranes and reducing efficiency. In addition, growth of microorganisms on the membranes can cause fouling.

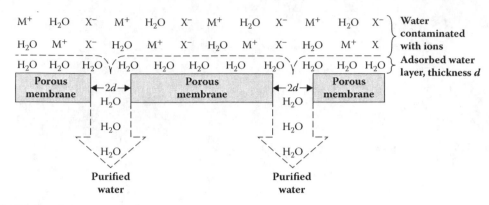

FIGURE 5.5   Solute removal from water by reverse osmosis.

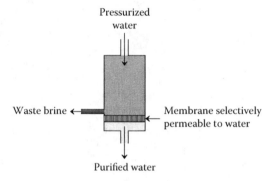

FIGURE 5.6   A reverse osmosis system for the removal of ions and other impurities from water. Highly pressurized water is forced through a membrane selectively permeable to water so that desalinated purified water penetrates the membrane and a waste brine is rejected. Membrane filtration processes that employ filters with larger pores are used to remove a variety of contaminants from water.

5.10.2   ELECTRODIALYSIS

Electrodialysis consists of applying a direct current across a body of water separated into vertical layers by membranes alternately permeable to cations and anions. Cations migrate toward the cathode and anions toward the anode. Cations and anions both enter one layer of water, and both leave the adjacent layer. Thus, layers of water enriched in salts alternate with those from which salts have been removed. The water in the brine-enriched layers is recirculated to a certain extent to prevent excessive accumulation of brine.

5.11   WATER DISINFECTION

5.11.1   PATHOGENS TREATED BY DISINFECTION

Water disinfection to kill or remove disease-causing organisms in water is discussed in this section. The most common waterborne diseases treated by disinfection are those in which pathogenic microorganisms of several types are ingested with drinking water, causing gastroenteritis, the irritation and inflammation of the digestive tract that usually is manifested by diarrhea and/or vomiting. In especially severe cases of dysentery, the form of diarrhea characterized by bloody stools, there can be a massive loss of water and electrolytes from the body resulting in death of the victim. Invasive gastroenteritis occurs when pathogens penetrate the epithelial layer of the gastrointestinal tract and afflict organs in the body. In such cases, there may be little or no diarrhea or vomiting, but organs, such as the kidneys, may be so badly affected that the results are fatal. The removal or killing of organisms in water responsible for gastroenteritis, or the prevention of their entry into water supplies, is one of the most important aspects of drinking water supply and treatment and arguably has been responsible for saving more lives since 1900 than any other public health measure. Even so, the number of children who die of diarrhea, much of it carried by impure drinking water, in predominantly tropical regions of Africa, Asia, and Latin America may number in the millions annually. Numerous species of microorganisms can be responsible for waterborne diseases, the main ones of which are discussed in the following paragraph.

Among bacteria that cause waterborne gastroenteritis, the most notorious is *Vibrio cholerae*, which grows in the small intestine producing an enterotoxin that causes explosive diarrhea and vomiting. Waves of eight cholera pandemics starting in India in the early 1800s and lasting into the 1900s killed millions of people. Several kinds of *Salmonella* bacteria are waterborne and cause gastrointestinal disease and other maladies. The most notorious of these is *Salmonella typhi*, which causes typhoid fever. Waterborne *Salmonella paratyphii* causes paratyphoid fever, which is similar to typhoid fever, though generally milder. Several kinds of *Shigella* bacteria can be waterborne. One of these that causes the most health problems is *Shigella dysenteriae*, which can cause dysentery. *S. dysenteriae* infections are especially common in Africa and may cause up to 600,000 deaths worldwide annually. *Escherichia coli* bacteria are normal residents of the intestinal tracts of animals where they can perform a useful function in synthesizing vitamins. Despite the generally benign and even beneficial nature of intestinal *E. coli* bacteria, there are pathogenic strains that can be carried by water and that can cause severe dysentery, loss of kidney function, and other effects that can be fatal. The most notable such strain is *E. coli*, 0157:H7, which has caused fatalities.[3] Other bacteria that can cause waterborne disease include *Yersinia enterocolitica* and *Campylobacter jejuni*.

Protozoa are unicellular animallike microorganisms that move about by means of flagella ("tails"), cilia ("hairs" on the cell walls), or by amoeboid locomotion in which the cell "oozes" its way about. Waterborne protozoa are responsible for several serious human diseases. *Entamoeba histolytica* is an amoeboid protozoan that can cause dysentery, sometimes with serious consequences. *Entamoeba dispar* causes a mild form of diarrhea. *Giardia lamblia* is a flagellated protozoan that invades the upper small intestine causing gastrointestinal maladies that usually go on for 1 or 2 weeks, but which may persist for months or even years if untreated. Animals, such as dogs, beavers, and bears, are reservoirs of this protozoan. Another important kind of waterborne pathogenic protozoa is *Cryptosporidium parvum*, a cause of diarrhea and other gastrointestinal misery. A fatal, though fortunately rare form of ameba that attacks human brain tissue is *Naegleria fowleri*, an organism that can be found in warmer water during summer months. Exposure to *N. fowleri* occurs when the victim inhales water through the nose from where it spreads to brain tissue that it destroys. Three deaths from this malady were reported in the United States during the hot, dry summer of 2011.

Viruses are much smaller than bacteria and protozoa. They do not have a cell structure, but consist of bundles of nucleic acid surrounded by a coat of protein. To reproduce, they must infect cells of other organisms and use the metabolic machinery of these cells to increase their own numbers. Viruses responsible for a number of diseases are thought to be spread by water. It is believed that poliomyelitis and viral hepatitis (inflammation of the liver accompanied by jaundice) can be spread by waterborne virus. Viruses of the rotavirus group are probably responsible for most waterborne viral gastroenteritis ailments.

5.11.2 DISINFECTION AGENTS

The agents that can be used to disinfect water include (1) chlorine, (2) chloramines, (3) ozone, (4) chlorine dioxide, (5) ultraviolet radiation, (6) membrane filtration to remove pathogens, and (7) miscellaneous agents including the evolving use of ferrate (iron(VI)). Of these, chlorine and chloramines have been the most popular, but are becoming less so because of the by-products they produce. Chlorine dioxide produces fewer by-products. Ozone is arguably the greenest of the disinfection agents because it can be made on-site with air as the raw material and produces few undesirable by-products.

The physical process of **filtration** is rather effective in removing pathogens from water. During the 1800s, before chlorine came into widespread use, filtration with simple sand filters employed in just a few cities cut down significantly on the incidence of waterborne cholera in those cities. With modern membrane technology (see Section 5.10), ultrafiltration can remove even viruses from water. Small amounts of chlorine or chloramines (see Section 5.11.3) can be added to maintain sterile water in distribution systems, but much less of these agents are required for membrane-filtered water than are required for total disinfection.

5.11.3 Disinfection with Chlorine and Chloramines

Chlorine is the most commonly used disinfectant employed for killing bacteria in water. When chlorine is added to water, it rapidly hydrolyzes

$$Cl_2 + H_2O \rightarrow H^+ + Cl^- + HOCl \tag{5.28}$$

to produce weakly acidic hypochlorous acid, HOCl. Salts of hypochlorite including calcium hypochlorite, $Ca(OCl)_2$, can be used as disinfectants and are safer to handle than gaseous chlorine.

The two chemical species formed by chlorine in water, HOCl and OCl^-, are known as **free available chlorine** and are very effective in killing bacteria and other pathogens. In the presence of ammonia, HOCl reacts with ammonium ion to produce monochloramine (NH_2Cl), dichloramine ($NHCl_2$), and trichloramine (NCl_3), three species collectively called **combined available chlorine**. Although weaker disinfectants than chlorine and hypochlorite, the chloramines persist in water distribution systems to provide residual disinfection.

A major problem with the use of chlorine as a disinfection agent is by-product production from the reaction of chlorine or bromine generated from chlorine with organics in water. The most common such by-products are the trihalomethanes including chloroform, $HCCl_3$, and dibromochloromethane, $HCBr_2Cl$, noted as water pollutants in Chapter 4, Section 4.13. Humic substances are common precursors to chlorinated by-products; their removal before water chlorination prevents the formation of organochlorine by-products.

Chlorine is used to treat water other than drinking water. It is employed to disinfect effluent from sewage treatment plants, as an additive to the water in electric power plant cooling towers, and to control microorganisms in food processing.

5.11.4 Chlorine Dioxide

Chlorine dioxide, ClO_2, is an effective water disinfectant that is of particular interest because, in the absence of impurity Cl_2, it does not produce impurity trihalomethanes in water treatment. In the neutral pH range, chlorine dioxide in water remains largely as molecular ClO_2 until it contacts a reducing agent with which to react. Chlorine dioxide is a gas that is violently reactive with organic matter and is explosive when exposed to light. For these reasons, it is not shipped, but is generated on-site by processes such as the reaction of chlorine gas with solid sodium chlorite:

$$2NaClO_2(s) + Cl_2(g) \leftrightarrow 2ClO_2(g) + 2NaCl(s) \tag{5.29}$$

A high content of elemental chlorine in the product may require its purification to prevent unwanted side reactions from Cl_2. As a water disinfectant, chlorine dioxide does not chlorinate or oxidize ammonia or other nitrogen-containing compounds.

5.11.5 TOXICITIES OF CHLORINE AND CHLORINE DIOXIDE

The toxicity of chlorine is a significant factor regarding its use for water disinfection. Chlorine is so toxic when it is inhaled that it was the first poison gas used in World War I. Elemental chlorine can be stored and transported as a liquid under pressure. Released to the atmosphere, it forms a dense layer of choking gas that reacts in moist respiratory tract tissue to produce acid and tissue-damaging oxidants. Levels of 10–20 ppm in air can cause immediate damage to the respiratory tract and a brief exposure to air containing 1000 ppm of chlorine can be fatal.

Like elemental chlorine, chlorine dioxide is a severe irritant to the respiratory tract and eye. It hydrolyzes in the body to produce chlorite ion, ClO_2^-, which is known to form methemoglobin, a product of hemoglobin in which the iron is in the +3 oxidation state and that cannot carry oxygen in the bloodstream. Chronic exposure to chlorine dioxide in workers has caused eye and throat irritation, respiratory symptoms, and bronchitis. At least one death has been reported of a worker exposed to 19 ppm of chlorine dioxide. Therefore, potentially dangerous chlorine dioxide is prepared only when needed, in the quantities needed, and where needed, which is in keeping with the best practice of green chemistry and technology. Several methods are available for the synthesis of chlorine dioxide without using toxic elemental chlorine as a reagent. One such method uses the reaction of sodium chlorite with hydrochloric acid:

$$5NaClO_2 + 4HCl \rightarrow 4ClO_2 + 5NaCl + 2H_2O \qquad (5.30)$$

5.11.6 GREEN OZONE FOR WATER DISINFECTION

A greener alternative to chlorine-based water disinfectants in many respects is ozone, O_3. Pumped into water, this form of oxygen kills pathogens without producing the undesirable by-products made by chlorine, and it is actually more effective than chlorine in killing viruses. Ozone is produced from oxygen in air by a high-voltage electrical discharge through dried air as illustrated in Figure 5.7. The lifetime of ozone in water is short, so a small amount of chlorine must usually be added to ozonated water to maintain disinfection in the water distribution system.

A major consideration with ozone is the rate at which it decomposes spontaneously in water, according to the overall reaction,

$$2O_3 \rightarrow 3O_2(g) \qquad (5.31)$$

Because of the decomposition of ozone in water, some chlorine must be added to maintain disinfectant throughout the water distribution system. Although ozone does not produce the undesirable halogenated organics that cause problems with the use of chlorine, it does produce some oxygenated by-products. These include aldehydes (acetaldehyde, $H_3CC(O)H$) and carboxylic acids and their salts (oxalic acid, HO_2CCO_2H).

With the caveat that release of toxic ozone to the ambient atmosphere must be avoided (see Section 5.11.7), disinfection of water by ozonation is a virtually ideal example of green chemical practice. The only raw material is universally available air, which is free. Ozone is produced only where it is needed as it is needed, without by-products. The ozone does not persist in water, where it decomposes to elemental oxygen, and there is very little likelihood of producing harmful disinfection by-products with ozone.

5.11.7 OZONE TOXICITY

The disinfection of water by ozone requires careful control of this gas because of its toxicity. Inhalation of ozone can be fatal. Ozone is a deep lung irritant and causes pulmonary edema, the accumulation of fluid in the lungs. Ozone is strongly irritating to the eyes and upper respiratory

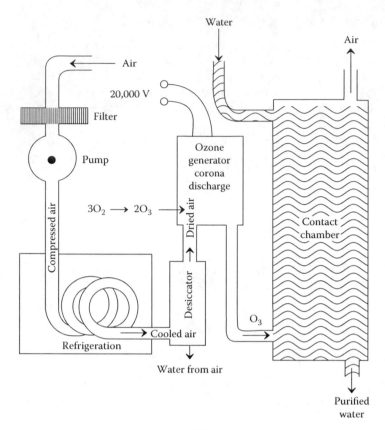

FIGURE 5.7 Ozone generator in which O_3 is produced from O_2 by an electrical discharge through chilled, dry air. Pumped into water, the ozone kills microorganisms including disease-causing pathogens.

tract. Ozone's toxic effects appear to be tied with its ability to produce reactive oxidant free radicals in an organism, which can lead to lipid peroxidation or reaction with sulfhydryl (–SH) groups. Ozone is also noted for being phytotoxic (toxic to plants). Because of ozone's toxicity, it is essential to avoid its release in applications such as water disinfection. Fortunately, it is relatively easy to contain ozone during water disinfection.

5.11.8 Miscellaneous Disinfection Agents

Several miscellaneous agents are used to disinfect water, especially in small quantities. One of these is elemental iodine, which is often carried by campers to disinfect small quantities of water taken from streams for drinking.

Iron(VI) in the form of ferrate ion, FeO_4^{2-}, is a strong oxidizing agent with excellent disinfectant properties. When it reacts in water, it produces a gelatinous precipitate of $Fe(OH)_3$, which acts as a coagulant that removes colloidal matter from water. It has the additional advantage of removing heavy metals, viruses, and phosphate. In a sense, ferrate is a universal water treatment chemical that disinfects, oxidizes organics, removes humic substances, and precipitates colloidal solid impurities.[4] Ferrate has commonly been used as the potassium salt prepared by reaction of a strong oxidant, such as Cl_2 or $NaClO$, with an iron(III) salt in KOH solution, from which K_2FeO_4 precipitates. This is a rather expensive process and the ferrate product does not store well, so efforts are under way to develop greener ferrate synthesis processes including electrochemical syntheses and solventless syntheses. At least one type of commercial unit is available for making ferrate on site as needed.

5.12 RESTORATION OF WASTEWATER QUALITY

One of the most important activities in water treatment is the treatment of wastewater from municipal sewage. This has generally been carried out to reduce the BOD in the wastewater effluent so that it does not deplete oxygen levels in receiving waters, to eliminate pathogens, and to remove nutrients that can promote algal growth and eutrophication in receiving waters. Increasingly, however, wastewater treatment is directed toward water recycling including even to the point of use as a drinking water source. Wastewater treatment is addressed here with emphasis on biological treatment processes.

5.12.1 PRIMARY WASTEWATER TREATMENT

Primary treatment of wastewater consists of the removal of insoluble matter such as grit, grease, and scum by processes such as skimming, screening, and grinding. **Primary sedimentation** removes both settleable and floatable solids and semisolids. Primary sedimentation is aided by aggregation of flocculent particles, a process that may be promoted by the addition of chemicals. The material that floats in the primary settling basin is collectively called grease and consists of fatty substances, oils, waxes, free fatty acids, and insoluble soaps containing calcium and magnesium. Primary sedimentation removes up to half of the contaminants in wastewater. These materials are commonly conveyed to an anoxic digester (see Figure 5.9 later in the chapter) where they degrade in the absence of air and produce combustible methane gas.

5.12.2 SECONDARY WASTE TREATMENT BY BIOLOGICAL PROCESSES

The most obvious harmful effect of biodegradable organic matter in wastewater is BOD, consisting of depletion of dissolved oxygen by microorganism-mediated degradation of the organic matter (see Chapter 3, Section 3.5). **Secondary wastewater treatment** is designed to remove BOD, usually by taking advantage of the same kind of biological processes that would otherwise consume oxygen in water receiving the wastewater. Secondary treatment by biological processes takes many forms, but consists basically of the action of microorganisms provided with added oxygen degrading organic material in solution or in suspension until the BOD of the waste has been reduced to acceptable levels. The waste is oxidized biologically under conditions controlled for optimum bacterial growth and at a site where this growth does not influence the environment.

One of the simplest biological waste treatment processes is the **trickling filter** (Figure 5.8) in which wastewater is sprayed over rocks or other solid support material covered with microorganisms.

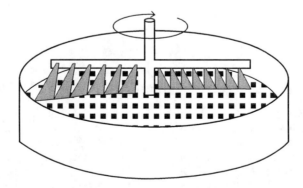

FIGURE 5.8 Trickling filter for secondary waste treatment. Wastewater is sprayed from holes in a rotating pipe onto a bed of rocks or other solid materials coated with microorganisms that metabolize biodegradable materials in the wastewater.

The structure of the trickling filter is such that contact of the wastewater with air is allowed and degradation of organic matter occurs by the action of the microorganisms.

Rotating biological reactors (**contactors**), another type of biological treatment system, consist of groups of large plastic discs mounted close together on a horizontal rotating shaft. The device is positioned so that at any particular instant, half of each disc is immersed in wastewater and half exposed to air. The shaft rotates constantly, so that the submerged portion of the discs is always changing. The discs, usually made of high-density polyethylene or polystyrene, accumulate thin layers of attached biomass, which degrades organic matter in the sewage. Oxygen is absorbed by the biomass and by the layer of wastewater adhering to it during the time that the biomass is exposed to air.

Both trickling filters and rotating biological reactors are examples of fixed-film biological (FFB) or attached growth processes. The greatest advantage of these processes is their low-energy consumption. The energy consumption is minimal because it is not necessary to pump air or oxygen into the water, as is the case with the popular activated sludge process described in the following paragraph. The trickling filter has long been a standard means of wastewater treatment, and a number of wastewater treatment plants still use trickling filters.

The **activated sludge process** (Figure 5.9) is generally acknowledged to be the most versatile and effective of all wastewater treatment processes. Microorganisms in the aeration tank convert organic material in wastewater to microbial biomass and CO_2. Organic nitrogen is converted to ammonium

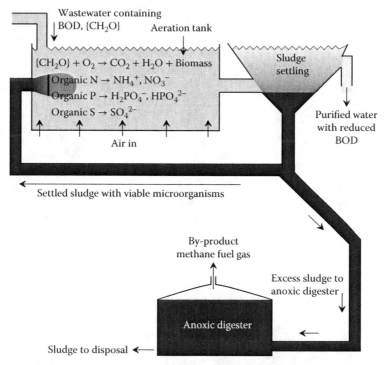

FIGURE 5.9 An activated sludge wastewater treatment facility. Wastewater containing degradable biomass is pumped into an aeration tank in which are suspended viable bacteria and protozoa capable of biodegrading the waste. The organic waste is converted to carbon dioxide and to additional viable microbial biomass. Organic N, P, and S are converted to simple inorganic forms. The treated wastewater is conveyed to a settling basin from which the microorganisms settle as sewage sludge (also called biosolids) and purified water is discharged. Much of the settled sludge is pumped back to the front of the aeration tank to degrade additional waste. Excess sludge is pumped to an anoxic digester where it undergoes fermentation to produce methane gas, a valuable fuel by-product of the process. Eventually, the sludge from the anoxic digester is dried and disposed, in some cases spread on land as a soil fertilizer.

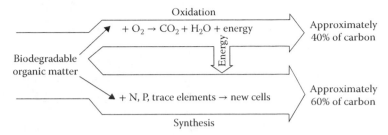

FIGURE 5.10   Pathways for the removal of BOD in biological wastewater treatment.

ion or nitrate. Organic phosphorus is converted to orthophosphate. The microbial cell matter formed as part of the waste degradation processes is normally kept in the aeration tank until the microorganisms are past the phase of growth in which they are increasing logarithmically, at which point the cells flocculate relatively well to form settleable solids. These solids settle out in a settler and a fraction of them is discarded. Some of the solids, the return sludge, are recycled to the head of the aeration tank and come into contact with fresh sewage. The combination of a high concentration of "hungry" cells in the return sludge and a rich food source in the influent sewage provides optimum conditions for the degradation of organic matter much more rapidly than would occur naturally if the wastes were discharged directly to a stream or body of water with a relatively small population of degrading organisms where such wastes would consume oxygen in receiving waters.

The activated sludge process provides two pathways for the removal of BOD, as illustrated schematically in Figure 5.10. BOD can be removed by (1) oxidation of organic matter to provide energy for the metabolic processes of the microorganisms and (2) synthesis, incorporation of the organic matter into cell mass. In the first pathway, carbon is removed in the gaseous form as CO_2 that does not present a disposal problem except for its small contribution to greenhouse gas atmospheric carbon dioxide. The second pathway provides for removal of carbon as a solid in biomass.

The excess solids consisting of biomass produced in the aeration tank along with unreacted material in the sewage is only about 1% solids and may contain undesirable components, so it must undergo further treatment. This is accomplished by placing the material in a large tank partially buried underground and protected from air called an **anoxic digester** (Figure 5.9). In the anoxic digester, methane-producing bacteria that function in the absence of air carry out the anoxic biodegradation of organic matter according to the following reaction:

$$2\{CH_2O\} \rightarrow CH_4 + CO_2 \qquad\qquad (5.32)$$

The anoxic digestion of these solids reduces both the volatile matter content and the sludge volume by about 60%. This process is very much in keeping with the practice of sustainable technology because it produces combustible methane gas, a premium fuel. A properly operating sewage treatment plant can generate enough methane by anoxic digestion to power all the pumps and produce all the electricity needed by the plant.

The sewage sludge (which is often referred to by the more palatable term "biosolids") produced in wastewater treatment processes requires disposal or use such as for soil conditioner. This issue is discussed in more detail in Section 5.14.

5.12.3   TERTIARY WASTE TREATMENT

Unpleasant as the thought may be, many people drink used water—water that has been discharged from a municipal sewage treatment plant or from some industrial process. This raises serious questions about the presence of pathogenic organisms or toxic substances in such water. Because of high population density and heavy industrial development, the problem is especially acute in

Europe, where some municipalities process 50% or more of their water from "used" sources. Obviously, there is a great need to treat wastewater in a manner that makes it amenable to reuse. This requires treatment beyond the secondary processes.

Tertiary waste treatment (sometimes called **advanced waste treatment**) is a term used to describe a variety of processes performed on the effluent from secondary waste treatment. The contaminants removed by tertiary waste treatment fall into the general categories of (1) suspended solids, (2) dissolved organic compounds, and (3) dissolved inorganic materials, including the important class of algal nutrients. Each of these categories presents its own problems with regard to water quality. Suspended solids are primarily responsible for residual biological oxygen demand in secondary sewage effluent waters. The dissolved organics are the most hazardous from the standpoint of potential toxicity. The major problem with dissolved inorganic materials is that presented by algal nutrients, primarily nitrates and phosphates. In addition, potentially hazardous toxic metals may occur among the dissolved inorganics. In addition to these chemical contaminants, secondary sewage effluent often contains a number of disease-causing microorganisms, requiring disinfection in cases where humans may later come into contact with the water. Ingestion of sewage still causes incidents of disease, even in more developed nations.

5.12.4 PHYSICAL–CHEMICAL TREATMENT OF MUNICIPAL WASTEWATER

Complete physical–chemical wastewater treatment systems offer both advantages and disadvantages relative to biological treatment systems. The capital costs of physical–chemical facilities can be less than those of biological treatment facilities, and they usually require less land. They are better able to cope with toxic materials and overloads. However, they require careful operator control and consume relatively large amounts of energy.

As illustrated in Figure 5.11, the basic steps of a complete physical–chemical wastewater treatment facility are as follows:

- Removal of scum and solid objects
- Clarification, generally with addition of a coagulant, and frequently with the addition of other chemicals (such as lime for phosphorus removal)
- Filtration to remove filterable solids
- Activated carbon adsorption
- Disinfection

During the early 1970s, it appeared likely that physical–chemical treatment would largely replace biological treatment, but this has not happened because of higher chemical and energy costs. However, as discussed in Section 5.15, physical–chemical processes are used in recycling water.

5.13 NATURAL WATER PURIFICATION PROCESSES

Virtually all of the materials that wastewater treatment processes are designed to eliminate can be absorbed by soil or degraded in soil. In fact, most of these materials can serve to add fertility to soil. Wastewater can provide the water that is essential to plant growth. The mineralization of biological wastes in wastewater provides phosphorus, nitrogen, and potassium usually added to soil by fertilizers. Wastewater also contains essential trace elements and vitamins. Stretching the point a bit, the degradation of organic wastes provides the CO_2 essential for photosynthetic production of plant biomass.

Soil may be viewed as a natural filter for wastes and one that rates very high in sustainability. Most organic matter is readily degraded in soil, and, in principle, soil constitutes an excellent primary, secondary, and tertiary treatment system for water. Soil has physical, chemical, and biological characteristics that can enable wastewater detoxification, biodegradation, chemical decomposition,

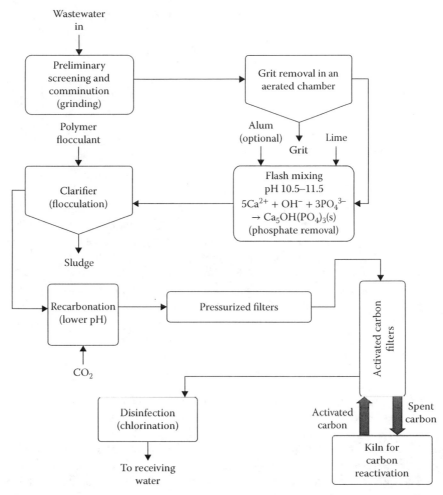

FIGURE 5.11 Major components of a complete physical–chemical treatment facility for municipal wastewater.

and physical and chemical fixation. A number of soil characteristics are important in determining its use for land treatment of wastes. These characteristics include physical form, ability to retain water, aeration, organic content, acid-base characteristics, and oxidation-reduction behavior. Soil is a natural medium for a number of living organisms that may have an effect on biodegradation of wastewaters, including those that contain industrial wastes. Of these, the most important are bacteria, including those from the genera *Agrobacterium*, *Arthrobacter*, *Bacillus*, *Flavobacterium*, and *Pseudomonas*. Actinomycetes and fungi are important in decay of vegetable matter and may be involved in biodegradation of wastes. Other unicellular organisms that may be present in or on soil are protozoa and algae. Soil animals, particularly earthworms, affect soil parameters such as soil texture. The growth of plants in soil may have an influence on its waste treatment potential in such aspects as uptake of soluble wastes and erosion control.

Early civilizations, such as the Chinese, used human organic wastes to increase soil fertility, and the practice continues today. The ability of soil to purify water was noted well over a century ago. In 1850 and 1852, J. Thomas Way, a consulting chemist to the Royal Agricultural Society in England, presented two papers to the Society entitled "Power of Soils to Absorb Manure." Mr. Way's experiments showed that soil is an ion exchanger. Much practical and theoretical information on the ion exchange process resulted from his work.

If soil treatment systems are not properly designed and operated, odor can become an overpowering problem. The author of this book is reminded of driving into a small town, recalled from some years before as a very pleasant place, and being assaulted with a virtually intolerable odor. The disgruntled residents pointed to a large spray irrigation system on a field in the distance—unfortunately upwind—spraying liquefied pig manure as part of an experimental feedlot waste treatment operation. The experiment was not deemed a success and was discontinued by the investigators, presumably before they met with violence from the local residents.

5.13.1 Industrial Wastewater Treatment by Soil

Wastes that are amenable to land treatment are biodegradable organic substances, particularly those contained in municipal sewage and in wastewater from some industrial operations, such as food processing. However, through acclimation over a long period of time, soil bacterial cultures may develop that are effective in degrading normally recalcitrant compounds that occur in industrial wastewater. Acclimated microorganisms are found particularly at contaminated sites, such as those where soil has been exposed to crude oil for many years.

Land treatment is most used for petroleum-refining wastes and is applicable to the treatment of fuels and wastes from leaking underground storage tanks. It can also be applied to biodegradable organic chemical wastes, including some organohalide compounds. Land treatment is not suitable for the treatment of wastes containing acids, bases, toxic inorganic compounds, salts, heavy metals, and organic compounds that are excessively soluble, volatile, or flammable.

5.14 SLUDGES AND RESIDUES FROM WATER TREATMENT

A challenge to the sustainability of water treatment processes has to do with sludges and generally semisolid residues collected or produced during water treatment. It is important to dispose of such material properly or preferably find uses for it.

As noted in Section 5.12.1, much sludge is present in wastewater before treatment and can be collected from it as part of primary wastewater treatment. There are two major kinds of sludge generated in a waste treatment plant. The first of these is organic sludge from activated sludge, trickling filter, or rotating biological reactors, commonly referred to as biosolids. The second is inorganic sludge from the addition of chemicals, such as in phosphorus removal (see Section 5.9).

As shown in Figure 5.9, sewage sludge is subjected to anoxic digestion in a digester designed to allow bacterial action to occur in the absence of air. This reduces the mass and volume of sludge, destroys pathogens, and ideally results in the formation of a stabilized humus.

Sludge normally requires treatment to reduce its volume and to get it into a form suitable for disposal or for some beneficial use, such as soil conditioning. **Conditioning** of sludge consists of the addition of polymers or lime to make the sludge more amenable to dewatering. **Thickening** of sludge is accomplished by settling, filtering, or centrifugation to give a relatively solid-free water phase that can be decanted from the solids fraction of the sludge. Thickened, conditioned sludge can be subjected to **dewatering** on beds consisting of layers of sand and gravel or to mechanical dewatering with filter presses, belt filter presses, or vacuum filters to provide a product that contains about 85% water, although it appears as a dry powder. Heat can also be used in the drying process.

Biosolid sewage sludge contains fertilizer materials required for plant growth and is typically around 5% N, 3% P, and 0.5% K on a dry-mass basis, making it useful to fertilize and condition soil. The humic material in the sludge improves the physical properties and cation-exchange capacity of the soil. Possible accumulation of heavy metals is of some concern insofar as the use of sludge on cropland is concerned. Sewage sludge is an efficient heavy metals scavenger and may contain elevated levels of zinc, copper, nickel, and cadmium. These and other metals tend to remain immobilized in soil by chelation with organic matter, adsorption on clay minerals, and precipitation as insoluble compounds such as oxides or carbonates. However, increased application of sludge on cropland has caused distinctly elevated levels of zinc and cadmium in both leaves and grain of corn. Therefore, caution has been advised in heavy or prolonged application of sewage sludge to soil. Prior control of heavy metal contamination from industrial sources has significantly reduced the heavy metal content of sludge and enabled it to be used more extensively on soil. One problem that has developed from soil application is accumulation of excessive levels of zinc, which is not very toxic to humans but can be detrimental to the growth of crops such as corn.

Incineration of sludge is effective in destroying all the organic matter, but does require relatively large amounts of fuel. Sludge can be pyrolyzed by heating in the absence of air to give a combustible gas and a mixture of liquid products, although most of the fuel thus produced is required by the pyrolysis process. A better alternative is to gasify sludge at relatively high temperatures in an atmosphere of oxygen at levels below those required to completely burn the sludge. Combustible gases produced are carbon monoxide, elemental hydrogen, and a significant fraction of methane.

An increasing problem in sewage treatment arises from sludge sidestreams. These consist of water removed from sludge by various treatment processes. Sewage treatment processes can be divided into mainstream treatment processes (primary clarification, trickling filter, activated sludge, and rotating biological reactor) and sidestream processes. During sidestream treatment, sludge is dewatered, degraded, and disinfected by a variety of processes, including gravity thickening, dissolved air flotation, anoxic digestion, oxic digestion, vacuum filtration, centrifugation, belt filter press filtration, sand drying bed treatment, sludge lagoon settling, wet air oxidation, pressure filtration, and Purifax treatment in which sludge is oxidized with heavy doses of chlorine. Each of these produces a liquid by-product sidestream that is circulated back to the mainstream. These add to the BOD and suspended solids of the mainstream.

A variety of chemical sludges are produced by various water treatment and industrial processes. Among the most abundant of such sludges is alum sludge produced by the hydrolysis of Al(III) salts used in the treatment of water, which creates gelatinous aluminum hydroxide:

$$Al^{3+} + 3OH^- (aq) \rightarrow Al(OH)_3 (s) \tag{5.33}$$

Alum sludges normally are 98% or more water and are very difficult to dewater.

Both iron(II) and iron(III) compounds are used for the removal of impurities from wastewater by precipitation of $Fe(OH)_3$. The sludge contains $Fe(OH)_3$ in the form of soft, fluffy precipitates that are difficult to dewater beyond 10% or 12% solids.

The addition of either lime, $Ca(OH)_2$, or quicklime, CaO, to water is used to raise the pH to about 11.5 and cause the precipitation of $CaCO_3$, along with metal hydroxides and phosphates. Calcium carbonate is readily recovered from lime sludges and can be recalcined to produce CaO, which can be recycled through the system.

Metal hydroxide sludges are produced in the removal of metals such as lead, chromium, nickel, and zinc from wastewater by raising the pH to such a level that the corresponding hydroxides or hydrated metal oxides are precipitated. The disposal of these sludges is a substantial problem because of their toxic heavy metal content. Reclamation of the metals is an attractive alternative for these sludges.

Pathogenic microorganisms may persist in the sludge left from the treatment of sewage, the most significant of which are (1) indicators of fecal pollution, including fecal and total coliform; (2) pathogenic bacteria, including *Salmonellae* and *Shigellae*; (3) enteric (intestinal) viruses, including enterovirus and poliovirus; and (4) parasites, such as *E. histolytica* and *Ascaris lumbricoides*. It is necessary both to be aware of pathogenic microorganisms in municipal wastewater treatment sludge and to find a means of reducing the hazards caused by their presence. Pathogen levels can be reduced by agitating the sludge for long periods with air, heating, composting, or raising the sludge pH to 12 or higher with lime.

5.15   WATER, THE GREENEST SUBSTANCE ON EARTH: REUSE AND RECYCLING

Water is the greenest substance imaginable, essential for life and totally recyclable. The hydrologic cycle (Chapter 3, Figure 3.2) is nature's recycling system for water. Unfortunately, water precipitation in the hydrologic cycle does not always fall where it is most needed. Water used in arid regions where many people live is usually largely lost to the atmosphere and may end up as precipitation thousands of miles away, often falling over the ocean. As a consequence, as demands for water exceed supply, water reuse and recycling are becoming much more common and promise to become even more so in the future.[5]

Reuse is a term applied when water discharged by one user is taken as a water source by another user. **Unplanned reuse** occurs as the result of waste effluents entering receiving waters or groundwater and subsequently being taken into a water distribution system. A typical example of unplanned water reuse occurs in London, which withdraws water from the Thames River that may have been through other water systems at least once, and which uses groundwater sources unintentionally recharged with sewage effluents from a number of municipalities. **Planned reuse** utilizes wastewater treatment systems deliberately designed to bring water up to standards required for subsequent applications. The term **direct reuse** refers to water that has retained its identity from a previous application; reuse of water that has lost its identity is termed **indirect reuse**. A distinction also needs to be made between recycling and reuse. **Recycling** commonly refers to reuse of water internally before it is ever discharged. An example is condensation of steam in a steam power plant followed by return of the steam to boilers.

Reuse of water continues to grow because of two major factors. The first of these is lack of supply of water. The second is that widespread deployment of modern water treatment processes significantly enhances the quality of water available for reuse. These two factors come into play in semiarid regions in countries with advanced technological bases. For example, Israel, which is dependent on irrigation for essentially all its agriculture, reuses about two-thirds of the country's sewage effluent for irrigation, whereas the United States, where water is relatively more available, uses only about 2–3% of its water for this purpose.

Since drinking water and water used for food processing require the highest quality of all large applications, intentional reuse for potable water traditionally has been regarded as less desirable, though widely practiced unintentionally or out of necessity. This leaves three applications within which water is most commonly reused at present:

1. Irrigation for cropland, golf courses, and other applications requiring water for plant and grass growth. This is the largest potential application for reused water and one that can take advantage of plant nutrients, particularly nitrogen and phosphorus, in water.
2. Cooling and process water in industrial applications. For some industrial applications, relatively low-quality water can be used and secondary sewage effluent is a suitable source.
3. Groundwater recharge. Groundwater can be recharged with reused water either by direct injection into an aquifer or by applying the water to land, followed by percolation into the aquifer.[6] The latter, especially, takes advantage of biodegradation and chemical sorption processes to further purify the water.

It is inevitable that water recycling and reuse will continue to increase, and growing pressures on water supplies will lead eventually to the widespread purification and recycling of water into municipal systems. This trend will increase the demand for water treatment, both qualitatively and quantitatively. In addition, it will require more careful consideration of the original uses of water to minimize water deterioration and enhance its suitability for reuse.

Figure 5.12 shows a system for planned reuse of water from wastewater sources. Most of the water treatment processes discussed in this chapter may be utilized in purifying water for reuse and

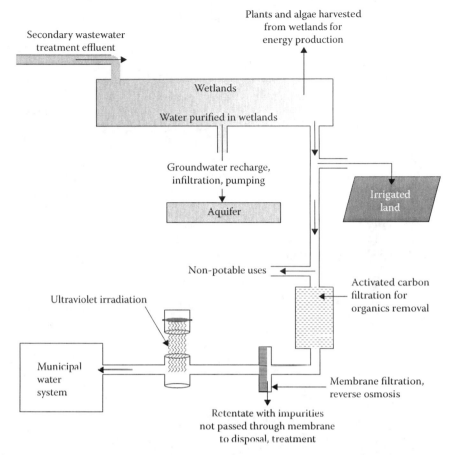

FIGURE 5.12 An advanced water treatment system capable of producing potable water from sewage. This system feeds secondary sewage waste treatment effluent into constructed wetlands where plant growth removes nutrients from the water reducing its potential to cause eutrophication. Plants grown in the wetlands are harvested and the biomass is used for energy production. Water from the wetlands infiltrates aquifers by gravity and by pumping, and some of the water is diverted to non-potable applications including irrigation and cooling water. The remainder of the water flowing from the wetlands is filtered over activated carbon to remove most organic impurities and then subjected to reverse osmosis to reduce the dissolved salt content. After treatment with ultraviolet radiation to destroy bioactive compounds including pharmaceuticals and their metabolites, the water is returned to the municipal water supply system.

employed in an operation such as the one shown (see especially the physical–chemical wastewater purification system, Figure 5.11). Figure 5.12 shows some of the main operations employed in a system for total water recycle capable of producing drinking water and water for other uses from sewage. It takes secondary wastewater effluent from which most of the BOD has been removed by the activated sludge process and discharges it into wetlands, taking advantage of nature's inherent ability to purify water. This water is nutrient-rich, and plants and algae grow profusely in the wetlands. This biomass has the potential to be harvested and utilized as a fuel or raw material using thermochemical gasification or related measures. Depending on geological conditions, significant amounts of water in the wetlands may infiltrate into groundwater, where it is purified further by natural processes underground. Water exiting the wetlands may be run over a bed of activated carbon to remove organics. The water can also be subjected to highly efficient membrane filtration processes including even reverse osmosis for removal of all salts. An important consideration with membrane purification is the production of streams of retentate enriched with impurities. Retentate can be pumped into deep underground saline aquifers in some areas, discarded to the ocean in coastal regions (often mixing with wastewater or cooling water to dilute the impurities in the retentate), or evaporated to produce a solid or sludge residue that can be disposed. Ultraviolet irradiation of the water can be used to destroy pharmaceuticals and their metabolites.

The world's largest water plant devoted to bringing sewage effluent up to a standard permitting potable uses began operation in the Orange County Water District in southern California in late 2007. Known as the Groundwater Replenishment System, this $481 million project is fed with sewage treated by secondary wastewater treatment processes and pumps its clean water product into underground aquifers to serve as future municipal water sources and to prevent saltwater intrusion from the nearby ocean. Around 275 million liters per day of sewage effluent is first forced through microfilters to remove solids, then treated by reverse osmosis to remove dissolved salts, and finally subjected to intense ultraviolet radiation to break down organic contaminants, especially pharmaceuticals and their metabolites before being discharged to underground aquifers. Although the purified water from the system is not currently used for municipal water supply, it certainly could be. It is expected to serve as a model for similar installations around the world.

5.16   WATER CONSERVATION

Water conservation, simply using less water, is one of the most effective, and clearly the fastest, means to ensure water supply. Water conservation falls into several major categories:

- Indoor and household water conservation practices
- Water-conserving devices and appliances
- Outdoor and landscaping water conservation
- Efficient irrigation practices
- Conservation in agricultural operations other than irrigation
- Water-efficient manufacturing

Each of these categories is summarized briefly in the following discussion.

Indoor and household water conservation practices are generally those simple measures that water consumers may implement to reduce use of water. Examples include sequential uses of water such as saving water from rinsing dishes to use for household cleaning, ensuring that there are no leaking faucets or toilet appliances, taking showers by first lathering one's wet body without water flowing followed by rinsing with water for the minimum length of time, minimizing the number of toilet flushes, minimizing use of kitchen sink disposal devices for food waste disposal, operating clothes washers and dishwashers only with full loads, and other common-sense measures to reduce water use. Implementation of these measures is largely a matter of education and good environmental citizenship. In cases of real water shortage, force may be applied through the painful mechanism of sharply higher water utility rates.

A number of water-saving devices and appliances are available. These can be quite simple and cheap, such as aerators with flow constrictors on faucets or showerheads, or they may be large and expensive such as low-water-consumption clothes washers that function well with cold water. Toilets that use minimal water per flush are now mandated by law, although the less effective of these may actually increase water consumption when multiple flushes are required. In public facilities, automatically flushing toilets and urinals are now commonplace, although some have the counterproductive tendency of performing multiple automatic flushes for each instance of use. Instant in-line water heaters installed near faucet outlets can reduce water use by making it unnecessary to run the hot water faucet for some time to obtain hot water.

Water used for lawns and landscaping often is one of the greatest consumers of water. In some areas, such as public parks, purified wastewater is used for this purpose. In principle, water used in some household applications can be recycled to water yard plants or grass. Some of the greater savings can be achieved in this area by growing ground cover and plants that require minimal water.

This may mean planting native grass species that can withstand drought without much watering. In extreme cases, yards have been converted to gravel cover. Mulching with materials such as tree bark or grass clippings can reduce water use significantly.

Irrigation of agricultural crops is one of the largest consumers of water and one that is amenable to conservation. Spray irrigation devices lose large quantities of water to evaporation and should be replaced by systems that do not spray the water into the air. The ultimate in irrigation efficiency is drip irrigation that applies minimum quantities of water directly to the plant roots. Application of excess water in irrigation should be minimized. However, particularly in areas that rarely receive enough rain to flush the soil, application of minimal amounts of irrigation water can result in harmful accumulation of salts in the soil.

Livestock production and processing agricultural products use large amounts of water. The production of 1 kg of beef requires around 20,000 L of water, including the water required to grow the grain that the animal eats. About 6000 L of water are required to produce 1 kg of chicken meat. Significantly less water is needed to produce an equivalent amount of vegetable protein. The production of fuel ethanol from the fermentation of sugar from corn requires relatively large amounts of water. Choices made in the kinds of agricultural products produced for market have important implications for water conservation.

Manufacturing can be very water-intensive. The manufacture of a single automobile requires around 150,000 L of water. Refining crude oil to gasoline requires approximately 44 times the volume of water as the gasoline produced. The conservation of water in manufacturing and commercial activities is an important aspect of industrial ecology (see Chapter 16).

5.16.1 RAINWATER HARVESTING

Rainwater harvesting refers to the practice of collecting and storing water before it can reach a river or infiltrate into an aquifer.[7] It has been practiced for centuries in various ways. Farm homes and rural school buildings in the U.S. Great Plains region used to have gutter and piping systems to collect rooftop water connected to underground cisterns for water storage. In some cases, these systems served as the main source of drinking water. When operated properly, when rain finally fell after a dry period, the first water running off the roof was supposed to be diverted to the ground to prevent bird droppings, dirt, radioactive fallout from aboveground nuclear weapons tests (late 1940s and 1950s), and other undesirable materials from getting into the stored water supply.

A form of rainwater harvesting involves collection on the soil surface. Soil can be prepared especially for this purpose, such as by planting grass varieties that produce a dense sod from which rainwater tends to wash off. Large parking lots and other paved areas are potential sources of rainwater runoff that can be harvested.

Although it is generally not advised as a source of drinking water for humans, harvested rainwater can be used for irrigation, for domestic applications other than drinking or cooking, for livestock, in car wash facilities, and in other applications where potable water is not required.

QUESTIONS AND PROBLEMS

1. During municipal water treatment, air is often mixed intimately with the water, that is, it is aerated. What kinds of undesirable contaminants would this procedure remove from water?
2. Phosphate in the form of $H_2PO_4^-$ and HPO_4^{2-} ions is the substance usually removed from secondary sewage effluent to prevent excessive algal growth and eutrophication in receiving waters. Of several possible algal nutrients, why is phosphate chosen? Show with a chemical reaction the most common means of removal.
3. Membrane filtration processes can be very effective in removing residual BOD from secondary wastewater effluent. What does this suggest regarding the nature of contaminants responsible for the BOD?

4. By doing some search on the Internet, gather information regarding the use of wastewater for irrigation. Is this a practice that is used and if so where does it usually take place? What are some of the benefits? What are some of the risks?

5. Look up proposals to restore the Hetch Hetchy Valley in Yosemite National Park to its former state. How might this affect water supply to parts of California? What might be some benefits of restoration of this valley to its former state?

6. What is the purpose of the return sludge step in the activated sludge process?

7. What are the two processes by which the activated sludge process removes soluble carbonaceous material from sewage?

8. How does reverse osmosis differ from a simple sieve separation or ultrafiltration process based on molecular size?

9. How many liters of methanol would be required daily to remove the nitrogen from a 200,000-L/day sewage treatment plant producing an effluent containing 50 mg/L of nitrogen? Assume that the nitrogen has been converted to NO_3^- in the plant and that the methanol is consumed in a denitrification reaction.

10. Discuss some of the advantages of physical–chemical treatment of sewage as opposed to biological wastewater treatment. What are some disadvantages?

11. Why is recarbonation necessary when water is softened by the lime–soda process?

12. Assume that a waste contains 300 mg/L of biodegradable {CH_2O} and is processed through a 200,000-L/day sewage treatment plant that converts 40% of the waste to CO_2 and H_2O. Calculate the volume of air (at 25°C, 1 atm) required for this conversion. Assume that the O_2 is transferred to the water with 20% efficiency.

13. If all of the {CH_2O} in the plant described in Question 12 could be converted to methane by anoxic digestion, how many liters of methane (STP) could be produced daily?

14. Assuming that aeration of water does not result in the precipitation of calcium carbonate, of the following, which one would not be removed by aeration: hydrogen sulfide, carbon dioxide, volatile odorous bacterial metabolites, alkalinity, and iron?

15. In which of the following water supplies would moderately high water hardness be most detrimental: municipal water, irrigation water, boiler feedwater, and drinking water (in regard to potential toxicity).

16. Which solute in water is commonly removed by the addition of sulfite or hydrazine?

17. Wastewater containing dissolved Cu^{2+} ion is to be treated to remove copper. Which of the following processes would *not* remove copper in an insoluble form: lime precipitation, cementation, treatment with NTA, ion exchange, and reaction with metallic Fe.

18. Match each water contaminant in the left column with its preferred method of removal in the right column.

A. Mn^{2+}	1. Activated carbon
B. Ca^{2+} and HCO_3^-	2. Raise pH by addition of Na_2CO_3
C. Trihalomethane compounds	3. Addition of lime
D. Mg^{2+}	4. Oxidation

19. A cementation reaction employs iron to remove Cd^{2+} present at a level of 350 mg/L from a wastewater stream. Given that the atomic mass of Cd is 112.4 amu and that of Fe is 55.8 amu, how many kg of Fe are consumed in removing all the Cd from 4.50×10^6 L of water?

20. Consider municipal drinking water from two different kinds of sources, one a flowing, well-aerated stream with a heavy load of particulate matter, and the other an anoxic groundwater. Describe possible differences in the water treatment strategies for these two sources of water.

21. Using appropriate chemical reactions for illustration, show how calcium present as the dissolved HCO_3^- salt in water is easier to remove than other forms of hardness, such as dissolved $CaCl_2$.

22. Suggest a source of microorganisms to use in a waste treatment process based on soil. Where should an investigator look for microorganisms to use in such an application? What are some kinds of wastes for which soil is particularly unsuitable as a treatment medium?

23. Biologically active masses of microorganisms are used in the secondary treatment of municipal wastewater. Describe three ways of supporting a growth of the biomass, contacting it with wastewater, and exposing it to air.

24. For water to be used commercially, label each of the following as external treatment (ex) or internal treatment (in): ( ) aeration, ( ) addition of inhibitors to prevent corrosion, ( ) adjustment of pH, ( ) filtration, ( ) clarification, ( ) removal of dissolved oxygen by reaction with hydrazine or sulfite, and ( ) disinfection for food processing

25. With respect to wastewater treatment, label each of the following as primary treatment (pr), secondary treatment (sec), or tertiary treatment (tert): ( ) screening, ( ) comminuting, ( ) grit removal, ( ) BOD removal, ( ) activated carbon filtration removal of dissolved organic compounds, and ( ) removal of dissolved inorganic materials.

26. Both activated-sludge waste treatment and natural processes in streams and bodies of water remove degradable material by biodegradation. Explain why activated-sludge treatment is so much more effective.

27. Of the following water treatment processes, the one that least belongs with the rest is (a) removal of scum and solid objects, (b) clarification, (c) filtration, (d) degradation with activated sludge, (e) activated carbon adsorption, and (f) disinfection.

28. Explain why complete physical–chemical wastewater treatment systems are better than biological systems in dealing with toxic substances and overloads.

29. What are the two major ways in which dissolved carbon (organic compounds) are removed from water in industrial wastewater treatment. How do these two approaches differ fundamentally?

30. What is the reaction for the hydrolysis of aluminum ion in water? How is this reaction used for water treatment?

31. Explain why coagulation is used with filtration.

32. What are two major problems that arise from the use of excessively hard water?

33. Show with chemical reactions how the removal of bicarbonate hardness with lime results in a net removal of ions from solution, whereas removal of non-bicarbonate hardness does not.

34. What two purposes are served by adding CO_2 to water that has been subjected to lime–soda softening?

35. Why is cation exchange normally used without anion exchange for softening water?

36. Show with chemical reactions how oxidation is used to remove soluble iron and manganese from water.

37. Show with chemical reactions how lime treatment, sulfide treatment, and cementation are used to remove heavy metals from water.

38. How is activated carbon prepared? What are the chemical reactions involved? What is remarkable about the surface area of activated carbon?

39. How is the surface of the membrane employed involved in the process of reverse osmosis?

40. Describe with a chemical reaction how lime is used to remove phosphate from water. What are some other chemicals that can be used for phosphate removal?

41. Why is nitrification required as a preliminary step in removal of nitrogen from water by biological denitrification?

42. What are some possible beneficial uses for sewage sludge? What are some of its characteristics that may make such uses feasible?

43. Give one major advantage and one major disadvantage of using chlorine dioxide for water disinfection.

44. Give one major advantage and one major disadvantage of using ozone for water disinfection.

45. Discuss how soil may be viewed as a natural filter for wastes. How does soil aid waste treatment? How can waste treatment be of benefit to soil in some cases?
46. In treating water for industrial use, consideration is often given to "sequential use of the water." What is meant by this term? Give some plausible examples of sequential use of water.

LITERATURE CITED

1. Manahan, Stanley E., *Water Chemistry: Green Science and Technology of Nature's Most Renewable Resource*, Taylor & Francis/CRC Press, Boca Raton, FL, 2010.
2. Wada, Yoshide, Ludovicus P. H. van Beek, Cheryl. M. van Kempen, Joseph. W. T. M. Reckman, Slavek Vasak, and Marc F. P. Bierkens, Global Depletion of Groundwater Resources, *Geophysical. Research Letters* **37**, 26–39, Preprint L20402 (2010)
3. Basic Information about *E. coli* 0157:H7 in Drinking Water, http://water.epa.gov/drink/contaminants/basicinformation/ecoli.cfm, 2012.
4. Sharma, Virender K, Ferrate: Green Chemistry Disinfectant and Oxidant in Water Treatment, Abstracts of Papers, 242nd American Chemical Society National Meeting & Exposition, Denver, CO, United States, August 28–September 1, 2011.
5. Miller, Wade, Integrated Concepts in Water Reuse: Managing Global Water Needs, *Desalination* **187**, 65–75 (2006).
6. Bhargav, J. S., K. Sivasankar, and V. Sambasiva Rao, Groundwater Quality in Artificial Recharge—Some Hypothetical Views, *Journal of Applied Geochemistry* **8**, 85–99 (2006).
7. Oweis, Theib Y., Dieter Prinz, and Ahmed Y. Hachum., *Rainwater Harvesting for Agriculture in Dry Areas*, Taylor & Francis/CRC Press, Boca Raton, FL, 2012.

SUPPLEMENTARY REFERENCES

Adin, Avner and Takashi Asano, The Role of Physical Chemical Treatment in Wastewater Reclamation and Reuse, *Water Science and Technology* **37**, 79–80 (1998).
American Water Works Association, *Reverse Osmosis and Nanofiltration*, American Water Works Association, Denver, CO, 1998.
American Water Works Association, *Water Treatment*, 3rd ed., American Water Works Association, Denver, CO, 2003.
Amjad, Sahid, *The Science and Technology of Industrial Water Treatment*, Taylor & Francis/CRC Press, Boca Raton, FL, 2010.
Baruth, Edward E., Ed., *Water Treatment Plant Design*, 4th ed, McGraw-Hill, New York, 2005.
Benjamin, Mark M., *Water Chemistry*, Waveland Press, Long Grove, IL, 2010.
Bergman, Robert, *Reverse Osmosis and Nanofiltration*, 2nd ed., American Water Works Association, Denver, CO, 2007.
Bitton, Gabriel, *Wastewater Microbiology*, Wiley-Liss, New York, 1999.
Brezonik, Patrick L., and William A. Arnold, *Water Chemistry: An Introduction to the Chemistry of Natural and Engineered Aquatic System*, Oxford University Press, Oxford, UK, 2011.
Cornwell, David A., *Water Treatment Residuals Engineering*, AWWA Research Foundation and American Water Works Association, Denver, CO, 2006.
Crittenden, John C., *Water Treatment Principles and Design*, 2nd ed., Wiley, Hoboken, NJ, 2005.
Drinan, Joanne E., and Frank Spellman, *Water and Wastewater Treatment*, 2nd ed., Taylor & Francis/CRC Press, Boca Raton, FL, 2012.
Faust, Samuel D. and Osman M. Aly, Eds., *Chemistry of Water Treatment*, 2nd ed., CRC Press, Boca Raton, FL, 1998.
Grady, C. P. Leslie, Glen T. Daigger, Nancy G. Love, and Carlos D. M. Filipe, *Biological Wastewater Treatment*, 3rd ed, Taylor & Francis/CRC Press, Boca Raton, FL, 2011.
Hammer, Mark, *Water and Wastewater Technology*, 6th ed., Prentice Hall, Upper Saddle River, NJ, 2007.
Hendricks, David, *Fundamentals of Water Treatment Unit Processes*, 2nd ed., Taylor & Francis/CRC Press, Boca Raton, FL, 2010.
Lauer, William C., *Desalination of Seawater and Brackish Water*, American Water Works Association, Denver, CO, 2006.

Mays, Larry W., *Water Distribution Systems Handbook*, McGraw-Hill, New York, 1999.

Norman, Terry and Gary Banuelos, *Phytoremediation of Contaminated Soil and Water*, CRC Press/Lewis Publishers, Boca Raton, FL, 1999.

Pizzi, Nicholas G., *Water Treatment Operator Handbook*, American Water Works Association, Denver, CO, 2005.

Polevoy, Savely, *Water Science and Engineering*, Blackie Academic & Professional, London, 1996.

Rao, D. G., R. Senthilkumar, J. Anthony Byrne, and S. Feroz, *Wastewater Treatment Advanced Processes and Technologies*, 2nd ed., Taylor & Francis/CRC Press, Boca Raton, FL, 2012.

Rice, Rip G., *Ozone Drinking Water Treatment Handbook*, CRC Press/Lewis Publishers, Boca Raton, FL, 1999.

Roques, Henri, *Chemical Water Treatment: Principles and Practice*, VCH, New York, 1996.

Sarai, Darshan Singh, *Water Treatment Made Simple for Operators*, Wiley, Hoboken, NJ, 2006.

Sincero, Arcadio P., *Physical-Chemical Treatment of Water and Wastewater*, CRC Press, Boca Raton, FL, 2003.

Spellman, Frank R., *The Science of Water: Concepts and Applications*, 2nd ed., Taylor & Francis/CRC Press, Boca Raton, FL, 2007.

Spellman, Frank R., *Handbook of Water and Wastewater Treatment Plant Operations*, 2nd ed., Taylor & Francis/CRC Press, Boca Raton, FL, 2008.

Steiner, V., UV Irradiation in Drinking Water and Wastewater Treatment for Disinfection, *Wasser Rohrbau* **49**, 22–31 (1998).

Stevenson, David G., *Water Treatment Unit Processes*, Imperial College Press, London, 1997.

Trimble, Stanley W., *Encyclopedia of Water Science*, 2nd ed., Taylor & Francis/CRC Press, Boca Raton, 2008.

Water Environment Federation, *Industrial Wastewater Management, Treatment, and Disposal*, 3rd ed., McGraw-Hill Professional, New York, 2008.

White, George C., *Handbook of Chlorination and Alternative Disinfectants*, John Wiley & Sons, New York, 1999.

6 Environmental and Toxicological Chemistry of the Atmosphere

6.1 ATMOSPHERE: AIR TO BREATHE AND MUCH MORE

We live and breathe in the atmosphere, a sea of gas composed mostly of elemental nitrogen gas (N_2), O_2, and water vapor. The fundamental properties of gases determine the properties of the atmosphere. Recall that gases consist of molecules and (in the case of noble gases) atoms with large amounts of space between them. The gas molecules are in constant, rapid motion, which causes gases to exert **pressure**. The motion of gas molecules becomes more rapid with increasing **temperature**. Due to their constant motion, gas molecules move by a process called **diffusion**. The relationships among the amount of a gas in moles and its volume, temperature, and pressure can be calculated by the gas laws discussed in Chapter 19.

Figure 6.1 shows some of the main features and aspects of the atmosphere and its relationship to other environmental spheres. Having a total mass of about 5.15×10^{15} t (only about one-millionth of Earth's total mass), the atmosphere is a layer of gases blanketing Earth, whose density diminishes rapidly with increasing altitude. More than 99% of the atmosphere's mass is within 40 km of Earth's surface, with the majority of the air lying below a 10 km altitude (compared to Earth's diameter of almost 13,000 km). A person exposed to air at the approximately 13,000 m altitude at which commercial jet aircrafts fly can remain conscious for only about 15 seconds without supplementary oxygen. There is no clearly defined upper limit to the atmosphere, which keeps getting thinner with increasing altitude. A practical upper limit may be considered to be an altitude of about 1000 km, above which air molecules can be lost to space (a region called the **exosphere**). If Earth were the size of a classroom globe, virtually all the mass of the atmosphere would be contained in a layer the thickness of the coat of varnish on the globe.

The atmosphere nurtures life on Earth in many important respects. Some of the main ones are as follows:

- The atmosphere constitutes much of Earth's natural capital because it serves as a source of materials, its regulation of climate, its protective function, and other attributes.
- The atmosphere is a source of molecular O_2 for all organisms that require it including humans and all other animals. In addition, pure oxygen, argon, and neon are extracted from the atmosphere for industrial uses.
- At approximately 0.039% of carbon dioxide, CO_2, the atmosphere is the source of carbon that plants and other photosynthetic organisms use to synthesize biomass.
- Consisting mostly of molecular N_2, the atmosphere serves as a source of nitrogen that is an essential component of protein and other biochemicals as well as a constituent of a variety of synthetic chemicals. Organisms "fix" this nitrogen in the biosphere chemically by the action of bacteria such as *Rhizobium*, and it is fixed synthetically in the anthrosphere under much more severe conditions of temperature and pressure.

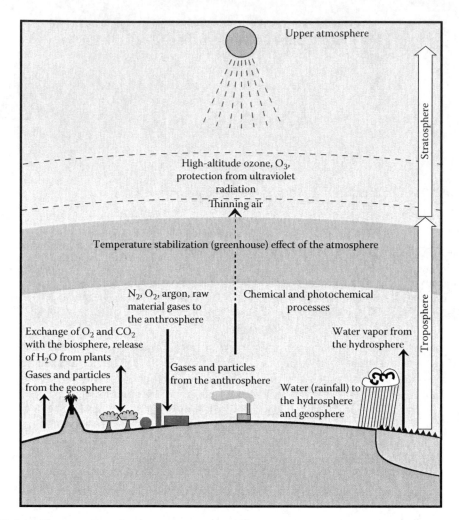

FIGURE 6.1 The atmosphere is a layer of gas, most of which is concentrated in a very thin region just above Earth's surface. It serves as a source of some essential materials; protects life on Earth from deadly electromagnetic radiation; transports water as part of the hydrologic cycle; participates in the carbon, oxygen, and sulfur cycles; and is an essential part of natural capital.

- The atmosphere acts as a blanket to keep Earth's surface at an average temperature of about 15°C at sea level and within a temperature range that enables life to exist (the good greenhouse effect).
- Earth's atmosphere absorbs very-short-wavelength ultraviolet (UV) radiation from the sun and space, which, if it reaches organisms on Earth's surface, would tear apart the complex biomolecules essential for life. In this respect, the stratospheric ozone layer is of particular importance.
- The atmosphere contains and carries water vapor evaporated from oceans that forms rain and other kinds of precipitation over land in the hydrologic cycle (Chapter 3, Figure 3.2).

6.2 REGIONS OF THE ATMOSPHERE

Figure 6.2 shows the stratification of the atmosphere and the influence of electromagnetic solar radiation in forming the layers of the atmosphere. Photochemical reactions in which energetic photons of UV solar radiation (represented as $h\nu$, where h is Planck's constant and ν is the frequency of the radiation) may break chemical bonds in atmospheric air molecules to produce O atoms as well as ions.

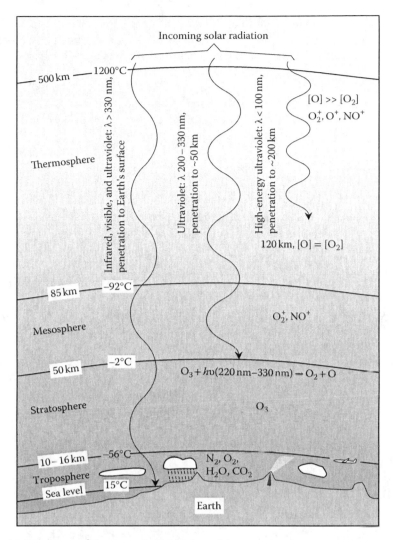

FIGURE 6.2   Major regions of the atmosphere (not to scale).

Except for a few aviators who fly briefly into the stratosphere, living organisms experience only the lowest layer of the atmosphere called the troposphere, which is characterized by decreasing temperature and density with increasing altitude. The troposphere extends from Earth's surface, where the average temperature is 15°C, to about 11 km (the approximate cruising altitude of commercial jet aircrafts), where the average temperature is –56°C. Most of the long-wavelength (high-frequency) UV solar radiation is filtered out before reaching the troposphere so that it has a relatively uniform composition of molecular N_2, O_2, and CO_2, and the noble gas argon. However, levels of H_2O vary appreciably with time, location, and altitude in the troposphere.

The very cold layer of air at the top of the troposphere, called the tropopause, serves as a barrier that causes water vapor to condense to ice so that it cannot reach altitudes at which it would photodissociate through the action of intense high-energy UV radiation. If this happens, the hydrogen produced would escape Earth's atmosphere and be lost. (Much of the hydrogen and helium gases originally present in Earth's atmosphere were lost by this process.)

Above the troposphere is the **stratosphere** in which the average temperature increases from about –56°C at its lower boundary that ranges between 10 and 16 km in altitude to about –2°C at its upper limit at around 50 km altitude. The stratosphere is warmed by the energy of intense solar radiation impinging on air molecules. Because this radiation can break the bonds holding O_2 molecules together, at higher altitudes the stratosphere maintains a significant level of O atoms and of ozone (O_3) molecules formed by the combination of O atoms with O_2 molecules (see the discussion on ozone layer in this section). The stratosphere is warmed by the UV energy absorbed by both O_2 and O_3 molecules.

A critical part of the stratosphere is the UV-absorbing ozone that it contains. Often referred to as the "ozone layer," stratospheric ozone is dispersed widely over several kilometers of altitude in the stratosphere. Stratospheric ozone actually constitutes only a minuscule fraction of the atmosphere's total mass, although it is absolutely essential in protecting humans and other organisms on Earth from deadly solar UV radiation. Stratospheric ozone and pollutant threats to it are discussed in more detail in Chapter 7, Section 7.9.

The absence of high levels of radiation-absorbing species in the rarified **mesosphere** immediately above the stratosphere results in a further decrease in temperature with increasing altitude to a low of about −92°C at altitudes around 85 km. Extending to the far outer reaches of the atmosphere is the **thermosphere**, in which the highly rarified gas reaches temperatures as high as 1200°C by the absorption of very energetic radiation of wavelengths less than approximately 200 nm by gas species in this region. Radiation energetic enough to tear electrons away from atmospheric molecules and atoms produces a region in the mesosphere and thermosphere containing ions, which is called the **ionosphere**. The upper regions of the mesosphere and higher define a region, called the exosphere, from which molecules and ions can completely escape the atmosphere.

Earth's atmosphere is crucial in absorbing, distributing, and radiating the enormous amount of electromagnetic energy that comes from the sun. A square meter of surface directly exposed to sunlight unfiltered by air would receive energy from the sun at a power level of 1340 W, the **solar flux** illustrated in Figure 6.3. The incoming radiation in the form of electromagnetic radiation centered in the visible wavelength region with a maximum intensity at a wavelength of 500 nm (1 nm = 10^{-9} m) is largely absorbed and converted to heat in the atmosphere and at Earth's surface, and the heat is reradiated back to space primarily as infrared radiation. The energy balance of the atmosphere is discussed in more detail in Section 6.5.

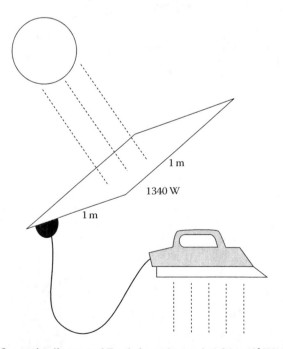

FIGURE 6.3   The solar flux at the distance of Earth from the sun is 1.34×10^3 W/m³.

6.3 ATMOSPHERIC COMPOSITION

At our level in the troposphere, air is a mixture of gases of uniform composition, except for water vapor, which composes 1%–3% of the atmosphere by volume, and some of the trace gases, such as pollutant sulfur dioxide. On a dry basis, air is 78.1% (by volume) N_2, 21.0% O_2, 0.9% argon, and 0.04% carbon dioxide. Trace gases at levels below 0.002% in air include ammonia, carbon monoxide, helium, hydrogen, krypton, methane, neon, nitrogen dioxide, nitrous oxide, ozone, sulfur dioxide, and xenon.

Two other important kinds of substances in the atmosphere are particles and free radicals. Atmospheric particles are discussed in more detail in Section 6.13. Free radicals are reactive fragments of molecules that are usually very reactive because they possess unpaired electrons (see Section 6.9).

By a wide margin, oxygen and nitrogen are the most abundant gases in the atmosphere. Because of the extremely high stability and low reactivity of the N_2 molecule, the chemistry of atmospheric elemental nitrogen is singularly unexciting, although nitrogen molecules are the most common "third bodies" that absorb excess energy from atmospheric chemical reactions, preventing the products of addition reactions in the atmosphere from falling apart. Oxides of nitrogen actively participate in atmospheric chemical reactions. Elemental nitrogen is an important commercial gas extracted from the atmosphere by nitrogen-fixing bacteria and in the industrial synthesis of ammonia. Atmospheric nitrogen is discussed in more detail in Section 6.11.

Oxygen is a reactive species in the atmosphere that reacts to produce oxidation products from oxidizable gases in the atmosphere. Two such species that are particularly important are sulfur dioxide gas, SO_2, and pollutant hydrocarbons. Molecular O_2 does not react with these substances directly; but it reacts with them indirectly through the action of reactive intermediates, especially **hydroxyl radical**, HO·.[1] Atmospheric oxygen is discussed in Section 6.10.

Under most conditions, the third most abundant constituent of the atmosphere is water, H_2O, which is discussed in more detail in Section 6.12. Normally within the range of 1%–3%, the water content of the atmosphere varies significantly with location, season, and time. Most of the atmosphere's water is in the gaseous molecular form, and some of it is in the form of very small liquid droplets that compose clouds.

Although they are usually present at levels well below 1 mg per liter of air, small suspended particles in the atmosphere (see Section 6.13), often called "particulates," are significant atmospheric constituents. Because of their ability to scatter light, particles are very important in determining visibility in the atmosphere. Atmospheric chemical reactions may occur on particle surfaces and within suspended water droplets. Particles are also very important with respect to health effects of the atmosphere, especially so in the case of very small respirable particles that can be drawn deep within the lungs.

6.4 NATURAL CAPITAL OF THE ATMOSPHERE

As discussed in Chapter 1, Section 1.4, Earth's natural capital is its ability to provide materials, protection, and conditions conducive to life including Earth's resources and its ecosystems. A large fraction of Earth's natural capital is in the atmosphere and includes materials, waste assimilative capacity, and esthetics, largely determining the degree to which our surroundings are pleasant and conducive to our existence. The atmosphere's natural capital is discussed in this section.

A huge part of the atmosphere's natural capital is its ability to absorb and protect organisms from destructive UV and other short-wavelength cosmic and solar electromagnetic radiation, which otherwise would make life on Earth impossible. The absorption of longer wavelength infrared radiation by which incoming solar energy is sent back into space leads to the second major protective function of the atmosphere, that is, its ability to maintain surface temperature at a level at which life can thrive.

The atmosphere is a source of essential raw materials, both for organisms and for industrial use, and has major applications in the practice of green chemistry. Plants that provide the foundation of food chains within which all organisms thrive extract the carbon dioxide that they use to build biomass from the atmosphere. Animals and other organisms that perform oxic respiration obtain the molecular O_2 they require from the atmosphere. The refractory N_2 in the atmosphere is converted to biomass and protein nitrogen by bacteria growing in soil and water.

Humans extract elemental oxygen, nitrogen, and other gases from the atmosphere for use in the anthrosphere. Historically, this has been done by distilling cold liquid air, a process that can also produce noble gases neon, krypton, and xenon, if desired. The initial step in air distillation is to compress air to about seven times atmospheric pressure and cool it to remove water vapor and carbon dioxide. Further compression and cooling yields a liquid air product that can then be fractionally distilled to give relatively pure oxygen, nitrogen, and other gases. These can be stored as cold liquids or as compressed gases. Adsorption and permeable membrane processes are now used to isolate nitrogen, oxygen, argon, and neon from air.

Essentially pure oxygen has a number of applications, such as for breathing by people with pulmonary insufficiencies. Although compressed oxygen in tanks used to be the most common source of this gas for breathing, it is now largely extracted from air as needed in apparatuses that use selective adsorption of different gases from air onto adsorbent materials. Huge amounts of elemental O_2 are consumed in steelmaking.

As a generally unreactive gas, pure nitrogen is used as an inert atmosphere that prevents fires and other chemical reactions. Boiling at a frigid −196°C, liquid nitrogen is the most widely used cryogenic liquid. One among its many applications is the preservation of viable human embryos for embryo implantation to produce "test-tube babies." Nitrogen extracted from air is an important raw material for the synthesis of ammonia, NH_3. Ammonia is then used to synthesize industrial chemicals, fertilizers, and explosives.

Noble gas argon from the atmosphere is totally chemically inert and is used industrially, such as in specialized welding processes.

As illustrated in Chapter 3, Figure 3.2, the atmosphere is the conduit by which water is evaporated from oceans and carried over land where it falls as precipitation in the hydrologic cycle. This ability of the atmosphere is an important component of its natural capital, and atmospheric conditions largely determine the quantity, quality, and distribution of water through the hydrologic cycle. Because of variations in atmospheric conditions, the distribution of rainfall is irregular, with excess in some locations and times and deficiencies in others. Hot drought conditions that cause great hardship and even starvation, especially in parts of Africa, are the result of climate conditions in the atmosphere. Sulfur dioxide and nitrogen oxides emitted to the atmosphere as air pollutants produce sulfuric acid and nitric acid, respectively, polluting the hydrosphere with strong acids, killing fish fingerlings, and harming vegetation.

Its ability to assimilate and process materials is an important part of the atmosphere's natural capital and a crucial component of nature's cycles. Transpiration of water from plant leaves is an important route for conveying water from soil to the atmosphere. Oxic respiration by humans and other organisms discharges carbon dioxide to the atmosphere as do forest fires and combustion processes in the anthrosphere. Photosynthetically produced elemental oxygen enters the atmosphere from the biosphere. Pollen and small particles such as smoke or fumes produced by anthrospheric processes are discharged to the atmosphere and are washed out by rain or deposited on Earth's surface. Hydrocarbons and nitrogen oxides from combustion are eventually purged from the atmosphere, often with the intermediate formation of oxidants, aldehydes, and particles characteristic of photochemical smog pollution.

The atmosphere's contribution to esthetics is a major facet of its natural capital. Clear, clean air that is free of visibility-obscuring particles, acidic gases, and ozone, which hinders breathing and irritates eyes, has genuine value including its contribution to good health. Whereas the water that humans use can be purified from muddy, even polluted, sources, air used for breathing must usually be taken as it comes. Humid, foggy air contaminated by acidic constituents and particles is unpleasant and even unhealthy to breathe as is air heated and dried to uncomfortable levels by greenhouse gas emissions.

6.5   ENERGY AND MASS TRANSFER IN THE ATMOSPHERE

As noted in Section 6.2, the flux of energy reaching Earth's atmosphere from the sun as sunlight is 1340 W/m² (Figure 6.3). This means that a square meter of area perpendicular to incoming sun-rays above Earth's atmosphere is receiving solar energy at a rate that is sufficient to power thirteen 100-W lightbulbs plus one 40-W bulb or enough to power an electric iron or a hair dryer set on high. This is an enormous amount of energy. As shown in Figure 6.4, some of the incoming energy reaches Earth's surface; some is absorbed in the atmosphere, warming it; and some is scattered back to space. The electromagnetic solar radiation energy that comes in primarily at a maximum intensity of 500 nm in the visible region must go out, which it does primarily as infrared radiation (with maximum intensity at about 10 μm, primarily between 2 and 40 μm). Water molecules, carbon dioxide, methane, and other minor species in the atmosphere absorb some of the outbound infrared, which eventually is all radiated to space. This temporary absorption of infrared radiation warms the atmosphere—a greenhouse effect.

The fraction of electromagnetic radiation from the sun that is reflected by Earth's surface, called **albedo**, varies with the nature of the surface, from freshly plowed black topsoil with a low albedo of only about 2.5% to fresh snow at about 90%. Human activities in the anthrosphere affect albedo by measures such as cultivating soil and paving large areas.

The maintenance of Earth's heat balance to keep temperatures within limits conducive to life is very complex and not well understood. Geological records show that in times past Earth was sometimes relatively warm and that at other times there were ice ages in which much of Earth's surface was covered by ice a kilometer or two thick. The differences in average Earth temperature between these extremes and the relatively temperate climate conditions that we now enjoy are a matter of only

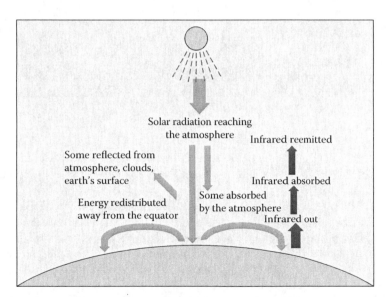

FIGURE 6.4   Some of the solar energy reaching the top of the atmosphere penetrates to Earth's surface; some is absorbed in the atmosphere, which warms it; and some is scattered by the atmosphere and by Earth's surface. Outgoing energy is in the infrared region. Some of it is temporarily absorbed by the atmosphere before being radiated to space, causing a warming effect (greenhouse effect). The equatorial regions receive the most energy, part of which is redistributed by warm air masses and latent heat in water vapor away from the equator.

a few degrees. It is also known that massive volcanic eruptions and almost certainly hits by large asteroids have caused cooling of the atmosphere that has lasted for a year or more. As addressed in more detail in Chapter 8, there is now concern that anthropogenic gas emissions, particularly of carbon dioxide from fossil fuel combustion, may be having a warming effect on the atmosphere.

Earth receives solar energy most directly at the equator, so equatorial regions are warmer than regions farther north and south. A significant fraction of this energy moves away from the equator. This is largely done by **convection** in which heat is carried by masses of air. Such heat can be in the form of **sensible heat** from the kinetic energy of rapidly moving air molecules (the faster their average velocities, the higher the temperature). Heat can also be carried as **latent heat** in the form of water vapor. The **heat of vaporization** of water is 2259 J/g, which means that 2259 J of heat energy are required to evaporate a gram of water without raising its temperature. This is a very high value, which means that the evaporation of ocean water by solar energy falling on it in warmer regions requires an enormous amount of heat. This vapor may be carried elsewhere and may condense to form rainfall. The heat energy thus released raises the temperature of the surrounding atmosphere.

6.6 METEOROLOGY, WEATHER, AND CLIMATE

The movement of air masses, cloud formation, and precipitation in the atmosphere are covered by the science of **meteorology**. In considering the environmental chemistry of the atmosphere, it is important to take meteorology into account because (1) meteorologic conditions strongly influence atmospheric chemistry and air pollution; and (2) emissions to the atmosphere, especially those responsible for atmospheric warming and particle formation, can significantly influence meteorology, weather, and climate. Meteorologic phenomena have a strong effect on atmospheric chemistry by processes such as the following:

- Movement of air pollutants from one place to another, such as the movement of air pollutant sulfur dioxide from the U.S. Ohio River valley to New England and southern Canada, where it forms acid rain.
- Conditions under which stagnant pollutant air masses remain in place so that secondary pollutants, such as photochemical smog, can form.
- Precipitation, which can carry acidic compounds from the atmosphere to Earth's surface in the form of acid rain.
- Active chlorine originally from chlorofluorocarbons (CFCs) is sequestered in stratospheric clouds during the dark winter above the South Pole and then released during the Antarctic spring (September and October), destroying protective stratospheric ozone and resulting in the Antarctic ozone hole.

Atmospheric chemical processes can influence meteorological phenomena. One of the most obvious examples of this influence is the formation of rain droplets around pollutant particles in the atmosphere.

Weather refers to relatively short-term variations in the state of the atmosphere as expressed by temperature, cloud cover, precipitation, relative humidity, atmospheric pressure, and wind. Long-term weather conditions are called **climate**. Weather is driven by redistribution of energy in the atmosphere. A particularly important aspect of this redistribution is the energy released when precipitation forms, which is the main driving force behind thunderstorms and tropical storms.

A very obvious manifestation of weather consists of very small droplets of liquid water composing **clouds**. These very small droplets may coalesce under appropriate conditions to form raindrops large enough to fall from the atmosphere. Clouds may absorb infrared radiation from Earth's surface, warming the atmosphere; but they also reflect visible light, which has a cooling effect. Pollutant particles are instrumental in forming clouds. Among the more active kinds of cloud-forming pollutants are atmospheric strong acids, particularly H_2SO_4.

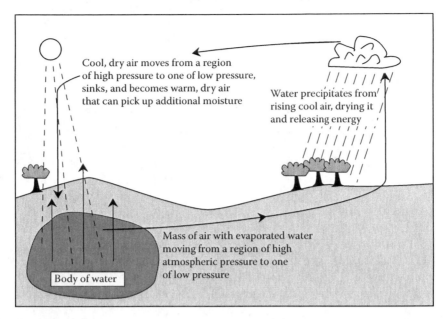

Cool, dry air moves from a region of high pressure to one of low pressure, sinks, and becomes warm, dry air that can pick up additional moisture

Water precipitates from rising cool air, drying it and releasing energy

Mass of air with evaporated water moving from a region of high atmospheric pressure to one of low pressure

Body of water

FIGURE 6.5 Circulation patterns involved in the movement of air masses and water; uptake and release of solar energy as latent heat in water vapor.

Air masses characterized by pressure, temperature, and moisture contents flow from regions of high atmospheric pressure to regions of low pressure, aspects that are illustrated in Figure 6.5. The boundaries between air masses are called **fronts**. The movement of air associated with moving air masses is **wind**, and vertically moving air is an **air current**. Wind is involved in the movement of pollutants from one place to another and is also responsible for dispersing pollutants to harmless levels. A tremendous amount of energy is contained in wind, and it is now being widely used to generate electricity sustainably.

6.6.1 Global Weather

The factors that determine and describe the movement of bodies of air are involved in the massive movement of air, moisture, and energy that occurs globally. The central feature of global weather is the redistribution of solar energy that falls unequally on Earth at different latitudes (relative distances from the equator and poles). Consider Figure 6.4. Sunlight and the energy flux from it are most intense at the equator because, averaged over the seasons, solar radiation falls perpendicular to Earth's surface at the equator. With increasing distance from the equator (higher latitudes), the angle is increasingly oblique and more of the energy-absorbing atmosphere must be traversed so that progressively less energy is received per unit area of Earth's surface. The net result is that equatorial regions receive a much greater share of solar radiation, progressively less is received farther from the equator, and the poles receive a comparatively minuscule amount. The excess heat energy in the equatorial regions causes the air to rise. The air ceases to rise when it reaches the stratosphere because in the stratosphere the air becomes warmer with higher elevation. As the hot equatorial air rises in the troposphere, it cools by expansion and loss of water and then sinks again. The air circulation patterns in which this occurs are called **Hadley cells**. As shown in Figure 6.6, there are three major groupings of these cells, which result in very distinct climatic regions on Earth's surface. The air in the Hadley cells does not move straight north and south but is deflected by Earth's rotation and by the atmosphere's contact with the rotating Earth; this is the **Coriolis effect**, which results in spiral-shaped air circulation patterns that are called cyclonic or anticyclonic depending on

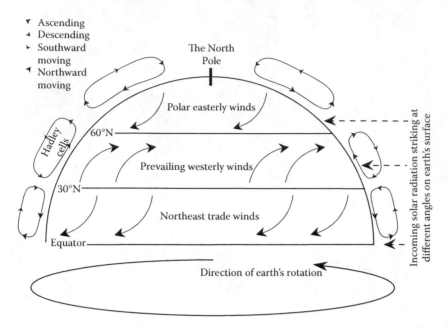

FIGURE 6.6 Global circulation of air in the Northern Hemisphere.

the direction of rotation. These give rise to different directions of prevailing winds depending on latitude. The boundaries between the massive bodies of circulating air shift markedly over time and season, resulting in significant weather instability.

The movement of air in Hadley cells interacts with other atmospheric phenomena to produce global **jet streams** that are, in a sense, shifting rivers of air that may be several kilometers deep and several tens of kilometers wide. Jet streams move through discontinuities in the tropopause (the boundary between the stratosphere and the troposphere) generally from west to east at velocities around 200 km/h (well over 100 mph); in so doing, they redistribute huge amounts of air and have a strong influence on weather patterns.

The aforementioned movement of air combines with other atmospheric phenomena to shift massive amounts of energy over long distances on Earth. If it was not for this effect, the equatorial regions would be unbearably hot and the regions closer to the poles intolerably cold. About half of the heat that is redistributed is carried as sensible heat by air circulation, almost one-third is carried by water vapor as latent heat, and the remaining approximately 20% is moved by ocean currents.

6.7 ATMOSPHERIC INVERSIONS AND ATMOSPHERIC CHEMICAL PHENOMENA

The complicated movement of air across Earth's surface is a crucial factor in the creation and dispersal of air pollutants and in atmospheric chemical phenomena generally. When air movement ceases, stagnation can occur, with the resultant buildup of atmospheric pollutants in localized regions. Although the temperature of air relatively near Earth's surface normally decreases with increasing altitude, certain atmospheric conditions can result in the opposite condition—increasing temperature with increasing altitude. Such conditions are characterized by high atmospheric stability and are known as **temperature inversions**, as illustrated in Figure 6.7. Because they limit the vertical circulation of air, temperature inversions result in air stagnation and the trapping of air pollutants in localized areas. Topographical features, such as a mountain range that limits horizontal

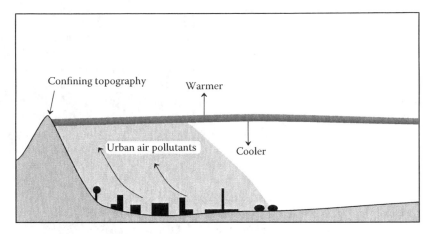

FIGURE 6.7 Illustration of pollutants trapped in a temperature inversion.

air movement, may make temperature inversions much more effective in trapping polluted masses of air. These conditions occur in the Los Angeles basin, noted for photochemical smog formation, which is discussed in Chapter 7.

Inversions can occur in several ways. In a sense, the whole atmosphere is inverted by the warm stratosphere, which floats atop the troposphere with relatively little mixing. An inversion can form from the collision of a warm air mass (warm front) with a cold air mass (cold front). The warm air mass overrides the cold air mass in the frontal area, producing the inversion. **Radiation inversions** are likely to form in still air at night when Earth is no longer receiving solar radiation. The air closest to Earth cools faster than the air higher in the atmosphere, which remains warm and thus less dense. Furthermore, the cooler surface air tends to flow into valleys at night, where it is overlaid by warmer, less dense air. **Subsidence inversions**, often accompanied by radiation inversions, can become very widespread. These inversions can form in the vicinity of a surface high-pressure area when high-level air subsides to take the place of surface air blowing out of the high-pressure zone. The subsiding air is warmed as it compresses and can remain as a warm layer several hundred meters above ground level. A **marine inversion** is produced during the summer months when cool air laden with moisture from the ocean blows onshore and under warm, dry inland air.

6.8 CLIMATE, MICROCLIMATE, AND MICROATMOSPHERE

Defined in Section 6.6 as consisting of long-term weather patterns over large geographical areas, perhaps the single most important influence on Earth's environment is climate. As a general rule, climatic conditions are characteristic of a particular region. This does not mean that climate remains the same throughout the year, of course, because it varies with season. One important example of such variations is the **monsoon**, that is, seasonal variations in wind patterns between oceans and continents. The climates of Africa and the Indian subcontinent are particularly influenced by monsoons. In the latter, for example, summer heating of the Indian landmass causes air to rise, thereby creating a low-pressure area that attracts warm, moist air from the ocean. This air rises on the slopes of the Himalayan mountains, which also block the flow of colder air from the north; moisture from the air condenses, and monsoon rains carrying enormous amounts of precipitation fall. Thus, from May until sometime into August, summer monsoon rains fall in India, Bangladesh, and Nepal. Reversal of the pattern of winds during the winter months causes these regions to have a dry season but produces winter monsoon rains in the Philippine Islands, Indonesia, New Guinea, and Australia.

Summer monsoon rains are responsible for the growth of tropical rain forests in Central Africa. Farther north in Africa is the arid Sahara desert. The interface between the Sahara desert and the wetter regions to the south encompasses the Sahel, a belt extending across the continent from the Atlantic Ocean in the west to the Red Sea in the east with a width that varies from a few hundred kilometers to a thousand kilometers and a total area of about 3 million square kilometers. The Sahel consists mainly of grasslands and savannas composed of grasslands interspersed with trees. The boundaries of the Sahel vary with time. Under good climate conditions, the Sahel supports an agricultural economy, a significant portion of which is based on grazing livestock. A periodic problem is drought in the region, which results in the shifting of the Sahel to the south with the northern belt of it overtaken by the Sahara desert, the result of diminished rainfall. This has caused starvation conditions in some years, and a major concern is that global warming may result in permanent desertification of large areas of the Sahel.

A significant feature of the Sahel is the Sahel dust zone that develops predominantly from December to April each year.[2] Characterized by frequent dust storms and high levels of atmospheric dust blown mostly from the Sahara desert, the Sahel dust zone is located in the latitudes of 10°N–16°N and extends for almost the entire width of Africa from east to west.

It is known that there are fluctuations, cycles, and cycles imposed on cycles in climate. The causes of these variations are not completely understood, but they are known to be substantial, even devastating to civilization. The last **ice age**, which ended only about 10,000 years ago and which was preceded by several similar ice ages, produced conditions under which much of the present landmass of the Northern Hemisphere was buried under thick layers of ice and, thus, uninhabitable. A "mini ice age" occurred during the 1300s, causing crop failures and severe hardship in northern Europe. In modern times, the El Niño Southern Oscillation occurs with a period of several years when a large, semipermanent tropical low-pressure area shifts into the Central Pacific region from its more common location in the vicinity of Indonesia. This shift modifies prevailing winds, changes the pattern of ocean currents, and affects the upwelling of ocean nutrients with profound effects on weather, rainfall, and fish and bird life over a vast area of the Pacific from Australia to the west coasts of South and North America.

6.8.1 HUMAN MODIFICATIONS OF THE ATMOSPHERE

Although Earth's atmosphere is huge and has an enormous ability to resist and correct for detrimental change, human activities have reached a point at which they are significantly affecting the atmosphere and Earth's climate. One such way is by the emission of large quantities of carbon dioxide and other greenhouse gases into the atmosphere, such that global warming may occur and cause substantial climatic change. Another way is through the release of gases, particularly CFCs (Freons) that may cause the destruction of vital stratospheric ozone.[3]

6.8.2 MICROCLIMATE

Section 6.6 describes climate on a large scale, ranging up to global dimensions. The climate that organisms and objects on the surface are exposed to close to the ground, under rocks, and surrounded by vegetation is often quite different from the surrounding macroclimate. Such highly localized climatic conditions are termed the **microclimate**. Microclimate effects are largely determined by the uptake and loss of solar energy very close to Earth's surface and by the fact that air circulation due to wind is much lower at the surface. During the day, solar energy absorbed by relatively bare soil heats the surface, but it is lost only slowly because of the very limited air circulation at the surface. This provides a warm blanket of surface air several centimeters thick, and an even thinner layer of warm soil. At night, radiative loss of heat from the surface of soil and vegetation can result in surface temperatures several degrees colder than the air about 2 m above ground level. These lower temperatures result in the condensation of **dew** on vegetation and the soil surface, thus providing a relatively more moist microclimate near ground level. Heat absorbed during early morning evaporation of dew tends to prolong the period of cold experienced right at the surface.

Vegetation substantially affects microclimate. In relatively dense growths, circulation can be virtually zero at the surface because vegetation severely limits convection and diffusion. The crown surface of vegetation intercepts most of the solar energy so that maximum solar heating in heavily vegetated areas may occur a significant distance up from Earth's surface. The region below the crown surface of vegetation thus becomes one of relatively stable temperature. In addition, in a dense growth of vegetation, most of the moisture loss is not from evaporation from the soil surface but rather from transpiration from plant leaves. The net result is the creation of temperature and humidity conditions that provide a favorable living environment for a number of organisms, such as insects and rodents. Some buildings are now constructed with roofs covered with vegetated soil that creates a favorable microclimate insulated by soil and vegetation from outside temperature extremes. Such roofs obviously have to be made very strong to prevent collapse. Also, the vegetation requires constant attention; if it dies, the roof readily becomes a layer of baked dirt during summer droughts and turns to mud during wet seasons.

Another factor influencing microclimate is the degree to which the slope of land faces north or south. South-facing slopes of land in the Northern Hemisphere receive greater solar energy. Advantage has been taken of this phenomenon in restoring land strips mined for brown coal in Germany by terracing the land such that the terraces have broad south slopes and very narrow, steep north slopes. On the south-sloping portions of the terrace, the net effect has been to extend the short summer growing season by several days, thereby significantly increasing crop productivity. In areas where the growing season is longer, better growing conditions may exist on a north slope because it is less subject to temperature extremes and to loss of water by evaporation and transpiration.

6.8.3　Effects of Urbanization on Microclimate

A particularly marked effect on microclimate is that induced by urbanization. In a rural setting, vegetation and bodies of water have a moderating effect, absorbing modest amounts of solar energy and releasing them slowly. The stone, concrete, and asphalt pavements of cities have an opposite effect, strongly absorbing solar energy and reradiating heat back to the urban microclimate. Rainfall is not allowed to accumulate in ponds, but it is drained away as rapidly and efficiently as possible. Human activities generate significant amounts of heat and produce large quantities of CO_2 and other greenhouse gases that retain heat. The net result of these effects is that a city is capped by a **heat dome** in which the temperature is as much as 5°C warmer than the surrounding rural areas such that large cities have been described as "heat islands." The rising warmer air over a city brings in a breeze from the surrounding area and causes a local greenhouse effect that probably is largely counterbalanced by the reflection of incoming solar energy by particulate matter above cities. Overall, compared with climatic conditions in nearby rural surroundings, the city microclimate is warmer, foggier, overlaid with more cloud cover for a greater percentage of the time, and subject to more precipitation, although generally the city microclimate is less humid.

6.8.4　Microatmosphere

The term microatmosphere has been used to describe enclosed bodies of gas as different as minuscule cavities produced in metals from radiation damage in nuclear reactors and oxygen-free bags for storing prepared food and probably other things as well. Here, **microatmosphere** is used as a term to describe a more or less isolated body of air or gas such as in a room, in a workplace, or at surface level of soil (the surface-level microclimate). A microatmosphere can be particularly important when it contains the air that humans must breathe. In rare cases, the microatmosphere may be depleted in oxygen, most commonly by purging with nitrogen. Usually, several deaths occur in the United States each year from nitrogen asphyxiation. Such an incident occurred in 1981 shortly before the first space shuttle mission when two technicians entered a compartment of the orbiter that had been filled with nitrogen as a precaution against fire; both lost consciousness and one died.

Improper handling of liquid nitrogen has led to microatmospheres deficient in oxygen. Such an incident occurred in 1999 in the Human Genetics Unit of the Western General Hospital in Edinburgh, Scotland, when an experienced laboratory worker was overcome and died while filling flasks with liquid nitrogen. Another fatality occurred at the Australian Animal Health Laboratories in Geelong, Australia, when the failure of a ventilation system in a sample storage room where samples were stored in liquid nitrogen led to the accumulation of a nitrogen atmosphere in the facility and the death by nitrogen asphyxiation of a laboratory staff worker who entered the area.

The microatmosphere to which humans are exposed can present hazards due to the presence of toxic substances. As an example, consider the tragic case of workers who sandblasted jeans in Turkey to give the jeans a "distressed" appearance.[4] In a study covering the period 2001–2009 of 32 young male textile workers who reported respiratory distress, it was found that 6 of the workers had died and 16 others had disabling lung damage likely to greatly shorten life expectancy because of silicosis resulting from the accumulation of finely divided sand particles in the lungs. The workers were employed in small-time operations and worked an average of 66 hours per week. Because of the possibility of silicosis, sandblasting jeans has long been banned in the United States and Europe.

The microatmosphere inside buildings is an important aspect of air pollution, for example, with respect to potentially toxic or allergenic vapors released from varnishes, wall paneling, and carpets. Indoor air pollution is discussed in more detail in Chapter 7, Section 7.10.

6.9   ATMOSPHERIC CHEMISTRY AND PHOTOCHEMICAL REACTIONS

Atmospheric chemistry describes chemical processes that occur in the atmosphere.[5] Atmospheric chemistry is a broad and complex science, covering a wide range of levels from global atmospheric chemistry extending tens of kilometers above Earth's surface down to room level. Indoor air chemistry may be particularly important with respect to human health, including occupational health. Much of what is known about atmospheric chemistry is of relatively recent origin. Although "smog" was long used as a term to describe the combination of smoke and fog (laced with choking sulfur dioxide), such as the smog that killed several thousand people in a severe air pollution episode in London in 1952, the term was applied to a much different form of air pollution characterized by lowered visibility in southern California after around 1950. The Dutch chemist Arie Jan Haagen-Smit, a Caltech faculty member, deduced that this photochemical smog, a brew of very fine particles along with ozone, nitric acid, peroxyacetyl nitrate, aldehydes, and other noxious chemicals, was produced by the action of sunlight on an atmosphere contaminated with hydrocarbons and nitrogen oxides, both emitted by automobile exhausts. (Appointed by Governor Ronald Reagan as the first chairman of the California Air Resources Board in 1968, Dr. Haagen-Smit was removed from this post in 1974 for refusing to obey the governor's directive to ease up on pollution control directives.)

It took until about 1970 for the importance of the hydroxyl radical, HO·, a very reactive intermediate because of its unpaired electron, to be recognized as a crucial intermediate in tropospheric chemistry and smog formation. As discussed later in this section, in the 1970s Paul Crutzen, Mario Molina, and F. S. Rowland (the three of whom received the Nobel Prize for their work in 1995) showed that CFCs, which are extremely stable and unreactive in the troposphere, could be photochemically dissociated by UV radiation in the stratosphere and the chlorine atom product could destroy stratospheric ozone, which is crucial for protection from UV radiation from the sun. One of the most ominous symptoms of ozone layer depletion was the Antarctic ozone hole discovered in 1985. This phenomenon was found to involve reactions on stratospheric ice cloud crystals and pointed to the importance of the challenging science of heterogeneous particle chemistry in the atmosphere. Another relatively recent development is the connection between atmospheric chemistry and global warming, a phenomenon generally attributed to emissions of carbon dioxide but now gaining recognition for the important contributions of methane, ozone, CFCs, nitrous oxide (N_2O), and particles.

One notable aspect of atmospheric chemistry is that it occurs largely in the gas phase where molecules are relatively far apart, so a molecule or a fragment of a molecule (a radical) may travel some distance before bumping into another species with which it reacts. This is especially true in the highly rarefied regions of the stratosphere and above.

A second major aspect of atmospheric chemistry is the occurrence of **photochemical reactions** that are initiated when a photon (essentially a packet of energy associated with electromagnetic radiation) of UV radiation is absorbed by a molecule. The energy of a photon, E, is given by $E = h\nu$, where h is Planck's constant and ν is the frequency of the radiation. Electromagnetic radiation of a sufficiently short wavelength breaks chemical bonds in molecules, leading to the formation of reactive species that can participate in reaction sequences called **chain reactions**.

An important example of a photochemical reaction is the photochemical dissociation of O_2 molecules in the stratosphere at altitudes above about 10 km that occurs when the molecules absorb highly energetic UV radiation from the sun

$$O_2 + h\nu \rightarrow 2O \tag{6.1}$$

followed by the reaction of the O atoms to produce ozone, O_3, where M is a third body, usually a molecule of N_2, that absorbs energy released by the reaction and prevents the ozone molecule from coming apart:

$$O + O_2 + M \rightarrow O_3 + M \tag{6.2}$$

These two reactions are responsible for the essential **stratospheric ozone layer**, which is discussed in more detail in Section 6.7.

An example of an important chain reaction sequence that begins with the photochemical dissociation of a molecule is the one that occurs when CFCs get into the stratosphere. **Chlorofluorocarbons**, trade name Freons, consist of carbon atoms to which are bonded fluorine and chlorine atoms. Noted for their extreme chemical stability and low toxicities, they were once widely used as refrigerant fluids in air conditioners, as aerosol propellants for products such as hair spray, and for foam blowing to make very porous plastic or rubber foams. Dichlorodifluoromethane, CCl_2F_2, was used in automobile air conditioners. When released into the atmosphere, this compound remained as a stable atmospheric gas until it reached very high altitudes in the stratosphere. In this region, UV radiation of sufficient energy ($h\nu$) is available to break the very strong C–Cl bonds

$$CCl_2F_2 + h\nu \rightarrow CClF_2 + Cl\cdot \tag{6.3}$$

releasing Cl atoms. The dot represents a single unpaired electron remaining with the Cl atom when the bond in the molecule breaks. Species with such unpaired electrons are very reactive and are called **free radicals**. As shown by Reactions 6.1 and 6.2, in addition to molecular O_2, there are oxygen atoms and molecules of ozone, O_3, also formed by photochemical processes in the stratosphere. A chlorine atom produced by the photochemical dissociation of CCl_2F_2 as shown in Reaction 6.3 can react with a molecule of O_3 to produce O_2 and another reactive free radical species, $ClO\cdot$. This species can react with free O atoms, which are present in the stratosphere along with the ozone (see Reaction 6.1), to regenerate Cl atoms, which in turn can react with more O_3 molecules. These reactions are shown as follows:

$$Cl + O_3 \rightarrow O_2 + ClO\cdot \tag{6.4}$$

$$ClO + O \rightarrow O_2 + Cl \tag{6.5}$$

These are chain reactions in which $ClO\cdot$ and $Cl\cdot$ are continually reacting and being regenerated, the net result of which is the conversion of O_3 and O in the atmosphere to O_2. One Cl atom can bring about the destruction of as many as 10,000 ozone molecules. Ozone serves a vital protective function in the atmosphere as a filter for damaging UV radiation, so its destruction is a very serious problem that has resulted in the banning of CFC manufacture.

As discussed in greater detail in Chapter 7 with respect to atmospheric reactions of air pollutants, hydroxyl radical, HO·, is the most important reactive intermediate in atmospheric chemical reactions.[6] Ultraviolet radiation is very much involved in the formation of hydroxyl radical through various photochemical reactions, and depletion of UV-absorbing stratospheric ozone may increase the levels of this species. The HO· radical is very active in determining the fates of atmospheric methane, carbon monoxide, hydrochlorofluorocarbons and hydrofluorocarbons that are substitutes for ozone-depleting CFCs, and other gases relevant to climate and ozone levels, and it is very much involved in the formation and dissipation of photochemical smog (see Chapter 7, Section 7.8).

Although it is a normal constituent present in an unpolluted atmosphere, hydroxyl radical may be an important consideration with respect to its toxicological chemistry (see Section 6.10). It is believed that hydroxyl radical (generally generated within the body) plays a role in oxidative stress in which reactive species such as hydroxyl attack tissue and DNA either directly or through their reaction products. Important biomolecules in the body may be altered by the introduction of oxygen atoms through hydroxyl radical. By reacting with disulfide bonds in proteins, hydroxyl radical can cause abnormal spatial configurations of some biomolecules, which may be a factor in some kinds of cancer, atherosclerosis, and neurological disorders.[7]

Very small particles of the size of a micrometer or less called **aerosols** are important in atmospheric chemical processes. Photochemical reactions often result in the production of particles. Particle surfaces can act to catalyze (bring about) atmospheric chemical reactions. Some particles in the atmosphere consist of water droplets with various solutes dissolved in them. Solution chemical reactions can occur in these droplets. One such process is believed to be the conversion of gaseous atmospheric sulfur dioxide (SO_2) to droplets of dilute sulfuric acid (H_2SO_4), which contribute to acid rain.

6.9.1 Atmospheric Ions and the Ionosphere

An important kind of photochemical reaction that occurs at altitudes generally above the stratosphere (50 km and higher) is the formation of ions by the action of UV and cosmic radiation energetic enough to remove electrons (e^-) from molecules as shown by the following example:

$$N_2 + h\nu \rightarrow N_2^+ + e^- \tag{6.6}$$

The ions formed are very reactive, but air is so rarefied at the altitudes at which they form that they persist for some time before reverting to neutral species. This results in an atmospheric layer called the **ionosphere** in which ions are constantly being formed and neutralized. At night when the solar radiation responsible for ion formation is shielded by Earth, the predominant process is recombination of positive ions with electrons, a phenomenon that proceeds most rapidly in the lower, denser regions of the ionosphere. The result is a lifting of the bottom boundary of the ionosphere, a phenomenon that was first hypothesized in 1901 when Marconi, attempting to bridge the Atlantic Ocean with shortwave radio transmissions, discovered that radio waves could be propagated over long distances, especially at night, making long-distance shortwave radio transmission possible as the signals bounced off the bottom layer of the ionosphere. For a time in the 1900s until it was made obsolete by satellite technology and fiber optics, the ionosphere was a useful part of the atmosphere's natural capital (see Section 6.2) by making possible long-distance shortwave radio broadcasts.

6.10 ATMOSPHERIC OXYGEN

Composing about one-fifth of the volume of gases in the atmosphere, elemental oxygen, O_2, is a crucial atmospheric component. Some of the primary features of the exchange of oxygen among the atmosphere, geosphere, hydrosphere, and biosphere are summarized in Figure 6.8. The oxygen cycle is critically important in atmospheric chemistry, geochemical transformations, and life processes.

A crucially important atmospheric chemical phenomenon involving oxygen is the stratospheric ozone layer, the formation of which is shown in Reactions 6.1 and 6.2. Stratospheric ozone is crucial for sustainability of life on Earth because it absorbs high-energy UV radiation from the sun that otherwise would seriously damage exposed biomolecules, including those on human skin. For all its benefits, there is not much ozone in the stratosphere. The ozone is distributed over a large range of altitudes of several kilometers. If it were distributed in a single pure layer of O_3 under the temperature and pressure conditions on Earth's surface, the ozone layer would be only 3 mm thick. Some classes of chemical species, especially the CFCs or Freons formerly used as refrigerants, are known to react in ways that destroy stratospheric ozone, and their elimination from commerce has been one of the major objectives of efforts in achieving sustainability. The preservation of the ozone layer is discussed in more detail in Chapter 8.

Oxygen in the atmosphere is consumed in the burning of hydrocarbons and other carbon-containing fuels. It is also consumed when oxidizable minerals undergo chemical weathering, such as

$$4FeO + O_2 \rightarrow 2Fe_2O_3 \tag{6.7}$$

All the oxygen in the atmosphere was originally placed there by the action of photosynthetic cyanobacteria in the early stages of life on Earth, as shown by

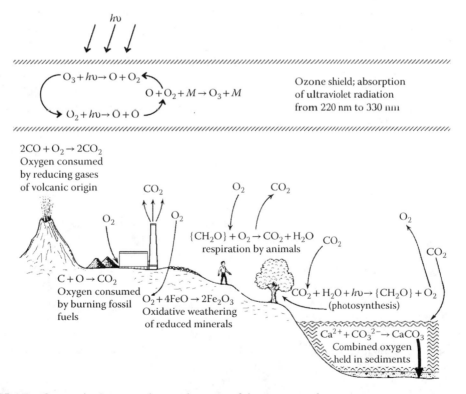

FIGURE 6.8 Oxygen in the atmosphere and aspects of the oxygen cycle.

$$CO_2 + H_2O + h\nu \rightarrow \{CH_2O\} + O_2 \tag{6.8}$$

where $\{CH_2O\}$ is a generic formula representing biomass.

The oxygen in the troposphere plays a strong role in the processes that occur on Earth's surface. Atmospheric oxygen takes part in energy-producing reactions, such as the burning of fossil fuels:

$$CH_4 \text{(in natural gas)} + 2O_2 \rightarrow CO_2 + 2H_2O \tag{6.9}$$

Atmospheric oxygen is utilized by oxic organisms in the degradation of organic material. Some oxidative weathering processes consume oxygen, such as

$$4FeO + O_2 \rightarrow 2Fe_2O_3 \tag{6.10}$$

Oxygen is returned to the atmosphere through plant photosynthesis:

$$CO_2 + H_2O + h\nu \rightarrow \{CH_2O\} + O_2 \tag{6.11}$$

The fact that air's oxygen was put there by photosynthetic organisms (Reaction 6.8) illustrates the importance of photosynthesis in the oxygen balance of the atmosphere. It can be shown that most of the carbon fixed by these photosynthetic processes is dispersed in mineral formations as humic material (see Chapter 10, Section 10.1); only a very small fraction is deposited in fossil fuel beds. Therefore, although combustion of fossil fuels consumes large amounts of O_2, there is no danger of running out of atmospheric oxygen.

Because of the extremely rarefied atmosphere and the effects of ionizing radiation, elemental oxygen in the upper atmosphere exists to a large extent in forms other than diatomic O_2. In addition to O_2, the upper atmosphere contains oxygen atoms, O; excited oxygen molecules that have been energized by the absorption of photons of UV radiation, O_2*; and ozone, O_3.

Atomic oxygen, O, is stable primarily in the thermosphere, where the atmosphere is so rarefied that the three-body collisions necessary for the chemical reaction of atomic oxygen seldom occur (the third body in this kind of three-body reaction absorbs energy to stabilize the products). As noted in Section 6.9, atomic oxygen is produced by a photochemical reaction:

$$O_2 + h\nu \rightarrow O + O$$

The oxygen–oxygen bond is strong (120 kcal/mol), and UV radiation in the wavelength regions 135–176 nm and 240–260 nm is most effective in causing dissociation of molecular oxygen. Because of photochemical dissociation, O_2 is virtually nonexistent at very high altitudes and less than 10% of the oxygen in the atmosphere at altitudes exceeding approximately 400 km is present in the molecular form. At altitudes exceeding about 80 km, the average molar mass of air is lower than the 28.97 g/mol observed at sea level because of the high concentration of atomic oxygen. The resulting division of the atmosphere into a lower section with a uniform molecular mass and a higher region with a nonuniform molar mass is the basis for classifying these two atmospheric regions as the **homosphere** and the **heterosphere**, respectively.

Oxygen atoms in the atmosphere can exist in the ground state (O) and in excited states (O*). Excited O* atoms are produced by the photolysis of ozone, which has a relatively weak bond energy of 26 kcal/mol, at wavelengths below 308 nm

$$O_3 + h\nu(\lambda < 308 \text{ nm}) \rightarrow O^* + O_2 \tag{6.12}$$

or by highly energetic chemical reactions such as

$$O + O + O \rightarrow O_2 + O* \tag{6.13}$$

Excited atomic oxygen emits visible light at wavelengths of 636, 630, and 558 nm. This emitted light is partially responsible for **airglow**, a very faint electromagnetic radiation that is continuously emitted by Earth's atmosphere. Although its visible component is extremely weak, airglow is more pronounced in the infrared region of the spectrum.

Oxygen ion, O^+, which may be produced by UV radiation acting upon oxygen atoms

$$O + h\nu \rightarrow O^+ + e^- \tag{6.14}$$

is the predominant positive ion in some regions of the ionosphere. Other oxygen-containing ions in the ionosphere are O_2^+ and NO^+.

Ozone, O_3, has an essential protective function because it absorbs harmful UV radiation in the stratosphere and serves as a radiation shield, protecting living beings on Earth from the effects of excessive amounts of such radiation. It is produced by a photochemical reaction

$$O_2 + h\nu \rightarrow O + O \tag{6.15}$$

(where the wavelength of the exciting radiation must be less than 242.4 nm), followed by a three-body reaction

$$O + O_2 + M \rightarrow O_3 + M \, (\text{increased energy}) \tag{6.16}$$

in which M is another species, such as a molecule of N_2 or O_2, which absorbs the excess energy given off by the reaction and enables the ozone molecule to stay together. The region of maximum ozone concentration is found within the range of 25–30 km high in the stratosphere where it may reach 10 ppm.

Ozone absorbs UV light very strongly in the region 220–330 nm. If this light were not absorbed by ozone, severe damage would result to exposed forms of life on Earth. Absorption of electromagnetic radiation by ozone converts the radiation's energy to heat and is responsible for the temperature maximum encountered at the boundary between the stratosphere and the mesosphere at an altitude of approximately 50 km. The reason that the temperature maximum occurs at a higher altitude than the altitude of maximum ozone concentration arises from the fact that ozone is such an effective absorber of UV light. Therefore, most of the UV radiation absorbed by ozone is absorbed in the upper stratosphere, where it generates heat, and only a small fraction reaches the lower altitudes, which remain relatively cool.

The overall reaction

$$2O_3 \rightarrow 3O_2 \tag{6.17}$$

is favored thermodynamically so that ozone is inherently unstable. Its decomposition in the stratosphere is catalyzed by a number of natural and pollutant trace constituents, including NO, NO_3, H, HO·, HOO·, ClO, Cl, Br, and BrO. Ozone decomposition also occurs on solid surfaces, such as metal oxides and salts produced by rocket exhausts.

Despite its essential protective role in the stratosphere, ozone is an undesirable pollutant in the troposphere. It is toxic to animals and plants and also damages materials, particularly rubber.

6.10.1 TOXICOLOGICAL CHEMISTRY OF OXYGEN

As noted in a very informative summary by McKersie,[8] although molecular oxygen is required by humans and other oxic organisms for energy metabolism and respiration, it plays an important role in a number of significant diseases and degenerative conditions. Such adverse health conditions are

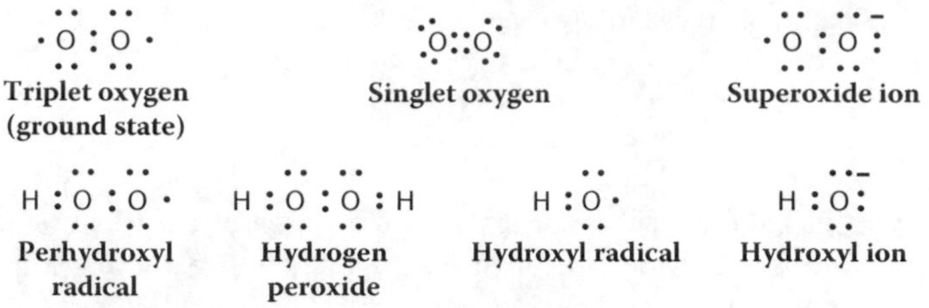

$$\cdot \overset{\cdot\cdot}{O} : \overset{\cdot\cdot}{O} \cdot$$

Triplet oxygen
(ground state)

$$\overset{\cdot\cdot}{O} :: \overset{\cdot\cdot}{O}$$

Singlet oxygen

$$\cdot \overset{\cdot\cdot}{O} : \overset{\cdot\cdot\,-}{O} :$$

Superoxide ion

$$H : \overset{\cdot\cdot}{O} : \overset{\cdot\cdot}{O} \cdot$$

Perhydroxyl
radical

$$H : \overset{\cdot\cdot}{O} : \overset{\cdot\cdot}{O} : H$$

Hydrogen
peroxide

$$H : \overset{\cdot\cdot}{O} \cdot$$

Hydroxyl radical

$$H : \overset{\cdot\cdot\,-}{O} :$$

Hydroxyl ion

FIGURE 6.9 Lewis structure representations of oxygen species involved in oxidative stress processes. Although triplet (ground state) molecular oxygen molecules are sometimes depicted by the simplified Lewis structural formulas shown, a complete description of the electronic behavior of these molecules requires use of molecular orbital theory, which is beyond the scope of this work.

commonly labeled **oxidative stress**, which in humans is associated with the aging process, arthritis, ALS syndrome (Lou Gehrig's disease), and other maladies. Oxidative stress is also associated with some toxic effects in plants (phytotoxicity).

Some important oxygen-containing species involved in oxidative stress are shown in Figure 6.9. Ground-state O_2 is an interesting molecule because it has two unpaired electrons with parallel spins, a triplet state that makes it relatively unreactive toward biomolecules. Energy absorbed by triplet-state oxygen can promote it to the singlet state in which the two electrons take on opposite spins, which makes singlet oxygen relatively much more reactive toward organic molecules and biomolecules. Molecular oxygen may be converted to superoxide ion, perhydroxyl radical, hydrogen peroxide, and hydroxyl radical, which are involved in the process of oxidative stress. The hydroxyl radical, HO·, is highly reactive and a potent oxidizing agent, which is ultimately responsible for most of the damage done by oxidative stress.

6.11 ATMOSPHERIC NITROGEN

The 78% by volume of nitrogen contained in the atmosphere constitutes an inexhaustible reservoir of this essential element. As shown in Figure 6.10, the nitrogen cycle is involved in all five environmental spheres. A small amount of nitrogen is fixed in the atmosphere by lightning, and some is also fixed by combustion processes, particularly in internal combustion and turbine engines. Figure 6.10 shows the major aspects of the nitrogen biogeochemical cycle through which nitrogen is interchanged among the atmosphere, organic matter, and inorganic compounds. It is one of nature's most vital dynamic processes.

Unlike oxygen, which is almost completely dissociated to the monatomic form in higher regions of the thermosphere, molecular nitrogen is not readily dissociated by UV radiation. However, at altitudes exceeding approximately 100 km, atomic nitrogen is produced by photochemical reactions:

$$N_2 + h\nu \rightarrow N + N \tag{6.18}$$

Several reactions of ionic species in the ionosphere may generate N atoms as well. The N_2^+ ion is generated by photoionization in the atmosphere

$$N_2 + h\nu \rightarrow N_2^+ + e^- \tag{6.19}$$

and may react to form other ions. The NO^+ ion is one of the predominant ionic species in the so-called E region of the ionosphere.

Pollutant oxides of nitrogen, particularly NO_2, are the key species involved in air pollution and the formation of photochemical smog. For example, NO_2 is readily dissociated photochemically to NO and reactive atomic oxygen:

$$NO_2 + h\nu \rightarrow NO + O \tag{6.20}$$

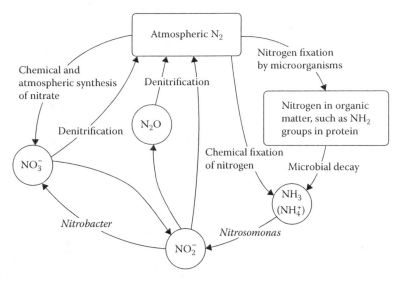

FIGURE 6.10 The nitrogen cycle.

This reaction is the most important primary photochemical process involved in smog formation. The roles played by nitrogen oxides in smog formation and other forms of air pollution are discussed in Chapter 7.

6.12 ATMOSPHERIC WATER

The water vapor content of the troposphere is normally within a range of 1%–3% by volume with a global average of about 1%. However, air can contain as little as 0.1% or as much as 5% water. The percentage of water in the atmosphere decreases rapidly with increasing altitude. Water circulates through the atmosphere in the hydrologic cycle, as shown in Chapter 3, Figure 3.2.

Water vapor absorbs infrared radiation even more strongly than does carbon dioxide, thus greatly influencing Earth's heat balance. Clouds formed from water vapor reflect light from the sun and have a temperature-lowering effect. On the other hand, water vapor in the atmosphere acts as a kind of "blanket" at night, retaining heat from Earth's surface by the absorption of infrared radiation.

As discussed in Section 6.5, water vapor and the heat released and absorbed by transitions of water between the vapor state and the liquid or solid state are strongly involved in atmospheric energy transfer. Condensed water vapor in the form of very small droplets is of considerable concern in atmospheric chemistry. The harmful effects of some air pollutants—for instance, the corrosion of metals by acid-forming gases—requires the presence of water, which may come from the atmosphere. Atmospheric water vapor has an important influence on pollution-induced fog formation under some circumstances. Water vapor interacting with pollutant particulate matter in the atmosphere may reduce visibility to undesirable levels through the formation of very small atmospheric aerosol particles.

In the very cold tropopause layer at the boundary between the troposphere and the stratosphere, water vapor in the atmosphere is condensed and forms ice crystals, a phenomenon that serves as a barrier to the movement of water into the stratosphere. Thus, little water is transferred from the troposphere to the stratosphere, and the main source of water in the stratosphere is the photochemical oxidation of methane:

$$CH_4 + 2O_2 + h\nu \xrightarrow{\text{Several steps}} CO_2 + 2H_2O \qquad (6.21)$$

TABLE 6.1

Important Terms Describing Atmospheric Particles

Term	Meaning
Aerosol	Colloidal-sized atmospheric particle
Condensation aerosol	Formed by condensation of vapors or reactions of gases
Dispersion aerosol	Formed by grinding of solids, atomization of liquids, or dispersion of dust
Fog	Term denoting a high level of water droplets
Haze	Denotes decreased visibility due to the presence of particles
Mists	Liquid particles
Smoke	Particles formed by the incomplete combustion of fuel

The water produced by this reaction serves as a source of stratospheric hydroxyl radical as shown by the following reaction:

$$H_2O + h\nu \rightarrow HO\cdot + H\cdot \qquad (6.22)$$

6.13 ATMOSPHERIC PARTICLES

Particles ranging in size from about one-half millimeter (the size of sand or drizzle) down to molecular dimensions are important atmospheric constituents. Atmospheric particles are made up of an amazing variety of materials and discrete objects that may consist of either solids or liquid droplets. A number of terms are commonly used to describe atmospheric particles; the more important of these are summarized in Table 6.1. **Particulates** is a term that has come to stand for particles in the atmosphere, although **particulate matter**, or simply **particles**, is the preferred usage.

As discussed in more detail in Chapters 7 and 8, particles can be very important air pollutants, the most obvious manifestation of which is reduced visibility in the atmosphere due to the ability of particles to scatter light. Particles are carriers of pollutants, such as lead, a property that makes particles an important consideration in environmental and occupational health. Particles can have direct health effects on the respiratory system and are important vectors by which toxic substances carried by particles are introduced into the body.

6.13.1 Physical Behavior of Atmospheric Particles

As shown in Figure 6.11, atmospheric particles undergo a number of processes in the atmosphere. Small colloidal particles are subject to **diffusion processes**. Smaller particles **coagulate** together to form larger particles. **Sedimentation** or **dry deposition** of particles, which have often reached sufficient size to settle by coagulation, is one of the two major mechanisms for particle removal from the atmosphere. In many areas, dry deposition on vegetation is a significant mechanism for particle removal. In addition to sedimentation, a major pathway for particle removal from the atmosphere is **scavenging** by raindrops and other forms of precipitation. Particles also react with atmospheric gases.

6.13.2 Atmospheric Chemical Reactions Involving Particles

In recent years, there has been an increasing recognition of the importance in atmospheric chemistry of chemical processes that occur on particle surfaces and in solution in liquid particles (Figure 6.12). Challenging as it is, gas-phase atmospheric chemistry is relatively straightforward compared to the heterogeneous chemistry that involves particles. Particles may serve as sources and sinks of atmospheric chemical reaction participant species and are very much involved in atmospheric

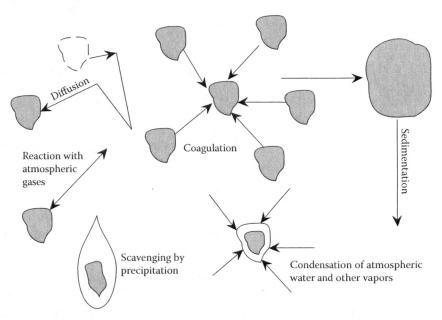

FIGURE 6.11 Processes that particles undergo in the atmosphere.

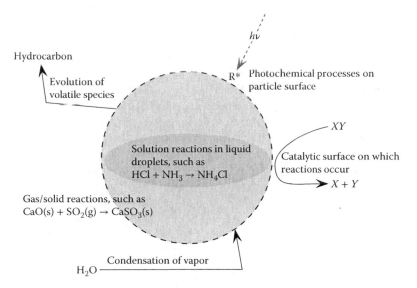

FIGURE 6.12 Particles provide sites for many important atmospheric chemical processes.

chemical processes.[9] Solid particle surfaces may adsorb reactants and products, serve a catalytic function, exchange electrical charge, and absorb photons of electromagnetic radiation, thus acting as photocatalytic surfaces. Liquid water droplets may act as media for solution reactions including photochemical reactions that occur in solution.

QUESTIONS AND PROBLEMS

Access to and use of the Internet is assumed in answering all questions, including general information, statistics, constants, and mathematical formulas required to solve problems. These questions are designed to promote inquiry and thought rather than just finding material in the chapter. So in some cases, there may be several "right" answers. Therefore, if your answer reflects intellectual effort and a search for information from available sources, it may be considered to be right.

1. Give two examples each of Earth's natural capital in terms of (1) protective function and (2) raw materials.
2. In what respect is the composition of gases in the troposphere not uniform (which atmospheric constituent varies widely in time and location)?
3. Other than avoiding turbulence due to lower altitude weather, suggest an advantage for commercial aircraft to cruise at a relatively high altitude of around 10 km.
4. Look up the tropopause on the Internet. Where is it located? What function does it serve in keeping Earth in a livable state?
5. What chemical species in the stratosphere is essential for life on Earth? What are the reactions by which it is made, and how may it be threatened?
6. What starts a photochemical reaction? What is it called when a series of reactions initiated by a photochemical event continues, often in a cyclical fashion?
7. What is a free radical?
8. In what two important respects may very small particles participate in atmospheric chemical processes?
9. In what respect does the radiation by which Earth loses energy differ from that by which it gets energy from the sun?
10. What are two catastrophic events that could cause a sudden cooling of Earth's atmosphere?
11. How is water vapor involved in moving energy through the atmosphere?
12. Distinguish among the terms meteorology, weather, and climate.
13. What do clouds consist of? What must happen before rain falls from clouds?
14. Why is there essentially no atmospheric chemistry involving elemental nitrogen gas in the atmosphere?
15. Cite an atmospheric chemical condition or phenomenon that shows that the O_2 molecule is easier to break apart than the N_2 molecule.
16. In what respect are elemental nitrogen and oxygen green elements?
17. After some search on the Internet, give a chemical reaction that produces oxygen that can be used for emergencies, especially in the case of loss of cabin pressure in aircrafts.
18. What are the two major classes of atmospheric particles based on how they are produced?

LITERATURE CITED

1. Platt, Ulrich, Atmospheric Gas Phase Reactions, *Geophysical Monograph* **187**, 7–15 (2009).
2. Klose, Martina, Yaping Shao, Melanie K. Karreman, and Andreas H. Fink, Sahel Dust Zone and Synoptic Background, *Geophysical Research Letters* **37**, L09802/1-L-09802/6 (2010).
3. Nielsen, Ole John, Depletion of the Ozone Layer, in *Air Pollution*, Jes Fenger and Jens Christian Tjell, Eds., Royal Society of Chemistry, Cambridge, UK, 389–414, 2009.
4. McNeil, Donald G., Turkey: Sandblasting Jeans for 'Distressed' Look Proved Harmful for Textile Workers, *New York Times*, November 1, 2011, p. D6.
5. Manning, Martin R., We Need to Know More About Atmospheric Chemistry, http://nzic.org.nz/CiNZ/articles/CiNZ%20April%202011%20Manning.pdf.
6. Tang, X., S. R. Wilson, K. R. Solomon, M. Shao, and S. Madronich, *Photobiological Sciences* **10**, 280–291 (2011).
7. Lupinski, Boguslaw, Hydroxyl Radical and its Scavengers in Health and Disease, *Oxidative Medicine and Cellular Longevity*, EPub 809696, 2011.
8. McKersie, Bryan D. Oxidative Stress, http://www.plantstress.com/Articles/Oxidative%20Stress.htm, 1996.
9. George, I. J., and J.P.D. Abbatt, Heterogeneous Oxidation of Atmospheric Aerosol Particles by Gas-Phase Radicals, *Nature Chemistry* **2**, 713–22 (2010).

SUPPLEMENTARY REFERENCES

Abhishek, Tiwary, and Jeremy Colls, *Air Pollution: Measurement, Modeling, and Mitigation*, 3rd ed., Routledge, London, 2010.

Ackerman, Steven A., and John Knox, *Meteorology: Understanding the Atmosphere*, 3rd ed., Jones and Bartlett Learning, Sudbury, MA, 2011.

Aguado, Edward, and James E. Burt, *Understanding Weather and Climate*, 5th ed., Pearson Education, Upper Saddle River, NJ, 2009.

Ahrens, C. Donald, *Meteorology Today: An Introduction to Weather, Climate, and the Environment*, 9th ed., Thomson Brooks/Cole, Belmont, CA, 2009.

Ahrens, C. Donald, and Perry Samson, *Extreme Weather and Climate*, Brooks/Cole, Belmont, CA, 2010.

Allaby, Michael, *Atmosphere: A Scientific History of Air, Weather, and Climate, Facts on File*, New York, 2009.

Barker, John. R., A Brief Introduction to Atmospheric Chemistry, *Advances Series in Physical Chemistry* **3**, 1–33 (1995).

Garratt, Richard, *Atmosphere: A Scientific History of Air, Weather, and Climate, Facts on File*, New York, 2009.

Hewitt, C. N., and Andrea Jackson, *Atmospheric Science for Environmental Scientists*, Wiley-Blackwell, Hoboken, NJ, 2009.

Hewitt, Nick C., and Andrea V. Jackson, *Atmospheric Science for Environmental Scientists*, Wiley-Blackwell, Hoboken, NJ, 2009.

Hewitt, Nick, and Andrea Jackson, Eds., *Handbook of Atmospheric Science*, Blackwell Publishing, Malden, MA, 2003.

Hidore, John J., John E. Oliver, Mary Snow, and Richard Snow, *Climatology: An Atmospheric Science*, 3rd ed., Prentice Hall, Upper Saddle River, NJ, 2009.

Hobbs, Peter V., *Introduction to Atmospheric Chemistry*, Cambridge University Press, New York, 2000.

Holloway, Ann M., and Richard P. Wayne, *Atmospheric Chemistry*, RSC Publishing, Cambridge, UK, 2010.

Jacob, Daniel J., *Introduction to Atmospheric Chemistry*, Princeton University Press, Princeton, NJ, 1999.

Jolliffe, Ian T., and David B. Stephenson, *Forecast Verification: A Practitioner's Guide in Atmospheric Science*, Wiley, Hoboken, NJ, 2012.

Lutgens, Frederick K., and Edward J. Tarbuck, *The Atmosphere: An Introduction to Meteorology*, 11th ed., Prentice Hall, New York, 2010.

Mankin, Mak, *Atmospheric Dynamics*, Cambridge University Press, New York, 2011.

Mohanakumar, K. *Stratosphere Troposphere Interactions: An Introduction*, Springer, New York, 2010.

Spellman, Frank R., *The Science of Air: Concepts and Applications*, 2nd ed., Taylor & Francis, Boca Raton, FL, 2009.

Wallace, John M., *Atmospheric Science: An Introductory Survey*, 2nd ed., Elsevier, Amsterdam, 2006.

7 Pollution of the Atmosphere

7.1 POLLUTION OF THE ATMOSPHERE AND AIR QUALITY

In Chapter 6, the atmosphere was introduced as one of the five main parts of the environment. In this chapter, pollution of the atmosphere is discussed. Atmospheric quality is extremely important for human welfare, in part because of the intimate relationship that we have with air in the atmosphere. It is almost impossible in a practical sense to avoid breathing air as we find it around us. Toxic air pollutants will get into our lungs. In the innermost alveoli of the lungs where gases are interchanged between air that is inhaled and blood in the bloodstream, a layer only the thickness of a single human cell separates the blood from the air. If the air that we breathe is polluted, we may be harmed by it.

In addition to human health effects, there are other reasons to be concerned about air pollution. Plants may be harmed by air pollutants. Acids in the atmosphere may attack materials and cause corrosion to occur. And the most damaging air pollutants of all may turn out to be the greenhouse gases, especially carbon dioxide and methane, that get into the atmosphere in part from natural sources, but which in excess from anthrospheric sources are causing global warming and may result in massive alteration of Earth's climate. Global climate change is discussed in Chapter 8, which deals with sustaining the atmosphere.

In discussing atmospheric pollution, it is important to make the distinction between primary and secondary air pollutants. **Primary air pollutants** are those that are pollutants in the form in which they are emitted into the atmosphere. An example would be light-scattering fine ash particles ejected from a smokestack. **Secondary air pollutants** are those that are formed from other substances by processes in the atmosphere. A prime example of a secondary pollutant develops when otherwise relatively innocuous levels of hydrocarbons (including terpenes from pine and citrus trees) and NO are emitted into the atmosphere and subjected to ultraviolet radiation from the sun, resulting in a noxious mixture of ozone, aldehydes, organic oxidants, and fine particles called photochemical smog.

As discussed in Chapter 6, Section 6.4, the atmosphere is a huge part of Earth's natural capital. Air pollution may be detrimental to Earth's natural capital. Pollutant particles in the atmosphere can reduce visibility, which is harmful to the atmosphere's esthetics, and may block some of the light that plants use in photosynthesis. Another detrimental effect of air pollution on Earth's natural capital has to do with the atmosphere's ability to absorb electromagnetic radiation. Depletion of the stratospheric ozone layer by chlorofluorocarbons (CFCs) can lead to greater penetration of damaging ultraviolet radiation into the troposphere, reducing this vital protective function of the atmosphere. Another very important part of the atmosphere's natural capital is its ability to keep Earth at a generally livable temperature. Greenhouse gases are shifting this delicate temperature balance, so that perhaps in the future, this part of the atmosphere's natural capital will not be as favorable to humans on the planet as it has been for the last several millennia.

Changing climatic conditions and atmospheric pollution that may result from such changes may have profound effects upon human health in the future.[1] The health implications of climate change are discussed in more detail in Chapter 8, Section 8.2.6

7.2 POLLUTANT PARTICLES IN THE ATMOSPHERE

In Chapter 6, Section 6.13, atmospheric particles were defined as solids or liquid droplets in a size range between the size of large molecules and ranging up to around half a millimeter. Finely divided dust blown by wind from topsoil is an example of solid particles and visibility-obscuring water droplets in fog are common liquid droplets. Particles in the atmosphere generally do not settle very rapidly, and those of colloidal size from about 0.001 to 1 μm, commonly called aerosols, tend to stay suspended in air and to scatter light. Terms applied to the major kinds of particles are given in Table 6.1. As noted in Chapter 6, Section 6.13, atmospheric particles are commonly called particulates, although particulate matter or simply particles is the preferred usage.

Atmospheric particles may consist of organic or inorganic materials or mixtures of both. Solid pollutant particles include very small combustion nuclei residues from fuel combustion, cement dust, silica dust from sandblasting, and soil dust mobilized by cultivation practices. Sulfuric acid droplets produced by oxidation of pollutant sulfur dioxide in the atmosphere are the most common kind of pollutant liquid droplets. Many kinds of particles are of biological origin and can be considered pollutants when they contribute to respiratory problems. These include bacteria, bacterial spores, fungal spores, and pollen.

7.2.1 Physical and Chemical Processes for Particle Formation: Dispersion and Condensation Aerosols

There are two major kinds of pollutant particles based on how they are formed. **Dispersion aerosols** are produced by breaking down bulk matter and include soil dust, dust emitted in rock crushing, silica dust from sandblasting, and formation of spray from cooling towers. Dispersion aerosols tend to be larger, usually above 1 μm in size, and settle readily from air. They are also generally less respirable and not very chemically reactive. Dispersion aerosols are usually largely free of toxic constituents such as heavy metals, although some such particles, notably silica (SiO_2) dust, which causes silicosis, and particulate asbestos, which can cause some forms of respiratory tract cancer, pose definite health hazards.

Although many dispersion aerosols originate from natural sources such as sea spray, windblown dust, and volcanic dust, a vast variety of human activities break up material and disperse it to the atmosphere. "All terrain" vehicles churn across desert lands, coating fragile desert plants with layers of dispersed dust. Quarries and rock crushers spew out plumes of ground rock. Spray from cooling towers produces liquid droplets from which the water may evaporate, leaving behind solid particles from minerals that were in the water. Cultivation of land has made it much more susceptible to dust-producing wind erosion. Areas of the world, especially in China, are now sometimes afflicted by plumes of dispersion particles stirred up by windstorms that disturb soil from land converted to desert by the climate effects of global warming, improper cultivation, and overgrazing, which tend to obscure visibility and reduce the intensity of sunlight reaching surface levels.

Condensation aerosols are produced by chemical reactions including combustion and reactions involving atmospheric gases. In contrast to dispersion aerosols, condensation aerosols are much smaller, hence more respirable, chemically reactive, likely to contain toxic substances, and of a size that is particularly effective in scattering light. Therefore, pollutant condensation aerosols are generally regarded as more harmful than dispersion aerosols. Before leaded gasoline was phased out of use beginning in the 1970s, large amounts of lead halide condensation aerosols were emitted into the atmosphere (not a good practice).

7.2.2 Chemical Processes for Inorganic Particle Formation

Some of the main aspects of the formation of inorganic particles are shown in Figure 7.1. Of particular importance are the toxic constituents, such as heavy metals, that may be incorporated into inorganic particles.

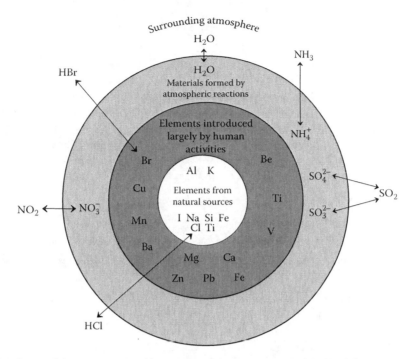

FIGURE 7.1 Some of the components of inorganic particulate matter and their origins.

Metal oxides constitute a major class of inorganic particles in the atmosphere. These are formed whenever fuels containing metals are burned. For example, particulate iron oxide is formed during the combustion of pyrite-containing coal:

$$3FeS_2 + 8O_2 \rightarrow Fe_3O_4 + 6SO_2 \tag{7.1}$$

Organic vanadium in residual fuel oil is converted to particulate vanadium oxide. Part of the calcium carbonate in the ash fraction of coal is converted to calcium oxide, some of which may be emitted into the atmosphere through the stack:

$$CaCO_3 + heat \rightarrow CaO + CO_2 \tag{7.2}$$

A common process for the formation of aerosol mists involves the oxidation of atmospheric sulfur dioxide to sulfuric acid, a hygroscopic substance that accumulates atmospheric water to form small liquid droplets:

$$2SO_2 + O_2 + 2H_2O \rightarrow 2H_2SO_4 \tag{7.3}$$

When basic air pollutants, such as ammonia or calcium oxide, are present, they react with sulfuric acid to form salts:

$$H_2SO_4 \left(droplet\right) + 2NH_3 \left(g\right) \rightarrow \left(NH_4\right)_2 SO_4 \left(droplet\right) \tag{7.4}$$

$$H_2SO_4 \left(droplet\right) + CaO\left(s\right) \rightarrow CaSO_4 \left(droplet\right) + H_2O \tag{7.5}$$

At low humidity, water is lost from these droplets and a solid salt aerosol is formed.

7.2.3 Composition of Inorganic Particles

As shown in Figure 7.1, the source of particulate matter is reflected in its elemental composition, taking into consideration chemical reactions that may change the composition. The three major sources of the elements in inorganic particles are (1) those from natural sources, the geosphere and sea salt; (2) those from pollutant sources including heavy metals, especially lead; and (3) those formed by the reaction of pollutant gases, such as sulfates from the oxidation of atmospheric SO_2. For example, particulate matter largely from ocean spray origin in a coastal area receiving sulfur dioxide pollution may show anomalously high sulfate and correspondingly low chloride content. The sulfate comes from atmospheric oxidation of sulfur dioxide to form nonvolatile ionic sulfate, whereas some chloride originally from the NaCl in the seawater may be lost from the solid aerosol as volatile HCl:

$$2SO_2 + O_2 + 2H_2O \rightarrow 2H_2SO_4 \tag{7.6}$$

$$H_2SO_4 + 2NaCl(\text{particulate}) \rightarrow Na_2SO_4(\text{particulate}) + 2HCl \tag{7.7}$$

Acids other than sulfuric acid can also be involved in the modification of sea-salt particles. The most common such acid is nitric acid formed by reactions of nitrogen oxides in the atmosphere. Traces of nitrate salts may be found among sea-salt particles.

Among the constituents of inorganic particulate matter found in polluted atmospheres are salts, oxides, nitrogen compounds, sulfur compounds, various metals, and radionuclides. In coastal areas, sodium and chlorine get into atmospheric particles as sodium chloride from sea spray. The trace elements that typically occur in particulate matter and their sources are the following:

- **Al, Fe, Ca, Si**: Soil erosion, rock dust, coal combustion
- **C**: Incomplete combustion of carbonaceous fuels
- **Na, Cl**: Marine aerosols, chloride from incineration of organohalide polymer wastes
- **Sb, Se**: Very volatile elements, possibly from the combustion of oil, coal, or refuse
- **V**: Combustion of residual petroleum (present at very high levels in residues from Venezuelan crude oil)
- **Zn**: Tends to occur in small particles, probably from combustion
- **Pb**: Combustion of fuels and wastes containing lead

Atmospheric particulate carbon, commonly called **black carbon**, produced as soot, carbon black, coke, and graphite originates from auto and truck exhausts, heating furnaces, incinerators, power plants, and steel and foundry operations and composes one of the more visible and troublesome particulate air pollutants.[2] Because of its good adsorbent properties, carbon can be a carrier of gaseous and other particulate pollutants. Both nitrogen and sulfur compounds in exhaust gases are adsorbed onto particulate carbon that is emitted by poorly controlled diesel engines. Particulate carbon surfaces may catalyze some heterogeneous atmospheric reactions, including the important conversion of SO_2 to sulfate. Black carbon is an effective absorber of heat both in the atmosphere and on snow surfaces and may be an important factor in global warming.

7.2.4 Fly Ash

Much of the mineral particulate matter in a polluted atmosphere is in the form of oxides and other compounds produced during the combustion of high-ash fossil fuel. Small particles in the exhaust from mineral-containing fossil fuels are called **fly ash**. Most fly ash enters furnace flues and is efficiently collected in a properly equipped stack system. However, some fly ash escapes through the stack and enters the atmosphere. Unfortunately, the fly ash thus released tends to consist of smaller particles that do the most damage to human health, plants, and visibility.

The composition of fly ash varies widely, depending upon the fuel. The predominant constituents are oxides of aluminum, calcium, iron, and silicon. Other elements that occur in fly ash are magnesium, sulfur, titanium, phosphorus, potassium, and sodium. Elemental carbon (soot, carbon black) is a significant fly ash constituent.

The size of fly ash particles is a very important factor in determining their removal from stack gas and their ability to enter the body through the respiratory tract. Although only a small percentage of the total fly ash mass is in the smaller size fraction of around 0.1 μm size, it includes the vast majority of the total number of particles and particle surface area. Submicrometer particles probably result from a volatilization—condensation process during combustion, as reflected in a higher concentration of more volatile elements such as As, Sb, Hg, and Zn. In addition to their being relatively much more respirable and potentially toxic, the very small particles are the most difficult to remove by electrostatic precipitators and bag houses (see Chapter 8, Section 8.4).

7.2.5 RADIOACTIVITY IN ATMOSPHERIC PARTICLES

The most significant source of radioactivity in the indoor microatmosphere is **radon**, a noble gas product of radium decay that is produced below ground and that may leak into the basements above. Radon may enter the atmosphere as either of two isotopes, ^{222}Rn (half-life 3.8 days) and ^{220}Rn (half-life 54.5 seconds). Both are alpha emitters in decay chains that terminate with stable isotopes of lead. The initial decay products, ^{218}Po and ^{216}Po, are nongaseous and adhere readily to atmospheric particulate matter. In some areas where radon is produced, homes have had to be fitted with ventilation systems to prevent radon infiltration.

When it was a common practice by the United States and Soviet Union from the end of World War II until the 1960s, above ground testing of nuclear weapons was a common source of atmospheric radioactive particulate matter. The catastrophic 1986 meltdown and fire at the Chernobyl nuclear reactor in the former Soviet Union spread large quantities of radioactive materials, mostly as particulate matter, over a wide area of Europe. The explosions and partial meltdown of nuclear reactors at Japan's Fukushima Daiichi power plant following the March 11, 2011, earthquake and tsunami in the area released significant quantities of radioactive particulate matter. Although the most immediate threat was from radioactive iodine-131, this isotope has a half-life of only 8 days and soon dissipated. The greater threat has been from cesium-137, a radioactive alkali metal that has a half-life of 30.1 years and behaves much like sodium and potassium in the body. Most of the radioactive particulate matter from the damaged power plant was rapidly washed from the atmosphere by rain, although some problems have resulted from contaminated soil, and potential contamination of rice grown on the soil has been a concern.

Although it is a gas not associated with particulate matter in the atmosphere, the radioactive noble gas ^{85}Kr (a beta and gamma emitter with a half-life of 10.3 years) is emitted into the atmosphere by the operation of nuclear reactors and the processing of spent reactor fuels. This radioisotope is largely contained in spent reactor fuel during reactor operation; nuclear fuel reprocessing releases most of this gas from the fuel elements, and its presence has been used to indicate clandestine plutonium production operations. Fortunately, living organisms cannot concentrate this chemically unreactive element and its toxicity is minimal.

7.2.6 ORGANIC POLLUTANT PARTICLES IN THE ATMOSPHERE

Organic particulate matter consists of a variety of different kinds of materials. Much of this matter is produced as secondary material that results from photochemical processes operating on volatile and semivolatile organic compounds emitted into the atmosphere. Such is the case with the organic particles with an approximate empirical formula of CH_2O, characteristic of photochemical smog (see Section 7.8). The compounds emitted into the atmosphere are predominantly hydrocarbon in nature, and the incorporation of oxygen and nitrogen through atmospheric chemical processes gives less volatile material in the form of organic particles.

Polycyclic aromatic hydrocarbons (PAH) in atmospheric particles have received a great deal of attention because of the known carcinogenic effects of some of these compounds. The most prominent of these compounds is benzo(a)pyrene and other examples are benz(a)anthracene, chrysene, benzo(e)pyrene, benz(e)acephenanthrylene, benzo(j)fluoranthene, and indenol. Some representative structures of PAH compounds are given as follows:

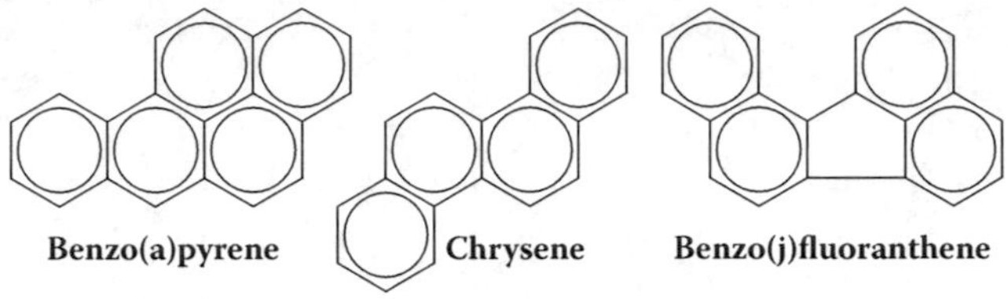

Benzo(a)pyrene   Chrysene   Benzo(j)fluoranthene

Elevated levels of PAH compounds of up to about 20 $\mu g/m^3$ are found in the atmosphere. Elevated levels of PAHs are most likely to be encountered in polluted urban atmospheres and in the vicinity of natural fires such as forest and prairie fires. Coal furnace stack gas may contain over 1000 $\mu g/m^3$ of PAH compounds and cigarette smoke contains almost 100 $\mu g/m^3$. Diesel engines can be prolific emitters of carbonaceous particulate matter including PAHs, and much of the effort to control particulate air pollution is now concentrated on this source.

Atmospheric polycyclic aromatic hydrocarbons are found almost exclusively in the solid phase, largely sorbed to soot particles (see black carbon in Section 7.2.3). Soot itself is a highly condensed product of PAHs. Soot contains 1%–3% hydrogen and 5%–10% oxygen, the latter due to partial surface oxidation. Benzo(a)pyrene adsorbed on soot disappears rapidly in the presence of light, yielding oxygenated products; the large surface area of the particle contributes to the high rate of reaction. Oxidation products of benzo(a)pyrene include epoxides, quinones, phenols, aldehydes, and carboxylic acids as illustrated by the composite structures shown in Figure 7.2.

7.2.7   Effects of Atmospheric Pollutant Particles

Pollutant particles in the atmosphere have both direct and indirect effects. The most obvious direct effects are reduction and distortion of visibility. The light scattering effects of particles in a size range of 0.1–1 μm are especially pronounced because of interference phenomena resulting from the particles being about the same size as the wavelengths of visible light. An indirect effect of particles is their ability to serve as reaction sites for atmospheric chemical reactions. They also act as nucleation bodies upon which water vapor condenses.

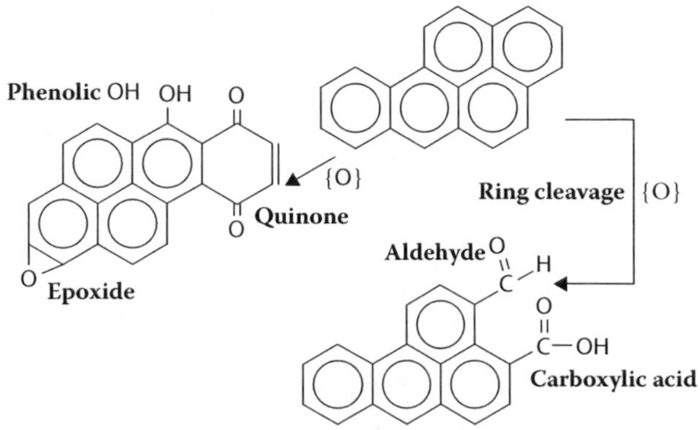

FIGURE 7.2 Oxidation of PAH compounds occurs in the atmosphere.

7.2.8 HEALTH EFFECTS AND TOXICOLOGY OF PARTICLES

Particles also have direct health effects when inhaled. This is especially true of very small particles that can be carried into the innermost parts (alveoli) of lungs. The particles that are most damaging are very small ones less than 2.5 µm in size (less than 1/30 the diameter of a human hair) designated $PM_{2.5}$ These are mostly condensation aerosols and may contain toxic elements, such as arsenic, acids, such as H_2SO_4, and carcinogenic polycyclic aromatic hydrocarbons including benzo(a)-pyrene. Because of their very small size, these particles have very high surface-to-volume ratios and biochemically active surfaces.

Figure 7.3 illustrates the pathway of exposure of humans to air pollutants including particles through the respiratory system, which is usually the most common route of exposure to toxicants. Because of the strong possibility for worker exposure to toxic respirable particles and other toxic substances in the atmosphere, **inhalation toxicology** is a very important aspect of industrial hygiene.

A common health condition caused by industrial exposure to particles is **silicosis** produced by the inhalation of silica (SiO_2) dust generated from finely ground rock including sand and granite. This material is used as an abrasive agent in applications such as cleaning rock surfaces (sand blasting) and is generated in the grinding and processing of rock and concrete. Usually, after many years exposure to silica dust, the lungs become inflamed and swollen and the victim becomes short of breath, is afflicted with a dry, nonproductive cough, and may eventually suffer respiratory failure and even death. In addition to bronchitis and chronic obstructive pulmonary disorder, silicosis has been associated with elevated levels of lung cancer, tuberculosis, and scleroderma, a disorder of skin, blood vessels, joints, and skeletal muscles.

One of the most notable kinds of particulate matter with known toxic effects is **asbestos**, one of several kinds of silicon-containing minerals with an approximate formula of $Mg_3P(Si_2O_5)(OH)_4$, which may be in the form of elongated fibers so thin that they can puncture individual cells. With extraordinary heat resistance and excellent insulating qualities, asbestos was once widely used for insulation, structural materials, brake linings, and pipe manufacture, although essentially all of these uses have now been eliminated. Asbestos has never been a pollution problem in the atmosphere as a whole, but it has been a troublesome indoor air pollutant in some cases and has posed particular health problems for workers who handled it. Now the major concern is with release of asbestos particles from demolishing and renovating old structures in which asbestos has been used, and programs have been undertaken to safely remove asbestos from buildings.

Inhalation of asbestos may causes **asbestosis** (a pneumonia condition), **mesothelioma** (a tumor of the mesothelial tissue lining the chest cavity around the lungs), and **bronchogenic carcinoma**

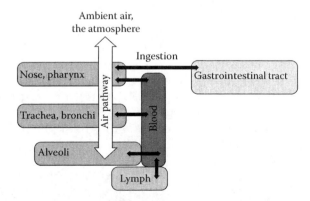

FIGURE 7.3 Exposure to toxicants from ambient air through the respiratory tract. Air is pulled into the alveoli, the innermost recesses of the lung where exchange occurs with the bloodstream. Toxicants not filtered out through the upper parts of the airway may enter blood through the alveoli. Particles filtered out by the upper parts of the airway may get into the throat and be ingested into the gastrointestinal tract.

(a cancer of the air passages in the lungs). Victims of asbestosis may suffer from lung inflammation associated with difficult breathing, coughing, and permanent lung damage, leading to premature death from lung failure. Mesothelioma has been one of the most notorious diseases associated with industrial exposure to toxicants, and this malady has firmly established asbestos as a known cause of human cancer.

When asbestos fibers are inhaled, they puncture individual cells in the respiratory tract acting as local irritants. The fibers undergo phagocytosis, meaning that they are engulfed by large white blood cells called macrophages. The bodies consisting of encapsulated asbestos fibers are then taken up by cellular organelles called lysosomes that secrete enzymes that digest the matter surrounding the fibers, releasing the fibers to start the process over again. The net result is that lymphoid tissue aggregates in the vicinity of the insult, forming fibrotic lesions from the synthesis of excess collagen.

There is a strong synergistic relationship between lung cancer from asbestos exposure and exposure to cigarette smoke. Typically, long-term exposure to asbestos increases the incidence of lung cancer about 5-fold, cigarette smoking increases lung cancer about 10-fold, but the two together cause an approximately 50-fold increase.

The very small $PM_{2.5}$ particles associated with photochemical smog (see Section 7.8) have been linked with adverse health effects.[3] Well documented in epidemiological studies, these effects are mostly cardiovascular stemming from cardiovascular inflammation. Lung inflammation from exposure to particulate matter can lead to systemic inflammation, a condition known to be associated with cardiovascular sudden death, resulting from inflammatory processes in atheromatous plaques, imbalance in coagulation factors that favor the propagation of thrombi (arterial blood clots) after thrombosis starts, and adverse effects on the autonomic nervous system involved in control of heart rhythm, potentially causing fatal dysrhythmia.

A significant health concern with particles, especially those from combustion sources, is their ability to carry toxic metals. Of these, lead is of the greatest concern because it usually comes closest to being at a toxic level. Problems with particulate lead in the atmosphere have been greatly reduced by the elimination of tetraethyl lead as a gasoline additive, an application that used to spew tons of lead into the atmosphere every day. Another heavy metal that causes considerable concern is mercury, which can enter the atmosphere bound to particles or as vapor-phase atomic mercury. Airborne mercury from coal combustion can become a serious water pollution problem, leading to unhealthy accumulations of this toxic element in some fish. Other metals that can cause health problems in particulate matter are beryllium, cadmium, chromium, vanadium, nickel, and arsenic (a metalloid).

7.2.9 ASIAN BROWN CLOUD: CLIMATE AND HEALTH EFFECTS

One of the most troublesome kinds of particulate air pollution consists of the **brown cloud** that afflicts parts of southern Africa, the Amazon basin, North America, and most prominently, large areas of Asia (Figure 7.4). The brown cloud is composed of a variety of kinds of particles, soot, black carbon, photochemical smog, and toxic chemicals. Clearly visible from airplanes and 1.6 km or more thick, the Asian brown cloud typically stretches from the Arabian Peninsula to the Yellow Sea and, in the spring, may reach over North and South Korea and Japan, even approaching the coast of California under some conditions. In addition to afflicting the pulmonary systems of millions of people, the brown cloud significantly decreases surface sunlight and is probably altering weather patterns in Asia, causing glaciers in the Himalayas to melt more rapidly, increasing the severity of dry season drought, and increasing the intensity of monsoon rains. Some authorities have estimated that there are more than 300,000 premature deaths in China and India each year, resulting from cardiovascular and respiratory diseases caused by inhalation of the brown cloud constituents. Although black carbon in the brown cloud and deposited on snow from the cloud tends to have a warming effect, overall, the brown cloud probably has a net cooling effect on regional climate, tending to offset the warming effects of greenhouse gases.

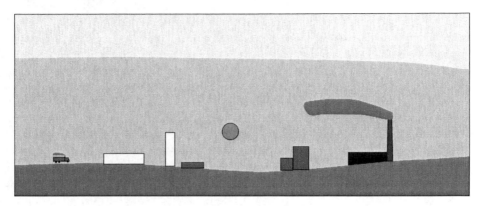

FIGURE 7.4   The brown cloud. Viewed through a layer of polluted air containing a high concentration of particles, the sun may appear as a dim red globe, its light attenuated by the particles. It will have a red hue because of the differential scattering of the shorter wavelength (blue) part of the spectrum.

A study of the Asian brown cloud has concluded that the greatest contribution to this phenomenon in Asia comes from biomass combustion.[4] Twigs, crop residues, and other biomass sources are widely used for cooking in parts of Asia, usually in inefficient modes of combustion that emit large quantities of particulate matter. Cow dung, which burns poorly, is a common source of cooking fuel in India. Direct inhalation of particle-laden smoke from burning biomass for cooking in the microatmosphere around cooking facilities probably contributes substantially to adverse respiratory health effects in people working around such facilities. Another major contributor to brown cloud emissions is the practice of slash-and-burn agriculture, in which trees are killed by slashing their bark, then burned to clear land for agriculture.

7.3   INORGANIC GAS POLLUTANTS

Several inorganic gases are significant air pollutants. Of these, sulfur dioxide and the oxides of nitrogen are of particular importance and are discussed separately in Sections 7.4 and 7.5, respectively. Carbon dioxide is also in a class by itself because, in excess, it is the most significant greenhouse gas responsible for global warming and climate change. It is discussed separately in Chapter 8, Section 8.2. Gases other than these that may be significant atmospheric pollutants are discussed in this section.

Ammonia, NH_3, is a common inorganic air pollutant released as a gas to the atmosphere from several sources. In addition to industrial pollution, such as from heating coal to make coke for steel making, ammonia can be added to the atmosphere by bacterial sources, from sewage treatment, and from the decay of animal wastes. Accidental releases can occur from liquid anhydrous ammonia used as an agricultural nitrogen fertilizer.

Ammonia is strongly attracted to water, and so it is normally present in the atmosphere in water droplets. It is the only significant gaseous base in the atmosphere, so that it reacts with atmospheric acids to produce corrosive ammonium salts as shown by the following reactions:

$$NH_3 + H_2SO_4 \rightarrow NH_4HSO_4 \tag{7.8}$$

$$NH_3 + HNO_3 \rightarrow NH_4NO_3 \tag{7.9}$$

Though not a concern as an atmospheric pollutant in the climate as a whole, hydrogen cyanide gas, HCN, can be deadly in a microatmosphere into which it has been released. The cyanide ion, CN^-, from HCN or cyanide salts in the body binds with ferricytochrome oxidase, the iron(III) form of the iron-containing metalloprotein, that cycles between the iron(III) form and the iron(II) form (ferrouscytochrome oxidase) in the metabolic process by which the body utilizes molecular oxygen in metabolism and thus acts as a chemical asphyxiant that prevents O_2 utilization.[5] Hydrogen cyanide is used in some chemical syntheses and is a by-product of some chemical reactions, both of which can lead to accidental release of HCN to the atmosphere. Its high toxicity and ability to penetrate minute places have led to HCN being used as a fumigant to kill pests such as rodents in warehouses, grain storage bins, greenhouses, and holds of ships. Numerous fatal accidental poisonings have occurred when people entered such areas while the HCN was still present. Cyanide salts are used to extract gold from ores in metal refining, in metal plating, and to salvage silver from photographic and x-ray films. Acidification of such solutions can lead to release of HCN. Hydrogen cyanide produced by the burning of nitrogen-containing compounds in fire has caused fatalities. As evidence of its extreme toxicity, hydrogen cyanide was once a favored agent for the execution of condemned criminals.

Gaseous chlorine, fluorine, and volatile fluorides are uncommon air pollutants, but very serious where they occur. Elemental chlorine, Cl_2, is widely produced and distributed as a water disinfectant, bleach, and industrial chemical. It is very reactive and so toxic that it was the first poisonous gas used as a military poison in World War I. Most toxic exposures of chlorine occur as the result of transportation accidents, leading to its release.

Hydrogen chloride, HCl, can get into the atmosphere by accidental releases of the gas from the reaction of reactive chlorine-containing chemicals, such as $SiCl_4$, with atmospheric water:

$$SiCl_4 + 2H_2O \rightarrow 2SiO_2 + 4HCl \tag{7.10}$$

and from the combustion of chlorine-containing polyvinylchloride (PVC) plastic. The strong affinity of HCl gas for water means that it exists as droplets of hydrochloric acid in the atmosphere. Atmospheric HCl is very irritating to mucous membrane tissue and damaging to corrodable materials.

Elemental fluorine (F_2) and hydrogen fluoride are both highly toxic. Fortunately, occurrences of these gases in the atmosphere are very rare. Gaseous silicon tetrafluoride, SiF_4, can be released during steel making and some metal smelting processes when fluorspar (CaF_2) reacts with sand (SiO_2):

$$2CaF_2 + 3SiO_2 \rightarrow 2CaSiO_3 + SiF_4 \tag{7.11}$$

Sulfur hexafluoride, SF_6, is an astoundingly unreactive gaseous compound used to blanket and degas molten aluminum and magnesium and in gas-insulated electrical equipment. It lasts essentially forever in the atmosphere. As noted in Chapter 8, Section 8.3, the greatest concern with its release is that it is a powerful greenhouse warming gas with an effect per molecule (radiative forcing, see Chapter 8, Section 8.2) of about 24,000 times that of carbon dioxide.

Carbonyl sulfide, COS, is another inorganic sulfur gas that can be detected in the atmosphere, though it is usually at very low levels. A related compound, carbon disulfide, CS_2, also occurs in the atmosphere. The toxic effects of carbon disulfide are relatively well known because of the widespread use of the compound as a solvent in the synthesis of rayon and cellophane from cellulose. Rayon plant workers have suffered blisters and other skin disorders from contact with carbon disulfide. Epidemiological studies of workers in the viscose rayon manufacturing industry and other industries where carbon disulfide is used have shown vascular atherosclerotic changes and increased mortalities, including those from cardiovascular causes. Very high levels of this volatile solvent of more than 10 parts per thousand by volume in the atmosphere can cause life-threatening central nervous system disorders.

Hydrogen sulfide, H_2S, enters the atmosphere from a number of natural sources, including geothermal sources, the microbial decay of organic sulfur compounds, and the microbial conversion of sulfate, SO_4^{2-}, to H_2S when sulfate acts as an oxidizing agent in the absence of O_2. Wood pulping processes can release hydrogen sulfide. Hydrogen sulfide commonly occurs as a contaminant of petroleum and natural gas. It also affects some kinds of materials, forming a black coating of copper sulfide, CuS, on copper roofing. This coating weathers to a rather attractive green layer (patina) of basic copper sulfate, $CuSO_4 \cdot 3Cu(OH)_2$, which protects the copper from further attack. Hydrogen sulfide in the atmosphere becomes oxidized to SO_2.

Atmospheric H_2S is phytotoxic, destroying immature plant tissue and reducing plant growth. With a toxicity to animals about the same as that of hydrogen cyanide, and a mode of toxicity resembling that gas, hydrogen sulfide is of particular concern for its toxic effects to humans. A tragic incident of hydrogen sulfide poisoning occurred in Poza Rica, Mexico, in 1950 as the result of a process to recover H_2S from natural gas. Incredibly, the hydrogen sulfide by-product was burned in a flare to produce sulfur dioxide. The flare became extinguished at night and the toxic hydrogen sulfide spread throughout the vicinity, killing 22 people and hospitalizing over 300. A 2003 blowout in a natural gas field in southwestern China released hydrogen sulfide that killed almost 200 people in the surrounding area. As an emergency measure, the escaping gas was set on fire producing sulfur dioxide, SO_2, still a toxic gas, but much less deadly than hydrogen sulfide.

Like cyanide, hydrogen sulfide inhibits the cytochrome oxidase system by which the body utilizes O_2 for respiration. Symptoms of hydrogen sulfide poisoning are largely in the central nervous system and include headache, dizziness, and excitement. Death follows rapidly upon inhalation of air containing 1000 parts per million (ppm) by volume of H_2S, and lower levels of exposure may cause death over a long time period. Exposure to hydrogen sulfide paralyzes the respiratory system, resulting in asphyxiation. Localized toxic effects of hydrogen sulfide may also occur at the point of contact; one such effect is pulmonary edema manifested by fluid accumulation in the lungs. Direct exposure of the eye to hydrogen sulfide gas can result in eye conjunctivitis, an inflammation of the thin transparent conjunctiva tissue lining the inner surface of the eyelid and covering the white part of the eye (familiar to most people as the result of infectious "pink eye"). Occupational occurrence of this condition from exposure to hydrogen sulfide is sometimes called "gas eye" after employees who developed it in gas works in which high-sulfur coal was gasified producing H_2S by-product.

Generally produced by partial combustion of fuels, largely in the internal combustion engine, **carbon monoxide**, CO, is an air pollutant of some concern because of its direct toxicity to humans. Carbon monoxide is toxic because it binds to blood hemoglobin and prevents the hemoglobin from transporting oxygen from the lungs to other tissues. Global and regional levels of atmospheric carbon monoxide are too low to be of concern. However, local levels in areas with heavy automobile traffic can become high enough to pose a health hazard and levels on some congested urban streets have reached concentrations of 50–100 ppm. The use of exhaust pollution control devices on automobiles has lowered these levels significantly during the last 30 years. The numerous fatal cases of carbon monoxide poisoning that occur each year are almost always the result of improperly vented heating devices in indoor areas. Carbon monoxide is removed from the atmosphere by reaction with hydroxyl radical, $HO\cdot$.

7.4   NITROGEN OXIDE AIR POLLUTANTS

Nitrous oxide (N_2O), colorless, odorless, **nitric oxide** (NO), and pungent-smelling, red-brown **nitrogen dioxide** (NO_2) occur in the atmosphere. They are discussed in this section.

Nitrous oxide, a commonly used anesthetic known as "laughing gas," is produced by some chemical syntheses and by microbiological processes and is one of the gaseous species by which nitrogen is returned to the atmosphere in the nitrogen biogeochemical cycle. Nitrous oxide is a component of the unpolluted troposphere at a level of approximately 0.3 ppm. This gas is relatively unreactive in the troposphere where it lasts for a long time and probably does not significantly influence important chemical reactions in the lower atmosphere. Its concentration decreases rapidly with altitude in the stratosphere owing to the photochemical reaction

$$N_2O + h\nu \rightarrow N_2 + O \qquad (7.12)$$

and some reaction with singlet atomic oxygen:

$$N_2O + O \rightarrow N_2 + O_2 \qquad (7.13)$$

$$N_2O + O \rightarrow 2NO \qquad (7.14)$$

These reactions are significant in terms of depletion of the stratospheric ozone layer. The NO produced from N_2O reacts with stratospheric ozone

$$NO + O_3 \rightarrow NO_2 + O_2 \qquad (7.15)$$

to produce O_2 and NO_2, the latter of which in turn reacts with O atoms present as part of the process by which stratospheric ozone is produced

$$NO_2 + O \rightarrow NO + O_2 \qquad (7.16)$$

to regenerate NO. The two preceding reactions constitute a repeating cycle in which NO originally produced in the stratosphere from N_2O destroys stratospheric ozone, with a single molecule of NO destroying many molecules of O_3. With the production of ozone-destroying CFCs now banned and the production of hydrochlorofluorocarbons now scheduled to cease by 2030 in accordance with the Montreal Protocol on Substances that Deplete the Ozone Layer, nitrous oxide may well emerge as the greatest threat to stratospheric ozone.[6] This problem may well be exacerbated by increased global fixation of nitrogen in the anthrosphere for use as crop fertilizer, accompanied by increased microbial production of N_2O by microorganisms in soil and in water that stands in rice paddies.

Both NO and NO_2, collectively designated as NO_x, are produced from natural sources, such as lightning and biological processes, and from pollutant sources. Pollutant concentrations of these gases can become too high locally and regionally, causing air pollution problems. A major pollutant source of these gases is the internal combustion engine, in which conditions are such that molecular elemental nitrogen and oxygen react

$$N_2 + O_2 \rightarrow 2NO \qquad (7.17)$$

to produce NO. Combustion of fuels that contain organically bound nitrogen, such as coal, also produces NO. Atmospheric chemical reactions convert some of the NO emitted into NO_2.

Exposed to electromagnetic radiation of wavelengths below 398 nm, nitrogen dioxide undergoes photodissociation

$$NO_2 + h\nu \rightarrow NO + O \qquad (7.18)$$

to produce highly reactive O atoms. The O atoms can participate in a series of chain reactions through which NO is converted back to NO_2, which can undergo photodissociation again to start the cycle over (see Figure 7.5). Nitrogen dioxide is very reactive, undergoing photodissociation within a minute or two in direct sunlight. The gas is a major player in tropospheric chemistry, particularly in its interactions involving sunlight and atmospheric hydrocarbons, leading to photochemical smog formation, discussed in more detail in Section 7.8.

Some of the nitrogen oxides in the atmosphere end up as nitric acid, HNO_3, or as nitrate salts, such as NH_4NO_3 or NH_4HSO_4, which typically are washed from the atmosphere with rainfall. Nitric acid contributes to acid rain (see Section 7.6). Figure 7.5 shows a cycle involving NO, NO_2, and HNO_3. This cycle is very much tied with the process of photochemical smog formation (see Section 7.8) under conditions of smog formation.

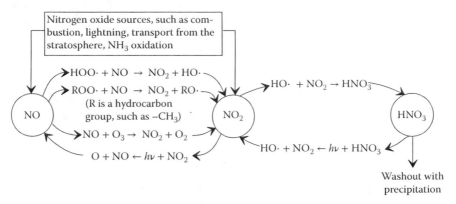

FIGURE 7.5 Principal reactions among NO, NO_2, and HNO_3 in the atmosphere. ROO· represents an organic peroxyl radical, such as the methylperoxyl radical, $CH_3OO·$.

7.4.1 TOXIC EFFECTS OF NITROGEN OXIDES

Nitrous oxide, N_2O, is not particularly toxic and is used as an anesthetic, an application in which it is generally regarded as being safe. The anesthetic effects of this gas may present some hazards from lowered levels of consciousness, and fatalities have even resulted from overexposure to this gas. Long-term exposures to relatively low levels of this gas are suspected of causing central nervous system, cardiovascular, and hepatic (liver) symptoms as well as effects on the organs that produce blood. Increased risk of adverse reproductive effects including spontaneous abortion, premature delivery, and infertility has been implied from epidemiological studies of operating room personnel.

Nitrogen dioxide, NO_2, is significantly more toxic than NO, although concentrations of NO_2 in the outdoor atmosphere rarely reach toxic levels. Accidental releases of NO_2 can be sufficient to cause toxic effects or even death. Brief exposures to 50–100 ppm of NO_2 in air inflames lung tissue for 6–8 weeks followed by recovery. Exposure to 500 ppm or more of NO_2 causes death within 2–10 days. Exposure to 100–500 ppm of NO_2 causes a lung condition with the ominous name of *bronchiolitis fibrosa obliterans* that is fatal within 3–5 weeks after exposure. Fatal incidents of NO_2 poisoning have resulted from accidental release of the gas used as an oxidant in rocket fuels and from burning of nitrogen-containing celluloid and nitrocellulose moving picture film (nitrocellulose has been used as an explosive and has long been banned in film because of catastrophic fires that killed numerous people). Plants exposed to nitrogen dioxide may suffer decreased photosynthesis, leaf spotting, and breakdown of plant tissue.

7.5 SULFUR DIOXIDE AIR POLLUTION

Sulfur dioxide enters the atmosphere as a result of the following:

- Direct emissions from volcanoes
- Atmospheric oxidation of H_2S emitted into the atmosphere by bacteria and from geothermal sources (volcanoes, hot springs, geysers)
- Atmospheric oxidation of dimethyl sulfide, $(CH_3)_2S$, emitted into the atmosphere from marine organisms
- Pollutant sources from the combustion of organic sulfur and iron pyrite, FeS_2, in fossil fuels

The pollutant sources are of most concern because of their contribution to local and regional air pollution problems and because they are sources that humans can do something about.

The fate of sulfur dioxide in the atmosphere is oxidation and reaction with water to produce sulfuric acid. The overall process is complex, but it can be described by the following reaction:

$$2SO_2 + O_2 + 2H_2O \rightarrow 2H_2SO_4 \tag{7.19}$$

This process is generally rather slow in the atmosphere, but it can be quite rapid under conditions of photochemical smog formation (see Section 7.8), in which highly reactive oxidizing species are present. It is very important because it is the main mechanism for forming acid rain (see Section 7.6), which can be directly harmful to vegetation, fish (especially fingerlings), and materials, such as building stone that can be attacked by acid. Sulfur dioxide forms aerosol droplets of sulfuric acid in the atmosphere. As a result, much of the Eastern United States is covered by a slight haze of sulfuric acid droplets during much of the year. In recent years, some volcanic eruptions have blasted enough sulfur dioxide into the atmosphere to produce a sufficient amount of sunlight-reflecting sulfuric acid aerosol to cause a perceptible cooling of the atmosphere.

In addition to indirect effects from the formation of acid rain, sulfur dioxide affects some plants directly, causing leaf necrosis (death of leaf tissue). Another symptom of sulfur dioxide phytotoxicity (toxicity to plants) is chlorosis, a bleaching or yellowing of green leaves.

7.5.1 Toxic Effects of Sulfur Dioxide

Although sulfur dioxide is not a particularly toxic material, episodes have occurred in which increased deaths have been associated with high levels of sulfur dioxide in the atmosphere. The most notable such incident occurred in London in December 1952 when a stagnant layer of dense fog and black coal smoke accompanied by high levels of sulfur dioxide descended on the city for days, a period during which thousands of excess deaths were recorded. Because of particles and fog in the atmosphere, visibility descended to near zero, and the fine particulate pollution was so bad that even the underwear of exposed people was soiled by it.

Sulfur dioxide is an irritant that affects eyes, the respiratory system, mucous membranes, and skin. Because of its water solubility, it is largely removed in the upper respiratory tract, which may be irritated by exposure to the sulfur dioxide, causing bronchioconstriction and increased airflow resistance. Some people are hypersensitive to sulfur dioxide and suffer allergic reactions to it. Included among such people are some asthmatics who are especially sensitive to exposure to the sulfur dioxide.

7.5.2 Toxic Effects of Atmospheric Sulfuric Acid

The toxic effects of sulfuric acid are discussed here because it is an oxidation product of atmospheric sulfur dioxide. Normally, sulfuric acid produced from sulfur dioxide is sufficiently dilute that the health effects are minimal, but the workplace atmosphere where it is used may develop unhealthy levels of sulfuric acid aerosol. The severely corrosive, dehydrating effects of exposure to concentrated H_2SO_4 are well known, resembling those of severe thermal burns, but slower to heal. Workers may become exposed to sulfuric acid fumes and mists in the workplace resulting in eye and respiratory tract irritation. Tooth erosion has been observed in workers exposed to sulfuric acid mists and fumes. The pulmonary tract may be damaged by exposure to sulfuric acid aerosols such as those responsible for acid rain, and sulfuric acid is a much more potent lung irritant than sulfur dioxide. Animal studies and limited data from human exposures to atmospheric sulfuric acid aerosols suggest increased airway resistance and inhibited bronchial clearance of inhaled particles. Asthmatics are susceptible to sulfuric acid inhalation, an effect that likely is synergistic with exposure to sulfur dioxide. In severe air pollution incidents, sulfuric acid particles, other kinds of particles, and sulfur dioxide are likely to occur together, adding up to a combined effect much greater than any one individually.

7.6 ACID-BASE REACTIONS IN THE ATMOSPHERE AND ACID RAIN

Acids are common constituents of the atmosphere. The acid that is always present in the atmosphere is CO_2, a very weak acid, which tends to make liquid water in the atmosphere slightly acidic because of the following reaction:

$$CO_2 + H_2O \rightarrow H^+ + HCO_3^- \tag{7.20}$$

It may be shown that at 25°C, water in equilibrium with unpolluted air containing current levels of about 390 ppm carbon dioxide, the value of dissolved $CO_2(aq)$ is 1.276×10^{-5} mol/L (M), resulting in a pH of 5.62 for the atmospheric water. Therefore, in the absence of pollutants that are stronger acids than CO_2, the pH of water from the atmosphere is just slightly acidic.

Along with hydrogen chloride, HCl, emitted into the atmosphere by the combustion of chlorine-containing organic compounds, sulfur dioxide and nitrogen oxides react in the atmosphere to produce strongly acidic H_2SO_4 (see Reaction 7.19) and HNO_3 (see Figure 7.5), respectively. Incorporated into rainwater, these acids fall to the ground as **acid rain**. A more general term **acid deposition** refers to the effects of atmospheric strong acids, acidic gases (SO_2), and acidic salts (NH_4NO_3 and NH_4HSO_4). Acid deposition is a major air pollution problem.

Figure 7.6 shows a typical distribution of acidic precipitation in the 48 contiguous U.S. states. This figure illustrates that acidic precipitation is a *regional* air pollution problem, not widespread enough to be a *global* problem, but spreading beyond *local* areas. (There have been some unfortunate cases where localized release of acidic emissions to the atmosphere, usually as sulfur dioxide from metal ore smelting operations, has affected local areas, often devastating vegetation within several kilometers of the source.)

Transport processes that move atmospheric acids and their precursor acid gases from their sources to downwind areas are very important in determining areas affected by acid rain. The northeastern United States and southeastern Canada are affected by acid originating from stack gas emissions carried by prevailing southwesterly winds from Missouri, Illinois, Kentucky, and other regions to the southwest. Southern Norway, Sweden, and Finland receive acid precipitation originating farther south in Europe.

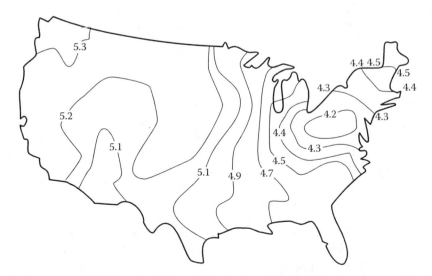

FIGURE 7.6 Isopleths of pH illustrating a hypothetical, but typical, precipitation-pH pattern in the lower 48 continental United States. Actual values found may vary with the time of year and climatic conditions, but generally show lower pH values (more acidic rainfall) in northeastern states.

Numerous adverse effects have been reported as the result of acidic precipitation. These can be divided into the following major categories:

- *Direct effects upon the atmosphere manifested by reduced and distorted visibility*: These effects are due to the presence of sulfuric acid droplets and solutions or solid particles of acidic salts, such as NH_4HSO_4.
- *Phytotoxicity (toxicity to plants) and destruction of sensitive forests*: These effects can be direct, resulting from exposure of plant leaves and roots to acidic precipitation and to acid-forming gases, particularly SO_2 and NO_2. They can also be indirect, primarily by the liberation of phytotoxic Al^{3+} ion by the action of acidic rainfall on soil.
- *Direct effects on humans and other animals*: These are usually respiratory effects, and asthmatics are especially vulnerable.
- *Effects upon plants and fish (especially fish fingerlings)*: In acidified lake water where the lake is not in contact with minerals, particularly $CaCO_3$, capable of neutralizing acid.
- *Damage to materials*: Stone (especially acid-soluble limestone and marble) and metal used in building construction can be corroded and etched by acidic precipitation. Electrical equipment, particularly relay contacts and springs, can be corroded by acidic precipitation.

7.7 ORGANIC AIR POLLUTANTS

7.7.1 ORGANICS IN THE ATMOSPHERE FROM NATURAL SOURCES

The quantities of organic chemicals emitted into the atmosphere from natural sources exceed those emitted as pollutants from the anthrosphere. This ratio is primarily the result of the huge quantities of methane produced by anoxic bacteria in the decomposition of organic matter (represented as $\{CH_2O\}$) in water, sediments, and soil:

$$2\{CH_2O\}(\text{bacterial action}) \rightarrow CO_2(g) + CH_4(g) \tag{7.21}$$

Flatulent emissions from domesticated animals, arising from bacterial decomposition of food in their digestive tracts, add about 85 million metric tons of methane to the atmosphere each year. Anoxic conditions in intensively cultivated rice fields produce large amounts of methane, perhaps as much as 100 million metric tons per year. Methane is a natural constituent of the atmosphere and is present at a level of about 1.8 ppm in the troposphere.

Methane in the troposphere contributes to the photochemical production of carbon monoxide and ozone. The photochemical oxidation of methane is a major source of water vapor in the stratosphere.

Natural sources of non-methane hydrocarbons emitted into the atmosphere include microorganisms, forest fires, animal wastes, volcanoes, and vegetation. Plants are the most significant source of non-methane **biogenic hydrocarbons** with several hundred different hydrocarbons released to the atmosphere from vegetation sources. One of the simplest organic compounds given off by plants is ethylene, C_2H_4, which is produced by a variety of plants and released to the atmosphere in its role as a messenger species regulating plant growth. Because of its double bond, ethylene from vegetation sources is highly reactive with hydroxyl radical, $HO\cdot$, and with oxidizing species in the atmosphere and is an active participant in atmospheric chemical processes.

Most of the hydrocarbons emitted by plants are **terpenes**, which are produced by conifers (evergreen trees and shrubs such as pine and cypress), plants of the genus *Myrtus*, and trees and shrubs of the genus *Citrus*. Three of the most common terpenes emitted by trees are α-pinene from pine trees, limonene from citrus trees and fruit, and isoprene (2-methyl-1,3-butadiene), a hemiterpene, from the *Hevea brasilinesis* tree, the main source of isoprene from which natural rubber is synthesized (see Figure 7.7). In many areas, terpenes from plants, which are highly reactive with atmospheric oxidants including ozone and hydroxyl radical, constitute the main source of hydrocarbons

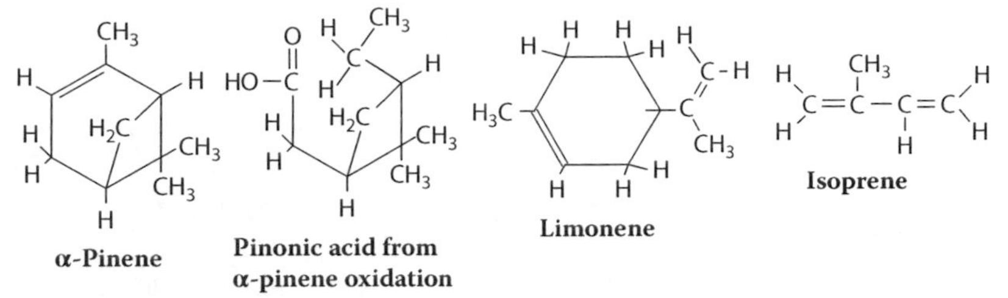

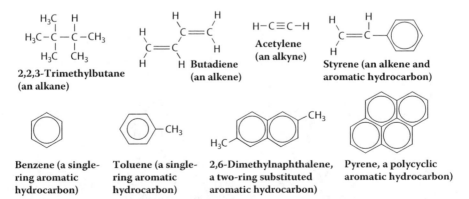

FIGURE 7.7   Some of the more important terpenes emitted by plants and an oxidation product of α-pinene.

that are oxidized by atmospheric chemical processes to produce **secondary organic aerosol**, often manifested as a haze around forests and citrus groves.[7] Such aerosols are particulate atmospheric oxidation products of terpenes and other reactive hydrocarbons in the atmosphere as shown by the example of pinonic acid in Figure 7.7.

Many different **ester** compounds are released to the atmosphere by plants as shown by the following example of citronellyl formate. Although the quantities of esters released are small, they are largely responsible for the fragrances of plants and their flowers and fruits.

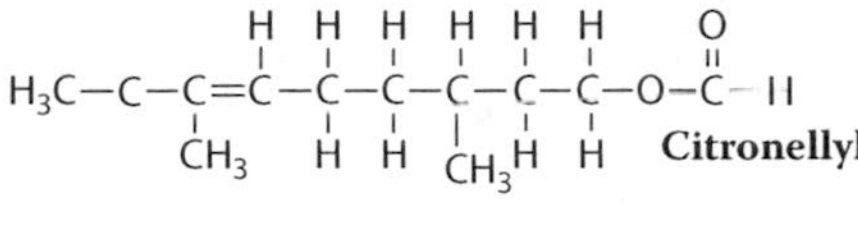

$$\underset{\underset{CH_3}{|}}{H_3C-C-C}=\underset{\underset{H}{|}}{\overset{\overset{H}{|}}{C}}-\underset{\underset{H}{|}}{\overset{\overset{H}{|}}{C}}-\underset{\underset{CH_3}{|}}{\overset{\overset{H}{|}}{C}}-\underset{\underset{H}{|}}{\overset{\overset{H}{|}}{C}}-\underset{\underset{H}{|}}{\overset{\overset{H}{|}}{C}}-O-\overset{\overset{O}{\|}}{C}-H \quad \textbf{Citronellyl formate}$$

7.7.2   POLLUTANT HYDROCARBONS FROM THE ANTHROSPHERE

Because of their widespread use in fuels and in petrochemical manufacture, hydrocarbons predominate among organic atmospheric pollutants released from anthrospheric activities, examples of which are shown in Figure 7.8. These include alkanes, of which methane, discussed earlier, and 2,2,3-trimethylbutane, typical of alkanes in gasoline, are examples; alkenes including ethylene (see above) and butadiene, which is used to synthesize synthetic rubber; the alkyne acetylene used as a fuel for cutting and welding torches; aromatic hydrocarbons with one or more rings; and substituted aromatic compounds. Alkenes are especially significant because of their high reactivity in the atmosphere, their widespread production as synthetic chemicals, and their generation in internal combustion engines from partial combustion of fuel.

2,2,3-Trimethylbutane (an alkane)

Butadiene (an alkene)

Acetylene (an alkyne)

Styrene (an alkene and aromatic hydrocarbon)

Benzene (a single-ring aromatic hydrocarbon)

Toluene (a single-ring aromatic hydrocarbon)

2,6-Dimethylnaphthalene, a two-ring substituted aromatic hydrocarbon)

Pyrene, a polycyclic aromatic hydrocarbon)

FIGURE 7.8   Structural formulas representative of classes of air pollutant hydrocarbons that may be introduced into the atmosphere from anthrospheric activities.

7.7.3 Nonhydrocarbon Organics in the Atmosphere

Many organic compounds that contain atoms in addition to hydrogen and carbon occur as atmospheric pollutants. Of these, the most significant are compounds containing oxygen, some of which are emitted as primary pollutants but are usually formed as secondary pollutants because of the tendency for atmospheric hydrocarbons to become oxidized. Important examples of organooxygen compounds encountered in the atmosphere are shown in Figure 7.9.

Two important organic **oxides**, ethylene oxide and propylene oxide, rank among the 50 most widely produced industrial chemicals and have a limited potential to enter the atmosphere as pollutants. Ethylene oxide is a moderately-to-highly toxic, sweet-smelling, colorless, flammable, explosive gas used as a chemical intermediate, sterilant, and fumigant. It is a mutagen and a carcinogen to experimental animals and classified as a hazardous substance because of both its toxicity and ignitability. Ethanol is an alcohol, now produced and distributed in large quantities as a gasoline additive. Although volatile, the lower alcohols, methanol and ethanol, are completely water soluble and are readily removed from the atmosphere with rain. Phenol is the most common aromatic alcohol. Phenol is toxic and may be of some concern as an indoor air pollutant in manufacturing operations and from construction materials.

Compounds that have a carbonyl moiety (group), C=O, on an end carbon (**aldehydes**) or middle carbon (**ketones**) are often the first species formed, other than unstable reaction intermediates, in the photochemical oxidation of atmospheric hydrocarbons. Aldehydes are important in atmospheric chemistry because they are second only to NO_2 as atmospheric sources of free radicals produced by the absorption of light. This is because the carbonyl group is a **chromophore**, a molecular group that readily absorbs light and it absorbs well in the near-ultraviolet region of the spectrum to produce active species that can take part in atmospheric chemical processes.

Formaldehyde is produced in the atmosphere as a product of the reaction of atmospheric hydrocarbons beginning with their reactions with hydroxyl radical, HO·. With annual global industrial production exceeding 1 billion kg, volatile formaldehyde is used in the manufacture of plastics, resins, lacquers, dyes, and explosives and is uniquely important because of its widespread distribution and toxicity, especially as a cause of hypersensitivity. Humans may be exposed to formaldehyde in manufacturing processes and as emissions from adhesives and resins used in indoor construction materials such as paneling and particle board. Because of potential exposure from such sources, formaldehyde is one of the indoor air pollutants of greatest concern. However, improved manufacturing processes have reduced the hazards from such materials. Acetaldehyde is a widely produced organic chemical used in the manufacture of acetic acid, plastics, and raw materials. Approximately, a billion kilograms of acetone are produced each year as solvents and for applications in the rubber, leather, and plastics industries.

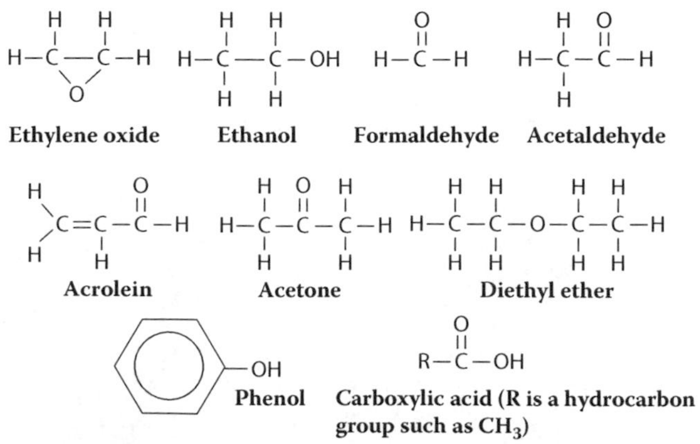

FIGURE 7.9 Structural formulas representative of classes of air pollutant oxygen-containing organic compounds.

Carboxylic acids, such as pinonic acid from the atmospheric oxidation of α-pinene shown in Figure 7.7, are often the final oxidation products of atmospheric hydrocarbon contaminants. Such acids with three or more carbon atoms are not very volatile and tend to be removed from the atmosphere with particulate matter.

7.7.4 ORGANOHALIDES

Organohalides consisting of halogen-substituted hydrocarbon molecules, each of which contains at least one atom of F, Cl, Br, or I, may be saturated (**alkyl halides** also called **haloalkanes**), unsaturated (**alkenyl halides**), or aromatic (**aromatic halides**). The organohalides of environmental and toxicological concern exhibit a wide range of physical and chemical properties. Although most organohalide compounds regarded as pollutants are from anthropogenic sources, it is now known that a large variety of such compounds are generated by organisms, particularly those in marine environments.

Structural formulas of several organohalides that are potentially important in the atmosphere are shown in Figure 7.10. **Dichloromethane** is a volatile liquid with excellent solvent properties and a number of industrial uses. It is commonly used as an extracting solvent for organic-soluble substances determined in chemical analysis. **Dichlorodifluoromethane** is one of the CFC compounds once widely manufactured as a refrigerant and subsequently found to cause stratospheric ozone depletion (see Section 7.9). **HCFC-124**, 2-chloro-1,1,1,2 tetrafluoroethane, was developed as a substitute for CFC compounds in refrigerants and propellants. It has a C–H bond that is acted upon by hydroxyl radical and, thus, is largely eliminated in the troposphere, but still has some capacity to destroy stratospheric ozone. **HFC-134a** has also been developed as a CFC substitute. It is destroyed by reaction with hydroxyl radical in the troposphere and does not deplete stratospheric ozone, but is a powerful greenhouse gas. **Vinyl chloride** is the monomer used to manufacture huge quantities of polyvinylchloride plastics. **Trichloroethylene** is a widely used solvent used in applications such as dry cleaning. The **polychlorinated biphenyl compounds** consist of a family of compounds containing two aromatic rings joined as shown in the example in Figure 7.10 and containing from 1 to 10 chlorine atoms. Brominated alkanes, commonly called **halons**, act to quench fires and have been the predominant materials for aircraft fire extinguishers. However, the halons are powerful ozone depleting compounds, which has led to restrictions on their use. Although the manufacture of halons has ceased, halon fire extinguishers are still used that employ recycled halons.

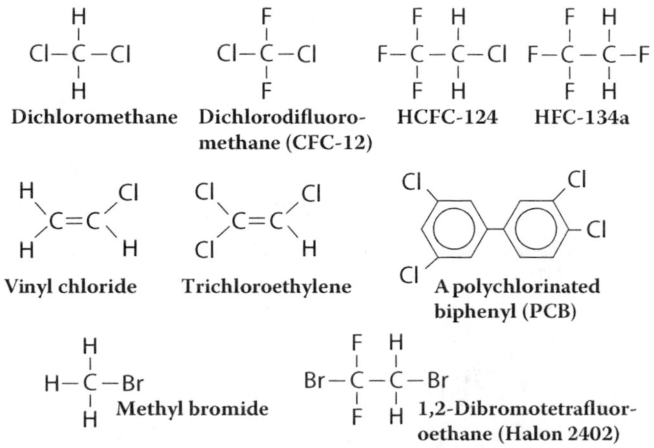

FIGURE 7.10 Some organohalide compounds that are, or have been in the past, potential air pollutants.

7.7.5 TOXICOLOGICAL CHEMISTRY OF ORGANOHALIDES

The levels of organohalides in ambient air are probably too low to cause any toxic effects. At higher air concentrations, as might be encountered in the workplace or around waste sites containing organohalides, a variety of toxic effects may be observed, resulting from inhalation exposure to organohalides. The once commonly used CFCs are notable for their low toxicities, and in the 1930s, their inventor used to do a demonstration in which he would fill his lungs with dichlorodifluoromethane vapor, then extinguish a candle flame by blowing his breath over it. The alkenyl halides tend to be relatively active metabolically because of the C=C double bond. Of this class of compounds, the most notorious is vinyl chloride, which has been shown to cause a rare angiosarcoma (cancer) of the liver in workers heavily exposed to the compound during polyvinylchloride polymer synthesis. Acute short-term inhalation of methyl bromide has caused lung damage in humans, and longer term inhalation has resulted in neurological damage. Such effects on workers applying methyl bromide as a soil fumigant have led to its being discontinued for general use. The less volatile and poorly metabolized organohalides, such as the PCBs, tend to bioaccumulate in lipid tissue.

7.7.6 ORGANOSULFUR COMPOUNDS

Substitution of alkyl or aryl hydrocarbon groups such as phenyl and methyl for H on hydrogen sulfide, H_2S, leads to a number of different **organosulfur thiols** (mercaptans, R–SH) and sulfides, also called thioethers (R–S–R). Structural formulas of examples of these compounds are shown in Figure 7.11. Methanethiol and other lighter alkyl thiols are fairly common air pollutants that have "ultragarlic" odors. The most significant atmospheric organosulfur compound is dimethylsulfide, produced in large quantities by marine organisms and introducing quantities of sulfur to the atmosphere comparable in magnitude to those introduced from pollution sources. Its oxidation produces most of the SO_2 in the marine atmosphere.

Although not highly significant as atmospheric contaminants on a large scale, organic sulfur compounds can cause local air pollution problems because of their bad odors. Major sources of organosulfur compounds in the atmosphere include microbial degradation, wood pulping, volatile matter evolved from plants, animal wastes, packing-house and rendering-plant wastes, starch manufacture, sewage treatment, and petroleum refining.

7.7.7 ORGANONITROGEN COMPOUNDS

Organic nitrogen compounds that may be found as atmospheric contaminants can be classified as **amines**, **amides**, **nitriles**, **nitro compounds**, or **heterocyclic nitrogen compounds**. Structures of common examples of each of these five classes of compounds reported as atmospheric contaminants

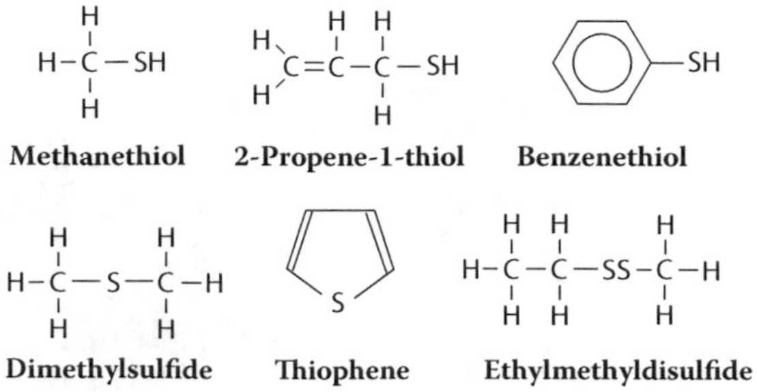

FIGURE 7.11 Common organosulfur compounds associated with air pollution. The organosulfur compounds are especially noted for their strong odors.

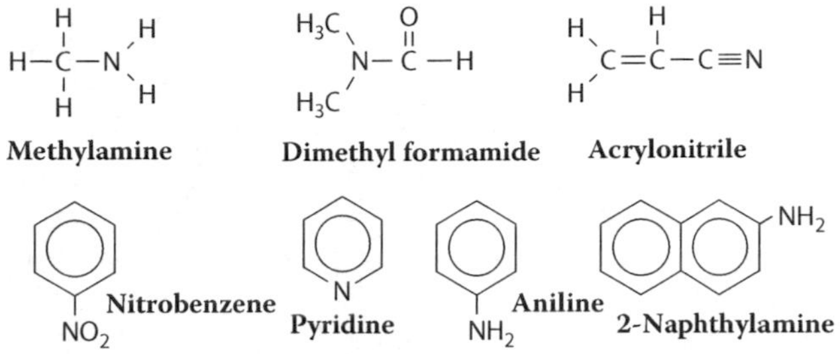

FIGURE 7.12 Organonitrogen compounds that may be encountered as air pollutants.

are shown in Figure 7.12. These organonitrogen compounds can come from anthropogenic pollution sources. Significant amounts of anthropogenic atmospheric organic nitrogen may also come from reactions of inorganic nitrogen (NO and NO_2) with reactive organic species.

Amines consist of compounds in which one or more of the hydrogen atoms in NH_3 has been replaced by a hydrocarbon moiety. Lower-molecular-mass amines are volatile. These are prominent among the compounds giving rotten fish their characteristic odor—an obvious reason that air contamination by amines is undesirable. A number of amines are widely used industrial chemicals and solvents, so industrial sources have the potential to contaminate the atmosphere with these chemicals. Decaying organic matter, especially protein wastes, produce amines, so rendering plants, packing houses, and sewage-treatment plants are potential sources of these substances. The simplest and most important aromatic amine is aniline, used in the manufacture of dyes, amides, photographic chemicals, and drugs.

The amide most likely to be encountered as an atmospheric pollutant is dimethylformamide. It is widely used commercially as a solvent for the synthetic polymer polyacrylonitrile (Orlon, Dacron). Most amides have relatively low vapor pressures, which limit their entry into the atmosphere. Nitriles, which are characterized by the C≡N group, have been reported as air contaminants, particularly from industrial sources. Both acrylonitrile and acetonitrile, CH_3CN, have been reported in the atmosphere as a result of emissions from synthetic rubber manufacture. Among the nitro compounds, RNO_2, reported as air contaminants are nitromethane, nitroethane, and nitrobenzene. These compounds are produced from industrial sources. Highly oxygenated compounds containing the NO_3 group, particularly peroxyacetyl nitrate (PAN, see Section 7.8), are end products of the photochemical oxidation of hydrocarbons in urban atmospheres.

7.7.8 TOXICOLOGICAL CHEMISTRY OF ORGANONITROGEN COMPOUNDS

Though probably not toxic at levels found in the ambient atmosphere, organonitrogen compounds can be harmful in the workplace atmosphere. The lower aliphatic amines are among the more toxic substances in routine, large-scale use. They are basic compounds that hydrolyze to raise the pH in tissue that they contact as shown below for monomethylamine:

$$H_3C - NH_2 + H_2O \rightarrow H_3C - NH_3^+ + OH^- \tag{7.22}$$

Lungs exposed to the lower amines can exhibit hemorrhage and edema (excessive fluid accumulation). Aniline and nitrobenzene may cause the iron(II) in blood hemoglobin to be oxidized to iron(III) so that the blood no longer transports oxygen. Some of the aromatic amines are known human carcinogens. Two important examples are 1-naphthylamine and 2-naphthylamine, which were shown to cause bladder cancer in German dye industry workers exposed to these compounds around 1900.

7.8 PHOTOCHEMICAL SMOG

One of the most common urban air pollution problems is the production of **photochemical smog**. As illustrated in Figure 7.13, this condition occurs usually under situations in which a confined mass of air is subjected to intense sunlight. Such an air mass normally is held in place by a temperature inversion in which warmer, less dense air overlays a mass of cooler, denser, stagnant air, which may be held in place by topographical features, especially a mountain range. A smoggy atmosphere contains ozone, O_3, organic oxidants, nitrogen oxides, aldehydes, and other noxious species. In latter stages of smog formation, visibility in the atmosphere is lowered by the presence of a haze of fine particles formed by the oxidation of organic compounds in smog.

The chemical ingredients of smog are nitrogen oxides and organic compounds, both released from the automobile, as well as from vegetation (see the discussion of biogenic hydrocarbons and terpenes in Section 7.7), and other sources. The driving energy force behind smog formation is electromagnetic radiation with a wavelength at around 400 nm or less, in the ultraviolet region, just shorter than the lower limit for visible light. Energy absorbed by a molecule from this radiation can result in the formation of active species, thus initiating *photochemical reactions*.

Although methane, CH_4, is one of the least active hydrocarbons in terms of forming smog, it will be used here to show the smog formation process, because it is the simplest hydrocarbon molecule. Smog is produced in a series of chain reactions. The first of these occurs when a photon of electromagnetic radiation (represented as $h\nu$) with a wavelength less than 398 nm is absorbed by a molecule of nitrogen dioxide

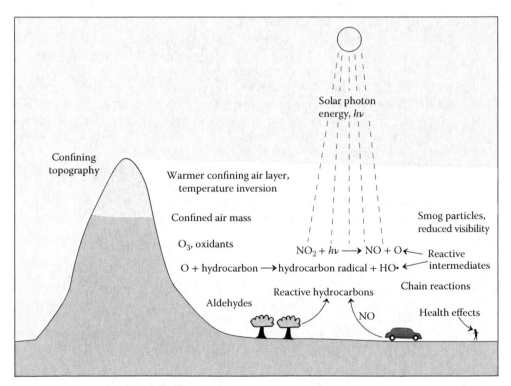

FIGURE 7.13 Photochemical smog formation. A mass of stagnant air receives reactive hydrocarbons from natural sources, such as pine trees, and from pollutant sources, such as automobile exhausts. Nitrogen oxides from combustion sources form NO_2, which is photochemically dissociated to produce reactive O atoms. These abstract H from hydrocarbons to produce the reactive HO· (hydroxyl) radical, starting sequences of chain reactions that produce the noxious products typical of photochemical smog, including oxidant ozone, organic oxidants, aldehydes, and visibility-obscuring fine particles.

$$NO_2 + h\nu \rightarrow NO + O \tag{7.23}$$

to produce an oxygen atom, O. The oxygen atom is a very reactive species that can abstract a hydrogen atom from methane

$$CH_4 + O \rightarrow HO\cdot + H_3C\cdot \tag{7.24}$$

to produce a methyl radical, $H_3C\cdot$, and a **hydroxyl radical**, $HO\cdot$. In these formulas, the dot shows a single unpaired electron. A chemical species with such a single electron is a **free radical**. The hydroxyl radical is especially important in the formation of smog and in a wide variety of other kinds of photochemical reactions. The methyl radical can react with an oxygen molecule

$$H_3C\cdot + O_2 \rightarrow H_3COO\cdot \tag{7.25}$$

to produce a methylperoxyl radical, $H_3COO\cdot$. This is a strongly oxidizing, reactive species. One of the very important reactions of peroxyl radicals is their reaction with NO, produced in the photochemical dissociation of NO_2 (see Reaction 7.23),

$$NO + H_3COO\cdot \rightarrow NO_2 + H_3CO\cdot \tag{7.26}$$

to regenerate NO_2, which can undergo photodissociation, reinitiating the series of chain reactions by which smog is formed.

Reaction 7.24 is illustrative of **abstraction reactions** in which an atom of H is abstracted from a hydrocarbon. The hydroxyl radical product of this reaction, $HO\cdot$, is very active in carrying out abstraction reactions. Even faster **addition reactions** are possible with alkene hydrocarbons that have unsaturated C=C bonds. For example, hydroxyl radical adds to a molecule of propene

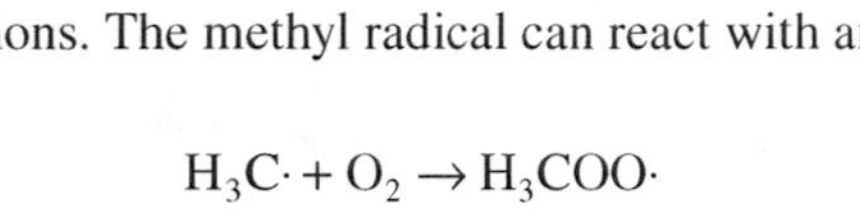

$$(7.27)$$

to produce a reactive radical species that can undergo additional chain reactions beginning with addition of O_2:

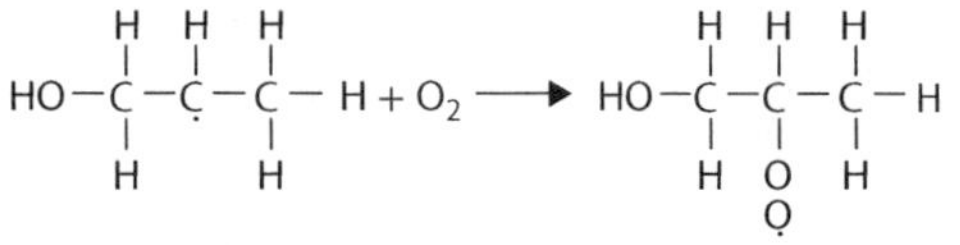

$$(7.28)$$

As the process of smog formation occurs (see Figure 7.14), numerous noxious intermediates are generated. One of the main ones of these is ozone, O_3, and it is the single species most characteristic of smog. Although ozone is an essential species in the stratosphere, where it filters out undesirable ultraviolet radiation, it is a toxic species in the troposphere, which is bad for both animals and plants. Another class of materials formed with smog consists of oxygen-rich organic compounds containing nitrogen of which PAN

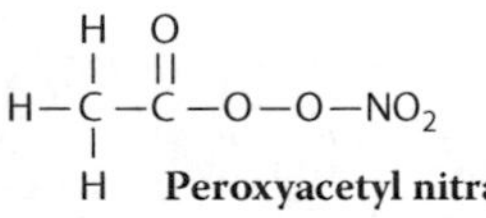

Peroxyacetyl nitrate (PAN)

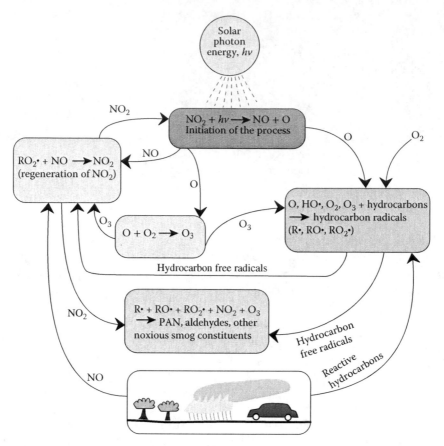

FIGURE 7.14 Overall scheme for the chemistry of the formation of photochemical smog. The key chemical ingredients are reactive hydrocarbons from vehicle engines, from combustion of biomass, such as from burning savannah grass, and from trees; nitric oxide (NO) from combustion sources; and atmospheric elemental oxygen, O_2. The initial energy input comes from the photochemical dissociation of NO_2 to produce reactive O atoms and the energy to sustain the process is from the thermodynamic tendency of elemental oxygen to react with hydrocarbons to produce oxidized organic matter. Free radicals are produced by the reaction of O and HO· with hydrocarbons to produce R·, RO·, and RO_2·, where R is a hydrocarbon group, in the simplest case the methyl group, CH_3, and the dot represents an unpaired electron. Hydroxyl radical, HO·, is a key intermediate in the chain reactions involved with smog formation. Ozone, O_3, and organic oxidants are produced during the formation of photochemical smog, as are aldehydes, which contribute substantially to the noxious character of smog. A key step is reaction of NO with RO_2· to regenerate NO_2, the photochemical dissociation of which initiates the whole process. An important end product consists of very small organic particles with an approximate empirical formula of CH_2O.

is the most common example. This compound and others with high nitrogen and oxygen contents are potent oxidizers and highly irritating to eyes and mucous membranes of the respiratory tract. Also associated with smog are aldehydes, which are irritants to eyes and the respiratory tract. The simplest aldehyde, and one commonly found in smoggy atmospheres, is formaldehyde:

Literally hundreds of reactions occur as photochemical smog is formed, leading eventually to oxidized organic matter that produces the small particulate matter characteristic of smog. In a typical daily cycle in which photochemical smog is produced, levels of hydrocarbons and NO are relatively high at dawn, but as the sunlight becomes more intense, the concentration of NO falls and that of NO_2 increases because of the conversion of NO to NO_2 by the action of pollutant oxidants. In the late afternoon, concentrations of ozone, organic oxidants, aldehydes, and particles, all harmful products of smog, reach high levels. At night, in the absence of sunlight, which is responsible for the photochemical dissociation of NO_2 that starts the smog-producing process (Reaction 7.23), the formation of smog is slowed, although reactions continue that are carried out by the nitrate radical, NO_3, formed by a reaction between NO_2 and O_3. (Although NO_3 forms in the daytime, it is rapidly dissociated to NO and O_2 by visible light with a wavelength less than 700 nm and to NO_2 and O by light less than 580 nm; artificial light in the urban atmosphere may be a factor in determining NO_3 concentrations at night as well.) Figure 7.14 summarizes the main chemical processes involved in photochemical smog formation.

The formation and reaction of compounds containing nitrogen and oxygen are very important in the smog-forming process. In addition to NO, NO_2, and NO_3, these include the inorganic species N_2O_5, HONO (nitrous acid), HNO_3 (nitric acid), and $HOONO_2$ (peroxynitric acid). Also important are organic compounds containing N and O, of which PAN was mentioned earlier. Other compounds of this type include the following (where R represents a hydrocarbon group, in the simplest case $-CH_3$): RONO (alkyl nitrite), $RONO_2$ (alkyl nitrate), and $ROONO_2$ (peroxyalkyl nitrate).

7.8.1 HARMFUL EFFECTS OF SMOG

Smog adversely affects human health and comfort, plants, materials, and atmospheric quality. Each of these aspects is addressed briefly here. Ozone is the smog constituent that is generally regarded as being most harmful to humans, plants, and materials, although very small particles, oxidants other than ozone, and some of the noxious organic materials, such as aldehydes, are harmful as well.

Visibility-reducing atmospheric aerosol particles are the most common manifestation of the harm done to atmospheric quality by smog. The smog-forming process occurs by the oxidation of organic materials in the atmosphere, and as a result, carbon-containing organic materials are the most common constituents of the aerosol particles in an atmosphere afflicted by smog. Conifer trees (pine and cypress) and citrus trees are major contributors to the organic hydrocarbons that are precursors to organic particle formation in smog.

Materials that are adversely affected by smog are generally those that are attacked by oxidants. The best example of such a material is rubber, especially natural rubber, which is attacked by ozone. Indeed, the hardening and cracking of exposed natural rubber has been used to test for the presence of atmospheric ozone.

Plants are harmed by exposure to nitrogen oxides, ozone, and PAN, all oxidants present in a smoggy atmosphere. PAN is the most harmful of these constituents, damaging younger plant leaves, especially. Ozone exposure causes formation of bleached or yellow spots on leaves, a condition called chlorotic stippling (Figure 7.15). Some plant species, including sword-leaf lettuce, black nightshade, quickweed, and double-fortune tomato, are extremely susceptible to damage by oxidant species in smog and are used as bioindicators of the presence of smog. Costs of crop and orchard damage by smog run into millions of dollars per year in areas prone to this kind of air pollution, such as southern California.

The adverse toxic effects of ozone on crops are a major sustainability concern. This is especially true when coupled with adverse climate effects from human activities such as increased greenhouse gas emissions, heat, drought, and sunlight-obscuring particle emissions from increased urbanization.[8] The areas most affected are those with rapidly growing populations and industrialization, with adverse effects on production of staple crops such as rice and wheat, giving rise to problems of food security.

FIGURE 7.15 The leaf of a lemon tree afflicted with chlorotic stippling because of exposure to oxidants in a smoggy atmosphere. This condition, which can seriously damage vegetation, is characterized by bleached or yellow spots on the leaves.

7.8.2 Toxic Effects of Smog and Its Constituents to Humans

People exposed to 0.15 ppm of ozone in air experience irritation to the respiratory mucous tissues accompanied by coughing, wheezing, and bronchial constriction. These effects may be especially pronounced for individuals undergoing vigorous exercise because of the large amounts of air that they inhale. On smoggy days, air pollution alerts may advise against exercise and outdoor activities. Because of these effects, the U.S. Environmental Protection Agency recommends an 8-hour standard limit for ozone of 0.075 ppm and is considering further lowering the standard. In a smoggy atmosphere, the adverse effects of ozone are aggravated by exposure to other oxidants and aldehydes.

Studies on the health effects of photochemical smog have concentrated largely on ozone, which is known to be toxic to humans. A statistical analysis has been performed on the long-term health effects of low-level exposure to ozone (a few parts per billion) and very small atmospheric particles less than 2.5 μm ($PM_{2.5}$) in size, both strongly associated with photochemical smog.[9] After correcting for confounding factors, such as smoking, the results showed that long-term exposure to both of these pollutants increased the risk of death from cardiopulmonary diseases by about 4%, a small, but significant increase. Both of these pollutants were associated with increased death from respiratory maladies. Ozone exposure, alone, increased the risk of death from respiratory maladies but, corrected for $PM_{2.5}$, did not increase the mortality rate owing to cardiovascular causes.

In addition to increased mortality, human exposure to ozone in smog decreases pulmonary functions and initiates inflammatory processes in the lung. The biochemical basis of these effects is not very well established. One possible mechanism is attack by ozone on endogenous lipids that act as surfactants in the lung.

Some of the more adverse health effects of photochemical smog are the result of inhalation of particles in the smog aerosol. The health effects of inhaled particles were discussed in Section 7.2.

7.9 CHLOROFLUOROCARBONS AND STRATOSPHERIC OZONE DEPLETION

As mentioned in Chapter 6, Section 6.2, and illustrated in Figure 6.1, stratospheric ozone, O_3, serves as a shield to absorb harmful ultraviolet radiation in the stratosphere, protecting living beings on the Earth from the effects of excessive amounts of such radiation. The two reactions by which stratospheric ozone are produced are

$$O_2 + h\nu \rightarrow O + O \ (\lambda < 242.4 \text{ nm}) \tag{7.29}$$

$$O + O_2 + M \rightarrow O_3 + M \ \left(\text{energy} - \text{absorbing } N_2 \text{ or } O_2\right) \tag{7.30}$$

and it is destroyed by photodissociation

$$O_3 + h\nu \rightarrow O_2 + O(\lambda < 325 \text{ nm}) \tag{7.31}$$

and a series of reactions from which the net result is the following:

$$O + O_3 \rightarrow 2O_2 \qquad (7.32)$$

Ozone in the stratosphere is present at a steady-state concentration resulting from the balance of ozone production and destruction by the above processes. The quantities of ozone involved are interesting. A total of about 350,000 metric tons of ozone are formed and destroyed daily. Ozone never makes up more than a minuscule fraction of the gases in the ozone layer. In fact, if all the atmosphere's ozone were in a single layer at surface temperature and pressure conditions of approximately 273 K and 1 atm, it would be only 3 mm thick!

Ozone absorbs ultraviolet radiation in the 220–330 nm region very strongly. Therefore, it is effective in filtering out dangerous UV-B radiation, 290 nm $< \lambda <$ 320 nm. (UV-A radiation, 320–400 nm, is relatively less harmful and UV-C radiation, <290 nm, does not penetrate to the troposphere.) If UV-B were not absorbed by ozone, severe damage would result to exposed forms of life on the Earth. Absorption of electromagnetic radiation by ozone converts the radiation's energy to heat and is responsible for the temperature maximum encountered at the boundary between the stratosphere and the mesosphere at an altitude of approximately 50 km. The reason that the temperature maximum occurs at a higher altitude than that of the highest ozone concentration is that ozone is so effective in absorbing ultraviolet radiation that most of this radiation is absorbed in the upper stratosphere, where it generates heat, and only a small fraction reaches the lower altitudes, which remain relatively cool.

Increased intensities of ground-level ultraviolet radiation caused by stratospheric ozone destruction would have some significant adverse consequences. One major effect would be on plants, including crops used for food. The destruction of microscopic plants that are the basis of the ocean's food chain (phytoplankton) could severely reduce the productivity of the world's seas. Human exposure would result in an increased incidence of cataracts. The effect of most concern to humans is the elevated occurrence of skin cancer in individuals exposed to ultraviolet radiation. This is because UV-B radiation is absorbed by cellular DNA (see Chapter 2, Section 2.6), resulting in photochemical reactions that alter the function of DNA so that the genetic code is improperly translated during cell division. This can result in uncontrolled cell division leading to skin cancer. People with light complexions lack protective melanin, which absorbs UV-B radiation, and are especially susceptible to its effects. The most common type of skin cancer resulting from ultraviolet exposure is squamous cell carcinoma, which forms lesions that are readily removed and has little tendency to spread (metastasize). Readily metastasized malignant melanoma caused by absorption of UV-B radiation is often fatal. Fortunately, this form of skin cancer is not very common, although becoming more so.

7.9.1 CHLOROFLUOROCARBONS AND STRATOSPHERIC OZONE DEPLETION

The major culprit in ozone depletion consists of CFC compounds, commonly known as "freons." Other kinds of compounds are implicated in ozone depletion as well, especially the brominated halons used in aircraft fire extinguishers, but never produced in quantities comparable to those of the CFCs. The volatile CFCs were manufactured, used, and released to a very large extent in the decades following their discovery in the 1930s. The major use associated with CFCs has been as refrigerant fluids. Other applications have included solvents, aerosol propellants, and blowing agents in the fabrication of foam plastics. The same extraordinarily high chemical stability that makes CFCs nontoxic enables them to persist for years in the atmosphere and to enter the stratosphere. In the stratosphere, the photochemical dissociation of CFCs by intense ultraviolet radiation

$$CF_2Cl_2 + h\nu \rightarrow Cl\cdot + CClF_2\cdot \qquad (7.33)$$

yields chlorine atoms, each of which can go through chain reactions involving first the reaction of atomic chlorine with ozone:

$$Cl \cdot + O_3 \rightarrow ClO \cdot + O_2 \qquad (7.34)$$

In the most common sequence of reactions involved with stratospheric ozone destruction, the ClO· radicals react to form a dimer, which then reacts to regenerate Cl atoms, which in turn react with ozone to regenerate ClO· in the following reaction sequence (where M is an energy-absorbing third body, such as an N_2 molecule):[7]

$$ClO \cdot + ClO \cdot \rightarrow ClOOCl \qquad (7.35)$$

$$ClOOCl + h\nu \rightarrow ClOO \cdot + Cl \qquad (7.36)$$

$$ClOO \cdot + M \rightarrow Cl \cdot + O_2 + M \qquad (7.37)$$

$$2Cl \cdot + 2O_3 \rightarrow 2ClO \cdot + 2O_2 \qquad (7.38)$$

$$2O_3 \rightarrow 3O_2 \,(\text{net reaction}) \qquad (7.39)$$

The net effect of these reactions is catalysis of the destruction of several thousand molecules of O_3 for each Cl atom produced. Because of their widespread use and persistency, the two CFCs of most concern in ozone destruction are CFC-11 and CFC-12, $CFCl_3$ and CF_2Cl_2, respectively. Even in the intense ultraviolet radiation of the stratosphere, the most persistent CFCs have lifetimes of the order of 100 years.

7.9.2 ANTARCTIC OZONE HOLE

The most prominent instance of ozone layer destruction is the so-called "Antarctic ozone hole" that was first firmly established in 1985 by the British Antarctic Survey and observed with great alarm in subsequent years. This phenomenon is manifested by the appearance during the Antarctic's late winter and early spring months of September and October of severely depleted stratospheric ozone (up to 50%) over the polar region. The reasons why this occurs are related to the normal effect of NO_2 in limiting Cl-atom-catalyzed destruction of ozone by combining with ClO:

$$ClO + NO_2 \rightarrow ClONO_2 \qquad (7.40)$$

During the winter in the polar regions, particularly Antarctica, at temperatures below $-70°C$, NO_x gases are removed along with water by freezing to produce ice crystals or aerosols composed of liquid supercooled ternary mixtures of HNO_3, H_2SO_4, and H_2O in which chlorine originally from CFCs is held in the form of $ClONO_2$ and HCl in polar stratospheric clouds. The reaction of HCl (which comes primarily from the reaction of stratospheric methane, CH_4, with Cl· atoms produced from CFCs) with $ClONO_2$

$$ClONO_2 + HCl \rightarrow Cl_2 + HNO_3 \qquad (7.41)$$

releases Cl_2. Under the conditions of low temperature and sunlight that prevail in the lower stratosphere above Antarctica in spring, the Cl_2 released and the HOCl produced by the reaction of Cl_2 with H_2O undergo photodissociation

$$Cl_2 + h\nu \rightarrow 2Cl \cdot \qquad (7.42)$$

$$HOCl + h\nu \rightarrow HO \cdot + Cl \cdot \qquad (7.43)$$

to produce Cl atoms that can undergo the sequence of chain reactions (Reactions 7.35 through 7.38) leading to ozone destruction. The preceding reactions are aided by the tendency of the HNO_3 product to become hydrogen-bonded with water in the cloud particles. The result of these processes is that over the winter months, photoreactive Cl_2 and HOCl accumulate in the Antarctic stratospheric region in the absence of sunlight, then undergo a burst of photochemical activity when spring arrives, leading to stratospheric ozone destruction and the formation of the Antarctic ozone hole.

The story of the discovery of the Antarctic ozone hole is an interesting one. Depletion of lower atmosphere ozone in Antarctica was first observed in the 1970s, and the first accurate measurements were taken in 1985. The drop in ozone levels during the Antarctic spring of 1985 was so dramatic that the scientists measuring it assumed that their instruments were faulty and had new instruments assembled and flown in, which confirmed the low ozone levels. The TOMS satellite designed to provide stratospheric ozone measurements did not pick up the ozone hole because the software analyzing the satellite data was designed to throw out very low readings as faulty! Subsequent analysis of the data showed that the Antarctic ozone hole did in fact occur, and it has been mapped accurately every year since 1985.

The largest Antarctic ozone hole on record was 29.8 million km^2 reached on September 10, 2000, and extending over the southern parts of South America, including the city of Ushuaia, Argentina. Again in 2006, the area of the Antarctic ozone hole was essentially as large at 29 million km^2. In 2012 the Antarctic ozone hole reached 21 million km^2 compared to 26 million km^2 in 2011, 22 million km^2 in 2010, and 24 million km^2 in 2009.

7.9.3 Nobel Prize in Environmental Chemistry

A richly deserved Nobel Prize in Chemistry, the first ever for environmental chemistry, was awarded in 1995 to three scientists, Paul J. Crutzen, Mario J. Molina, and F. Sherwood Rowland, for their work on the role of CFCs in ozone depletion. In 1970, Dr. Crutzen, Director of the Department of Atmospheric Chemistry at Max Planck Institute for Chemistry, Mainz, Germany, showed that nitrogen oxides are involved with the balance of levels of upper atmospheric ozone, suggesting that catalytic substances from anthropogenic sources, such as NO emitted by high-flying supersonic aircraft, could accelerate the natural destruction of stratospheric ozone, lowering the levels of this essential substance. Drs. Molina and Rowland working at the University of California Irvine established that photodissociation of stratospheric CFC contaminants could put catalytic amounts of atomic Cl into the stratosphere, which would be extraordinarily effective in destroying ozone. Their work provided the basis upon which the United Nations Environment Programme (UNEP) arrived at the Montreal Protocol of 1987, through which production and use of CFCs were to be phased out.

7.10 INDOOR AIR POLLUTION AND THE MICROATMOSPHERE

The indoor microatmosphere (see Chapter 6, Section 6.8) presents some special problems for air pollution and health.[10] Generally, whatever is in outside air comes inside as well. And additional air pollutants may be generated inside from sources such as outgassing of volatile organic compounds (VOCs) from paints and resins, combustion by-products from cooking stove burners or incense burning, and ozone from laser printers and electrostatic air cleaners. Cigarette smoking produces a variety of indoor air pollutants that affect not only the smoker, but other people in the room as well with emissions that include formaldehyde, benzene, and fine particles including carcinogenic polycyclic aromatic hydrocarbons.

VOCs can present special problems in the indoor microatmosphere. A common source of such emissions consists of oil-based paints from which the volatile paint vehicles evaporate as the paint dries. Water-based latex paints even have the potential to emit air pollutants. One such source consists of coalescents added to these paints that enable the colloidal particles in the paint to spread, giving a uniform painted surface. A Presidential Green Chemistry Challenge award was given to the Archer Daniels Midland Company in 2005 for the development of vegetable-oil-based coalescents that do not evaporate but react with oxygen in the air to become part of the paint coating.

Cases have been reported of "sick building syndrome," in which people occupying certain buildings have been afflicted with respiratory distress, allergenic reactions, and other adverse health effects attributed to the microatmospheres in the buildings in which they work. In such cases, the culprit has often been attributed to emissions such as VOCs emitted from synthetic materials used in wall paneling, furniture, and carpets in the building.

Some of the most harmful indoor atmospheres are created in less developed regions where poorly ventilated open fires are used for cooking. These may be in regions where cooking fuel is scarce and wood, brush, and dried animal dung are used for fuel.

As discussed in Section 7.2, radioactive radon emitted from the geosphere below dwellings can be a serious indoor air pollutant. In some areas where radon can be a problem, testing for this indoor air pollutant is required when a dwelling is sold, and it may be necessary to install ventilation facilities around the foundations of afflicted buildings.

The workplace can often suffer from unhealthy indoor air quality due to emissions of materials as the result of activities in the workplace. In Chapter 6, Section 6.8, the unfortunate case was discussed of workers in Turkey who developed lung silicosis resulting from their sandblasting jeans to give them a "distressed" appearance. Because of emissions from primers and paints, the microatmospheres in auto body shops may contain high levels of VOCs and particles produced in spray painting.

An important device for controlling indoor air pollutants is the HEPA (high-efficiency particulate air) air filtration system that removes at least 99.97% of particles greater than 0.3 μm in size from the air passing through the filtration system. Air pollutant molecules such as those of ozone, SO_2, or NO_2 are not removed by particle filtration, but may be reduced by reaction with the filter material and particulate matter held on the filtration system.

QUESTIONS AND PROBLEMS

Access to and use of the Internet is assumed in answering all questions including general information, statistics, constants, and mathematical formulas required to solve problems. These questions are designed to promote inquiry and thought rather than just finding material in the text. So in some cases, there may be several "right" answers. Therefore, if your answer reflects intellectual effort and a search for information from available sources, it may be considered to be "right."

1. What starts a photochemical reaction? What is it called when a series of reactions initiated by a photochemical event continues?
2. What is a free radical?
3. In what two important respects may very small particles participate in atmospheric chemical processes and why are their health effects particularly important?
4. What are the two major classes of atmospheric particles based upon how they are produced?
5. In the earlier days of coal utilization, fly ash was not a major problem. What has changed that has resulted in much greater production of fly ash? What modern mode of coal combustion significantly reduces the production of fly ash and acid gases from combustion?
6. Suggest why lead has become less of a problem as an atmospheric pollutant in recent years.
7. What is the radioactive element that can get into indoor spaces from underground sources?
8. What is the atmospheric phenomenon caused most prominently by particles 0.1–1 μm in size? Why are very small particles especially dangerous to breathe?
9. What is the major adverse health effect of carbon monoxide? Using an Internet search, see if there is any evidence for the production of carbon monoxide in the human body as part of normal metabolic processes.

10. In what sense is acid rain a secondary air pollutant? What is the chemistry behind the formation of most acid rain constituents? Explain how particles in the atmosphere may be either primary or secondary air pollutants?

11. Explain why condensation aerosols are generally regarded as of greater health risks than are dispersion aerosols. Give two examples of dispersion aerosols that have caused health problems.

12. What is black carbon in the atmosphere? How have sources of this air pollutant changed over the last century? Why may it be particularly important in Arctic regions?

13. It is believed that some victims of residential or industrial fires have been killed by a toxic gas released when acrylonitriles, polyurethanes, wool, or silk are burned. What is the likely toxic agent in such a case and why is it produced from these materials?

14. Two small-molecule gaseous air pollutants are emitted from rice paddies to the atmosphere. What are these chemicals, and what are their major air pollution effects?

15. What is the condition called *bronchiolitis fibrosa obliterans*? What common air pollutant that may be present at harmful levels in localized atmosphere may cause this toxic effect?

16. If Yellowstone National Park in the United State was an industrial facility, parts of it might be put off limits by regulatory authorities to unprotected workers due to a toxic air contaminant. Explain.

17. A search of the Internet will bring up stories that during the Civil War in the United States, the atmospheres of battlefields were often afflicted by two organonitrogen pollutants when bodies of dead soldiers and animals were left in the field for several days. These compounds had sickening odors and could even be toxic in the atmosphere. What are they and how are they produced?

18. What is the serious air pollution phenomenon resulting from an atmospheric reaction of sulfur dioxide?

19. In what form may approximately half of the sulfur in coal be physically separated before combustion?

20. What is an important health effect of nitrogen dioxide? Why is NO_2 particularly important in atmospheric chemistry?

21. In 2008/2009, indoor air pollution problems arose in newer houses due to toxic dry wall. What was the cause of this problem and how does it relate to material covered in this chapter?

22. What are five categories of adverse effects from acid precipitation?

23. Chemically, what is distinctive about ammonia in the atmosphere?

24. What is the historic evidence for the toxicity of elemental chlorine in air?

25. What are some sources of atmospheric hydrogen sulfide? Is it a health concern?

26. In what respect is atmospheric carbon dioxide essential to life on Earth? Why may it end up being the "ultimate air pollutant?"

27. What are the ingredients and conditions leading to the formation of photochemical smog? Are all the ingredients from what are commonly regarded as pollution sources?

28. Can microorganisms be involved in the destruction of atmospheric carbon monoxide? Explain.

LITERATURE CITED

1. Epstein, Paul R., and Dan Ferber, *Changing Planet, Changing Health: How the Climate Crisis Threatens Our Health and What We Can Do About It*, University of California Press, Berkeley, CA, 2011.
2. Skele, R. B., T. Berntsen, G. Myhrel, C. A. Pedersen, J. Strom, S. Gerland, and J. A. Ogren, Black Carbon in the Atmosphere and Snow, from Pre-Industrial Times Until Present, *Atmospheric Chemistry and Physics* **11**, 6809–6836 (2011).

3. Donaldson, Ken, Nicholas Mills, William MacNee, Simon Robinson, and David Newby, Role of Inflammation in Cardiopulmonary Health Effects of $PM_{2.5}$, *Toxicology and Applied Pharmacology* **207**, S483–S488 (2005).

4. Gustafson, Örjan, Martin Krusa, Zdenek Zencak, Rebecca J. Sheesley, Lennart Granat, Erik Engström, P.S.P. Rao, Caroline Leck, and Henning Rodhe, Brown Clouds over South Asia: Biomass or Fossil Fuel Combustion? *Science* **323**, 495–498 (2009).

5. Manahan, Stanley E., Toxic Inorganic Compounds, Chapter 11 in *Toxicological Chemistry*, Taylor & Francis/CRC Press, Boca Raton, FL, 2002.

6. Kemsley, Jillian M., Nitrous Oxide Threat to Ozone, *Chemical and Engineering News* **87**(35), 8 (August 31, 2009).

7. Noziere, Barbara, Nelida J. D. González, Anna-Karin Borg-Karlson, Yuxing Pei, Johan Petterson Redeby, Radovan Krejci, Joseph Dommen, Andre S. H. Prevot, and Thorleif Anthonsen, Atmospheric Chemistry in Stereo: A New Look at Secondary Organic Aerosols from Isoprene, *Geophysical Research Letters* **38**, L11807 (2011).

8. Fuhrer, Jurg, Ozone Risks for Crops and Pastures in Present and Future Climates, *Naturwissenschaften* **96**, 173–194 (2009).

9. Jerrett, Michael, Long-Term Ozone Exposure and Mortality, *New England Journal of Medicine* **360**(11), 1085–1095 (2009).

10. Salthammer, Tunga, and Erik Uhde, *Organic Indoor Air Pollutants*, 2nd ed., Wiley-VCH, Weinheim, Germany, 2009.

SUPPLEMENTARY REFERENCES

Abhishek, Tiwary, and Jeremy Colls, *Air Pollution: Measurement, Modelling, and Mitigation*, 3rd ed., Routledge, London, 2010.

Ackerman, Steven A., and John Knox, *Meteorology: Understanding the Atmosphere*, 3rd ed., Jones and Bartlett Learning, Sudbury, MA, 2011.

Cassee, Fleming R., Nicholas Mills, and David E. Newby, *Cardiovascular Effects of Inhaled Ultrafine and Nano-Sized Particles*, Wiley, Hoboken, NJ, 2011.

Cooper, C. David, and F. C. Alley, *Air Pollution Control: A Design Approach*, 4th ed., Waveland Press, Long Grove, IL, 2010.

De Nevers, Noel, *Air Pollution Engineering*, 2nd Reissue ed., Waveland Press, Long Grove, IL, 2010.

Desonie, Dana, *Atmosphere: Air Pollution and its Effects*, Chelsea House Publishers, New York, 2007.

Gurjar, Bhola R., Luisa T. Molina, and C.S.P. Ojha, Eds., *Air Pollution: Health and Environmental Impacts*, Taylor & Francis/CRC Press, Boca Raton, FL, 2010.

Heck, Ronald M., Robert J. Farrauto, and Suresh T. Gulati, *Catalytic Air Pollution Control: Commercial Technology*, 3rd ed., Wiley, Hoboken, NJ, 2009.

Hewitt, C. N., and Andrea Jackson, *Atmospheric Science for Environmental Scientists*, Wiley-Blackwell, Hoboken, NJ, 2009.

Holloway, Ann M., and Richard P. Wayne, *Atmospheric Chemistry*, RSC Publishing, Cambridge, UK, 2010.

Jacobs, Andrew, Haunting Asia, a Brown Cloud Blots Out Sun, *New York Times Health Science* http://www.nytimes.com/2008/11/13/health/13iht-cloud.4.17808200.html, 2008.

Kessel, Anthony, *Air, the Environment and Public Health*, Cambridge University Press, Cambridge, UK, 2011.

Lazaridis, Mihalis, and Ian Colbeck, *Human Exposure to Pollutants Via Dermal Exposure and Inhalation*, Springer, Dordrecht, Netherlands, 2010.

Saponaro, Sabrina, Elena Sezenna, and Luca Bonomo, Eds., *Vapor Emission to Outdoor Air and Enclosed Spaces for Human Risk Assessment: Site Characterization, Monitoring and Modeling*, Gazelle Distribution, Lancaster, UK, 2011.

Sheffield, Mark E., Ed., *Encyclopedia of Air Pollution*, Nova Science Publishers, Hauppage, NY, 2010.

Spellman, Frank R., *The Science of Air: Concepts and Applications*, 2nd ed., Taylor & Francis, Boca Raton, FL, 2009.

Sportisse, Bruno, *Fundamentals in Air Pollution: From Processes to Modelling*, Springer, New York, 2010.

Tammemagi, Hans, *Air: Our Ailing Planet's Atmosphere*, Oxford University Press, New York, 2009.

Vallero, Daniel, *Fundamentals of Air Pollution*, 4th ed., Elsevier, Amsterdam, 2007.

Wallace, John M., *Atmospheric Science: An Introductory Survey*, 2nd ed., Elsevier, Amsterdam, 2006.

Wang, Lawrence K., Norman C. Pereira, and Yung-Tse Hung, *Air Pollution Control Engineering (Handbook of Environmental Engineering)*, Humana Press, Totowa, NJ, 2010.

8 Sustaining the Atmosphere
Blue Skies for a Green Earth

8.1 PRESERVING THE ATMOSPHERE

Nothing is more important to the sustenance of life on Earth than the sustainability of the atmosphere. At the beginning of Chapter 7, it was noted that humans have a most intimate relationship with the atmosphere through the air that they breathe. If the atmosphere contains toxic pollutants, human health will be harmed. Toxic substances in the atmosphere may also harm plants and stunt the growth of food crops and forest trees. The atmosphere also plays a vital protective role for the Earth System, defined and discussed in Chapter 1, Section 1.6, especially in regulating Earth's temperature and in filtering out harmful electromagnetic radiation.

This chapter is about sustaining the atmosphere in a condition most conducive to life on Earth and the quality of the Earth System. It emphasizes means of controlling the release of atmospheric pollutants, including measures to destroy or sequester such pollutants before they can be released into the atmosphere. It is important to emphasize practices that prevent the production of pollutants in the first place, including green chemistry and industrial ecology (see Chapters 14 through 16). Included is a discussion of threats to the stratospheric ozone layer, which is so essential in filtering out damaging ultraviolet solar radiation and measures that may be taken to preserve this vital filter. Of overwhelming importance is the influence of the atmosphere in regulating Earth's temperature and the threat of global climate change resulting from release of greenhouse gases to the atmosphere. In addition to minimizing the release of these gases, it is important to adopt measures to deal with the effects of global climate change that are already under way.

In considering sustainability of the atmosphere, it is very important to consider the interactions of the atmosphere with the other global spheres. The anthrosphere is obviously important both as a source of atmospheric pollutants and for the effects of atmospheric conditions on it. One of the most prominent aspects of the anthrosphere/atmosphere interaction is the production and utilization of energy in the anthrosphere, particularly with respect to burning fossil fuels that release greenhouse gas carbon dioxide. The biosphere is obviously influenced strongly by the atmosphere with respect to such things as the toxic effects of atmospheric pollutants on organisms and the influence of meteorological conditions on bioproductivity. The biosphere also has strong influence on the atmosphere, including the ability of photosynthetic organisms to remove atmospheric carbon dioxide from the atmosphere and the release of biogenic hydrocarbons that are involved in photochemical smog formation. The hydrosphere is strongly affected by the atmosphere, especially with respect to the influence of atmospheric conditions on rainfall, evaporation of liquid water, melting of snow cover accelerated by the deposition of light-absorbing matter (black carbon) from the atmosphere, and the hydrologic cycle in general. The geosphere is strongly affected by the atmosphere, especially in conversion of large areas of the geosphere to desert resulting from atmospheric climate change. In addition, the geosphere influences the atmosphere through volcanic emissions of sulfur dioxide and particles and by the topography of the geosphere surface that helps to confine air masses in which photochemical smog is formed.

8.1.1 Preservation of the Atmosphere's Natural Capital

In Chapter 6, Section 6.4, the atmosphere's natural capital was described. It includes the raw materials that are taken from the atmosphere such as oxygen used by organisms, for combustion, and as an industrial chemical; nitrogen that in the elemental form serves as an inert atmosphere and as a cryogenic liquid that is fixed biologically and chemically to serve as an essential plant nutrient and industrial chemical; carbon dioxide required for plant photosynthesis; protection from damaging solar ultraviolet radiation; temperature stabilization (greenhouse effect); esthetics (visibility); and capacity to absorb and process wastes. Sustaining the atmosphere requires preservation of its natural capital.

There is no danger of diminishing the atmosphere's store of essential oxygen, nitrogen, or carbon dioxide. However, human activities are threatening other parts of the atmosphere's natural capital. The emission of chemical species that have the capacity to diminish the stratospheric ozone layer has posed some threat to this valuable ultraviolet radiation filter. Esthetics and visibility are marred by emission of particles and gases that react to form particles. The capacity of the atmosphere to process wastes is taxed by emissions of acid-forming gases, particles, and organics. And the greatest threat of all is posed by increased levels of carbon dioxide and other greenhouse gases that can disturb the delicate equilibrium that keeps Earth's temperature within limits suitable for life on the planet. These and other sustainability matters related to the atmosphere's natural capital are addressed in this chapter.

8.2 GREATEST THREAT: GLOBAL CLIMATE WARMING

Arguably, the greatest threat posed to the atmosphere and to the Earth System as a whole is the emission of carbon dioxide and other gases that have a warming effect on the atmosphere. Figure 6.4, in Chapter 6, shows the heat balance of Earth's atmosphere, primarily in the troposphere. The electromagnetic solar energy that impinges on Earth comes in primarily as light in a wavelength range of 400–800 nm with maximum intensity at a wavelength of about 500 nm. An exactly equivalent amount of energy must leave the Earth System, which it does primarily as infrared radiation with maximum intensity at about 10 μm, mostly within the range between 2 and 40 μm. Infrared-absorbing species, the greenhouse gases that include water vapor (H_2O), carbon dioxide (CO_2), methane (CH_4), ozone (O_3), and other minor species in the atmosphere, absorb some of the outbound infrared radiation, which which is eventually all radiated to space. The temporary absorption of infrared radiation warms the atmosphere—a greenhouse effect—without which Earth's average surface temperature would be about −18°C and unsuitable for life instead of the relatively comfortable average level now of about 14°C.

An increase in greenhouse gas levels tends to raise the altitude of the region from which infrared radiation is emitted into space and to increase the temperature of the troposphere below. Compared to preindustrial climate conditions, the extra radiant energy that must be emitted to maintain the heat balance of the Earth System is now about 2 W/m², which is small compared to the solar flux of 1340 W/m², but sufficient to raise atmospheric temperature somewhat more than 0.5°C. The Intergovernmental Panel on Climate Change (IPCC) has called this effect **radiative forcing**, which the group defined in 2001 as, "The radiative forcing of the surface-troposphere system due to the perturbation in or the introduction of an agent (say, a change in greenhouse gas concentrations) is the change in net (down minus up) irradiance (solar plus long-wave; in W/m²) at the tropopause **after** allowing for stratospheric temperatures to readjust to radiative equilibrium, but with surface and tropospheric temperatures and state held fixed at the unperturbed values."[1] Basically, radiative forcing is a concept that enables assignment of values of temperature increase or decrease to factors that influence global temperature—as examples, a positive value for an increase in atmospheric levels of greenhouse gases (carbon dioxide) that have a warming effect and a negative value for factors that tend to decrease atmospheric temperature, such as increased levels of radiation-reflecting aerosols.

Atmospheric gases tend to absorb outbound infrared radiation largely throughout its wavelength range except for a gap between 8 and 13 μm. Small increases in levels of carbon dioxide, especially, have a relatively large effect on infrared absorption because they fill gaps in the infrared absorption spectrum where other gases, especially water vapor, absorb less strongly. Figure 8.1 shows the infrared absorption spectrum of carbon dioxide. A strong absorption band spanning about 13–18 μm is of particular significance because it is in an area where the main greenhouse gas, water vapor, absorbs relatively weakly. Increases in atmospheric carbon dioxide levels effectively widening this band fill in the "wings" of the band and increase the absorption of outbound infrared radiation. Sloan and Wolfendale[2] have called this phenomenon "closing the blinds" on the "window" through which infrared radiation leaves the Earth System. Methane, which is present at much lower levels in the atmosphere than carbon dioxide, and some other trace-level gases from the anthrosphere likewise have a large impact per molecule on the absorption of infrared radiation, although their overall impacts are relatively small because of their low atmospheric concentrations.

As shown in Figure 8.2, the atmospheric concentration of carbon dioxide has been increasing. During the past several decades, the increase has been around 1 ppm/year and now may be approaching 2 ppm/year. This increase has been so significant that climatologists have used it as a

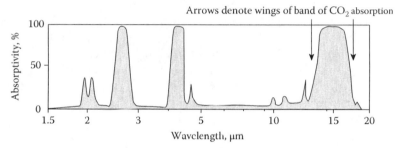

FIGURE 8.1 The infrared absorption spectrum of atmospheric carbon dioxide. The "wings" on the large band on the right (see vertical arrows) are particularly significant with respect to atmospheric warming because the increase in carbon dioxide levels has the effect of widening the band into these areas of the infrared spectrum and filling in a gap in the wavelength region in which atmospheric gases absorb infrared radiation leaving the Earth System.

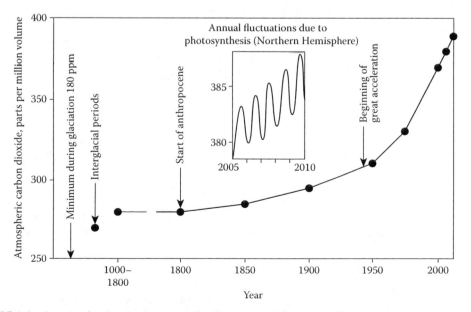

FIGURE 8.2 Levels of atmospheric carbon dioxide as a function of time. The inset shows the annual variation in the Northern Hemisphere due to photosynthesis. The "great acceleration" marks a point at which atmospheric carbon dioxide levels began rising at a very rapid rate.

means of measuring Earth's evolution into the Anthropocene epoch, which, as noted in Chapter 1, Section 1.6, many scientists believe is taking the place of the Holocene epoch during which civilization has existed since the end of the last ice age. The beginnings of the Anthropocene are commonly dated to around 1800 after which the first perceptible human-caused increases in atmospheric carbon dioxide levels began. From around 280 ppm in 1800, the concentration of atmospheric carbon had reached 310 ppm in 1950. This date marks the beginning of what has been called "the great acceleration" in which levels of atmospheric carbon dioxide have increased at an accelerating pace until the present. As shown in the plot, the level of atmospheric carbon dioxide was around 270 ppm in past interglacial periods and went down to as low as 180 ppm at the peaks of past glaciations.

Most of the recent increase in atmospheric carbon dioxide has been attributed to combustion of fossil fuels and to a lesser extent loss of rain forests. The latter adds carbon dioxide to the atmosphere from the burning of biomass and the decay of biomass forest litter, which is accelerated by tree removal. The destruction of vegetation slows the removal of carbon dioxide from the atmosphere by photosynthesis. This phenomenon is shown by the inset in Figure 8.2, which exhibits an annual fluctuation of several ppm in atmospheric CO_2. This plot is for the Northern Hemisphere where abundant vegetation, especially in forests thriving in middle latitudes, is particularly active in photosynthesis. The minimum in the annual cycle of atmospheric carbon dioxide in the Northern Hemisphere occurs in late September and early October as the growing season ends and the maximum is observed in April following a winter season with relatively little photosynthesis during which carbon dioxide has been entering the atmosphere from combustion and respiration of organisms including biomass decay.

8.2.1 INCREASING TEMPERATURE

The likely consequences of global climate change are many and profound. Global warming of just 1°C or 2°C would seriously impact the Earth System and it is possible that warming over the next century will be more than 2°C. Climate models predict an average global temperature increase of 1.5–5°C with a doubling of atmospheric carbon dioxide levels. That amount of warming is about as much again as the temperature increase that occurred from the last ice age until now. Especially if the warming is toward the high side of the projected range, it would greatly affect climate and rainfall.

It is virtually certain that global warming is already under way as Earth enters the Anthropocene epoch. There are some uncertainties in measurements of Earth's temperature with differences between land surface temperature and those of the oceans and variations with geographic area. Earlier data from a century or more ago are especially uncertain, because they are based on readings of surface temperature taken in scattered locations over the globe. However, since 1979, very accurate global temperature readings have been obtained by satellite measurements of Earth's irradiance. Figure 8.3 shows Earth surface temperature anomalies relative to the 1960–1990 average as measured from surface monitoring locations. The plot shows a definite upward trend since 1975.

Global temperatures analyzed by the Goddard Institute for Space Studies (GISS) in New York City show that the 1980s were the warmest decade on record globally since reasonably accurate global temperatures have been measured. The 1980s were followed by a warmer decade in the 1990s and the warmest decade of all from the beginning of 2000 to the end of 2009. The hottest years ever documented historically were 2005 and 2010 with 1998, 2002, 2003, 2006, 2007, and 2009 all essentially tied for third.

8.2.2 PASSING THE TIPPING POINTS

The increase in global temperature during recent years shown in Figure 2.5 is small compared to normal short-term fluctuations in weather temperature. The Earth System has huge inertia that will tend to delay temperature increases. A major factor contributing to this inertia is the great mass of the oceans, which average about 4 km deep around the globe and are filled with water, which has

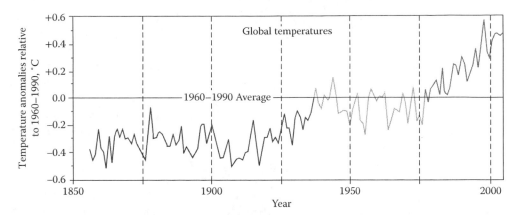

FIGURE 8.3 Global surface temperature variations compared to the 1960–1990 average as measured at Earth's surface indicating an upward trend since 1975. (Data from U.S. National Oceanic and Atmospheric Administration National Climatic Data Center, http://www.ncdc.noaa.gov/sotc/global.)

the highest heat capacity of all common substances. The second major factor contributing to Earth's vast inertia consists of great ice sheets, which may be 2–3 km thick in Antarctica and Greenland. Ice has a very high heat of fusion, which means that a tremendous amount of energy must be absorbed to melt it. The Earth System heat inertia has the effect of masking global warming for some time. The danger is that climate **tipping points** will be exceeded before it is obvious that they even exist. Tipping points are factors with positive feedbacks, which beyond a certain level feed upon themselves. A common example discussed in Section 8.2.3 is melting of surface ice that exposes heat-absorbing rock and soil leading to additional ice melting.

8.2.3 LOSS OF ICE COVER

Ice, especially when covered by layers of snow, reflects incoming solar energy. As the ice melts, surfaces are exposed, which absorb solar energy much more effectively. As an example, the melting of Arctic Ocean sea ice, which is now taking place during summer months, exposes seawater, which absorbs solar energy much more effectively than does ice, which leads to increased melting of the ice and additional absorption of heat. A second example is provided by the melting of ice that has covered Greenland's land mass for millennia. The melting exposes surface rock and soil that absorbs solar energy much more effectively than ice with the result that more heat is absorbed and more of the ice melts. Furthermore, the liquid water produced by the melting flows to a low spot on the ice where it burrows and creates a vertical shaft called a moulin that penetrates to the base of the ice mass. This infiltrating water lubricates the interface between the ice and the surface below, which enables large masses of ice to move physically into oceans, further contributing to ice loss. Of particular concern is the possible melting and loss of the West Antarctic ice sheet now resting on bedrock below sea level.

Satellite measurements of Arctic sea ice since the late 1970s show a shrinkage of about 30% in the late summer ice cap during that period. The loss of Antarctic ice is now around 100 km³/year. During the past several decades, the melt area of the Greenland ice cap has increased by 50% with an annual loss of about 200 km³/year. (Indications are that snowfall in Greenland above an altitude of 2000 m has been increasing as global warming has pumped more water into the atmosphere.)

8.2.4 GLACIERS AND WATER SUPPLY

Important collateral damage from the shrinkage of mountain glaciers in the Himalayas, Andes, Alps, and the U.S. and Canadian Rockies is the adverse effect on water supply. Some of the world's most notable rivers, crucial as sources of irrigation water and water supply as well as waterways for navigation, are fed by glaciers. One of the most important of these is the Brahmaputra River, which goes by several names in its 2900 km course from glacial origins in Tibet, through India, and which discharges through Bangladesh into the Bay of Bengal. Prone to catastrophic spring-time flooding, this river is an essential source of irrigation water for crops that feed millions of people and is widely used for navigation. Another great glacier-fed river is the 3200 km long Indus, which originates on the Tibetan Plateau (the world's third largest store of ice after Antarctica and Greenland), flows through parts of India, and then across Pakistan to empty into the Arabian Sea near Karachi. The major concern with these and some other glacier-fed rivers is that with global warming, the glaciers will stop producing liquid water during the late summer causing severe shortages of irrigation water and disruptions in navigation. Another possibility is that with increased water vapor in the atmosphere due to global warming, huge quantities of snow will accumulate in high-altitude glacial regions followed by rapid melting and catastrophic down-stream flooding in the warmer spring and early summer. This could lead to an annual cycle of devastating flooding in part of the year followed by severe drought and water shortages later in the year.

8.2.5 Expansion of Subtropical Arid Regions and Drought

An effect of global warming with special implications for food supply is the poleward spread of the subtropical dry regions. This occurs as air rises in the tropics and sinks in the subtropics causing the subtropics to become drier. This effect is now estimated to have caused movement of the subtropic margin toward the poles averaged over all longitudes (east/west around Earth) of about 4 degrees of latitude. Areas of Africa, parts of the southern United States, regions of Australia, and the Mediterranean region have become drier as a consequence. In 2011, the worst drought in 60 years hit African regions of Ethiopia, Somalia, and Northern Kenya putting as many as 10 million people at risk of starvation in the region and killing hundreds of thousands of farm animals.

Although a warmer Earth with increased water evaporation and subsequent precipitation will cause some areas to become wetter and more prone to storms with severe rainfall events, other areas, such as the subtropical arid regions discussed in the preceding paragraph, will experi-ence less rainfall and severe drought. It is anticipated that in the United States, the problem will be especially severe in the Colorado River basin where water sources are already stressed by population pressures. Increased evaporation of water will intensify water shortages brought on by drought by reducing runoff and water available for agricultural, industrial, and municipal applications. The demand for irrigation water will increase leading to pumping and depletion of underground water aquifers. The irrigation water and runoff water will tend to be more saline and of generally lower quality.

8.2.6 Some Other Effects of Global Climate Change

It is likely that there will be a number of environmental surprises other than increasing tempera-tures associated with global warming. These include infestations of plant and animal pests such as weeds, insects, and rodents likely to thrive better under warmer climate conditions. It is possible that the geographic distribution of some diseases now associated with tropical regions will spread along with the expansion of warmer areas. One of these of great concern is mosquito-borne malaria. Another likely effect of global warming is deterioration of coral reefs resulting from higher ocean levels and increased levels of dissolved carbon dioxide in seawater, which has the effect of dissolv-ing the calcium carbonate that is the structural material of the reefs.

8.3 DEALING WITH GLOBAL CLIMATE CHANGE

Possible means of coping with global warming may be divided into the three categories of (1) **mitigation** or minimization of emissions of greenhouse gases, (2) **counteracting measures** consisting of geoengineering measures to counteract the effects of greenhouse gases by increasing the fraction of solar radiation reflected back into space, and (3) **adaptation** to global warming and climate change. These possibilities are discussed in Section 8.3.

8.3.1 MITIGATION AND MINIMIZATION OF GREENHOUSE GAS EMISSIONS

The minimization of greenhouse gas emissions is strongly connected to the utilization of fossil fuels for energy because of the carbon dioxide that is emitted when these fuels are burned. While this source of atmospheric carbon dioxide realistically cannot be eliminated, much can be done to minimize it with existing technology.

8.3.1.1 Less Carbon Dioxide from Internal Combustion Engines

Internal combustion engines used in transportation are major sources of greenhouse gas carbon dioxide emissions and a logical place to look for reductions in such emissions. In keeping with that goal, on July 29, 2011, the U.S. government and major automakers announced that the U.S. automobile fleet is targeted to reach an average of 54.5 miles per gallon fuel economy by 2025. This would double the mileage of 27 mpg, the average for 2011 cars. One area in which greenhouse gas emissions from automobiles is being significantly reduced is with increasingly popular hybrid vehicles that combine batteries with highly efficient internal combustion engines that can readily cut in half the amount of carbon dioxide emitted to the atmosphere in moving equivalent quantities of passengers and freight. By using batteries that can be recharged from an electrical current source, as much as 50 km of driving range can be provided before the internal combustion engine is required to maintain propulsion. Such a driving range would enable at least half of routine commuting and local freight deliveries with energy taken from the electrical grid, which ultimately can be powered primarily with energy sources other than fossil fuels. Provision of charging stations in parking facilities can further extend the driving range of hybrids that can be charged from external electricity sources. Charging stations are provided by some employers as a fringe benefit.

The use of methane (natural gas, CH_4) to power the internal combustion engines in both conventional and hybrid vehicles can significantly reduce greenhouse gas emissions from transportation sources. Compared to other fossil fuels, methane contains a much higher proportion of hydrogen to carbon so that its combustion produces relatively less carbon dioxide and more water. A problem with methane fuel is the limited amount that can be carried as the pressurized gas. That is less of a problem with vehicles that are locally driven, and that thus never need be far from a methane refueling station. One of the most common examples consists of methane-fueled buses, which now make up a large proportion of the buses used around many airport terminals. Long-distance truck transportation is also a good candidate for use of methane fuel dispensed from truck stops. Methane should make a good fuel for locomotives with the compressed fuel contained in several tank cars attached immediately behind the locomotives that could readily be exchanged for filled cars as the fuel becomes exhausted. The exceptional fuel economy achieved with hybrid vehicles means that a methane-powered hybrid would contain enough fuel to provide acceptable driving range. Efforts are being made to develop solid adsorbent materials, generally based on specialized activated carbon, that can hold large quantities of methane or elemental hydrogen fuel.

A possible option for trucks and locomotives is the use of engines running primarily on methane but compression-ignited with diesel fuel.[3] With such an engine, air mixed with methane is pulled into the engine cylinder on the intake stroke and highly compressed during compression (see

Figure 8.8). The mixture is too lean in methane fuel to ignite, but at the peak of the compression stroke, a small amount of diesel fuel is injected, which ignites and causes the methane to burn. This is an application of the stratified-charge ignition concept that enables internal combustion engines to run on a fuel/air mixture that is too lean (fuel-deficient) to ignite with a spark or under compression but can be ignited by a small fuel-rich ignition zone created by injecting fuel directly into a small region of the combustion chamber (in a spark-ignited engine directly onto the spark plug). With this kind of engine, about 90% of the energy could come from clean-burning methane with the remainder from diesel fuel. It should also be possible to design a stratified charge engine that could temporarily switch over completely to diesel fuel if the methane tank were depleted.

8.3.2 Transportation Alternatives to the Internal Combustion Engine

Greenhouse gas emissions from the transportation sector can be greatly reduced by switching from trucks for freight transportation to electrically powered rail using electric power generated by means that do not use fossil fuels. Unlike Europe and some Asian countries where electrically powered trains are the rule, relatively little of the U.S. rail system is powered with electricity. Since about 1990, there has been substantial growth in the shipment of containers and trucks by rail, and this less energy-intensive means of transport has already saved large amounts of fuel and prevented release of much carbon dioxide. Increasing reliance on rail freight transport in the United States would benefit from updating the rail network with connections to some locations not now served by rail. In some cases, interstate highway rights of way can be used for construction of such connections.

8.3.3 Heating and Cooling

Fossil fuels can be used much more efficiently for heating and cooling. Because of its low carbon dioxide emissions, methane is the best fossil fuel for heating buildings. Significant savings in fuel and greenhouse gas emissions can be achieved by using methane in hybrid heating and cooling systems. In such a system, natural gas would be employed in an internal combustion engine coupled to a heat pump and to an electrical generator to produce electricity. For heating, all of the engine heat normally dispersed in cooling the engine and in the exhaust would be reclaimed. The heat pump would pump in heat energy from the outside for heating and reverse the process for cooling during hot periods. Numerous such units attached to dwellings and to commercial buildings could serve to provide electricity to an electrical grid reducing pressures on centralized electricity distribution systems and providing diversity in such systems.

For relatively high-density residential and commercial areas, **district heating and cooling** systems can provide significant economies and reduced greenhouse emissions. Such systems are common in Europe and on university campuses in the United States. Basically, they consist of centralized power stations from which steam or hot water are distributed in pipes for heating and chilled water is distributed for cooling. The outline of such a system is shown in Figure 8.4.

8.3.4 Carbon Capture

Carbon capture refers to measures that at least temporarily remove carbon dioxide from the atmosphere or prevent its release to the atmosphere. One way in which net release of carbon dioxide to the atmosphere is prevented is the use of biomass for fuel and raw material in place of petroleum. When biomass is burned, carbon dioxide is released, but exactly the same amount of carbon dioxide was removed from the atmosphere for the photosynthetic production of the biomass. Therefore, on balance, using biomass as fuel does not add carbon dioxide to the atmosphere. If biomass is not burned or does not decay, its production amounts to a net loss of atmospheric carbon dioxide.

A tantalizing possibility with respect to the use of biomass fuels arises from the fact that photosynthesis in most plants is only about 0.5% efficient. Increasing this value to even just 1% would

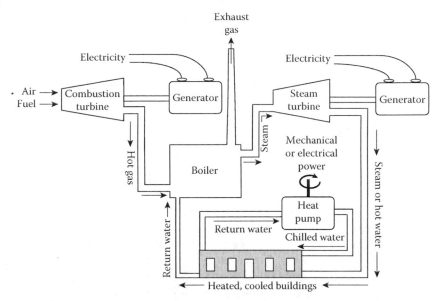

FIGURE 8.4   A combined power cycle in which natural gas or oil is first used to fire a gas turbine connected to an electrical generator. The hot exhaust gases from this turbine are fed to a boiler to produce steam, which drives a steam turbine, also connected to a generator. The still hot exhaust steam from the steam turbine is distributed to residential and commercial buildings for heating. The water condensed from the steam used to heat buildings is returned to the power plant to generate more steam, which conserves water and also prevents the necessity of treating more water to the high standards required for a boiler. A heat pump powered by electricity generated in the facility or linked mechanically to one of the turbines produces chilled water used to cool buildings.

make photosynthetic production of biomass much more attractive. Such an increase may be possible using genetically engineered plants. Another possibility is to use algae to produce biomass. Algae typically make biomass at a rate several times that of terrestrial plants.

Biomass added to soil in the form of photosynthetically generated stalk or straw by-products of crop production has the effect of temporarily removing carbon dioxide from the atmosphere. Eventually, most of the biomass incorporated into soil decays and releases its carbon back to the atmosphere as carbon dioxide. However, biomass can be pyrolyzed to release volatile matter leaving a pure carbon residue called **biochar**. Biochar does not decay and the elemental carbon of which it is composed remains in soil. Biochar is being advocated as a soil additive that retains nutrients and water and sequesters within soil the carbon dioxide produced in the decay of plant matter. The volatile matter released in making biochar contains liquids and gases that can be used as fuel and as feedstocks for chemical synthesis.

Carbon sequestration is employed to capture carbon dioxide generated in combustion, fossil fuel gasification, and fermentation (such as in the production of bioethanol) and to place the carbon dioxide where it cannot be released into the atmosphere.[4] Carbon sequestration works best with concentrated sources of carbon dioxide. The carbon dioxide from conventional sources where fossil fuels are burned in air is so dilute that its capture for sequestration is challenging and expensive. In coal gasification to produce elemental hydrogen, however, carbon in coal is reacted at high temperatures with oxygen and steam in several reactions for which the overall process is

$$2C(coal) + O_2 + 2H_2O \rightarrow 2CO_2 + 2H_2 \tag{8.1}$$

The net result is the production of a mixture of equal volume percentage carbon dioxide and elemental hydrogen. The hydrogen product is a premium nonpolluting fuel in applications such as fuel cells and as a feedstock that can be used to make synthetic hydrocarbon fuels and a variety of chemical products such as ammonia, NH_3.

Carbon dioxide can be sequestered in several ways. One approach is to pump the gas deep into cold ocean water, although there is concern that the slight lowering of pH of ocean water will be detrimental to marine life, particularly by interfering with the formation and settling of calcium carbonate shells in some marine organisms (see the discussion of coral in Section 8.2.6). The carbon dioxide can be pumped into secure mineral formations far underground. Saline water in deep aquifers is very effective in permanently sequestering carbon dioxide. Installations for pumping contaminant carbon dioxide from some natural gas sources have been operated successfully for many years in several locations, most notably in the Sleipner gas field off the coast of Norway where about a million metric tons per year of carbon dioxide that composes around 9% of the gas is separated and pumped into the Ultsira aquifer formation about 800 m below the bed of the North Sea. In some cases, contaminant toxic hydrogen sulfide in natural gas sources is being sequestered underground along with the carbon dioxide. Another practice is to pump carbon dioxide into petroleum-bearing formations where it serves to stimulate release of crude oil in secondary petroleum recovery operations.

An interesting use of carbon dioxide is its conversion to methane or synthetic petroleum fuel. For such a system to work, it is necessary to have an abundant and inexpensive source of elemental hydrogen, H_2, made by the electrolysis of water using renewable wind power or (especially in Iceland) geothermal energy. The key chemical reaction is a reverse water gas shift reaction

$$CO_2 + H_2 \rightarrow CO + H_2O \qquad (8.2)$$

followed by further reaction of CO with H_2 to produce methane, CH_4

$$CO + 3H_2 \rightarrow CH_4 + H_2O \qquad (8.3)$$

or through the Fischer–Tropsch synthesis reaction of CO and H_2 to make higher alkanes such as isomers of octane, C_8H_{18}. The carbon dioxide used in this way is discharged to the atmosphere when the hydrocarbon is burned, but the fuel produced substitutes for fossil hydrocarbon fuels that otherwise would be burned giving a net saving of carbon dioxide emissions.

8.3.5 Avoiding Fossil Fuels

The most effective means of avoiding carbon dioxide emissions to the atmosphere is to not use fossil fuels. Substantial progress has been made in this area such as in the application of solar energy for direct home heating. Another possibility is the development of more efficient photovoltaic cells that convert solar energy directly to electricity. Photovoltaics are now marginally competitive with conventional ways of producing electricity and improved efficiencies would make them more so. Development of processes for efficient production of hydrogen and oxygen by direct photodissociation of water

$$2H_2O + h\nu \rightarrow 2H_2 + O_2 \qquad (8.4)$$

followed by reaction of H_2 and O_2 in a fuel cell to produce electricity could also reduce consumption of fossil fuels and lower the release of carbon dioxide to the atmosphere.

8.3.6 Avoiding Greenhouse Gases Other than Carbon Dioxide

It is important to prevent the release of greenhouse gases other than carbon dioxide, especially ultrastable volatile compounds that have a high greenhouse gas potential. An excellent example of green chemistry has been the replacement of chlorofluorocarbons (CFCs) (Freons such as

CF_2Cl_2) with analogous compounds having at least one C–H bond, which are rather quickly destroyed in the troposphere by chain reactions that begin by the abstraction of the H atom by tropospheric hydroxyl radical (HO·). Although this was done to prevent destruction of stratospheric ozone by CFCs, it has been useful to reduce greenhouse warming. Although the compounds that have a C–H bond act as greenhouse gases, their lifetimes during which they are available to absorb infrared radiation are much shorter than those of the CFCs. The production of extremely stable sulfur hexafluoride, SF_6, and completely fluorinated hydrocarbons, such as CF_4, should be avoided.

Another measure to avoid atmospheric warming is to limit the emissions of methane, CH_4. Large quantities of methane are released by anoxic bacteria growing in flooded rice paddies. By developing strains of rice and means of cultivation that enable the crop to be grown in unflooded soil, this source of methane can be greatly reduced. Methane collection systems placed in municipal waste landfills can prevent the release of methane generated by anoxic decay of biomass in the landfills and provide a source of methane fuel.

8.3.7 ECONOMIC AND POLITICAL MEASURES

In addition to appropriate technologies, economic, regulatory, and political measures are required to ensure that such technologies are employed to reduce carbon dioxide emissions. **Mileage standards** mandating automotive fuel economy have been effective in reducing use of petroleum and lowering amounts of carbon dioxide emitted. Such standards were implemented in the United States following shortages of gasoline in the 1970s. Unfortunately, they were not updated during the late 1900s and early 2000s, a time when the growing popularity of behemoth "sport utility" vehicles contributed to increased fuel consumption and it was not until 2007 that renewed mileage standards were implemented. As noted in Section 8.3.1.1, in 2011, regulations were announced that would require the U.S. automobile fleet to achieve an average of 54.5 miles per gallon fuel economy by 2025.

Carbon taxes on fossil fuel use can be very effective in reducing carbon dioxide emissions to the atmosphere. Very high taxes on gasoline and diesel fuel in Europe have resulted in significantly reduced automotive fuel use compared to the United States. One result has been a much higher percentage of diesel engines in vehicles in Europe; such engines are thermodynamically much more efficient than conventional gasoline engines because of their higher peak combustion temperatures. Cap and trade systems can be effective in limiting carbon dioxide emissions. Cap and trade regulations applied in this manner allocate maximum amounts of carbon dioxide that firms may emit. A firm may buy rights to exceed its cap from another firm that is below its limit. An advantage of cap and trade is that it does not mandate the technologies used leaving that aspect to the private sector and to the ingenuity of entrepreneurs.

A political initiative to lower carbon dioxide emissions was the Kyoto treaty that resulted from a 1997 meeting in Kyoto, Japan, of 160 nations representing most of the world's population. Under the terms of this treaty, greenhouse gas emissions were to be reduced to 1990 levels over the 2008–2012 time period, which would have resulted in a reduction by 23% compared to levels that would have been in effect without any action. In addition to carbon dioxide, other greenhouse gases restricted by the treaty include methane, nitrous oxide, sulfur dioxide, sulfur hexafluoride, hydrofluorocarbons (HFCs), and perfluorocarbons. The treaty was implemented in 2005 with virtually all the nations that participated signing it except for the United States. The refusal of the United States to ratify the agreement has resulted from the provision that exempts developing nations from the treaty's terms for economic reasons. India and China are of particular concern because even though their emissions per capita are much less than those of countries such as the United States and Australia, their huge and growing populations combined with their rapidly developing economies mean that they are emitting increasing quantities of greenhouse gases.

8.3.8 COUNTERACTING MEASURES

Counteracting measures, sometimes called geoengineering, are those that potentially could be taken to slow greenhouse warming. Most of the measures suggested would operate by increasing aerosol particulate matter in the atmosphere, which would reflect incoming solar energy back into space. Most of these are based on production of light-reflecting haze produced by the atmospheric oxidation of atmospheric sulfur dioxide. The sulfuric acid and sulfates produced from it act as condensation nuclei around which atmospheric water vapor condenses to produce the reflective aerosol particles. The net effect is to increase the coverage, density, and brightness of the cloud cover that reflects incoming solar radiation. Evidence that this approach works in principle is provided by documented cooling effects of past massive volcanic emissions that were particularly rich in sulfur dioxide. There is also evidence to suggest that pollutant sulfur dioxide emissions cause a cooling effect in central Europe and the eastern United States during the summer as well as in urban areas of Asia. Some concerns have been expressed that successful efforts to curb sulfur dioxide emissions will result in additional atmospheric warming because of a reduction in the aerosol produced by sulfur dioxide.

A geoengineering approach based on production of light-reflecting atmospheric sulfate aerosols would entail discharging large quantities of sulfur dioxide from the anthrosphere into the atmosphere. The enormous quantities of sulfur dioxide required and the collateral consequences, particularly increased acid rain, make it unlikely that this solution will ever be employed.

Alterations of surface albedo might be employed to reduce the amount of incoming solar radiation absorbed at Earth's surface. Freshly exposed dark soil absorbs most of the radiation that reaches it, so turning over soil by plowing increases heat absorption. Modern practices of conservation tillage in which crop cover is left on top of the soil can reduce surface warming in agricultural areas. Conversion of cropland to forests and grasslands would also tend to have a cooling effect. Urban surfaces may also be modified to reduce absorption of solar radiation. Concrete pavement tends to reflect more solar radiation than does asphalt. One approach that has been tried is to cover roofs of buildings, typically covered by black roofing, with aluminum roofing or even with soil and energy-reflecting vegetation.

8.3.9 ADAPTATION

Given the inevitability of at least some (and perhaps a lot) of global warming and resultant climate change, adaptation in various forms will be required. Adaptation must deal with temperature increases and other climate changes and will be among the main endeavors of the next generation of scientists and engineers.

8.3.10 HEAT

One of the greater challenges of global warming is simply dealing with the effects of heat on people. Excessive heat kills more people each year than any other kind of weather event, with the elderly especially vulnerable. The August 2003 heat wave that afflicted much of Europe killed over 1000 people in the United Kingdom. In that same deadly month, 15,000 people, mostly elderly, died from the heat in France where refrigerated warehouses were put into service to store bodies until they could be identified and disposed. Increased global temperatures will lead to demands for more air conditioning of dwellings, stores, and workplaces. Such will be the case in areas of Europe where air conditioning has not been common. There will also be increased demand for cooling installations in developing countries in hotter climates, such as India and southern China. Such installations will need more electricity, especially from renewable sources. There will also be a need for facilities to temporarily meet peak demand for electricity, probably from methane-fueled gas turbines.

8.3.11   DROUGHT

One of the most troublesome aspects of global warming is likely to be drought and the shortage of water. Current shortages of water around the world are likely to become worse as areas of drought spread. The worst effects of drought will be reduction in the capacity to produce food. Drought-resistant crops, some developed by genetic engineering, will have to be grown. Irrigation practices will need to be as efficient as possible. Trickle irrigation systems that deliver the minimum amount of water required for growth will have to be implemented. To prevent excessive salt buildup in soil, irrigation water may need to be treated by reverse osmosis to remove ionic solutes. World resources of groundwater suitable for irrigation are now being rapidly depleted and will be unavailable as global warming progresses. To help alleviate groundwater deficiencies, it will be necessary to establish groundwater recharge systems to take advantage of periodic wet periods. To enable utilization of brackish groundwater, it will be necessary to use reverse osmosis to remove salts employing renewable wind energy to power the desalination processes.

A promising approach to food production in arid regions near ocean shores is to grow salt-tolerant plants called halophytes irrigated with seawater. Among the halophytes that can grow in saltwater and produce biomass are salt grass, saltbush, glasswort, and sea blight. Some of these plants are capable of producing 1–2 kg of dry biomass per square meter of area, productivity similar to that of alfalfa grown in soil. Most of these plants do not produce grain, but their biomass can serve as forage suitable for animal consumption. Because of the high salt content of this forage, animals may require more drinking water.

Salicornia bigelovii is a seed-producing saltwater plant that is often the first to colonize coastal mud flats. With a salt content of only about 3%, the seeds contain about 35% protein. With about 30% content of polyunsaturated oil, similar to safflower oil, the seeds can serve as a rich source of lipids for nutrition. One disadvantage of this plant is that the seed residue remaining after extraction of oil contains bitter saponins, which limit the amounts that can be fed to animals. It may be possible to genetically engineer this plant to eliminate the saponins.

The most productive photosynthesizers that can be grown in seawater are algae. Some algae are especially productive of oils. There is a significant potential to genetically engineer saltwater algae free of toxins and rich in protein and oils making the biomass suitable for animal and even human consumption. A promising approach is to grow saltwater algae in impoundments along with fish that use the algae as food. This could provide an abundant source of protein for human consumption.

8.3.12   WATER BANKING

Water banking is now widely employed across the western United States to store and allocate water relying on a market system for the most efficient utilization of water. Water banks enable interests with water rights to store water and sell it to users including agricultural and industrial concerns and municipalities. The idea is to store water, usually in underground aquifers, during times when it is in surplus, and release it to buyers when water is in short supply.

The Kern County Water Agency located in California's Central Valley operates one of the largest water banks in the United States. The county is very large with an area almost equal to that of New Jersey and includes the city of Bakersfield. Gravel and sand deposited in the area from the Sierra Nevada Mountains by the Kern River constitutes a porous layer through which surface water may percolate into aquifers below, which make up the water storage capacity of the water bank. Water is allowed to infiltrate the aquifer during periods of heavy runoff from mountain snows in the springtime and the stored water is sold for use during times of the year when drought conditions prevail. A surplus of water is accumulated in wet years and the reservoir of groundwater is drawn down during drought years.

As global climate changes due to greenhouse warming, a possible effect will be earlier and heavier springtime water runoff from mountain snows increased by greater transport of atmospheric

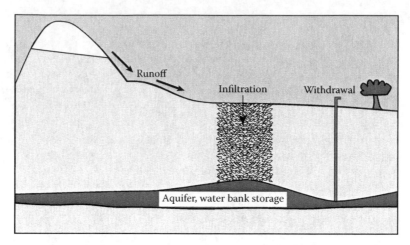

FIGURE 8.5 The concept of the water bank in which water is accumulated in underground aquifers during times of runoff and sold later to interests needing it.

water at warmer temperatures, followed by longer and more intense drought periods the rest of the year. Such conditions should increase the importance of water banking as a means of allocating the water resource. If the net result of global warming in an area such as the Sierra Nevada snowshed is simply less snowfall, however, water banking will suffer due to lack of water for recharge (Figure 8.5).

8.4 CONTROL OF PARTICLE EMISSIONS

The removal of particulate matter from gas streams is the most widely practiced means of air pollution control and is very important in sustaining the quality of the atmosphere. A number of devices that differ widely in effectiveness, complexity, and cost have been developed for this purpose. The selection of a particle removal system for a gaseous waste stream depends on the particle loading, nature of particles (size distribution), and type of gas scrubbing system used.

8.4.1 Particle Removal by Sedimentation and Inertia

The simplest means of particulate matter removal is **sedimentation**, a phenomenon that occurs continuously in nature. Gravitational settling chambers can be employed for the removal of particles from gas streams by simply settling under the influence of gravity. These chambers take up large amounts of space and have low collection efficiencies, particularly for small particles.

Gravitational settling of particles is enhanced by increased particle size, which occurs spontaneously by coagulation (see Chapter 6, Figure 6.11). Thus, over time, the size of particles increases and the number of particles decreases in a mass of air that contains particles. Brownian motion of particles less than about 0.1 μm in size is primarily responsible for their contact, enabling coagulation to occur. Particles greater than about 0.3 μm in radius do not diffuse appreciably and serve primarily as receptors of smaller particles.

Inertial mechanisms used for particle removal depend on the fact that the radius of the path of a particle in a rapidly moving, curving airstream is larger than the path of the stream as a whole. Therefore, when a gas stream is spun by vanes, a fan, or a tangential gas inlet, the particulate matter may be collected on a separator wall because the particles are forced outward by centrifugal force. Devices utilizing this mode of operation are called **dry centrifugal collectors**.

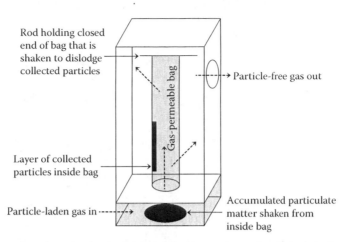

Rod holding closed end of bag that is shaken to dislodge collected particles

Gas-permeable bag

Particle-free gas out

Layer of collected particles inside bag

Particle-laden gas in

Accumulated particulate matter shaken from inside bag

FIGURE 8.6 Illustration of a unit of a single bag in a baghouse for removal of particulate matter from a gas stream. Particle-laden gas is indicated by gray shading and the flow of gas is illustrated by the dashed lines. A bag is mounted on an opening in a plate such that particle-laden gas is forced up through the inside of the cylindrical bag and is filtered through the bag walls and then flows out as filtered gas. The particulate matter accumulates on the inside of the bag. Periodically, the gas flow through the bag is stopped and the rod suspending the closed upper end of the bag is shaken to dislodge particulate matter from the bag walls, which is collected at the bottom of the apparatus.

8.4.2 PARTICLE FILTRATION

Fabric filters, as their name implies, consist of fabrics that allow the passage of gas but retain particulate matter. These are used to collect dust in bags contained in structures called **baghouses**. Periodically, the fabric composing the filter is shaken to remove the particles and to reduce back-pressure to acceptable levels. Typically, the bag is in a tubular configuration, as shown in Figure 8.6. Numerous other configurations are possible. Collected particulate matter is removed from bags by mechanical agitation, blowing air on the fabric, or rapid expansion and contraction of the bags.

Although simple, baghouses are generally effective in removing particles from exhaust gas. Particles as small as 0.01 μm in diameter are removed, and removal efficiency is relatively high for particles down to 0.5 μm in diameter.

8.4.3 SCRUBBERS

A venturi scrubber passes gas through a duct in which the gas is first constricted and then allowed to expand. Injection of the scrubbing liquid at right angles to incoming gas in the constricted part of the scrubber breaks the liquid into very small droplets, which are ideal for scavenging particles from the gas stream. In the reduced-pressure (expanding) region of the venturi, some condensation can occur, adding to the scrubbing efficiency. In addition to removing particles, venturis can serve as quenchers to cool exhaust gas and as scrubbers for pollutant gases.

8.4.4 ELECTROSTATIC PRECIPITATION

Aerosol particles may acquire electrical charges. In an electric field, such particles are subjected to a force that is proportional to the voltage gradient in the electric field and the electrostatic charge on the particle. This phenomenon has been widely used in highly efficient **electrostatic precipitators**, as shown in Figure 8.7. The particles acquire a charge when the gas stream is passed through a high-voltage, direct-current corona. Because of the charge, the particles are attracted to a grounded

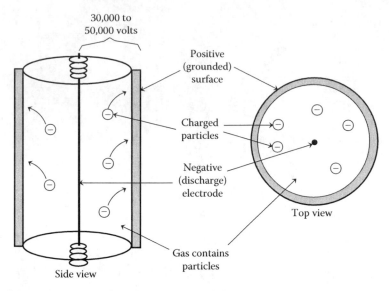

FIGURE 8.7   Schematic diagram of an electrostatic precipitator.

surface, from which they may be later removed. Ozone may be produced by the corona discharge. Similar devices used as household dust collectors may produce toxic ozone if not operated properly.

8.4.5   WHERE DOES IT ALL GO?

It is often the case that solving one pollution problem tends to create another. This is very much so with the dusts and sludges produced in cleaning up gas streams, such as those from power plants. These materials may be undesirable and potential sources of pollutants such as heavy metals. Disposed in landfills, they may have a tendency to leach water pollutants into groundwater.

Kilns used to heat minerals to make lime or Portland cement are major sources of particulate matter that must be removed from the kiln exhaust gas before its release. These kilns consist of enormous rotating drums inclined at a slight angle and fed with raw solid materials at the upper end and fuel at the bottom end where the calcined solids are discharged. For the manufacture of quicklime (CaO), the raw material is limestone ($CaCO_3$), and for the manufacture of cement, the raw materials are limestone, sand, and clay. Both processes generate finely divided dust that is removed from the gas stream. The greenest means of dealing with this material is to simply recycle it into the kiln. Lime kiln dust can be used in place of calcium carbonate in a cement kiln. Kiln dusts can also be applied to agricultural land to treat acidic soils. In the case of cement kiln dust, care must be taken in agricultural applications because of the potential to accumulate excess aluminum, which is toxic to plants, and because many cement kilns now burn hazardous wastes as a means of waste disposal and as a source of fuel, so the possibility of heavy metals in the dust must be considered.

The combustion of fossil fuels, particularly coal, generates large quantities of solid by-product, fly ash, blown out of furnaces with combustion gas. The best way to deal with fly ash is to not make it, but that is not possible for most commonly used combustion processes with fuels other than ash-free natural gas. Petroleum fuels produce much less fly ash than coal, but the petroleum fractions that are burned are the heavier ones containing ash-generating mineral matter. One special case is that of vanadium-rich Venezuelan crude oils that release an ash rich in vanadium oxide.

Coal typically contains several percent mineral matter and is a huge source of ash in power generation. The most efficient means of burning coal is to grind it to a powder and blast it into a burner much like oil injected into a furnace. As a result, the fly ash product consisting of metal oxides, silicates, and some unburned elemental carbon is ejected from the furnace and must be collected before going up the stack. Much effort has gone into finding uses for fly ash, including as a raw material for cement manufacture, in ceramics, as a filler for concrete, as a soil amendment agent in agriculture, as a sorbent for liquid wastes, and in sludge conditioning. Some coal fly ash is rich in aluminum that can be extracted with HCl for aluminum metal manufacture. This process is not yet competitive with preparation of aluminum with bauxite ore (Al_2O_3), but may become more so if bauxite supplies become tight. It must always be kept in mind that coal is a potential source of toxic elements, with arsenic in ash being a particular concern.

8.5 CONTROL OF CARBON MONOXIDE EMISSIONS

Toxic carbon monoxide gas resulting from incomplete combustion of carbon-based fuels is not a global or even a regional air pollution problem, but can accumulate to dangerous levels in localized situations, such as urban areas with heavy motor vehicle traffic. It is a potentially deadly indoor pollutant, and each year, significant numbers of people meet their demise by the action of this silent killer generated in improperly vented furnaces or by acts of bad judgment, such as using charcoal grills indoors. Until natural gas (CH_4) became widely available around 1920, carbon monoxide, mixed with elemental hydrogen (H_2), was the fuel generated by thousands of municipal gas plants from coal and piped into houses for lighting and cooking (it was too expensive to use for heating). During the late 1800s and early 1900s, many people were killed in their homes by exposure to this lethal gas mixture.

Since the internal combustion engine is the primary source of localized pollutant carbon monoxide emissions, control measures have been concentrated on automobiles and have been very successful in reducing carbon monoxide emissions. Carbon monoxide emissions may be lowered by employing a leaner air–fuel mixture, that is, one in which the mass ratio of air to fuel is relatively high. At air–fuel (mass:mass) ratios exceeding approximately 16:1, an internal combustion engine emits virtually no carbon monoxide. Modern automobiles use catalytic exhaust reactors and precise computerized control of engine operation to cut down on carbon monoxide emissions.

8.6 CONTROL OF NITROGEN OXIDE EMISSIONS

Collectively designated NO_X, colorless, odorless nitric oxide (NO), and pungent red-brown nitrogen dioxide (NO_2) gases enter the atmosphere from natural sources, such as lightning and biological processes, and from pollutant sources. Most NO_X enters the atmosphere as NO, which is then converted to NO_2 by atmospheric chemical processes. As shown in Figure 7.5, Chapter 7, there is continuous exchange between NO and NO_2 in the atmosphere. Pollutant sources of NO_X can lead to regionally high NO_X concentrations that can cause severe air quality deterioration because of the ability of NO_2 to undergo photochemical dissociation to produce O atoms that lead to the formation of ozone, photochemical smog, and nitric acid, which is a contributor to acid rain. Practically all anthropogenic NO_2 enters the atmosphere as a result of the combustion of fossil fuels in both stationary and mobile sources, and NO_X control has been concentrated on these sources.

At very high combustion temperatures, the following reaction occurs:

$$N_2 + O_2 \rightarrow 2NO \tag{8.5}$$

High temperatures favor both a high equilibrium concentration and a rapid rate of formation of NO. Especially in internal combustion engines, rapid cooling of the exhaust gas from combustion "freezes" NO at a relatively high concentration because equilibrium is not maintained. Thus, by its nature, the combustion process both in the internal combustion engine and in furnaces produces high levels of NO in the combustion products.

The level of NO_X emitted from stationary sources such as power plant furnaces generally falls within the range of 50–1000 ppm. Generation of NO is favored both kinetically and thermodynamically by high temperatures and by high excess oxygen concentrations, and these factors must be considered in reducing NO emissions from combustion sources. Reduction of combustion temperature to prevent NO formation in an internal combustion engine is commonly accomplished by adding recirculated exhaust gas.

Low-excess-air firing is effective in reducing NO_x emissions during the combustion of fossil fuels. As the term implies, low-excess-air firing uses the minimum amount of excess air required for oxidation of the fuel, so that less oxygen is available for the production of NO in the high-temperature region of the flame shown in Reaction 8.5. Problems with incomplete fuel burnout as well as elevated emissions of hydrocarbons, soot, and CO that occur with low-excess-air firing may be overcome by a two-stage combustion process. In the first stage, fuel is fired at a relatively high temperature with a substoichiometric amount of air, and NO formation is limited by the absence of excess oxygen. In the second stage, fuel burnout is completed at low temperatures in excess air; the low temperature prevents formation of NO.

Removal of NO_x from stack gas presents some formidable problems. Use of liquid scrubbers such as those employed to remove SO_2 from stack gas is not very effective for NO_x removal because of the low solubilities of nitrogen oxides. Sorption onto solids followed by destruction of the sorbed gases has been tried. Catalytic reduction and decomposition of nitrogen oxides are employed in automobile catalytic converters (see Section 8.8) and may be applicable to stack gas, although sulfur gases and particles in stack gas may interfere and poison the catalysts. Another possibility is the use of biofilters in which microorganisms held on support media metabolize NO_x (see Section 8.9).

8.7 CONTROL OF SULFUR DIOXIDE EMISSIONS

A number of processes are being used to remove sulfur and sulfur oxides from fuel before combustion and from stack gas after combustion. Most of these efforts concentrate on coal, since it is the major source of sulfur oxides pollution. Physical separation techniques can be used to remove discrete particles of pyritic sulfur from coal. Chemical methods can also be employed for removal of sulfur from coal.

Fluidized bed combustion of coal can be used to eliminate SO_2 emissions at the point of combustion. The process consists of burning granular coal in a bed of finely divided limestone or dolomite maintained in a fluidlike condition by air injection. Heat calcines the limestone to produce CaO, which absorbs SO_2 as shown by the following two reactions:

$$CaCO_3 \rightarrow CaO + CO_2 \tag{8.6}$$

$$CaO + SO_2 (+ O_2) \rightarrow CaSO_3 (\text{and } CaSO_4) \tag{8.7}$$

Many processes have been proposed or studied for the removal of sulfur dioxide from stack gas. Several of these are in widespread use. They include throwaway and recovery systems as well as wet and dry systems.

Slurries of either lime $(Ca(OH)_2)$ or limestone can be injected into stack gas scrubbers downstream from the boilers. With lime, the reaction is

$$Ca(OH)_2 + SO_2 \rightarrow CaSO_3 (CaSO_4) + H_2O \tag{8.8}$$

and with limestone, the reaction is as follows:

$$CaCO_3 + SO_2 \rightarrow CaSO_3 + CO_2 \tag{8.9}$$

The reaction with lime is the more efficient because the pH of the slurry is higher. These scrubbers can remove well over 90% of both SO_2 and fly ash when operating properly. In addition to corrosion and scaling problems, disposal of lime sludge poses formidable obstacles (about 200 kg of lime are required for each metric ton of coal burned). However, the $CaSO_3$ may be oxidized to $CaSO_4$ and the gypsum product, $CaSO_4 \cdot 2H_2O$, made into wallboard as is described in the Kalundborg, Denmark system of industrial ecology discussed in Chapter 14, Section 14.2.

Recovery systems in which sulfur dioxide or elemental sulfur is removed from the spent sorbing material, which is recycled, are much more desirable from an environmental and sustainability viewpoint than are throwaway systems. Many kinds of recovery processes have been investigated, including those that involve scrubbing with magnesium oxide slurry, sodium sulfite solution, ammonia solution, or sodium citrate solution. One type of recovery system uses a solution of sodium sulfite to react with sulfur dioxide in the flue gas

$$Na_2SO_3 + H_2O + SO_2 \rightarrow 2NaHSO_3 \tag{8.10}$$

followed by heating the sodium hydrogen sulfite product to produce a stream of pure SO_2 gas and regenerated sodium sulfite:

$$2NaHSO_3 \rightarrow Na_2SO_3 + H_2O + SO_2 \tag{8.11}$$

A portion of the sulfur dioxide can be reduced to hydrogen sulfide by reaction with natural gas (methane, CH_4) or with synthesis gas made from coal (a mixture of CO and H_2)

$$SO_2 + reducing\ gas \rightarrow H_2S + CO_2 \tag{8.12}$$

and the H_2S product subjected to the Claus reaction with SO_2 to produce elemental sulfur:

$$SO_2 + 2H_2S \rightarrow 3S + 2H_2O \tag{8.13}$$

Sulfur is in demand for the synthesis of sulfuric acid and other industrial applications. Recovery processes such as the one just described are much greener and in keeping with sustainability than are throwaway processes that generate large quantities of waste lime.

8.8 CONTROL OF HYDROCARBON EMISSIONS AND PHOTOCHEMICAL SMOG

Hydrocarbon emission control is particularly important in limiting photochemical smog discussed as an air pollution phenomenon in Section 7.8. Significant progress has been made in limiting emissions of volatile hydrocarbons to the atmosphere by measures such as gasoline delivery hoses that are equipped with fittings to prevent release of gasoline vapors as the tank is filled. A greater challenge has been in limiting emissions from internal combustion engines including those in vehicles.

To understand the production and control of automotive hydrocarbon exhaust products, it is helpful to understand the basic principles of the internal combustion engine. As shown in Figure 8.8, the four steps involved in one complete cycle of the four-cycle gasoline-fueled engine used in most vehicles in the United States are as follows: (1) **intake** of air or a gasoline/air mixture as the piston moves down with the intake valve open; (2) **compression** at a ratio of around 7:1 as the piston moves upward with both valves closed; (3) **ignition** with a spark plug near the top of the compression stroke resulting in a temperature of about 2500°C and a pressure up to 40 times atmospheric pressure, which forces the piston downward in the **power stroke**; and (4) **exhaust** of gases largely composed

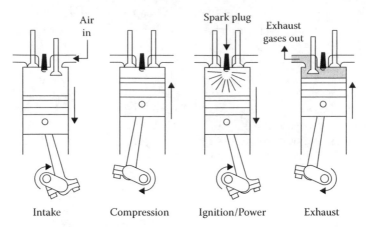

FIGURE 8.8 Steps in one complete cycle of a four-cycle internal combustion engine. Fuel is mixed with the intake air or injected separately into each cylinder.

of N_2, CO_2, and H_2O vapor with traces of hydrocarbons, CO, and NO as the piston moves upward with the exhaust valve open.

Several aspects of the cycle described in the preceding paragraph are related to hydrocarbon and other emissions from the engine. Higher compression ratios favor thermal efficiency and complete combustion, but may result in "pinging" and increased NO emissions. As the gas volume increases with downward movement of the piston in the power stroke, the temperature decreases in a few milliseconds. This rapid cooling "freezes" nitric oxide in the form of NO without allowing it time to dissociate to N_2 and O_2, which are thermodynamically favored at the normal temperatures and pressures of the atmosphere. The primary cause of unburned hydrocarbons in the engine cylinder is wall quench, wherein the relatively cool wall in the combustion chamber of the internal combustion engine causes the flame to be extinguished within several thousandths of a centimeter from the wall.

Several engine design and operational characteristics favor lower exhaust hydrocarbon emissions. The production of unburned hydrocarbons from wall quench is diminished by design that decreases the combustion chamber surface/volume ratio through reduction of compression ratio, more nearly spherical combustion chamber shape, increased displacement per engine cylinder, and increased ratio of stroke relative to bore. Spark retard also reduces exhaust hydrocarbon emissions. For optimum engine power and economy, the spark should be set to fire appreciably before the piston reaches the top of the compression stroke and begins the power stroke. Retarding the spark to a point closer to top-dead-center reduces the hydrocarbon emissions markedly. One reason for this reduction is that the effective surface to volume ratio of the combustion chamber is reduced, thus cutting down on wall quench. Second, when the spark is retarded, the combustion products are purged from the cylinders sooner after combustion. Therefore, the exhaust gas is hotter, and reactions consuming hydrocarbons are promoted in the exhaust system.

As shown in Figure 8.9, the air–fuel ratio in the internal combustion engine has a marked effect on the emission of hydrocarbons. As the air–fuel ratio becomes richer in fuel than the stoichiometric ratio, the emission of hydrocarbons increases significantly. There is a moderate decrease in hydrocarbon emissions when the mixture becomes appreciably leaner in fuel than the stoichiometric ratio requires. The lowest level of hydrocarbon emissions occurs at an air–fuel ratio somewhat leaner in fuel than the stoichiometric ratio. This behavior is the result of a combination of factors, including minimum quench layer thickness at an air–fuel ratio somewhat richer in fuel than the stoichiometric ratio, decreasing hydrocarbon concentration in the quench layer with a leaner mixture, increasing oxygen concentration in the exhaust with a leaner mixture, and a higher peak exhaust temperature at a ratio slightly leaner in fuel than the stoichiometric ratio.

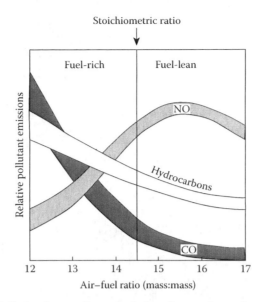

FIGURE 8.9 Effects of air/fuel ratio on pollutant emissions from an internal combustion piston engine.

8.8.1 Compression-Fired Engines

The preceding discussion applies to engines in which ignition occurs as a result of a spark set to fire near the peak of compression. Such engines are used in most automobiles and light trucks in the United States. However, because of their significantly higher peak temperatures and thus greater efficiency and fuel economy, compression-fired engines, commonly known as diesel engines, are widely used in heavier trucks in the United States and in automobiles in many countries. These engines tend to emit particulate carbon and higher levels of NO than do gasoline engines. Diesel engines are the major sources of atmospheric particle emissions in many urban areas. Particles from diesel engine emissions are very small and fit into the category of nanoparticles (those with at least one dimension less than 100 nm), and their inhalation probably has significant pulmonary, cardiovascular, and other health effects.[5] In addition to measures commonly applied to control NO_x, CO, and hydrocarbon emissions from gasoline engines, diesel engines may also be equipped with filters that filter out particles and then burn them off from the ceramic filter material. One exhaust emission control system that is used on diesel engines employs injection of urea into the exhaust that produces ammonia to react with and destroy NO. Since commonly used pollution control equipment is readily rendered ineffective by exposure to sulfur, the performance of diesel engine exhaust emission controls has required use of low-sulfur diesel fuel, the production of which has proven to be a challenge for producers of the fuel.

8.8.2 Catalytic Converters for Exhaust Gas Control

Catalytic converters are now used to destroy pollutants in exhaust gases. Currently, the most commonly used automotive catalytic converter is the three-way conversion catalyst, so called because a single catalytic unit destroys all three of the main classes of automobile exhaust pollutants—hydrocarbons, carbon monoxide, and nitrogen oxides. This catalyst depends on accurate sensing of oxygen levels in the exhaust combined with computerized engine control that cycles the air–fuel mixture several times per second back and forth between slightly lean and slightly rich relative to the stoichiometric ratio. In the first stage of the catalytic converter, a reduction catalyst in which the active catalytic metals are platinum and rhodium acts to reduce NO to N_2 and O_2. In the second stage, a platinum and palladium oxidation catalyst uses any excess oxygen in the exhaust gas and oxygen released in the reduction stage to oxidize carbon monoxide, hydrogen, and hydrocarbons (C_cH_h):

$$CO + 1/2 O_2 \rightarrow CO_2 \tag{8.14}$$

$$H_2 + 1/2 O_2 \rightarrow H_2O \tag{8.15}$$

$$C_cH_h + (c + h/4)O_2 \rightarrow cCO_2 + h/2\ H_2O \tag{8.16}$$

Automotive exhaust catalysts are dispersed on a high surface area substrate, most commonly consisting of cordierite, a ceramic composed of alumina (Al_2O_3), silica, and magnesium oxide. The substrate is formed as a honeycomb-type structure providing maximum surface area to contact exhaust gases. The support needs to be mechanically strong to withstand vibrational stresses from the automobile, and it must resist severe thermal stresses in which the temperature may rise from ambient temperatures to approximately 900°C over an approximately 2-minute period during "light-off" when the engine is started (during which time the catalyst is ineffective). The catalytic material, which composes only about 0.10–0.15% of the catalyst body, consists of precious metal rhodium, platinum, and palladium. These are very expensive and catalytic converters are sometimes stolen off from vehicles and sold by thieves for their precious metal content.

Since lead can poison auto exhaust catalysts, automobiles equipped with catalytic exhaust-control devices require lead-free gasoline, which has become the standard motor fuel. Sulfur in gasoline is also detrimental to catalyst performance, so sulfur levels in gasoline are kept very low.

8.8.3 PHOTOCHEMICAL SMOG AND VEGETATION

Not all photochemical smog originates from the anthrosphere. Vegetation can be a source of the main smog-forming ingredients. As noted in Section 7.7, terpenes from pine and citrus trees along with other vegetation sources have double C=C bonds and are very active in smog formation. Burning vegetation, a practice in destructive "slash-and-burn" agriculture, produces nitrogen oxides, which are essential ingredients for photochemical smog formation. For these reasons, some largely rural areas afflicted with reactive hydrocarbons and nitrogen oxides may have problems with photochemical smog.

Another consideration related to photochemical smog and plants is the sensitivity of some crops to ozone and other smog ingredients. In some areas afflicted by smog, consideration may need to be given to growing crops that are relatively less susceptible to the effects of smog.

8.8.4 PREVENTING SMOG WITH GREEN CHEMISTRY

Smog is basically a chemical problem, which would indicate that it should be amenable to chemical solutions. Indeed, the practice of green chemistry and the application of the principles of industrial ecology can help to reduce smog. This is due in large part to the fact that a basic premise of green chemistry is to avoid the generation and release of chemical species with the potential to harm the environment. The best way to prevent smog formation is to avoid the release of nitrogen oxides and organic vapors that enable smog to form. At an even more fundamental level, measures can be taken to avoid the use of technologies likely to release such substances, for example, by using alternatives to polluting automobiles for transportation.

The evolution of automotive pollution control devices to reduce smog provides an example of how green chemistry can be used to reduce pollution. The first measures taken to reduce hydrocarbon and nitrogen oxide emissions from automobiles were very much command-and-control and "end-of-pipe" measures. These primitive measures implemented in the early 1970s did reduce emissions, but with a steep penalty in fuel consumption and in driving performance of vehicles. However, since about 1980, the internal combustion automobile engine has evolved into a highly sophisticated computer-controlled machine that generally performs well, emits few air pollutants, and is highly efficient. (And it would be much more efficient if those drivers who feel that they must drive "sport utility" behemoths would switch to vehicles of a more sensible size.) This change has required an integrated approach involving reformulation of gasoline. The first major change was elimination of tetraethyllead from gasoline, an organometallic compound that poisoned automotive exhaust catalysts (and certainly was not good for people). Gasoline was also reformulated to eliminate excessively volatile hydrocarbons and unsaturated hydrocarbons (those with double bonds between carbon atoms) that are especially reactive in forming photochemical smog.

An even more drastic approach to eliminating smog-forming emissions is the use of electric automobiles that do not burn gasoline. These vehicles certainly do not pollute as they are being driven, but they suffer from the challenging problem of a limited range between charges and the need for relatively heavy batteries. However, hybrid automobiles using a small gasoline or diesel engine that provides electricity to drive electric motors propelling the automobile and to recharge relatively smaller batteries can largely remedy the emission and fuel economy problems with automobiles. The internal combustion engine on these vehicles runs only as it is needed to provide power and, in so doing, can run at a relatively uniform speed that provides maximum economy with minimum emissions.

Another approach that is being used on vehicles as large as buses that have convenient and frequent access to refueling stations is the use of fuel cells that can generate electricity directly from the catalytic combination of elemental hydrogen and oxygen, producing only harmless water as a product. There are also catalytic processes that can generate hydrogen from liquid fuels, such as methanol, so that vehicles carrying such a fuel can be powered by electricity generated in fuel cells.

Green chemistry can be applied to devices and processes other than automobiles to reduce smog-forming emissions. This is especially true in the area of organic solvents used for parts cleaning and other industrial operations, vapors of which are often released into the atmosphere. The substitution of water with proper additives or even the use of supercritical carbon dioxide fluid can eliminate such emissions.

8.9 BIOLOGICAL CONTROL OF AIR POLLUTION

Living organisms have a large effect on the atmosphere. The oxygen now in the atmosphere was put there by the action of photosynthetic cyanobacteria. Carbon dioxide is removed from the atmosphere by the action of plants performing photosynthesis. Organisms may act effectively to purify air. There are two major categories of such purification. The first of these uses microorganisms in biofilters through which polluted air and gases are forced to filter the air and remove pollutants from it. The second uses plants growing in areas afflicted by air pollution to sequester and remove pollutant gases, vapors, and particulate matter from air. These two approaches are addressed in this section.

8.9.1 Bioreactors for Air Pollutant Removal

Bioreactors that contain microorganisms that degrade pollutants, primarily bacteria and fungi, can be used to purify contaminated air from a variety of sources.[6] The first such reactors to be used were simple biofilters, which removed odorous compounds from wastewater treatment plants, solid waste treatment plants, and composting facilities by forcing the air up through a porous bed of biomass material, such as wood chips or peat, that held biodegrading organisms and provided nutrients. Although the first materials to be removed from air by bioreactors were generally biodegradable organic compounds, later research has shown that a variety of substances usually regarded as poorly biodegradable, such as chlorinated ethylene compounds, could be removed by bioreactors as well. In addition, inorganic gases including NO_X, SO_2, H_2S, and CO can also be removed.

The kinds of contaminants most commonly removed from air by bioreactors are hydrocarbons, such as toluene, and organooxygen compounds, such as formaldehyde and methanol vapors. Such substances are mineralized to inorganic carbon dioxide and water and to a certain extent are used by microorganisms to produce biomass. For the biodegradation of hydrocarbons and organooxygen compounds, it may be necessary to add other nutrients including phosphorus, sulfur, and especially nitrogen. One of the most commonly studied compounds is toluene for which the biodegradation reaction in a medium supplemented by NH_4Cl as a nitrogen source may be represented as follows:

$$1.55C_7H_8 + 12.9O_2 + 0.2NH_4Cl \rightarrow CH_{1.8}N_{0.2}O_{0.5}\,(biomass) + 9.85CO_2 + 5.6H_2O + 0.2HCl$$

(8.17)

The production of acidic HCl in this reaction may require addition of base to keep the microbial medium from becoming too acidic.

Initially, the filter beds used on biofilters were composed exclusively of natural materials including wood, peat, compost, and even soil. Such materials served as a source of microorganisms and nutrients and were readily disposed after use. However, the nutrients in them could become depleted and compaction over time that reduced air flow could occur. More recently, nonorganic materials including plastic rings and saddles; silicate-based celite, perlite, or porous lava rock; polyurethane; and activated carbon have been used as biomass supports. Activated carbon has the advantage that it absorbs vapors and holds them in place for microbial biodegradation to occur. These materials provide no nutrients, which normally must be added for optimum biotreatment. They do not degrade, but can be rejuvenated by, for example, scrubbing processes to remove excess biomass.

A relatively recent development in bioreactors is that of the **biotrickling filter** similar to trickling filters used in wastewater treatment (see Chapter 5, Section 5.12 and Figure 5.8). The configuration for a biotrickling filter for air treatment is shown in Figure 8.10. The major advantage of a

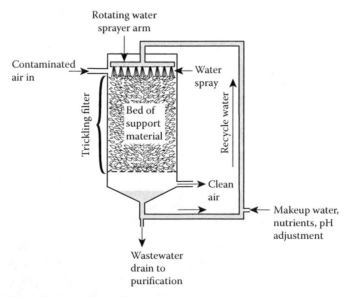

FIGURE 8.10 Configuration of a biotrickling filter for air purification. As is the case with a trickling filter for wastewater treatment, the microorganisms responsible for removing air pollutants are held on pieces of solids in the filter bed. The contaminated air is first run through a filter to remove particles and, if necessary, its temperature lowered to a level compatible with the microorganisms in the trickling filter.

biotrickling filter over a fixed bed reactor is the flowing water phase, which enables pH control, removal of metabolites, temperature control, and facile addition of nutrients.

A **bioscrubber** for air treatment consists of two separate units: (1) a scrubbing tower into which water is sprayed, often over a medium that disperses the scrubbing solution, and air is blown counter to the flow of the scrubbing water, and (2) a biological wastewater treatment unit in which the impurities absorbed from the air are degraded, often with an activated sludge treatment process (see Chapter 5, Section 5.12 and Figure 5.9). This process is most suitable for removal of water-soluble air pollutants.

8.9.2 REMOVING AIR POLLUTION WITH VEGETATION

Vegetation in contact with a plume of polluted air can act to attenuate air pollutants by intercepting, absorbing, and metabolizing air pollutants and can divert plumes of polluted air carried by wind. Advantage has been taken of these capabilities by planting greenbelts of trees and shrubs around industrial facilities from which air pollutants may be emitted. A study of such a greenbelt 500 m in width planted around a petroleum refinery showed that it was about 40% effective in removing SO_2, NO_X, and small particulate matter and around 90% effective in removing total hydrocarbons, volatile organic compounds, and CO.[7]

For a greenbelt to be effective in controlling air pollution, it must be located such that the pollutant plume flows through it at ground level. Proper location of the greenbelt requires knowledge of meteorological conditions, including wind patterns and temperature profiles. Topography must also be considered. For example, locating a greenbelt on the upslope of an elevated area in the plume means that the polluted air is forced into the vegetation canopy increasing contact of the pollutants with the vegetation.

It is important to consider the types of vegetation in the greenbelt. Obviously, plants have to be robust and tolerant of the air pollutants. Plant height, flowering and production potential, foliage form, and canopy structure should be considered. Another consideration is the possibility of air pollutant emissions from the plants themselves. As discussed in Chapter 7, Section 7.7, plants may release biogenic compounds. Of these, the ones of most concern for potential contributions to air pollution are alkenes, including gaseous ethylene and the terpenes, which have C=C bonds. These hydrocarbons react very readily with hydroxyl radical and ozone and in the presence of pollutant NO_X have the potential to contribute to photochemical smog conditions, which would be counterproductive to the purpose of a greenbelt for air pollution control.

Although greenbelts around industrial facilities have the potential to control air pollution at the local level, on a larger scale, vegetation, especially forest trees, planted over large areas susceptible to air pollution may have beneficial effects in limiting air pollution regionally. A concern with respect to global warming is reduction of forest growth due to heat and drought and the consequent loss of the atmospheric cleansing ability of trees.

8.10 CONTROLLING ACID RAIN

Acid rain and its effects were discussed in Chapter 7, Section 7.6. One of the two major contributors to acid rain are sulfur dioxide, which produces sulfurous acid, H_2SO_3, but, more importantly, sulfuric acid, H_2SO_4, and the acidic salt, NH_4HSO_4. The other major contributor is NO_X, which is oxidized to HNO_3 and may also contribute the acidic salt NH_4NO_3 to the atmosphere. There is some contribution to acid rain by HCl, primarily released in the burning of organochlorine materials, especially polyvinylchloride plastic.

Acid rain is best controlled by limiting emissions to the atmosphere of sulfur dioxide and nitrogen oxides, discussed in Sections 8.6 and 8.7. Combustion of coal containing sulfur is a major source of atmospheric sulfur dioxide, the release of which can be limited by stack gas controls, fluidized bed combustion in a sulfur-sequestering medium, and removal of sulfur from coal before combustion (removal of pyrite, FeS_2). A more sustainable approach is to use fuels that do not contain sulfur (and most sustainable of all are energy sources in which fossil fuels are not burned at all). The most promising low-sulfur fossil fuel is natural gas, which is now being extracted from abundant shale resources of methane. Some natural gas contains sulfur as toxic H_2S, but it is readily removed from the methane and disposed below ground or used as a raw material to make industrial sulfuric acid. It is important to control sulfur emissions from metal refining operations where sulfide ores, including CuS, ZnS, and PbS, are used as sources of metal. Some areas around smelters that used sulfide ores have been devastated of vegetation killed by sulfur dioxide and sulfuric acid released from the smelters.

Limiting NO_X emissions prevents production of acidic HNO_3 in the atmosphere. Nitric acid is the major inorganic product of NO_X in the process of photochemical smog formation so that curtailment of smog also prevents production of atmospheric HNO_3. This involves not only control of NO_X emissions but also reactive hydrocarbons. Sulfur dioxide is rapidly oxidized to sulfuric acid under photochemical smog–forming conditions and curtailment of smog reduces this source of acid as well.

Potential sources of atmospheric HCl can be limited by scrubbing stack gas from facilities where organochlorine compounds, especially polyvinylchloride plastic, are incinerated. To the extent possible, it is best to avoid incineration of such materials.

8.10.1 Dealing with Toxic and Other Adverse Effects of Acid Rain

Acid rain in the form of rainfall poses little threat to human health; the pH is not low enough to affect exposed tissue. However, droplets composing fog may be quite acidic and are respirable, which can cause asthma attacks and other adverse health effects to exposed individuals. In addition, inhalation of particles containing acidic salts (ammonium sulfates and nitrates) can have adverse pulmonary effects.

Acidic precipitation adversely affects vegetation and crops and is now regarded as a major biological stressor around the world.[8] The growth, nutrition, and physiology of crop plants are adversely affected by acid rain. The acidification of soil by acidic precipitation leaches essential cations from the rhizosphere region of soil where plant roots are located. A significant problem is leaching of aluminum from soil minerals, which competes with nutrient metal ions for entry into root systems and has the effect of starving plants of other essential cations.

Acid rain adversely affects aquatic organisms in bodies of water that lack a pH buffering capacity because of lack of contact with minerals that can neutralize acid, especially limestone, $CaCO_3$. The greatest adverse effect is on fish fingerlings that do not thrive in acidic water.

The best way to deal with acid rain is to eliminate its production by limiting emissions of sulfur and nitrogen oxides. Consideration may need to be given to growing crops that are relatively less susceptible to the adverse effects of acid rain and the nitrogen and sulfur oxides that form it. Acid in excessively acidic bodies of water can be neutralized by treatment with $CaCO_3$.

8.11 LIMITING STRATOSPHERIC OZONE DEPLETION

As mentioned in Chapter 6, Section 6.2, stratospheric ozone, O_3, serves as a shield against harmful ultraviolet radiation from the sun. Threats to this essential shield, especially from CFC compounds released into the atmosphere, were discussed in Chapter 7, Section 7.9. This section discusses measures being taken to protect stratospheric ozone, especially with the development of substitutes for CFCs that are much less likely to harm stratospheric ozone.

In a sense, the CFCs now blamed for stratospheric ozone depletion were an example of green chemistry, developed in the 1930s long before the concept of green technology was even imagined. The fluids that they replaced, sulfur dioxide and ammonia, are quite toxic and had even caused fatalities when leaked from refrigerators in homes. The CFC replacements performed ideally and were remarkably nontoxic. Several related compounds, such as halothane, 2-bromo-2-chloro-1,1,1-trifluoroethane, have been used as anesthetics. It was not until the 1970s and later that the analytical capability became available to show that CFCs had persisted throughout the global atmosphere and are far from green when considering the global environment as a whole.

The solutions to the problem of stratospheric ozone depletion posed by CFCs provide a good example of green chemistry and green technology, taking advantage of fundamental knowledge regarding the properties and behavior of chemicals. The reason that CFCs are so stable and do not break down at all until they have entered the stratosphere—and then only slowly—is the extreme stability of the C–Cl and C–F bonds. Essentially, all anthropogenic chemical species that are broken down in the troposphere are attacked by hydroxyl radical, HO·, which is abundant in the troposphere. This reactive species attacks and breaks C–H bonds, but is not reactive enough to break C–Cl and C–F bonds. So, the solution to the problems posed by CFCs has been to develop **hydrohaloalkanes** that possess at least one C–H bond per molecule that is susceptible to attack by HO· radical in the troposphere, thereby eliminating the compound with its potential to produce ozone-depleting Cl atoms before it reaches the stratosphere.

The first substitutes for CFCs widely used were hydrochlorofluorocarbons (HCFCs), of which one of the most popular was HCFC-22 ($CHClF_2$). However, because it is a carrier for the chlorine atom, this compound still has an ozone-depletion potential of 0.030 relative to a value of 1.0 for CFC-11, a non-hydrogen-containing CFC with a formula of $CFCl_3$, and the use of HCFC-22 and that of other HCFCs is being discontinued. The substitutes that were favored as replacements are HFCs, including the very popular HFC-134a (CH_2FCF_3) and HFC-410A (CHF_2CF_3). The HFCs do not deplete ozone because they have no chlorine atoms, the fluorine atom does not affect ozone, and they are largely destroyed in the troposphere by attack of hydroxyl radical on the H atom. Unfortunately, HFC-134a has a relatively high global warming potential, so replacements are being sought for it.

Several classes of compounds are considered as replacements for HCFCs. Ironically, one of the practical replacements is carbon dioxide, the substance most commonly mentioned as a greenhouse gas! Hydrocarbon gases can be used as refrigerants. One such replacement is designated HC-12a, a mixture of propane and isobutene. It could be used as a "drop-in" replacement for dichlorodifluoromethane in automobile air conditioners that did not require any modification of seals or other components of the system, but was banned in 1995 because of its flammability. The latest kind of fluid to gain favor consists of hydrofluoroolefins (HFOs) that have a C=C double bond. The most popular of these is HFO-1234yf, a fluorinated alkene with a chemical formula of $CF_3CF=CH_2$, with physical properties very similar to those of HFC-134a, which it readily replaces. Like HFC-134a, HFO-1234yf has no ability to deplete stratospheric ozone. Furthermore, HFO-1234yf has virtually no capacity to cause climate warming, unlike HFC-134a, which is a potent greenhouse gas. The reason for the very low potential of HFO-1234yf to cause global warming is that in addition to its C–H bonds that are susceptible to attack by hydroxyl radical in the troposphere it has a C=C double bond, which undergoes extremely rapid addition reactions with HO·, leading to very short atmospheric lifetimes for the HFO molecule.

QUESTIONS AND PROBLEMS

1. What is the Gaia hypothesis or theory? What does it have to do with sustaining the atmosphere? Who first proposed this hypothesis and for what discovery related to sustaining the atmosphere is this scientist noted?

2. Critique the statement that (Greenhouse gases in the atmosphere prevent a significant fraction of the energy that Earth receives from the sun from ever leaving the Earth System thus resulting in a steady increase in the atmosphere's temperature, which causes global warming.) Is that statement true and, if not, in what sense is it false? If you think it is false, suggest a better explanation for global warming.

3. Black carbon is a term that is applied both to a material in the atmosphere and to a material commonly found in the geosphere (soil). Explain how black carbon in or carried by the atmosphere may contribute to global warming. How does the production of black carbon that ends up in soil (biochar) tend to reduce global warming?

4. What is meant by the concept of the Anthropocene and how is it related to sustainability of the atmosphere? Who is commonly credited with coining the term "Anthropocene" and with which other major environmental issue is this person commonly associated?

5. Some of the most important adverse effects of global warming are mentioned in this chapter as "collateral damage." One such example of collateral damage from global warming is the bark beetle infestation that is devastating forests in the western United States and Canada. What has the bark beetle done that is so damaging? How might the infestation be related to global warming?

6. What is carbon capture and how is it used to reduce atmospheric warming? What are some specific examples of the practice of carbon capture? What are the best places to put captured carbon and why is it not such a good idea to pump it deep into oceans?

7. How does the cultivation of rice contribute to greenhouse warming? Does rice itself contribute to greenhouse gas emissions? How may changes in the way that rice is grown reduce greenhouse gas emissions?

8. In what sense is the "brown cloud" a counteracting measure against global warming? Does that mean that measures should not be taken to counteract the brown cloud?

9. Are there any examples in which volcanic eruptions have acted to counteract global warming? If so, what is the historical evidence for it?

10. Look up the construction of a venturi gas scrubber. What are the converging section, throat, and diverging sections? How does the venturi scrubber work? What air pollutants does it remove?

11. What did Frederick G. Cottrell invent that pertains to air pollution control? When was this device first used for? Is there any possibility that the production of ozone could be a problem with this device and, if so, how?

12. Chemically, what is nature's way of dealing with carbon monoxide emissions so that this gas does not accumulate in the atmosphere?

13. In what sense is nitrogen dioxide, NO_2, a secondary air pollutant? Explain. Why is NO_2 especially harmful in the atmosphere?

14. Suggest a series of reactions by which sulfur dioxide can be removed from stack gas and converted to a commercially valuable product that ranks high in annual production of inorganic chemicals.

15. What pollution problem does a lean mixture aggravate when employed to control hydrocarbon emissions from an internal combustion engine?

16. The Blue Ridge Mountains in the United States are part of the Appalachian Mountain chain extending from Georgia into Pennsylvania. They are noted for the blue haze that envelops the mountains. What is the chemical process that makes these mountains "blue" and why were they blue even in times when anthropogenic air pollution was not a factor?

17. Vehicles powered by fuel cells are uniquely well suited for Iceland. What is it about the geography and resources of Iceland that make it particularly well suited for fuel cells in transportation?

18. Explain chemically why pine and citrus trees would not be good choices for a greenbelt to control air pollution around an industrial facility from which one of the major air pollutants emitted consists of NO_X.

19. Explain why deforestation associated with the heat and drought conditions that occur with global warming may contribute to air pollution.

20. Explain how the formation of photochemical smog may contribute to acid rain formation. What is the chemistry involved?

21. If the concept of green chemistry had been around in the 1930s when CFCs were developed and used as refrigerants, how would the practitioners of green chemistry probably have reacted to this development, given the state of knowledge regarding atmospheric chemistry at that time? What happened about 40 years later that would have changed that view?

LITERATURE CITED

1. Intergovernmental Panel on Climate Change (IPCC), 2001, IPCC Third Assessment Report—Climate Change 2001—Complete online versions, http://www.grida.no/publications/other/ipcc_tar/?srcc =/climate/ipcc_tar/wg1/214.htm#611.
2. Sloan, T., and A. W. Wolfendale, Man-Made Global Warming Explained—Closing the Blinds (2010), http://arxiv.org/pdf/1001.4988. e-Print Archive, Physics Pages 1–7, 2010.
3. Chan, E. C., M. H. Davy, G. deSimone, and V. Mulone, Numerical and Experimental Characterization of a Natural Gas Engine with Partially Stratified Charge Spark Ignition, *Journal of Engineering for Gas Turbines and Power* **133**, 22801 (2011).
4. Wilson, Elizabeth J., and David Gerard, *Carbon Capture and Sequestration: Integrating Technology, Monitoring and Regulation*, Blackwell Publishing, Ames, IA, 2007.
5. Hesterberg, Thomas W., Christopher M. Long, Charles A. Lapin, Ali K. Hamade, and Peter A. Valberg, Diesel Particulate (DEP) and Nanoparticle Exposures: What do DEP Human Clinical Studies Tell Us About Potential Health Hazards of Nanoparticles?, *Inhalation Toxicology* **22**, 679–694 (2010).
6. Kennes, Christian, Eldon R. Rene, and María C. Veig, Bioprocesses for Air Pollution Control, *Journal of Chemical Technology and Biotechnology* **84**, 1419–1436 (2009), online at http://onlinelibrary.wiley.com/doi/10.1002/jctb.2216/pdf.
7. Gupta, Rakhi B., P. R. Chaudhari, and S. R. Wate, Overview on Attenuation of Industrial Air Pollution by Greenbelt, *Journal of Industrial Pollution Control* **24**, 1–8 (2008).
8. Balasubramanian, G., C. Udayasoorian, P. C. Prabu, and G. Senthil Kumar, Impact of Acid Rain on Agro-eco Systems, *Journal of Ecobiology* **21,** 1–38 (2007).

SUPPLEMENTARY REFERENCES

Archer, David, *Global Warming: Understanding the Forecast*, Blackwell Publishing Co., Malden, MA, 2007.

Archer, David, and Stefan Rahmstorf, *The Climate Crisis: An Introductory Guide to Climate Change*, Cambridge University Press, Cambridge, UK, 2010.

Balduino, Sergio P., Ed., *Progress in Air Pollution Research*, Nova Science Publishers, New York, 2007.

Beerling, David, *The Emerald Planet: How Plants Changed Earth's History*, Oxford University Press, Oxford, UK, 2007.

Bily, Cynthia A., Ed., *Global Warming*, Greenhaven Press, San Diego, CA, 2006.

Bodine, Corin G., Ed., *Air Pollution Research Advances*, Nova Science Publishers, New York, 2007.

Brown, Paul, *Global Warning: The Last Chance for Change*, Guardian Books, London, 2006.

Chambers, Frank, and Michael Ogle, Eds., *Climate Change: Critical Concepts in the Environment*, Routledge, London, 2002.

Chehoski, Robert, Ed., *Critical Perspectives on Climate Disruption*, Rosen Publishing Group, New York, 2006.

Cheremisinoff, Nicholas P., *Handbook of Air Pollution Control and Prevention Technologies*, Butterworth-Heinemann, Woburn, MA, 2002.

Colls, Jeremy, and Abhishek Tiwary, *Air Pollution: Measurement, Modelling, and Mitigation*, 3rd ed., Routledge, London, 2010.

Cooper, C. David, and F. C. Alley, *Air Pollution Control: A Design Approach*, 4th ed., Waveland Press, Long Grove, IL, 2010.

De Nevers, Noel, *Air Pollution Control Engineering*, 2nd ed., Waveland Press, Long Grove, IL, 2010.

Gore, Al, *An Inconvenient Truth: The Planetary Emergency of Global Warming and What We Can Do About It*, Rodale Press, Emmaus, PA, 2006.

Heck, Ronald M., Robert J. Farrauto, and Suresh T. Gulati, *Catalytic Air Pollution Control: Commercial Technology*, 3rd ed., Wiley, Hoboken, NJ, 2009.

Johansen, Bruce E., *Global Warming in the 21st Century*, Praeger Publishers, Westport, CN, 2006.

Lehrer, Jonathan, Ed., *Technical Report On Ozone Exposure, Risk, and Impact Assessments for Vegetation*, U.S. Environmental Protection Agency, Office of Air Quality Planning and Standards, Health and Environmental Impacts Division, Ambient Standards Group, Research Triangle Park, NC, 2007.

Livingston, James V., Ed., *Air Pollution: New Research*, Nova Science Publishers, New York, 2007.

McCaffrey, Paul, *Global Climate Change*, H. W. Wilson Company, Bronx, NY, 2006.

Mohanakumar, K., *Stratosphere Troposphere Interactions: An Introduction*, Springer, Berlin, 2010.

Muzio, L. J., G. C. Quartucy, and J. E. Cichanowicz, Overview and Status of Post-Combustion NO_X Control: SNCR, SCR and Hybrid Technologies, *International Journal of Environment and Pollution* **17**, 4–30 (2002).

Parker, Larry, and Wayne A. Morrissey, *Stratospheric Ozone Depletion*, Novinka Books, New York, 2003.

Parson, Edward, *Protecting the Ozone Layer: Science and Strategy*, Oxford University Press, New York, 2003.

Ramachandra, T. V., and S. P. Mahajan, *Air Pollution Control, Tata Energy Research Institute*, New Delhi, 2011.

Shareefdeen, Zarouk, and Ajay Singh, *Biotechnology for Odor and Air Pollution Control*, Springer, Berlin, 2008.

Srivastava, R. K., W. Neuffer, D. Grano, S. Khan, J. E. Staudt, and W. Jozewicz, Controlling NO_X Emission from Industrial Sources, *Environmental Progress* **24**, 181–197 (2005).

Wang, Lawrence K., Norman C. Pereira, Yung-Tse Hung, and Kathleen Hung, Eds., *Air Pollution Control Engineering*, Humana Press, Totowa, NJ, 2004.

9 Environmental and Toxicological Chemistry of the Geosphere

9.1 GEOSPHERE

As illustrated in Figure 9.1, the **geosphere** is the solid Earth (which sometimes is not so solid when earthquakes or volcanic eruptions occur). The geosphere is an enormous source of natural capital, the management and preservation of which are of utmost importance to sustainability.[1] It provides the platform upon which most food is grown and is the source of plant fertilizers, construction materials, and fossil fuels that humans use. As part of its natural capital, the geosphere receives large quantities of consumer and industrial wastes, although past and current practices of using the geosphere as the anthrosphere's waste dump are ultimately unsustainable. As shown in Figure 9.1, the geosphere interacts strongly with the hydrosphere, atmosphere, biosphere, and anthrosphere.

The geosphere is a layered structure, most of which is hot enough to melt rock. Earth's core is a huge ball of iron at a temperature above the normal melting point of iron, but solid because of the enormous pressure that it is under. Above this core is the **mantle** composed of rock and ranging in depth between 300 and 1890 km. The deeper inner mantle, though hot enough for the rock to be liquid under ordinary pressures, is solid because of the enormous pressure to which it is subjected. On top of the inner mantle is the outer mantle at a depth between 10 and 300 km composed of hot molten rock called **magma**. Floating on the magma is the solid **lithosphere** composed of relatively strong rock, varying in thickness from just a few to as much as 400 km, averaging about 100 km. The transition layer between the molten magma and the lithosphere is the **athenosphere** composed of hot rock that is relatively weak and plastic. Earth's **crust** is the outer layer of the lithosphere, which is only 5–40 km thick.

9.1.1 Geosphere Related to the Other Environmental Spheres

Virtually all things and creatures commonly regarded as parts of Earth's environment are located on, in, or just above the geosphere. Major segments of the hydrosphere including the oceans, rivers, and lakes rest on the geosphere, and groundwater exists in aquifers underground. Water dissolves minerals from the geosphere that nourish aquatic life. These minerals and rock particles eroded by moving water from the geosphere are deposited in layers and transformed into rock again. The atmosphere exchanges gases with the geosphere. For example, organic carbon produced by photosynthetic plants from atmospheric carbon dioxide may end up as soil organic matter in the geosphere, and the photosynthetic processes of plants growing on the geosphere put elemental oxygen back into the atmosphere. The majority of biomass of organisms in the biosphere is located on or just below the surface of the geosphere. Most structures that are parts of the anthrosphere are located on the geosphere, and a variety of wastes from human activities are discarded to the geosphere.

Modifications and excavations of the geosphere to accommodate the anthrosphere have major effects on the geosphere. Human activities have a tremendous influence on the geosphere as evidenced by hills leveled, valleys filled in, and vast areas paved to make freeways, parking lots, and shopping centers. One such influence is on **surface albedo**, defined as the percentage of impinging

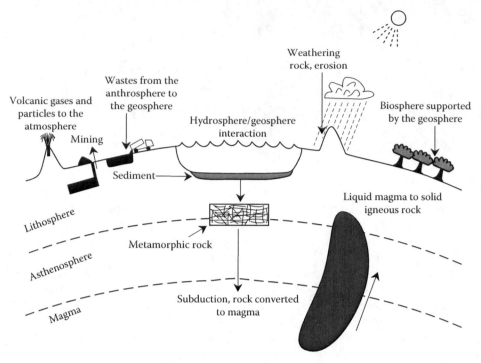

FIGURE 9.1 The geosphere. The "solid earth" consists of a surface layer of rock about 100 km thick on average floating on top of molten magma. Rock cycles occur within the geosphere. It is an enormous source of natural capital including essential minerals and provides a thin surface of soil upon which most food is grown. The geosphere interacts strongly with the other environmental spheres.

solar energy reflected back from Earth's surface. The surface albedo of an asphalt-paved surface is only about 8%. A more alarming effect is desertification, in which normally productive soil is converted to unproductive desert in areas where rainfall is marginal. This phenomenon is discussed in more detail in Chapter 10.

9.1.2 PLATE TECTONICS

Introduced to much controversy in the mid-1900s, the theory of **plate tectonics** views Earth's surface as consisting of huge lithospheric plates upon which the continents and Pacific Ocean rest, behaving as units.[2] Earth's crust is a dynamic system in which the lithospheric plates move relative to each other by, typically, a few centimeters per year. When abrupt plate movement occurs, an earthquake results. Magma coming to the surface along plate boundaries results in emissions of hot and molten rock, ash, and gases in the form of volcanoes. The major aspects of plate tectonics are shown in Figure 9.2.

9.1.3 ROCK CYCLE

The rock that composes the geosphere circulates through the geosphere in the **rock cycle**. Figure 9.3 illustrates the rock cycle in which rock circulates among liquid magma and solid igneous, sedimentary, and metamorphic rock. As molten magma penetrates near the top of Earth's crust then cools and solidifies, it forms **igneous rock**. Exposed to water and the atmosphere, igneous rock undergoes physical and chemical changes in a process called **weathering**. Weathered rock material carried by water and deposited as sediment layers may be compressed to produce **secondary minerals**, of which clays are an important example. The action of high pressure and elevated temperatures converts sedimentary rock to metamorphic rock.

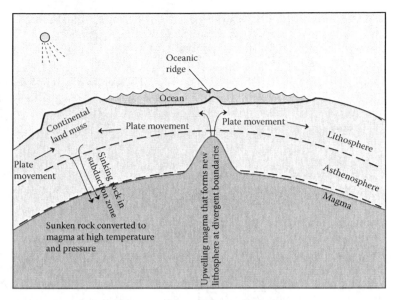

FIGURE 9.2 Illustration of the tectonic cycle in the geosphere. Divergent boundaries form on ocean floors between tectonic plates that are moving apart and are where hot magma undergoes upwelling and cooling to form new solid lithospheric rock, creating ocean ridges. Convergent boundaries are where plates move toward each other forcing matter downward into the asthenosphere in subduction zones, eventually to form new molten magma and, in some cases, forcing matter upward to produce mountain ranges. Two plates moving laterally relative to each other create fault lines along which earthquakes occur.

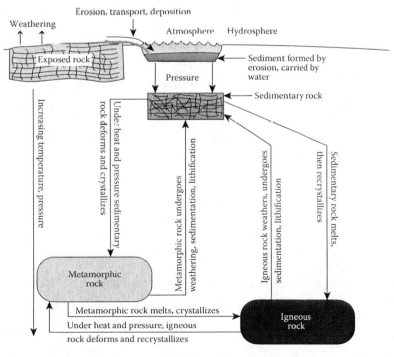

FIGURE 9.3 The rock cycle. Solid rock is formed initially when liquid magma reaches the surface and solidifies to produce igneous rock. The igneous rock undergoes weathering, a slow process in which it is broken down physically and chemically. The products of weathering are carried by water and deposited as sediments. Buried and subjected to heat and pressure, sediments form metamorphic rock, which can melt to produce magma and start the cycle over again.

A part of the crust crucial for the existence of humans and most other nonaquatic life forms is the thin layer of weathered rock, partially decayed organic matter, air spaces, and water composing **soil** that supports plant life. Were Earth the size of a geography classroom globe, the average thickness of the soil layer on it would be only about the size of a human cell! The top layer of soil that is most productive of plants is topsoil, which is often only a few centimeters thick in many locations or even nonexistent where poor cultivation practices and adverse climatic conditions have led to its loss by wind and water erosion. The conservation of soil and enhancement of soil productivity are key aspects of sustainability. Soil is discussed in detail in Chapter 10.

9.2 CHEMICAL COMPOSITION OF THE GEOSPHERE AND GEOCHEMISTRY

For the most part, the crust consists of **rocks**, which in turn are made up of **minerals** characterized by a definite chemical composition and crystal structure. Only about 25 of the approximately 2000 known minerals compose most rocks. Because most of the crust consists of chemically combined oxygen (49.5%) and silicon (25.7%), the most abundant minerals are **silicates** composed of various silicon oxides, examples of which are quartz, SiO_2, and potassium feldspar, $KAlSi_3O_8$. Other elements in Earth's crust are aluminum (7.4%, commonly occurring as Al_2O_3), iron (4.7% as Fe_3O_4 and other iron oxides), calcium (3.6% in limestone, $CaCO_3$, and dolomite, $CaCO_3 \cdot MgCO_3$), sodium (2.8%), potassium (2.6%), and magnesium (2.1%). That leaves only 1.6% of the crust to serve as a source of other important mineral substances, including metals other than iron and aluminum, phosphorus required for plant growth, and sulfur widely used in industrial applications. Careful management of this natural capital resource of scarce essential minerals is one of the primary requirements for sustainability.

The rocks that compose Earth's crust participate in the **rock cycle** shown in Figure 9.3, a process that alters the chemical nature of rocks and forms new minerals. As shown in Figure 9.3, igneous rock is produced from the solidification of liquid magma. Exposed to water, atmospheric oxygen, and various organisms, igneous rock undergoes weathering and becomes highly altered, reaching a state of greater physical and chemical equilibrium with the atmosphere. Weathering products end up as soil and sediments in bodies of water and are carried by water to be deposited as sediments. Sediments that become buried and compressed turn into secondary minerals, among the most abundant of which are **clays**, consisting of hydrated silicon and aluminum oxides, produced by the weathering of minerals such as potassium feldspar, $KAlSi_3O_8$. A common clay is **kaolinite**, $Al_2Si_2O_5(OH)_4$, structural aspects of which are shown in Figure 9.4.

Geochemistry is the branch of chemistry that deals with rocks and minerals and the chemical interactions of the geosphere with other environmental spheres.[3] The specialized branch of geochemistry relating to environmental influences and interactions of the geosphere is **environmental geochemistry**. Weathering by chemical processes is a particularly important aspect of geochemistry and largely determines the chemical nature of minerals near Earth's surface. Almost imperceptible under dry conditions, weathering proceeds at a much more rapid rate in the presence of water. The rate of weathering is also increased by the action of microorganisms, some of which secrete chemical species that attack rock and leach nutrients from it. Particularly important to weathering are **lichens**, which are algae and fungi living together synergistically. The algae utilize solar energy to convert atmospheric carbon dioxide to plant biomass and the fungi utilize the biomass and anchor the organisms to the rock surface and extract nutrients from it.

Weathering enables the rock/water/mineral system to attain equilibrium through chemical mechanisms of dissolution or precipitation, acid-base reactions, complexation, hydrolysis, and oxidation-reduction. Water plays a key role in weathering, enabling weathering agents to come into intimate chemical contact with rock, removing weathered material from the rock surface, and participating

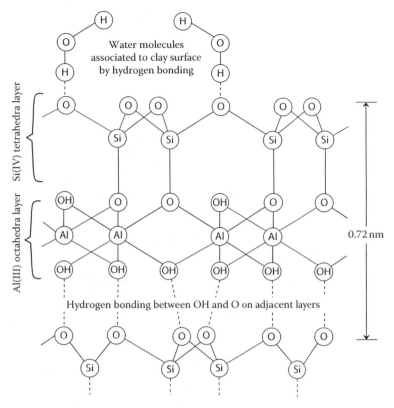

FIGURE 9.4 Representation of the structure of kaolinite, a two-layer clay, in which layers of hydrated Si(IV) oxide alternate with layers of Al(III) oxide.

as a reactant in the chemical reactions involved in weathering. Acid-forming gases such as CO_2 and SO_2 produce H^+ ion, a key weathering agent, by reactions with water:

$$CO_2 + H_2O \rightarrow H^+ + HCO_3^-$$ (9.1)

Rainwater, which is free of alkalinity and hardness ions, but is somewhat acidic due to dissolved CO_2 and sometimes other acidic constituents, such as those characteristic of acid rain (see Section 4.8), is an aggressive weathering agent.

There are several important processes in which water is involved in chemical weathering. Water acts as a solvent to dissolve minerals,

$$CaSO_4 \cdot 2H_2O(s)(water) \rightarrow Ca^{2+}(aq) + SO_4^{2-}(aq) + 2H_2O$$ (9.2)

may add to minerals in **hydration** reactions,

$$CaSO_4(s) + 2H_2O \rightarrow CaSO_4 \cdot 2H_2O(s)$$ (9.3)

or be lost in **dehydration** processes:

$$2Fe(OH)_3 \cdot xH_2O(s) \rightarrow Fe_2O_3(s) + (3 + 2x)H_2O$$ (9.4)

The reaction

$$FeCO_3(s) + H_2O \rightarrow FeOH^+(aq) + HCO_3^-(aq) \qquad (9.5)$$

illustrates dissolution accompanied by hydrolysis. **Acid hydrolysis** is an important process by which $CaCO_3$ and $CaCO_3 \cdot MgCO_3$ are dissolved

$$CaCO_3(s) + H_2O + CO_2(aq) \rightarrow Ca^{2+}(aq) + 2HCO_3^-(aq) \qquad (9.6)$$

and the dissolved carbonates may be transported some distance in solution in water and deposited as the solid carbonates as CO_2 is lost to the atmosphere or taken up by algal photosynthesis, reversing Reaction 9.6. The dissolution of pyrite through the action of bacterially mediated reactions (see Section 3.3) is an example of weathering by **oxidation**:

$$4FeS_2(s) + 15O_2(g) + (8 + 2x)H_2O \rightarrow 2Fe_2O_3 \cdot xH_2O + 8SO_4^{2-}(aq) + 16H^+(aq) \qquad (9.7)$$

Complexation of metal ion may enhance weathering. An example is the dissolution of iron(II) silicate minerals, such as fayalite, Fe_2SiO_4, by the action of fulvic acid chelating agents (represented as H_2FA; see Section 3.11 and Figure 3.14):

$$Fe_2SiO_4(s) + 2H_2FA(aq) + H_2O \rightarrow 2FeFA(aq) + H_4SiO_4(aq) \qquad (9.8)$$

The formation of soluble iron or fulvic acid chelates, represented in general as FeFA (Equation 9.8), has been shown to be responsible for soluble iron in runoff from humic-substance-rich peat bog leachate carrying iron into coastal seawater off North Scotland.[4]

The kinds and concentrations of solute species in natural waters are largely determined by weathering processes. Acid hydrolysis of silicate minerals is especially important in adding Na^+, K^+, and Ca^{2+} to water in contact with these minerals.

9.2.1 Biological Aspects of Weathering

As noted previously, organisms, especially lichens, are responsible for much weathering followed by soil formation. Miniature ecosystems develop in cavities on rock surfaces where they accumulate water and inorganic and organic debris, including weathering products (Figure 9.5). In addition to lichens, such systems support nitrogen-fixing cyanobacteria, green algae, fungi, bacteria, and insects. The organisms release organic acids and produce humic material from the partial biodegradation of plant matter that act to weather the rock, enlarging the cavities. As this process occurs, small crystals of minerals are released from the rock and may undergo further weathering. Eventually, the weathering products, windblown dust, and organic matter produce soil in the cavities. This enables vascular plants to develop, which may contribute to further weathering of rock through, for example, the action of bacteria associated with the plant roots.

Weathering is an important process for releasing trace levels of elements, including nutrients and toxic substances, into water and soil. As noted previously, fulvic acids chelate iron in the weathering of iron-rich minerals and may be responsible for maintaining levels of this micronutrient high enough in water to support the growth of algae. Also discussed previously, bacterially mediated weathering of mineral pyrite, FeS_2, can release sulfuric acid and excessive levels of soluble iron into water. The acid is toxic to aquatic life as is soluble iron at higher levels. The dissolution of aluminum-containing minerals by acidic water, such as that from acid rainwater, can raise levels of

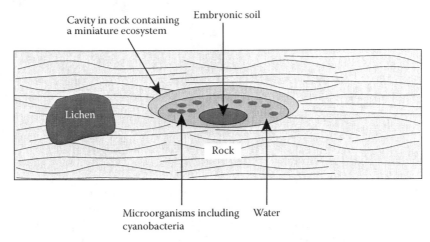

FIGURE 9.5 Water, solids of various kinds, and organisms accumulate in cavities on the surface of rocks, contributing to weathering. These structures may host miniature ecosystems in which the organisms contribute to the weathering of rock and make use of materials produced by weathering.

dissolved Al^{3+} high enough to be phytotoxic (toxic to plants). Some toxic elements are released by weathering. One of these is radioactive radium released in the weathering of some kinds of shale, discussed as a water pollutant in Chapter 4, Section 4.14. Another element is arsenic, commonly associated with iron-rich minerals (see Chapter 4, Section 4.4 and Figure 4.1). Lead leached from mineral galena, PbS, can be toxic in water consumed by humans and animals.

Weathering can release essential plant nutrients into the soil solution, which is taken up by plant roots carrying nutrients with it. One of the most important of these is K^+ ion, which goes into solution from the weathering of minerals such as potassium feldspar, $KAlSi_3O_8$. Nutrient phosphorus is released in the weathering of hydroxyapatite, $Ca_5OH(PO_4)_3$. Selenium is leached from minerals that contain this element, too much of which can be toxic and a deficiency of which can cause adverse health effects, especially in livestock, as well (see Chapter 10, Section 10.5).

9.3 GEOSPHERE AS A SOURCE OF NATURAL CAPITAL

The geosphere is an immense source of natural capital, providing a living environment for most humans, minerals required by modern civilizations, and room for disposal of wastes. One of the greatest concerns with regard to sustainability is the acquisition of essential elements from the geosphere in ways that sustain to the maximum extent possible supplies of these irreplaceable resources. As technology advances, priorities for specific geospheric resources change. In recent years, numerous uses have emerged for the **rare earth elements** consisting of the 15 lanthanides (elements with atomic numbers 58 through 71 in the periodic table shown in Chapter 19) plus scandium and yttrium, transition elements with atomic numbers 21 and 39, respectively. The chemical properties of the lanthanides are generally quite similar to each other, making their separation difficult. The chemical properties of scandium and yttrium are similar to those of the lanthanides, so they are commonly classified as rare earths.

The uses that have emerged for the rare earths are varied. The applications of various rare earths include making metal alloys, superconductors, phosphors that glow various colors in light-emitting diodes (LEDs), electrodes, electrolytes, electronic filters, lasers, specialty (colored) glasses, x-ray tubes, mercury vapor lamps, computer memory, oxidizing agents, and reducing agents. Rare earths are widely used in hybrid automobiles and in wind turbines. Each Toyota Prius hybrid automobile reportedly requires 1 kg of neodymium for its electric motor with terbium and dysprosium added in smaller quantities to preserve magnetic qualities and 10–15 kg of lanthanum for its battery.

Lithium has emerged as an important element because of the development of high-powered lithium storage batteries capable of storing and releasing large quantities of energy per unit mass of battery. These have largely been used in computers and other electronic devices, but will certainly find growing applications in electric and hybrid automobiles.

With the rather sudden development of new applications for rare earth elements and lithium, questions of supply have become important. China has had a near monopoly on rare earth elements and, with the advancement of high-tech industries in China, which may consume available supply, other countries have become alarmed regarding availability. China is also a source of lithium, although Bolivia is the main supplier. Fortunately, rare earths are not very rare, and the vast U.S. deposit in Mountain Pass, California, was the largest supplier until the facility was closed because of competition from China where labor costs are very low. Now the mine is being modernized in preparation for production to resume. Rare earth deposits also occur in Canada and even Vietnam.

In June 2010 U.S. military officials and geologists revealed that war-torn Afghanistan was a treasure trove of desired minerals including rare earths, with a total value of all mineral resources estimated at around 1 trillion dollars. The most abundant and valuable of these is iron (estimated at \$420 billion), copper (\$274 billion), niobium (\$81 billion), cobalt (\$51 billion), gold (\$25 billion), molybdenum (\$24 billion), and rare earths (\$7.4 billion). Other minerals of commercial value in Afghanistan likely include silver, potash, aluminum, graphite, fluorite, phosphorus, lead, zinc, mercury, strontium, sulfur, talc, magnesite, and kaolin clay. There are also believed to be lithium deposits in dry lakebeds of Afghanistan's eastern province of Ghazni. The lithium deposits may in fact be equal to those of Bolivia, which currently produces most of the lithium used in battery manufacture. Development of these mineral sources has the potential to help move the economy of the troubled country of Afghanistan from dependence on the opium trade (and U.S. military expenditures) to an economy based upon mineral resources.

9.4 ENVIRONMENTAL HAZARDS OF THE GEOSPHERE

Volcanoes and earthquakes are manifestations of the awesome, potentially destructive forces that reside in the geosphere to which modern civilization is very vulnerable, beyond the power of humans to prevent or even accurately predict. Although humans cannot predict or prevent these natural disasters, human activities can significantly influence the degree of damage that they cause. As examples, structures constructed on poorly consolidated fill dirt are much more susceptible to earthquake damage than are those attached firmly to bedrock, and the construction of dwellings in areas known to be subject to periodic volcanic eruptions simply means that unstoppable lava flows and other volcanic effects will be much more damaging when they occur. Other less spectacular but very destructive geospheric phenomena can be greatly aggravated by human activities. Destructive and sometimes life-threatening landslides, for example, often result from human alteration of surface soil and vegetation.

9.4.1 VOLCANOES

On March 5, 2011, the Kilauea volcano in Hawaii's Volcanoes National Park suddenly erupted, spewing thousands of tons of molten lava into the ocean. The spectacle immediately increased the tourist trade to the area, although people were kept away for some distance because of the plume of volcanic gas including more than 10,000 tons per day of sulfur dioxide released to the atmosphere. A much more serious eruption took place almost exactly a year earlier when, having lain dormant for almost two centuries, the Eyjafjallajokull volcano, one of Iceland's largest, began to ooze lava on March 20, 2010, visible as a red glow above the huge glacier covering the volcano. Initially, the eruption was nothing more than an interesting tourist attraction and the volcano appeared to revert to its normal state after a few days. However, on April 14, an enormous explosion sent volcanic ash as far as 11,000 m into the atmosphere, followed by days in which the volcano continued to spew ash high into the sky. This presented a significant problem for commercial aviation as the plume of volcanic ash spread eastward across the British Isles and northern and central Europe, because volcanic ash can damage jet engines and even cause them to stop running. (In 1982, all four engines of a British Airways 747 stopped when it was inadvertently flown into an ash cloud from Indonesia's Mount Galunggung and for several terrifying minutes, what suddenly became the world's largest glider, descended from 11,000 to 4100 m before the engines restarted, enabling an emergency landing in Jakarta.) The result of the Eyjafjallajokull eruption was that within 2 days, most of Europe's major airports were closed, canceling thousands of flights. Because of ripple effects across the world, this incident became the worst peacetime travel disruption in history, stranding millions of travelers, many with diminished financial resources from limited travel budgets. The result was a period of many days of travel chaos as flight bookings were rescheduled to eventually get travelers to their destinations. Airlines estimated financial losses of about $1.7 billion, resulting from the cancellation of more than 100,000 flights.

Illustrated in Figure 9.6, a volcano results from the presence of liquid rock magma near the surface.[5] In addition to liquid rock lava at temperatures ranging from 500°C to 1400°C that flows from volcanoes, these often very destructive phenomena are manifested by discharges of gases, steam, ash, and particles. Volcanic disasters have always plagued humankind. The 79 AD eruption of Mount Vesuvius in ancient Rome buried the city of Pompei in ash, preserving a snapshot of life in Rome at that time. The astoundingly massive eruption of Indonesia's Tambora volcano in 1815 was caused when water infiltrated the hot magma beneath the volcano, resulting in an explosion equivalent to 100 million tons of TNT explosive and blasting an estimated 30 km^3 of solid material into the atmosphere. The May 18, 1980, Mount St. Helens eruption in Washington State blew about 1 km^3 of material into the atmosphere, killed 62 people, and caused about $1 billion in damage.

In addition to their immediate effects upon surrounding areas, volcanoes can affect the atmosphere and climate. The Tambora volcano blasted enough particulate matter into the atmosphere to produce a very pronounced cooling effect. The following "year without a summer" caused global crop failures and starvation, and perceptible global cooling was observed for the next 10 years. Huge quantities of water vapor, dense carbon dioxide gas, carbon monoxide, hydrogen sulfide, sulfur dioxide, and hydrogen chloride may be emitted into the atmosphere in volcanic eruptions. Hydrogen chloride along with hydrogen sulfide and sulfur dioxide oxidized in the atmosphere to sulfuric acid can contribute to acidic rainfall. Volcanic emissions differ in their atmospheric chemical effects. The 1982 El Chichón eruption in Mexico generated little particulate mineral matter, but emitted vast amounts of sulfur oxides that were oxidized to sulfuric acid in the atmosphere. The tiny droplets of sulfuric acid suspended into the atmosphere effectively reflected enough sunlight to cause a perceptible cooling in climate.

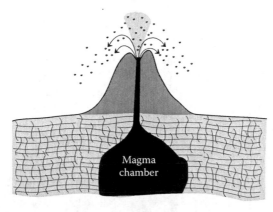

FIGURE 9.6 A volcano in a classic cone shape produced when molten lava and ash are ejected from a magma chamber underground.

9.4.2 TOXICOLOGICAL AND PUBLIC HEALTH ASPECTS OF VOLCANOES

Volcanic eruptions can cause some harmful health effects. Tourists were kept at some distance from the 2011 eruption of the Kilauea volcano (see Section 9.4.1) because of its emissions of sulfur dioxide gas. People may suffocate in the carbon dioxide or be poisoned by the toxic carbon monoxide and hydrogen sulfide. (If the geothermally active Yellowstone National Park in the United States was an industrial installation, it is likely that authorities would consider placing some areas off limits because of emissions of hydrogen sulfide, readily detected by visitors by its foul odor.) Fine particles blown into the atmosphere from volcanic eruptions can cause respiratory problems when inhaled. Plants may be killed or their growth stunted by exposure to sulfur dioxide and hydrogen sulfide from volcanoes.

Massive, atmospheric-damaging eruptions of volcanoes in recorded history have caused catastrophic crop failures. These will happen again. And since the world as a whole carries little food surplus from year to year, the certainty of food supply disruptions due to volcanic activity points to the desirability of storing substantial amounts of food for emergency use.

9.4.3 EARTHQUAKES

Earthquakes consist of violent horizontal and vertical movement of Earth's surface resulting from relative movements of tectonic plates. Huge masses of rock in the plates may be locked relative to each other for as long as centuries, then suddenly move along fault lines. This movement and the elastic rebound of rocks that occur as a result cause the earth to shake, often violently and with catastrophic damage.

History provides many examples of astoundingly damaging earthquakes. Over 1 million lives (out of a much lower global population than now) were lost by an earthquake in Egypt and Syria in 1201 AD. The May 12, 2008, 7.9 magnitude Wenchuan earthquake in Sichuan Province, China, left 80,000 people dead or missing. Financial costs of earthquakes in highly developed areas are enormous; the 1989 Loma Prieta earthquake in California cost about 7 billion dollars. Phenomena caused by earthquakes can add to their destructiveness. In addition to their direct shaking effects, earthquakes can cause ground to rupture, subside, or rise. **Liquefaction** of poorly consolidated ground, especially where groundwater levels are shallow, occurs when soil particles disturbed by an earthquake separate, become mixed with water, and behave like a liquid, causing structures to sink and collapse. Soil liquefaction was a major cause of destruction in the 6.3 magnitude Christchurch, New Zealand, earthquake in February 2011 that killed as many as 240 people and caused billions of dollars in property damage. Liquefaction damaged large numbers of the city's sewers, and even 2 weeks after the event, as many as 158,000 residents were left without functioning toilets; authorities were advising some residents to dig holes in their backyards for temporary toilet facilities! One of the more terrifying effects of earthquakes is a **tsunami**, a giant ocean wave that can reach heights of as much as 30 m. On December 26, 2004, a huge earthquake off the coast of Sumatra generated a tsunami up to 30 m high, killing more than 150,000 people in countries around the Indian Ocean. On March 11, 2011, a magnitude 8.9 earthquake, the largest ever recorded in the country, occurred off the coast of Sendai, Japan, northeast of Tokyo. It was followed by a massive tsunami that swept boats, automobiles, wreckage of buildings, and other debris far inland along some coastal regions of Japan. At least 13,800 people were known to be killed, 14,000 were missing and probably dead a month later, and the property damage came to about $300 billion.

Earthquakes have defied efforts to predict them, a fact that makes them all the more frightening. (An alarm of an impending earthquake did reach a number of residents in the vicinity of the March 11, 2011, quake in Japan, a warning of *30 seconds* before the quake hit.) Warnings of impending tsunamis are possible, and many areas have established alarm systems including sirens. Areas thousands of miles from the epicenter of a quake may receive word of an impending tsunami hours before it hits; such was the case in Hawaii after Japan's great quake. However, the epicenter of this earthquake was so close to the shore that residents of Japan's coastal areas, especially in the vicinity of Sendai, had very little warning before the massive tsunami generated by the quake hit the coast. Earthquake-prone areas, such as southern California, are well known, and loss of life and property can be minimized by taking appropriate measures. Buildings can be constructed to resist the effects of earthquakes using practices that have been known for some time. For example, some buildings in Niigata, Japan, were constructed to be earthquake-resistant in the 1950s. When a destructive earthquake hit that city in 1964, some buildings tipped over on the liquefied soil but remained structurally intact! (Current practice calls for the construction of more flexible structures designed to dissipate the energy imparted to them by an earthquake.) The construction of buildings, roadways, railroads, and other structures to withstand the destructive effects of earthquakes provides an excellent example of designing the anthrosphere in a manner that is as compatible as possible with the geosphere and the natural hazards it poses.

Although humans can do nothing to prevent earthquakes, there is some evidence that anthrospheric activities have helped cause them. Some seismologists have suggested that the pressure of water from newly constructed reservoirs in China provided lubrication between underground rock formations that enabled earth movement in the Wenchuan earthquake mentioned previously. One experiment near Basel, Switzerland, an area known to be close to a significant geological fault, that involved injecting water into hot rock formations to produce steam for power had to be stopped because it was believed to have caused a number of very small quakes. The head geologist of the company conducting the experiment was put on trial in 2009 for allegedly causing the tremors, but was acquitted. A 4.7 magnitude earthquake was recorded near Greenbrier, Arkansas, in February 2007, the largest in the state in 35 years. It was preceded by a number of smaller quakes that coincided in location and time with drilling and hydrofracking (fracturing by a suspension of highly pressurized sand in water containing additives) of underground shale formations to extract natural gas, a practice that some authorities have suspected of causing the tremors.

9.4.4 Toxicological and Public Health Aspects of Earthquakes

Direct toxicological effects of earthquakes are relatively unlikely. One possibility is the release of poisonous hydrogen sulfide gas from underground deposits or from coastal ocean sediments stirred up by sediments. Some concern has been expressed over the potential release of methane gas. Another possibility is the release into the atmosphere of suffocating levels of carbon dioxide gas produced by subterranean volcanic activity (see the discussion of such a devastating release from Africa's Lake Nyos in Section 3.7).

There can be significant indirect toxicological and health effects from earthquakes. Following the great 2011 Japanese earthquake, destruction and fires at petroleum refining facilities released toxic substances, possibly including toxic vapors of organic liquids such as benzene as well as carcinogenic polycyclic aromatic hydrocarbons from the combustion of hydrocarbons in petroleum refinery fires. Of much greater concern was the release of radioactive nuclear fission products from the damaged Fukushima Daiichi nuclear power reactor complex following partial meltdown of some of the fuel elements.

Significant public health problems are generated by earthquakes' destruction of housing, water supplies, and waste disposal systems. These were especially severe after the January 2010 magnitude 7.0 quake that devastated Port-au-Prince in Haiti and are suspected of contributing to an epidemic of cholera late in 2010 that killed a number of people.

9.4.5 Surface Effects

Though less spectacular than major earthquakes or volcanic eruptions, surface earth movement causes enormous damage and significant loss of life. Furthermore, surface earth movement is often strongly influenced by human activities. Surface phenomena result from the interaction of forces that act to thrust earth upward countered by weathering and erosion processes (see Section 9.2) that tend to bring earth material down. Both of these phenomena are influenced by the exposure of earth masses to water, oxygen, freeze–thaw cycles, alternate saturation with water and drying, and organisms and human influences.

Landslides occur when finely divided (unconsolidated) earthen material slides down a slope often with devastating results.[6] The 1970 earthquake-initiated landslide of dirt, mud, and rocks on the slopes of Mt. Huascaran in Peru may have killed 20,000 people. A 1963 landslide on slopes surrounding a reservoir held by the Vaiont Dam in Italy suddenly filled the reservoir, causing a huge wall of water to overflow the dam, killing 2600 people and destroying everything in its path.

Along with weather and climate, human activities can influence the likelihood and destructiveness of landslides. Roads and structures constructed on sloping land can weaken the integrity of earthen material or add mass to it, increasing its tendency to slide. In some cases, strong root structures of trees and brush anchor sloping land in place. However, some plant roots destabilize and add mass to soil, increase the accumulation of water underground, and cause earth to slide. Fortunately, predicting a tendency for landslides to occur is relatively straightforward based upon the nature and slope of geological strata, climate conditions, and observations of evidence of a tendency toward landslides, such as movement of earth and appearance of cracked foundations in buildings built on slopes. In some cases, remedial actions may be taken, but more important are the indications that structures should not be built on slide-susceptible slopes.

Less spectacular than landslides is **creep**, which is characterized by a slow, gradual movement of earth. Creep is especially common in areas where the upper layers of earth undergo freeze or thaw cycles. A special challenge is **permafrost**, which occurs in northern Scandinavia, Siberia, and Alaska. Permafrost refers to a condition in which ground at a certain depth never thaws, and thawing occurs only on a relatively thin surface layer. Structures built on permafrost may end up on a pool of water-saturated muck resting on a mixture of frozen ice and soil (Figure 9.7). One of the greater challenges posed by permafrost in recent times has been the construction of the Trans-Alaska pipeline in Alaska on a permafrost surface. Global warming is causing thawing of permafrost in Arctic regions such as parts of Siberia and is resulting in significant structural damage.

Expansive clays that alternately expand and contract when saturated with water, then become dried out, can cause enormous damage to structures, making the construction of basements virtually impossible in some areas. **Sinkholes** occur in areas where rock formations are dissolved by

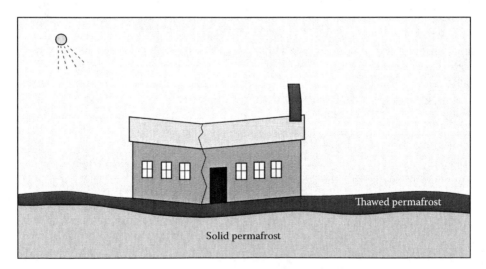

FIGURE 9.7 The thawing of permafrost due to global warming is causing structural damage in Arctic regions such as the northern parts of Siberia.

chemical action of water (particularly dissolved carbon dioxide acting on limestone). Earth can fall into a cavity generated by this phenomenon, causing huge holes in the ground that can swallow several houses at a time.

9.4.6 RADON, A TOXIC GAS FROM THE GEOSPHERE

A toxicological hazard from the geosphere is radon gas, a radioactive alpha particle emitter that may be released into the atmosphere and inhaled, potentially causing lung cancer. Geospheric radon originates with uranium, which decays to radium, which in turn decays to radon gas. This noble gas has a brief lifetime before it decays to other products, during which it may infiltrate dwellings through cracks in basement floors. The radioactive decay products of radon attach to particles in indoor air and are taken into the lungs, where they may damage lung tissue and cause cancer. Most of the cases of lung cancer attributed to radon have also involved synergistic effects with smoking.

9.5 WATER IN AND ON THE GEOSPHERE

The geosphere is the repository of virtually all the world's freshwater. As shown in Figure 9.8, this water may be in underground aquifers as groundwater; on the surface as streams, rivers, lakes, and impoundments; or as deposits of ice (glaciers) resting on Earth's surface. Water collected by the geosphere constitutes virtually all Earth's freshwater resources. This water is susceptible to pollution, which in extreme cases can render the water sources virtually useless. One of the greater water pollution problems is when water in underground aquifers becomes contaminated with hazardous waste material improperly discarded in the geosphere.

Water commonly moves on the geosphere in **streams** or **rivers** consisting of channels through which water flows. Rivers collect water from drainage basins or watersheds. To protect water quality in rivers, pollution and pollution-causing agricultural practices in drainage basins must be avoided.

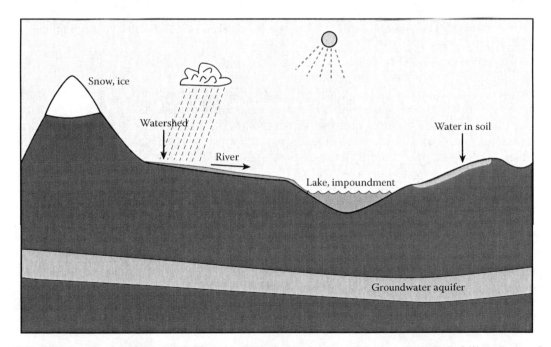

FIGURE 9.8 Water is generally abundant on and below the surface of the geosphere. Rain falling as part of the hydrologic cycle is spread on a watershed from which it may flow to produce rivers and seep into porous geological formations called aquifers as groundwater. Water is an important constituent of soils.

Rivers continually erode the geosphere over which they flow and leave deposits of sediments. Over time, a river will erode earth away and create valleys. An undisturbed river continually cuts curving patterns known as **meanders** in a river valley. The flat area of a valley formed by erosion and sediment deposition in the valley and susceptible to periodic flooding is the river's floodplain.

Floods are the phenomena associated with river flow, which are most likely to cause damage to anthrospheric structures. Despite their destructiveness, floods are normal phenomena by which a river does much of its work of shaping the surface of the geosphere. However, by unwisely building in floodplains, humans have made themselves susceptible to the damaging effects of floods. This was illustrated most tragically in 2005 by the deadly flood of the city of New Orleans following Hurricane Katrina, in which many of the areas flooded had been constructed below sea level! Human activities on the geosphere surface can make the effects of floods much worse. For example, flash floods following intense rainfall in urban areas are made much worse by the removal of vegetation from watersheds and its replacement with paving. Concrete and asphalt surfaces do not slow down the flow of water like well-rooted plants do and such surfaces prevent the infiltration of water into the ground.

Attempts to control water flow and flooding provide interesting examples of how humans interact with their natural environment. Control measures have concentrated on the downstream end on the rivers themselves by constructing levees to confine rivers to their banks, by straightening and deepening river channels to increase the velocity and the flow of the water in an effort to move it quickly downstream away from the potentially flooded area, and by building dams to contain floodwater until it can be safely released. Such measures can be deceptively successful, sometimes for many decades, until a massive flood overwhelms them. When a contained river carrying vast amounts of water flowing at a high velocity eventually breaks through the levees and dams designed to contain it, the resulting damage can be catastrophic. During the record-breaking May 2011 Mississippi River flood, the U.S. Army Corps of Engineers used explosives to create a 2-mile breach in the river levee to enable water to flow into the Birds Point-New Madrid Floodway in Missouri in order to save Cairo, Illinois, and Hickman, Kentucky, from flooding. These towns were saved, but the 200-square-mile lake that was created inundated 130,000 acres of rich farmland and around 100 homes.

An approach to flood control based upon the best practice of sustainability provides a means of minimizing flood damage. Such an approach tends to concentrate more on the upstream end, the watersheds from which water produced by rainfall flows into the river. With the proper kind of vegetation cover, such as forests, and with terraces and small dams designed to temporarily slow the flow of water into the river from the watershed, extremes of high water (flood crests) can be greatly reduced. With regard to protection of dwellings and agricultural land in the river's flood plain, a fundamental question has to be asked whether houses should even be located in these areas and whether the land should be cultivated. In many cases, the answer is no, and the least costly alternative overall is to pay for removal of the structures and conversion of the land back to an uncultivated state, simply allowing the flooding that comes naturally to the river.

9.5.1 GEOSPHERIC WATER AND HEALTH EFFECTS

Water on and beneath the surface of the geosphere plays a strong role in pollution and the distribution of toxic substances. Toxic substances from wastes improperly disposed to the geosphere can leach into groundwater and contaminate water supplies. Radioactive radium resulting from the decay of uranium in aquifer formations has caused some groundwater sources of drinking water to be abandoned.

One of the most tragic examples of water as a vector for toxic substances in the geosphere happened in Bangladesh, where tube wells drilled to supply pathogen-free water (Figure 9.9) became contaminated with arsenic leached from the aquifers into which they were drilled and caused numerous cases of arsenic poisoning. Characterized by the World Health Organization as

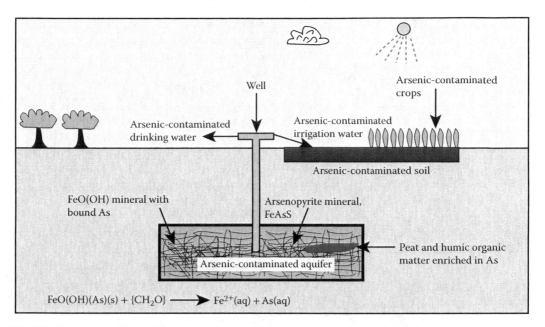

FIGURE 9.9 Tube wells sunk into arsenic-containing aquifer formations may yield water contaminated by toxic soluble arsenic. Use of the water for irrigation can contaminate soil with arsenic and provide a pathway for arsenic to get into food crops such as rice. The problem has been especially acute in Bangladesh, but has occurred in other parts of the world as well.

"the largest mass poisoning of a population in history," the arsenic-contaminated water may have exposed as many as 77 million people, half the population of the country, to toxic levels of arsenic. In a study of approximately 12,000 people in Bangladesh over a 10-year period,[7] it was concluded that more than 20% of the deaths of this group over the course of the study were due to arsenic poisoning. The people in the top 25% of arsenic exposures had almost a 70% higher likelihood of dying over a 6-year period compared to those exposed to the less than 10 µg/L arsenic, generally accepted as an exposure limit. The maladies commonly attributed to arsenic intake include bladder, kidney, liver, and skin cancers as well as elevated levels of heart disease. It is believed that excessive exposure to arsenic, primarily from drinking water, occurs in approximately 70 countries including India, Mexico, and the United States.

There are several possible mechanisms for arsenic to get into water from the geosphere. Probably the most common of these is the release of arsenic that has accumulated on iron(III) oxide (goethite, FeO(OH)). As shown in Figure 9.9, when the insoluble iron(III) is reduced to soluble Fe^{2+} by the action of anoxic bacteria on organic matter, $\{CH_2O\}$, arsenic is released as soluble As(III) or As(V) species. The oxidation of arsenopyrite, FeAsS, can release arsenic as can the decomposition of arsenic-contaminated organic matter (peat) in underground deposits.

9.6 ANTHROSPHERIC INFLUENCES ON THE GEOSPHERE

The urge to "dig in the dirt" and alter Earth's surfaces seems to be innate in humans. During recent decades, the potential of humans to alter the geosphere has been greatly increased by the development of massive earth-moving equipment. Flooding of rivers caused by human activities was discussed in Section 9.5. Other geospheric disturbances that can be detrimental include landslides on mounds of waste mine tailings, adverse effects resulting from exposure of minerals during mining (production of acid mine water from exposure of pyrite, FeS_2, in coal mining), and filling and destruction of wetlands upon which many forms of wildlife depend for breeding grounds.

Human effects upon the geosphere can be both direct and indirect. Constructing dams and reservoirs, flattening whole mountaintops to get to underground coal seams, and plowing natural prairies to grow crops are obvious direct effects. Indirect effects include pumping so much water from underground aquifers that the ground subsides or exposing minerals to the atmosphere by strip mining so that the minerals undergo weathering to produce polluted acidic water. Mineral extraction can result in earthen material being disturbed and rearranged in ways that can cause almost irreversible damage to the environment. A major objective of the practice of green chemistry and industrial ecology is to minimize these detrimental effects and, to the extent possible, eliminate them entirely.

Many of the effects of human activities on the geosphere have to do with the extraction of resources of various kinds from Earth's crust. These may range from gravel simply scooped from pits on Earth's surface to precious metals at such low concentrations that tons of ore must be processed to get a gram or less of the metal. The most straightforward means of obtaining materials from Earth's crust is surface mining. This often involves removing unusable material in the form of the overburden of soil and rock that covers the desired resource. Such surface mining may leave a pit that fills with water alongside a pile of the overburden. This kind of mining practice caused many environmental problems in the past. However, with the modern practice of surface mining, topsoil is first removed and stored, rock removed to get to the resource is either placed back in the pit or on contoured piles, and the topsoil is placed over it for revegetation. In favorable cases, the result can be attractive lakes that support fish life and vegetated, gently sloping artificial hills.

Underground mining usually does not leave the visible scars that may be inflicted by surface mining. However, it can have profound environmental effects. Collapse of underground mines can cause surface subsidence. Water flowing through and from underground mines can pick up water pollutants. Most ores require a degree of beneficiation in which the usable portion of the ore is concentrated, leaving piles of tailings. These may collapse, and materials leached from them can pollute water. Examples of the latter include acidic water produced by the action of bacteria on iron pyrite, FeS_2, removed from coal and radium leached from the tailings remaining from uranium mining operations.

9.7   GEOSPHERE AS A WASTE REPOSITORY

As discussed previously, mineral processing produces large quantities of waste solids. Other sources of waste solids include ash from coal combustion, municipal garbage, and solid wastes from various industrial processes. Ultimately, these wastes are placed on or in the geosphere. Such measures have an obvious potential for pollution.

One of the most common waste materials that ends up as part of the geosphere is **municipal refuse**, the "garbage" generated by human activities. This material is largely disposed in **sanitary landfills** made by placing the solid wastes on top of the ground or in depressions in the ground and covering it with soil to minimize effects such as windblown waste paper and plastic, emission of odorous materials to the atmosphere, and water pollution (Figure 9.10). Although "garbage dumps" used to be notably unsightly and polluting, modern practice of sanitary landfilling can result in areas that can be used as parkland, golf courses, or relatively attractive open space. The unconsolidated nature of decaying garbage and the soil used to cover it make municipal landfills generally unsuitable for building construction. Biological decay of degradable organic material ($\{CH_2O\}$) in the absence of oxygen generates methane gas by a process represented as

$$2\{CH_2O\} \rightarrow CO_2 + CH_4 \tag{9.9}$$

Methane is a powerful greenhouse gas in the atmosphere, much more effective per molecule at absorbing infrared radiation than is CO_2, so it is undesirable to release CH_4 to the atmosphere. However, modern sanitary landfills may be equipped with pipes and collection systems so that the methane can be collected and used as a fuel.

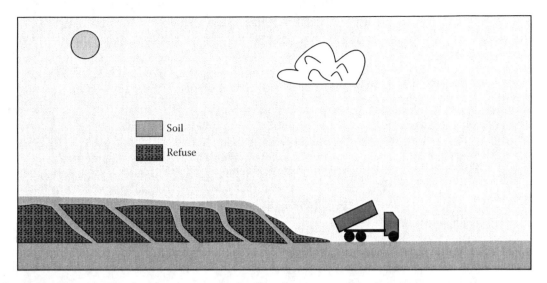

FIGURE 9.10 A sanitary landfill. The refuse is placed in compartments that may be separated with poorly permeable fill dirt, typically after each day's disposal, then covered with dirt. In larger, more advanced landfills, the methane produced by anoxic fermentation of biodegradable biomass is collected and used as fuel.

Although the release of gases, particularly methane, to the atmosphere is a potential air pollution problem with sanitary landfills, contaminated **leachate** consisting of water seeping through the landfilled wastes can pollute water, especially groundwater. This water may contain heavy metals, organic acids, odor-causing organics, and other undesirable pollutants. There are two general approaches to minimizing problems from contaminated landfill leachate. One of these is to construct the landfill in a manner that minimizes water infiltration, thus reducing the amount of leachate produced to lowest possible levels. To prevent the leachate from getting into groundwater, it is desirable to locate the landfill on a layer of poorly permeable clay. In some cases, the bottom of the landfill may be lined with impermeable synthetic polymer liners that prevent leachate from getting into groundwater. In cases where significant quantities of badly polluted leachate are generated, it is best to collect and treat the leachate, usually by biological treatment processes.

It is highly desirable to minimize quantities of materials requiring sanitary landfill disposal using the best practice of industrial ecology and green chemistry, preferably by reducing quantities of materials at the source, using less material that ultimately will require disposal. Wherever possible, materials, such as packing materials, that ultimately get into landfills should be biodegradable. Recycling programs in which glass, plastic, paper, and food cans are removed from refuse before disposal are effective in reducing quantities of material requiring disposal in landfills. Burning of garbage with proper pollution control measures can reduce it to a low-volume ash that can be placed in a landfill. Although not practiced to a significant extent, anoxic digestion of macerated wet refuse in an oxygen-free digester has the potential to produce methane for fuel use and significantly reduce the mass of the degradable wastes.

Sanitary landfills are not suitable for the disposal of hazardous substances. These materials must be placed in special **secure landfills**, which are designed to contain the wastes and leachate, thus preventing pollution of water, air, and the geosphere. One way in which this is accomplished is with impermeable synthetic membranes that do not allow water to seep into the fill and that prevent leachate from draining into groundwater. These landfills are often equipped with water treatment systems to treat leachate before it is released from the system. Unfortunately, many hazardous chemicals **never** degrade and a "secure" chemical landfill leaves problems for future generations to handle. One of the major objectives of green chemistry is to prevent the generation of any hazardous materials that would require disposal on land. The best way to do that is to avoid making or using such materials. In cases where that is not possible and hazardous materials are generated, they should be treated in a way that renders them nonhazardous before disposal.

QUESTIONS AND PROBLEMS

Access to and use of the Internet is assumed in answering all questions including general information, statistics, constants, and mathematical formulas required to solve problems. These questions are designed to promote inquiry and thought rather than just finding material in the text. So in some cases, there may be several "right" answers. Therefore, if your answer reflects intellectual effort and a search for information from available sources, it can be considered to be "right."

1. Do some search on the Internet regarding potential shortages of rare earths. What are the current major sources of these elements? Are some alternate sources being developed? What would be the consequences of drastic shortages of these elements.

2. What is channelization as it applies to the geosphere and hydrosphere (the term has another meaning as applied to communication technology)? Can there be adverse effects of channelization on the geosphere?

3. Suggest the main contributions made by the geosphere to the biosphere.

4. Do an Internet search of melting permafrost as related to global warming. What are some of the major detrimental effects? Which areas are the most impacted?

5. What are watersheds located on the geosphere? What are some of the threats to watersheds? How may watersheds be sustained?

6. Look up the current status of biogas production from landfills. Where has collection of landfill biogas been implemented? What is likely to be its future development?

7. Distinguish between rocks and minerals.

8. How does igneous rock turn into secondary minerals?

9. What is the branch of chemistry that deals with rocks and minerals and their chemical characteristics and interactions?

10. Give an example of an air pollutant and of a water pollutant that may be generated by sanitary landfills.

11. Why do silicates and oxides predominate among earth's minerals?

12. Explain how the following are related: weathering, igneous rock, sedimentary rock, soil.

13. What is the distinction between weathering and erosion? Suggest ways in which air pollution may contribute to both phenomena.

14. In what sense may volcanoes contribute to air pollution? What possible effects may this have on climate?

15. Large areas of central Kansas have vast deposits of halite. What is halite? What does this observation say about the geological history of the area?

16. One way in which coal and other fossil fuels may be used without contributing to higher levels of greenhouse gas carbon dioxide in the atmosphere is through carbon sequestration by pumping carbon dioxide into mineral strata. Explain with a chemical reaction how formations of limestone (calcium carbonate) might be used for this purpose. Suggest how this might cause problems on the surface.

17. Rust-colored deposits of iron oxides have been observed on the soil surface where leachate from sanitary landfills leaks onto the surface. Given that iron oxides are not at all water-soluble and that the iron in the leachate is carried in a soluble form, suggest a pathway by which these deposits may be formed. Some review of water chemistry may be required to answer this question.

18. Suggest ways in which improved materials, some made by green chemical processes, can reduce the effects of earthquakes.

19. In what respect do volcanoes have the potential to drastically affect global climate? Is there any evidence for such an effect?
20. How may human activities lead to landslides?
21. What are the formations called that contain water underground? What is a major threat to groundwater in such formations?
22. What is FeS_2? Why is the exposure of this material from mining a potential problem?

LITERATURE CITED

1. Desonie, Dana, *Geosphere: The Land and its Uses (Our Fragile Planet)*, Chelsea House Publications, New York, 2008.
2. Frisch, Wolfgang, Martin Meschede, and Ronald C. Blakey, *Plate Tectonics: Continental Drift and Mountain Building*, Springer, Berlin, 2010.
3. Albarede, Francis, *Geochemistry: An Introduction*, Cambridge University Press, Cambridge, UK, 2009.
4. Krachler, Regina, Relevance of Peat-Draining Rivers for the Riverine Input of Dissolved Iron into the Ocean, *Science of the Total Environment* **408**, 2402–2408 (2010).
5. Siebert, Lee, Tom Simkin, and Paul Kimberly, *Volcanoes of the World*, 3rd ed., University of California Press, Berkeley, CA, 2011.
6. Sassa, Kyoji, Badaoui Rouhban, Salvano Briceno, and Bin He, *Landslides: Global Risk Preparedness*, Springer, New York, 2012.
7. Ahsan, Habibul, et al, Arsenic Exposure from Drinking Water, and All-Cause and Chronic-Disease Mortalities in Bangladesh (HEALS); A Prospective Cohort Study, *Lancet* **376**, 252–258 (2010).

SUPPLEMENTARY REFERENCES

Anderson, Michael, Ed., *Investigating Plate Tectonics, Earthquakes, and Volcanoes*, Britannica Educational Publications, New York, 2012.

Coch, Nicholas K., *Geohazards: Natural and Human*, Prentice Hall, Upper Saddle River, NJ, 1995.

Coyne, Mark S., and James A. Thompson, *Fundamental Soil Science*, Thomson Delmar Learning, Clifton Park, NY, 2006.

Eby, G. Nelson, *Principles of Environmental Geochemistry*, Thomson-Brooks/Cole, Pacific Grove, CA, 2004.

Essington, Michael E., *Soil and Water Chemistry: An Integrative Approach*, Taylor & Francis/CRC Press, Boca Raton, FL, 2004.

Gates, Alexander E., and David Ritchie, *Encyclopedia of Earthquakes and Volcanoes*, 3rd ed., Facts on File, New York, 2007.

Hyndman, Donald, and David Hyndman, *Natural Hazards and Disasters*, 2nd ed., Thomson Brooks/Cole, Belmont, CA, 2009.

Kearey, Philip, Keith A. Klepeis, and Frederick J. Vine, *Global Tectonics*, 3rd ed., Wiley Blackwell, Hoboken, NJ, 2009.

Keller, Edward A., *Environmental Geology*, 9th ed., Pearson/Prentice Hall, Upper Saddle River, NJ, 2010.

Keller, Edward A., *Introduction to Environmental Geology*, 5th ed., Pearson/Prentice Hall, Upper Saddle River, NJ, 2011.

Lutgens, Frederick K., and Edward J. Tarbuck, *Foundations of Earth Science*, Pearson Prentice Hall, Upper Saddle River, NJ, 2011.

Marti, Joan, and Gerald Ernst, Eds., *Volcanoes and the Environment*, Cambridge University Press, Cambridge, UK, 2005.

McCollum, Sean, *Volcanic Eruptions, Earthquakes, and Tsunamis*, Chelsea House, New York, 2007.

Montgomery, Carla W., *Environmental Geology*, 9th ed., McGraw-Hill, Boston, 2010.

Pipkin, Bernard W., *Geology and the Environment*, 5th ed., Thomson Brooks/Cole, Belmont, CA, 2008.

Sammonds, P. R., and J. M. T. Thompson, Eds., *Advances in Earth Science: From Earthquakes to Global Warming*, Imperial College Press, London, 2007.

Savino, John, and Marie D. Jones, *Supervolcano: The Catastrophic Event that Changed the Course of Human History (Could Yellowstone be Next?)*, New Page Books, Franklin Lakes, NJ, 2007.

Stone, George W., *Raging Forces: Life on a Violent Planet*, National Geographic, Washington, DC, 2007.

Waltham, Tony, *Foundations of Engineering Geology*, Taylor & Francis, London, 2009.

10 Soil
A Critical Part of the Geosphere

10.1 HAVE YOU THANKED A CLOD TODAY?

A common bumper sticker is one that asks the question, "Have you thanked a green plant today?" This is an obvious reference to plants whose photosynthesis produces the food that we and most other animals depend on for our existence. An even more fundamental question is whether we have thanked the soil—the clods of dirt—on which green plants depend for their existence. Good, productive soil combined with a suitable climate and adequate water is the most valuable asset that a nation can have. Vast areas of the world lack this fundamental asset, and the people living in areas with poor soil often suffer poverty and malnutrition as a result. Furthermore, areas that once had adequate soil have seen it abused and degraded to the extent that it is no longer productive. One of the central challenges faced by the practice of green chemistry and industrial ecology is to retain and enhance the productive qualities of soil. The remainder of this chapter addresses soil and the aspects of agriculture related specifically to soil.

10.1.1 WHAT IS SOIL?

Soil is a term that actually describes a wide range of finely divided mineral matter containing various levels of organic matter and water that can sustain and nourish the root systems of plants growing on it. Soil is largely a product of the weathering of rock by physical, chemical, and biochemical processes that produce a medium amenable to the support of plant growth. A healthy soil contains water that is available to plants, has a somewhat loose structure with air spaces, and supports an active population of soil-dwelling organisms, including fungi and bacteria that degrade dead plant biomass and animals, such as earthworms. Although the solids in a typical soil are composed of about 95% inorganic matter, some soils contain up to 95% organic matter and some sandy soils may have only about 1% organic matter.

Figure 10.1 shows the major aspects of the physical structure of soil. Soil is divided into layers called **horizons** formed by the weathering of parent rock, chemical processes, biological processes, and the action of water including leaching of colloidal matter to lower horizons. The most important of these for plant growth is **topsoil**. Plant roots permeate the topsoil and take water and plant nutrients from it. Topsoil is the layer of maximum biological activity. The **rhizosphere** is the part of topsoil in which plant roots are especially active and in which the elevated levels of biomass are composed of plant roots and microorganisms associated with them. There are strong synergistic relationships between plant root systems and microorganisms in the rhizosphere. The surfaces of root hairs are commonly colonized by microorganisms, which thrive on carbohydrates, amino acids, and root-growth-lubricant mucigel secreted from the roots. The microorganisms in turn aid in the uptake of nutrients by plant root hairs, and in the case of legume plants bacteria growing in nodules on the roots convert N_2 gas to chemically bound nitrogen that the plants can utilize as a nutrient.

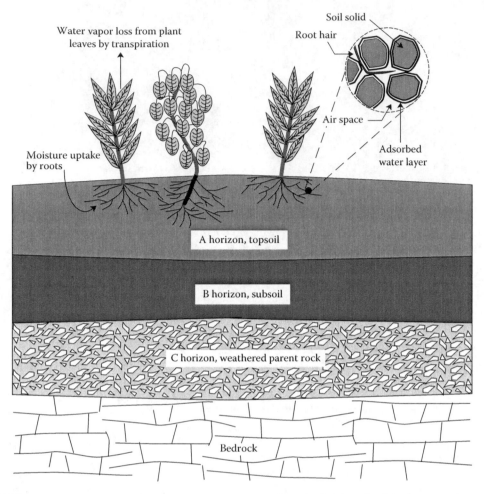

FIGURE 10.1 Major aspects of soil structure showing a typical distribution of soil horizons: resting on the parent rock is the C horizon consisting of weathered parent rock, above which is the B horizon or subsoil. The A horizon or topsoil is the top layer and is the most important part of the soil where plants are rooted. The inset shows aspects of soil microstructure including solid soil particles, water bound to soil particles, and air spaces.

10.1.2 INORGANIC SOLIDS IN SOIL

Reflecting the fact that the two most common elements in Earth's crust are oxygen and silicon (see Chapter 9, Section 9.2), silicates are the most common mineral constituents of soil. These include finely divided quartz (SiO_2), orthoclase ($KAlSi_3O_8$), and albite ($NaAlSi_3O_8$). Other elements that are relatively abundant in Earth's crust are aluminum, iron, calcium, sodium, potassium, and magnesium; their abundance is reflected by various minerals such as epidote ($4CaO \cdot 3(AlFe)_2O_3 \cdot 6SiO_2 \cdot H_2O$), geothite ($FeO(OH)$), magnetite ($Fe_3O_4$), calcium and magnesium carbonates ($CaCO_3$, $CaCO_3 \cdot MgCO_3$), and oxides of manganese and titanium in soil. Soil parent rocks undergo weathering processes to produce finely divided colloidal particles, by far the most abundant of which are clays. These secondary minerals hold moisture and mineral nutrients, such as K^+ required for plant growth, that are accessible by plant roots and are repositories of plant nutrients. Inorganic soil colloids can absorb toxic substances in soil, thus reducing the toxicity of substances that would harm plants. It is obvious that the abundance and nature of inorganic colloidal material in soil are important factors in determining soil productivity.

10.1.3 Soil Organic Matter

The small percentage of soil mass consisting of organic matter has a strong influence on the physical, chemical, and biological characteristics of soil. Among its important effects in soil, organic matter is effective in holding soil moisture, and it holds and exchanges with plant roots some of the ions that are required as plant nutrients. Temperature, moisture, and climatic conditions significantly affect the kinds and levels of soil organic matter. Cold, wet conditions in which soil stays saturated with moisture preventing the access of microorganisms to oxygen tend to prevent complete biodegradation of plant residues that compose soil organic matter, allowing it to accumulate. This is clearly illustrated by the accumulation of peat in Ireland and other locales with similar climatic conditions such that most of the solid soil is composed of organic matter. Tropical conditions, especially with alternate wet and dry seasons, can result in loss of soil organic matter. One reason that the soil supporting tropical rain forests degrades so quickly when the trees are removed is that the organic matter in the soil undergoes rapid biodegradation when the forest cover is removed.

The plant biomass residues that form soil organic matter undergo a biodegradation process by the action of soil bacteria and fungi in which the cellulose in the biomass is readily degraded, leaving modified residues of the lignin material that binds the cellulose to the plant matter. This is the process of **humification** and the residue is **soil humus**, a black organic material of highly varied chemical structure.[1] A fraction of soil humus is soluble in water (see the discussion on humic substances in water in Chapter 3, Section 3.11), especially when a base is present in the water. Another fraction called **humin** does not dissolve and stays in the solid soil.

Although usually composing not more than a small percentage of soil mass, soil humus has a very strong influence on the characteristics of soil. It has a strong affinity for water and holds much of the water in a typical soil. Primarily because of their carboxylic acid, $-CO_2H$, groups, soil humic molecules exchange H^+ ion and act to buffer the pH of water in soil (the soil solution). Humic substances bind metal ions and other ionic plant nutrients. Soil humus also binds and immobilizes organic materials, such as herbicides applied to soil.

10.1.4 Water in Soil and the Soil Solution

Water in soil is required for plants. This water is taken up by plant root hairs, transferred through the plant, and evaporated from the leaves, a process called **transpiration**. The quantities of water involved are enormous; for example, the water transpired to produce a kilogram of dry hay can amount to several hundred kilograms. Most of the water in normal soils is not present as visible liquid but is absorbed to various degrees on the soil solids. In fact, a condition in which all the spaces in soil are filled with water—waterlogging—slows the growth of most plants. The water that is available in soil is called the **soil solution** and contains a number of dissolved materials, including plant nutrients. It plays an essential role in transferring substances, such as dissolved metal cations, between roots and the soil solid. Cations commonly present in the soil solution include H^+, Ca^{2+}, Mg^{2+}, Na^+, K^+, and NH_4^+, along with very low levels of Fe^{2+}, Mn^{2+}, and Al^{3+}. Common anions present are HCO_3^-, CO_3^{2-}, HSO_4^-, SO_4^{2-}, Cl^-, and F^-.

10.1.5 Chemical Exchange Processes in Soil

Soil holds and releases a number of substances including ions and organics. The most commonly exchanged species are cations, which are held by negatively charged sites on the soil. These cation exchange sites are usually associated with clays, which have negatively charged sites due to the substitution of Al(III) for Si(IV) in the clay mineral, and with organic matter, usually humic materials with negatively charged carboxylate, $-CO_2^-$, functional groups. Negatively charged anions can also be exchanged by soil, but they are less strongly held than cations because of the lack of positive binding sites on the soil solid.

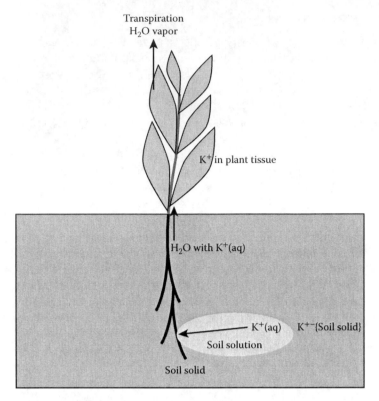

FIGURE 10.2   Soil cation exchange and uptake by plants: in the example shown, nutrient K^+ is desorbed from soil into the soil solution, in which it is absorbed by a plant root and transported upward into the plant tissue by the osmotic flow of water. The water eventually enters the atmosphere as water vapor through the process of transpiration, leaving the K^+ in the plant, where it participates in essential metabolic processes.

Figure 10.2 shows the pathway of a cation, in this case nutrient K^+, being desorbed from the soil solid and, accompanied by an anion, most likely HCO_3^-, entering the soil solution. The soil solution containing the cation is absorbed by a plant root and transported upward into the plant tissue by the osmotic flow of water. The water eventually enters the atmosphere as water vapor through the process of transpiration, leaving the K^+ in the plant where it participates in essential metabolic processes. Other species, including toxic heavy metal ions such as Cd^{2+}, enter the plant tissue by the same mechanism.

Soil acidity in the form of H^+ ion builds up as plant roots exchange H^+ for other cationic nutrients in soil. When acidity reaches excessive levels, the soil is no longer productive. Acidity can be neutralized by the addition of lime, $CaCO_3$, which neutralizes acidity according to the following reaction:

$$\text{Soil}\}(H^+)_2 + CaCO_3 \rightarrow \text{Soil}\}Ca^{2+} + CO_2 + H_2O \qquad (10.1)$$

This process also adds calcium to soil.

The essential plant nutrient nitrogen is very much involved with nature's nitrogen cycle, which is significantly modified by human activities. Major aspects of this cycle are as follows:

- At 79% N_2, Earth's atmosphere constitutes an inexhaustible nitrogen resource, although, because of the extreme stability of the N_2 molecule, it is difficult to extract nitrogen from air in a chemically combined form.

- *Rhizobium* bacteria growing on the roots of leguminous plants, such as clover and soybeans, convert atmospheric nitrogen to nitrogen that is chemically bound in biomolecules. This nitrogen is converted to ammonium ion, NH_4^+, when plant residues and animal feces, urine, and carcasses undergo microbial decay.
- Lightning and combustion processes convert atmospheric nitrogen to nitrogen oxides, and ammonia-manufacturing plants produce NH_3 from atmospheric elemental nitrogen and elemental hydrogen produced from natural gas (methane).
- Soil microbial processes oxidize ammoniacal nitrogen, NH_4^+, to nitrate ion, NO_3^-, which is the form of nitrogen that is most readily used by plants. Microbial processes also produce gaseous N_2 and N_2O, which are released into the atmosphere, a process called denitrification that completes the nitrogen cycle.

10.2 PLANT NUTRIENTS AND FERTILIZERS IN SOIL

Plant biomass is composed largely of carbon, hydrogen, and oxygen, which plants extract from water, and atmospheric carbon dioxide. Other nutrients that plants require in relatively large quantities are calcium, magnesium, and sulfur, which are usually present in sufficient abundance in soil, and nitrogen, phosphorus, and potassium, which are commonly added to soil as fertilizers.

Natural processes usually do not produce sufficient nitrogen to allow maximum plant growth so that artificial means are used to extract nitrogen in a chemically combined form from the atmosphere. This is done by the Haber process, which combines elemental N_2 and H_2 over a catalyst at very high pressures of about 1000 times atmospheric pressure and an elevated temperature of 500°C. The reaction is

$$N_2 + 3H_2 \rightarrow 2NH_3 \qquad\qquad (10.2)$$

producing ammonia that is 82% chemically bound N. This anhydrous ammonia can be applied directly below the soil surface where its tremendous attraction to soil moisture binds it to the soil. It can also be applied as a 30% solution of NH_3 in water and is sometimes added directly to irrigation water. Ammonia, which is held in soil as ammonium ion, NH_4^+, is not well assimilated directly by most plants. But it is slowly oxidized by the action of soil bacteria using atmospheric O_2 as oxidant to nitrate ion, NO_3^-, which is used directly by plants.

Anhydrous ammonia is a corrosive toxic substance. It can cause frostbite when it evaporates from exposed skin. Ammonia is a potent skin corrosive and harms exposed eye tissue. Because of its high water solubility, inhaled ammonia is absorbed by the moist tissues of the upper respiratory tract. Inhalation of ammonia causes constriction of the bronchioles and can cause lung edema (fluid accumulation) and changes in lung permeability. Anhydrous liquid ammonia stored in tanks has been a favorite target of operators of illicit "meth labs" in rural areas who use it in the synthesis of methamphetamines. Certainly, many of the thieves are harmed by exposure to ammonia, but they rarely seek treatment for the injury.

A solid form of nitrogen fertilizer can be made by reacting ammonia with oxygen over a platinum catalyst to make nitric acid, HNO_3, and reacting the acid with basic ammonia to make ammonium nitrate, NH_4NO_3. This molten material is solidified into small pellets that can be applied to soil as fertilizer. Ammonium nitrate mixed with fuel oil is used for blasting to quarry rock, and it was the explosive used in the bombing of the Oklahoma City federal building in 1995. A safer alternative to ammonium nitrate as a solid nitrogen fertilizer is urea which is made by a process that, overall, involves the reaction of carbon dioxide and ammonia:

$$CO_2 + 2NH_3 \rightarrow CO(NH_2)_2 + H_2O \tag{10.3}$$

Phosphorus is an essential plant nutrient required for cellular DNA and other biomolecules. It is utilized by plants as $H_2PO_4^-$ and HPO_4^{2-} ions. Phosphate minerals that can be used to manufacture phosphorus-containing fertilizers occur in a number of places throughout the world. In the United States, Florida has especially abundant phosphate resources, largely as fluorapatite, $Ca_5(PO_4)_3F$, as well as hydroxyapatite, $Ca_5(PO_4)_3OH$. These phosphate minerals are too insoluble to serve directly as fertilizers and are treated with phosphoric acid and sulfuric acid to make superphosphates that are much more soluble and available to plants:

$$2Ca_5(PO_4)_3F(s) + 14H_3PO_4 + 10H_2O \rightarrow 2HF(g) + 10Ca(H_2PO_4)_2 \cdot H_2O \tag{10.4}$$

$$2Ca_5(PO_4)_3F(s) + 7H_2SO_4 + 3H_2O \rightarrow 2HF(g) + 3Ca(H_2PO_4)_2 \cdot H_2O + 7CaSO_4 \tag{10.5}$$

It may be noted here that along with the production of phosphate fertilizers, elemental white phosphorus is often produced as an industrial chemical intermediate and food additive. For the latter application, white phosphorus, P_4, is oxidized to P(V) oxide and reacted with water to produce orthophosphoric acid, H_3PO_4. By going through the elemental phosphorus intermediate in making food-grade orthophosphates, toxic arsenic, which occurs with phosphate minerals, can be removed. Elemental white phosphorus is highly toxic by exposure through inhalation, the oral route, and dermal routes. Its systemic effects include anemia, gastrointestinal system dysfunction, and bone brittleness. A characteristic symptom is phossy jaw in which the jawbone deteriorates and becomes brittle, the bone may break, and the teeth fall from the jaw. There have been fatalities from exposure to white phosphorus, including one case from the 1920s in which a child ate a firecracker containing white phosphorus.

Potassium as the potassium ion, K^+, is required by plants to regulate water balance, activate some enzymes, and enable some transformations of carbohydrates. Potassium is one of the most abundant elements in Earth's crust, of which it makes up 2.6%; however, much of this potassium is not easily available to plants. For example, some silicate minerals such as leucite, $K_2O \cdot Al_2O_3 \cdot 4SiO_2$, contain strongly bound potassium. Exchangeable potassium held by clay minerals is relatively more available to plants. Potassium for fertilizer is simply mined from the ground as salts, particularly KCl, or pumped from beneath the ground as potassium-rich brines. Large potassium deposits occur in the Canadian province of Saskatchewan.

Plants require several **micronutrients**, largely elements that occur only at trace levels, for their growth. These include boron, chlorine, copper, iron, manganese, molybdenum (for N fixation), and zinc. Some of these are toxic at levels above those required for optimum plant growth. Most of the micronutrients are required for adequate function of essential enzymes. Photosynthetic processes use manganese, iron, chlorine, and zinc. Since the micronutrients are required at such low levels, soil normally provides sufficient amounts.

10.3 SOIL AND PLANTS RELATED TO WASTES AND POLLUTANTS

Soil is a repository of large quantities of wastes and pollutants, and plants act as filters to remove significant quantities of pollutants from the atmosphere. Sulfates and nitrates from the atmosphere, including acid-rain-causing H_2SO_4 and HNO_3, deposit largely on land and the plants growing on it. Gaseous atmospheric SO_2, NO, and NO_2 are absorbed by soil and oxidized to sulfates and nitrates. Soil bacteria and fungi are known to convert atmospheric CO to CO_2. When leaded gasoline was widely used, soil along highways became contaminated with lead, and lead mines and smelters were significant sources of this toxic element. Organic materials, such as those involved in photochemical smog formation, are removed by contact with plants and are especially attracted by the waxy organic-like surfaces of the needles of pine trees.

A number of materials that can be considered as pollutants are deliberately added to soil. The most obvious of these consist of insecticides and herbicides added to soil for pest and weed control. Chemicals from hazardous waste disposal sites can get onto soil or below the soil surface by leaching from landfills or drainage from waste lagoons. Some kinds of wastes, especially petroleum hydrocarbons, are disposed on soil where adsorption and microbial processes immobilize and degrade the wastes. Soil can be effective for the treatment of sewage. Leakages from underground storage tanks of organic liquids, such as gasoline and diesel fuel, have created major soil contamination problems.

Soils in parts of New York State have been contaminated with polychlorinated biphenyls (PCBs) discarded from the manufacture of industrial capacitors (see Chapter 4, Figure 4.13 for the structural formula of one of the many PCB compounds). Analyses of PCBs in soils of the United Kingdom archived for several decades have shown levels of these pollutants that parallel their production. Starting with very low levels around 1940 before PCBs were manufactured in large quantities, concentrations of PCBs increased markedly, peaking around 1970, when PCB manufacture was ceased. More recent soil samples have shown PCB concentrations near the pre-1940 levels. It is believed that these results reflect evaporation of PCBs and their condensation onto soil. They are consistent with observations of high PCB levels in remote Arctic and sub-Arctic regions, which is believed to be due to the condensation of these compounds from the atmosphere onto soil in very cold regions.

The degradation and eventual fate of the enormous quantities of herbicides and other pesticides applied to soil are very important in understanding the environmental effects of these substances. Many factors are involved in determining the fate of pesticides. One of the main factors is the degree of adsorption of pesticides to soil, which is strongly influenced by the nature and organic content of the soil surface as well as the solubility, volatility, charge, polarity, and molecular structure and size of the pesticides. Strongly adsorbed molecules are less likely to be released and thus harm organisms, but they are less biodegradable in the adsorbed form. The leaching of adsorbed pesticides into water is important in determining their water pollution potential. The effects and potential toxicities of pesticides to soil bacteria, fungi, and other organisms have to be considered. It must be kept in mind that pesticides may be converted to more toxic products by microbial action.

10.4 SOIL LOSS: DESERTIFICATION AND DEFORESTATION

Soil erosion refers to the loss and relocation of topsoil by water and wind action.[2] About a third of U.S. topsoil has been lost to erosion since cultivation began on the continent, and at present about a third of U.S. cropland is eroding at a rate that is sufficient to lower productivity. About 10% of U.S. land is eroding at an unacceptable rate in excess of 14 t of topsoil per acre annually. Soil erosion is largely a product of cultivation. Except in cases of extreme slopes, very high winds, and torrential rains, uncultivated soils undergo little erosion. Erosion was recognized as a problem in the central United States within a few years after forests and prairie grasslands were first plowed to plant crops, particularly in the latter 1800s. The observation that precious topsoil was being lost at an unsustainable rate led to soil conservation measures going back to 1900, or even earlier. In that sense, soil conservation was the first environmental movement, predating efforts to alleviate water and air pollution by many decades.

Water erosion is responsible for greater loss of soil than wind erosion. Whereas wind erosion tends to move soil around and deposit it in areas where it can still be used for growing crops, water erosion normally moves greater quantities of soil and carries them into streams and rivers and ultimately to oceans. The overall pattern of soil erosion in the central continental United States is shown in Figure 10.3. This figure shows that erosion is especially bad in agricultural areas draining

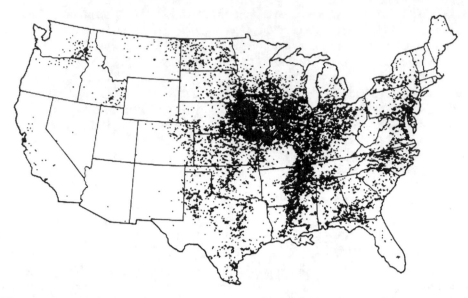

FIGURE 10.3   Pattern of soil erosion (dark areas) in the major agricultural states of the central continental United States: it is seen that erosion is particularly severe in the highly cultivated Missouri and Mississippi River watersheds.

into the Missouri and the Mississippi Rivers; millions of tons of soil are carried by these rivers into the Gulf of Mexico each year. These are areas of relatively high rainfall, which can sometimes come as very intense rainstorms, especially during the spring. A high proportion of the farmlands in these areas are devoted to row crops, which are crops such as corn, soybeans, and sorghum grains planted in rows with bare soil in between. This mode of cultivation leaves soil that is especially susceptible to water erosion.

The ultimate result of soil erosion and other unsustainable agricultural practices in relatively dry areas is **desertification**. This condition occurs when permanent plant cover is lost from soil so that the soil loses its capacity to retain moisture, dries out, and loses fertility so that plants no longer grow on it. Among the interrelated factors involved in desertification are wind erosion, water erosion (which occurs during sporadic cloudbursts even in arid areas), development of adverse climate conditions, depletion of underground water aquifers, lack of water for irrigation, accumulation of salt in water supplies, loss of soil organic matter, and deterioration of soil physical and chemical properties. Eventually, the land becomes unable to support agriculture, grazing, or even significant human populations. Desertification is one of the most troublesome results of global warming caused by greenhouse gases. It is actually a very old problem and is a serious concern in many parts of the world, such as the Middle East, the southern boundary of Africa's Sahara, and regions of the southwestern United States. Formerly productive areas of the Middle East and North Africa, "lands of milk and honey" described in biblical terms, have turned into desert, largely due to human agricultural activities. The growth of domestic grazing animals on these areas—especially goats, which tend to pull vegetation up by its roots—is a particularly strong contributor to desertification. Much of the productive capacity of arid grasslands in the western and southwestern United States has been drastically diminished by overgrazing.

Fortunately, human ingenuity and technological tools can be used to prevent or reverse desertification. For example, water, which on occasion falls as torrential rain on normally dry desert lands, can be collected and used to recharge underground water aquifers. Advanced cultivation and irrigation techniques can be used to establish perennial plant cover on erosion-prone desert soils. Potentially, plants can be genetically engineered to grow under severe conditions of temperature, drought, and salinity. Environmentally friendly mining practices can be employed to obtain minerals, and land surfaces damaged by harmful strip mining practices can be restored.

The loss of forest growth to cultivated land—**deforestation**—has occurred extensively in the United States.[3] However, much of the colonial United States, particularly New England, which was deforested for cultivation of crops, is now undergoing largely spontaneous reforestation as unprofitable farmlands are abandoned and trees become established again. Deforestation is a particularly severe problem in tropical regions. Rich tropical forests contain most known plant and animal species, many of which are becoming extinct as the forests are destroyed. Once destroyed, tropical forests are almost impossible to restore. This is because tropical forest soil has been leached of nutrients by the high annual rainfalls in tropical regions. When forest cover is removed, the soil erodes rapidly, loses the plant roots and other biomass that tend to hold it together, loses nutrients, and becomes unable to sustain either useful crops or the kinds of forests it formerly supported.

10.5 TOXICOLOGICAL AND PUBLIC HEALTH ASPECTS OF SOIL

Soil, the plants that grow on it, and the animals that consume these plants constitute an important connection between the geosphere and the biosphere and may influence human and animal health. In considering such a connection, it is important to keep in mind that the health status of people is influenced by several factors. A commonly accepted estimate of these factors is about 50% due to lifestyle, 10%–20% from genetic factors, 10%–20% from level of health care, and 20% from environment. The environment becomes much more important in highly polluted areas.

There are some striking geographic correlations with the occurrence of cancer. For example, breast cancer in women is five to six times more likely in Western Europe and North America than in parts of Africa. The difference is even more striking for lung cancer in men in which there are very high rates in Eastern Europe compared to very low rates in parts of Africa. Although most of these differences are probably associated with diet and lifestyle (the high incidence of lung cancer in Eastern Europe is almost certainly due to cigarette smoking), the possibility exists that some of the variation may be due to soil type and its influence on food. Such a correlation would be more likely in less developed countries where the connection between local soil and the food consumed is relatively more direct.

Medical geochemistry addresses the medical effects of deficiencies or excesses of environmental chemicals, usually trace elements. In addition to trace elements, some specific chemical species can be important. One that has received much attention is nitrate ion, NO_3^-. Chemically bound nitrogen gets into soil from the action of nitrogen-fixing bacteria, fertilizers added to soil, and decay of nitrogenous biomass, most commonly as ammonium ion, NH_4^+. Although this species is bound strongly to soil by cation-binding groups and is not toxic at any reasonable levels, it can be converted to NO_3^- by oxic bacteria exposed to air in soil. Nitrate is not bound well by soil and can get into drinking water. Nitrate does not harm adult humans, but it may be converted to toxic nitrite, NO_2^-, in the digestive systems of ruminant animals (including cattle) and human infants. Nitrite oxidizes the iron(II) in hemoglobin to iron(III), producing methemoglobin, which is useless for transporting oxygen in the blood. This condition can be very detrimental to the health of ruminant animals and can cause a condition called "blue babies" that is potentially fatal in infants due to insufficient oxygen reaching tissues through the blood.

An important connection between soil and the biosphere is the incorporation into food from soil of micronutrient elements essential for human health. Excesses or deficiencies of certain chemical elements can cause various illnesses. Although it is rare in the United States, but more common in China where much of the soil lacks selenium, a deficiency of this element has been related to hypothyroidism, weakened immune system, and heart disease. Selenium poisoning from ingestion of too much of the element can be manifested by symptoms that include nausea; vomiting; hair loss; discoloration, brittleness, and loss of nails; irritability; fatigue; joint pain; and a foul breath odor commonly described as "garlic breath." (Maladies due to too little or too much selenium have been commonly observed in livestock.) Goiter is caused by iodine deficiency. There is also a condition caused by too much of the wrong kind of iodine, the radioactive iodine-131 isotope produced as a fission product in nuclear power reactors, which accumulates in the thyroid and may cause thyroid cancer. As an antidote, children exposed to this isotope in the 1986 Chernobyl reactor fire in the former Soviet Union were given large doses of nonradioactive iodide salts, which resulted in the elimination of most of the radioactive form through the mass action effect; doses of iodide were also made available for administration to affected populations following the partial meltdown of power reactor cores in Japan resulting from the massive March 11, 2011, earthquake there. Fluorosis characterized by mottled teeth and bone abnormalities results from too much fluoride (F^-), and arseniasis is the name given to arsenic poisoning. Other trace elements that may cause illness when consumed in excess include cadmium, antimony, lead, zinc, copper, and mercury. Concern has long been expressed regarding a possible relationship between the lack of essential micronutrient trace element ingestion and cardiovascular disease.

A study published quite some time ago reviewed the soil–cancer relationship for gastric, esophageal, urinary, breast, bone, and bronchial cancers as well as pleural mesothelioma.[4] In addition to general factors, the study considered trace elements including selenium, cesium, and rubidium; potassium; and natural radioactivity. Although some correlations between soil constituents and cancer have been observed, the cause and effect relationship between soil characteristics and occurrence of cancer could not be proved definitively.

A study of health and population has been conducted of the Spissko-Gemerske Rudohorie Mountains, Slovakia, an area in which metal mining and smelting activities have been conducted for more than 1000 years.[5] Of the metals studied, chromium, mercury, lead, and antimony showed slight accumulation from contaminated soil into plants, whereas medium accumulation occurred for arsenic, cadmium, and copper. The area showed a dearth of older residents, which is suspected to be due to deaths from environmental exposure to toxic substances, a high percentage of low-birth-mass infants, and elevated levels of neoplasms (cancer) and cardiovascular diseases. Elevated levels of arsenic and antimony in human tissues reflected the high concentrations of these elements in contaminated soils in the area.

It may be concluded that the role of soil in environmental health is not definitively known and has not been as extensively studied as it deserves. The amount of research on the influence of soil in producing foods that are more nutritious and lower in content of naturally occurring toxic substances is quite small compared to the research on higher soil productivity (greater crop yields). It is to be hoped that the environmental health aspects of soil and its products will receive greater emphasis in the future.

10.5.1 TOXICOLOGICAL ASPECTS OF SOIL HERBICIDES

Many chemicals are applied to soil and crops to control insects, fungi, and weeds. The potential toxicities of these substances to humans, wildlife, and beneficial plants is of course a concern.[6] A comprehensive coverage of the toxicological chemistry of agricultural pesticides is beyond the scope of this chapter. However, three important herbicides, 2,4-dichlorophenoxyacetic acid (2,4-D), atrazine, and glyphosate, are mentioned here; their structural formulas are shown in Figure 10.4. Herbicides are mentioned because they are so widely used in agriculture and are spread over wide areas.

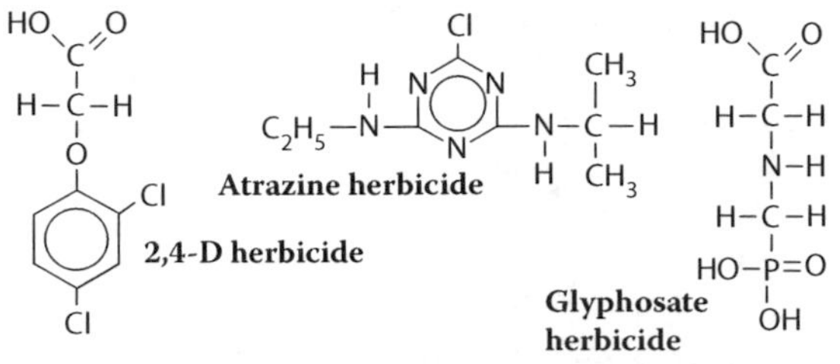

FIGURE 10.4   Structural formulas of three of the most widely used herbicides.

Herbicidal 2,4-D selectively kills broadleaf weeds and is the most widely used of all herbicides worldwide on wheat, corn, rice, and other cereal grass crops. It is a synthetic auxin, or plant hormone that is absorbed through leaves, is transferred throughout the plant, and causes unsustainable growth that results in plant death. Fortunately, given its widespread use and human exposure, it has very low acute toxicity. It is "possibly carcinogenic," a status shared with many other pesticides, and is a suspected endocrine disruptor (see Chapter 2, Section 2.16).

Atrazine is one of a large class of triazine herbicides (those with three N atoms in a ring) and is a broad-spectrum herbicide for preemergent and postemergent control of both broadleaf and grassy weeds. Inexpensive and effective, atrazine is one of the most widely used herbicides in the world, especially in conservation tillage (see Chapter 11, Section 11.11) to eliminate competing weeds prior to crop growth. Because of its widespread use, it is a common water contaminant and appears in drinking water in some agricultural areas. Its acute toxicity is low, although at least one study has associated atrazine exposure with the occurrence of cancer, birth defects, and menstrual problems. The European Union has banned the reregistration of atrazine, but it is still widely used in North America and elsewhere throughout the world.

Glyphosate (N-(phosphonomethyl)glycine) is a broad-spectrum herbicide of particular interest because crops have been developed that are immune to its herbicidal action so that it can be applied directly to the crop to kill competing weeds and grasses. Originally patented by the Monsanto Company in the 1970s and marketed under the brand name Roundup, glyphosate is the most widely used herbicide in the United States, with around 40 million kilograms being applied to crops each year and another approximately 3 million kilograms being used on lawns and yards. It acts upon target plants by absorption through foliage and translocation to the growth regions of plants where it inhibits the synthesis of the aromatic amino acids tyrosine, tryptophan, and phenylalanine. It does not act as a pre-emergent herbicide. The acute toxicity of glyphosate is slight, it is not likely carcinogenic, and its potential for endocrine disruption is uncertain.

10.6 TOXICOLOGICAL CONSIDERATIONS IN LIVESTOCK PRODUCTION

Livestock production practices may result in the distribution of toxic substances to the environment. This has been especially true with concentrated livestock feeding operations that produce large amounts of waste products in small areas. The crowding of animals in these kinds of operations is conducive to the spread of disease so that antibiotics and other agents are often fed to the animals, and these materials and their metabolites appear in livestock wastes.

Since the 1940s, organic arsenic compounds have been fed to chickens to control parasitic infections, promote weight gain, and improve meat color. Two such additives are roxarsone and nitarsone, structural formulas of which are shown in Figure 10.5. On June 9, 2011, the subsidiary of Pfizer that had been making roxarsone announced that it would cease the production of roxarsone because of concerns raised about elevated levels of potentially carcinogenic inorganic arsenic found in the livers of broiler chickens that had been fed the additive.

Antibiotics are another class of feed additives used in concentrated livestock feeding operations to control disease and increase weight gain. The greatest concern with the long-term subtherapeutic use of antibiotics in livestock feed is the evolution of antibiotic-resistant bacteria.

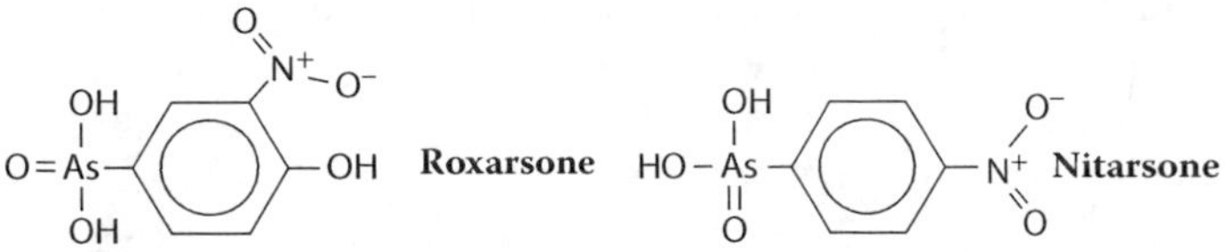

FIGURE 10.5 Two organoarsenic compounds that were widely used as additives in chicken feed to control parasites and augment weight gain.

QUESTIONS AND PROBLEMS

Access to and use of the Internet is assumed in answering all questions, including general information, statistics, constants, and mathematical formulas required to solve problems. These questions are designed to promote inquiry and thought rather than just finding material in the chapter. So in some cases, there may be several "right" answers. Therefore, if your answer reflects intellectual effort and a search for information from available sources, it can be considered to be right.

1. What is the system called in which plants are grown without soil? How does such a system work? For what kinds of plants is it used?
2. A common soil mineral is orthoclase ($KAlSi_3O_8$). Suggest the kind of secondary mineral found in soil that would be formed by the weathering of orthoclase.
3. An important chemical characteristic of soil is its pH buffering ability. What is in soil that acts as a buffer? Which chemical group is commonly involved?
4. Which kinds of ions are commonly held and exchanged by soil? What is largely responsible for this ion-exchanging ability?
5. The nitrogen cycle is very important with respect to soil and its productivity. Discuss how the demand for plant fertilizer nitrogen is affecting the nitrogen cycle.
6. What is a major advantage of using anhydrous ammonia as a source of nitrogen fertilizer in soil? What role is played by bacteria that enables anhydrous ammonia to be an effective plant fertilizer?
7. Rock phosphate (hydroxyapatite) used to be applied to soil as a source of phosphorus fertilizer. Why is this source of phosphorus relatively ineffective? What is done to it to make it a good source of phosphorus fertilizer?
8. Which event or series of events in the United States during the 1900s led to serious concerns regarding soil erosion? What measures have been taken by the U.S. government to limit soil erosion and conserve soil?
9. Suggest ways in which global climate change is likely to exacerbate problems with soil erosion. Is climate change likely to increase both wind erosion and water erosion? Explain.
10. What are two human health conditions associated with human exposure to iodine, or particular kinds of iodine?
11. Explain why glyphosate has become a particularly important herbicide in recent times.
12. In late 2011, concern was raised about elevated levels of arsenic in fruit juices, especially apple juice and grape juice. What is the current status of that controversy? Given orchard management practices from several decades ago, why might particular attention be paid to the possibility of arsenic contamination of apple juice from old orchards?
13. How is soil divided? Which is the top one of these divisions?
14. What is humification, and what does it have to do with soil?
15. What is water in soil called? Give the name of the process by which this water enters the atmosphere by way of plants.
16. Name a gaseous form and two solid forms of fixed nitrogen used as fertilizer.
17. Explain what is meant by desertification.
18. What is the good news in the United States regarding deforestation?
19. Of the following, the one that is **not** a manifestation of desertification is (explain) (a) declining groundwater tables, (b) salinization of topsoil and water, (c) increased organic matter in soil, (d) reduction of surface waters, and (e) unnaturally high soil erosion.
20. Explain how the following are related: weathering, igneous rock, sedimentary rock, and soil.
21. What is the distinction between weathering and erosion? Suggest ways in which air pollution may contribute to both phenomena.

22. In what respect is biochar a material that gets into soil by natural processes? Where can soils with significant levels of naturally occurring biochar likely be found?
23. In what respects do humus and biochar perform similar functions in soil? What are the main differences between these two kinds of materials?

LITERATURE CITED

1. Tan, Kim H., *Humic Matter in Soil and the Environment: Principles and Controversies*, Amazon Kindle edition, Marcel Dekker, New York, 2009.
2. Morgan, R. P. C., and M. A. Nearing, Eds., *Handbook of Erosion Modelling*, Wiley, Hoboken, NJ, 2011.
3. Lamont, Grant, *Effects of Deforestation-The Global Impact on the Environment*, Lamont Hall Publishing/Amazon Kindle Direct, Seattle, 2011.
4. Peeters, E.G., The Possible Influence of the Components of the Soil and the Lithosphere on the Development and Growth of Neoplasms, *Experientia* **43**, 74–81 (1987).
5. Rapant, S., V. Cveckova, Z. Dietzová, M. Khun, and L. Letkovicová, Medical Geochemistry Research in Spissko-Gemerske Rudohorie Mountains, Slovakia, *Environmental Geochemistry and Health* **31**, 11–25 (2009).
6. Matthews, Graham, *Pesticides: Health, Safety and the Environment*, Wiley-Blackwell, Hoboken, NJ, 2006.

SUPPLEMENTARY REFERENCES

Bleam, William F., *Soil and Environmental Chemistry*, Academic Press, Boston, 2012.
Brady, Nyle C., and Ray R. Weil, *Elements of The Nature and Properties of Soils*, 3rd ed., Pearson Education, Upper Saddle River, NJ, 2010.
Chesworth, Ward, Ed., *Encyclopedia of Soil Science*, Springer, Dordrecht, Netherlands, 2008.
Coyne, Mark S., and James A. Thompson, *Fundamental Soil Science*, Thomson Delmar Learning, Clifton Park, NY, 2006.
Eash, Neal S., *Soil Science Simplified*, 5th ed., Blackwell Publishing, Ames, IA, 2008.
Essington, Michael E., *Soil and Water Chemistry: An Integrative Approach*, Taylor & Francis/CRC Press, Boca Raton, FL, 2004.
Gardiner, Duane T., and Raymond W. Miller, *Soils in Our Environment*, 11th ed., Pearson/Prentice Hall, Upper Saddle River, NJ, 2008.
Ikerd, John E., *Crisis and Opportunity: Sustainability in American Agriculture (Our Sustainable Future)*, Bison Books, Winnipeg, 2008.
Lal, Rattan, *Soil Degradation in the United States: Extent, Severity, and Trends*, CRC Press/Lewis Publishers, Boca Raton, FL, 2003.
Morgan, R. P. C., *Soil Erosion and Conservation*, 3rd ed., Blackwell Publishing, Malden, MA, 2005.
Paul, Eldor A., *Soil Microbiology, Ecology, and Biochemistry*, 3rd ed., Academic Press, Boston, 2007.
Plaster, Edward J., *Soil Science and Management*, 5th ed., Delmar Cengage Learning, Clifton Park, NJ, 2009.
Singer, Michael J. and Donald N. Munns, *Soils: An Introduction*, 6th ed., Pearson Prentice Hall, Upper Saddle River, NJ, 2006.
Sposito, Garrison, *The Chemistry of Soils*, 2nd ed., Oxford, UK, 2008.
Tabatabai, M. A., and D. L. Sparks, Eds., *Chemical Processes in Soils*, Soil Science Society of America, Madison, WI, 2005.
Tan, Kim H., *Principles of Soil Chemistry*, 4th ed., Taylor & Francis/CRC Press, Boca Raton, FL, 2010.
Van Elsas, Jan Dirk, Janet K. Jansson, and Jack T. Trevors, Eds., *Modern Soil Microbiology*, 2nd ed., Taylor & Francis/CRC Press, Boca Raton, FL, 2007.
Waltham, Tony, *Foundations of Engineering Geology*, Taylor & Francis, London, 2009.
White, Robert E., *Principles and Practice of Soil Science: The Soil as a Natural Resource*, 4th ed., Blackwell Publishing, Malden, MA, 2006.
Yong, Raymond N., Masashi Nakano, and Roland Pusch, *Environmental Soil Properties and Behaviour*, Taylor & Francis/CRC Press, Boca Raton, FL, 2012.

11 Sustaining the Geosphere

11.1 MANAGING THE GEOSPHERE FOR SUSTAINABILITY

Humankind often has an uneasy relationship with the geosphere upon which it dwells. Frequently, there is a need to modify the geosphere to construct dwellings and other structures, to build roads and railroads, to impound water, or for a number of other purposes. Humans may be unpleasantly surprised by the results of their efforts as destructive landslides form on sloping grounds, dams collapse releasing devastating floods, and other unforeseen consequences result. Some natural geospheric processes are quite destructive of property and even human life. The two most dangerous of these are earthquakes and volcanoes. These are internal phenomena that result from changes taking place deep underground. Surface processes, including landslides and ground subsidence, can be very destructive to property. Although they are usually not threatening to human life, surface processes can cause fatalities. Perhaps the most dangerous of such processes are mudslides following extremely high rainfalls that sometimes bury whole villages, entombing their residents. Cases have occurred in which vast amounts of earthen material have slid into reservoirs, causing them to overflow and violently drowning many people downstream.

Human activities have profound effects on the geosphere and its sustainability. Such effects may be obvious and direct, such as strip mining or rearranging vast areas for construction projects, such as roads and dams. Or the effects may be indirect, such as pumping so much water from underground aquifers that the ground subsides or abusing soil such that it no longer supports plant life well and it erodes. As the source of minerals and other resources used by humans, the geosphere is dug up, tunneled, stripped bare, rearranged, and subjected to many other kinds of indignities. The land is often severely disturbed, air can be polluted with dust particles during mining, and water may be polluted. Many of these effects, such as soil erosion caused by human activities, are addressed in Section 9.4.

This chapter discusses several major related aspects of the geosphere. It considers destructive geospheric phenomena, especially volcanic eruptions, earthquakes, and landslides, and what can be done to cope with these often catastrophic phenomena. This chapter discusses preservation of the geosphere and modifications to it that can preserve and enhance geospheric quality, such as conversion of contaminated areas back to safe and productive uses. In addition, the geosphere is considered in this chapter as a source of raw materials, especially essential elements and metals and the environmental and sustainability challenges posed by the acquisition of these materials from Earth.

11.2 SUSTAINING THE GEOSPHERE IN THE FACE OF NATURAL HAZARDS

In Chapter 9, Section 9.4, earthquakes and volcanoes are discussed as two great natural hazards involving the geosphere. Both of these phenomena are inevitable; they have happened throughout Earth's lifetime and will continue to occur long after Earth becomes uninhabitable by humans. In sustaining the geosphere, nothing can be done to prevent these destructive phenomena. Although individual earthquakes cannot be forecast, areas that are earthquake-prone are relatively well known. The condition of the geosphere in such areas is important in limiting the damage that they cause. Structures should be located on terrain that is not prone to liquefaction during earthquakes, for example.

Areas of the geosphere have been rendered uninhabitable by volcanic activity due to the explosive power of volcanic eruptions, the physical hazards of molten lava, and the toxicological hazards of sulfur dioxide and finely divided ash. This happened in 2000 when an eruption of the Mount Oyama volcano on the small Japanese volcanic island of Miyake, a granitic composite volcanic cone located in the Pacific Ocean 180 km south of Tokyo, released an astounding total of 18 Mt of sulfur dioxide at a peak rate of about 54 kt/day, and the resulting toxic atmosphere forced the temporary evacuation of all the residents of the island. Although the residents of the island were allowed to return in 2005, the continued emissions of sulfur dioxide from the volcano since then along with the continued threat of further eruptions have caused many residents to stay away from the island, and its population had dropped below 2700 and was still shrinking by 2012.

11.2.1 Vulnerable Coasts

Coastal areas are among the most vulnerable regions with respect to natural disasters. This vulnerability is highly increased by the tendencies of people to live near coasts and to place structures as close as possible to water. About 75% of the U.S. population lives in coastal states, and the total coastline (including the Great Lakes) is about 150,000 km. Furthermore, tropical storms, which are some of nature's most damaging phenomena, are at their most destructive along coasts and lose their destructive powers rather abruptly as they move inland. The infamous Hurricane Katrina in 2005, followed by Hurricane Rita, virtually wiped out New Orleans and leveled structures along the coasts of Mississippi, Louisiana, and Texas that had stood for more than a century, emphasizing the vulnerability of these areas to natural disasters. Insurance costs to replace structures destroyed by coastal storms have reached prohibitive levels. Intelligent land use regulations are needed for these regions.

There are several main areas of coastal vulnerability. In some regions, the U.S. Gulf Coast, for example, the greatest potential for damage is from **tropical cyclones**, which are called **hurricanes** in the Atlantic and the Gulf of Mexico and **typhoons** in the Pacific and the Indian Oceans. Commonly developing between 15° south and 8° north of the equator, tropical cyclones release enormous amounts of energy as their water vapor condenses to form rain. This energy release creates air currents moving vertically and strong winds exceeding 100 km/h, often approaching 150 km/h, and getting as high as 300 km/h. The most devastating effects of tropical cyclones occur when they hit coastal areas at high tide accompanied by a **storm surge**. Storm surges develop as the consequence of low pressures in cyclones that raise water levels by several meters. In addition to damage from storm surges and high winds, coastal areas hit by tropical cyclones are subject to flooding from the extreme rainfall from the storms. Rainfall amounts approaching a meter over two or three days have been reported, causing great problems from conventional flooding.

The coastal geosphere may also be damaged by tsunamis resulting from earthquakes. The giant breakers produced by tsunamis can reach heights of 10–15 m, or even higher. The geosphere along coasts can be eroded and damaged by tsunamis. Tsunamis carry large quantities of saltwater inland, contaminating soil such that its capacity to support plant life is impaired.

The geosphere in coastal regions takes a severe beating from coastal storms and from normal water and wind processes aggravated by unwise human management of the shoreline. As shown in Figure 11.1, a coastline typically consists of a bank of earthen material composed of a seacliff of some height below which is a sandy beach. In many cases, such as much of the Gulf Coast region that was ravished by Hurricane Katrina in 2005 or the coastal areas of Bangladesh that were devastated by the destructive tropical cyclone that hit that area in November 2007, the coastal land is not much higher than the sea that it borders, which makes it susceptible to flooding by tidal surges.

Shorelines are subject to destructive forces that erode the banks. Houses originally constructed at a comfortable distance from the edge may eventually end up perched precariously above the beach, often supported by poles extending downslope. Such structures frequently collapse following an extreme weather event or normal coastline erosion. Usually during winter, heavy storm waves

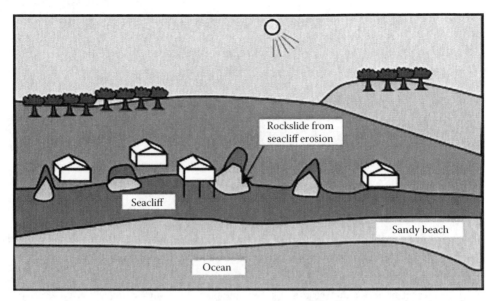

FIGURE 11.1 Coastal regions are subject to damage from storm surges, normal erosive processes, and human influences. Much of the damage and loss stems from unwise construction practices along the coastline.

pound the base of a seacliff, washing sand away from it and exposing it to damage from the waves. In addition to wave action, the seacliff may be eroded by a number of other factors including rain wash, weathering, the action of some organisms such as boring mollusks, and tree roots that penetrate the seacliff rock and force it apart. Anthrospheric constructs can exacerbate seacliff erosion. Paving of surface areas without proper diversion of rainwater runoff can cause erosive flows of water that erode the seacliff.

The most valued part of a coastline is often the sandy beach. Sand is washed into the ocean as the result of weathering of silicon-rich quartz and feldspar rocks upstream and distributed along the shoreline by the action of ocean currents and waves. Human intervention has reduced the amount of sand available. Dams built on streams trap sand as sediment, depriving the ocean of its source of sand. Improper beach management can result in loss of beach sand. The term **littoral cell** applies to a region of shoreline that includes inflow of sand from a stream or from seacliff erosion, transport of sand generally parallel to the coast by ocean currents and wave action, and eventual loss of sand from the coastal region.

A number of measures may be taken to preserve the coastline. Foremost among these is to avoid doing those things that result in damage to the shoreline and destruction along it. Dwellings and other structures that may be subject to damage by natural forces along the coast simply should not be placed where they are likely to be harmed. Damaging rain runoff can be diverted to drainage pipes extending to the base of a seacliff. Water infiltrating from excessive watering of vegetation along the edge of a seacliff can infiltrate and flow out of the base as seeps and springs that weaken the shoreline rock.

Areas that have lost beach sand are sometimes restored by expensive programs of **beach nourishment** to haul in more sand. This is usually a losing proposition in that the forces that caused the loss of beach sand originally will simply wash away the new sand. Walls composed of large rock or concrete constructed along a shoreline can be used in attempts to enhance beach areas. A common such structure consists of **groins** composed of dams perpendicular to the coastline that intercept the flow of water and suspended sediment in a littoral cell (see earlier in this section). These are generally constructed in groups constituting a groin field in a littoral cell. Sand accumulates in the upflow direction from each groin, leading eventually to a segment of beach extending out to the end of the cell. However, erosion of the beach occurs downflow from the groin, leading to an irregular beach.

Perhaps the most massive undertakings to prevent coastline damage are high **seawalls** that are designed to intercept ocean surges and prevent damage onshore. Following the deadly 1900 Galveston, Texas, hurricane that killed around 8000 people, the city undertook the construction of a concrete seawall 17 ft high, 16 ft thick at the base, and 7 mi long. In addition, sand was pumped into the city to raise the grade substantially. For a century following the construction of the seawall, Galveston avoided major damage from seawater surges until Hurricane Ike breached the seawall and flooded the city in 2008. A disadvantage of seawalls is the loss of beach sand that their construction often causes. Hurricane Ike caused massive loss of sand along the Galveston seawall, which was subsequently replaced by 400,000 m^3 of sand hauled in from land.

About 40% of Japan's 32,000-km coastline is protected by seawalls designed primarily to hold back tsunamis caused by earthquakes. Much of the damage from the March 11 earthquake off the coast of Sendai, Japan, occurred when seawalls were breached by the tsunami that followed. Large areas of farmland were contaminated by seawater as a result.

11.2.2 Threat of Rising Sea Levels

Large numbers of people live at a level near, or in some cases below, sea level. As a result, any significant temporary or permanent rise in sea level poses significant risk to lives and property. Such an event occurred on February 1, 1953, when high tides and strong winds combined to breach the system of dikes protecting much of the Netherlands from seawater. About one-sixth of the country was flooded as far inland as 64 km from the coast, killing about 2,000 people and leaving approximately 100,000 without homes.

Although isolated instances of flooding by seawater caused by combinations of tidal and weather phenomena described earlier in this section will continue to occur, a much more long-lasting threat is posed by long-term increases in sea level. These could result from global warming due to the greenhouse gas emissions discussed in Chapter 8. Several factors could raise ocean levels to destructive highs, which is also a result of greenhouse warming. Simple expansion of warmed oceanic water could raise sea levels by about 1/3 meter over the next century. The melting of glaciers, such as those in the Himalayas, has probably raised ocean levels by about 5 cm during the last century, and the process is continuing. The greatest concern, however, is that global warming could cause the great West Antarctic Ice Sheet to melt, which would raise sea levels by as much as 6 m.

The measurement of sea levels has proved to be a difficult task because the levels of the surface of land keep changing. The land most recently covered with ice age glaciers in areas such as Scandinavia is still "springing back" from the immense mass of the glaciers, so that sea levels measured by gauges fixed on land actually appear to be dropping by several millimeters per year in such locations. An opposite situation exists on the east coast of North America where land that was pushed outward and raised around the edge of the enormous sheet of ice that covered Canada and the northern United States about 20,000 years ago is now settling back. Factors such as these illustrate the advantages of remarkably accurate satellite technology now used in the determination of sea levels.

11.3 SUSTAINABLE DEVELOPMENT ON THE GEOSPHERE'S SURFACE

The most direct interaction between the geosphere and the anthrosphere occurs when humans build structures on the geosphere or below its surface. Such endeavors involve **engineering geology**, which addresses the ways in which geological materials and structures are used and dealt with technologically with respect to structures, materials science, and other engineering aspects.[1] Practically anything that is built on or near the surface of Earth requires a consideration of geological engineering aspects. For example, if a structure is to be built on a slope, it is important to consider the geological properties of the surface on which it is to be built. The degree of the slope and the nature of the material on which it is located can be used to predict the likelihood of earth slides and to design structures to avoid these natural disasters. Engineering geology is used to design rock quarries and to determine the most efficient means of removing rocks from the ground.

Large public works projects are the human endeavors most likely to affect the geosphere because they entail earthmoving, digging, boring, and other operations performed on the geosphere. In turn, the nature, costs, and safety of structures constructed as part of huge public works projects are all highly dependent on the characteristics of the geological formations on which they are built. Therefore, public works projects require a high degree of sophisticated geological engineering throughout their planning and construction stages, and their maintenance and operation require consideration of geological engineering as well. Large public works projects include dams, roads, railroads, airports, pipelines, canals, tunnels, and other structures. Much of the engineering that goes into such projects involves evaluation of the geologic strata upon which the structures are located and development of measures to prevent problems. For example, dams should be located on and anchored to strong formations of rock, preferably igneous or metamorphic rock. Fractures in the rock formations along which leaks may develop should be detected and filled with a wet mixture of cement and sediment called grout. Numerous factors must be considered in highway construction. To a greater extent than with most structures, highways have to use surrounding geological materials, which must be evaluated for their suitability for fill and roadbed. Topography (surface configuration) and slope are crucial for grading and drainage. Geologic engineering is a crucial consideration in the siting and construction of large structures, such as buildings or nuclear or fossil-fueled power plants. The underlying strata must be carefully evaluated for its load-bearing capacity and to discover unexpected features, such as faults or fractures that might shift or caverns that might cause subsidence and sinkholes.

11.3.1 Site Evaluation

Before placing a structure on or below the surface of Earth, it is important to consider a number of factors regarding the suitability of the construction, a process called **site evaluation**. A number of factors must be considered in site evaluation. Site topography is important; for some applications, steep slopes may be detrimental. The present and former use of the site must be considered. For example, if the site was formerly employed in an application that might have involved leakage and infiltration of wastes, extensive site remediation must be considered. The physical, chemical, and engineering properties of the earthen material must be known. In many cases, structures should rest on bedrock, so depth to bedrock should be measured. Natural hazards have to be evaluated. These include possibilities of earthquake (locating a critical structure directly on a seismic fault can be particularly troublesome) or volcanic activity. The potential for less spectacular but much more common hazards should be evaluated, including tendency to landslides and erosion. Potential interactions with the hydrosphere are crucial. These include surface runoff characteristics, surface water infiltration underground, flooding potential, depth of the water table, and groundwater flow.

11.3.2 Kinds of Structures on the Geosphere

Numerous kinds of structures are located on or below the surface of the geosphere, most requiring some degree of excavation. The most common of these are as follows:

Highways: Important considerations are topography, considering limits to permissible gradients, excavation required such as in constructing cuts through hills, stability of slopes related to the likelihood of landslides onto the highway, strength of base rock, potential for flooding, and availability of construction materials including rock and sand to use in making paving.

Railroads: In general, the same considerations given to highway construction apply to railroads. Gradient is much more important with railroads and must be much lower for them than for highways. This requirement leads to the consideration of the need for cuts through hills and tunnels.

Bridges: Both highways and railroads require bridges to cross rivers, estuaries, and deep valleys. In planning a bridge, it is essential to assess the earthen material upon which it rests. In some cases pilings must be driven to bedrock, and in other cases grout must be pumped underground to fill voids and provide stability.

Airports: Important considerations are topography including the presence of, or potential for, essentially level areas of sufficient size to construct long runways, soil characteristics including good load-bearing capacity, surface drainage and absence of flooding potential, and availability of rock and sand to use in construction.

Tunnels: An important consideration is whether a tunnel is constructed through solid rock or through poorly cohesive earth, thus requiring structures to prevent the flow of earth into the tunnel. It is common for both types of material to be encountered in the same tunnel. Rock structure and rock fractures must be considered. Also important is the potential for water infiltration and for means to remove water.

Buildings: The larger the building, the more important the considerations given to siting it. Often, extensive drilling and coring are performed to evaluate the suitability of the material on which the building will rest. In areas where underground voids may exist, it is crucial to make sure that none are in the vicinity of the building. In some areas, the potential to make structures earthquake resistant is crucial. Soil that can undergo liquefaction (see Chapter 9, Section 9.4) during earthquakes must be avoided if at all possible. Another consideration is the potential for soil to expand and contract, which may crack or even destroy basement walls. The likelihood of groundwater infiltration must be considered; in some unfortunate cases, building excavations have intercepted springs or aquifers with a high flow potential.

Dams: Both concrete and earth dams require careful assessment of the geological strata involved. A dam is almost always placed across a valley, which then fills upstream with water. The susceptibility of the reservoir slopes to landslides must be considered; such slides may be more likely as the bottom parts of the slopes become constantly saturated with water from the reservoir. The rate of sediment accumulation in the reservoir should be evaluated. The strata upon which the dam rests are very important, and its strength, stability, and potential to develop routes for leakage of water must be evaluated. Especially in the case of earthen dams, the suitability of earthen material in the vicinity for construction is important.

Mines: There is not much choice of locations of mines because they must be placed exactly where the desired minerals are found. However, an evaluation of the geosphere at the location of a mine can be used to determine its construction. A major issue in designing a mine is the choice between an underground mine and an open pit. The type of mine construction depends strongly on the type and stability of the earthen material that must be excavated. Water infiltration from groundwater sources is a very important consideration and may require elaborate measures for pumping and drainage. Especially for coal mines, the potential for the infiltration of explosive methane must be known. Also important is the consideration of locations for the placement of mine spoils (wastes).

11.4 DIGGING IN THE DIRT

One of the most common human interactions with Earth is its excavation to extract materials, construct structures such as roadways, construct basements for buildings, alter surface slope, or simply put holes in the ground for uses such as harbors. The ease with which earthen material is excavated and moved and the measures required to do so depend strongly on the nature of the material excavated. Earthen material may vary in hardness from dirt through rock of varying degrees of hardness and consolidation to extremely hard rock. The easiest means of excavation is simply scooping up unconsolidated material composed of relatively small particles. When this is not possible, the earthen material can be loosened by a process called ripping. A ripper consists of a strong, vertical blade mounted on the back of a heavy tracked tractor that extends a meter or more below the ground and is pulled through the earth so that it breaks up the earthen material. Material that cannot be broken down by ripping may require blasting, in which holes are drilled at intervals and packed with explosives that are detonated to break the hard rock into small pieces that can be removed with a scoop.

The properties of an excavation may deteriorate with time. One way in which this occurs is by slumping of walls, which is more likely and more severe with more steeply angled walls. The removal of earthen material reduces the mass resting on the floor of an excavation so that it may have a tendency to rise. Some of the greater challenges in maintaining an excavation occur when it extends to below the water table. Tapping a large, highly permeable aquifer can result in an excavation rapidly filling with water, which can be a major problem unless the purpose of the excavation is to construct a water reservoir. In some cases, problems with water can be solved by surrounding the excavation with wells that are constantly pumped, which not only prevents water infiltration but also stabilizes the floor and walls of the excavation.

Numerous means are employed to stabilize slopes to prevent slumping. Rock bolts and rock anchors can be installed in holes drilled into a slope to hold it in place. Anchors embedded deep within a slope can be used to fasten concrete panels to the sloping surface, preventing material from sliding down. The loss of the integrity of a slope face due to water infiltration can be mitigated to a large extent by applying cement pneumatically to the surface, which cures to provide a relatively impervious material that resists weathering.

11.4.1 Subsurface Excavations

The most challenging excavations are those that are made subsurface for tunnels, mines, subways, and underground caverns. Among the factors to be considered in such excavations are the integrity and hardness of the earthen material, presence of faults, and groundwater infiltration. These conditions can be assessed prior to construction by indirect methods from the surface, such as seismic testing, or by direct methods, including drilling, coring, and pilot tunnels.

Seismic testing that uses vibrational waves generated on the surface to assess subsurface conditions is the most commonly used indirect method of examining subsurface strata. Seismic testing involves the generation of shock waves on the surface by an explosive charge or by mechanically dropping a heavy weight and monitoring the resulting seismic P-waves and S-waves (both observed during earthquakes). The waves generated are monitored by geophones placed appropriately relative to the shot point where the waves are initiated. Times at which the waves reflected by the underlying strata are received are very accurately monitored along with their intensities by the geophones. Computer analysis of the data enables evaluation of the geometry and lithology (hardness, integrity, and strength of the underlying strata). Faults, joints, synclines, and anticlines can also be mapped by seismic testing.

Figure 11.2 illustrates a subsurface excavation for a water tunnel used for water supply to a large urban area. In addition to a tunnel to conduct water, such an installation has shafts extending from the surface to the tunnel. Shafts are excavated from the surface; when they are excavated upward from below the surface, they are called raises. A large water system may also have caverns consisting of large voids hollowed out of the underground strata for water storage.

11.4.2 Green Underground Storage

Underground storage facilities offer a number of advantages and rank high in sustainability. Such facilities must be located in suitable strata free from major water infiltration. They will maintain

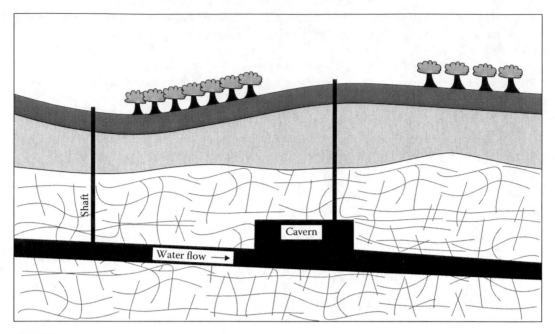

FIGURE 11.2 An underground tunnel for conveying water: underground caverns can be used for water storage. Access to the tunnel and caverns is through vertical shafts.

a particular temperature within a relatively narrow range regardless of outside conditions. With suitable insulation, they can be maintained at almost any reasonable temperature including low temperatures for cold storage.

The most commonly used underground storage facilities are located in limestone formations. In some cases natural caves have been used, but underground limestone quarries are usually superior. One of the largest such installations is the Atchison Storage Facility located along the Missouri River about 2 mi southeast of Atchison, Kansas. This huge facility was mined from limestone in a river bluff. It covers 127 acres and consists of a series of caverns with a limestone ceiling supported by 78 pillars of undisturbed limestone, each around 10 m in diameter. During World War II, the U.S. War Food Administration leased the Atchison Storage Facility and converted part of it to a refrigerated food storage facility maintained at 0°C for the storage of sides of beef, eggs, vegetables, fruits, butter, lard, and salt pork. An initial delivery of 12 railcars of dried eggs in 1944 expanded to 8,900 t of eggs along with 48 t of dried milk; 1,000 t of raisins; and 20,000 t of prunes by 1949. Considering that the war ended in 1945 and that dried eggs and prunes were very low on the soldiers' dietary preferences, it is likely that much of this food was eventually discarded.

During the 1950s, parts of the Atchison Storage Facility were converted to warehouse space for specialized military hardware. An ammonia-to-brine dehumidification system was installed to reduce the humidity to 42% relative humidity and to maintain the temperature of the facility at a range of 18°C–22°C. Later, the facility was used for record storage.

11.4.3 SALT DOME STORAGE

Salt domes were formed when deposits of salts (predominantly sodium chloride) were left from the evaporation of water from vast inland seas. Eventually, the salt deposits became covered with sediments. Being less dense than the sedimentary material that covered them, the salt deposits rose over time, forming domes 1–10 km in diameter and up to 6.5 km in thickness. The salt in salt domes is generally dry and impermeable.

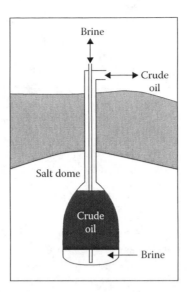

FIGURE 11.3   Salt dome storage of petroleum such as that used for the U.S. National Strategic Petroleum Reserve: a cavity is formed in the dome by the action of water. Petroleum floats on top of a layer of brine and is pumped out of the storage cavern by pumping in more brine.

Initial interest in salt domes was as sources of salt that could be extracted by forcing water through wells into the domes, pumping out the resulting brine, and evaporating the water solvent to obtain a solid salt residue. It was also found that petroleum deposits are often encountered above the domes, and such deposits constitute the dominant source of petroleum along the Gulf of Mexico. More recently, salt domes have come into use for petroleum storage, especially for the U.S. Strategic Petroleum Reserve,[2] which is located in salt domes in Texas and Louisiana. The Strategic Petroleum Reserve has a capacity of about 727 million barrels (116 million cubic meters) of petroleum. Figure 11.3 shows a basic salt dome facility with a cavern in the dome for petroleum storage. In such a facility, the crude oil floats on top of a layer of brine (salt solution). When the crude oil is to be removed, brine is pumped in from a surface reservoir and the oil is forced out. The process is reversed to place petroleum back into the storage cavity.

11.5   EXTRACTION OF MATERIALS FROM EARTH

A large number of minerals are extracted from the geosphere. As an example, the following is an impressively long list of minerals published by the World Mineral Exchange:[3] alunite, andalusite, andradite, apatite, aragonite, arsenic, asbestos, asphalt, ball clay, barites, bauxite, bentonite, beryl, beryllium, biotite, bismuth, borates, borax, bornite, boron, brines, cadmium sulfide, cadmium telluride, calcite, carbonatite, cassiterite, celestite, chalcopyrite, china clay, chromite, chrysoprase, coal, cobalt, columbite, copper, corundum, cryolite, diatomite, diamond, dimension stone, diopside, dolomite, epidote, fayalite, feldspar, fire clay, flint, fluorite, galena, gallium, garnet, gold, granite, graphite, gypsum, halite (sodium chloride), hornblende, ilmenite, iodine, iridium, iron ore, iron oxide pigments, kaolinite, kyanite, lazulite, lead, limestone, lithium, magnesite, manganese, marble, mercury, mica, molybdenite, monazite, nickel, ochre, opal, palladium, palygorskite, perlite, phosphate, platinum, potash, pumice, pyrite, pyrochlore, pyrophyllite, quartz, rare earths, selenium, siderite, silica, sillimanite, silver, slate, sodium, sodium bicarbonate, sodium carbonate, sodium sulfate, staurolite, strontium, sulfur, talc, tin, tourmaline, ulexite, uranium, vanadinite, vermiculite, wollastonite, wulfenite, xenotime, zeolites, zinc, zircon, and zoisite. Many human effects on the geosphere result from the extraction of multitudes of minerals such as these from Earth's crust.

The most potentially damaging method of mineral resource extraction is **surface mining**, employed in the United States to extract virtually all of the rock and gravel that is mined, well over half of the coal, and numerous other resources. In earlier times, before strict reclamation laws were in effect, surface mining, particularly of coal, left large areas of land scarred, devoid of vegetation, and subject to erosion. However, with appropriate excavation and restoration practices, the damage from surface mining can be minimized and surface mining may even be used to improve surface quality, such as by the construction of reservoirs where rock or gravel has been extracted.

Several approaches are employed in surface mining. Sand and gravel located under water are extracted by **dredging** with draglines or chain buckets attached to large conveyers. In most cases, resources are in deposits beneath an **overburden** of earthen material that does not contain any of the resource that is being sought. This material must be removed as **spoil**. As the name implies, **open-pit mining** is a procedure in which gravel, building stone, iron ore, and other materials are simply dug from a big hole in the ground. Some of these pits, such as several pits from which copper ore has been taken in the United States, are truly enormous in size.

The best known (sometimes infamous) method of surface mining is **strip mining**, in which strips of overburden are removed by draglines and other heavy earthmoving equipment to expose seams of coal, phosphate rock, or other materials. Heavy equipment is used to remove a strip of overburden, and the exposed mineral resource is removed and hauled away. Overburden from a parallel strip is then removed and placed over the previously mined strip, and the procedure is repeated numerous times. On highly sloping terrain, overburden is removed on progressively higher terraces and placed on the terrace immediately below.

Because of past mining practices, surface mining got a well-deserved bad name. With strip mining, the common practice was to dump waste overburden in rather randomly constructed **spoil banks** consisting of poorly compacted, steeply sloped piles of finely divided material. The rock and soil on spoil banks was highly susceptible to erosion and physical and chemical weathering. No effort was made to replace topsoil on the surface of the spoil banks, and the topsoil that was there quickly eroded away, so that vegetation on these unsightly piles was sparse. Now, however, with modern reclamation practices, topsoil is first removed and later placed on top of overburden that is replaced such that it has gentle slopes and proper drainage. The topsoil spread over the top of the replaced spoil, which is often carefully terraced to prevent erosion, is seeded with indigenous grass and other plants, fertilized, and watered, if necessary, to provide vegetation. The end result of carefully done **mine reclamation** projects is a well-vegetated area suitable for wildlife habitat, recreation, forestry, and other beneficial purposes.

A controversial practice that has developed in recent years is **mountaintop removal strip mining** that is being practiced in West Virginia and to a lesser extent in Kentucky and Virginia.[4] This procedure involves blasting the tops off mountains and pushing the overburden into valleys to get to coal seams that are then dug up with huge draglines and shipped for use in power plants. Proponents contend that the practice does minimal harm and even provides flatland in areas notably short of level areas. Opponents cite destruction of hardwood forests and damage to water sources as major problems.

The extraction of minerals from placer deposits formed by deposition from water has obvious environmental implications. Placer deposits can be mined by dredging from a boom-equipped barge. Another means that can be used is hydraulic mining with large streams of water. One interesting approach for more coherent deposits is to cut the ore with intense water jets and then suck up the resulting small particles with a pumping system.

For many minerals, underground mining is the only practical means of extraction. An underground mine can be very complex and sophisticated. The structure of the mine depends on the nature of the deposit. It is of course necessary to have a shaft that reaches to the ore deposit. Horizontal tunnels extend out into the deposit, and provision must be made for sumps to remove water and for ventilation. Factors that must be considered in designing an underground mine include the depth, shape, and orientation of the ore body, as well as the nature and strength of the rock in and around it; thickness of overburden; and depth below the surface.

Usually, significant amounts of processing are required before a mined product is used or even moved from the mine site. Such processing, and its by-products, can have significant environmental effects. Even rock to be used for aggregate in cement road construction must be crushed and sized, a process that has the potential to emit air-polluting dust particles to the atmosphere. Crushing is also a necessary first step for further processing of ores. Some minerals occur to an extent of a small percentage or even less in the rock taken from the mine and must be concentrated on site so that the residue does not have to be hauled far. For metals mining, these processes, as well as roasting, extraction, and similar operations, are covered under the category of **extractive metallurgy**.

11.5.1 ENVIRONMENTAL EFFECTS OF MINING AND MINERAL EXTRACTION

Some of the environmental effects of surface mining have been mentioned in Section 11.5. Although surface mining is most often considered for its environmental effects, subsurface mining can also have a number of effects, some of which are not immediately apparent and may be delayed for decades. Underground mines have a tendency to collapse, leading to severe subsidence. Mining disturbs groundwater aquifers. Water seeping through mines and mine tailings may become polluted. One of the more common and damaging effects of mining on water occurs when pyrite, FeS_2, commonly associated with coal, is exposed to air and becomes oxidized to sulfuric acid by bacterial action to produce acid mine water (see Chapter 3, Section 3.5). Some of the more damaging environmental effects of mining are the result of the processing of mined materials. Usually, ore is only part—often a small part—of the material that must be excavated. Various **beneficiation** processes are employed to separate the useful fraction of ore, leaving a residue of **tailings**. A number of adverse effects can result from environmental exposure of tailings. For example, residues left from the beneficiation of coal are often rich in pyrite, FeS_2, which produces acidic water pollutants as described in Section 3.5. Uranium ore tailings unwisely used as fill material have contaminated buildings with radioactive radon gas.

One of the more environmentally troublesome by-products of mineral refining consists of waste **tailings**. By the nature of the mineral processing operations employed, tailings are usually finely divided and prone to chemical weathering processes. Heavy metals associated with metal ores can be leached from tailings, producing water runoff contaminated with cadmium, lead, and other pollutants. Adding to the problem are some of the processes used to refine ores. Large quantities of cyanide solution are used to extract low levels of gold from its ore, posing obvious toxicological hazards.

11.6 SUSTAINABLE UTILIZATION OF GEOSPHERIC MINERAL RESOURCES

An essential part of Earth's natural capital consists of minerals containing metals and other essential raw materials that the geosphere provides for use by humans. Important considerations in utilizing mineral resources include a finite supply of essential minerals, the need to utilize less rich sources of minerals as better sources are depleted, and environmental damage associated with mineral exploitation. There are numerous kinds of mineral deposits that are used in various ways. The most available deposits have already been exploited. As the more available mineral resources are exploited, a major challenge is to find sustainable ways to utilize available resources in a cost-effective manner that is consistent with maximum conservation of the resource, environmental protection, and material recycling.

In order to make their extraction worthwhile, sources of minerals must be those that are enriched at a particular location in Earth's crust relative to their average percentage abundance in Earth's crust. Such an enriched deposit is called an ore, a term usually applied to metal deposits. The value of an ore is expressed in terms of a **concentration factor**:

$$\text{Concentration factor} = \frac{\text{concentration of material in ore}}{\text{average crustal concentration}} \tag{11.1}$$

Obviously, higher concentration factors are always desirable. Required concentration factors decrease with average crustal concentrations and with the value of the commodity extracted. A concentration factor of 4 might be adequate for iron, which makes up a relatively high percentage of Earth's crust. Concentration factors must be several hundred or even several thousand for relatively low-value metals that are not present at very high percentages in Earth's crust. However, for an extremely valuable metal, such as platinum, a relatively low concentration factor is acceptable because of the high financial return obtained from extracting the metal.

In addition to large variations in the concentration factors of various ores, there are extremes in the geographic distribution of mineral resources. With some exceptions, the United States is reasonably well endowed with mineral resources, possessing in terms of its mineral resources, possessing significant resources of copper, lead, iron, gold, and molybdenum but virtually no resources of some important strategic metals, including chromium, tin, and platinum-group metals. For its size and population, South Africa is particularly blessed with some important metal mineral resources.

11.6.1 Metals

With an adequate supply of all of the important elements and energy, almost any needed material can be manufactured. Most of the elements, including practically all of those likely to be in short supply, are metals. Some metals are considered especially crucial because of their importance to industrialized societies, uncertain sources of supply, and price volatility in world markets. One of these is antimony, which is used in automobile batteries, fire-resistant fabrics, and rubber. Chromium, another crucial metal, is used to manufacture stainless steel (especially for parts exposed to high temperatures and corrosive gases), jet aircraft, automobiles, hospital equipment, and mining equipment. The platinum-group metals (platinum, palladium, iridium, rhodium) are used as catalysts in the chemical industry, in petroleum refining, and in automobile exhaust anti-pollution devices. Around 90% of these metals used in the United States are imported with the remainder obtained from recycling.

Mining and processing of metal ores involve major environmental concerns, including disturbance of land, air pollution from dust and smelter emissions, and water pollution from disrupted aquifers. These problems are aggravated by the fact that the general trend in mining involves the utilization of less rich ores. This is illustrated in Figure 11.4, which shows the average percentage of copper in copper ore mined since 1900. The average percentage of copper in ore mined in 1900 was about 4%, but by 1982 it was about 0.6% in domestic ores and 1.4% in richer foreign ores. Ores as low as 0.1% copper may eventually be processed. Increased demand for a particular metal, coupled with the necessity to utilize lower grade ores, has a multiplying effect on the amount of ore that must be mined and processed and on accompanying environmental consequences.

Metals have a wide variety of properties and uses. The metals in ores occur in a number of different compounds; in some cases, two or more compounds are significant mineral sources of the same metal. Usually, these compounds are oxides or sulfides. However, other kinds of compounds and, in the cases of gold and platinum-group metals, the elemental (native) metals themselves serve as metal ores. Table 11.1 lists the important metals and their properties, major uses, and sources.

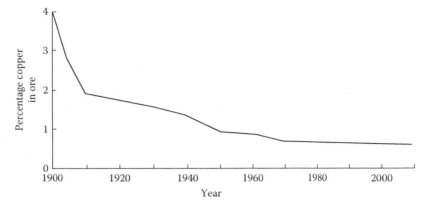

FIGURE 11.4   Average percentage of copper in ore that has been mined.

TABLE 11.1
Geospheric Metal Resources

Metals and Major Uses	Aspects of Resources
Aluminum: metal products, autos, electrical transmission lines	From bauxite (Al_2O_3), limited U.S resources, ample world resources
Chromium: metal plating, stainless steel, specialized alloys, chemicals	From chromite, an oxide mineral; virtually none in the United States, most in South Africa, Zimbabwe, Russia
Cobalt: alloys, magnet alloys, pigments	Co_3S_4 and other minerals, abundant resources
Copper: ductile, excellent electrical conductor, valuable metal with many uses	In low percentages in sulfides, oxides, carbonates, increasingly short supply
Gold: jewelry, currency, electronics, industrial and high technology uses	In various minerals as low as several parts per million, greatly increased prices in recent years, limited supply
Iron: most widely used metal, usually as steel containing 0.3%–1.7% carbon	Abundant U.S. and global resources of hematite (Fe_2O_3), goethite ($Fe_2O_3 \cdot H_2O$), magnetite (Fe_3O_4)
Lead: storage batteries, electronics, chemicals (limited because of toxicity)	Galena (PbS), limited U.S. and marginal global resources, heavily recycled
Manganese: alloys, dry cells, chemicals	Few U.S. but adequate world resources
Mercury: instruments, electronics, chemicals, use curtailed because of toxicity	From cinnabar (HgS), no U.S. production, from China, much is recycled
Molybdenum: alloys, chemicals	Large U.S. and global resources
Nickel: stainless steel, alloys, rechargeable batteries, catalysts	Associated with iron ores, tight supplies with most from Russia, Australia, Canada
Silver: jewelry, bearings, solder, electronics, no longer used much in photography	Sulfide minerals along with Cu, Pb, Zn, relatively short supply
Tin: coatings, solders, bronze, alloys, chemicals including organometallics	No U.S. production, largely from China, Indonesia, Peru, much is recycled
Titanium: strong, corrosion resistant, used in alloys, white paint pigment	TiO_2, abundant in Earth's crust, produced in the United States and worldwide
Tungsten: very strong, high-melting alloys, drill bits, turbines, nuclear reactors	As scheelite ($CaWO_4$), abundant U.S. and world resources
Vanadium: used for strong steel alloys	Largely from China, Russia, South Africa
Zinc: alloys, galvanized steel, paint pigments, chemicals, ranks fourth in the world	In many ore minerals in the United States and other countries, relatively short supply

11.6.2 Nonmetal Mineral Resources

A number of minerals other than those used to produce metals are important resources. There are so many of them that it is impossible to discuss them all in this chapter; however, mention will be made of the major ones. As with metals, the environmental aspects of mining many of these minerals are quite important. Typically, even the extraction of ordinary rock and gravel can have important environmental effects.

Clays are secondary minerals composed predominantly of aluminum, silicon, and oxygen, which occur widely in the geosphere (see Chapter 9, Section 9.2). Other than stone tools, clays were the first minerals used by humans, especially in making pottery containers. Various clays are used for clarifying oils; as catalysts in petroleum processing; as fillers and coatings for paper; and in the manufacture of firebrick, pottery, sewer pipe, and floor tile. A typical clay mineral is kaolin, chemical formula $Al_2(OH)_4Si_2O_5$, which is used in numerous applications such as paper filler, refractories, pottery, dinnerware, and petroleum-cracking catalysts. The U.S. production of clay is several tens of million metric tons per year, and global and domestic U.S. resources are abundant.

Fluorine compounds are widely used in industry. Large quantities of fluorspar, CaF_2, are required as flux in steel manufacture. Synthetic and natural cryolite, Na_3AlF_6, is used as a solvent for aluminum oxide in the electrolytic preparation of aluminum metal. Sodium fluoride is added to water to help prevent tooth decay, a measure that is commonly called water fluoridation. World reserves of high-grade fluorspar are adequate for several decades with about 13% of the production in the United States. A great deal of by-product fluorine is recovered from the processing of fluorapatite, $Ca_5(PO_4)_3F$, which is used as a source of phosphorus.

Micas are complex aluminum silicate minerals, which are transparent, tough, flexible, and elastic. Muscovite, $K_2O \cdot 3Al_2O_3 \cdot 6SiO_2 \cdot 2H_2O$, is a major type of mica. Better grades of mica are cut into sheets and used in electronic apparatus, capacitors, generators, transformers, and motors. Finely divided mica is widely used in roofing, paint, welding rods, and many other applications. Sheet mica is imported into the United States, and finely divided "scrap" mica is recycled domestically. Shortages of this mineral are unlikely.

In addition to its consumption in fertilizer manufacture (Chapter 10, Section 10.2), phosphorus is used for supplementation of animal feeds, synthesis of detergent builders, and preparation of chemicals such as pesticides and medicines. The most common phosphate minerals are hydroxy-apatite, $Ca_5(PO_4)_3(OH)$, and fluorapatite, $Ca_5(PO_4)_3F$. Ions of Na, Sr, Th, and U are found substituted for calcium in apatite minerals. Small amounts of PO_4^{3-} may be replaced by AsO_4^{3-}, and the arsenic must be removed for food applications. Approximately 17% of world phosphate production is from igneous minerals, primarily fluorapatites. About three-fourths of world phosphate production is from sedimentary deposits, generally of marine origin. Vast deposits of phosphate, accounting for approximately 5% of world phosphate production, are derived from guano droppings of seabirds and bats. World production of phosphate is several tens of millions of tons annually with the United States, Morocco, and China among the largest producers. Although current phosphate reserves are adequate for the immediate future, their eventual exhaustion is of concern for the long-term outlook for food production.

Pigments and fillers of various kinds are used in large quantities. The only naturally occurring pigments still in wide use are those containing iron. These minerals are colored by limonite, an amorphous brown-yellow compound with the formula $2Fe_2O_3 \cdot 3H_2O$, and hematite, which is composed of gray-black Fe_2O_3. Along with varying quantities of clay and manganese oxides, these compounds are found in ocher, sienna, and umber. Manufactured pigments include carbon black, titanium dioxide, and zinc pigments. About 1.5 million metric tons of carbon black, which is manufactured by the partial combustion of natural gas, are used in the United States each year, primarily as a reinforcing agent in tire rubber.

Several million metric tons of minerals are used in the United States each year as fillers for paper, rubber, roofing, battery boxes, and many other products. In addition to the aforementioned clays, some of the minerals used as fillers are carbon black, diatomite, barite, fuller's earth, kaolin, mica, limestone, pyrophyllite, and wollastonite ($CaSiO_3$).

Although sand and gravel are the cheapest of mineral commodities per ton, the average annual dollar value of these materials is greater than all but a few mineral products because of the huge quantities involved. In tonnage, sand and gravel production is by far the greatest of nonfuel minerals. Almost 1 billion tons of sand and gravel are employed in construction in the United States each year, largely to make concrete structures, road paving, and dams. Slightly more than that amount is used to manufacture portland cement and as construction fill. Although ordinary sand is predominantly silica, SiO_2, about 30 million tons of a more pure grade of silica are consumed in the United States each year to make glass, high-purity silica, silicon semiconductors, and abrasives.

At present, old river channels and glacial deposits are used as sources of sand and gravel. Many valuable deposits of sand and gravel are covered by construction and lost to development. Transportation and distance from source to use are especially crucial for this resource. Environmental problems involved with the defacement of land can be severe, although bodies of water used for fishing and other recreational activities frequently are formed by the removal of sand and gravel.

The biggest single use for sulfur is in the manufacture of sulfuric acid. However, the element is employed in a wide variety of other industrial and agricultural products. Several tens of millions of metric tons of sulfur are produced worldwide each year. The most important sources of sulfur are sulfur recovered from sour natural gas (as H_2S) and petroleum and sulfur recovered during the processing of mineral ores, such as galena, PbS. Deposits of elemental sulfur are mostly mined with hot water (Frasch process). Supply of sulfur is no problem either in the United States or worldwide. The United States has abundant deposits of elemental sulfur, and sulfur recovery from fossil fuels as a pollution control measure provides a large source of this element.

Sodium chloride, gypsum, and potassium salts are all important minerals that are recovered as evaporites remaining from the evaporation of seawater and from brines pumped from below the ground. Sodium chloride in the form of mineral halite is used as a raw material for the production of industrially important sodium, chlorine, and their compounds. It is used directly to melt ice on roads, in foods, and in other applications. Potassium salts are, of course, essential ingredients of fertilizers and have some industrial applications as well. Gypsum, hydrated calcium sulfate, is used to make plaster and wallboard and is an ingredient in the manufacture of portland cement.

Graphite, a form of elemental carbon in flat sheetlike structures, is mined in a number of countries and is also made synthetically from carbonaceous feedstocks. Worldwide production of mined graphite is about 1 million metric tons per year. High-quality flake graphite is the most desirable natural form of graphite and has a number of applications. The use of this form of graphite is expected to rise significantly as high-energy-density lithium-ion batteries come to dominate the market for hybrid vehicles.

11.6.3 How Long Will Essential Minerals Last?

During about a 30-year period following World War II, demand for most important mineral commodities increased at a very rapid rate. Coinciding roughly with the "energy crisis" of the early to mid-1970s, demand slowed. Now, however, the emergence of newly developing economies, particularly those in the highly populous countries of China and India, is causing a greatly increasing demand for strategic minerals.

To a degree, the economic demand for and the price of a resource determine its availability. Higher prices lead to greater exploration, exploitation of less available resources, and often spectacular increases in supply. This phenomenon has led to the misinterpretation of the resource supply and demand equation by authorities whose understanding of economics is limited to only conventional monetary supply and demand models, without due consideration of sustainability aspects. The fact is that the total available amounts of most resources are limited, painfully so in terms of the time span over which they will be needed by humankind. Although higher prices and improved technologies can increase supplies of critical resources significantly, the ultimate result will be the same: the resource will run out. Furthermore, exploitation of lower and lower grades of resources results in ever-increasing environmental disruption, adding significantly to the cost considering environmental economics.

Mineral resources may be divided into several categories based on current production and consumption and known reserves. In the first category are those minerals that are in relatively comfortable supply, with supply of at least 100 to several hundred years. Minerals in this category include bauxite, the source of aluminum; iron ore; platinum-group metals; and potassium salts. In an intermediate category are minerals with a current projected lifetime supply of 25–100 years. These include chromium, cobalt, copper, manganese, nickel, gypsum, phosphate minerals, and sulfur. The most critical group consists of minerals for which the supply based on current rates of consumption and known reserves is 25 years or less. Among these minerals are sources of lead, tin, zinc, gold, and silver.

The United States is essentially without economic reserves of a number of essential minerals. These include aluminum, antimony, chromium, cobalt, manganese, tantalum, niobium, platinum, nickel, and tin. Domestic reserves of fluorine, gold, potash, silver, tungsten, sulfur, vanadium, and zinc are limited. As far as the United States is concerned, metals of most concern are chromium, manganese, and cobalt. These substances are essential for a modern industrialized economy. Although global supplies are adequate for the immediate future, they are threatened by the potential instability of the countries from which they come—Zaire, Zambia, South Africa, and Russia and other countries in the former Soviet Union.

As discussed in Chapter 9, Section 9.3, there is current concern regarding the availability of rare earth elements and the monopoly that China has developed for supplies of these elements.[5] Although they are not really very rare and potentially can be produced in several places around the world including the United States, there are some serious environmental challenges involved in the development of rare earth resources.

The world economy will never totally run out of any of the aforementioned minerals. However, severely constrained supplies of any one or several of them will have some marked effects. For example, world food production now depends on fertilizers, which require phosphorus, of which resources are limited. Within the next century, a food crisis related to phosphate shortages may be anticipated.

11.6.4 Green Sources of Minerals

One of the most crucial aspects of green technology is the sustainable utilization of minerals. In a sense, the concept of sustainable mineral utilization is an oxymoron because minerals removed from the geosphere are not replaced. However, the idea of sustainability can greatly extend supplies of minerals. This section addresses the approach to sustainability in obtaining minerals. The broader questions of green utilization of materials, substitution of materials, and recycling are discussed in more detail in Chapters 14 and 16.

As discussed in Chapters 14–16, modern technology and human ingenuity are very effective in alleviating shortages of important minerals. Applications of materials science (see Chapter 14) continue to produce substances made from readily available materials that provide good substitutes for more scarce resources. For example, concrete covered by strong layers of composite materials can often be substituted for iron in construction. Ceramics with special heat- and abrasion-resistant qualities are being used where high-temperature metal alloys were formerly required.

As minerals become less available, one of the measures to be taken is one that has been taken historically: find more. Modern technology provides a number of useful tools for finding new mineral deposits. Arguably the most useful approach to finding new mineral deposits is through the applications of geology. Recent advances in plate tectonics, for example, have contributed to the understanding of likely locations of significant mineral deposits.

Slight differences in magnetic field, gravity, and electrical conductivity can be detected very sensitively, and they reflect differences in density, magnetic properties, and electrical properties that indicate the presence of remote mineral deposits. These techniques are in the realm of **geophysical prospecting**. Another useful technique for finding mineral deposits is **geochemical prospecting**. As its name implies, this method depends on detecting the presence of chemical species, usually specific metals. In addition to finding such substances directly in rock, geochemical prospecting can reveal evidence of minerals in water some distance from the mineral sources. Even gas analysis can be indicative of some minerals, such as volatile mercury or sulfur compounds from sulfur deposits. Plants, particularly those that concentrate some elements, such as copper, can be analyzed to indicate the presence of minerals, a process called **biogeochemical prospecting**.

Photography and the measurement of light and infrared radiation from aircraft and from satellites have greatly increased human understanding of Earth and its resources. These **remote sensing** techniques make it unnecessary to go to remote, poorly accessible, and dangerous regions. With satellite measurements, it is possible to cover huge areas of Earth's surface in a reasonable period of time. The most ambitious program of remote sensing for mineral exploration is the Landsat satellite system, which was first launched in 1972 and subsequently followed by other launches from the United States and other countries. The sensors on these satellites measure visible and infrared radiation. The Landsat images reveal numerous features of Earth's surface, including abundance and types of vegetation, soil and rock type, and moisture. Such features in turn may reflect the presence of various kinds of mineral deposits with potential for exploitation. An interesting aspect of satellite-based minerals exploration is geobotany in which, for example, trees poisoned by heavy metals indicative of metal deposits may be observed. More subtle changes, such as the timing by which deciduous trees start to produce leaves in the spring and change color and lose leaves in the fall, may provide clues to mineral deposits.

11.6.5 EXPLOITATION OF LOWER GRADE ORES

Modern technology enables the exploitation of lower grade ores, thus significantly increasing supplies. A striking example of this phenomenon has been provided by copper. About 100 years ago, the average copper content of ore mined in the United States was around 5%; now, it is only about one-tenth that figure or less for copper ore mined globally. Despite the decline in copper ore quality, during the latter half of the 1900s known copper reserves increased about fivefold and, adjusted for inflation, the price of copper became less than it was a century earlier (although for about the last decade it has increased rapidly because of strong demand from developing economies).

The ability to exploit much less rich sources of ores has resulted from improved technologies. Of particular importance are advances in the means of moving huge quantities of rock, which is essential for the exploitation of lower grade ores. Earthmoving equipment has greatly increased in size and versatility during the last several decades. There has been an environmental cost, of course, for these advances. As an approximation, for each 10-fold decrease in mineral content, it is necessary to move 10 times as much material to obtain the same amount of metal. In addition to disruption of land, disturbed material is more prone to erosion, landslides, and water pollution. Much more energy is required, as is more water for those mining operations that use large quantities of water. Not the least of the factors required for exploiting lower grade resources is the need for additional capital and operating investment, which may be in short supply.

All the rich mineral ores in readily accessible areas have already been found and exploited. Therefore, any rich deposits will be found in remote locations and hostile environments, such as locations deep under the ocean. One major possibility is Antarctica, a remote continent noted for its ice, wind, and generally hostile conditions. It is very likely that rich mineral deposits are buried beneath the thick Antarctic ice sheet. However, the probability of severe environmental damage from extracting minerals there is very high even if the extreme climate conditions can be overcome. In recognition of this concern, 26 nations involved in Antarctic exploration signed a treaty in 1991 banning mineral extraction for 50 years. However, if shortages of crucial minerals become severe, it is likely that efforts will be made to find and extract them in Antarctica.

Exciting possibilities exist for the extraction of minerals from very hostile places, especially at great depths. One possibility is ultradeep mining under conditions too severe to enable human participation. It may one day be possible to use robots to mine deposits several kilometers deep, where extreme pressures and heat would make it impossible for humans to work.

11.6.6 MINING THE OCEAN FLOORS

Another potentially abundant remote source of minerals is the ocean floor, which remains largely unexplored to this day. Compared to space exploration, only small amounts of money and time have been devoted to the exploration of the ocean floor. Despite their vast unrealized potential as a source of natural capital, relatively little has been done to take advantage of ocean-floor resources.

The potential for developing deep sea resources has been greatly increased in recent years by the development of remotely operated vehicles. With probes, robots, and sensors controlled by computers from the surface, these unmanned submarines are capable of operating in the cold, extreme-pressure regions of ocean depths, opening new possibilities for exploration and resource utilization. It may be anticipated that in the future remotely operated mining equipment will be developed that will permit the mining of minerals on the ocean floor.

Large areas of the ocean are covered with manganese-rich lumps called **manganese nodules**.[6] In addition to manganese, these lumps also contain other metals, including valuable platinum, copper, and nickel. Extraction of these metals as by-products adds to the economic attractiveness of mining manganese nodules.

11.6.7 WASTE MINING

Waste mining is the term given to the extraction of useful materials from waste streams. Sulfur is one of the best examples of waste mining. Technology has been developed for the removal of sulfur from flue gas in nonferrous metal smelters. The recovery of sulfur from the smelting of lead, copper, zinc, and other metals also provides an example of policy-driven waste mining because it is mandated by law in most developed countries. Waste mining is employed to recover some metals as part of the production of other metals. By-product metals can be recovered from the gangue remaining from beneficiation of ore, slag from smelting, or dust collected from flues in metal refining operations. Arsenic and cadmium are recovered from the production of copper and zinc, respectively. Coal ash is a huge untapped resource for the waste mining of aluminum and ferrosilicon. Factoring in the costs of waste disposal and potential environmental degradation can make the economics of waste mining relatively more attractive. Some caution is suggested in that policy-driven waste mining of some substances creates a need to market them, sometimes to the detriment of the environment. Cadmium and arsenic are both examples of substances recovered from waste mining that should not be used any more than necessary because of their toxicities.

Related to waste mining is the utilization of scrap materials, especially scrap metal. There are two major categories of scrap: **new scrap** consists of materials that are reclaimed during the manufacture of an item, such as metal shavings from machining. **Old scrap** consists of material that has been in products used in the consumer market and reclaimed as scrap material. The quality of new scrap can be carefully controlled, and it can be reclaimed very quickly and efficiently. However, old scrap has a recycling time that depends on the life of the product in which it is contained, and it is difficult to control its quality. Furthermore, the percentage return of material is lower from old scrap because of the products that are discarded and not recycled.

The anthrosphere provides a large reservoir of materials that eventually can be recycled. A prime example of such a material is that of copper contained in copper wiring and plumbing in buildings, electrical lines, and other anthrospheric structures. The relatively high price of copper metal makes it attractive to reclaim the metal from these sources, and copper is commonly stolen from unoccupied buildings, including even buildings under construction. Another example is the large amount of iron contained in vehicles, machinery, rail lines, bridges, and other artifacts.

Landfills constitute another anthrospheric construct that contains large amounts of materials, including metals. Unfortunately, the usable materials in landfills are usually too dispersed and mixed with materials of no value to be reclaimable. Any metals that are put in landfills should be put in segregated areas from which they may later be reclaimed.

11.6.8 RECYCLING

Recycling should be practiced for all major mineral commodities. Fortunately, both economic and environmental concerns have resulted in vastly increased efforts to recycle materials in recent years. The largest quantity of metal that is recycled consists of ferrous metals (iron). During the last 30 years, electric arc furnaces for iron processing have become commonplace. Fortunately, these devices require scrap iron as feedstock and have resulted in a continuing market for recyclable iron scrap. Aluminum ranks next to iron in quantity of metal recycled; at least one-third of aluminum is recycled in the United States and globally. Particular success has been achieved in the recycling of aluminum beverage cans. The refining of aluminum metal from bauxite ore is particularly energy consumptive, so a big advantage of aluminum recycling is reduced energy consumption. In addition, recycling produces only about 5% the amount of wastes and potential pollutants as are generated by refining aluminum from its ore. Cost savings are huge as well. Other metals that are largely recycled are copper and copper alloys; cadmium; lead; tin; mercury; zinc; silver; and, of course, gold and platinum.

A crucial consideration in recycling is the nature of source material. Copper is relatively easy to recycle because it is often found in a relatively pure form in wire, pipe, and electrical apparatus. The lead in lead storage batteries can simply be melted down and recast into battery electrodes. Large amounts of aluminum are available from waste cans and structural materials. Although iron is largely recycled, it often occurs as specialized alloys containing varying contents of other metals, such as titanium or tungsten. The contents of these elements complicate the utilization of scrap iron.

11.7 TOXICOLOGICAL IMPLICATIONS OF MINERAL MINING AND PROCESSING

A potential toxicological hazard from fluoride may be caused by the processing of fluorapatite mineral, $Ca_5(PO_4)_3F$, to make superphosphate fertilizer and to produce phosphorus compounds. The reaction of fluorapatite with sulfuric acid to produce superphosphate is as follows:

$$2Ca_5(PO_4)_3F + 7H_2SO_4 + 3H_2O \rightarrow 7CaSO_4 + 3Ca(H_2PO_4)_2 \cdot H_2O + 2HF \qquad (11.2)$$

Toxic hydrogen fluoride, HF, is a product of this reaction as a gas or an aqueous solution of hydrofluoric acid. Hydrogen fluoride is so reactive that it is used to etch glass. Both the gas and the acid are extremely corrosive irritants, causing potentially gangrenous lesions on exposed tissue and ulcers in the upper respiratory tract. Although beneficial in preventing tooth decay when added to water at levels of around 1 ppm, fluoride ion, F^-, at higher levels can cause **fluorosis**, a condition that is characterized by bone abnormalities and mottled teeth. Pollution of grazing land by fluoride fallout can harm livestock, causing lameness and even death in extreme cases. Obviously, the control of HF and fluoride exposure is very important in the processing of fluorapatite and other fluoride minerals.

11.7.1 PNEUMOCONIOSIS FROM EXPOSURE TO MINERAL DUST

Pneumoconiosis is the term given to lung disease caused by the inhalation of mineral dust. The cause of this malady is often exposure to dusts in mining and processing minerals.

The most common cause of pneumoconiosis is **silicosis**, a chronic fibronodular lung disease manifested by fluid buildup and scar tissue in lungs accompanied by reduced pulmonary function resulting from exposure to silica, SiO_2. Sand, sandstone, slate, and granite all contain silica, and occupational exposure to dusts from these minerals has caused silicosis. Silicosis has been known for centuries and was called potters rot in Victorian England. The introduction of mechanical mining that produced copious quantities of silica dust from drilling and rock-cutting operations greatly increased the incidence of silicosis, but effective industrial hygiene measures have reduced the number of workers afflicted by this disease.

Black lung disease is a form of pneumoconiosis caused by inhalation of coal dust that has afflicted many coal miners. Although the silica in coal dust contributes to black lung disease, it is a separate condition resulting from exposure to carbonaceous material in coal. Black lung disease is also caused by exposure to other carbonaceous dusts including graphite dust in the mining and processing of natural graphite. Early-stage black lung disease, simple pneumoconiosis, does not seriously impair those who are afflicted. It can progress to a debilitating form called complicated pneumoconiosis or progressive massive fibrosis. There is no specific treatment for black lung disease, and the condition is not reversible.

11.7.2 HEAVY METAL POISONING

There are numerous potential toxicological effects of mining, processing, and recycling metals. The materials of concern include heavy metals, especially lead, and other materials such as silica dust and coal dust. Most exposures are through inhalation. Lead has adverse effects on many parts of the body including the central nervous system, kidney, and blood and may cause developmental problems in children. Due to lead's toxicity to the nervous system, a condition called lead palsy used to be a common affliction of lead miners and lead smelter workers.

There is serious concern about lead poisoning among workers and others exposed to the metal in recycling operations in Mexico. Because of the less stringent regulations pertaining to occupational lead exposure in Mexico compared to the United States, it has become a common practice to send old lead storage batteries to Mexico from the United States for recycling. In the United States, lead battery recycling operations are carried out in highly mechanized sequestered plants in which the stacks from lead smelting are fitted with effective scrubbers and lead emission monitors, but as of 2011 such safeguards allegedly had not been in place in Mexico. According to a report in the *New York Times*,[7] recycling operations commonly occur in poor neighborhoods where workers dismantle old batteries by beating on their cases with hammers and the scrap lead is melted in furnaces lacking adequate stack emission controls. A major concern is lead fallout onto nearby residential areas and schools. In 2011, about 20 million old lead batteries, each containing about 18 kg of lead, were sent from the United States to Mexico for recycling.

Occupational exposure to zinc oxide fumes can cause **metal fume fever**, which is manifested by fever, chills, fatigue, and other flu-like symptoms. The condition is reversible with recovery taking place usually within a few hours after exposure to the fumes has stopped. Although the most common cause of zinc metal fume fever is from exposure during welding and metal cutting of zinc-coated (galvanized) steel, it may also occur during processing of zinc ore and zinc recycling operations. Mining, processing, and recycling of cadmium and mercury also presents hazards from these highly toxic metals, but they are used much less than lead and zinc.

11.8 SUSTAINING THE GEOSPHERE TO MANAGE WATER

Some of the most extensive modifications of the geosphere are those undertaken to manage water. The interaction of water with the geosphere is important and is addressed by the science of **hydrogeology**. Water is encountered both on top of the geosphere in streams, lakes, and reservoirs and beneath the geosphere surface as groundwater contained in aquifers. This section addresses human efforts to manage water in the geosphere other than saline water in oceans and seas.

Most water falls on the geosphere as precipitation (rain, snow, sleet) from the atmosphere. Much of this water flows along the surface in streams; it may be held in soil, evaporate into the atmosphere, or infiltrate into the ground. Water falling on the surface of the geosphere may infiltrate downward; if it reaches a zone of saturation, the process is called percolation. Percolation is an important process by which groundwater recharge of underground aquifers occurs. Groundwater recharge is essential to maintaining crucial underground water resources. Paving land in urban areas is detrimental to groundwater recharge, and in some cases artificial recharge is employed to pump water into aquifers. In some places, paving materials are now made of permeable solids that allow surface water to penetrate and eventually recharge groundwater reservoirs.

An important effect of the hydrosphere on the geosphere consists of flooding from streams and rivers. A flood occurs when a stream develops a high flow such that it leaves its banks and spills out onto the floodplain of the stream. Floods are arguably the most common and damaging of surface phenomena in the geosphere. Although they are natural and in many respects beneficial occurrences, floods cause damage to structures located in their paths, and the severity of their effects is greatly increased by human activities.

Floods are made more intense by higher fractions and higher rates of runoff, both of which may be aggravated by human activities. This can be understood by comparing a vegetated drainage basin to one that has been largely denuded of vegetation and paved over. In the former case, rainfall is retained by vegetation, such as grass cover. Thus, the potential floodwater is delayed, the time span over which it enters a stream is extended, and a higher proportion of the water infiltrates into the ground. In the latter case, less rainfall infiltrates, and the runoff tends to reach the stream quickly and to be discharged over a shorter time period, thus leading to more severe flooding. These factors are illustrated in Figure 11.5.

The conventional response to the threat of flooding is to control a river, particularly by the construction of raised banks called levees. In addition to raising the banks to contain a stream, the

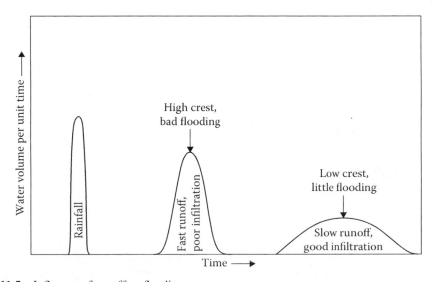

FIGURE 11.5 Influence of runoff on flooding.

stream channel may be straightened and deepened to increase the volume and velocity of water flow, a process called channelization. Although effective for common floods, these measures may exacerbate extreme floods by confining and increasing the flow of water upstream such that the capacity to handle water downstream is overwhelmed.

One of the most common human modifications of the geosphere to manage water is the construction of reservoirs in which flowing water is retained for water supply, for recreational areas, to generate hydroelectric power, and for flood control. Figure 11.6 shows the basic components of a reservoir. It is seen that a reservoir is constructed when a dam is placed across a stream. The streambed and parts of the valley occupied by the stream compose the bed of the reservoir. The dam contains a sluiceway through which water from the reservoir can be drained to a very low volume, or the sluiceway may serve as a conduit to hydroelectric turbines attached to generators for hydroelectric power. Somewhat below the top of the dam is a spillway through which excess water automatically overflows when the water level in the reservoir reaches that of the spillway. Water flowing over the top of a dam, particularly an earthen one, can destroy it rather quickly and is to be avoided. In addition to water held behind the dam, water in bank storage in rock formations along the edge of the reservoir may flow into it as the water level is lowered, adding to the useful storage volume of the reservoir.

Many factors must be considered in constructing a dam to make a reservoir. These include stream flow, rainfall, geologic conditions, and topographic conditions. Ideally, dams are constructed in locations where the stream is constricted and where the banks are high and steep so that a small dam will impound a large reservoir volume. It is particularly important to have relatively watertight reservoir walls so that there is no excessive loss of water. One of the major sources of dam failure has been dam construction in regions where underground channels exist through which leakage can take place from the reservoir to below the dam. Eventually, such leakage can erode the earth on which the dam rests, leading to its collapse. In some cases, channels and voids can be filled with fluid cement (grout) prior to dam construction.

An often-cited case of reservoir failure occurred with the Baldwin Hills Reservoir near Los Angeles, California, in 1963. This rectangular basin had been carved on top of a hill in 1951 and rested on poorly consolidated earthen material consisting primarily of silt, sand, and clay. Pumping of oil in the vicinity had led to ongoing settling of the land. Movement of faults beneath the reservoir dam led to an abrupt loss of water, erosion of material from beneath the dam, and an abrupt breach of the dam. The sudden rush of water from the reservoir that resulted killed 5 people and caused $15 million of damage.

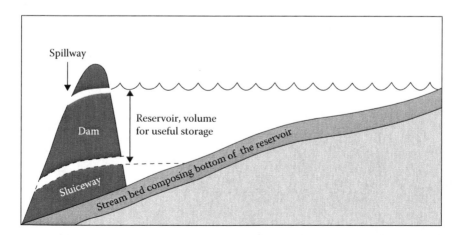

FIGURE 11.6 A dam placed across a stream impounds water in the stream valley above the dam, producing a reservoir. The maximum level of water in the reservoir is at the level of the spillway, and it can be lowered to the level of the sluiceway.

As discussed in Chapter 3, Section 3.3, bodies of water tend to become stratified, with significant effects on aquatic chemistry. Impoundment of water in reservoirs leads to stratification such that reduced chemical species predominate in the bottom hypolimnion layer of the water. This can cause the production of relatively high concentrations of soluble Fe^{2+} and Mn^{2+} species, which can be undesirable water contaminants. Toxic hydrogen sulfide, H_2S, and methane, CH_4, a gas that can contribute to global warming, are also produced in reservoir sediments.

11.8.1 CHINA'S THREE GORGES DAM PROJECT

One of humankind's most ambitious efforts to control water and a truly vast project of construction on the geosphere is China's Three Gorges Dam project on the Yangtze River. The dam was completed in 2006 followed by construction of additional power stations associated with it. This project has set numerous world records including largest dam and largest single power plant project in one location. Its construction involved the greatest consumption of dirt, concrete, and stone of any construction project ever completed. About 1.13 million people were displaced by the reservoir created by the dam. Since its construction, the project has controlled floods that annually caused much misery and destruction along the Yangtze and has added an enormous amount of clean hydroelectric power–generating capacity to China's electrical power grid. But it has caused significant environmental problems. Water pressure from the 175-m-deep reservoir is thought to have affected surrounding geological strata and contributed to landslides around the shoreline. Sediment accumulation in the reservoir is a problem. Many of the environmental problems have resulted from the activities of the largely rural population that was displaced to higher ground. Farmers have attempted to cultivate the steeply sloping land and have stripped it of forests. This has increased problems from landslides, causing the Chinese government to undertake reforestation programs.

11.8.2 WATER POLLUTION AND THE GEOSPHERE

Water pollution is addressed in detail in Chapter 4. Much water pollution arises from interactions of groundwater and surface water with the geosphere. These aspects are addressed briefly here.

The relationship between water and the geosphere is twofold. The geosphere may be severely damaged by water pollution. This occurs, for example, when water pollutants produce contaminated sediments, such as those polluted by heavy metals or polychlorinated biphenyls. In some cases, the geosphere serves as a source of water pollutants. Examples include acids produced by exposed metal sulfides in the geosphere or synthetic chemicals improperly discarded in landfills.

The sources of water pollution are divided into two main categories. The first of these consists of **point sources**, which enter the environment at a single, readily identified entry point. An example of a point source is a sewage-water outflow. Point sources tend to be those directly identified as to their origins from human activities. **Nonpoint sources** of pollution are sources from broader areas. Such a source is water contaminated by fertilizer from fertilized agricultural land or water contaminated with excess alkali leached from alkaline soils. Nonpoint sources are relatively harder to identify and monitor. Pollutants associated with the geosphere are usually nonpoint sources.

An especially common and damaging geospheric source of water pollutants consists of sediments carried by water from land into the bottoms of bodies of water. Most such sediments originate from agricultural land that has been disturbed such that soil particles are eroded from land into water. The most common manifestation of sedimentary material in water is opacity, which seriously detracts from the esthetics of the water and inhibits the growth of light-requiring algae. Sedimentary material deposited in reservoirs or canals can clog them and eventually make them unsuitable for water supply, flood control, navigation, and recreation. Suspended sediments in water used for water supply can clog filters and add significantly to the cost of treating the water. Sedimentary material can devastate wildlife habitats by reducing food supplies and ruining nesting sites. Turbidity in water can severely curtail photosynthesis, thus reducing primary productivity necessary to sustain the food chains of aquatic ecosystems.

11.9 WASTE DISPOSAL AND THE GEOSPHERE

The geosphere receives many kinds and large amounts of wastes. Its ability to cope with such wastes with minimal damage is an important part of the geosphere's natural capital. A variety of wastes, ranging from large quantities of relatively innocuous municipal refuse to much smaller quantities of potentially lethal radioactive wastes, are deposited on land or in landfills. These are addressed briefly in Sections 11.9.1 and 11.9.2.

11.9.1 MUNICIPAL REFUSE

Currently, the most widely practiced method for disposing of municipal solid wastes—household garbage—is in **sanitary landfills** (Chapter 9, Section 9.7), which consist of refuse piled on top of the ground or into a depression such as a valley, compacted and covered at frequent intervals by soil (see Figure 9.10). An important aspect of sustaining the geosphere is to construct and operate municipal landfills so that they are of net environmental benefit. Some landfills are equipped for the collection of methane gas fuel from the anoxic biodegradation of wastes, a practice that is certainly in keeping with sustainability. Because of production of combustible gases, settling, and surface instability, closed landfill sites are generally not suitable for building construction but can be attractive sites for parks, forests, or even gardens and agricultural land.

11.9.2 HAZARDOUS WASTE DISPOSAL

Hazardous chemical wastes are disposed of in so-called **secure landfills**. Equipped with a variety of measures to prevent contamination of groundwater and the surrounding geosphere, they are designed to prevent leakage and geospheric contamination of toxic chemicals disposed in them. The base of the landfill is made of compacted clay that is largely impermeable to leachates. An impermeable polymer liner is placed over the clay liner. The surface of the landfill is covered with material designed to reduce water infiltration, and the surface is contoured with slopes that also minimize the amount of water running in. Elaborate drainage systems are installed to collect and treat leachates.

The best way to manage hazardous waste disposal to the geosphere is to render all wastes nonhazardous before disposal. Unlike radioactive wastes that eventually decay (see later in this section), some hazardous chemical wastes never degrade and no secure landfill can be assumed to be secure forever. Organic wastes, such as refractory organochlorine compounds, should be destroyed by thermal process, and toxic heavy metals should be separated and recycled.

One of the more pressing matters pertaining to geospheric disposal of wastes involves radioactive wastes. Most of these wastes are **low-level** wastes, including discarded radioactive laboratory chemicals and pharmaceuticals, filters used in nuclear reactors, and ion-exchange resins used to remove small quantities of radionuclides from nuclear reactor cooler water. When disposed of in properly designed landfills, such wastes pose minimal hazards.

Of greater concern are **high-level** radioactive wastes, primarily fission products of nuclear power reactors and by-products of nuclear weapons manufacture. Many of these wastes are currently stored as solutions in tanks, many of which have outlived their useful lifetimes and pose leakage hazards, at sites such as the federal nuclear facility at Hanford, Washington, where plutonium was generated in large quantities during the post–World War II years. Eventually, such wastes must be placed in the geosphere such that they will pose no hazards. Numerous proposals have been advanced for their disposal, including disposal in salt formations, subduction zones in the seafloor, and ice sheets. The most promising sites appear to be those in poorly permeable formations of igneous rock. Among these are basalts, which are strong, glassy igneous types of rock found in the Columbia River plateau. Granite and pyroclastic welded tuffs fused by past high-temperature volcanic eruptions are also likely sites for disposing of nuclear wastes and keeping them isolated for tens of thousands of years.

11.10   DERELICT LANDS AND BROWNFIELDS

Derelict lands, commonly called **brownfields**, consist of properties that have been damaged by anthrospheric activities and are generally unsuitable for further use without restoration.[8] Often, such lands are contaminated with potentially hazardous substances and require cleanup and decontamination. Generally, brownfields are the result of abandoned industrial enterprises and mining. In the latter case, subsidence into voids excavated underground can be a major problem. Chemical contamination can result from industrial activity and mining.

Vast amounts of land are covered by brownfields that were formerly sites of enterprises such as factories, mills, quarries, petroleum refineries, trucking depots, and rail yards. The Environmental Protection Agency estimates that there are 450,000 brownfield sites in the United States, with many more in other countries. It has been estimated that England has approximately 66,000 ha of brownfield land. Some of the most challenging brownfield sites are located in nations of the former Soviet Union where the collapse of former state-owned manufacturing enterprises combined with a disregard of environmental considerations have given rise to many impaired areas with severe contamination problems. Brownfields present both problems and opportunities. Abandoned industrial sites at best are eyesores and at worst pose real problems with respect to pollution and dispersion of toxic substances to groundwater and surrounding areas. However, they are usually centrally located near population centers with good access to rail lines, highways, and utilities, thereby providing excellent potential development opportunities. In recognition of these factors, in 1995 the U.S. Environmental Protection Agency initiated a brownfield program to encourage cleanup and conversion to beneficial purposes of abandoned contaminated sites. This program provides funds for cleanup and limits liability, particularly in the event of discovery of additional hazardous contamination during cleanup. The program was strengthened by legislation in 2002, which provided additional funds for cleanup and further limited liability.

Restoration of brownfields may entail physical remediation as well as treatment or removal of chemical contamination. Often, particularly where they are located at abandoned mine sites, brownfields are afflicted by subsidence. In some cases, concrete and stone salvaged from old structures and their foundations can be used as fill to treat subsidence. Some kinds of chemical contamination can be treated in place. If excavation and disposal of contaminated soil is required, the costs of brownfield restoration may increase dramatically.

In recent years, a strong driving force behind brownfields development has been greatly increased real estate prices. Therefore, it has become attractive to construct dwellings and commercial developments on brownfield sites. Housing developments for private homes have been constructed on renovated brownfields, although such sites are usually more attractive for higher density developments such as condominiums and apartment complexes. Many people are willing to trade the open spaces and greenery of suburbs for the convenience and much shorter commutes offered by more concentrated developments closer to urban centers. Ideally, such developments should conform to the standards of **smart growth** defined by the Urban Land Institute as development that is environmentally sensitive, economically viable, community oriented, and sustainable.

11.10.1   LAND RESTORATION FROM THE FUKUSHIMA DAIICHI NUCLEAR ACCIDENT

As of 2012, one of the greater land cleanups ever undertaken was that associated with the release of radioactivity from the Fukushima Daiichi Nuclear Power Station, which was the consequence of the March 11, 2011, earthquake off the coast of Honshu, Japan, and the resulting tsunami that destroyed the power station (see Chapter 4, Section 4.14). The greatest problem with radioactivity from this incident has come from cesium-137, a fission product with a 30-year half-life. As a result of the Fukushima disaster, an area within a radius of 20 km from the damaged reactors was evacuated as was another area beyond this radius afflicted by a plume of radioactive fallout from the reactors. Radiation levels within the evacuated zone have been measured at up to 510 mSv/year, about 25 times the level at which evacuations are required. (A millisievert is a measure of radiation damage to tissue. The average person in the United States receives about 6 mSv/year radiation dose from natural sources, and a limit of 1 mSv/year is recommended from artificial sources.) The cleanup in Japan includes removal and burial of topsoil in an area potentially as large as that of the U.S. state of Connecticut. A major barrier to this cleanup has been the lack of places willing to accept the contaminated soil.

11.11 SUSTAINING SOIL

The most crucial part of the geosphere to sustain is soil, the thin layer on the surface of much of the geosphere that supports the plant life on which humans depend for their sustenance. As discussed in Chapter 10, Section 10.4, the thin layer of soil is fragile and much of it has been lost to erosion (see Figure 10.3). In this section, we discuss some means of sustaining this precious resource, preventing erosion, reversing the desertification process by which soil is degraded to unproductive desert, and reversing deforestation.

The key to preventing soil loss from erosion as well as preventing desertification lies in a group of practices that agriculturists term **soil conservation**.[9] A number of different approaches are used to retain soil and enhance its quality. Some of these are old, long-established techniques such as construction of terraces and planting crops on the contour of the land (see Figure 11.7). Crop rotation and occasional planting of fields to cover crops, such as clover, are also old practices. A relatively new practice involves minimum cultivation and planting crops through the residue of crops from the previous year. This practice, now commonly called **conservation tillage**, is very effective in reducing erosion because of the soil cover of previous crops and the roots that are left in place. Conservation tillage makes minimum use of herbicides to kill competing weeds until the desired crop is established enough to shade out competing plants. There is some concern that fungi (molds) will thrive in old crop residues and cause problems with new crops.

The ultimate in no-till agriculture is the use of perennial plants that do not have to be planted each year. Trees in orchards and grapevines in vineyards are certainly such plants. The roots of perennial plants are very effective in holding plants in place. Efforts to develop perennial plants that produce grain have had only limited success so far. This is because a productive grain-bearing plant

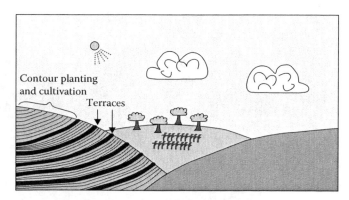

FIGURE 11.7 Construction of terraces on the contour of land and planting crops on the contour are practices that have been very effective in reducing soil erosion.

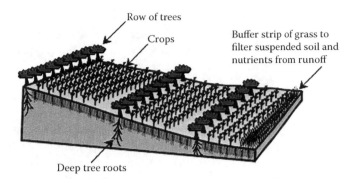

FIGURE 11.8   Alley cropping of crops between rows of trees running across sloping land can be an effective means of practicing agroforestry sustainably.

is one that dedicates its metabolic processes to the production of large quantities of seed that can be used for grain, whereas perennial plants put their energy into the development of large, bulbous root structures that store food for the next growing system. It is possible that sometime in the future genetic engineering may be applied to the development of perennial plants capable of producing high grain yields.

An interesting application of perennial plants (trees) to sustainable agriculture is **agroforestry** in which crops are grown in strips between rows of trees, "alley cropping across the slope" as shown in Figure 11.8. The trees stabilize the soil, particularly on a sloping terrain. By choosing trees with the capability to fix nitrogen, the system can be partially self-sufficient in this essential nutrient. Fast-growing, nitrogen-fixing trees hold the sloping soil in place. In between crop seasons, the trees are pruned and the nutrient-rich prunings are spread on the soil where crops are grown, fertilizing the soil, adding organic matter, and holding the soil in place. The trees potentially have economic value in providing wood for construction, firewood for cooking, fruits, and nuts. At the bottom of the slope, a buffer strip of grass can serve to filter nutrients and suspended soil from runoff from the fields. Potentially, rich topsoil collected by the buffer strip can be returned to higher levels to enrich the soil.

11.11.1   BIOCHAR FOR SOIL CONSERVATION AND ENRICHMENT

An interesting measure that may be taken to enhance soil quality is the use of biochar as a soil amendment. **Biochar** is made by the pyrolysis of organic matter such as crop residues. It is a natural constituent of soil as the result of forest and prairie fires, but it is now produced artificially for addition to soil.[10] The two major advantages of biochar are its high affinity for nutrients by adsorption and its extremely high persistence; unlike humic material, it never degrades. Furthermore, the production of elemental carbon in biochar has the net effect of permanently sequestering atmospheric carbon dioxide fixed by photosynthesis, thus helping to alleviate global warming.

11.11.2   REVERSING DESERTIFICATION

As discussed in Chapter 10, Section 10.4, **desertification** is a process in which vegetation is removed from once-fertile land; streams and groundwater sources dry up; and the atmospheric, terrestrial, and living environments assume characteristics of desert conditions. There are several causes of desertification. The most troubling potential cause is greenhouse warming. "Slash-and-burn" agriculture by which trees and other vegetation are stripped from rain forests for short-term production of pasture and agricultural land is currently a major contributor to desertification. Excessive grazing is ruinous to soil and pushes land along the path to desert formation. Desertification has a strong tendency toward positive feedback, which means that it feeds on itself. Decreased plant cover leads to erosion and rapid loss of water from soil, which in turn further decreases the capacity of land to support plant life. Obviously, desertification is a major environmental problem that must be dealt with firmly and vigorously if Earth is to sustain its present populations.

Fortunately, human ingenuity and technological tools can be used to prevent or reverse desertification. One of the first points to recognize with respect to desertification is that deserts are the natural state of much of Earth's surface area. Some substantial problems have occurred because of humankind's insistence on trying to convert desert lands to cropland, lawns, golf courses, and other applications unsuitable for regions in which deserts occur naturally.

In areas where land has been converted to desert from grassland or cropland, it is important to stop doing the things that cause desertification. This means, particularly, not cultivating land that should not be cultivated and not overgrazing land with ruminant animals.

In the United States alone, several million acres of once-productive rangeland have deteriorated to desert conditions by overgrazing so that **rangeland restoration** is the most important countermeasure to desertification. A key aspect of rangeland restoration is to remove cattle and other grazing animals from rangeland for periods of up to several years to give grasses and other native plants the opportunity to be reestablished. Another helpful measure is to plant riparian (stream bank) areas with hardy trees, such as aspens, cottonwoods, willows, and fast-growing hybrid willows, in order to stabilize streams and reduce erosion.

Once rangeland restoration is underway, it is critical to manage the grazing for maximum sustainability. This requires consideration of the feeding and social habits of livestock. For example, allowing pasturage from May to mid-July enables utilization of bunchgrass when it is most palatable. After mid-July, cattle tend to congregate in the cool shade under trees and to feed on new willow and aspen sprouts and seedlings, which has the undesirable effect of preventing the propagation of these trees. By removing cattle in mid-July, new growth of trees is allowed and bunchgrass growth is restored by fall.

Perennial grasses are preferable to annual grasses for rangeland restoration. In general, perennial grasses are more productive, allow for longer grazing and haying seasons, and are better for weed control than annual grasses. Prolonged growth of perennial grasses generally improves soil quality. Whereas the root zones of annual grasses normally extend to depths of only about 20 cm, the roots of perennial grasses can extend to 10 times this depth. This enables perennial grasses to recapture nutrients and water that have leached to greater depths.

Broad-leaved flowering plants called **forbs** are also important in rangeland restoration and are important indicators of rangeland health. Although forbs are often considered to be weeds, many are legumes that harbor nitrogen-fixing bacteria and many species serve as forage plants for range animals. Of the many common forbs, one interesting one is goats rue (*Tephrosia virginiana*), a legume that is a source of the natural pesticide rotenone used by Native Americans to kill fish.

In the United States, a program that can significantly reduce the deterioration of marginal agricultural areas tending toward desertification is the U.S. Department of Agriculture's Conservation Reserve Program. At its peak, this program has included 34 million acres—an area equal to that of New York State—of land in Kansas, North Dakota, eastern Washington, and other agricultural states in which landowners are paid to simply leave usually highly erodable land alone to revert to natural vegetation. The program costs about $2 billion per year, which is about 8% of total agricultural subsidies in the United States but less than what the subsidies would be on the same land for crop production. The program has been highly beneficial to wildlife, and it is estimated that it produces 2 million additional ducks per year by providing habitats and safe nesting regions. Farmers in the program have made additional income by selling hunting rights on their reserve lands. One criticism has been that prohibitions on grazing have led to the predominance of just a few kinds of very tall grass, whereas grazing would allow much more diverse populations of shorter plants. Another problem is with invasive trees, and it has been proposed that such lands should be burned every 3 years to prevent the growth of undesirable trees.

In addition to more "natural" approaches to combat desertification, such as rangeland restoration described earlier in this section, there are potentially more proactive measures that humans can take to restore desert regions. Modifications of terrain to construct terraces and water impoundments can be performed to conserve water. Mulching with biomass can be used to improve soil quality. One interesting approach has been to spread biosolids from municipal wastewater treatment onto desertified land to add biomass and promote plant growth, although the generally large distances from biosolids production sources to receptor lands are a deterrent to this measure. Nitrogen, phosphorus, and potassium fertilizers can be applied to promote the growth of ground cover. It may be possible to genetically engineer plant species that are particularly well adapted to colonizing impaired desert soil in order to establish conditions conducive to the later growth of other plant species.

11.11.3 REFORESTATION

Reforestation refers to the establishment of tree growth in areas from which trees have been removed for lumber or to clear agricultural land. Reforestation can occur naturally or by human intervention. A common practice in forest areas in which trees have been harvested is to leave some trees that can provide seeds to start new trees. A more proactive approach may be applied in which seedlings are planted deliberately. In such cases, small amounts of plant fertilizer may be added to the soil around the seedlings' roots to promote early tree growth and establishment of a strong root system that can enable the young trees to survive later drought conditions. In some cases, it may be desirable to plant improved varieties of trees including even fast-growing, hardy hybrids. One criticism of deliberate tree planting has been that it may contribute to a lack of biodiversity in the new forest stands. This may be overcome by planting several varieties of trees. Also, many decades of forest fire prevention has a tendency to reduce biodiversity because of the predominance of just one or two kinds of very large, old trees, and reforestation provides the opportunity to introduce additional biodiversity.

Reforestation is most advantageous when performed as soon as possible after trees have been harvested. There are several reasons why this is true. Trees planted on freshly harvested land can take advantage of nitrogen fixed in soil by organisms before it is lost by denitrification processes. Another possible advantage is that there is a relatively rapid rate of biodegradation in newly cleared soil and seedlings can benefit from the carbon dioxide given off near the soil surface. Also, particularly in rapidly growing tropical forests, the rate of atmospheric carbon dioxide sequestration is more rapid during the early years of tree growth.

Reforestation is important in preserving the geosphere, especially topsoil. Tree root systems anchor topsoil in place and prevent its loss by erosion due to running water. Tree cover also tends to break up rainfall and slow the impact of driving rain, which can be erosive.

11.11.4 WATER AND SOIL CONSERVATION

Conservation of soil and conservation of water go together very closely. Soil is normally the first part of the geosphere that water contacts, and contaminated soil yields contaminated water. Most freshwater falls initially on soil, and the condition of the soil largely determines the fate of the water and how much water is retained in a usable condition. Soil in a condition that retains water allows rainwater to infiltrate into groundwater. If water drains too rapidly from soil, the soil erodes and the water runoff is badly contaminated with soil sediments. Measures taken to conserve soil usually conserve water as well. Terraces, contour cultivation, constructed waterways, and water-retaining ponds prevent water from washing the soil away, but they also retain water and help prevent flash floods. Some of these measures involve modification of the contour of the soil, particularly terracing, construction of waterways, and construction of water-retaining ponds. Bands of trees can be planted on the contour to retain both soil and water (Figure 11.7). Avoiding practices, such as overgrazing, that tend to lead to desertification and reforestation of land unsuitable for growing crops conserves water as well as land.

QUESTIONS AND PROBLEMS

Access to and use of the Internet is assumed in answering all questions, including general information, statistics, constants, and mathematical formulas required to solve problems. These questions are designed to promote inquiry and thought rather than just finding material in the chapter. So in some cases there may be several "right" answers. Therefore, if your answer reflects intellectual effort and a search for information from available sources, it can be considered to be right.

1. Some island nations are particularly concerned about the potential loss of their geosphere. How might this occur, perhaps within the next century? Which are some of these nations?

2. Find and describe at least one specific example of a beach restoration project. Why is simply transporting sand to a depleted beach usually not an effective means of beach restoration? What other measures may be employed?

3. Adequate amounts of land are often not available around congested cities for the construction of new airports. What is the solution to this problem in some coastal cities? Cite at least one specific example.

4. Some of the larger projects involving subsurface excavations are those that involve construction of tunnels to move water or wastewater. See if you can find at least one such project that either is ongoing or has just been completed, and describe it.

5. The construction of pipelines is a common kind of large-scale project that involves the geosphere. Explain why pipelines are in many respects a desirable form of transport. What are some potential hazards of pipelines? Describe one pipeline project that is underway or in the planning stages and the controversies it has caused.

6. Section 11.5 contains a long list of minerals extracted from the geosphere for commercial purposes. In a number of cases, the name of the mineral does not reveal the element or compound of which it is a source. For what specific materials are andalusite, chrysoprase, epidote, hornblende, pyrochlor, staurolite, and wollastonite sources?

7. One of the factors involved in starting World War II was the concern of some nations for secure supplies of essential minerals. Find one specific example. Discuss the possibility that mineral resources may be a source of disagreement among nations in the future.

8. Has there ever been a source of platinum-group metals in the United States? What are the prospects for developing such a source in the future? Answer the same questions for rare earths.

9. For what purpose is cryolite used? Is this substance mined from the geosphere? What are the feedstocks and chemical processes for making cyrolite synthetically?

10. Explain why demand for graphite is expected to increase in the future. What are some natural sources of graphite? How is it made synthetically, and what is at least one disadvantage of the synthetic production of graphite?

11. A source of elemental sulfur for chemical synthesis is that from "sour" natural gas. What makes some natural gas sour? Give a sequence of chemical reactions leading from the sulfur-containing component of natural gas to its major chemical product.

12. Explain how sustainable energy is very much tied with the provision of sustainable supplies of essential minerals.

13. Explain how the practice of industrial ecology is tied with the provision of essential materials such as some metals.

14. Which element is a toxic by-product from the mining and processing of phosphorus minerals? Why is this element often found with phosphorus-containing minerals? For which application of phosphorus must it be removed?

15. What is black lung disease? Explain why black lung disease is probably a major problem in China, and see if there are any specific recent reports on this problem in China.

16. How are watersheds involved in the conservation of both soil and water?

17. What is the largest dam and reservoir project completed in the world to date? How is it related to energy and food production? What have been some of the environmental implications of this massive project?

18. One way of reclaiming something of value from municipal refuse is harvesting biogenic methane from landfills. Another way is to burn the refuse, including wastepaper, to generate electrical power and for district heating. How does the latter option substantially reduce the amount of solid waste disposed to the geosphere? What are the main advantages of burning municipal wastes? Cite some specific examples of where it is being practiced.

19. What is the brownfields program? Cite at least one specific example of a successful brownfields project.

20. What is conservation tillage? In what sense is it a "natural" means for growing crops? When did the practice become widespread? What is the current state of conservation tillage?

21. Describe the production of a common commodity (other than wood products) that is grown through the practice of agroforestry?

22. Although the production of biochar to enhance soil fertility is the subject of much current interest, it is actually a very old process. In what sense is it older than civilization itself? See if there are any examples that occurred before the development of modern agriculture.

23. How are tree plantations an example of reforestation? What are the advantages and disadvantages of tree plantations? How widespread is the practice of growing trees on such plantations?

24. Explain how bamboo might be used for restoring degraded soil. How does bamboo prevent erosion?

LITERATURE CITED

1. Price, David George, and Michael de Freitas, *Engineering Geology: Principles and Practice*, Springer, New York, 2010.
2. Strait, Albert L., *Strategic Petroleum Reserve*, Nova Science Publishers, Hauppauge, NY, 2009.
3. World Mineral Exchange: http://www.mineralszone.com/minerals/
4. Hall, Madison A., *Mountaintop Mining: Background and Issues*, Nova Science Publishers, Hauppage, NY, 2010.
5. Eliseeva, Svetlana V., and Jean-Claude G. Buenzli, Rare Earths: Jewels for Functional Materials of the Future, *New Journal of Chemistry* **35**, 1165–1176 (2011).
6. Sen, P. K., Metals and Materials from Deep Sea Nodules: An Outlook for the Future, *International Materials Reviews* **55**, 364–391 (2010).
7. Rosenthal, Elisabeth, Lead from Old U.S. Batteries Sent to Mexico Raises Risks, *New York Times* December 8, 2011, pp. A1.
8. Kliucininkas, L., and D. Velykiene, Environmental Health Damage Factors Assessment in Brownfield Redevelopment, *WIT Transactions on Biomedicine and Health* **14**, 179–186 (2006).
9. Blanco-Canqui, Humberto, and Rattan Lal, *Principles of Soil Conservation and Management*, Springer, New York, 2009.
10. Lehmann, Johannes, and Stephen Joseph, *Biochar for Environmental Management: Science and Technology*, Earthscan, London, 2009.

SUPPLEMENTARY REFERENCES

Anderson, Michael, Ed., *Investigating Plate Tectonics, Earthquakes, and Volcanoes*, Britannica Educational Publications, New York, 2012.

Bleam, William F., *Soil and Environmental Chemistry*, Academic Press, Burlington, MA, 2011.

Brady, Nyle C., and Ray R. Weil, *Elements of The Nature and Properties of Soils*, 3rd ed., Pearson Education, Upper Saddle River, NJ, 2010.

Chesworth, Ward, Ed., *Encyclopedia of Soil Science*, Springer, Dordrecht, Netherlands, 2008.

Coch, Nicholas K., *Geohazards: Natural and Human*, Prentice Hall, Upper Saddle River, NJ, 1995.

Coyne, Mark S., and James A. Thompson, *Fundamental Soil Science*, Thomson Delmar Learning, Clifton Park, NY, 2006.

Eash, Neal S., *Soil Science Simplified*, 5th ed., Blackwell Publishing, Ames, IA, 2008.

Eby, G. Nelson, *Principles of Environmental Geochemistry*, Thomson-Brooks/Cole, Pacific Grove, CA, 2004.

Essington, Michael E., *Soil and Water Chemistry: An Integrative Approach*, Taylor & Francis/CRC Press, Boca Raton, FL, 2004.

Gardiner, Duane T., and Raymond W. Miller, *Soils in Our Environment*, 11th ed., Pearson/Prentice Hall, Upper Saddle River, NJ, 2008.

Gates, Alexander E., and David Ritchie, *Encyclopedia of Earthquakes and Volcanoes*, 3rd ed., Facts on File, New York, 2007.

Hyndman, Donald, and David Hyndman, *Natural Hazards and Disasters*, 2nd ed., Thomson Brooks/Cole, Belmont, CA, 2009.

Ikerd, John E., *Crisis and Opportunity: Sustainability in American Agriculture (Our Sustainable Future*, Bison Books, Winnipeg, 2008.

Keller, Edward A., *Environmental Geology*, 9th ed., Pearson/Prentice Hall, Upper Saddle River, NJ, 2010.

Keller, Edward A., *Introduction to Environmental Geology*, 5th ed., Pearson/Prentice Hall, Upper Saddle River, NJ, 2011.

Lal, Rattan, S*oil Degradation in the United States: Extent, Severity, and Trends*, CRC Press/Lewis Publishers, Boca Raton, FL, 2003.

Lutgens, Frederick K., and Edward J. Tarbuck, *Foundations of Earth Science*, Pearson Prentice Hall, Upper Saddle River, NJ, 2011.

Marti, Joan, and Gerald Ernst, Eds., *Volcanoes and the Environment*, Cambridge University Press, Cambridge, UK, 2005.

McCollum, Sean, *Volcanic Eruptions, Earthquakes, and Tsunamis*, Chelsea House, New York, 2007.

Montgomery, Carla W., *Environmental Geology*, 9th ed., McGraw-Hill, Boston, MA, 2010.

Morgan, R. P. C., *Soil Erosion and Conservation*, 3rd ed., Blackwell Publishing, Malden, MA, 2005.

Nash, David J., and Sue J. McLaren, Eds., *Geochemical Sediments and Landscapes*, Blackwell Publishing, Malden, MA, 2007.

Paul, Eldor A., *Soil Microbiology, Ecology, and Biochemistry*, 3rd ed., Academic Press, Boston, MA, 2007.

Pipkin, Bernard W., *Geology and the Environment*, 5th ed., Thomson Brooks/Cole, Belmont, CA, 2008.

Plaster, Edward J., *Soil Science and Management*, 5th ed., Delmar Cengage Learning, Clifton Park, NJ, 2009.

Sammonds, P. R., and J. M. T. Thompson, Eds., *Advances in Earth Science: From Earthquakes to Global Warming*, Imperial College Press, London, 2007.

Savino, John, and Marie D. Jones, *Supervolcano: The Catastrophic Event that Changed the Course of Human History (Could Yellowstone be Next?)*, New Page Books, Franklin Lakes, NJ, 2007.

Singer, Michael J., and Donald N. Munns, *Soils: An Introduction*, 6th ed., Pearson Prentice Hall, Upper Saddle River, NJ, 2006.

Sposito, Garrison, *The Chemistry of Soils*, 2nd ed., Oxford, UK, 2008.

Stone, George W., *Raging Forces: Life on a Violent Planet*, National Geographic, Washington, DC, 2007.

Tabatabai, M. A., and D. L. Sparks, Eds., *Chemical Processes in Soils*, Soil Science Society of America, Madison, WI, 2005.

Van Elsas, Jan Dirk, Janet K. Jansson, and Jack T. Trevors, Eds., *Modern Soil Microbiology*, 2nd ed., Taylor & Francis/CRC Press, Boca Raton, FL, 2007.

Waltham, Tony, *Foundations of Engineering Geology*, Taylor and Francis, London, 2009.

White, Robert E., *Principles and Practice of Soil Science: The Soil as a Natural Resource*, 4th ed., Blackwell Publishing, Malden, MA, 2006.

12 Environmental and Toxicological Chemistry of the Biosphere

12.1 LIFE AND THE BIOSPHERE

The water-rich boundary region at the interface of Earth's surface with the atmosphere, a paper-thin skin compared to the dimensions of Earth or its atmosphere, is the **biosphere** where life exists.[1,2] The biosphere includes soil on which plants grow, a small bit of the atmosphere into which trees extend and in which birds fly, the oceans, and various other bodies of water. Although the numbers and kinds of organisms decrease very rapidly with distance above Earth's surface, the atmosphere as a whole, extending many kilometers upward, is essential for life as a source of oxygen, medium for water transport, blanket to retain heat by absorbing outgoing infrared radiation, and protective filter against high-energy ultraviolet radiation. Indeed, if not for the ultraviolet-absorbing layer of ozone in the stratosphere, life on Earth could not exist in its present form (see Chapter 7, Section 7.9).

Understanding life requires defining what life really is and an understanding of the chemistry of life, a topic introduced in Chapter 2. Living organisms are constituted of cells that are bound by a membrane, contain nucleic acid genetic material (DNA), and possess specialized structures that enable the cell to perform its functions. A living organism may consist of only one cell or of billions of cells of many specialized types. All living organisms have two characteristics: (1) they process matter and energy through metabolic processes and (2) they reproduce. The ability of an organism to process matter and energy is called **metabolism**. Another important characteristic of living organisms is their ability to maintain an internal environment that is favorable to metabolic processes and that may be quite different from the external environment. Warm-blooded animals, for example, maintain internal temperatures that may be much warmer or even cooler than their surroundings. Finally, through succeeding generations, living organisms can undergo fundamental changes in their genetic composition that enable them to adapt better to their environment.

Living species are present in the biosphere because they have evolved with the capability to survive and to reproduce. Every single species in the biosphere has become an expert in these two things; otherwise it would not be here. The key factors for existing—at least long enough to reproduce—are the ability to utilize energy and to process matter. In so doing, life systems and processes are governed by the principles of thermodynamics and the law of conservation of matter. Organisms handle energy and matter in various ways. Plants, for example, process solar energy by photosynthesis and utilize atmospheric carbon dioxide and other simple inorganic nutrients to make their biomass. Herbivores are animals that eat the matter produced by plants, deriving energy and matter for their own bodies from it. Carnivores in turn feed on the herbivores.

Life forms require several things to exist. The appropriate chemical elements must be present and available. For most organisms, energy for photosynthesis is required in the form of adequate sunlight. Temperatures must stay within a suitable range and preferably should not be subject to large, sudden fluctuations. Liquid water must be available. And, as noted in the preceding discussion, a sheltering atmosphere is required. The atmosphere should be relatively free of toxic substances. This is an area in which human influence can be quite damaging, through release of air pollutants that are directly toxic or that react to form toxic products, such as life-damaging ozone produced through the photochemical smog–forming process.

The nature of life is determined by the surroundings in which the life forms must exist. Much of the environment in which organisms live is described by physical factors, including whether or not the surroundings are primarily aquatic or terrestrial. For a terrestrial environment, important physical factors are the nature of accessible soil and availability of water and nutrients. These are **abiotic factors**. There are also important **biotic factors** relating to the life forms present, their wastes and decomposition products, their availability as food sources, and their tendencies to be predatory or parasitic.

12.1.1 Biosphere in Stabilizing the Earth System: Gaia Hypothesis

There is a very strong connection between life forms on earth and the nature of earth's climate, which determines its suitability for life. As proposed by James Lovelock, a British chemist, this forms the basis of the **Gaia hypothesis**, which contends that the atmospheric O_2/CO_2 balance established and sustained by organisms determines and maintains Earth's climate and other environmental conditions, largely through photosynthesis.[3] For about 3.5 billion years, stabilizing feedback mechanisms have maintained the Earth/atmosphere boundary region within a narrow range conducive to the presence of liquid water and in which life can exist. It is incumbent on humankind to avoid upsetting this delicate balance, which could take place within just a few years period of time, something that may be happening with ongoing global warming.

12.2 ORGANISMS AND SUSTAINABLE SCIENCE AND TECHNOLOGY

There is an extremely strong connection between organisms and green science and technology that is sustainable and environmentally friendly. The most fundamental reason to practice green science and technology as well as environmental chemistry is to maintain an environment on Earth that is hospitable to life. Humans, of course, have a vital self-interest in this endeavor. In general, an environment that is conducive to life and a high diversity of life forms is by nature sustainable. Some of the main aspects in which green science and technology are related to organisms and life are as follows:

- Conditions under which life thrives are generally by nature mild and consistent with green science and technology.
- One of the main characteristics of green science and technology is the absence of toxic substances that harm or kill organisms. Unhealthy or dying organisms are indicative of unsustainable conditions.
- Living organisms and their ecosystems provide models for sustainable anthrospheric systems. Sustainable systems of industrial ecology (Chapter 14) can be largely modeled on natural ecosystems.
- Green science and technology conserve matter and energy to a maximum extent and are characterized by a high-degree recycling. Organisms and their ecosystems have evolved to a maximum degree of efficient energy and matter utilization and provide excellent models for anthrospheric systems.
- Through photosynthesis, plants are outstanding sources of renewable materials in the form of cellulose, lignocellulose, wood, and other materials. Therefore, plants are important and growing sources of materials and fuels for anthrospheric systems.
- Organisms have sophisticated enzyme systems that can perform chemical syntheses and transitions that are either impossible in anthrospheric systems or possible only under extreme conditions.

12.3 LIFE SYSTEMS

To consider the biosphere and its ecology in their entirety, it is necessary to look at several levels in which life exists. The unimaginably huge numbers of individual organisms in the biosphere belong to **species**, or kinds of organisms. Groups of organisms of the same species living together and occupying a specified area over a particular period of time constitute a **population**, and that part of Earth on which they dwell is their **habitat**. In turn, various populations coexist in **biological communities**. Members of a biological community interact with each other and with their atmospheric, aquatic, and terrestrial environments to constitute an **ecosystem**. An ecosystem describes the complex manner in which energy and matter are taken in, cycled, and utilized; the foundation on which an ecosystem rests is the production of organic matter by photosynthesis. Assemblies of organisms living in generally similar surroundings over a large geographic area constitute a **biome**. Each biome may contain many ecosystems. Examples of biomes include tropical rain forests, temperate deciduous forests in which trees grow new leaves in the spring and shed them in the fall, grasslands, deserts, and others.

To sustain life, an ecosystem must provide energy and nutrients. Energy enters an ecosystem as sunlight. Part of the solar energy is captured by photosynthesis, and part is absorbed to keep organisms warm, which enables their metabolic processes to occur faster. In addition to capturing energy, an ecosystem must provide for recycling essential nutrients, including carbon, oxygen, phosphorus, sulfur, and trace-level metal nutrients such as iron.

Virtually all food upon which organisms depend is produced by the fixation of carbon from carbon dioxide and energy from light in the form of energy-rich, carbon-rich biomass through the process of photosynthesis represented by

$$CO_2 + H_2O + h\nu \rightarrow \{CH_2O\} + O_2 \tag{12.1}$$

where $h\nu$ stands for light energy absorbed in photosynthesis and $\{CH_2O\}$ represents biomass. Thus, the photosynthetic plants in the biosphere are the basic **producers** upon which all other members of the community, the **consumers**, depend for food and for their existence. The rate of biomass production is called **productivity**. The sequence of food utilization, starting with biomass synthesized by photosynthetic producers, is called the **food chain**; food chains in turn are interconnected to form often intricate **food webs**.

A biological community is the biological component of an ecosystem and consists of an assembly of organisms that occupy a defined space in the environment. There are many interactions of organisms in a community, many of which are mutually advantageous and others of which are highly competitive. Biological communities are subject to constant change. Some of these changes are relatively short term and cyclical, following daily and seasonal patterns, whereas others are long term. Many transitions are the result of human activities, such as those that take place when agricultural land is taken out of production for a number of years. In the past, major transitions in biological communities have occurred with changes in climate as discussed in the following paragraph.

An important characteristic of a biological community is its hierarchy in which more abundant members lower in the community provide support for those at higher levels. As shown in Figure 12.1, a dominant plant species anchors the community as its major producer of biomass. In addition to providing most of the food through photosynthesis that the rest of the community uses, the dominant plant species often acts to modify the physical environment of the community in ways that enable the other species to exist in it. For example, the trees in a forest community provide the physical habitat in which birds can nest, relatively safe from predators. In addition, the trees provide shade that significantly modifies the habitat at ground level and prevents the growth of most kinds of low-growing plants.

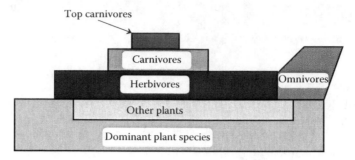

FIGURE 12.1 A dominant plant species typically provides most of the food for a biological community. The biomass that it produces is consumed primarily by herbivores, which are fed upon by carnivores, of which there may be more than one level. Omnivores feed on both plants and animals.

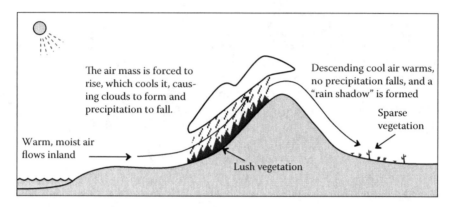

FIGURE 12.2 Illustration of the effects of topography on biological communities. A mass of warm moist air forced to rise over a mountain range deposits rain and snow providing water that enables a productive community to thrive. On the other side of the range, a "rain shadow" creates drought conditions in which the productivity is low, and a much different community exists.

The nature of the geosphere on which it rests is very important in determining the type and function of a biological community. Topography can have a strong influence. The population of a community on a south-facing slope (in the Northern Hemisphere) is influenced by the generally warmer temperatures and greater intensity of sunlight compared to that of a north-facing slope. Marked contrasts in communities may exist on either side of a mountain range over which moisture-laden air flows, as illustrated in Figure 12.2. As warm moist air flows over the upwind slope, it is forced to rise, causing it to cool and release precipitation. On the downwind slope, the cool air sinks, becomes warmer, and has a drying effect on the vegetation and terrain that it contacts. The absence of rain in this region is called a "rain shadow."

12.3.1 Biosphere/Atmosphere Interface and the Crucial Importance of Climate

There are strong interactions between the biosphere and the atmosphere, each greatly influencing the other. As discussed in Chapter 2, Section 2.2, the atmosphere as it exists today was largely made by cyanobacteria in the biosphere, which released through photosynthesis all of the elemental O_2 in the atmosphere. Figure 12.3 indicates another biosphere/atmosphere interaction in which a plant takes water in through its roots and releases it as water vapor to the atmosphere through the process of **transpiration**.

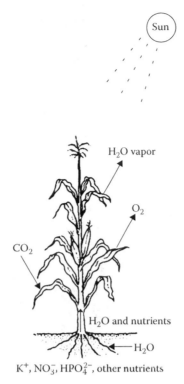

FIGURE 12.3 Transpiration in a corn plant. Water from soil is carried by capillary action to the leaf surfaces of the plant from where it evaporates into the atmosphere. The water carries nutrients with it. On a hot summer's day, a field of corn transfers vast quantities of water from soil to the atmosphere.

The nature and productivity of ecosystems is a function of **climate**, weather conditions over prolonged periods of time. The most productive ecosystems are those in the tropics. Productivity is lower in the temperate zones, and much less near the poles. Seasonal cold has a tremendous effect on plants and hence on productivity. Annual plants are killed by freezing, whereas deciduous trees lose their leaves and stop photosynthesis when it freezes. Thus, cold weather and freezing temperatures tremendously decrease primary productivity.

Specific species of plants and animals thrive only within certain temperature ranges. There are several important ways in which animals maintain internal temperatures conducive to their existence. The temperatures of plants are largely those of the surroundings. (An interesting exception is the "warm-blooded" skunk cabbage plant that uses respiratory processes to metabolize energy-producing materials stored in its large root to keep its temperature as much as 30°C higher than the surrounding atmosphere on cold spring days.) However, for most plants, temperature extremes cause enzymes to become inactivated and proteins to become denatured. Plant membranes are injured by very high or very low temperatures.

All life forms require water. Undesirable levels of moisture, usually moisture deficiency, constitute one of the most common environmental stresses faced by organisms and one often exacerbated by human activities. Drought conditions can be very harmful to biological communities. For a terrestrial community, a lack of water combined with high temperatures may be devastating to plants and animals, although organisms have various ways to respond to moisture deficiency. Plant mechanisms for dealing with moisture deficiency include shedding leaves and reducing water loss to the atmosphere by transpiration. Animals have evolved that can tolerate some dehydration. One interesting mechanism of coping with short water supplies is that of producing very concentrated urine. Such urine, which usually has a very pungent odor, is characteristic of the cat family, which is largely adapted to dry conditions.

12.4 METABOLISM AND CONTROL IN ORGANISMS

As discussed in Section 2.8, in respect to the processing of biochemicals, living organisms continually process materials and energy, a process called *metabolism*. *Photosynthesis*, which is mentioned earlier, is the metabolic process that provides the base of the food chain for most organisms. Animals break down complex food materials to smaller molecules through the process of **digestion**. **Respiration** occurs as nutrients are metabolized to yield energy:

$$C_6H_{12}O_6 \text{ (glucose)} + 6O_2 \rightarrow 6CO_2 + 6H_2O + \text{energy} \qquad (12.2)$$

Organisms assemble small molecules to produce biomolecules, such as proteins, by a **synthesis** process.

In addition to viewing metabolism as a phenomenon within an individual organism, it can be viewed as occurring within groups of organisms living in an ecosystem. Consider, for example, the metabolism of nitrogen within an ecosystem (the nitrogen cycle). Elemental nitrogen from the atmosphere may be fixed as organic nitrogen by bacteria living symbiotically on the roots of leguminous plants and then converted to nitrate when the nitrogen-containing biomass decays. The nitrate may be taken up by other plants and incorporated into protein. The protein may be ingested by animals and the nitrogen excreted as urea in their urine to undergo biological decay and return to the atmosphere as elemental nitrogen. Carbon from carbon dioxide in the atmosphere may be incorporated into biomass by plant photosynthesis and then eventually returned to the atmosphere as carbon dioxide as the biomass is used as a food source by animals.

12.4.1 Enzymes in Metabolism

Discussed in Section 2.7, **enzymes** are biochemical catalysts that are very important in metabolism. Enzymes were discussed in Chapter 2, Section 2.7, and their action illustrated in Figure 2.8 with respect to their processing of biochemicals. Enzymes speed up metabolic reactions by as much as almost a billion-fold. In addition to making reactions go much more rapidly, enzymes are often highly specific in the reactions that they catalyze. The reason for the specificity of enzymes is that they have very specific structures that fit with the substances upon which they act.

A number of factors can affect enzyme action. One important factor is temperature. Organisms without temperature-regulating mechanisms have enzymes that increase in activity as temperature increases up to the point where the heat damages the enzyme, after which the activity declines precipitously with increasing temperature. Enzymes in mammals function optimally at body temperature (37°C for humans) and are permanently destroyed by about 60°C. There is particular interest in enzymes that function in bacteria that live in hot springs and other thermal areas where the water is at or near boiling. These enzymes may turn out to be very useful in commercial biosynthesis operations where the higher temperature enables reactions to occur faster. Acid concentration also affects enzymes, such as those that function well in the acidic environment of the stomach, but stop working when discharged into the slightly basic environment of the small intestine (were this not the case, they would tend to digest the intestine walls).

Enzymes play an important role in toxicological chemistry. A significant concern with potentially toxic substances is their adverse effects on enzymes. As an example, organophosphate compounds, such as insecticidal parathion and military poison sarin "nerve gas," bind with acetylcholinesterase required for nerve function, causing it not to act and stopping proper nerve action. Some substances cause the intricately wound protein structures of enzymes to come apart, a process called denaturation, which stops enzyme action. The active sites of enzymes at which substrates are recognized have a high population of –SH groups. Heavy metals, such as lead and cadmium, have a strong affinity for –SH groups and may bind at enzyme active sites, thus destroying the function of the enzymes.[4] A second area in which enzymes are very much involved in toxicological chemistry is the enzymatic modification of toxicants and protoxicants. In some cases, toxic substances are detoxified by enzyme action. In other cases, enzymes act to convert the precursors to toxicants (protoxicants) to metabolically active toxic forms.

Enzymes are of significant concern in the practice of green chemistry and sustainability science and technology. One obvious relationship is that between enzymes and chemicals that are toxic to them. In carrying out green chemical processes, such chemicals should be avoided wherever possible. Another obvious relationship has to do with the use of biological processes to perform chemical operations, which are usually done under much milder and environmentally friendly conditions biologically than chemically. Biochemical processes are all carried out by enzymes. For example, several enzymes, starting with hexokinase, are involved in the multistepped biochemical fermentation synthesis of ethyl alcohol from carbohydrate glucose. With recombinant DNA technology, it is now possible to invest bacteria with enzyme systems from other organisms designed to carry out desired biochemical processes. Bacteria are much more amenable to handling and usually much more efficient than the organisms from which the genes for the desired enzyme systems are taken. Another approach is to use isolated enzymes immobilized on a solid support to carry out biochemical processes without the direct involvement of an organism.

12.4.2 Nutrients

The raw materials that organisms require for their metabolism are **nutrients**. Those required in larger quantities include oxygen, hydrogen, carbon, nitrogen, phosphorus, sulfur, potassium, calcium, and magnesium and are called macronutrients. Plants and other autotrophic organisms use these nutrients in the form of simple inorganic species, such as H_2O and CO_2, which they obtain from soil, water, and the atmosphere. Heterotrophic organisms obtain much of the macronutrients that they need as carbohydrates, proteins, and lipids (see Chapter 2) from organic food material. An important consideration in plant nutrition is the provision of fertilizers consisting of sources of nutrient nitrogen, phosphorus, and potassium as discussed in Chapter 10, Section 10.2.

Organisms also require very low levels of a number of **micronutrients**, which are usually used by essential enzymes that enable metabolic reactions to occur. For plants, essential micronutrients include the elements boron, chlorine, copper, iron, manganese, sodium, vanadium, and zinc. The bacteria that fix atmospheric nitrogen required by plants use trace levels of molybdenum. Animals require in their diet elemental micronutrients including iron and selenium as well as micronutrient vitamins consisting of small organic molecules.

12.4.3 Control in Organisms

Organisms must be carefully regulated and controlled to function properly. A major function of these regulatory functions is the maintenance of the organism's **homeostasis**, its crucial internal environment. The most obvious means of control in animals is through the **nervous system** in which messages are conducted very rapidly to various parts of the animal as **nerve impulses**. More advanced animals have a brain and spinal cord that function as a **central nervous system** (CNS). This sophisticated system receives, processes, and sends nerve impulses that regulate the behavior and function of the animal. Effects on the nervous system are always a concern with toxic substances. For example, exposure to organic solvents that dissolve some of the protective lipids around nerve fibers can lead to a condition in which limbs do not function properly called **peripheral neuropathy**.[5] Therefore, a major objective of green chemistry is to limit the use of and human exposure to such solvents.

Both animals and plants employ molecular messengers that move from one part of the organism to another to carry messages by which regulation occurs. Messages sent by these means are much slower than those conveyed by nerve impulses. Molecular messengers are often *hormones* discussed as lipids in Section 2.5. In animals, regulatory hormones are commonly released by endocrine glands including in humans the anterior pituitary gland that releases human growth hormone and

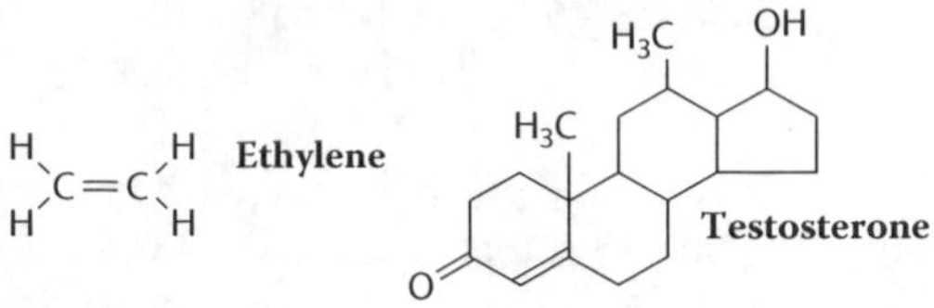

FIGURE 12.4 A simple molecule that acts as a plant hormone to promote maturation processes (ethylene) and a common animal hormone, testosterone, the male sex hormone. See Chapter 20, Section 20.3, for the meaning of line formulas, such as those shown for testosterone.

the pancreas that releases insulin to stimulate glucose uptake from blood. Hormones are carried by a fluid medium in the organism, such as the bloodstream, to cells where they bind to **receptor proteins** causing some sort of desired response. For example, the process may cause the cell to synthesize a protein to counteract an imbalance in homeostasis (see discussion of homeostasis in Section 12.6). Some hormones called **pheromones** carry messages from one organism to another. They commonly serve as sex attractants. Some biological means of pest control use sex pheromones to cause sexual confusion in pesticidal insects, thus preventing their reproduction. Figure 12.4 shows a common plant hormone and a common animal hormone.

12.5　REPRODUCTION AND INHERITED TRAITS

As noted in the preceding section, one of the major activities of organisms is metabolism by which organisms process materials and energy. The other major activity of all organisms is reproduction. Most organisms are capable of reproducing a large excess of their species because throughout time predators and hostile conditions have required large numbers of juveniles to ensure survival of enough members to continue the species. Unrestrained reproduction, especially by humans, poses a strong threat of overpopulation that will outstrip Earth's resources and is a major concern related to reproduction and the environment. A second major concern is the potential effect of environmental chemicals on reproduction and the threat of such chemicals to cause birth defects. Therefore, chemicals that may affect reproduction are given strong consideration in the practice of green chemistry.

Primitive single-celled organisms, particularly bacteria, undergo **asexual reproduction** in which a cell simply splits to form two cells. Humans and most other multicelled organisms undergo **sexual reproduction** requiring that male sperm cells fertilize female egg cells to produce cells capable of dividing and producing new individuals.

Reproduction is directed by **genes** that occur in molecules of DNA, discussed in Chapter 2, Section 2.6. The DNA of an individual, which in sexual reproduction has contributions from both parents, determines the physical, biochemical, and behavioral traits of the organism. The DNA can be altered resulting in changes called **mutations**. A minuscule fraction of mutations are desirable and convey advantages to an individual that are passed along as heritable characteristics in offspring. This is the process of natural selection that has resulted in literally millions of different species of organisms.

Some chemicals are capable of producing mutations. Control of production and exposure to these **mutagens** is a major thrust of green chemistry. This is particularly so because substances that cause mutations are generally regarded as being capable of causing cancer as well and substances that give positive tests for mutagenicity are suspect carcinogens.

12.6　STABILITY AND EQUILIBRIUM OF THE BIOSPHERE

For an organism to survive and thrive, it must reach a state of stability and equilibrium with its environment. The term given to such a state is **homeostasis** ("same status"). In maintaining homeostasis, an organism must interact with its surroundings and other organisms in its surroundings and must balance flows and processing of matter (including nutrients) and energy. On an individual basis, organisms do a remarkably good job of keeping their internal levels of water, materials such as calcium in blood, and temperature within a range conducive to their well-being. Mammals have developed extraordinary capabilities of homeostasis; a healthy individual maintains its internal temperature within a few tenths of a degree. The concept of homeostasis applies to entire groups of organisms living together in ecosystems and, ultimately, to the entire biosphere. Therefore, a major objective of environmental science, including the practice of green chemistry, is to maintain and enhance conditions of homeostasis in the biosphere.

Various concepts pertaining to life systems were defined in Section 12.3 and are reviewed in this section as they apply to the overall stability and equilibrium of the biosphere. **Ecology** describes the interaction of organisms with their surroundings and each other. An important consideration in ecology is the manner in which organisms process matter and energy. An ecosystem describes a segment of the environment and the organisms in it with all of the interactions and relationships that implies. An ecosystem has means of capturing energy, almost always by plants or algae that perform photosynthesis. Light, temperature, moisture, and nutrient supplies are critical aspects of an ecosystem. Ecosystems recycle essential nutrient carbon, oxygen, nitrogen, phosphorus, sulfur, and trace elements. An important part of any ecosystem is the **food chain**, or more complicated **food webs**, in which food generated by photosynthesis is utilized by different organisms at different levels. An important aspect of the food chain in respect to persistent, poorly degradable organic chemicals that are soluble in lipid (fat) tissue occurs through the sequence of animals eaten in the food chain (small creatures in water are fed upon by small fish that are eaten by large fish that are eaten by large birds). Thus, aquatic pollutants become more concentrated in lipid tissue at the top of the food chain, a process called **biomagnification** (discussed with respect to organisms in the hydrosphere in Chapter 3, Section 3.13).[6] An objective of the practice of green chemistry is to avoid the generation and use of chemicals capable of biomagnification in the environment.

The surroundings over a relatively large geographic area in which a group of organisms live constitute a **biome**. There are a number of different kinds of biomes. Regions near the equator may support **tropical rain forest** biomes that stay warm all of the year and in which nutrients remain largely in the organisms (rain forest soil is often notably poor in nutrients, which are mostly held in forest biomass). Temperate regions may support **temperate deciduous forests** in which the trees grow new leaves for a warm, wet summer season and shed them for cold winters. Temperate regions may also have **grassland biomes** in which grass grows from a tough mass of dense roots called **sod**. **Tundra** are treeless arctic regions in which during summer only a layer of wet soil thaws above a permanently frozen foundation of permafrost. Global warming is causing some profound changes in tundra biomes.

The hydrosphere provides several examples of important biomes. One of these consists of **coral reefs**, a specialized type of seashore structure supporting unique marine ecosystems built up by calcium carbonate deposits produced by coral and other marine organisms. Coral reefs form in tropical regions where a firm geological formation is available at shallow depths, conditions that often exist around volcanic islands. These structures provide habitats for the coral itself, associated algae, crustaceans, echinoderms, mollusks, sponges, and fish. Many coral reefs around the world are threatened by the effects of global warming, which raises sea levels, thus reducing the intensity of sunlight reaching the coral for photosynthesis. Coral reefs are also threatened by elevated dissolved carbon dioxide levels in seawater, and the slightly lowered pH resulting from the dissolved CO_2 tends to dissolve the calcium carbonate composing the coral skeleton.

Estuaries, locations where freshwater from rivers mixes with seawater from the ocean, are another kind of biome in the hydrosphere (see Figure 12.5). Estuaries exhibit gradations of salinity, which vary with the ebb and flow of tides. The food chain in estuaries is based on both detrital food sources and phytoplankton. They are especially important as nurseries for marine fish and shellfish, in part because potential predators from the ocean are intolerant of the lower salinity of estuarine waters.

Different kinds of biomes pose a variety of environmental challenges. Some of these have come about from the conversion of biomes to cropland. Grasslands in which the sod has been broken to support wheat and other crops have proven susceptible to wind erosion, which gave rise to the catastrophic Dust Bowl that caused such great hardship on the U. S. Great Plains during the 1930s.

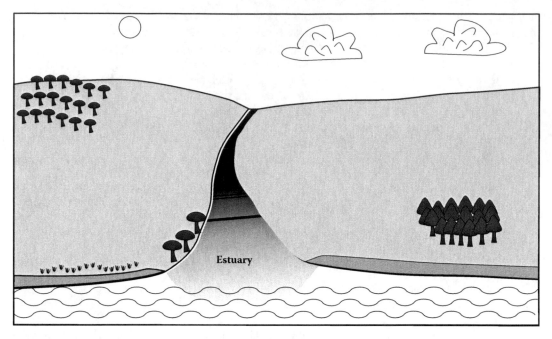

FIGURE 12.5 Estuaries, biomes where freshwater mixes with ocean water, are important nurseries for a number of aquatic species.

Climate changes resulting from global warming could change the distribution of biomes, giving rise to much larger areas of hot deserts that humans might have to learn how to utilize.

12.6.1 BIOMES IN UNEXPECTED PLACES

The conventional thinking in the past was that biomes would occur only in those areas where sunlight enabled conversion of inorganic carbon to biomass that could sustain a food web. It came as a surprise to marine scientists in 1977 that thousands of meters below the surface of the Pacific Ocean, far too deep for light to penetrate and without significant amounts of fallout from biomass generated in surface waters, biomes existed that teemed with tubeworms, clams, and mussels. It is now known that this abundance of life is nourished by microorganisms that thrive in hot volcanic springs and that get their energy through chemosynthesis by mediating reactions of hydrogen sulfide and other substances often toxic to more familiar organisms.

A new kind of habitat was found in 1984 with the discovery of organisms, including tube worms over 2 m long that may be centuries old, that thrive on petroleum seepage on relatively cold ocean floors. These colonies, which may contain hundreds of different species, are especially abundant on the seabed of the Gulf of Mexico, where Spanish accounts from the 1500s noted oil slicks from natural petroleum leakage.

12.6.2 RESPONSE OF LIFE SYSTEMS TO STRESS

Organisms and the ecosystems in which they exist are subject to a number of threats that can result in loss of populations and even total destruction of the system. Natural threats include drought, flooding, fire, landslide, and volcanic eruption. Humans threaten life systems with cultivation, deforestation, mining, and severe pollution. The ability of a community of organisms to resist alteration and damage from such threats, sometimes called **inertia**, depends on several factors and provides important lessons for the survival of the human community in the face of environmental threats.

One of the basic factors involved in providing resistance of a community to damage is its overall rate of photosynthesis, its **productivity** (see Section 12.3). important factor is **diversity** of species so that if one species is destroyed or seriously depleted, another species may take its place. **Constancy** of numbers of various organisms is desirable; wide variations in populations can be very disruptive to a biological community. Finally, **resilience** is the ability of populations to recover from large losses.

The ability of a biological system to maintain high levels of the desirable factors of productivity, diversity, constancy, and resilience is commonly determined by factors other than the organisms present. This is clearly true of productivity, which is a function of available moisture, suitable climate, and nutrient-rich soil. Since all organisms depend on the availability of good food sources, diversity, constancy, and resilience tend to follow high productivity.

12.6.3 RELATIONSHIPS AMONG ORGANISMS

In a healthy, diverse ecosystem, there are numerous, often complex relationships among the organisms involved (Figure 12.6). Species of organisms strongly influence each other. And organisms may greatly alter the physical portion of the system in which they live. An example of such an influence is the tough, soil-anchoring sod that develops in grassland biomes.

In most ecosystems, there is a **dominant plant species** that provides a large fraction of the biomass anchoring the food chain in the ecosystem (Figure 12.1). This might be a species of grass, such as the bluestem grass that thrives in the Kansas Flint Hills grasslands. Herbivores feed on the dominant plant species and other plants and, in turn, are eaten by carnivores. At the end of the food cycle are organisms that degrade biomass and convert it to nutrients that can nourish growth of additional plants. These organisms include earthworms that live in soil and bacteria and fungi that degrade biological material.

In a healthy ecosystem, different species compete for space, light, nutrients, and moisture. In an undisturbed ecosystem, the **principle of competitive exclusion** applies in which two or more potential competitors exist in ways that minimize competition for nutrients, space, and other factors required for growth. Much of agricultural chemistry is devoted to trying to regulate the competition of weeds with crop plants. Large quantities of herbicides are applied to cropland each year to kill competing weeds. In this never-ending contest, green chemistry has an important role in areas such as the synthesis of herbicides that have maximum impact on target pests with minimum impact on the environment.

Within ecosystems, there are large numbers of **symbiotic relationships** between organisms that exist together to their mutual advantage. The classic case of such a relationship is that of **lichen** consisting of algae and fungi growing together. The fungi anchor the system to a rock surface and produce substances that slowly degrade the rock and extract nutrients from it. The algae are photosynthetic, so they produce the biomass required by the system, which is utilized in part by the fungi. Another important symbiotic relationship is that in which nitrogen-fixing bacteria grow in nodules on leguminous plant roots. The bacteria receive nutrients from the plants in exchange for chemically fixed nitrogen required for plant nutrition.

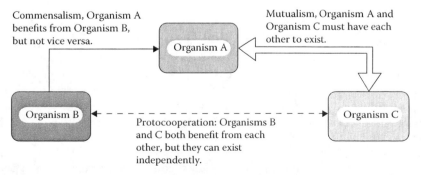

FIGURE 12.6 Types of beneficial relationships among organisms.

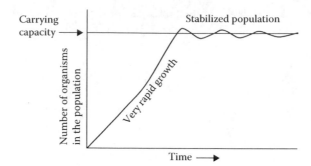

FIGURE 12.7 Very rapid growth of a population newly introduced into an environment suitable for its survival, followed by stabilization of numbers around the carrying capacity. This is an idealized picture subject to perturbation by many factors such as predation, disruption of habitat, or disease, and it applies to humans and other organisms.

12.6.4 POPULATIONS

Recall from Section 12.3 that populations consist of groups of the same species of organisms living together and occupying a specified area over a particular period of time. Populations have numerous characteristics, including numbers, genetic composition, birth and death rates, and age and sex distribution. Here are addressed some of the important factors involved with populations in biological communities.

All existing species are present on Earth because they have developed exceptional abilities to reproduce and to survive long enough to do so. The potential for reproduction always greatly exceeds the ultimate capacity of an environment to support a population. This capacity is known as the **carrying capacity**. Introduction of members of a population into an area that is amenable to their growth causes their numbers to increase very rapidly, often at a rate approaching exponential growth. Eventually, numbers reach a level around the carrying capacity at which point rapid growth ceases abruptly because of some limiting factor, such as limited food, nutrients, water, air, or shelter, or because of stress from crowding. After the very rapid growth phase ceases, the numbers of organisms tend to fluctuate somewhat around the carrying capacity as shown in Figure 12.7. In some cases, a large growth in population can cause it to temporarily exceed the carrying capacity by a significant margin, or the carrying capacity can be suddenly reduced by circumstances such as droughts. This can result in a temporary overpopulation until reduced reproduction rates and increased death rates can adjust the population to accord with the carrying capacity. The result can be an abrupt decrease in population, or population crash. If the carrying capacity has been altered by overpopulation, such as destruction of grassland by overgrazing, or if it has been reduced by external factors, the new population will stabilize at a figure that reflects the new carrying capacity.

12.7 DNA AND THE HUMAN GENOME

In Chapter 2, Section 2.6, DNA was discussed and it was noted that this macromolecule stores and passes on genetic information that organisms need to reproduce and synthesize proteins. Recall that DNA is composed of repeating units called **nucleotides**, each consisting of a molecule of the sugar 2-deoxy-β-D-ribofuranose, a phosphate ion, and one of the four nitrogen-containing bases, adenine, cytosine, guanine, and, thymine (conventionally represented by the letters A, C, G, and T respectively). DNA is one of two **nucleic acids**, the other one of which is RNA. Like DNA, RNA consists of repeating nucleotides but the sugar in RNA is β-D-ribofuranose and it contains uracil instead of thymine in its bases. The structural formulas of segments of DNA and RNA are shown in Figure 2.6.

The structure of DNA is a key aspect of its function, and its elucidation by Watson and Crick in 1953 was a scientific insight that set off a revolution in biology that is going on to this day.

The huge DNA molecules consist of two strands counterwound with each other and held together by hydrogen bonds; a representation of this structure is shown in Figure 2.7. In this structure, the hydrogen bonds connecting complementary bases on the two strands are represented by dashed lines. Because of their structures that make hydrogen bonding possible, adenine on one strand is always hydrogen bonded to thymine on the opposite strand and guanine to cytosine. During cell division, the two strands of DNA unwind and each generates a complementary strand for the DNA of each new cell.

In organisms with eukaryotic cells, DNA is divided into units associated with protein molecules called **chromosomes**. The number of these varies with the organism; humans have 23 pairs of chromosomes, a total of 46. The strands of DNA in chromosomes, in turn, are divided into sequences of nucleotides, each distinguished by the nitrogen-containing base in it. These sequences of nucleotides give directions for the synthesis of a specific kind of protein or polypeptide. (See the discussion of proteins in Chapter 2, Section 2.4.) These specific groups of nucleotides, each of which has a specific function, are called **genes**. When a particular protein is made, DNA produces a nucleic acid segment designated mRNA, which goes out into the cell and causes the protein to be formed through a process called transcription and translation (the gene is said to be expressed).

As the units that give the directions for protein synthesis, genes are obviously of the utmost importance in living organisms. As discussed in Chapter 13, genes can now be transferred between different kinds of organisms and will direct the synthesis of the protein for which they are designed in the recipient organism. It is now known that a number of human diseases are the result of defective genes, and there is a genetic tendency toward getting other kinds of diseases.[7] For example, certain gene characteristics are involved in susceptibility to breast cancer.

Because of the known relationship of gene characteristics to disease, the decision was made in the mid-1980s to map all the genes in the human body. This collective body of genes is called the human genome and the project to map it is called the Human Genome Project. The original impetus for this project in the United States arose because of interest in the damage to human DNA by radiation, such as that from nuclear weapons. But, from the beginning, it was recognized that the project had enormous commercial potential, especially in the pharmaceutical industry, and could be very valuable in human health.[8]

The announcement in 2001 that mapping of the human genome was complete promised great progress in biology, especially in medicine. Genes function by directing the synthesis of specific proteins, and the action of most pharmaceuticals is to alter the activities of proteins, the drug's target. In some cases, proteins are made more active and in others their activity is diminished. Knowledge of the human genome enables a better understanding of protein activity and is resulting in the development of more specific drugs. For example, the gene responsible for cystic fibrosis was discovered in 1989 by examination of family histories of the disease, and about 10 years later, pharmaceutical dornase alfa, brand name Pulmozyme, made by recombinant DNA technology became available to treat people afflicted by cystic fibrosis.

12.8   BIOLOGICAL INTERACTION WITH ENVIRONMENTAL CHEMICALS

Organisms in the environment interact significantly with xenobiotic materials (those foreign to living systems) in their surroundings. The uptake of such materials by organisms is discussed in this section. The biological interactions of organisms with xenobiotic substances in the hydrosphere are discussed in Sections 3.13 and 3.14.

Bioaccumulation is the term given to the uptake and concentration of xenobiotic materials by living organisms. The materials may be present in water in streams or bodies of water, sediments in bodies of water, drinking water, soil, food, or even the atmosphere. Bioaccumulation can lead to **biomagnification** in which xenobiotic substances become successively more concentrated in the tissues of organisms higher in the food chain. This usually occurs with poorly degradable, lipid-soluble organic compounds. Suppose, for example, that such a compound contacts lake water, accumulates in solid detritus in the water, sinks to the sediment, and is eaten by small burrowing creatures in the sediment, which are eaten by small fish. The small fish may be eaten by larger fish, which in turn are consumed as food by birds. At each step, the xenobiotic substance may become more concentrated in the organism and may reach harmful concentrations in the birds at the top of the food chain. This is basically what happened with DDT, which almost caused the extinction of eagles and hawks.

Fish that bioaccumulate poorly degradable, lipid-soluble organic compounds from water will lose them back to water if they are placed in an unpolluted environment. The process by which this occurs is called **depuration**. The time required to lose half of the bioaccumulated xenobiotic material is called the **half-life** of the substance.

12.8.1 BIODEGRADATION

Biodegradation is the process by which biomass from deceased organisms is broken down to simple inorganic constituents, thus completing the cycle in which biomass is produced from atmospheric carbon dioxide and from water by photosynthesis. Important aspects of biodegradation are discussed in Chapter 3, Section 3.14, with respect to biodegradation in the hydrosphere, and the concepts introduced there including biodegradability, cometabolism, mineralization, and detoxification apply in general to organisms in the biosphere.

12.9 EFFECTS OF THE ANTHROSPHERE ON THE BIOSPHERE

Human intervention, both subtle and drastic, has a large potential effect on biological communities and the biosphere as a whole. One of the greatest such influences took place with the cultivation of forests and grasslands as Europeans settled in North and South America. An important way in which humans may influence biological communities is through the introduction of new species. If it is successful in a community, a new species affects those already there and may significantly modify the physical nature of the habitat. New species may prey upon those already present or serve as prey that attracts predatory species from outside the community. When forests are cut and grasslands established, larger numbers of herbivores and representatives of species not previously present are attracted. These animals in turn attract carnivores that feed on them. Parasites usually accompany newly introduced species that can serve as their hosts.

Some introduced species are particularly destructive to biological communities and habitat. One of the worst of these is the goat, which has a well-earned reputation for indiscriminate consumption of vegetation, destruction of plant life, and damage to sod with its hard hooves. Rats introduced into islands have wreaked havoc with indigenous species. Domestic house cats reverting to a wild state have wiped out whole populations of birds. Aggressive bird species, particularly house sparrows and starlings, have displaced more desirable native species.

The human species has become inextricably linked with technology such that in a sense *Homo sapiens* are not "natural animals." Much of what is known about the effects of humans on biological communities is negative—destruction of habitat, emission of pollutants to the environment, and a potential permanent change in climate from greenhouse gases. These kinds of influences are unfortunate and very harmful to biological communities. However, humans are linked to technology irreversibly, and it will be necessary to adapt themselves and their technologies to the biological communities upon which ultimately humankind depends for its existence.

12.9.1 BENEFICIAL EFFECTS OF HUMANS ON THE BIOSPHERE

Not all human activities are detrimental to the biosphere. In fact, the anthrosphere can be designed and operated in ways that improve the quality of the biosphere. For example, the productivity of the biosphere can be very much improved by applications of fertilizers, advanced cultivation techniques, and other measures in the agricultural sector. Technology can be applied to the construction and enhancement of habitats, such as wetlands. The potential beneficial effects of humans on the biosphere are discussed in more detail in Section 13.4.

QUESTIONS AND PROBLEMS

1. Look up information pertaining to the career of James Lovelock, author of *The Vanishing Face of Gaia: A Final Warning*. What is the scientific discovery early in his career that pertains to environmental chemistry? What is Gaia? What is the "final warning?"

2. Look up biomarkers or biomarkers of exposure. What are biomarkers telling us about the state of the biosphere? Are they providing warning signs?

3. What two things essential for sustaining the biosphere are captured by photosynthesis? Explain.

4. Biomass (other than that contained in bone) is represented in this chapter by the simplified formula of $\{CH_2O\}$, which shows the three most abundant elements in biomass. What are the next three most abundant elements in biomass? See if you can find on the Internet an empirical formula that takes these elements into account and write a chemical reaction representing photosynthesis that includes all these elements.

5. What are thermogenic plants? How do they relate to material in this chapter? What are some examples of thermogenic plant species?

6. What are the distinctions among digestion, respiration, and photosynthesis? Which of these predominantly involve transitions of energy?

7. What are the major aspects of the metabolism of nitrogen within ecosystems as a whole? How is it now significantly perturbed by activities in the anthrosphere?

8. Look up the use of pheromones for insect control. To what general class of biological materials do pheromones belong? Are pheromones regarded as toxic to insects? In what sense is their use to control insects a green technology?

9. How do lichen involve a symbiotic relationship? What do lichen have to do with the geosphere (weathering)?

10. What is the Human Genome Project? Why is it important? What is the current status of the project?

11. Explain why poorly degradable, lipid-soluble compounds are the most likely to undergo biomagnification. Name a class of compounds that are likely to undergo biomagnification and a detrimental ecosystem effect that has resulted from biomagnification processes in the past.

12. Look up the Earth System on the Internet. How is it defined? Is there unanimous agreement regarding the definition? Are there programs of study for Earth System science?

13. Explain why glucose is very important in both plants and animals.

14. In the decades preceding about 1950, there was considerable concern that world populations were exceeding levels that could be supported by the biosphere. What happened to change that picture and what was this "revolution" called? What are some of the environmental and sustainability factors that may make dire predictions of the adverse effects of overpopulation come true in coming decades?

15. Are there examples from human history in which human populations have exceeded the carrying capacity of the ecosystems that support them? (The works of Jared Diamond may be useful in answering this question).

16. Look up current threats to coral reefs posed by atmospheric carbon dioxide levels. Is there evidence to suggest that coral reefs may be threatened? Explain.

LITERATURE CITED

1. Voet, Donald, Judith G. Voet, and Charlotte W. Pratt, *Fundamentals of Biochemistry: Life at the Molecular Level*, 4th ed., Wiley, Hoboken, NJ, 2012.
2. Kaufman, Donald G., *The Biosphere: Protecting Our Global Environment*, 4th ed., Kendall Hunt Publishing, Dubuque, IA, 2011.
3. Lovelock, James, *The Vanishing Face of Gaia: A Final Warning*, Basic Books, New York, 2009.
4. Gasser, Gilles, and N. Metzler-Nolte, Metal Compounds as Enzyme Inhibitors, *Bioinorganic Medicinal Chemistry*, Enzo Alessio, Ed., pp. 351–382. Wiley-VCH, Weinheim, Germany, 2011.
5. Watkins, John B., and Curtis Klaassen, *Casarett and Doulls Essentials of Toxicology*, 2nd ed., Kindle Edition, McGraw Hill Professional, New York, 2010 (Kindle Locations 10759–10761).
6. Jonker, Michiel, T. O., What is Causing Biomagnification of Persistent Hydrophobic Organic Chemicals in the Aquatic Environment?, *Environmental Science and Technology* **46**, 110–111 (2012).
7. Gilham, Nicholas Wright, *Genes, Chromosomes, and Disease: From Simple Traits, to Complex Traits, To Personalized Medicine*, Pearson Education/FT Press, Upper Saddle River, NJ, 2011.
8. McElheny, Victor, *Drawing the Map of Life: Inside the Human Genome Project*, Basic Books, New York, 2010.

SUPPLEMENTARY REFERENCES

Brooker, Robert, Eric Widmaier, Linda Graham, and Peter Stiling, *Biology*, McGraw Hill, New York, 2010.

Connelly, Matthew, *Fatal Misconception: The Struggle to Control World Population*, Belknap Press of Harvard University Press, Cambridge, MA, 2008.

Dickinson, Gordon, and Kevin Murphy, *Ecosystems*, 2nd ed., Routledge, New York, 2007.

Dotretsov, Nicolay, *Biosphere Origin and Evolution*, Springer, Berlin, 2007.

Huggett, Richard J., *The Natural History of the Earth: Debating Long Term Change in the Geosphere and the Biosphere*, Routledge, New York, 2006.

Kaufman, Don, and Cecilia Franz, *The Biosphere: Protecting our Global Environment*, 4th ed., Kendall Hunt Publishing, Dubuque, IA, 2005.

Kunitz, Stephen J., *The Health of Populations: General Theories and Particular Realities*, Oxford University Press, Oxford, UK, 2007.

León, Rosa, Aurora Gaván, and Emilio Fernández, *Transgenic Microalgae as Green Cell Factories*, Springer, Berlin, 2006.

McNeil, Brian, and Linda Harvey, *Practical Fermentation Technology*, Wiley, New York, 2008.

Molles, Manuel C., *Ecology: Concepts and Applications*, 4th ed., McGraw-Hill Higher Education, Boston, MA, 2008.

Raven, Peter H., Linda R. Berg, and David M. Hassenzahl, *Environment*, 6th ed., Wiley, Hoboken, NJ, 2008.

Schmitz, Oswald J., *Ecology and Ecosystem Conservation*, Island Press, Washington DC, 2007.

Smil, Vaclav, *The Earth's Biosphere: Evolution, Dynamics, and Change*, MIT Press, Cambridge, MA, 2003.

Spilsbury, Louise, and Richard Spilsbury, *Food Chains and Webs: From Producers to Decomposers*, Heinemann Library, Chicago, IL, 2004.

Stevens, Christian V., and Roland Verhé, Eds., *Renewable Bioresources: Scope and Modification for Non-food Applications*, Wiley, Hoboken, NJ, 2004.

UNESCO, *Knowledge for Sustainable Development: An Insight into the Encyclopedia of Life Support Systems*, UNESCO, Paris, 2002.

Zikov, G. E., Ed., *Biotechnology and Industry*, Nova Science Publishers, New York, 2004.

13 Sustaining the Biosphere and Its Natural Capital

13.1 KEEPING LIFE ALIVE

Nothing is more important in the effort to achieve sustainability than keeping the biosphere and the organisms that compose it healthy. In the past century, especially, human activities have significantly threatened the biosphere, and there is evidence to suggest that a mass extinction event associated with the beginning of the Anthropocene epoch is under way. Fossil records clearly show five major extinction events millions of years in the past as illustrated in Figure 13.1.[1] The first of these took place at the end of the Ordovician period 434 million years ago with a loss of an estimated 60% of all marine and terrestrial organisms. The second mass extinction took place in the late Devonian period. The end of the Permian mass extinction 251 million years ago resulted in the extinction of up to 95% of marine species. The mass extinction at the end of the Triassic period extinguished about 80% of all terrestrial quadrupeds. The dinosaurs were wiped out in the end of the Cretaceous mass extinction 65 million years ago.

The causes of past mass extinction events are not known with certainty, although there is convincing evidence that asteroid impacts caused the last two of these. The fossil record indicates dramatic climate changes and increased atmospheric carbon dioxide levels associated with past events. Now, a sixth mass extinction is under way associated with the beginning of the Anthropocene epoch (see Chapter 1, Section 1.6). This Anthropocene extinction event is associated with what humans have done to the Earth System as their activities affect the biosphere. Habitat has been significantly altered, especially with the development of modern agriculture and construction of buildings, highways, and parking lots on the surface of the geosphere. Air and water pollution have harmed many species of animal life and plant life. And the greatest effect is likely to be from global climate change associated especially with increased atmospheric carbon dioxide levels. The diversity of the gene pool of domesticated plants and animals is being severely reduced as plant and animal breeding programs concentrate on a relative few strains.

Humans, themselves, are of the biosphere and are totally dependent on it for their existence, especially in respect to provision of adequate food. So, a central sustainability challenge is to maintain and enhance the biosphere upon which we all depend. This chapter is about this challenge. It also addresses, for lack of a better word, the exploitation of the biosphere to meet human needs. Especially, as usable supplies of petroleum become scarcer, the biosphere will be called upon to fulfill increased demand for materials and fuel. To meet those needs while maintaining and even enhancing the state of the biosphere is a major challenge for those involved in environmental science and sustainability.

13.2 NATURAL CAPITAL OF THE BIOSPHERE

The biosphere is an immense source of natural capital in many respects. The most obvious of these is production of food, beginning with plant photosynthesis. In addition, the biosphere generates large quantities of raw materials and feedstocks such as wood. As noted in the discussion of the Gaia hypothesis in Section 12.1.1, Chapter 12, the biosphere is very much involved in maintaining the Earth System in a state compatible with life and human existence on the planet. Another important aspect of the biosphere's natural capital is its ability to process and detoxify wastes.

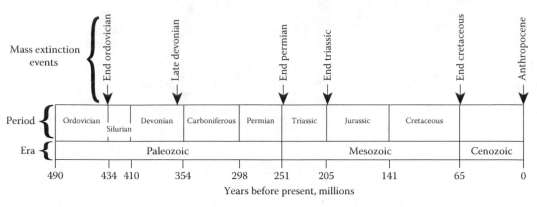

FIGURE 13.1 Mass extinction events from the past as indicated by fossil records.

Organisms break down waste materials such as crop residues and even potentially toxic synthetic substances. By converting organic wastes to simple inorganic compounds, the process of **mineralization**, organisms in the biosphere recycle essential nutrients including chemically bound nitrogen and phosphorus, an important aspect of natural biogeochemical cycles.

One of the most important aspects of the biosphere's natural capital is its ability to sequester atmospheric carbon dioxide and convert it to biomass. As discussed in Chapter 8, Section 8.2, and illustrated in Figure 8.2, photosynthesis by plants and algae in the biosphere removes significant amounts of carbon dioxide from the atmosphere, a process that helps to control global warming.

13.2.1 Types of Biomaterials from the Biosphere

The biosphere provides a vast source of biomaterials including those that can be employed for fuel.[2] This section presents the various alternatives for biomaterials.

By photosynthetic processes, plants generate carbohydrate, for which the approximate simple formula is CH_2O. The most fundamental of the carbohydrates, and the one generated directly by photosynthesis, is glucose, chemical formula $C_6H_{12}O_6$, a simple sugar or monosaccharide. Two molecules of simple sugars with chemical formulas of $C_6H_{12}O_6$ can bond together chemically with the elimination of a molecule of H_2O to produce a disaccharide, chemical formula $C_{12}H_{22}O_{11}$. Sucrose used as a sweetener in foods and drinks is a disaccharide. Many molecules of simple sugars can bond together with the elimination of a molecule of H_2O for each one bonded to produce polysaccharides consisting of huge macromolecules. Starch and cellulose are the most prominent such polysaccharides.

Glucose is being studied intensely as a feedstock for making a variety of substances. It is most readily obtained by breaking down starch produced by grains. Cornstarch is the most abundant source of glucose. Cellulose found in woody plant material and in substances such as cotton is a very abundant potential source of glucose, although it is more difficult to extract glucose from cellulose than from starch. Sucrose squeezed as sap from sugarcane is a ready source of glucose and of the similar sugar fructose. Starch, itself, is an excellent raw material for many applications, as is cellulose. Chemically modified starch and chemically altered cellulose both have a variety of applications.

Lignin is a biological polymer with a complex structure (see Figure 13.2), which is associated with cellulose in woody parts of plants, binding fibers of cellulose together in lignocellulose. In a number of applications, most prominently paper making, it is necessary to separate the lignin from the cellulose. Unfortunately, lignin is a rather refractory material of variable molecular composition that has few uses. It would be of great use if plants could be developed that produced copious quantities of cellulose without lignin. The cotton plant does this, producing a pure form of cellulose in the cotton fiber, but the small ball of cotton is only a small fraction of the cotton plant biomass.

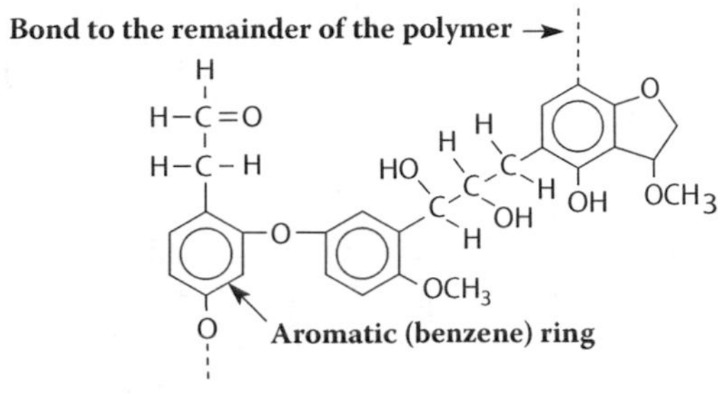

FIGURE 13.2 Segment of the lignin polymer showing the major kinds of functionality in this material. There is a lot of variability in the molecular structure of lignin, which tends to make it difficult to use as a feedstock.

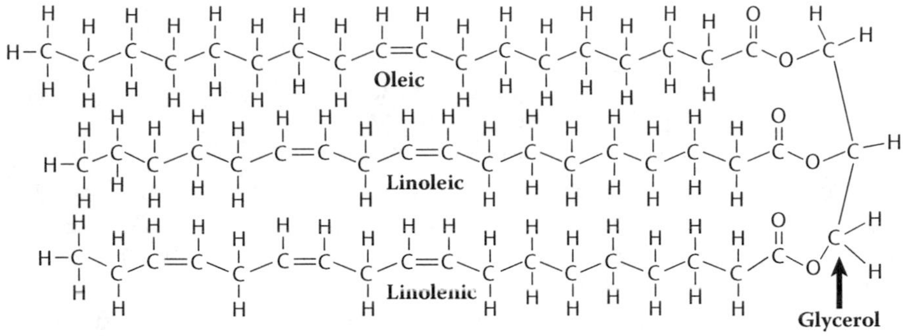

FIGURE 13.3 Structural formula of a typical vegetable oil containing three of the most common fatty acids (oleic, linoleic, and linolenic) in canola oil, which is extracted from seeds of the rape plant (*Brassica napus*).

Many useful **lipid oils** are extracted from a variety of plant seeds including rapeseed, soybeans, sunflowers, and corn. In addition to their food uses, these oils are used in a large variety of applications including raw materials for making other chemical products, lubricants, and as biodiesel fuels. Part of the usefulness of lipid oils in many applications is due to their similarity to petroleum hydrocarbons. Volatile solvents, most commonly the 6-carbon straight-chain alkane *n*-hexane, C_6H_{14}, are used to extract oils from plant sources. In this process, the solvents are distilled off from the extract and recirculated through the process.

The seeds of the *Brassica napus* plant, commonly called rape, are an excellent source of vegetable oil. The structural formula of a vegetable oil that would be typical of rapeseed oil is shown in Figure 13.3. Rapeseed oil is a triglyceride with three fatty acid groups esterified with the alcohol glycerol, which has three –OH groups available for bonding on each molecule. Triglycerides with unsaturated fatty acid groups (those with C=C double bonds) are oils rather than semisolids that are common in animal fats. The seeds of *B. napus* are high in both oil (40%) and protein (23%) compared to about 20% and 40%, respectively, for soybeans. Varieties of rapeseed grown before about 1970 were used as sources of lubricants but were not well suited for food and animal feed production because the oil contained high levels of erucic acid, a long-chain fatty acid suspected of adverse effects on the heart when ingested at high levels, and in addition, the protein-rich residues of rapeseed meal left from oil extraction were not suitable for animal feed because of high levels of sulfur-containing glucosinolate compounds that made the meal unpalatable and potentially toxic to livestock. In 1979, the Western Canadian Oilseed Crushers Association registered the name of canola to designate rapeseed oil from varieties of the plant that had been developed that had both very low levels of erucic acid in the oil and low levels of glucosinolates in the reside left from oil extraction. This enabled production of both a healthy oil product for human and animal consumption and a meal by-product that could safely be used as a protein source for animals, thus enabling economical production of rapeseed oil as a food source.

An attractive aspect of vegetable oils is that there are significant sources in oil-bearing plants in arid and semiarid regions where other crops do not grow well. Many of these oils contain fatty acids that are desirable industrially, for nutrition, and even medicinally including oleic (from almond and olive), linoleic (grape seed and walnut), erucic (crambe), α-linolenic (chia), γ-linolenic (evening primrose), wax esters (jojoba), and hydroxy fatty acids (lesquerella). In some cases, these crops produce oils for which markets are well established and in others markets may well be developed in the future. An oil plant with interesting possibilities is crambe (*Crambe abyssinica*), an annual plant from the Mediterranean area with seeds that contain up to one-third oil, significantly higher than the oil content of soybeans. Crambe oil is up to 60% erucic acid

Erucic acid

with a long hydrocarbon chain. Crambe oil is unsuitable for human consumption, but has a number of potential industrial uses in lubricants, plastics, and coatings.

Some plants produce hydrocarbons directly in the form of **terpenes**, hydrocarbon molecules containing at least one C=C double bond. Terpenoids are oxygen-containing analogs of terpenes. Most of the plants that produce terpenes are conifers (evergreen trees and shrubs such as pine and cypress), plants of the genus *Myrtus*, and trees and shrubs of the genus *Citrus*. One of the most common terpenes emitted by trees is α-pinene, a principal component of turpentine, a material formerly widely used in paint formulations because it reacts with atmospheric oxygen to form a peroxide, then a hard resin that produces a durable painted surface. The only commercial source of natural rubber, at the moment, is the Brazilian rubber tree *Hevea brasiliensis*. The hydrocarbon terpenes that occur in rubber trees can be tapped from the trees as a latex suspension in tree sap. Steam treatment and distillation can be employed to extract terpenes from sources such as pine or citrus tree biomass. The production of terpenes by plants is of particular interest because of the hydrocarbon nature of terpenes.

Isoprene

α-Pinene

Grain seeds are rich sources of protein, almost always used for food, but potentially useful as chemical feedstocks for specialty applications. An exciting possibility just now coming to fruition in a practical sense is to transplant genes into plants so that they will make specialty proteins such as medicinal agents.

Biological materials used as sources of feedstocks are usually complex mixtures, which make separation of desired materials difficult. However, some compensation is made for that disadvantage in that in some biological starting materials' nature has done much of the synthesis of the final product. Most biomass materials are partially oxidized, as is the case with carbohydrates, which contain approximately one oxygen atom per carbon atom (compared to petroleum hydrocarbons that have no oxygen). This can avoid expensive, sometimes difficult oxidation steps, which may involve potentially hazardous reagents and conditions. The complexity of biomass sources can make the separation and isolation of desired constituents relatively difficult.

There are several main pathways by which feedstocks can be obtained from biomass. The most straightforward of these is a simple physical separation of biological materials, such as squeezing oil from oil-bearing biomass or tapping latex from rubber trees. Only slightly more drastic treatment consists of extraction of oils by organic solvents as is done for lipid oils extraction. Physical and chemical processes can be employed to remove useful biomass from the structural materials of plants, which consist of lignocellulose composed of cellulose bound together by lignin "glue."

13.2.2 BIOREFINERIES

Biorefineries are facilities in which materials taken from the biosphere are extracted, isolated, refined, and processed to produce useful products.[3] Figure 13.4 is an outline of a biorefinery. It shows several pathways by which biomass can be utilized. The simplest of these is extraction using a volatile organic solvent that is recycled or simply squeezing biomaterials from biomass, such as is done in the extraction of oil from oilseeds or removal of sucrose-rich sap from sugarcane or sugar beets. Biomass consists largely of complex biopolymers, such as lignocellulose, which under favorable circumstances can be broken down by enzymes or microorganisms to yield sugars and other small molecules. Chemical processing can be employed to convert biomass to usable products, including pyrolysis, which yields liquids and gases plus a char residue (biochar), and hydrogenation by high-temperature treatment of biomass with H_2. The most drastic treatment is gasification in which biomass is subjected to high temperatures in a sub-stoichiometric atmosphere of O_2 to yield a synthesis gas product consisting of CO_2 and combustible CO and H_2 that can be burned for fuel or used to chemically synthesize methane (CH_4), higher hydrocarbons including gasoline or diesel fuel, and alcohols.

13.2.3 USING THE BIOSPHERE THROUGH AGRICULTURE

Manipulated by humans, the biosphere produces enormous quantities of food and fiber through the practice of **agriculture**. In addition to food, organic biomass produced by plants can be used as a renewable source of raw material and fuel. Some plants are now being genetically engineered to produce specific chemicals. The production of materials from the biosphere is very closely connected with the practice of green chemistry and sustainability and has numerous implications with respect to environmental and toxicological chemistry. Agricultural chemicals, including fertilizers,

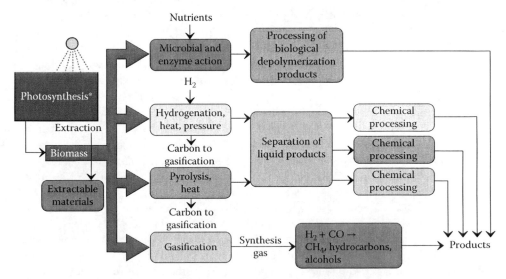

FIGURE 13.4 A biorefinery for processing materials taken from the biosphere. *Production of biomass from terrestrial plants, algae, elipse and cyanobacteria by photosynthesis.

herbicides, and insecticides, are produced and applied to crops and land in enormous quantities. Annual production of millions of kilograms of these chemicals demands the proper practice of green chemistry and engineering and an understanding of environmental chemistry. The judicious use of relatively small quantities of herbicides enables planting of crops in soil covered with residues of the previous year's crops with little or no cultivation of soil. This practice of low-tillage agriculture, now called conservation tillage (see Chapter 11, Section 11.11), is in keeping with the best practice of green chemistry and sustainability.

The managed production of biomass by the biosphere is absolutely essential for the survival of humankind. To continue to feed growing world populations while maintaining and even enhancing the ability of the biosphere to produce food, it is necessary that the practice of agriculture be as green as possible. In the past and still today, this has often not been true. Cultivation of soil by humans has displaced native plants, destroyed wildlife habitat, contaminated soil with pesticides, filled rivers and bodies of water with sediments, and otherwise perturbed and damaged the environment and significantly reduced the natural capital of the biosphere. Agricultural practices arguably represent the greatest incursion of the anthrosphere into the other environmental spheres. On the positive side, growth of domestic crops temporarily removes greenhouse gas carbon dioxide from the atmosphere and provides organic raw materials and biomass fuel without any net addition of carbon dioxide to the atmosphere.

The basis of agriculture is the development of domestic plants from their wild ancestors.[4] (The same can be said of animals, but only a handful of animal species have been domesticated, although each consists of many different breeds.) Our prehistoric ancestors learned to select plants with desired characteristics for the production of food and fiber, developing new species that often require the careful efforts of expert botanists or DNA testing to relate them to their wild ancestors. Only around 1900 were the scientific principles of heredity applied to plant breeding, eventually with excellent results. Using scientific methods, agriculturists accomplished a "green revolution" in the 1950s and 1960s that resulted in varieties of rice and wheat, especially, that had vastly increased yields. The techniques used included selective breeding, hybridization, cross-pollination, and backcrossing to develop grain varieties, which, combined with chemical fertilizers and pesticides, lead to much higher crop yields. India, a country on the verge of starvation in the 1940s, increased its grain output by 50%. Developments such as higher yielding and faster maturing dwarf varieties of rice enabled better nutrition for an increasing world population, at least postponing the inevitable problems that will result from excessive population growth. By breeding plants resistant to cold, drought, and insects, overall crop productivity has been further increased. Increased nutritional values for grain have been achieved, such as the development of corn varieties that have higher levels of lysine amino acid.

One of the major advances in plant breeding has been the development of **hybrids** produced by crossing true-breeding strains of plants. So-called hybrid vigor is well known, and many hybrids have vastly greater yields than their parent strains. Corn (maize), a remarkably productive photosynthesizer, has proven most amenable to the production of hybrids, in part because of the separation of male flowers that grow on the tops of plants from female flowers attached to the budding corn ears. By planting rows of corn that alternate between two different strains and cutting the tassels from the tops of the plants that are to produce the corn seed, hybrid corn varieties are readily produced. More recently, techniques have been developed for growing hybrids of other kinds of plants.

There are, of course, many factors other than the genetic strains of plants that are involved in high crop productivity. The effects of weather have been mitigated by the development of crop varieties that resist heat, cold, and drought. The provision of water by artificial irrigation has greatly increased crop productivity and is essential for crop productivity in some regions, such as the vegetable-growing areas of California. Irrigation practices continue to become more efficient with the replacement of wasteful spray irrigators by systems that apply water directly to soil or even directly to plant roots. Computerized control of irrigation can make it much more efficient. However, currently, one of the greater concerns related to sustainability is the rapid depletion of groundwater resources (see discussion of groundwater depletion in Chapter 3, Section 3.4).

13.2.4 Genome Sequencing and Green Chemistry

The Human Genome Project (see Chapter 12, Section 12.7)[5] and related genome sequencing of other organisms have a number of implications for green chemistry and sustainability science and technology. One of the key goals of green chemistry is to use chemicals that have maximum effectiveness for their stated purpose with minimum side effects. This certainly applies to pharmaceuticals in which knowledge of the human genome may enable development of drugs that do exactly what they are supposed to do without affecting nontarget systems. This means that drugs can be made very efficiently with little waste material.

Some of the most important effects of DNA sequencing as it relates to green chemistry has to do with a wide variety of organisms other than humans. With an exact knowledge of DNA and the genes that it contains, it is possible to deal with organisms on a highly scientific basis in areas such as pest control and the biosynthesis of raw materials. An accurate map of the genetic makeup of insects, for example, should result in the synthesis of precisely targeted insecticides that kill target pests without affecting other organisms. Such insecticides should be effective at very low doses, thus minimizing the amount of insecticide that has to be synthesized and applied, consistent with the goals of green chemistry.

An exact knowledge of the genomes of organisms is extremely helpful in the practice of genetic engineering in which genes are transferred between species to enable production of desired proteins and to give organisms desirable characteristics such as pest resistance. A number of medically useful proteins and polypeptides are now produced by genetically engineered microorganisms, most commonly genetically modified *Escherichia coli* bacteria. Perhaps, the greatest success with this technology has been the biosynthesis of human insulin, a lack of which causes diabetes in humans. Two genes are required to make this relatively short polypeptide that consists of only 51 amino acids. Other medically useful substances produced by genetically engineered organisms include human growth hormone, tissue plasmogen activator that dissolves blood clots formed in heart attacks and strokes, and various vaccine proteins to inoculate against diseases such as meningitis, hepatitis B, and influenza. Genetic engineering is discussed in more detail in Section 13.3.

13.3 GENETIC ENGINEERING

13.3.1 Recombinant DNA and Genetic Engineering

Ever since humans started raising crops for food and fiber (and later animals), they have modified the genetic makeup of the organisms that they use. This is particularly evident in the cultivation of domestic corn, which is physically not at all like its wild ancestors. Until now, breeding has been a slow process. Starting with domestication of wild species, selection and controlled breeding have been used to provide desired properties such as higher yield, heat and drought tolerance, cold resistance, and resistance to microbial or insect pests. For some domesticated species, these changes have occurred over thousands of years. As discussed in Section 13.2.3, during the 1900s, increased understanding of genetics greatly accelerated the process of breeding different varieties. The development of high-yielding varieties of wheat and rice during the "green revolution" of the 1950s has prevented (or at least postponed) the starvation of millions of people. A technology that enabled a quantum leap in productivity of domestic crops was the development of **hydrids** from crossing of two distinct lines of the same crop, dating in a practical sense from the mid-1900s.

Traditional breeding normally takes a long time and depends largely on random mutations to generate desirable characteristics. One of its greatest limitations has been that it is essentially confined to the same species, whereas more often that not, desired characteristics occur in species other than those being bred. Since about the 1970s, however, humans have developed the ability to alter DNA so that organisms synthesize proteins and perform other metabolic feats that would otherwise be impossible. Such alteration of DNA is commonly known as **genetic engineering** and

recombinant DNA technology. Organisms produced by recombinant DNA techniques that contain DNA from other organisms are called transgenic organisms. With recombinant DNA technology, segments of DNA that contain information for the specific syntheses of particular proteins are transferred between organisms. Most often, the recipient organisms are bacteria, which can be reproduced (cloned) over many orders of magnitude from a cell that has acquired the desired qualities. Therefore, to synthesize a particular substance such as human insulin or growth hormone, the required genetic information can be transferred from a human source to bacterial cells, which then produce the substance as part of their metabolic processes.

The mechanics of recombinant DNA gene manipulation is a complex and sophisticated operation. The first step involves lyzing (opening up) a cell that has the genetic material needed and removal of this material from the cell. Through enzyme action, the sought-after genes are cut from the donor DNA chain. These are next spliced into small DNA molecules. These molecules, called cloning vehicles, are capable of penetrating the host cell and becoming incorporated into its genetic material. The modified host cell is then reproduced many times to carry out the desired biosynthesis.

Recombinant DNA technology is a rapidly growing area that is having profound effects, especially in agriculture and medicine. It is being used increasingly to produce crops with unique characteristics, to synthesize pharmaceuticals, and to make a variety of useful raw materials as renewable feedstocks. Recombinant DNA technology has a lot of potential in the development of green chemistry and sustainability, such as in the sustainable production of chemical feedstocks and products of various kinds. An example is synthesis of polylactic acid using lactic acid produced enzymatically with corn and polymerized by standard chemical processes. In the environmental area, genetic engineering offers the potential for the production of bacteria engineered to safely destroy troublesome wastes and to produce biological substitutes for environmentally damaging synthetic pesticides.

Plants are particularly amenable to recombinant DNA manipulation. In part, this is because huge numbers of plant cells can be grown in appropriate media and mutants can be selected from billions of cells that have desired properties such as virus resistance. Individual plant cells are capable of generating whole plants, so cells with desired qualities can be selected and allowed to grow into plants that may have the qualities desired. Ideally, this accomplishes in weeks what conventional plant breeding techniques would require decades to do.

Once plants containing desired transgenes have been produced, an exhaustive evaluation process occurs. This process has several objectives. The most obvious of these is an evaluation of the transplanted gene's activity to see if it produces adequate quantities of the protein for which it is designed. Another important characteristic is whether or not the gene is passed on reliably to the plant's progeny through successive generations. It is also important to determine whether the modified plant grows and yields well and if the quality of its products is high.

Only a few strains of plants are amenable to the insertion of transgenes, and normally, their direct descendants do not have desired productivity or other characteristics required for a commercial crop. Therefore, transgenic crops are crossbred with high-yielding varieties. The objective is to develop a cross that retains the transgene while having desired characteristics of a commercially viable crop. The improved variety is subjected to extensive performance tests in greenhouses and fields for several years and in a number of locations. Finally, large numbers of genetically identical plants are grown to produce seed for commercial use. Many kinds of genetically modified plants have been developed and more are being marketed commercially every year.

Early concerns about the potential of genetic engineering to produce "monster organisms" or new and horrible diseases have been largely allayed, although not entirely so, and resistance to the application of recombinant DNA technology is strong in some quarters. Opposition has been especially strong in Europe, and the European Commission, the executive body of the European Union, has disallowed a number of transgenic crops. However, caution is still required with this technology. One example of a problem has been the emergence of weeds resistant to the widely used herbicide glyphosate.

Despite these concerns, transgenic crops are growing in importance and they have become widely utilized in highly populated countries, particularly China, where they are seen as a means of feeding very large populations.

13.3.2 Major Transgenic Crops and Their Characteristics

The two characteristics most commonly developed in transgenic crops is tolerance for herbicides that kill competing weeds and resistance to pests, especially insects, but including microbial pests (viruses) as well. In the earlier years of transgenic crop plantings, most of the crops had traits for only one of these characteristics, but in more recent years, so-called stacked varieties with two or more characteristics have become more common and now comprise the fastest growing sector of transgenic crops. As of 2010, the land areas planted to transgenic crops in the eight leading countries were the following (millions of hectares in parentheses):[6] United States (66.8), Brazil (25.5), Argentina (22.9), India (9.4), Canada (7.6), China (3.5), Paraguay (2.6), and South Africa (2.2). A total of 150 million hectares of genetically modified (GM) crops were planted throughout the world in 2010. The most common biotech crops are as follows (hectares planted in 2010 in parentheses): soybeans (74), maize or corn (45), cotton (20), and canola (8). Other crops for which transgenic varieties have been developed are tomato, poplar, papaya, sweet pepper, sugar beet, alfalfa, and squash. Herbicide tolerance, which enables crops to be unharmed by spraying with herbicide (especially glyphosate, see Section 13.3.3) to kill competing weeds, has been the predominant biotech trait introduced with 91 million hectares (of a global total of 150 million hectares). Next were stacked traits including both herbicide tolerance and insect resistance (32 million hectares), then insect resistance (27 million hectares).

In 2010, Monsanto and Dow Agrosciences introduced stacked transgenic corn with 8 traits including resistance to insects above and below the soil as well as tolerance to some common herbicides. It is claimed that this variety will reduce the refuge area for corn planting from 20% to 5%. (The refuge is a fraction of the area of a crop that is planted to non-transgenic crops to generate enough insecticide-susceptible insects to dilute the resistant ones that eventually develop in the transgenic areas. The rationale for this approach is that insects growing in refuge areas without any incentive to develop resistance will crossbreed with resistant strains, preventing them from becoming dominant.) In addition to pest resistance and tolerance to herbicides, future stacked transgenic crops are expected to have characteristics such as drought resistance, high omega-3 lipid production in soybeans, and elevated levels of pro-Vitamin A (substance that is converted to Vitamin A by metabolic processes after consumption) in "golden rice."

13.3.3 Crops versus Pests

The disruption of natural ecosystems by cultivation of land and planting agricultural crops provides an excellent opportunity for opportunistic plants—weeds—to grow in competition with the desired crops. To combat weeds, farmers use large quantities of a variety of herbicides. The heavy use of herbicides poses a set of challenging problems. In many cases, to be effective without causing undue environmental damage, herbicides must be applied in specified ways and at particular times. Collateral damage to crop plants, environmental harm, and poor biodegradation leading to accumulation of herbicide residues and contamination of water supplies are all problems with herbicides.[7] A number of these problems can be diminished by planting transgenic crops that are resistant to particular herbicides.[7] The most common such plants are those resistant to Monsanto's Roundup herbicide glyphosate (structural formula shown in Chapter 4, Figure 4.11). This widely used compound is a broad-spectrum herbicide, meaning that it kills most plants that it contacts. One of its advantages from an environmental standpoint is that it rapidly breaks down to harmless products in soil, minimizing its environmental impact and problems with residue carryover. By using "Roundup Ready" crops, of which by far the most common are transgenic soybeans, the herbicide can be applied directly to the crop, killing competing weeds. Application when the crop plants are relatively small, but after weeds have had a chance to start growing, kills weeds and enables the crop to get a head start. After the crop has developed a significant size, it deters the growth of competing weeds by shade that deprives the weeds of sunlight.

Unfortunately, not long after the introduction of Roundup-ready seeds, glyphosate-resistant weeds began appearing, including horseweed, giant ragweed, and several of a number of species in the genus *Amaranthus* (pigweed) that by 2010 had afflicted 7–10 million acres of the approximately 170 million acres of corn, soybeans, and cotton planted in the United States. Pigweed has been an especially bad actor that can grow 7 or 8 cm in a day, reach heights of 2 m, and with stalks so thick and strong that they can damage harvesting machinery.

Aside from weeds, the other major class of pests that afflict crops consists of a variety of insects. Two of the most harmful of these are the European corn borer and the cotton bollworm, which cost millions of dollars in damage and control measures each year and can even threaten an entire year's crop production. Even before transgenic crops were available, *Bacillus thuringiensis* (Bt) was used to control insects. This soil-dwelling bacterium produces a protein called delta-endotoxin. Ingested by insects, delta-endotoxin partially digests the intestinal walls of insects causing ion imbalance, paralyzing the system, and eventually killing the insects. Fortunately, the toxin does not affect mammals or birds. Bt has been a popular insecticide because as a natural product it degrades readily and has gained the acceptance often accorded to "natural" materials (many of which are deadly).

Genetic engineering techniques have enabled transplanting genes into field crops that produce Bt. This is an ideal circumstance in that the crop being protected is generating its own insecticide, and the insecticide is not spread over a wide area. There are several varieties of insecticidal Bt, each produced by a unique gene. Several insect pests are well controlled by transgenic Bt including the European corn borer, the Southwestern corn borer, and the corn earworm. Cotton varieties that produce Bt are resistant to cotton bollworm. Bt-producing tobacco resists the tobacco budworm. Potato varieties have been developed that produce Bt to kill the Colorado potato beetle, although this crop has been limited because of concerns regarding Bt in the potato product consumed directly by humans. Although human digestive systems are not affected adversely by Bt, there is concern over it being an allergen because of its proteinaceous nature.

Virus resistance in transgenic crops has concentrated on papaya. This tropical fruit is an excellent source of Vitamins A and C and is an important nutritional plant in tropical regions. The papaya ringspot virus is a devastating pest for papaya, and transgenic varieties resistant to this virus are now grown in Hawaii. One concern with virus-resistant transgenic crops is the possibility of transfer of genes responsible for the resistance to wild relatives of the plants that are regarded as weeds, but are now kept in check by the viruses. For example, it is possible that virus-resistant genes in transgenic squash may transfer to competing gourds, which would crowd out the squash grown for food.

13.3.4 FUTURE CROPS

The early years of transgenic crops can be rather well summarized by soybeans, corn, and cotton resistant to herbicides and insects. In retrospect, these crops will almost certainly seem rather crude and unsophisticated. In part, this lack of sophistication is due to the fact that the genes producing the desired qualities are largely expressed by all tissues of the plants and throughout their growth cycle, giving rise to problems such as the Bt-contaminated corn pollen that may threaten Monarch butterflies or Bt-containing potatoes that may not be suitable for human consumption. It is anticipated that increasingly sophisticated techniques will overcome these kinds of problems and will lead to much improved crop varieties in the future.

There are several potential green chemistry benefits from genetic engineering of agricultural crops. One promising possibility is to increase the efficiency of photosynthesis, which is only a few tenths of a percent in most plants. Doubling this efficiency should be possible with recombinant DNA techniques, which might significantly increase the production of food and biomass by plants. For example, with some of the more productive plant species, such as fast-growing hybrid poplar trees and sugarcane, biomass is competitive with fossil fuels as an energy source. A genetically engineered increase in photosynthesis efficiency could enable biomass to economically replace expensive petroleum and natural gas for fuel and raw material. A second possibility with genetic engineering is the development of the ability to support nitrogen-fixing bacteria on plant roots in plants that cannot do so now. If corn, rice, wheat, and cotton could be developed with this capability, it could save enormous amounts of energy and natural gas (a source of elemental hydrogen) now consumed to make ammonia synthetically.

A wide range of other transgenic crops are under development. One widely publicized crop is "golden rice" incorporating β-carotene in the grain, which is therefore yellow, rather than the normal white color of rice.[8] The human body processes β-carotene to Vitamin A, the lack of which impairs vision and increases susceptibility to maladies including respiratory diseases, measles, and diarrhea. Since rice is the main diet staple in many Asian countries, the widespread distribution of golden rice could substantially improve health. As an example of the intricacies of transgenic crops, two of the genes used to breed golden rice were taken from daffodil and one from a bacterium! Some investigators contend that humans cannot consume enough of this rice to provide a significant amount of Vitamin A.

As of 2010, transgenic alfalfa and sugar beets resistant to glyphosate herbicide were being promoted for agricultural use. Alfalfa is a nutritious forage crop for animal feed, a legume that grows synergistically with nitrogen-fixing *Rhizobium* bacteria growing in nodules attached to its roots. In 2010, the U.S. Supreme Court overturned a lower court decision that had prevented widespread distribution of these crops because of the possibility that their glyphosate-resistant qualities might spread to other plants and violate restrictions on foods designated as "organic" and that some countries have put in place against all transgenic foods.

Work continues on improved transgenic oilseed crops. The one getting the most attention is canola, the source of canola oil. Efforts are under way to modify the distribution of oils in canola to improve the nutritional value of the oil. Another possibility is increased Vitamin E content in transgenic canola. Sunflower, another source of vegetable oils, is the subject of research designed to produce improved transgenic varieties. Herbicide tolerance and resistance to white mold are among the properties that are being developed in transgenic sunflowers.

Decaffeinated coffee and tea have become important beverages. Unfortunately, the processes that remove caffeine from coffee beans and tea leaves also remove flavor, and some such processes use organic solvents that may leave undesirable residues. The genes that produce caffeine in coffee and tea leaves have now been identified, and it is possible that they may be removed or turned off in the plants to produce coffee beans and tea leaves that give full-flavored products without the caffeine. Additional efforts are under way to genetically engineer coffee trees in which all the beans ripen at once, thereby eliminating the multiple harvests that are now required because of the beans ripening at different times.

Although turf grass for lawns would not be regarded as an essential crop, enormous amounts of water and fertilizers are consumed in maintaining lawns and grass on golf courses and other locations. Healthy grass certainly contributes to the "green" esthetics of a community and can be regarded as part of the natural capital of the biosphere. Furthermore, herbicides, insecticides, and fungicides applied to turf grass leave residues that can be environmentally harmful. So, the development of improved transgenic varieties of grass and other groundcover crops can be quite useful. There are many desirable properties that can benefit grass. Included are tolerances for adverse conditions of water and temperature, especially resistance to heat and drought. Disease and insect resistance are desirable. Reduced growth rates can mean less mowing, saving energy. For grass used on waterways constructed to drain excess rain runoff from terraced areas (see Chapter 11, Figure 11.7), a tough, erosion-resistant sod composed of masses of grass roots is very desirable. Research is under way to breed transgenic varieties of grass with some of these properties. Also, grass is being genetically engineered for immunity to the effects of glyphosate herbicide, which is environmentally more benign than some of the herbicides such as 2,4-D currently used on grass.

An interesting possibility for transgenic foods is to produce foods that contain vaccines against disease. This is possible because genes produce proteins that resemble the proteins in infectious agents, causing the body to produce antibodies to such agents. Diseases for which such vaccines may be possible include cholera, hepatitis B, and various kinds of diarrhea. The leading candidate as a carrier for such vaccines is the banana. This is because children generally like bananas and this fruit is readily grown in some of the tropical regions where the need for vaccines is the greatest.

13.4 ROLE OF HUMAN ACTIVITIES IN PRESERVING AND ENHANCING THE BIOSPHERE

Technology can be harnessed to preserve and improve the condition of life on Earth, in a sense, the anthrosphere working with and for the biosphere. A prime example of how that is done is provided by agriculture discussed in Section 13.3. With the application of plant genetics, herbicides, fertilizers, and advanced cultivation and harvesting techniques, agricultural interests can vastly increase the productivity of a plot of soil. By building terraces and waterways planted to grass that forms a tough, erosion-resistant sod, the productivity of land may be increased while erosion is slowed to a negligible level.

Human intervention can be used to create and enhance habitats that are not maintained for agricultural production. For example, although not all reservoirs formed by damming streams are desirable, many provide a welcome variety of habitat for species that live in or around bodies of water. Impounding water can cause suspended material to settle from streams, thus improving stream quality below the dams. In a limited, but encouraging number of cases, human intervention is being applied to reverse damage done to habitats by human activities in the past. Once meandering streams straightened and turned into ugly, erosive ditches by channelization have been restored in some cases to provide the bending channels that make the stream hospitable to life and tolerant of flooding. Productive wetlands are being restored, or even constructed where none existed before, usually as a means to aid wastewater treatment. Badly used, eroded farmland has been converted to forests and grassland. Special structures can be made and sunk in shallow coastal areas to provide shelter and habitat for marine life; even old ship hulls and airplane fuselages have been used for this purpose.

The restoration of ecosystems by human intervention is called **restoration ecology**.[9] Restoration ecology has become a significant area of human endeavor, and it may be hoped that it will increase in importance as technology is used increasingly to benefit the natural environment. The restoration ecologist needs to be familiar with basic ecology as well as with the kinds of technology used to rebuild ecosystems. Knowledge of related areas, such as geology, hydrology, limnology, and soil science, is also required. After catastrophic floods along the Missouri and Mississippi Rivers in the United States in 1993, the deliberate decision was made to forego reconstruction of some river dikes destroyed by the flood and restoration ecology was applied to some limited areas to restore wetlands and river bottomlands for wildlife habitat.

Much of the work that has been done to preserve wildlife and to restore ecosystems in which wild species exist has been the result of efforts to maintain and increase numbers of game animals. Enlightened hunting and fishing laws have reduced the harvest of many species to sustainable levels. In some cases, these have brought species back from very low numbers or even the brink of extinction. Important examples in the United States are American bison, wood ducks, wild turkeys, snowy egrets, and white-tailed deer. In addition to hunting and fishing restrictions, habitat restoration has been very important in increasing numbers of game animals. Restoration of wetland breeding areas has enabled significant increases in numbers of waterfowl.

Information is essential to understand, preserve, and enhance biological systems. The capabilities of technology to gather and process information are enormous. Sophisticated chemical analysis techniques provide detailed profiles of the chemical characteristics of the environment in which organisms live, including both nutrients and pollutants. Sensors for temperature, wind, moisture, and sunlight can be used to give a continuous picture of the physical environment. This and other information can be subjected to sophisticated computer analysis to provide a profile of the life system and to direct human intervention in constructive ways.

One of the more useful relatively recent technologies employed to study life systems consists of satellite images of Earth, such as those provided by the Landsat satellite. Such images can be gathered by infrared measurements, digitized, and processed by computer to provide profiles of geological features, water on Earth's surface, and vegetation. By remotely sensing the absorption of electromagnetic radiation at specific wavelengths, instruments mounted on satellites can monitor gases or reactive chemical species in the atmosphere. One example of the latter is ClO, a reactive intermediate produced during the photochemical processes that occur as part of ozone depletion from stratospheric chlorofluorocarbons. The levels of greenhouse gases, including carbon dioxide, nitrous oxide, and methane, can also be monitored. This information can be used to predict the effects of atmospheric species on life forms that may occur from global warming or ozone depletion.

13.4.1 ARTIFICIAL HABITATS AND HABITAT RESTORATION

To a degree, plants can be preserved artificially by seed banks in which seeds are stored for long periods of time under appropriate conditions for their preservation. Botanical gardens and arboreta enable growth of plants under artificial conditions that can prevent at least some species from becoming extinct.

The number of animal species that can be maintained in zoos is limited, but zoos are still important for protecting various kinds of animals from extinction. Zoos are being used to a greater degree for wildlife preservation, in some cases with the goal of introducing animal species back into the wild. **Captive breeding** programs have been established to salvage individuals of endangered species from the wild, increase their population by breeding in captivity, and reintroduce them into the wild state. The numbers of endangered bird species have been increased by taking eggs from nests of birds in the wild and hatching them in captivity, sometimes with surrogate parents from other bird species. On a much larger scale, fish hatcheries have been in use for many decades to ensure a steady supply of fingerlings, particularly of trout and salmon species. There have been some tentative successes in captive breeding programs to restore species to the wild. In the United States, captive peregrine falcon and blackfooted ferret have been reintroduced to some areas. The Arabian oryx (a large species of antelope) has been restored to some of its former habitats in the Middle East. Golden lion tamarins have been reintroduced to rain forests in Brazil. The widely publicized reintroduction of the California condor, a large carrion-eating bird, from individuals bred in captivity, has been difficult because of the deaths of many of the specimens released, but has achieved some success in reestablishing this species in its natural habitats.

A major problem with captive breeding programs has been the vulnerability of limited numbers of any species population to loss. When only a few individuals remain, the sudden onset of disease can be devastating. Not the least of the problems is the limited genetic diversity of a small population and the adverse effects of inbreeding. An intriguing possibility is the collection of DNA from threatened species with the goal of cloning the animals at a later date when that becomes feasible.

13.5 PRESERVING THE BIOSPHERE BY PRESERVING THE ATMOSPHERE

Measures taken to preserve and enhance the quality of the other environmental spheres may be effective in maintaining the quality of the biosphere. This is certainly true of the atmosphere. A polluted atmosphere is generally harmful to the biosphere. Particulate matter that blocks out sunlight reduces the ability of the atmosphere to carry out photosynthesis. Particles in air that humans and other animals breathe can harm the lungs and have adverse health effects. Nitrogen oxides and sulfur oxides are phytotoxic.

Ground-level atmospheric ozone is toxic to plants including important crop plants. It has been estimated that atmospheric ozone has now reduced yields of corn (maize) by 2.2–5.5%, 3.9–15% for wheat, and 8.5–14% for soybean.[10] Ozone causes visible damage to foliage including early senescence (aging) and premature abscission (detachment and dropping) of leaves. Typical of the phytotoxicity of O_3, ozone damage to a lemon leaf is manifested by chlorotic stippling (characteristic yellow spots on a green leaf), as represented in Figure 13.5. Ponderosa and Jeffrey pines exposed to ozone and smog in California's San Bernardino Mountains have suffered chlorotic mottle and premature needle death. Reduction in plant growth may occur without visible lesions on the plant. Leaves exposed to ozone may exhibit reduced stomatal aperture. Carbon availability to plants and decreased translocation of fixed carbon to edible fruits, grains, pods, or roots may result from ozone exposure. Of particular importance given global climate change, ozone exposure may reduce plant resistance to heat and drought stress.

Some plant species, including sword-leaf lettuce, black nightshade, quickweed, and double-fortune tomato, are so susceptible to the effects of ozone and other photochemical oxidants in the atmosphere that they are used as bioindicators of the presence of smog. Brief exposure to approximately 0.06 ppm of ozone may temporarily cut photosynthesis rates in some plants in half. Crop damage from ozone and other photochemical air pollutants in California alone is estimated to cost millions of dollars each year. The geographic distribution of damage to plants in California is illustrated in Figure 13.6.

Air pollution control measures taken to reduce photochemical smog and associated ozone (see Chapters 7 and 8) have certainly reduced the damage that otherwise would have been done to the

FIGURE 13.5   Representation of ozone damage to a lemon leaf. In color, the spots appear as yellow chlorotic stippling on the green upper surface caused by ozone exposure.

FIGURE 13.6   Geographic distribution of plant damage from smog in California.

biosphere from tropospheric ozone and other oxidants. Adverse effects on both human health and crop productivity have been lessened by such measures.

A second consideration pertaining to atmospheric ozone and the health of the biosphere has to do with the "good ozone" in the stratosphere. As discussed in Chapter 7, Section 7.9, the relatively small amount of ozone in the stratosphere is crucial in filtering out ultraviolet radiation that would be very damaging to organisms including humans on Earth's surface. The Nobel-Prize-winning discovery that this ozone is destroyed by the products of chlorofluorocarbons and their subsequent ban have prevented much damage to the biosphere.

13.6   PRESERVING THE BIOSPHERE BY PRESERVING THE HYDROSPHERE

Water is essential for life. Therefore, preservation of the hydrosphere is an important aspect of preserving the biosphere. Efforts in water conservation have the effect of conserving the biosphere.

One of the important aspects of conserving water is preservation and enhancement of **watersheds**, areas in which rainwater collects, runoff is slowed, and infiltration into groundwater is maximized (see Chapter 3, Section 3.3). Effective watershed areas are generally those with significant vegetation that catches rainwater, slows its runoff, and maximizes infiltration to groundwater. Good watersheds are likely to have healthy biomes with diverse and healthy vegetation.

Agricultural practices designed to conserve rainwater are generally beneficial to the biosphere. One such practice in semiarid regions is **summer fallowing** in which crop residues are left to stand in place for periods of time to allow for buildup of soil moisture to be available for crops that are planted later. In many areas such as those of the high plains where summer fallowing is practiced, winter snowfall is an important source of water, and stubble left in fields is effective in catching snow and holding it in place so that it can melt and add water to the soil. The soil in summer-fallowed regions supports healthy populations of microorganisms, earthworms, and other organisms that live in the biosphere.

As discussed in Chapter 5, Section 5.15, constructed wetlands are useful for purifying and recycling water. Wetlands can also be highly bioproductive and, as indicated in Figure 5.12, can be used as sources of biomass, such as for fuel production. Wetlands also support populations of waterfowl.

Constructed lakes and reservoirs are used to conserve water and increase water infiltration to groundwater aquifers. These bodies of water can serve as sources of irrigation water for crops. Potentially, such bodies of water can support high populations of fish and shellfish. As discussed in Chapter 17, they also have the potential to support heavy growths of algal biomass that can be harvested for synthetic fuels and biomaterials.

13.7   PRESERVING THE BIOSPHERE BY PRESERVING THE GEOSPHERE

As discussed in Chapter 10, Section 10.4, damage to the geosphere's top layer of soil can lead to erosion and desertification. One characteristic of desert formation is the loss of vegetation from the soil surface. Therefore, measures to prevent desert formation or to reverse it add to the population of organisms supported on land surfaces.

There are strong synergisms between preserving the geosphere and preserving the biosphere. Plants established on arid lands tend to anchor soil in place and are especially effective in preventing wind erosion, a strong contributor to desertification. Decayed plant biomass can add to soil organic matter, an important soil quality characteristic.

Measures taken to prevent the release of toxic substances from the geosphere can aid in increasing the quality and productivity of the biosphere. An important example is the prevention of acid mine water formation (see Chapter 3, Section 3.5.1).

13.7.1   CONSTRUCTING THE GEOSPHERE TO SUPPORT THE BIOSPHERE: WHAT THE ANCIENT INCAS KNEW

The vast Native American Inca Empire that existed along the west coast of South America and centered in modern Peru reached its peak of development from 1438 until the coming of the Spanish conquerors in 1533. This highly developed society numbered perhaps 6 million people until decimated by European diseases to which the native population had no resistance. Second only to the Himalayas, the Andes mountains of the Inca Empire are the world's tallest and most formidable. But by altering this inhospitable geosphere, the Incas succeeded in building an agricultural system that fed their large population. Incan agriculture was based on a system of terraces covering about a million acres that were cut into the steeply sloping terrain.[11] The terraces were much like stairway steps with walls of carefully laid stone between which were deposited mixtures of gravel and soil upon which crops were grown. The stone walls of the terraces had the additional advantage of retaining heat from sunshine during the day, preventing damage to plant roots from cold night temperatures. An elaborate system of irrigation canals and cisterns provided water for growing crops. The Incas grew crops especially adapted to the conditions on the terraced systems including corn, potatoes, and quinoa (a gluten-free grain, especially high in protein, which has become a popular item in modern-day health food stores).

The exploitation of the Inca population by the Spanish and the drastic reduction in numbers from European diseases led to a decline in the ingenious system of agriculture built on terraces. Now, however, especially in light of the agricultural problems being brought on by climate change, there is renewed interest in the Inca agricultural system and, based on knowledge gleaned from archaeological research, some of the old systems of terraces and irrigation canals are being rebuilt and the traditional Inca crops are being grown again in systems that are relatively more productive and efficient in water use. The biologically productive Inca system of agriculture built on terraces on steeply sloping land, now undergoing reincarnation, stands as an example of the positive effects that modifications of the geosphere by humans may have on the biosphere.

QUESTIONS AND PROBLEMS

1. Look up information on the Internet pertaining to current rates of species extinction. What are reliable estimates of current species extinction? How is extinction related to the Anthropocene? What may be the effects of global climate change?

2. What is the Gaia hypothesis? What happened in the earlier stages of life on Earth that was arguably the best example of Gaia? How may humans have a potentially beneficial effect on Gaia in the future?

3. Can lipid oils be used directly as fuel for internal combustion engines? If so, for which type of engine? Can you see any problems with such use in cold climates (try putting olive oil into your refrigerator's freezer and see what happens)? How are lipid oils currently used to make engine fuel?

4. Look up the history of canola oil. Has it been used as food oil for a long time? What has plant breeding had to do with canola oil?

5. Some medicinal agents, particularly proteinaceous substances such as insulin, are now made by fermentation using microorganisms. Is there any possibility of using plants instead? If so, what would be the advantages of plants?

6. Look up biorefineries on the Internet? Are there any biorefineries currently operating at least on a pilot scale? If so, what do they make? Some authorities suggest that in the future there will be large biorefineries operating on algal biomass. Why might this be the case?

7. How large are world food reserves? Is food production vulnerable to interruptions that might result in widespread food shortages? Are there any examples of such interruptions in recorded history?

8. When was hybrid corn developed? When did it come into widespread use? Why is corn especially amenable to hybridization? What has been the impact of hybrid corn on overall bioproductivity?

9. Plants that are hybrids and those that are produced by recombinant DNA are both important in the production of agricultural crops. What is the distinction between the two? May a crop variety be both a hybrid and a product of recombinant DNA?

10. What is the most common environmental problem associated with herbicides? From a search of the Internet, see if there are any concerns regarding herbicides and the Ogallala aquifer?

11. Look up the controversy involving Monarch butterflies and Bt herbicides. What was the suspected problem? Was anything done to fix it? Is it still regarded as a problem?

12. Look up "landraces." What do landraces have to do with sustainability, especially in respect to agricultural sustainability?

13. What are some of the major pitfalls involved with trying to reestablish viable populations of a species when numbers have fallen to very low values? Name and describe two species of large birds in the United States where attempts have been made to build up a breeding population in the wild from only a small group of individuals. What is the degree of success with these efforts?

14. What is biomonitoring and what are bioindicators? How is biomonitoring done for tropospheric ozone pollution and what are some specific bioindicators of such pollution?

15. What is the distinction between "good ozone" and "bad ozone" in the atmosphere? Where is each found?

16. What is the history of quinoa as a food? How might more extensive cultivation of quinoa add to the biosphere and lead to increased human nutrition in the future?

17. Where did potatoes originate as a cultivated crop? What happened sometime after potatoes were introduced into Europe that tragically illustrated the vulnerability of the biosphere and food supply?

18. What are biofuels? What is the conflict of "food versus fuel?"

19. Suggest why the production of useful organic compounds from lignocellulose is considerably more complex than the acquisition of oils from oil seeds such as soybeans.

20. In the decades preceding about 1950, there was considerable concern that world populations were exceeding levels that could be supported by the biosphere. What happened to change that picture and what was this "revolution" called? What are some of the environmental and sustainability factors that may make dire predictions of the adverse effects of overpopulation come true in coming decades?

21. What are two advantages of producing Bt insecticide in transgenic crops compared to making the insecticide using bacterial processes and simply spreading it onto crops?

22. What are seed banks and how are they related to preservation and enhancement of the biosphere? Can you find any specific examples of seed banks?

LITERATURE CITED

1. Barnosky, Anthony D., Has Earth's Sixth Mass Extinction Already Arrived? *Nature* **471**, 51–57 (2011).

2. Rosillo-Calle, Frank, and Francis X. Johnson, Eds., *Food versus Fuel: An Informed Introduction to Biofuels*, Zed Books, London, 2011.

3. Aresta, Michele, Angela Dibenedetto, and Frank Dumeignil, Eds., *Biorefinery: From Biomass to Chemicals and Fuels*, Walter de Gruyter, Berlin, 2012.

4. Kingsbury, Noel, *Hybrid: The History and Science of Plant Breeding*, University of Chicago Press, Chicago, IL, 2011.

5. McElheny, Victor, *Drawing the Map of Life: Inside the Human Genome Project*, Basic Books, New York, 2010.

6. James, Clive, 2010 ISAAA Report on Global Status of Biotech/GM Crops, *International Service for the Acquisition of Agri-biotech Applications* (ISAAA), http://www.isaaa.org.
7. Dewar, Alan M., Weed Control in Glyphosate-Tolerant Maize in Europe, *Pest Management Science* **65**, 1047–1058 (2009).
8. Tang, Guangwen, Jian Qin, Gregory G. Dolnikoswki, Robert M. Russell, and Michael A. Grusak, Golden Rice is an Effective Source of Vitamin A, *American Journal of Clinical Nutrition* **89**, 1776–1783 (2009).
9. Van Andel, Jelte, and James Aronson, *Restoration Ecology: The New Frontier*, 2nd ed., Wiley/Blackwell, Hoboken, NJ, 2012.
10. Wilkinson, Sally, Gina Mills, Rosemary Illidge, and William J. Davies, How is Ozone Pollution Reducing Our Food Supply?, *Journal of Experimental Botany* **63**, 527–326 (2012).
11. Graber, Cynthia, Farming Like the Incas, Smithsonian, September 7, 2011, http://www.smithsonianmag .com/history-archaeology/Farming-Like-the-Incas.html#.

SUPPLEMENTARY REFERENCES

Brand, Stewart, *Whole Earth Discipline: Why Dense Cities, Nuclear Power, Transgenic Crops, Restored Wildlands, and Geoengineering are Necessary*, Penguin, New York, 2010.
Brooker, Robert, Eric Widmaier, Linda Graham, and Peter Stiling, *Biology*, McGraw Hill, New York, 2010.
Connelly, Matthew, *Fatal Misconception: The Struggle to Control World Population*, Belknap Press of Harvard University Press, Cambridge, MA, 2008.
Curtis, Ian S., Ed., *Transgenic Crops of the World: Essential Protocols*, Kluwer Academic Publishers, Boston, MA, 2004.
Dickinson, Gordon, and Kevin Murphy, *Ecosystems*, 2nd ed., Routledge, New York, 2007.
Dotretsov, Nicolay, *Biosphere Origin and Evolution*, Springer, Berlin, 2007.
Gimelli, Salvatore Paul, *Carbohydrates: Fundamentals and Applications*, Micelle Press, Port Washington, NY, 2006.
Gressel, Jonathan, *Genetic Glass Ceilings: Transgenics for Crop Biodiversity*, Johns Hopkins University Press, Baltimore, MD, 2008.
Hu, Thomas Q., Ed., *Chemical Modification, Properties, and Usage of Lignin*, Kluwer Academic Publishers, New York, 2002.
Huggett, Richard J., *The Natural History of the Earth: Debating Long Term Change in the Geosphere and the Biosphere*, Routledge, New York, 2006.
Ikerd, John E., *Crisis and Opportunity: Sustainability in American Agriculture Our Sustainable Future*, Bison Books, Winnipeg, 2008.
Kamide, Kenji, *Cellulose and Cellulose Derivatives: Molecular Characterization and its Applications*, Elsevier, Amsterdam, 2005.
Kaufman, Don, and Cecilia Franz, *The Biosphere: Protecting our Global Environment*, 4th ed., Kendall Hunt Publishing, Dubuque, IA, 2005.
Kole, Chittaranjan, Charles Michler, Albert G. Abbott, and Timothy C. Hall, *Transgenic Crop Plants*, Springer, Berlin, 2010.
Kunitz, Stephen J., *The Health of Populations: General Theories and Particular Realities*, Oxford University Press, Oxford, UK, 2007.
Kuo, Tsung Min, and Harold W. Gardner, Eds., *Lipid Biotechnology*, Marcel Dekker, New York, 2002.
León, Rosa, Aurora Gaván, and Emilio Fernández, *Transgenic Microalgae as Green Cell Factories*, Springer, Berlin, 2006.
Manning, Richard, *Against the Grain: How Agriculture Has Hijacked Civilization*, North Point Press, New York, 2005.
Mazoyer, Marcel, and Laurence Roudart, *A History of World Agriculture: From the Neolithic Age to the Current Crisis*, Monthly Review Press, New York, 2006.
McNeil, Brian, and Linda Harvey, *Practical Fermentation Technology*, Wiley, New York, 2008.
Molles, Manuel C., *Ecology: Concepts and Applications*, 4th ed., McGraw-Hill Higher Education, Boston, MA, 2008.
Pua, Eng Chong, and Michael R. Davey, Eds., *Transgenic Crops VI*, Springer, Berlin, 2007.
Raven, Peter H., Linda R. Berg, and David M. Hassenzahl, *Environment*, 6th ed., Wiley, Hoboken, NJ, 2008.
Schmitz, Oswald J., *Ecology and Ecosystem Conservation*, Island Press, Washington DC, 2007.
Smil, Vaclav, *The Earth's Biosphere: Evolution, Dynamics, and Change*, MIT Press, Cambridge, MA, 2002.

Spilsbury, Louise, and Richard Spilsbury, *Food Chains and Webs: From Producers to Decomposers*, Heinemann Library, Chicago, IL, 2004.

Stevens, Christian V., and Roland Verhé, Eds., *Renewable Bioresources: Scope and Modification for Non-Food Applications*, Wiley, Hoboken, NJ, 2004.

UNESCO, *Knowledge for Sustainable Development: An Insight into the Encyclopedia of Life Support Systems*, UNESCO, Paris, 2002.

Zikov, Gennady E., Ed., *Biotechnology and Industry*, Nova Science Publishers, New York, 2004.

14 Environmental and Toxicological Chemistry of the Anthrosphere

14.1 ANTHROSPHERE

Shown as one of the five environmental spheres in Chapter 1, Figure 1.2, the **anthrosphere** is the part of the environment made and operated by humans. The anthrosphere is where pollutants are made and from which they are released with profound effects on all the other environmental spheres. It is also strongly affected by pollutants, for example, acid rain that causes deterioration of stone structures and corrosion of metal components.

It is essential to view the anthrosphere as a distinct environmental sphere when considering environmental chemistry and sustainability. Just a look around us shows the dwellings, buildings, roads, airports, factories, power lines, and numerous other things constructed and operated by humans as visible evidence of the existence of the anthrosphere on Earth (see Figure 14.1). The anthrosphere and its influences are so obvious and even intrusive that the Nobel Prize–winning atmospheric chemist Paul Crutzen has argued convincingly that Earth is undergoing a transition from the Holocene geological epoch to a new one, the **Anthropocene**. This is occurring because human activities are now quite significant compared to nature in their impact on Earth's environment and are changing Earth's fundamental physics, chemistry, and biology. There is concern that, especially through changes in global climate, activity in the anthrosphere will detrimentally alter Earth's relatively stable, nurturing environment and produce one that is much more challenging to human existence.

There are many distinct segments of the anthrosphere as determined by a number of factors including where and how humans dwell; the movement and distribution of goods; the provision of services; the utilization of nonrenewable materials; the provision of renewable food, fiber, and wood; the collection, conversion, and distribution of energy; and the collection, treatment, and disposal of wastes. With these factors in mind, it is possible to list a number of specific things that are parts of the anthrosphere, as shown in Figure 14.1. These include dwellings as well as other structures used for manufacturing, commerce, education, and government functions. Utilities include facilities for the distribution of water, electricity, and fuel; systems for the collection and disposal of municipal wastes and wastewater (sewers); and—of particular importance to sustainability—systems for materials recycle. Transportation systems include roads, railroads, and airports, as well as waterways constructed or modified for transport on water. The anthrospheric segments used in food production include cultivated fields for growing crops and water systems for irrigation. A variety of machines, including automobiles, trains, construction machinery, and airplanes, are part of the anthrosphere. The communications sector of the anthrosphere includes radio transmitter towers, satellite dishes, and fiber optics networks. Oil and gas wells are employed for extracting fuels from the geosphere, and mines are excavated into the geosphere for removing coal and minerals.

The boundaries between the anthrosphere and the other environmental spheres tend to be blurred. Most of the anthrosphere including buildings, highways, and railroads is anchored to the geosphere. Gardens that adorn the anthrosphere are planted on geospheric soil, and the flowering plants in them are part of the biosphere. Farm fields are modifications of the geosphere, but the crops raised on

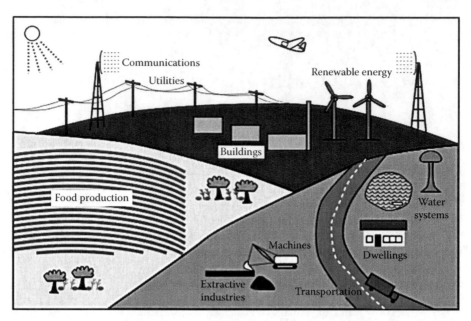

FIGURE 14.1   There are many aspects to the anthrosphere as illustrated by the examples in this figure. It is tied closely with the other environmental spheres.

them are part of the biosphere. Coal mines are burrowed into the geosphere. Ships move over ocean waters in the hydrosphere, and airplanes fly through the atmosphere.

14.1.1 CRUCIAL ANTHROSPHERIC INFRASTRUCTURE

A critical part of the anthrosphere may be classified as **infrastructure**.[1] Infrastructure is generally considered to be parts of the anthrosphere used by large numbers of people in common. It consists of utilities, facilities, and systems essential to a properly operating society. The physical components of infrastructure include electrical power–generating facilities and distribution grids, communications systems, roads, railroads, air transport systems, airports, buildings, water supply and distribution systems, and waste collection and disposal systems. A very important part of infrastructure is nonphysical, composed of laws, regulations, instructions, and operational procedures (see the discussion on sociosphere in Section 14.1.2). Because they are used in common by many people, crucial parts of infrastructure are in the public sector; other segments are privately owned and operated. For example, major airports are almost always publicly owned, whereas the aircraft that serve them are generally owned by private corporations.

An outdated, cumbersome, poorly designed, and worn-out anthrospheric infrastructure causes economic systems and societies to operate in a very inefficient manner that is inconsistent with sustainability. Catastrophic failure can result as has occurred with cascading breakdowns of electrical power grids or failure of wastewater treatment systems resulting in the discharge of sewage to streams. The devastating magnitude 9.0 earthquake that occurred on March 11, 2011, off the coast of Honshu, Japan, about 373 km (231 mi) northeast of Tokyo and the devastating tsunami that followed illustrated the catastrophic effects of infrastructure failure that resulted ultimately in the greatest nuclear emergency since the 1986 Chernobyl nuclear meltdown and fire. This occurred at the Fukushima Daiichi Nuclear Power Station. The three reactors that were running at the plant shut down automatically when the quake hit, as they were designed to do. However, the quake destroyed the electrical power infrastructure leading to the plant so that water could not be pumped to cool the reactor core and the spent fuel rods normally stored under water close to the core. Zirconium metal alloy cladding the hot fuel rods reacted with water to generate hydrogen gas, which exploded, destroying the structures that sheltered the reactors. Significant amounts of radioactivity were released.

The deterioration of infrastructure is a continuing concern. One of the greatest problems is corrosion, a chemical process in which metals, such as the steel that composes bridge girders, tend to revert to the state in which they occur in nature (in the case of steel, rust). Human negligence, misuse, and vandalism can all cause premature loss of infrastructure function. A major concern with terrorism is potential damage to infrastructure, including cybercrime that could cripple electrical distribution systems. Infrastructure problems frequently begin with improper design. Sustainability requires that elements of infrastructure be properly designed, maintained, and protected to avoid the expense and material and energy required to rebuild infrastructure if it fails before its expected lifetime is up.

Careful attention must be given to sustainability and the maintenance of environmental quality in the development of infrastructure. Examples include highly effective waste treatment systems with recovery of materials and energy from wastes, high-speed rail systems to replace inefficient movement of people and freight by private cars and by trucks, and electrical systems that use wind and solar power to the maximum extent possible.

Critical infrastructure is a term applied to those parts of the infrastructure that, if destroyed or harmed, could have an immediate and devastating effect on the function of a society.[2] The U.S. Department of Homeland Security defines critical infrastructure as including (but not limited to) telecommunications, energy, banking and finance, transportation, water systems, and emergency services, both governmental and private. Critical infrastructure is obviously of concern with respect to terrorist threats.

14.1.2 SOCIOSPHERE

The **sociosphere** is the societal organization of people including their governments, laws, cultures, religions, families, and social traditions and is a critical part of infrastructure. A well-functioning sociosphere enables people to lead good lives within a sustainable environment and economic system. Largely because of dysfunctional social systems, the quality of life and the environment in some countries with substantial resources, especially of petroleum, is often substandard. Societies in countries with dictatorial, corrupt governments that do not nurture human rights are not conducive to the maintenance of sustainability. Sustainability and quality of life are also not served well by antigovernment creeds that reject the role of well-functioning governments in implementing sensible, well-administered laws and regulations designed to protect the environment, promote human well-being, and maintain sustainability.

An important consideration in the sociosphere is the science of **economics**, which describes the production, distribution, and use of income, wealth, and materials (commodities). Much of economics as it has been traditionally practiced is inconsistent with the development of sustainability on which functional economic systems must ultimately depend. Economic value has traditionally been measured in terms of financial and material possessions with emphasis on growth and with a narrow view of the environment. Earth has been largely regarded as a part of the economic system from which materials may be extracted, which is to be "developed" with structures and other artifacts of the anthrosphere, and into which wastes are to be discarded. Such an approach is putting a severe strain on environmental support systems and Earth's natural capital (see Chapter 1, Section 1.4). A more enlightened economic view regards Earth's natural capital as an endowment. As with financial endowments, Earth's store of natural capital should be nurtured, with only a portion of its income spent for immediate needs and the rest devoted to enhancing the natural capital. Therefore, it is essential for sustainability that economics is viewed as a part of Earth's greater environmental system, rather than viewing the environment as a subsection of a world economic system. Instead of defining wealth in material possessions, it should be measured in terms of well-being and satisfaction with life, operating within rules that promote and require sustainability.

14.2 INDUSTRIAL ECOLOGY AND INDUSTRIAL ECOSYSTEMS

Industrial ecology, which is an important aspect of the anthrosphere, describes a web of industrial concerns, distributors, and other enterprises functioning to mutual advantage, using each other's products, recycling each other's potential waste materials, and utilizing materials and energy as efficiently as possible.[3] Above all, it is a **sustainable** means of providing goods and services; the practice of industrial ecology is important in maintaining the anthrosphere in a state of compatibility with the other environmental spheres and with Earth. In so doing, industrial ecology considers every aspect of the provision of goods and services from concept, through production, and to the final fate of products remaining after they have been used. Industrial ecology considers industrial systems in a closed-loop model rather than a linear one, thereby emulating natural biological ecosystems, which are sustainable by nature.

By analogy with natural ecosystems, an **industrial ecosystem** consists of a group of mutually interacting enterprises practicing industrial ecology. A key measure of the success of such a system can be given by the following relationship:

$$\frac{\text{Market value of products and services}}{\text{Consumption of material and energy}}$$

As has been the case with natural ecosystems, the best means of assembling industrial ecosystems is through natural selection in which the various interests involved work out mutually advantageous relationships. However, with a knowledge of the feasibility of such systems, external input and various kinds of incentives can be applied to facilitate the establishment of industrial ecosystems.

Just as organisms in natural ecosystems develop strong symbiotic relationships—the inseparable union of algae and fungi in lichens growing on rock surfaces, for example—concerns operating in industrial ecosystems develop a high degree of **industrial symbiosis**. It is the development of such mutually advantageous interactions between two or more industrial enterprises that results in the self-assembly of an industrial ecosystem in the first place. The recycling components of an industrial ecosystem are absolutely dependent on symbiotic relationships with their sources of supply and with their markets for recycled products.

Figure 14.2 outlines a general industrial ecosystem. The major inputs to such a system are energy and virgin raw materials. A successful system minimizes use of virgin raw materials and maximizes efficiency of energy utilization. The materials processing sector produces processed materials such as sheet steel or synthetic organic polymers. These in turn go to a goods fabrication sector in which the processed materials are formed and assembled or, in the case of consumables such as detergents, formulated to give the desired product. Scrap materials, rejected components, and off-specification consumables generated during goods fabrication may go to the recycling and remanufacturing sector. From goods fabrication, manufactured items or formulated substances are taken to a user sector, which includes consumers and industrial users. In a successful system of industrial ecology, waste materials from the user sector are minimized and, ideally, totally eliminated. Spent goods from the user sector are taken to the recycling and remanufacturing sector to be introduced back into the materials flow of the system. Examples of such items include automobile components that are cleaned, have bearings replaced, and are otherwise refurbished for the rebuilt automobile parts market. Another typical item is paper, which is converted back to pulp that is made into paper again. In some cases, the recycling and remanufacturing sector salvages materials that go back to materials processing to start the whole cycle over. An example of such a material is scrap aluminum that is melted down and recast into aluminum for goods fabrication. Communications are essential to a successful industrial ecosystem, as is a reliable, rapid transportation system. It is especially important that these two sectors work well in modern manufacturing practice, which calls for "just-in-time" delivery of materials and components to avoid the costs of storing such items.

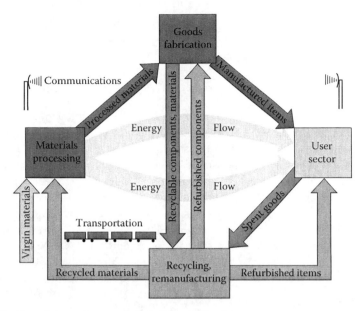

FIGURE 14.2 Outline of the major components of an industrial ecosystem.

An important characteristic of an industrial ecosystem is its **scope**. A regional scope large enough to encompass several industrial enterprises but small enough for them to interact with each other on a constant basis is probably the most satisfactory scale to consider. Frequently, such systems are based around transportation systems. Webs of interstate highways over which goods and materials move between enterprises by truck may constitute industrial ecosystems.

14.2.1 KALUNDBORG INDUSTRIAL ECOSYSTEM

The most commonly cited example of a specific industrial ecosystem is one that occupies a relatively small region around Kalundborg, Denmark. The Kalundborg industrial ecosystem is often cited for the spontaneous way in which it developed, beginning in the 1960s with steam and electricity provided to the petroleum refinery from the power plant then branching out to a large variety of other enterprises in the vicinity, as shown in Figure 14.3. The wallboard-manufacturing facility

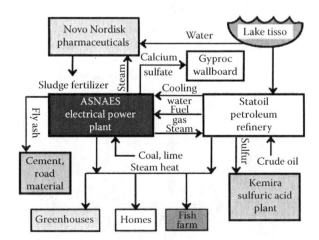

FIGURE 14.3 The system of industrial ecology that has developed spontaneously around Kalundborg, Denmark.

developed from the regulatory requirement for the removal of sulfur dioxide from stack gas in the coal-fired power plant and sulfuric acid manufacture was the result of the requirement to remove sulfur from petroleum.

14.3   METABOLIC PROCESSES IN INDUSTRIAL ECOSYSTEMS

Industrial metabolism refers to the processes to which materials and components are subjected in industrial ecosystems. It is analogous to the metabolic processes that occur with food and nutrients in biological systems. Like biological metabolism, industrial metabolism may be addressed at several levels. A level of industrial metabolism at which green chemistry especially comes into play is at the molecular level where substances are changed chemically to give desired materials or to generate energy. Industrial metabolism can be addressed within individual unit processes in a factory, at the factory level, at the industrial ecosystem level, and even globally.

A significant difference between industrial production as it is now largely practiced and natural metabolic processes relates to the wastes that these systems generate. Natural ecosystems have developed such that true wastes are virtually nonexistent. For example, even those parts of plants that remain after biodegradation of plant materials form soil humus that improves the conditions of soil on which plants grow. Anthropogenic industrial systems, however, have developed in ways that generate large quantities of wastes, where a waste may be defined as the *dissipative use of natural resources*. Furthermore, human use of materials has a tendency to dilute and dissipate the materials and disperse them to the environment. Materials may end up in a physical or chemical form from which reclamation becomes impractical because of the energy and effort required. A successful industrial ecosystem overcomes such tendencies.

Organisms performing their metabolic processes degrade materials to extract energy (catabolism) and synthesize new substances (anabolism). Industrial ecosystems perform analogous functions. The objective of industrial metabolism in a successful industrial ecosystem is to make desired goods with the least amount of by-product and waste. This can pose a significant challenge. For example, to produce lead from lead ore for the manufacture of lead/acid storage batteries requires mining large quantities of ore, extracting the relatively small fraction of the ore consisting of lead sulfide mineral, and roasting and reducing the mineral to get lead metal. The whole process generates large quantities of lead-contaminated tailings left over from mineral extraction and significant quantities of by-product sulfur dioxide, which must be reclaimed to make sulfuric acid and not released to the environment. The recycling pathway, by way of contrast, takes essentially pure lead from recycled batteries and simply melts it down to produce lead for new batteries. The advantages of recycling in this case are obvious, although there can be significant pollution and health problems connected with lead recycling, such as some operations in Mexico (see Chapter 11, Section 11.7.2).

There are some interesting comparisons between natural ecosystems and industrial systems as they now operate. The basic unit of a natural ecosystem is the organism, whereas that of an industrial system is the firm or, in the case of large corporations, the branch of a firm. Natural ecosystems handle materials in closed loops; in current practice, materials traverse an essentially one-way path through industrial systems. It follows that natural systems completely recycle materials, whereas in industrial systems the level of recycling is often very low. Organisms have a tendency to concentrate materials. For example, carbon in carbon dioxide that is only about 0.04% of atmospheric air is transferred to the biosphere and becomes concentrated in organic carbon through photosynthesis. Industrial systems in contrast tend to dilute materials to a level where they cannot be economically recycled but still have the potential to pollute. Aside from maintaining themselves during their limited lifetime, the major function of organisms is reproduction. Industrial enterprises do not have reproduction of themselves as a primary objective; their main function is to generate goods and services in a manner that maximizes monetary income.

Unlike natural ecosystems in which reservoirs of needed materials are essentially constant (oxygen, carbon dioxide, and nitrogen from air as examples), industrial systems are faced with largely depleting reservoirs of materials. For example, the aforementioned lead ore is a depleting resource; more will be found, but only a finite amount is ultimately available. Fossil energy resources are also finite. For example, much more fossil energy from coal may be available, but its utilization as the world's main source of energy over the long term would come at an unacceptable cost of global warming from carbon dioxide emissions. Again, industrial metabolic processes that emphasize recycling are desirable because recycling gives partially renewable reservoirs of materials in the recycling loop. Ideally, even in the case of energy, renewable energy resources such as wind and solar energy are continually replenished.

The metabolism that occurs in an entire natural ecosystem is **self-regulating**. If herbivores that consume plant biomass become too abundant and diminish the stock of the biomass, their numbers cannot be sustained, the population dies back, and their food source rebounds. The most successful ecosystems are those in which this self-regulating mechanism operates continuously without wide variations in populations. In contrast, industrial systems do not inherently operate in a self-regulating manner that is advantageous to their surroundings, or even to themselves in the long run. Examples of the failure of self-regulation of industrial systems abound in which enterprises have wastefully produced large quantities of goods of marginal value, running through limited resources in a short time, dissipating materials to their surroundings, polluting the environment in the process, and adding to economic instability. Despite these bad experiences, within a proper framework of laws and regulations designed to avoid wastes and excesses, industrial ecosystems can be designed to operate in a self-regulating manner. Such self-regulation operates best under conditions of maximum recycling in which the system is not dependent on a depleting resource of raw materials or energy.

Obviously, recycling is the key to the successful function of industrial metabolism. Figure 14.4 illustrates the importance of the **level of recycling**. In low-level recycling, a material or component is taken back to near the beginning of the steps through which it is made. For example, an automobile engine block might be melted down to produce molten metal from which new blocks are then cast. With high-level recycling, the item or material is recycled as close to the final product as possible. In the case of the automobile engine block, it may be cleaned, the cylinder walls rehoned, the flat surfaces replaned, and the block used as the platform for assembling a rebuilt engine. In this example and many others that can be cited, high-level recycling uses much less energy and materials and is inherently more efficient. The term given to the value attributed to an item or material recycled near the top of the energy/materials pyramid shown in Figure 14.4 is called its **embedded utility**.

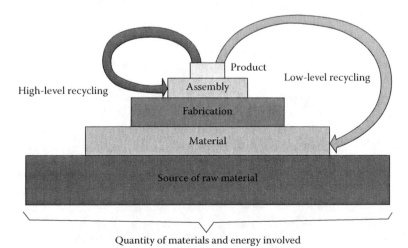

FIGURE 14.4 The level at which recycling occurs in the energy/materials pyramid strongly influences the amount of energy required and the quantity of materials that must be processed.

14.3.1 ATTRIBUTES OF SUCCESSFUL INDUSTRIAL ECOSYSTEMS

An important characteristic of successful biological ecosystems is their **inertia**, which is their resistance to alteration and damage, the key factors of which are **productivity** of basic food materials, **diversity** of species, **constancy** of numbers of various organisms, and **resilience** in the ability of populations to recover from loss. Industrial ecosystems likewise have key attributes that are required for their welfare. These include **energy**, **materials**, and **diversity**.

With enough usable energy, almost anything is possible. Therefore, the provision of adequate amounts of energy that can be used without damaging the environment too much is essential for the function of industrial ecosystems. The provision of such energy is a central challenge facing humankind. Fossil fuels are depleting resources, and their use contributes greenhouse-warming gases to the atmosphere. Wind and solar energy are disperse and intermittent sources. The 2011 nuclear disaster at Japan's Fukushima Daiichi Nuclear Power Station has given pause to the development of nuclear energy sources.

There are several approaches to providing materials. These can be classified as **dematerialization** in which less material is used for a specific purpose, **substitution** of abundant materials for scarce ones, **recycling** materials, and **waste mining** in which needed materials are extracted from wastes.

Examples abound in recent decades in which the need for materials from depleting resources has been reduced by dematerialization and material substitution. The switch from 6-V to 12-V electrical systems in automobiles has enabled lighter wiring in automotive electrical systems. The switch to digital photography has essentially eliminated the use of silver in photography. The most spectacular advances have been made in electronics where material substitution has provided electronic circuits with many orders of magnitude more capability than the circuits that they replaced. The glowing, electricity-consuming vacuum tubes, capacitors, resistors, and transformers of the receiver circuit of a 1950s tabletop radio have been replaced with a tiny circuit almost invisible to the human eye. The huge numbers of copper wires that carried telegraph and telephone messages in the 1940s have now been replaced by fiber optic signal conductors that carry unimaginably more information per unit mass of carrier. Polyvinylchloride (PVC) pipe, synthesized from inexhaustible world resources of chlorine and potentially from biomass hydrocarbon sources, has replaced copper and steel for water and wastewater transmissions.

Recycling is of course one of the major objectives of a system of industrial ecology, one in which significant progress is being made. There are some consumable items that are not practical to recycle, and for them the raw materials are abundant enough that recycling is not required. Household detergents are in this category. A second group of recyclables consists of those items that are not particularly scarce but for which recycling is feasible and desirable. Wood and paper fall into this category. A third category of recyclable materials consists of metals, particularly the more valuable and scarce ones, such as chromium, platinum, and palladium. These metals definitely should be recycled. A fourth category of recyclables consists of parts and apparatus that can be refurbished and reused.

Waste mining, the extraction of useful materials from wastes, provides more materials while benefitting the environment. One of several important examples of waste mining is the extraction of combustible methane gas, a low-polluting premium fossil fuel, from municipal refuse landfills in which the biodegradation of organic matter in the absence of oxygen generates the gas. Sulfur in sulfur dioxide extracted from the flue gases generated in burning coal that contains sulfur can be reclaimed and used to make sulfuric acid. Methods have been developed to extract aluminum from finely divided coal fly ash generated in coal combustion. In this case, the finely divided, homogeneous, dry nature of the fly ash is a definite advantage in processing it. Until 1992, Japan's Mitsubishi operated a refining operation in Malaysia to recover the rare earths, essential raw materials for modern technological applications, left in huge piles of tailings from tin ore refining that had been carried out in the country since the 1820s. As of 2011, the operation was being refurbished to process rare earth ores from Australia.

14.3.2 DIVERSITY

Diversity in industrial ecosystems tends to impart a **robust** character to them, which means that if one part of the system is diminished other parts will take its place and keep the system functioning well. Many communities that have become dependent on one enterprise or just a few major enterprises have suffered painful economic crises when a major employer leaves or cuts back. The fouling of beaches in Louisiana, Mississippi, Alabama, and Florida from the 2010 BP Deepwater Horizon oil well blowout in the Gulf of Mexico devastated the tourist trade in that region for some time and forced painful economic adjustments. In many parts of the world, water supply from a single vulnerable source threatens diversity.

14.4 LIFE CYCLES IN INDUSTRIAL ECOSYSTEMS

In conventional industrial systems, a product is manufactured and marketed after which the vendor forgets about it (unless some product defect, such as sticking accelerator pedals on an automobile, forces a recall). In a system of industrial ecology, however, the entire **life cycle** of the product is considered. An important aspect of such a consideration is **life cycle assessment**.[4] The overall goal of a life cycle assessment is to determine, measure, and minimize environmental and resource impacts of products and services.

An important decision that must be made at the beginning of a life cycle assessment is determination of the **scope** of the assessment. Evaluation of the scope includes the time period to be considered; the area (space) to be considered; and the kinds of materials, processes, and products that will go into the assessment. As an example, consider the chemical synthesis of an insecticide that releases harmful vapors and generates significant quantities of waste material. A narrowly focused life cycle assessment might take into account control measures in the synthesis process to capture released vapors and the best means of disposal of the waste by-products. A broader scope would consider a different synthetic process that might not cause the problems mentioned. An even broader scope might consider whether or not the insecticide even needs to be made and used; perhaps there are more acceptable alternatives to its use.

Life cycle assessment involves an **inventory analysis** to provide information about the consumption of material and release of wastes from the point that raw material is obtained to make a product to the time of its ultimate fate, an **impact analysis** to consider the environmental and other impacts of the product, and an **improvement analysis** to determine the measures that can be taken to reduce impacts. A life cycle assessment gives a high priority to the choice of materials in a way that minimizes wastes. It considers which materials and whole components can be used or recycled. And it considers alternate pathways for manufacturing processes or, in the case of chemical manufacture, alternate synthesis routes.

In doing life cycle assessments, it is useful to consider the three major categories of **products**, **processes**, and **facilities**, all of which have environmental and resource impacts (Figure 14.5). Products are obviously the things and commodities that consumers use. They are discussed further in Section 14.5. Processes refer to the ways in which products are made. Facilities consist of the infrastructural elements in which products are made and distributed.

Some of the greater environmental impacts from commerce result from the processes by which items are made. An example of this is paper manufacture. The environmental impact of the paper **product** tends to be relatively low. Even when paper is discarded improperly, it does eventually degrade without permanent effect. But the **process** of making paper, beginning with the harvesting of wood and continuing through the chemically intensive pulping process and final fabrication, has significant environmental impact. In addition to potential air and water pollution, papermaking consumes energy and requires large amounts of water. Processes can be made much more environmentally friendly by the application of the principles of industrial ecology, enabling maximum recycling of materials that otherwise would have significant pollution potential.

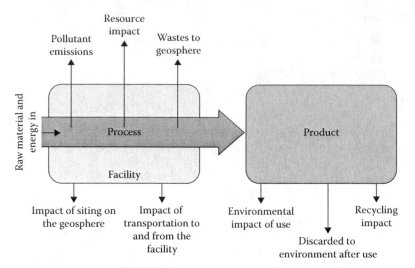

FIGURE 14.5   A life cycle assessment must consider the environmental and resource impacts of processes, the facilities in which processes are carried out, and the products themselves.

The impact of facilities can vary over a wide range. A specialized facility such as a steel mill or a petroleum refinery can have a significant environmental impact. Abandoned sites of these facilities can be blighted and difficult and expensive to restore for some other use. (As noted in Chapter 11, Section 11.10, the term "brownfields" is sometimes used to describe sites of abandoned industrial facilities, and restoration of blighted brownfields is often a major goal of urban renewal projects.) One of the more challenging kinds of facilities to decommission are sites of nuclear power reactors in which there is a significant amount of radioactivity to deal with in dismantling and disposing of some of the reactor components. The impact of facilities can be minimized by designing them with future use and eventual decommissioning in mind. Typically, well-designed commercial buildings may have a number of lives during which they are used by a sequence of enterprises. A key aspect of a building destined for multiple uses is structure flexibility so that it can easily be rearranged for new uses.

14.4.1 Product Stewardship

An important consideration with respect to the resource, environmental, and toxicological impacts of products is **product stewardship**, basically the control exercised in their use and ultimate fate. Product stewardship is relatively facile for products used within an enterprise or for those that are leased with responsibility for their fate retained by the owner rather than the user. Stewardship becomes more problematic for products sold to the consumer market. Refundable deposits such as those placed on beverage containers in some localities help ensure the recycling of products. Recycling is also increased by the ready availability of recycling centers or bins where spent items or materials may be deposited.

14.5 KINDS OF PRODUCTS

In considering life cycle assessments, it is useful to divide products into three major categories as illustrated in Figure 14.6. **Consumable products**, such as gasoline burned in an automobile engine, are used up or dispersed to the environment with no possibility of recovery. **Recyclable commodities** are potentially reclaimed, reprocessed, and reused; engine oil is such a material. **Service products** (sometimes called durable products) are devices that can be used multiple times and have a long lifetime. An automobile is a service product.

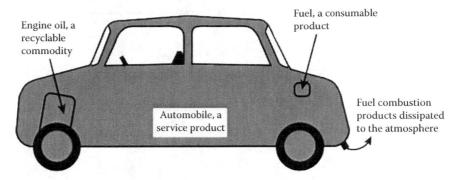

FIGURE 14.6 An automobile is a service product that is used multiple times. The fuel that it burns is a consumable material, the combustion products of which are dissipated to the atmosphere. The engine oil is a recyclable commodity that can be reprocessed and used again.

Since consumable products are dispersed to the environment, it is important that they have environmentally and toxicologically friendly characteristics. They should first of all be **nontoxic** at the levels and manner in which organisms are exposed to them. In addition to not causing acute toxicity, they should not be carcinogenic or mutagenic and should not cause birth defects. Another characteristic that consumable products and recyclable commodities should have is that they should not be **bioaccumulative**. As discussed in Chapter 3, Section 3.12, and illustrated in Figure 3.15, **bioaccumulation** is the term given to the uptake and concentration of xenobiotic materials by living organisms. Poorly biodegradable, lipid-soluble materials such as polychlorinated biphenyl (PCB) compounds have a strong tendency to bioaccumulate, and such substances should be avoided in consumable products. Consumable products should also be **degradable**. The most common type of degradation is biodegradation, which occurs primarily through the action of microorganisms. The practice of green chemistry can aid in making biodegradable products by, for example, avoiding branched-chain hydrocarbon structures in organic compounds and by attaching functional groups, such as the organic carboxylic acid group, $-CO_2H$, that are amenable to microbial attack.

Recyclable commodities should be designed with durability and recycling in mind. In order for them to last through a normal life cycle, such commodities should not be as degradable as consumables. An example of making a product more amenable to recycling is the use of bleachable and degradable inks on newsprint, which makes it easier to recycle the newsprint to produce a grade of recycled paper that meets acceptable color standards.

Although service products are designed to last relatively long, they do reach a stage requiring disposal or recycling. Service products should be designed and constructed to facilitate disassembly so that various materials can be separated for recycling. A key factor in recycling is the availability of channels through which such products can be recycled. Proposals have been made for "de-shopping" centers where items such as old computers and broken small appliances can be returned for recycling.

14.6 ENVIRONMENTAL IMPACTS OF THE ANTHROSPHERE

The anthrosphere certainly has profound effects on the atmosphere, hydrosphere, geosphere, and biosphere (Figure 14.7). Most environmental pollutants and toxicants dispersed to the environment originate in the anthrosphere. Anthrospheric influences may range from highly localized effects to global effects, such as greenhouse warming or stratospheric ozone depletion. The magnitude of the effects may be minor, or they may be catastrophic. Until recently, the effects of human activities on the surrounding environment were of relatively little concern, resulting in neglect that is the cause of many of the environmental problems that exist even today. However, the proper practice

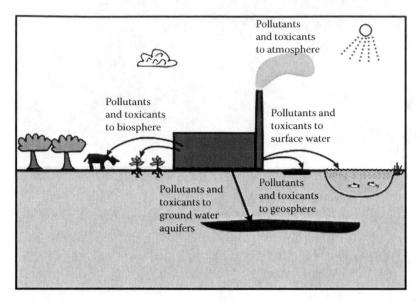

FIGURE 14.7   The anthrosphere is the source of many pollutants and toxicants that get into the other environmental spheres. These include particles, inorganic gases, and organics emitted to the atmosphere; inorganic and organic pollutants and toxicants that are discharged to surface water and that seep into groundwater; wastes disposed to the geosphere, a source of groundwater contamination; and pollutants and toxicants that affect plants and animals.

of industrial ecology requires that consideration be given to the various influences that anthrospheric activities have on the surrounding environment. Of particular concern are the toxic effects on humans and other living things of xenobiotic substances produced in and discharged from the anthrosphere.

Consider the kinds of effects that industrial activities may have on the natural environmental spheres. One of the most obvious influences is on the atmosphere because of the emission to the atmosphere of pollutant gases, vapors from volatile compounds, and particles. Released carbon dioxide and vapors such as those of fluorinated hydrocarbons have a high potential to cause greenhouse warming. Particles obscure visibility and cause adverse health effects in people who must breathe the air contaminated by particles. Chlorofluorocarbons lead to stratospheric ozone depletion and hydrocarbons and nitrogen oxides released to the atmosphere can cause formation of photochemical smog. Industrial activities often utilize large quantities of water for cooling and other purposes. Water may become polluted or be warmed excessively when used for cooling (thermal pollution). The biosphere is most affected by industrial activity when toxic substances are released. Other effects on the biosphere may be indirect as the result of adverse effects on the atmosphere, hydrosphere, or geosphere.

Many industries require large quantities of materials that are taken from Earth by the extractive industries. This may result in the disruption of the geosphere from mining and dredging and from pumping petroleum. The other major effect on the geosphere results from the need to dispose of wastes. Scarce land may be required for waste disposal dumps, and the geosphere may become contaminated with pollutants from disposed wastes.

Industrial systems are largely dependent on the utilization of fossil fuels, so many environmental effects are due to fossil fuel extraction and combustion. Greenhouse-warming carbon dioxide emissions, acid gas emissions, smog-forming hydrocarbons and nitrogen oxides, and deterioration of atmospheric quality from particles released from fossil fuel combustion are all atmospheric effects associated with fossil fuel combustion. Coal mining activities have the potential to release acid mine water to the hydrosphere, petroleum production can release brines or result in ocean oil spills, acid precipitation may acidify isolated lakes, and water used as cooling water in power plants may become thermally polluted. The geosphere may be disrupted by fossil fuel extraction, especially in the surface mining of coal. Coal is extracted from some areas of West Virginia by cutting off entire mountaintops overlying coal seams and dumping the overburden into the valleys below in order to get to the coal. Effects on the biosphere from fossil fuel utilization may be direct (birds coated with tar from oil spills come to mind); but they are more commonly indirect, such as acidified bodies of water from acid rain resulting from sulfur dioxide emissions from coal combustion.

14.6.1 IMPACT OF AGRICULTURAL PRODUCTION

Agricultural activities certainly have to be considered as parts of the anthrosphere, and modern agricultural practices are part of vast agriculturally based industrial systems. Large quantities of greenhouse-warming methane are released to the atmosphere from the action of anoxic bacteria in rice paddies and in the intestines of ruminant animals. "Slash-and-burn" agricultural techniques practiced in some tropical countries release greenhouse gas carbon dioxide to the atmosphere and destroy the capacity of forests to sequester atmospheric carbon dioxide by photosynthesis. Enormous quantities of water are run through irrigation systems. Some of this water is evaporated and lost from the hydrosphere. The water that returns to the hydrosphere from irrigated fields picks up significant amounts of salt from the land and fertilizers applied to the land, so water salinity can become a problem. Underground aquifers become severely depleted by pumping large quantities of water for irrigation.

The production of protein from livestock requires much more water overall than the production of an equivalent amount of protein from grain. Animal wastes from huge livestock feedlots are notorious water polluters, adding oxygen-depleting biochemical oxygen demand (BOD) (see Chapter 3, Section 3.5) and potentially toxic inorganic nitrogen compounds to water. The disturbance of the geosphere from crop cultivation is enormous. Raising livestock for food entails a much greater degree of land cultivation than the cultivation of cereal grains. Agricultural production replaces entire, diverse biological ecosystems with artificial ecosystems, which causes a severe disturbance in the natural state of the biosphere. Another agricultural activity that affects the biosphere is the loss of species diversity in the raising of crops and livestock. In addition to the loss of entire species of organisms, the number of strains or breeds of organisms grown within species tends to become severely diminished in modern agricultural practice. Obviously, those varieties of crops and livestock that are most productive are the ones that will be used to produce grain, meat, and dairy products. However, if a problem develops, such as a particular variety of crop becoming susceptible to a newly mutated virus, alternative resistant varieties may no longer be available. Finally, the raising of transgenic crops and livestock poses some danger of profound and potentially unforeseen effects on the biosphere.

Figure 14.8 illustrates an environmental/toxicological effect that has been observed in some concentrated livestock feeding operations. Livestock feces and urine contain organically bound nitrogen, including urea in urine, that is deposited on soil in the feedlot. Bacterial biodegradation of the nitrogenous wastes initially produces ammonium ion, NH_4^+. This ion is bound by cation exchange processes to the soil and does not move readily with water in soil. However, in the top layers of the waste exposed to air, the ammonium nitrogen is oxidized to nitrate ion, NO_3^-. As an anionic species, the nitrate is not bound well to soil and can be carried into groundwater by infiltrating surface water. The groundwater pumped from a well located close to the feedlot may be contaminated with the nitrate. The nitrate in water consumed by cattle or other ruminant animals (those with complex digestive systems in which ruminant bacteria convert cellulose to organic acids that the animals absorb in their intestinal tracts and utilize as food) can be converted in the animal digestive tracts to nitrite ion, NO_2^-. The nitrite ion converts the iron(II) in the animal blood hemoglobin to iron(III) producing methemoglobin, which does not carry oxygen in the bloodstream, with the result that the animal may suffer oxygen deprivation and become ill or even die. Adult humans are

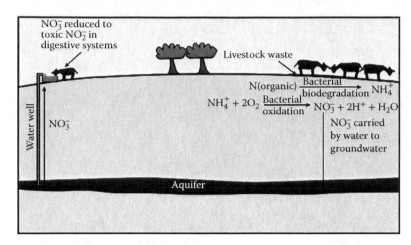

FIGURE 14.8 Organic nitrogen in livestock wastes in concentrated livestock feeding operations is mineralized to ammonium ion, which in turn is oxidized to nitrate by oxic bacteria. Nitrate ion is not retained well by soil and gets into groundwater by the seepage of surface water. Nitrate in contaminated well water gets converted to nitrite in the digestive systems of ruminant animals, converting blood hemoglobin to methemoglobin, which prevents the blood from transporting oxygen to tissues in such animals.

relatively tolerant of nitrate in drinking water, but nitrate may be reduced to nitrite in the stomachs of infants resulting in the production of methemoglobin and in oxygen deprivation. In a few cases, this methemoglobinemia condition ("blue baby") that results from mixing infant formula using nitrate-contaminated water has been fatal to infants.

14.6.2   Design of Industrial Ecosystems to Minimize Environmental Impact

From the discussion in Section 14.6, it is obvious that industrial activity, broadly defined to include agriculture also, has a high potential to adversely affect the atmosphere, hydrosphere, biosphere, and geosphere. Inherent to the ways in which properly designed industrial ecosystems operate, however, are measures and systems designed to minimize such impacts.

Several measures may be taken to minimize the effects of industrial ecosystems on the geosphere. Since most of the raw materials required for manufacturing originally have to be extracted from the geosphere, the recycling of materials within well-designed industrial ecosystems minimizes their impact on the geosphere. The selection of materials can also be important. As an example, the mining of copper to make copper wire, once widely used to carry communications signals, involves digging large holes in the ground and exposing minerals that tend to release toxic heavy metals and acidic pollutants. The silica used in the fiber optic cables that now largely substitute for copper is simply obtained from sand. The impacts of disturbing the geosphere for food and fiber production can be minimized by some of the conservation methods and agricultural practices discussed in Chapter 11, Section 11.11.

Well-designed industrial ecosystems emit much less harmful material to the atmosphere than conventional industrial systems. Industrial atmospheric emissions have been decreasing markedly in recent years as the result of improved technology, more stringent regulation, and requirements to release to the public information about atmospheric emissions. One of the main classes of industrial atmospheric pollutants has consisted of the vapors of volatile organic compounds (VOCs). These have been significantly reduced by modifying the conditions under which they are used to lower emissions and by using measures such as activated carbon filters to trap the vapors. The practice of industrial ecology goes beyond these kinds of measures and attempts to find substitutes, such as water-based formulations, so that VOCs need not even be used.

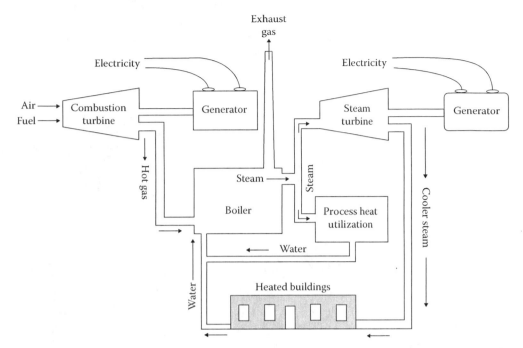

FIGURE 14.9 A combined power cycle system in which energy is used at several levels: the combustion of fuel in a turbine linked to a generator produces electricity. The heat from this turbine raises steam in a boiler that can drive a steam turbine linked to another generator, and the exhaust steam from this turbine can be used to heat buildings (district heating). Steam from the boiler can also be used for process heat in manufacturing.

Years of regulation have greatly lowered releases of water pollutants from industrial operations due largely to sophisticated water treatment operations that are applied to water before it is released from a plant. Desirable as these "end-of-pipe" measures are, the practice of industrial ecology goes beyond such pollution control, minimizing the use of water and preventing its pollution in the first place. One way to ensure that water pollutants are not released from an industrial operation is to completely recycle the water in the system—no water out, no water pollutants.

In past years, many hazardous solid and liquid wastes have been improperly disposed of to sites in the geosphere, giving rise to a large number of "hazardous waste sites," which is the subject of Superfund activity in the United States. The practice of industrial ecology seeks to totally eliminate any such wastes that would require disposal. Ideally, such wastes simply represent material resources that are not properly utilized, a fact that can serve as a guideline for the prevention of such wastes.

The expenditure of energy entails the potential to cause environmental harm to the various spheres of the environment. A prime goal in the proper practice of industrial ecology is the most efficient use of the least polluting sources of energy possible. More efficient electric motors in industrial operations can significantly reduce electricity consumption. The proper design of buildings to reduce heating and cooling costs can also reduce energy consumption. Many industrial operations require heat (process heat in industrial parlance) and steam. Rather than generating them separately, they can be produced in combined power cycles along with the generation of electricity, thereby greatly increasing the overall efficiency of energy utilization (Figure 14.9).

14.7 GREEN CHEMISTRY AND THE ANTHROSPHERE

Green chemistry (defined in Chapter 1, Section 1.5, and illustrated in Figure 1.5) has an essential role to play in the development of successful industrial ecosystems, especially in making industrial metabolism as efficient, nonpolluting, and safe as possible. The practice of green chemistry has become an essential part of the operation of the anthrosphere in an environmentally favorable manner. A major advantage of the practice of green chemistry to reduce environmental impact is that, ideally, it is inherently safe and clean. By using nontoxic chemicals and processes that do not threaten the environment, green chemistry avoids posing threats to the people who practice it and to the surrounding environment. Of course, these are ideals that can never be completely realized in practice; but by having these ideals as goals and making constant incremental improvements, the practice of green chemistry can become increasingly safe, environmentally friendly, and sustainable. This reduces dependence on the command and control measures that require constant vigilance to maintain. Rather than depending on regulations imposed from the outside to maintain its safe operation, green chemistry is inherently self-regulating.

Green chemistry is defined as the practice of chemical science and manufacturing in a manner that is sustainable, safe, and nonpolluting, and it consumes minimum amounts of materials and energy while producing little or no waste material. The major aspects of green chemistry include the following:

- Efficient use of matter with minimum production of wastes
- Catalysis for maximum efficiency in chemical processing
- Utilization of biological processes
- Maximization of renewable raw materials
- Green product design
- Minimization or elimination of solvents, use of water where possible
- Process intensification

Green chemistry gives prime consideration to the chemical reactions and processes by which chemicals are manufactured. One approach to making chemical synthesis greener is to use an existing chemical synthesis process, but make the process itself safer and less polluting while also making the reagents required for it by greener processes. An example of the former might be to substitute a less volatile, less toxic solvent as a reaction medium for a chemical synthesis reaction. In some cases, a reagent may be made more safely by using biological processes for its preparation in place of chemical processes. A second general approach to making chemical preparations greener is to use different reagents for a synthesis that are safer and less likely to pollute.

The practice of green chemistry is largely applied to the synthesis of organic chemicals. The history of organic synthesis abounds with examples of processes that are emphatically not green. One example that is sometimes cited is the synthesis beginning with explosive trinitrotoluene (TNT) of phloroglucinol, a chemical used in relatively small quantities in the fine chemicals industry. The synthesis began with oxidation by dichromate (a carcinogenic substance) in fuming sulfuric acid (a highly corrosive material that causes horrid lesions to skin) followed by reduction with iron in hydrochloric acid and heating to isolate the product. Although the quantity of product made was only about 100 t/year, the process generated about 4000 t/year of solid waste containing $Cr_2(SO_4)_3$, NH_4Cl, $FeCl_2$, and $KHSO_4$. Clearly, this was not an environmentally friendly process, and a major objective of green chemistry has been to find substitute pathways for syntheses such as this one.

Several key parameters are calculated in quantifying green chemistry. As discussed in Section 14.9, one of these is **atom economy**, defined as the fraction of reactant material that actually ends up in the final product. The higher the atom economy—ideally 100%—the greener the process. Other important parameters include the E factor, which measures wastes (Section 14.9), and oxygen and hydrogen availability (Section 14.14).

14.7.1 Presidential Green Chemistry Challenge Awards

The U.S. Presidential Green Chemistry Challenge Awards, administered by the Environmental Protection Agency with partial sponsorship by the American Chemical Society, are given annually to recognize efforts to reduce hazards and wastes from chemical processes and to help meet pollution prevention goals. The 77 winners of these awards since 1995 are estimated to have eliminated 97 million kilograms of hazardous chemicals from use, prevented the release of 26 million kilograms of greenhouse-warming carbon dioxide into the atmosphere, and saved 80 million liters of water. The winners of these awards and the processes for which they are given are published each June in the ACS publication *Chemical and Engineering News.* An example is a 2010 award for the design of a biomaterials refinery as a substitute for a petroleum refinery that uses transgenic microorganisms to convert sugars to hydrocarbon alkanes and alkenes, long-chain fatty acids, and fatty esters.

14.8 PREDICTING AND REDUCING HAZARDS WITH GREEN CHEMISTRY

The conventional approach to making chemical processes less dangerous to workers and less harmful to the environment has emphasized **exposure reduction** in which the hazard is still present, but workers are protected from it. In the arena of worker safety, this has involved measures such as wearing protective gear to prevent contact with hazardous chemicals. For the environment as a whole, it has largely consisted of end-of-pipe measures to prevent the release of pollutants once they are generated.

In contrast to exposure reduction, green chemistry relies on **hazard reduction**. The first step in hazard reduction is to know what the hazards are and where they originate. Hazards may arise from the raw materials used, the media (solvents) in which chemical processes are carried out, the catalysts that enable chemical reactions to occur, and by-products. The direct hazards posed to workers in a chemical process fall into the two main categories of toxicity hazards and hazards associated with uncontrolled events such as fires and explosions.

14.9 ATOM ECONOMY AND THE E FACTOR IN GREEN CHEMISTRY

14.9.1 YIELD AND ATOM ECONOMY

Traditionally, synthetic chemists have used **yield**, which is defined as a percentage of the degree to which a chemical reaction or synthesis goes to completion, to measure the success of a chemical synthesis. For example, if the stoichiometry of a chemical reaction shows that 100 g of a product should be produced, but only 85 g is produced, the yield is 85%. A synthesis with a high yield may still generate significant quantities of useless by-products if the reaction does so as part of the synthesis process. Instead of yield, green chemistry emphasizes **atom economy**, the fraction of reactant material that actually ends up in the final product. With 100% atom economy, all of the material that goes into the synthesis process is incorporated into the product. For efficient utilization of raw materials, a 100% atom economy process is most desirable. Figure 14.10 illustrates the concepts of yield and atom economy.

Although atom economy is a useful concept, a more accurate measurement of the environmental acceptability of a chemical manufacturing process is the **E factor**, defined as follows:[5]

$$\text{E factor} = \frac{\text{Total mass of waste from process}}{\text{Total mass of product}} \tag{14.1}$$

The E factor takes into account waste by-products, leftover reactants, solvent losses, spent catalysts and catalyst supports, and anything else that can be regarded as a waste. Its calculation depends on what is defined as waste. For example, water is a significant by-product of many chemical processes and is generally harmless, so its mass is usually omitted from the total mass of waste in the calculation. However, it may be included in those processes in which it is severely contaminated and difficult to reclaim in a form pure enough to use or discharge to a publicly owned wastewater treatment facility. A leftover reactant that can be easily reclaimed and recycled to the process is not included as waste, whereas a reactant that cannot be salvaged is counted in the waste.

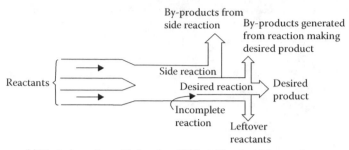

(a) Typical reaction with less than 100% yield and with by-products

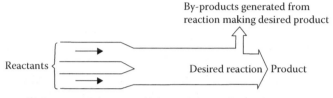

(b) Reaction with 100% yield, but with by-products inherent
    to the reaction

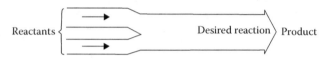

(c) Reaction with 100% atom, economy, no leftover reactants, no
    by-products

FIGURE 14.10 Illustration of the concept of atom economy compared to yield: a high atom economy, ideally 100%, is very desirable.

The ideal E factor is 0, and higher E factors are relatively less desirable. E factors that can be tolerated depend on the value of the product and the amount of the product produced. Larger E factors may be tolerated in the production of small quantities of relatively high-value chemicals.

Until recently, little attention had been given to the amounts of wastes produced in pharmaceutical manufacturing because the prices of the products were so high and the total amounts of wastes produced were so low. However, with the realization that even the generation of a few hundred tons per year of waste can be undesirable and costly, the pharmaceutical industry is becoming a leader in the implementation of green chemical practice. It should be noted that, although they are not considered in the calculation of E factors for pharmaceutical manufacture, annual releases of postconsumer pharmaceuticals and their metabolites are not insignificant. Of greatest concern is the contamination of wastewater, some of which gets back into drinking water supplies, by pharmaceuticals and their metabolites discharged with urine or simply flushed down the drain when no longer needed. By their nature, pharmaceuticals are metabolically active, and their presence in drinking water can be a concern.

14.9.2 NATURE OF WASTES

There are wastes, and then there are wastes. Production of a few thousand tons of carbon dioxide per year in a chemical synthesis process may be of little concern because, although it can be discharged to the atmosphere, contributing to the atmosphere's burden of greenhouse gases, it is negligible compared to the millions of tons released by burning fossil fuels. However, the generation of a few kilograms of heavy metal wastes can be problematic because of heavy metal toxicity. So it matters what kinds of wastes are produced. Attempts have been made to assign an **environmental quotient**

(**EQ**) to wastes where Q is a number assigned to a particular kind of waste that, when multiplied by the E factor (Equation 14.1), provides in principle a means of weighting the potential harm of various kinds of wastes. The toxicological chemistry of specific wastes is very important in the assignment of environmental quotients. Whereas, E is easily measured by simple weighing, Q is a much more arbitrary number and subject to change as information is obtained regarding the potential harm of particular kinds of wastes.

14.10   CATALYSTS AND CATALYSIS IN GREEN CHEMISTRY

The main components of a chemical process can be divided into the categories of catalysts, media, feedstocks, and reagents, all of which are important in the practice of green chemistry. Catalysts and catalysis are addressed in this section, and the other aspects of chemical production are addressed in Section 14.12.

An ideal green chemical reaction occurs with 100% atom efficiency using only reactants and no other reagents under mild conditions with only moderate input of thermal energy and without any catalysts. Unfortunately, few chemical processes meet these criteria.

In the past, especially in the synthesis of fine chemicals and pharmaceuticals where the objective has been to simply make the desired product without much consideration of waste, so-called stoichiometric reagents have been used. These have included inorganic oxidants such as MnO_2, $KMnO_4$, and $K_2Cr_2O_7$; metal reductants including zinc, magnesium, sodium, and iron; and metal hydride reducing agents, especially $LiAlH_4$ and $NaBH_4$. Various organic reactions including sulfonations and nitrations employ Lewis acids (BF_3, $AlCl_3$, and $ZnCl_2$) and mineral acids (HF, H_2SO_4, and H_3PO_4). Many of these processes are indirect methods of adding to organic molecules hydrogen (reduction), oxygen (oxidation), carbon, and nitrogen. A problem with these reagents is the large amount of inorganic wastes produced. Where possible, it is much more desirable to employ catalysts to attach the simplest possible forms of the elements including H_2 for reduction, O_2 or H_2O_2 for oxidation, CO or CO_2 for attachment of carbon, and NH_3 for attachment of N.

Instead of relying only on stoichiometric reagents, catalysts are commonly employed. **Catalysts** are substances that enable reactions to proceed at significant rates without themselves being consumed. Catalysts are of great importance in the practice of green chemistry for a number of reasons including their ability to facilitate reactions and to reduce the energy required to enable reactions to proceed. There are two major approaches to contacting a catalyst with a reaction mixture, as shown in Figure 14.11. In **heterogeneous catalysis**, the catalyst is held on a support and the reactants flow over it. In **homogeneous catalysis**, the catalyst is placed in the reaction mixture and either remains in the product or is separated from the product in a separate step. In general, homogeneous catalysts have high activity and selectivity, whereas heterogeneous catalysts are more easily recovered and recycled. Much of the activity in green chemistry has been in the area of catalysis, especially in the development of heterogeneous catalysts that do not contaminate the product and that can be reused multiple times. Large numbers of different catalysts are used in chemical processes, and their potential toxicities, production of by-products and contaminants, recycling, and disposal are matters of considerable importance in the chemical industry.

An important area of endeavor in the development of improved catalysts with respect to green chemistry is **selectivity enhancement**. Basically, this means developing a catalyst that is very selective in what it does, ideally making the right product and nothing else. A highly selective catalyst increases the percentage utilization of raw material (increased percent yield) and decreases the amount of waste by-products from undesired side reactions (smaller E factor).

Another important attribute of a good catalyst is related to the basic way in which a catalyst works, which is by lowering the activation energy that is required to make a reaction proceed at a significant rate. As a consequence, catalysts lower the total amount of energy that must be put into

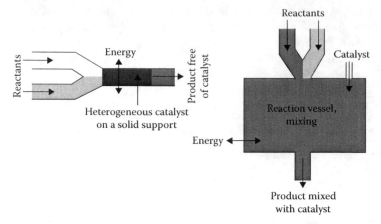

FIGURE 14.11   A heterogeneous catalyst (left) is held on a solid support and the reaction mixture flows over it. In homogeneous catalysis (right), the catalyst is mixed with the reactants and either remains in the product or is removed in a separation step. Some reactions require addition of heat energy, whereas some exothermic reactions require cooling. In the practice of green chemistry, energy is sometimes added as electrical current, ultraviolet radiation, microwave radiation, or ultrasound.

a chemical process for it to occur. Lowered energy requirements are a basic part of the practice of green chemistry, and in this respect highly efficient catalysts can be extremely beneficial in reducing energy consumption and, in so doing, lowering costs and environmental impact.

Numerous green industrial chemical reactions have been developed in recent years, most employing some sort of catalyst. An example is the synthesis of the widely used industrial chemical propylene oxide starting with elemental H_2 and O_2. Around 1985, it was discovered that propylene could be oxidized to propylene oxide with 30% hydrogen peroxide using a titanium silicate catalyst, but the process was not economic because of the high cost of hydrogen peroxide. A Presidential Green Chemistry Challenge prize was awarded in 2007 for an economical process to make hydrogen peroxide by combining H_2 and O_2 in a gas mixture at levels of H_2 directly below the lower flammability limit of H_2 using a catalyst composed of palladium–platinum nanoparticles:

$$H_2 + O_2 \rightarrow 2H_2O_2 \tag{14.2}$$

As a second step in the process, the hydrogen peroxide is reacted with propylene to produce propylene oxide with water as the only by-product:

$$ \tag{14.3}$$

Propylene   **Propylene oxide**

As shown in Figure 14.12, an approach that is being tried to obtain the selectivity and high catalytic activity of homogeneous catalysts with the high recovery and recycling possible with heterogeneous catalysts is **liquid–liquid biphasic catalysis** in which the catalyst is contained in one phase and the reactants and products in a second phase. During the reaction, the two phases are normally subjected to intense mixing to enable contact of reactants with the catalyst. With this approach, when the reaction is complete the catalyst is recovered by simply separating the phases and, in ideal circumstances, can be directly recycled for further reaction without further processing.

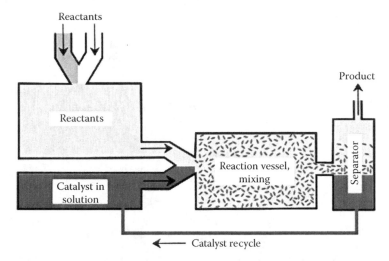

FIGURE 14.12   Illustration of a biphasic catalyst system: reactants in one solution contact the catalyst in another immiscible solution. After the reaction has had a chance to go to completion, the two immiscible phases are separated and the catalyst solution is recycled. In the reaction vessel, this kind of system provides the superior qualities of a homogeneous catalyst but the product/catalyst separability of a heterogeneous catalyst.

14.11 BIOCATALYSIS WITH ENZYMES

Nature has provided some ideal catalysts in the form of enzymes (see Chapter 2, Section 2.7), the use of which offers some substantial advantages in green chemistry. The most obvious advantage is that enzymes have evolved to work under the mild conditions in which organisms function, especially moderate temperatures and physiological pH. Unlike the precious metal catalysts that are commonly used, enzymes are made renewably from biomaterials. Enzymes have generally high activities and are highly selective in the chemical processes that they carry out. Whereas conventional organic syntheses often require activation or protection of functional groups, both of which may consume reagents and hence produce more wastes, these measures are often not needed with enzymes. The result is that biocatalyzed reactions can frequently be carried out with relatively fewer steps and less wastes, making them more attractive environmentally and economically.

Although some biocatalyzed reactions have been used for the production of chemicals for centuries—the production of ethanol by yeast fermentation of sugar comes immediately to mind—relatively recent advances in biotechnology have greatly increased their versatility and utility. One such advance has been with recombinant DNA in which enzymes that perform specific functions may be transferred between organisms. The other major advance has been with **directed evolution** in which the amino acid sequences in enzymes produced by genes are shuffled randomly and large numbers of the products are sampled for their activity, particularly for carrying out a particular biochemical synthetic step. This may be done within living cells, but this can be carried out on a much larger scale outside the cells, a process called **in vitro evolution**. Obviously, most of the enzymes produced by this technique are not superior, or are even useless, but out of the enormous numbers generated, some will be superior. In vitro evolution is being carried out to provide enzymes with properties such as improved or different catalytic activity, catalytic specificity, thermostability, and pH optima that can be used in industrial, medical, or agricultural applications. It is proving particularly useful in developing enzymes that act upon compounds that do not occur in nature and for which enzymes have not evolved through natural evolution.

14.11.1 Immobilized Enzyme Catalysts

Disadvantages of catalytic enzymes often include low stability, limited storage times, difficult recovery, and product contamination, almost always a consideration with homogeneous catalysts. The stability and recyclability of enzymes may often be enhanced by immobilizing them on solid supports. A common means of immobilizing enzymes is by precipitating them from the fermentation broth in which they are generated by a buffer, such as ammonium sulfate, and stabilizing the precipitated aggregates with a reagent possessing two bonding functional groups, commonly glutaraldehyde, a reagent that is widely used for cross-linking proteins:

$$
\begin{array}{ccccc}
O & H & H & H & O \\
\parallel & | & | & | & \parallel \\
H-C-&C-&C-&C-&C-H \\
& | & | & | & \\
& H & H & H &
\end{array}
$$

Glutaraldehyde

The precipitation and cross-linking of an enzyme combines its isolation and immobilization into a single step. Enzymes prepared in this way are usually highly productive, stable, and resistant to denaturation (loss of enzyme function by structural alteration) due to exposure to organic solvents, heating, or breakdown to shorter peptide chains or amino acids by the action of proteolytic enzymes (proteolysis).

14.11.2 Reduction in Synthesis Steps with Enzyme Catalysts

Typical synthesis of organic compounds, especially ones as complicated as many pharmaceutical agents, may involve multiple steps. Because of factors such as product loss, the need to protect and deprotect functional groups, and generation of wastes from each step, these multistep syntheses tend to build high E factors overall. The ideal synthesis is a "one-pot" process in which all steps are carried out in the same operation without the need to isolate intermediates. Living cells are often one-pot synthetic factories, so it is natural to look to enzymes to accomplish the same thing in the laboratory and in chemical production using multiple enzymes in a single container and employing a single multiple-step operation. Accomplishment of such a process can be complicated by the incompatibility of different enzymes and the different conditions under which enzymes operate, although in general they operate in water under ambient conditions compatible with life. (The relatively recent discovery of organisms that live under hot, extreme conditions on the deep ocean floor raises some interesting possibilities for the isolation of enzymes that might function under unusual conditions, particularly elevated temperatures.)

14.11.3 Enzyme Catalysts and Chirality

As shown by the example of the herbicide mecoprop in Figure 14.13, **chiral molecules** are three-dimensional molecules with structures such that a molecule cannot be directly imposed on its mirror image. Chiral molecules have different groups arranged around an atom, usually of carbon, that constitutes a chiral center. Two chiral molecules of the same compound are called **enantiomers**, commonly designated as R and S. The physical and chemical properties of enantiomers are generally identical: they have the same melting points, boiling points, and solubilities. However, one enantiomer of a compound may fit exactly with an enzyme active site, whereas the other does not. This often results in markedly different biochemical properties of enantiomers and consequently completely different environmental and toxicological behaviors. One enantiomer of a chiral pharmaceutical may function very well, whereas the other does not work at all or may even be toxic. Biochemical differences between enantiomers may be especially pronounced for pesticides.

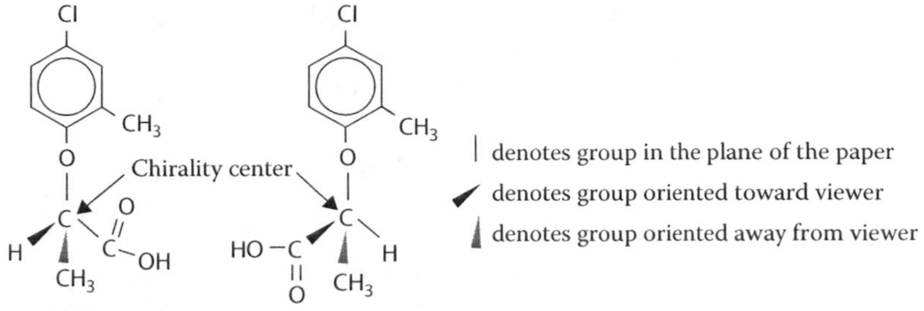

The two enantiomers of the herbicide mecoprop

FIGURE 14.13 Illustration of the chiral herbicide mecoprop: only the R enantiomer has herbicidal activity and is now marketed in the enantiomeric pure form.

For example, the R enantiomer of herbicidal mecoprop (Figure 14.13) kills weeds very effectively, whereas the S enantiomer is inactive; therefore, the pure R enantiomer is now marketed as an herbicide in place of the racemic mixture with the S enantiomer.

Normally, when a chiral compound is synthesized by conventional chemical means, a racemic mixture of the two enantiomers is produced. Because of their essentially identical chemical properties, they are very hard to separate. However, it is possible to produce enantiomerically pure chemical compounds with appropriate enzymatic catalysts. Furthermore, it is possible to use enzymatic catalysts to convert racemic mixtures of compounds to enantiomerically pure forms. In one of the larger such industrial operations, the German chemical firm BASF now uses enzymatically catalyzed processes to prepare enantiomerically pure amines in thousand-ton quantities.

14.12   ENERGIZING CHEMICAL REACTIONS AND PROCESS INTENSIFICATION

One of the most important aspects of green chemistry is the enhancement of the speed and degree of completion of chemical reactions. One of the ways in which this is done is by lowering the activation energy required to enable a reaction to proceed. That is what catalysts do, as discussed in Section 14.11. The other way to enhance a reaction is by adding energy as discussed in this section.

The most straightforward means to add energy to a reaction is by heating the reaction mixture. On an industrial scale, this is commonly accomplished with coils of tubing immersed in the reaction mixture that are heated with steam passing through the coils. Heating by passing a current of electricity through electrically resistant coils is also a means of adding energy to a chemical system. Much of the effort in green chemistry has been devoted to finding more efficient and sophisticated ways of energizing chemical systems.

Microwaves can be used to add energy to reactions to enhance reaction rates.[6] Microwaves are electromagnetic radiation with wavelengths of 1 cm to 1 m (frequencies of 30 GHz to 300 Hz). To avoid interference with the microwave bands used in communication, industrial and household microwave generators commonly operate at 2.45 GHz. Microwaves are absorbed by polar molecules, such as those of water, causing rapidly repeating reorientation of the molecules in a microwave field. The result is a high input of energy directly into substances subjected to microwaves, thereby adding energy to and speeding up reactions. Microwave energy can be put directly into relatively small volumes of reaction media, reducing material requirements and minimizing wastes. Microwaves can be used to enhance reactions in (1) water media; (2) polar organic solvents such as dimethylformamide; and (3) media-free reactions, such as mixed solid reactants.

Sonochemistry adds energy to a chemical reaction by subjecting a reaction medium to ultrasound energy at frequencies between 20 and 100 KHz, which introduces very high energy pulses into the medium.[7] Commonly, the ultrasound is produced by the piezoelectric effect through which crystals of substances such as ceramic-impregnated barium titanate are subjected to rapidly reversing electrical fields, converting the electrical energy to sound energy with an efficiency that can reach 95%. An advantage of sonochemistry is that it can introduce high energy into microscopic regions, enabling reactions to occur without appreciably heating the reaction medium.

Electrochemistry by the passage of a direct current of electricity through a reaction medium can cause both reductions and oxidations to occur. Reduction, the addition of electrons, e^-, occurs at the relatively negatively charged cathode, and oxidation, the loss of electrons, occurs at the relatively positively charged anode. Electrochemical oxidation and reduction can be controlled by the electrical potentials applied, by the reaction media, and by the electrodes used. Because the addition of electrons to a reaction medium (reduction) and the accompanying removal of electrons (oxidation) do not add matter, electrolytic syntheses meet the goal of minimal material in green chemistry. Electrochemistry as a means of carrying out reagent-free oxidation and reduction is discussed in more detail in Section 14.14, and the electrolytic production of oxygen and of hydrogen, a nonpolluting fuel and valuable raw material, is shown later in the chapter in Figure 14.15.

Photochemical reactions use the energy of photons of light or ultraviolet radiation to cause reactions to occur. The energy, E, of a photon of electromagnetic radiation of frequency v is $E = hv$, where h is Planck's constant. Since a photon can be absorbed directly by a molecule or a functional group on a molecule, the application of electromagnetic radiation of appropriate energy to a reaction medium can introduce a high amount of energy into a reactant species without significantly heating the medium. Photochemical energy can be used to cause synthesis reactions to occur more efficiently with less production of waste by-products than nonphotochemical processes.[8]

A reaction participant does not have to absorb a photon directly to undergo a photochemically induced reaction. In some cases, photochemically reactive species are added to a reaction mixture to absorb photons and then produce reactive excited species or free radicals that carry out additional reactions. An example of this is provided with hydrogen peroxide, which absorbs photons

$$H_2O_2 + hv \rightarrow HO \cdot + HO \cdot \qquad (14.4)$$

to produce reactive hydroxyl radicals that react with a number of other species.

14.12.1 PROCESS INTENSIFICATION AND INCREASED SAFETY WITH SMALLER SIZE

Process intensification can be employed with continuous-flow reactors (Figure 14.14) used to intensify chemical processes and enable increased output of products with a smaller footprint of apparatus. This is especially the case when continuous flow is combined with heterogeneous catalysis and energy input. A big benefit to such reactors from the green chemistry viewpoint is increased safety. If something goes wrong in a large batch reactor, an accident such as an explosion or a fire with a large amount of material may occur in the worst case. With a continuous-flow reactor, the problem can be confined to the small volume of the reactor and the process can be shut down immediately by stopping the inflow of reactants.

14.13   SOLVENTS AND ALTERNATE REACTION MEDIA

Chemical reactions are often carried out in **media**, usually organic **solvents** or water, to provide a medium in which feedstocks and reagents can dissolve and come into close, rapid contact at the molecular level. A good solvent for chemical synthesis is one that enables facile product separation and is amenable to purification and reuse with minimum loss. Substances dissolved in a solvent are **solvated** by binding the solvent to the molecules or ions of the dissolved substance, the **solute**. The solvation of reactants and products often plays an important role in determining the kinds and rates of reactions. Many organic feedstocks and reagents are not soluble in water or are decomposed by it; so organic solvents including hydrocarbons, chlorinated hydrocarbons, and ethers have to be used as reaction media.

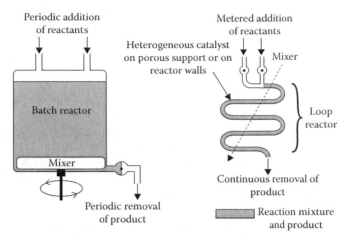

FIGURE 14.14 A batch reactor (left) involves mixing a large amount of reactants, often with a homogeneous catalyst, and allowing the reaction to take place for the required amount of time with the result that if anything goes wrong a large amount of material is spoiled and a considerable hazard is developed. In a continuous-flow reactor (right), only a small amount of material is reacting at any given time, heterogeneous catalysts can be used, energy input can be accurately regulated, and the process can be stopped at any time if something goes wrong.

Organic solvents cause several problems in chemical synthesis. Particularly because of problems associated with their containment, recovery, and reuse, organic solvents especially are major contributors to undesirably high E factors. Many of the environmental and occupational health problems associated with making chemicals are the result of the use of organic solvents as media. Hydrocarbon solvents will burn, and hydrocarbon vapors in air are explosive. Although many hydrocarbon solvents are not particularly toxic, some can cause the condition of peripheral neuropathy (damage to peripheral nerves such as those in feet and legs), and benzene is regarded as a carcinogen that is thought to cause leukemia. Released to the atmosphere, hydrocarbons can also participate in photochemical processes leading to the formation of photochemical smog (see Chapter 7, Section 7.8).

One approach to making chemical synthesis processes greener is to replace specific solvents with less hazardous ones. For this reason, toxic benzene solvent is replaced by toluene wherever possible. As shown by their structural formulas later in this section, toluene has a methyl group, $-CH_3$, that benzene does not possess. The methyl group in toluene can be acted upon by human metabolic systems to produce a harmless metabolite (hippuric acid) that is eliminated in the urine, whereas metabolic processes acting upon benzene convert it to a toxic intermediate that can react with cellular DNA and cause blood abnormalities including leukemia.

Benzene   **Toluene**

As another example of solvent replacement, *n*-hexane, which can cause peripheral neuropathy in exposed individuals, can be replaced with 2,5-dimethylhexane, which does not cause this condition, for reactions in which the higher boiling temperature of the latter compound is not a problem.

There is significant interest in reaction media other than organic solvents. The ultimate approach to eliminating problems with solvents in chemical synthesis is to do reactions without solvents of any kind. Some reactions can be performed in which the reactants are simply mixed together or are held on solid supports, such as clays. Microwave heating of such reaction mixtures has proved to be effective in providing energy to enable reactions to occur rapidly. However, in many cases, this is not possible and solvents are required. Some alternative solvents are discussed in Sections 14.13.1 through 14.13.3.

14.13.1 WATER SOLVENT

Many reagents are reactive with water, preventing its use as a reaction medium. However, where applicable, water is the greenest solvent for green chemical processes. Water is abundant, cheap, and nontoxic, and it does not burn. Because of its polar nature and ability to form hydrogen bonds (see Chapter 3, Section 3.1 and Figure 3.1), water is an especially good solvent for ionic compounds—acids, bases, and salts. Water is particularly useful as one of the solvents in biphasic catalysis, which is described in Section 14.10. Normally, the catalyst is held in the water phase and the product in a water-immiscible organic solvent, which allows facile separation of the catalyst after the reaction is complete.

Because of water's many advantages, significant efforts have been made in using it to replace organic solvents employed for reaction media. Although water does not appreciably dissolve many nonpolar organic compounds, in some cases these materials may be suspended as very small colloidal particles in water, enabling close enough contact of organic materials to undergo reactions. Water is a good solvent for some of the biological materials, such as glucose, that are now favored as chemical feedstocks where they can be used.

14.13.2 CARBON DIOXIDE SOLVENT

At a high pressure above 73.8 atm (73.8 times normal atmospheric pressure at sea level) and a temperature exceeding 31.1°C, carbon dioxide becomes a **supercritical fluid**, a relatively dense state of matter in which there is no longer a distinction between liquid and gas. A good solvent for organic compounds, supercritical carbon dioxide can be used as a reaction medium for organic chemical reactions.[9] An advantage of supercritical carbon dioxide in this application is that its viscosity is only about 1/30 that of common liquid organic solvents, which enables reactant species to move much faster through the fluid, thus speeding the reactions that they undergo. At temperatures and pressures somewhat below those at which carbon dioxide becomes critical, it exists as separate gas and liquid phases while retaining many of the solvent properties of supercritical carbon dioxide. Under these conditions, carbon dioxide is called a **dense phase fluid**, a term that also encompasses supercritical fluids.

The adjustment of the composition of dense phase fluid carbon dioxide and the conditions under which it is maintained can provide significant variations in its solvent properties and its ability to act as a reaction medium. In addition to variations in temperature and pressure, dense phase fluid carbon dioxide may be mixed with small quantities of other solvents (modifiers), such as methanol, to further vary its solvent properties.

In addition to its solvent properties, dense phase fluid carbon dioxide offers the advantage of low toxicity and low potential for environmental harm (the small amounts of greenhouse gas carbon dioxide released from its application as a solvent are negligible compared to the quantities released from the combustion of fossil fuels). A big advantage of dense phase fluid carbon dioxide is its volatility, which means that it separates readily from reaction products when pressure is released. Furthermore, carbon dioxide released from a reaction mixture can be captured and recycled for the same application. Carbon dioxide can be obtained at low cost from biological fermentation processes.

14.13.3 IONIC LIQUID SOLVENTS

Ionic liquids present another alternative to organic solvents for use as media for chemical synthesis.[10] Inorganic salts consisting of ions, such as NaCl composed of Na^+ and Cl^- ions, are normally hard, high-boiling solids. However, when one or both of the ions consist of large charged organic molecules, as is the case with the cation in the compound

1-Butyl-3-methylimidazolium hexafluorophosphate

the salts can be liquids at room temperature and are called ionic liquids. These materials have the potential to act as suitable media in which substances can be dissolved and undergo reactions, and active research is underway to explore this possibility. There is an enormous variety of such ionic liquids with widely varying solvent properties because of the large number of kinds of ions that can be combined, leading to almost limitless possibilities for various ionic liquids.

14.14 FEEDSTOCKS AND REAGENTS

14.14.1 FEEDSTOCKS

Feedstocks are the main ingredients that go into the production of chemical products. Feedstocks are addressed here and further in Chapter 16.

As discussed here, feedstocks may be acted upon by reagents, and there is often some overlap between the two categories of materials. Feedstocks are very important in the practice of green chemistry. The acquisition of feedstocks has some profound environmental and sustainability implications.

The **source** of a feedstock may consist of a depleting resource, such as petroleum, in which case the lifetime of the resource and the environmental implications of obtaining it must be considered. From the standpoint of sustainability, it is preferable to use recycled materials as feedstocks, although the availability of recycled materials suitable for this purpose is usually limited. A third source that is very desirable consists of renewable resources, particularly from materials made by photosynthesis and biological processes.

A second major aspect of converting feedstocks to final products is **separation and isolation** of the desired substance. An example of this step is the isolation of specific organic compounds from crude oil to provide a feedstock for organic chemical synthesis. It may be necessary to process raw materials from a source to convert it to the specific material used as a feedstock for a chemical process. Often, most of the environmental harm in providing feedstocks comes during the isolation process, largely because of the relatively large amount of waste material that often is generated in obtaining the needed feedstock.

The world chemical industry has been built primarily on fossil carbon feedstocks. Much of the impetus for the organic chemical industry was built during the late 1800s and early 1900s on the basis of organic chemicals isolated from the coal tar by-product of coal coking. Later, petroleum and natural gas became the basis of the petrochemicals industry, which has produced enormous quantities of polymers, plastics, synthetic rubber, and thousands of other kinds of chemicals. Eventually, this reliance on depleting fossil carbon resources must end. Therefore, one of the main goals of green chemistry has become a shift toward renewable feedstocks—biomass produced by photosynthesis—and renewable reaction media, specifically water and supercritical and pressurized liquid carbon dioxide. In addition to being renewable, such feedstocks offer advantages over petroleum in that they are usually not toxic and they, and most of their products and intermediates, are biodegradable.

14.14.2 REAGENTS

The term **reagents** is used here to describe the substances that act upon basic chemical feedstocks to convert them to new chemicals in synthetic processes. The kinds of reagents used have a very strong effect on the acceptability of a chemical process with respect to green chemical aspects. Much of the work that has been done in developing and using green reagents has involved organic chemical processes, many of which are beyond the scope of this book. However, some of the general aspects of chemical reagents from a green chemical perspective are discussed here.

The most obvious characteristic required of a good chemical reagent is that it does what it is supposed to do completely at an acceptable rate. A reagent with a high **product selectivity** produces a high percentage of the desired product with a low percentage of undesired by-products. Another desirable characteristic of a good reagent is high **product yield**, which means that most of the feedstocks are converted to products. The use of reagents that provide high selectivity and yield means that less unreacted feedstock and by-product material have to be handled, recycled, or disposed of.

One of the most common measures taken in implementing green chemical processes is selection of alternative reagents. The criteria used in selecting a reagent include what it does, how efficient it is, and its cost and availability. Important considerations with a chemical transformation are whether it is stoichiometric or catalytic, the degree to which it is atom economical, and the quantities and characteristics of any wastes produced.

14.14.3 Reagents for Oxidation and Reduction

One of the main kinds of reactions for which reagents are used is **oxidation**, which usually consists of the addition of oxygen to a chemical compound or a functional group on a compound and a loss of electrons. An example of an oxidation reaction is the conversion of ethanol to acetic acid

$$H-\underset{\underset{H}{|}}{\overset{\overset{H}{|}}{C}}-\underset{\underset{H}{|}}{\overset{\overset{H}{|}}{C}}-OH + 2\{O\} \rightarrow H-\underset{\underset{H}{|}}{\overset{\overset{H}{|}}{C}}-\overset{\overset{O}{||}}{C}-OH + H_2O$$

Ethanol   Acetic acid

where {O} is used to represent oxygen from some unspecified oxidant. Oxidation is one of the most common steps in chemical synthesis. A number of reagents are used as oxidants. Some of these reagents, such as potassium dichromate, $K_2Cr_2O_7$, are dangerous (dichromate salts are considered to be carcinogenic when inhaled for prolonged periods of time) and leave troublesome residues that require disposal.

Because of problems with oxidants that are commonly used, a major objective in the practice of green chemistry is to use more benign oxidants. Alternatives to the more traditional oxidant reagents include molecular oxygen, O_2, and hydrogen peroxide, H_2O_2, usually used with a suitable catalyst that enables the oxidation reaction to occur. Ozone, O_3, may also be regarded as a green oxidant in the sense that it is made by an electrical discharge through O_2; however, ozone presents hazards due to its tendency to form explosive ozonides with organic molecules. Under the right conditions, hydrogen peroxide can be used as an alternative to elemental chlorine, Cl_2, a strong oxidant used to bleach colored materials such as paper pulp and cloth. Since chlorine is toxic (it was used as a poison gas in World War I) and has a tendency to react with organic compounds to produce undesirable chlorinated organic compounds, hydrogen peroxide is a much preferable bleaching agent.

Sodium percarbonate serves as a solid source of hydrogen peroxide and oxidant. This compound is made by adding hydrogen peroxide to solid sodium carbonate to give a material with the formula $2Na_2CO_3 \cdot 3H_2O_2$. Although it is stable in the dry solid form and is used in household cleaning formulations, it can be a very reactive oxidant and a potent source of hydrogen peroxide in nonaqueous media.

In contrast to the usually harsh conditions under which chemical oxidations are carried out, organisms carry out biochemical oxidations under mild conditions. In so doing, they use monooxygenase and peroxidase enzymes that catalyze the oxidizing action of molecular oxygen or hydrogen peroxide. An area of significant interest in green chemistry is to perform such oxidations in biological systems or to attempt the use of catalysts that mimic the action of enzymes in catalyzing oxidations with molecular oxygen or hydrogen peroxide.

Oxidants are rated according to **oxygen availability**, the percentage of the mass of the oxidant molecule that is available oxygen, which is represented here as {O}. In some reactions that introduce all of the oxygen from O_2 into an oxidation product, the oxygen availability is 100%. Theoretically, the molecule with the greatest oxygen availability is molecular O_2, which has an oxygen availability of 100% if both O atoms add to the species being oxidized. In practice, one of the two O atoms usually ends up as water, H_2O, such that the oxygen availability is 50%. Hydrogen peroxide, H_2O_2, adds oxygen to an organic molecule, represented as R, according to the following reaction:

$$H_2O_2 + R \rightarrow R\{O\} + H_2O \qquad (14.5)$$

Since the O atom is 47% of the mass of H_2O_2, the oxygen availability of hydrogen peroxide is 47%.

Reduction, which consists of loss of O, gain of H, or gain of electrons by a chemical species, is also a common operation in chemical synthesis. As is the case with oxidants, the reagents used to accomplish reduction can pose hazards and produce undesirable by-products. Such reductants include lithium aluminum hydride, $LiAlH_4$, and tributyl tin hydride.

For reduction involving the addition of H atoms to a molecule, the concept of **hydrogen availability** may be used. When lithium hydride, molecular mass 7.94, is employed as a reducing agent as in the synthesis of silane

$$SiCl_4 + 4LiH \rightarrow SiH_4 + 4LiCl \qquad (14.6)$$

all of its hydrogen is used and its hydrogen availability is 12.6%, which (because of the low atomic mass of Li, the formula mass of LiH is low) is the highest hydrogen availability of all the metal hydrides. The hydrogen availability of H_2 is 100% when molecular hydrogen is added to C=C double bonds in some organic chemical reductions.

14.14.4 ELECTRONS AS REAGENTS FOR OXIDATION AND REDUCTION

As an alternative to using potentially troublesome reagents, **electrochemistry** provides a reagent-free means of doing oxidation and reduction. This is possible because an electrical current consists of moving electrons, oxidation consists of electron removal from a chemical species, and reduction is the addition of an electron. The passage of an electrical current between metal or carbon graphite electrodes through a solution resulting in oxidation and reduction reactions is called **electrolysis**. Consider the simplest possible case of electrolysis, that of water containing a nonreactive salt, such as Na_2SO_4, shown in Figure 14.15.

At the **cathode**, where electrons (e^-) are pumped into the system and where reduction occurs, reduction of water occurs, releasing H_2

$$2H_2O + 2e^- \rightarrow H_2 + 2OH^- \qquad (14.7)$$

and at the **anode** where electrons are removed, O_2 is released as the water is oxidized:

$$2H_2O \rightarrow O_2 + 4H^+ + 4e^- \qquad (14.8)$$

In the setup shown, H^+ ion generated at the anode and OH^- ion generated at the cathode diffuse through the solution and react on contact to produce water again. At the cathode, a dissolved chemical species can be reduced directly or the hydrogen generated can add to a species, reducing it. And at the anode, another species can be oxidized directly by the loss of electrons or the oxygen generated can add to a species, oxidizing it.

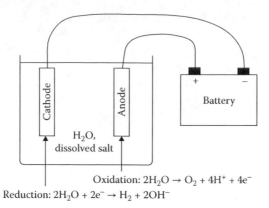

Oxidation: $2H_2O \rightarrow O_2 + 4H^+ + 4e^-$

Reduction: $2H_2O + 2e^- \rightarrow H_2 + 2OH^-$

FIGURE 14.15 Apparatus for electrolysis in which a direct current of electricity is passed through a reaction medium, in this case water with a dissolved salt to make it electrically conducting: reduction occurs when electrons are added to the medium at the cathode, and oxidation occurs when they are removed at the anode. Electrolysis is a reagent-free way of doing oxidation and reduction.

The aforementioned electrolytic production of elemental hydrogen has important implications for sustainability and green chemistry. Hydrogen has many crucial industrial uses, such as in the manufacture of ammonia, NH_3, and in some important respects it is an ideal fuel because its combustion produces only water and it can be used in fuel cells to generate electricity directly and efficiently. Elemental hydrogen feedstock currently is produced primarily by the reaction of fossil fuels with water at a high temperature. Coal can be gasified by reaction with hot steam (see Chapter 8, Section 8.3 and Reaction 8.1) to generate H_2, and currently the main source of elemental hydrogen is the reaction of steam with natural gas, CH_4. These sources produce large quantities of greenhouse gas CO_2 and are not very sustainable means of generating H_2. However, the electrolysis of water to generate H_2 is pollution free and uses water, which is in unlimited supply, as a raw material. Furthermore, pure elemental oxygen, a material with numerous uses (such as an oxidant in fuel cells), is generated as a by-product.

Wind-powered electricity potentially is an ideal source of H_2. In a sense, wind energy is free, and it is nonpolluting. By providing storage for the hydrogen product, the intermittent nature of wind energy in this application is not a problem. A proposed system for wind-generated H_2 is shown in Figure 14.16. This system uses an electrolysis system located deep underground at a depth of 1 or 2 km and pressurized by a column of water. The hydrogen gas is produced at a high pressure by passing a direct current through the water[11] and can be stored under pressure in a suitable geological formation. As H_2 is released, it goes through a turbine where the pressurized gas generates power, or the pressure can be maintained for the utilization of H_2 in high-pressure chemical syntheses or in pressurized fuel cells to generate electricity. In the latter application, high-pressure O_2 produced by the electrolysis of water can be utilized as well.

14.15 ANTHROSPHERE AND OCCUPATIONAL HEALTH

The anthrosphere, its activities, and its products are major potential sources of hazards from dangerous and toxic materials. Various countries have laws and regulations to protect workers and people in surrounding areas from such hazards. In the United States, the 1970 Occupational Safety and Health Act is designed to maintain a work environment free from safety and health hazards. Such hazards may include exposure to toxicants, ionizing radiation, and microwave radiation; fire hazards; explosives hazards; unsanitary conditions; excessive noise; temperature stresses; and mechanical dangers.

Industrial hygiene is the discipline devoted to recognizing and preventing conditions in the anthrospheric workplace that may have detrimental effects on workers and on people in the

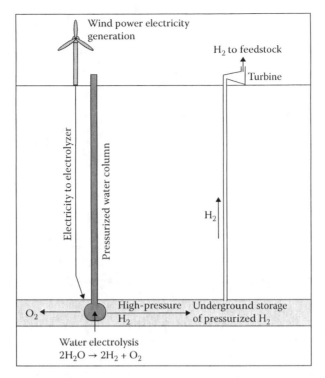

FIGURE 14.16 Pressurized hydrogen produced by the electrolysis of high-pressure water in an underground electrolysis apparatus: the electricity required is provided by generators powered by wind, and the hydrogen pressure is used to power a turbine.

immediate surroundings. Such effects obviously include those from exposure to toxic substances, and much of the efforts of the more than 10,000 industrial hygienists in the United States and thousands more of their counterparts from around the world are devoted to minimizing exposure to, and adverse effects from, such agents. Industrial hygiene activities are closely tied with safety and environmental control in any industrial operation.

Chemical hazards in the workplace may be classified into several major categories including the following:

- **Acute poisons** such as hydrogen cyanide, HCN
- **Chronic systemic poisons** such as benzene, long-term exposure to which may cause reduction in the number of blood cell erythrocytes, leukocytes, and thrombocytes (platelets)
- **Carcinogens** such as vinyl chloride
- **Corrosive substances** such as strong acids or bases
- **Pneumoconiosis agents** such as silica dust and coal dust that cause inflammation and fibrosis of lungs and may even lead to the destruction (necrosis) of lung tissue
- **Reproductive hazards** that may cause infertility, miscarriage, and birth defects, for example, ethylene oxide used to sterilize surgical instruments
- **Neurotoxins** such as some organophosphate esters
- **Nephrotoxins** such as lead and carbon tetrachloride that damage kidneys or impair kidney function

The toxicological chemistry of chemicals that may be encountered in the anthrosphere is obviously important in assessing their hazards. In a number of important cases, toxicological hazards are well known from the unfortunate history of human maladies resulting from their exposure. Benzene, lead, carcinogenic asbestos, and carcinogenic vinyl chloride are in this category, and their toxic effects were revealed from epidemiologic studies of exposed workers. Especially since about the mid-1900s, a wide variety of new organic chemicals have been synthesized and have entered materials flow in the anthrosphere. With awareness of possible detrimental effects of exposure to these substances, these substances have been subjected to animal tests from which toxicity to humans has been inferred. Typical of compounds that have been subjected to exhaustive animal studies are the nitrosamines or N-nitroso compounds containing the N–N=O functional group, most of which have been found to be carcinogens. Dimethylnitrosamine was once widely used as an industrial solvent, and liver damage and jaundice in exposed workers led to studies dating to the 1950s of its carcinogenicity and the subsequent ban on its use for most purposes:

$$
\begin{array}{c}
\text{H} \\
| \\
\text{H} - \text{C} - \text{H} \\
| \\
\text{N} - \text{N} = \text{O} \\
| \\
\text{H} - \text{C} - \text{H} \quad \textbf{Dimethylnitrosamine} \\
| \qquad\qquad \textbf{(N-nitrosodimethylamine)} \\
\text{H}
\end{array}
$$

There are numerous potential avenues for worker exposure in industrial operations. These include the following:

- **Fugitive emissions** such as those around pipe flanges and pump seals
- **Process operations**, in modern chemical operations, automation and computer control have greatly reduced contact with feedstock, intermediates, and products
- **Extrusion** of solid products such as a solid polymer that normally is hot when released through an extruder head and may give off vapors of residual monomers, oligomers, and additives that may be inhaled
- **Maintenance and repair** that requires closed systems to be opened and may result in the disturbance of deposits and release of toxic substances from them
- **Sandblasting** in which abrasive agents such as finely divided silica (SiO_2) may be inhaled, which may damage lungs
- **Chemical cleaning** that may involve exposure to acids, caustic substances, and solvents and their vapors
- **Welding** that may release metal fumes such as those of zinc that can cause a condition of metal fume fever characterized by elevated temperature and chills
- **Painting** and **coating**, especially in surface preparation where cleaning old surfaces may release airborne lead, chromium, or other toxic substances
- **Waste handling** that may expose workers to toxic waste materials

The most common route of worker exposure in the anthrosphere is through inhalation of gases, vapors from liquids, and suspended particles. Whereas it is relatively easy to protect them from dermal exposure and ingestion, workers normally must breathe ambient air, which may contain toxicants.

An important activity in industrial hygiene is workplace monitoring and analysis of chemical hazards. Since air is the most likely route of exposure, it is usually the medium analyzed. Careful consideration must be given to the sampling protocol to make sure that it follows potential exposure as closely as possible. Various sampling devices are used, including those in which air is pumped over an adsorbent solid. Passive samplers that absorb toxic substances from surrounding air are also used. Sampling should be done in a location as close as possible to where workers may be exposed (close to the breathing zone). Normally, results are expressed as a time-weighted exposure or time-weighted concentration. A time-weighted concentration is calculated by measuring the concentration of a toxicant over several time intervals multiplied by the duration of each interval and divided by the sum of the times for all the intervals. Occupational exposure is commonly expressed for a working shift of 8 hours.

14.15.1 ROLE OF GREEN CHEMISTRY IN OCCUPATIONAL HEALTH

From the preceding discussion in Section 14.15, it is seen that protection of worker health is a top priority in the operation of the anthrosphere. This may involve workplace monitoring for hazards, preventative measures, protective gear for workers, diligent attention to equipment maintenance, and proper operating procedures to prevent exposure. The proper practice of green chemistry can aid substantially in this task. Green chemistry seeks to avoid the use and production of toxic and otherwise hazardous substances such that their effects never become an issue. Although not invariably safe (consider "brown lung disease" from exposure to cotton dust and explosions of grain dusts), the use of biomaterials for feedstocks tends to reduce hazards from materials. Green chemistry seeks to minimize the volume in which reactions are carried out, such as by using small-volume flow-through reactors rather than batch reactor processes, thus reducing the volume of potential releases (see Figure 14.14). Green chemistry seeks to carry out processes under relatively mild conditions of temperature and pressure, thus reducing hazards. By their nature, biological and enzymatic chemical transformations are carried out under mild conditions, which reduces hazards when such processes are feasible.

QUESTIONS AND PROBLEMS

Access to and use of the Internet is assumed in answering all questions, including general information, statistics, constants, and mathematical formulas required to solve problems. These questions are designed to promote inquiry and thought rather than just finding material in the chapter. So in some cases, there may be several "right" answers. Therefore, if your answer reflects intellectual effort and a search for information from available sources, it can be considered to be right.

1. What is meant by "level of recycling," and how is it related to embedded utility?
2. Given that an abundant source of energy can make almost anything possible in an industrial ecosystem, in what respects do vast reserves of coal, wind power, and solar energy fall short of being ideal energy sources?
3. What is your part of the anthrosphere? How is it tied to each of the other four environmental spheres?
4. What is the status of infrastructure in the United States? How does it compare to that of China? What are the long-term implications of infrastructure development, or lack thereof?
5. Look up cascading failure. To what part of the infrastructure does it generally apply? Are there other parts of the anthrosphere or infrastructure to which cascading failure may apply?
6. Look up information pertaining to critical infrastructure. Why is it a concern regarding security?
7. Growth is commonly mentioned as a concern in political discussions? How do most politicians regard growth? Is growth necessarily always a good thing? What are the alternatives to growth as it is commonly regarded?
8. Look up microsociology. What does it have to do with the anthrosphere and sustainability? Do you think there is such a thing as the microsociosphere? If so, of what does it consist?
9. Look up on the Internet recent or upcoming international conferences on industrial ecology. What are some of the topics covered? What do these topics tell you about the nature of industrial ecology?
10. What is meant by just-in-time manufacturing processes? How does it contribute to sustainability?
11. See if you can find any information pertaining to lead battery recycling in countries other than the United States. Have there been any problems with such operations? What are

some of the occupational health problems that have been documented associated with the processing of lead?

12. Explain how conditions in Cuba over the last several decades have provided an example (though painful in some respects) of at least one kind of recyclable service product.

13. Explain how a laser printer provides examples of consumable products, recyclable commodities, and service products.

14. Some power plants provide an example of how a toxic substance released from the anthrosphere may get into water, then into the biosphere, and finally adversely affect human health. What is this substance? Has there been any recent regulatory activity pertaining to it?

15. Look up on the Internet recent or upcoming international conferences on green chemistry. What are some of the topics covered? What do these topics tell you about the nature of green chemistry?

16. What are the environmental problems with MnO_2, $KMnO_4$, and $K_2Cr_2O_7$ as oxidants in chemical processing? What are some of the more sustainable alternatives to the use of these reagents?

17. Perchloric acid, $HClO_4$, is a potent oxidizing agent when heated. See if you can find an example (from a long time ago) in which perchloric acid was involved in a devastating industrial accident with a number of fatalities.

18. What influence do catalysts have on rate of reaction? What is meant by the specificity (selectivity) of catalysts and why is it important?

19. The chapter mentions "an economical process to make hydrogen peroxide by combining H_2 and O_2 in a gas mixture at levels of H_2 directly below the lower flammability limit of H_2 using a catalyst" (Section 14.10). Why not just mix H_2 and O_2 in a 1/1 mole ratio and let them undergo a simple addition reaction to produce H_2O_2?

20. In what respect is a biphasic catalyst like a heterogeneous catalyst and in what respect is it like a homogeneous catalyst?

21. Suggest why enzyme catalysts may be particularly advantageous in producing pharmaceuticals or insecticides where specific biological activity is important.

22. Name and describe at least one large-scale commodity chemical made by electrolysis. What chemical product is produced at the cathode and what at the anode in producing this chemical? How is electrolysis used in the preparation or purification of some metals?

23. Find and describe a photochemical process that is used in the chemical analysis of organic chemical compounds in water.

24. A relatively recent development in the synthetic fuels industry has made carbon dioxide much more available as a supercritical fluid that can be used as a medium in chemical synthesis. Explain.

25. Around 1900, organic compounds from coal tar became widely available and were used in chemical synthesis, especially in making synthetic dyes. What was the source of the coal tar? What were some of the unfortunate occupational health results of this industry?

26. A popular cleaning agent is marketed under the brand name of Oxi-Clean. What is the chemical nature of this agent, and how does it function as a cleaning and bleaching agent? What other substance is commonly used as a laundry bleaching agent?

27. In what sense is electrolysis a reagent-free chemical synthesis process?

28. What are some examples in which dusts composed of biological materials have caused accidents that have led to fatalities?

29. Name two desirable characteristics of reagents insofar as their ability to generate products is concerned.

30. What are the two oxidants used by organisms? What catalysts are used with these oxidants?

31. What is a nonchemical alternative to the use of oxidant and reductant reagents?

32. What are the two most common media used for chemical reactions?

33. Why is toluene preferred to benzene as a solvent medium for organic reactions?
34. What is a common toxicity problem with some lighter hydrocarbons, such as hexane?
35. Name an environmental effect of industrial activities in the anthrosphere on each of the other four environmental spheres.
36. Name an environmental effect of agricultural activities in the anthrosphere on each of the other four environmental spheres.
37. For each of the other four environmental spheres, name a measure that may be taken in anthrospheric activities to reduce the environmental impact of these activities.
38. Why should the use of lipid-soluble organics be reduced in the practice of green chemistry?
39. What is a characteristic of VOCs that both makes them useful in industrial applications and increases their hazards?
40. Distinguish between reactive and corrosive substances. Can a compound be both?

LITERATURE CITED

1. Petroski, Henry, Infrastructure, *American Scientist* **5**, 370–374 (2009).
2. Radvanovsky, Robert, and Allan McDougal, *Critical Infrastructure: Homeland Security and Emergency Preparedness*, 2nd ed., Taylor & Francis/CRC Press, Boca Raton, FL, 2009.
3. Graedel, Thomas E., and Braden. R. Allenby, *Industrial Ecology and Sustainable Engineering*, Prentice Hall, Upper Saddle River, NJ, 2009.
4. Crawford, Robert, *Life Cycle Assessment in the Built Environment*, Routledge, New York, 2011.
5. Sheldon, Roger A., E Factors, Green Chemistry and Catalysis: An Odyssey, *Chemical Communications* **29**, 3352–3365 (2008).
6. Baar, Marsha R., Greener Organic Reactions under Microwave Heating, in *Green Organic Chemistry in Lecture and Laboratory*, Andrew P. Dicks, Ed., CRC Press, Boca Raton, FL, 2011, pp. 225–256.
7. Cravotto, Giancarlo, and Pedro Cintas, Introduction to Sonochemistry: A Historical and Conceptual Overview, in *Handbook on Applications of Ultrasound*, Dong Chen, Sanjay K. Sharma, and Ackmez Mudhoo, Eds., CRC Press, Boca Raton, FL, 2012, pp. 23–40.
8. Protti, Stefano, Simone Manzini, Maurizio Fagnoni, and Angelo Albini, Angelo, The Contribution of Photochemistry to Green Chemistry, *RSC Green Chemistry Series* **3**, 80–111 (2009).
9. Machida, Hiroshi, Masafumi Takesue, and Richard L. Smith, Green Chemical Processes with Supercritical Fluids: Properties, Materials, Separations and Energy, *Journal of Supercritical Fluids* **60**, 2–15 (2011).
10. Jenkins, Harry, and Donald Brooke, Ionic Liquids-An Overview, *Science Progress* **94**, 265–297 (2011).
11. Kelly, Nelson A., Thomas L. Gibson, and David B. Ouwerkerk, Generation of High-Pressure Hydrogen for Fuel Cell Electric Vehicles Using Photovoltaic-Powered Water Electrolysis, *International Journal of Hydrogen Energy* **36**, 15803–15825 (2011).

SUPPLEMENTARY REFERENCES

Ahluwalia, V. K., M. Kidwai, *New Trends in Green Chemistry*, Kluwer Academic, Boston, MA, 2004.
Allen, David T., and David R. Shonnard, *Green Engineering: Environmentally Conscious Design of Chemical Processes*, Prentice Hall, Upper Saddle River, NH, 2002.
Anastas, Paul, Ed., *Handbook of Green Chemistry*, Wiley-VCH, New York, 2010.
Ayres, Robert U., and Leslie W. Ayres, Eds., *A Handbook of Industrial Ecology*, Edward Elgar Publishing, Cheltenham, UK, 2002.
Ayres, Robert U., and Benjamin Warr, *The Economic Growth Engine: How Energy and Work Drive Material Prosperity*, Edward Elgar, Northampton, MA, 2009.
Clark, James, and Duncan MacQuarrie, *Handbook of Green Chemistry and Technology*, Blackwell Science, Malden, MA, 2002.
Cote, Ray, James Tansey, and Ann Dale, Eds., *Linking Industry and Ecology: A Question of Design*, University of British Columbia Press, Vancouver, 2006.
Doble, Mukesh, and Anil Kumar Kruthiventi, *Green Chemistry and Processes*, Elsevier, Amsterdam, 2007.
Doxsee, Kenneth M., and James E. Hutchison, *Green Organic Chemistry: Strategies, Tools, and Laboratory Experiments*, Thomson-Brooks/Cole, Monterey, CA, 2004.
El-Haggar, Salah M., *Sustainable Industrial Design and Waste Management: Cradle-to-Cradle for Sustainable Development*, Elsevier Academic Press, Amsterdam, 2007.

Graedel, Thomas E., and Braden R. Allenby, *Industrial Ecology*, 2nd ed., Prentice Hall, Upper Saddle River, NJ, 2003.

Graedel, Thomas E., and Jennifer A. Howard-Grenville, *Greening the Industrial Facility: Perspectives, Approaches, and Tools*, Springer, Berlin, 2005.

Grossman, Elizabeth, *Chasing Molecules: Poisonous Products, Human Health, and the Promise of Green Chemistry*, Shearwater, Washington, DC, 2009.

Gupta, Surendra M., and A.J.D. Lambert, Eds., *Environment Conscious Manufacturing*, Taylor & Francis/CRC Press, Boca Raton, FL, 2008.

Hawken, Paul, Amory Lovins, and L. Hunter Lovins, *Natural Capitalism: Creating the Next Industrial Revolution*, Back Bay Books, Boston, 2008.

Hendrickson, Chris T., Lester B. Lave, and H. Scott Matthews, *Environmental Life Cycle Assessment of Goods and Services: An Input-Output Approach*, Resources for the Future, Washington, DC, 2006.

Horvath, Istvan T., and Paul T. Anastas, Innovations and Green Chemistry, *Chemical Reviews* **107**, 2169–2173 (2007).

Hunkeler, David, Kerstin Lichtenvort, and Gerald Rebitzer, Eds., *Environmental Life Cycle Costing*, Taylor & Francis/CRC Press, Boca Raton, FL, 2008.

Islam, M. Rafiqul, Ed., *Nature Science and Sustainable Technology*, Nova Science Publishers, New York, 2007.

Kronenberg, Jakub, *Ecological Economics and Industrial Ecology: A Case Study of the Integrated Product Policy of the European Union*, Routledge, New York, 2007.

Kutz, Myer, Ed., *Environmentally Conscious Transportation*, Wiley, Hoboken, NJ, 2008.

Lankey, Rebecca L., and Paul T. Anastas, Eds., *Advancing Sustainability through Green Chemistry and Engineering*, American Chemical Society, Washington, DC, 2002.

Li, Chao-Jun, and Barry M. Trost, Green Chemistry for Chemical Synthesis, *Proceedings of the National Academy of Sciences of the United States of America* **105**, 13197–13202 (2008).

Matlack, Albert, *Introduction to Green Chemistry*, 2nd ed., Taylor & Francis/CRC Press, Boca Raton, FL, 2010.

McDonough, William, and Michael Braungart, *Cradle to Cradle: Remaking The Way We Make Things*, North Point Press, New York, 2002.

Nelson, William M., *Green Solvents for Chemistry: Perspectives and Practice*, Oxford University Press, New York, 2003.

Nelson, William M., Ed., *Agricultural Applications in Green Chemistry*, Oxford University Press, New York, 2004.

Roesky, Herbert W. Dietmar Kennepohl, and Jean-Marie Lehn, Eds., *Experiments in Green and Sustainable Chemistry*, Wiley-VCH, New York, 2009.

Rosenfeld, Paul E., and Lydia Feng, *Risks of Hazardous Wastes*, Elsevier, Amsterdam, 2011.

Sheldon, Roger A., Isabel Arends, and Ulf Hanefield, *Green Chemistry and Catalysis*, Wiley-VCH, New York, 2007.

Simpson, R. David, Michael A. Toman, and Robert U. Ayres, Eds., *Scarcity and Growth Revisited: Natural Resources and the Environment in the New Millennium*, Resources for the Future, Washington, DC, 2005.

Tundo, Pietro, Alvise Perosa, and Fulvio Zecchini, *Methods and Reagents for Green Chemistry*, Wiley-Interscience, Hoboken, NJ, 2007.

Tundo, Pietro, and Vittorio Esposito, Eds., *Green Chemical Reactions*, Springer, Dordrecht, Netherlands, 2008.

Vallero, Daniel A., *Sustainable Design: The Science of Sustainability and Green Engineering*, Wiley, Hoboken, NJ, 2008.

15 Anthrosphere, Pollution, and Wastes

15.1 WASTES FROM THE ANTHROSPHERE

In its modern form, the anthrosphere has a tendency to produce wastes that can harm the environment and be dangerous to humans. Therefore, one of the most important activities of humans is to deal with the potential waste products that their activities generate. Of such wastes, hazardous wastes that pose particular dangers are of most concern.

A **hazardous substance** is a material that may pose a danger to living organisms, materials, structures, or the environment by explosion or fire hazards, corrosion, toxicity to organisms, or other detrimental effects. What, then, is a hazardous waste? A simple definition of a **hazardous waste** is that it is a hazardous substance that has been discarded, abandoned, neglected, released, or designated as a waste material or one that might interact with other substances to be hazardous. The definition of hazardous waste is addressed in greater detail in Section 15.2, but, in a simple sense, it is a material that has been left where it should not be and that could cause harm to living creatures or its surroundings.

15.1.1 HISTORY OF HAZARDOUS SUBSTANCES

Humans have always been exposed to hazardous substances, going back to prehistoric times when people inhaled noxious volcanic gases or succumbed to carbon monoxide from inadequately vented fires in cave dwellings sealed too well against ice-age cold. Slaves in Ancient Greece developed lung disease from weaving mineral asbestos fibers into cloth to make it more degradation-resistant. Some archaeological and historical studies have concluded that lead wine containers were a leading cause of lead poisoning in the more affluent ruling class of the Roman Empire, leading to erratic behavior such as fixation on spectacular sporting events, chronic unmanageable budget deficits, speculative purchases of overvalued real estate, illicit trysts by high officials in governmental offices, and ill-conceived, overly ambitious military ventures in remote foreign lands. Alchemists who worked during the Middle Ages often suffered debilitating injuries and illnesses resulting from the hazards of their explosive and toxic chemicals. During the 1700s, runoff from mine spoil piles began to create serious contamination problems in Europe. As the production of dyes and other organic chemicals developed from the coal tar industry in Germany during the 1800s, pollution and poisoning from coal tar by-products was observed. By around 1900, the quantity and the variety of chemical wastes produced each year were increasing sharply with the addition of wastes such as spent steel and iron pickling liquor, lead battery wastes, chromic wastes, petroleum refinery wastes, radium wastes, and fluoride wastes from aluminum ore refining. As the century progressed into the World War II era, the wastes and hazardous by-products of manufacturing increased markedly from sources such as chlorinated solvent manufacture, pesticide synthesis, polymer manufacture, plastics, paints, and wood preservatives.

The Love Canal affair of the 1970s and 1980s brought hazardous wastes to public attention as a major political issue in the United States. Starting around 1940, the Hooker Chemical Company used this site of an abandoned canal in Niagara Falls, New York, to dispose of about 20,000 metric tons (Mg) of chemical wastes containing at least 80 different chemicals, including benzene, chlorinated hydrocarbons, alkaline wastes, fatty acids, solvents, and other materials from the manufacture of synthetic resins, rubber, dyes, and perfumes. The U.S. Army and the city of Niagara Falls had also used the site for waste disposal. Given the technology at the time, the 1.6-km-long, 15-m-wide, and 3–12-m-deep canal excavation appeared to be ideal for waste disposal because of its supposedly impermeable clay walls. Eventually, the dumping ceased and the wastes were covered by a layer of clay. Eventually, the town of Love Canal was developed at the site and construction of buildings and plumbing breached the clay barriers. In 1953, the Niagara Falls Board of Education purchased land at the site for $1.00 and built an elementary school on it (the school's architect raised questions about the site after excavation uncovered chemicals, but his main concern was over the integrity of the school building's foundations). Strange odors had been noted at the site in the 1950s, but by the 1970s, health problems including skin irritation had become common in the area.[1] By 1978, the waste problems associated with Love Canal had become so obvious that the U.S. Federal Government became involved and the site became the first on the Superfund list under legislation establishing the Comprehensive Environmental Response, Compensation, and Liability Act (CERCLA) of 1980 (see Section 15.1.3). After more than 2 decades of remediation effort, relocation of about 800 residents in the Love Canal area, and expenditure of $400 million in public and private funds spent by Occidental Chemical Corp., which had purchased Hooker Chemical, in 2004, the remediated site was declared safe and removed from the Superfund list.

Other areas in the United States containing hazardous wastes that received attention included an industrial site in Woburn, Massachusetts that had been contaminated by wastes from tanneries, glue-making factories, and chemical companies dating back to about 1850; the Stringfellow Acid Pits near Riverside, California; the Valley of the Drums in Kentucky; and Times Beach, Missouri, an entire town that was abandoned because of contamination by 2,3,7,8-tetrachlorodibenzo-p-dioxin (TCDD) (dioxin). Many sites around the world have been contaminated with hazardous wastes. Many areas of the former Soviet Union suffer from improperly discarded wastes, the management of which was not a high priority in state-run economies. (The largest area ever abandoned as a result of waste contamination is that around the site of the 1986 Chernobyl nuclear reactor fire in the former Soviet Union.) A major problem has occurred with dumping of hazardous wastes in poorer countries by industrial concerns from wealthier nations.[2] The problem with hazardous wastes has become particularly acute in countries that are undergoing rapid industrialization, such as China and India.

15.1.2 Pesticide Burial Grounds

An example of problems from improper disposition of hazardous wastes is provided by "pesticide burial grounds" or "pesticide tombs."[3] In Poland, these consisted of about 20,000 Mg of obsolete pesticides, primarily organochlorine compounds, especially DDT, but also including organophosphates, carbamates, dinitrophenols, phenoxy acids, and inorganic compounds. An inventory of these sites was taken by the National Waste Management Plan in 1993 with site remediation that begun in 1999. Including the hazardous substances themselves, contaminated soils, and construction materials, the total amount of material in these sites in Poland amounted to about 100,000 Mg.

15.1.3 Legislation

Governments in a number of nations have passed legislation to deal with hazardous substances and wastes. In the United States, such legislation has included the following:[4]

- Toxic Substances Control Act of 1976
- Resource Conservation and Recovery Act (RCRA) of 1976 (amended and reauthorized by the Hazardous and Solid Wastes Amendments Act [HSWA] of 1984)

- CERCLA of 1980
- Pollution Prevention Act of 1990

RCRA legislation charged the U.S. Environmental Protection Agency (EPA) with protecting human health and the environment from improper management and disposal of hazardous wastes by issuing and enforcing regulations pertaining to such wastes. RCRA requires that hazardous wastes and their characteristics be listed and controlled from the time of their origin until their proper disposal or destruction. Regulations pertaining to firms generating and transporting hazardous wastes require that they keep detailed records, including reports on their activities and manifests to ensure proper tracking of hazardous wastes through transportation systems. Approved containers and labels must be used, and wastes can only be delivered to facilities approved for treatment, storage, and disposal.

CERCLA (Superfund) legislation deals with actual or potential releases of hazardous materials that have the potential to endanger people or the surrounding environment at uncontrolled or abandoned hazardous waste sites in the United States. The act requires responsible parties or the government to clean up waste sites. Among CERCLA's major purposes are the following:

- Site identification
- Evaluation of danger from waste sites
- Evaluation of damages to natural resources
- Monitoring of release of hazardous substances from sites
- Removal or cleanup of wastes by responsible parties or government

CERCLA was extended for 5 years by the passage of the Superfund Amendments and Reauthorization Act (SARA) of 1986, legislation with greatly increased scope and additional funding. Actually longer than CERCLA, SARA encouraged the development of alternatives to land disposal that favor permanent solutions reducing volume, mobility, and toxicity of wastes; increased emphasis on public health, research, training, and state and citizen involvement; and establishment of a new program for leaking underground (petroleum) storage tanks. After 1986, few new legislative initiatives dealing with hazardous wastes were introduced in the United States.

In its earlier years, Superfund was supported by taxes on petroleum and chemical manufacture. However, in the 1990s, the U.S. Congress refused to reinstate these taxes, and the fund that they supported ran out in 2003. Since that time, superfund cleanup efforts costing around $1 billion per year have been financed from general revenues and from legal settlements with parties judged responsible for improper waste disposal. As of January, 2012, a total of 1296 waste sites were on the U.S. EPA's National Priorities List designated for cleanup under the Superfund program. Periodically, sites that have been cleaned up are deleted from the list, and others are added to the list.

The Pollution Prevention Act of 1990 was passed in recognition of the fact that existing regulations, emphasis on treatment and disposal, and industrial resources diverted to compliance can hinder pollution prevention. This act is designed to change production, operation, and raw materials use in ways that reduce pollution. It attempts to focus industry, government, and public attention on pollution reduction. This act recognizes that source reduction is more desirable than waste management or pollution control, so it seeks to reduce the amount of pollution through cost-effective changes in production, operation, and raw materials use. The objectives of the Pollution Prevention Act are met very well by the proper practice of green chemistry and industrial ecology.

15.2   CLASSIFICATION OF HAZARDOUS SUBSTANCES AND WASTES

Many specific chemicals in widespread use are hazardous because of their chemical reactivities, fire hazards, toxicities, and other properties. There are numerous kinds of hazardous substances, usually consisting of mixtures of specific chemicals. These include such things as explosives; flammable liquids; flammable solids such as magnesium metal and sodium hydride; oxidizing materials such as peroxides; corrosive materials such as strong acids; etiologic agents that cause disease; and radioactive materials.

Three basic approaches to defining hazardous wastes are (1) a qualitative description by origin, type, and constituents; (2) classification by characteristics largely based on testing procedures; and (3) by means of concentrations of specific hazardous substances. Wastes may be classified by general type such as "spent halogenated solvents" or by industrial sources such as "pickling liquor from steel manufacturing."

15.2.1 CHARACTERISTICS AND LISTED WASTES

For regulatory and legal purposes in the United States, hazardous substances are listed specifically and are defined according to general characteristics. Under the authority of the RCRA, the U.S. EPA defines hazardous substances in terms of the following **characteristics**:

- **Ignitability**: Characteristic of substances that are liquids, the vapors of which are likely to ignite in the presence of ignition sources; nonliquids that may catch fire from friction or contact with water and that burn vigorously or persistently; ignitable compressed gases; and oxidizers
- **Corrosivity**: Characteristic of substances that exhibit extremes of acidity or basicity or a tendency to corrode steel
- **Reactivity**: Characteristic of substances that have a tendency to undergo violent chemical change (examples are explosives, pyrophoric materials, water-reactive substances, or cyanide- or sulfide-bearing wastes)
- **Toxicity**: Defined in terms of a standard extraction procedure followed by chemical analysis for specific substances

In addition to classification by characteristics, EPA designates more than 450 **listed wastes** that are specific substances or classes of substances known to be hazardous. Each such substance is assigned an EPA **hazardous waste number** in the format of a letter followed by three numerals, where a different letter is assigned to substances from each of the four following lists:

1. **F-type wastes from nonspecific sources**: For example, quenching wastewater treatment sludges from metal heat treating operations where cyanides are used in the process (F012).
2. **K-type wastes from specific sources**: For example, heavy ends from the distillation of ethylene dichloride in ethylene dichloride production (K019).
3. **P-type acute hazardous wastes**: Wastes that have been found to be fatal to humans in low doses or capable of causing or significantly contributing to an increase in serious irreversible or incapacitating reversible illness. These are mostly specific chemical species such as fluorine (P056) or 3-chloropropane nitrile (P027).
4. **U-Type miscellaneous hazardous wastes**: These are predominantly specific compounds such as calcium chromate (U032) or phthalic anhydride (U190).

15.2.2 HAZARDOUS WASTES AND AIR AND WATER POLLUTION CONTROL

Somewhat paradoxically, measures taken to reduce air and water pollution (Figure 15.1) have had a tendency to increase production of hazardous wastes. Most water treatment processes yield sludges or concentrated liquors that require stabilization and disposal. Air scrubbing processes likewise produce sludges. Baghouses and precipitators used to control air pollution all yield significant quantities of solids, some of which are hazardous.

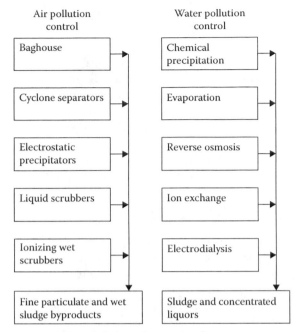

Air pollution control

Water pollution control

FIGURE 15.1 Potential contributions of air and water pollution control measures for hazardous wastes production.

15.3 SOURCES OF WASTES

Quantities of hazardous wastes produced each year are not known with certainty and depend on the definitions used for such materials. In the United States, there are around 17,000 RCRA-regulated sites that generate approximately 30 million tons of wastes. However, most of this material is water, with only a few million tons consisting of solids. Some high-water-content wastes are generated directly by processes that require large quantities of water in waste treatment, and other aqueous wastes are produced by mixing hazardous wastes with wastewater.

Some wastes that might exhibit a degree of hazard are exempt from RCRA regulation by legislation. These exempt wastes include fuel ash and scrubber sludge from power generation by utilities, oil and gas drilling muds, by-product brine from petroleum production, cement kiln dust, waste and sludge from phosphate mining and beneficiation, mining wastes from uranium and other minerals, and household wastes.

15.3.1 Types of Hazardous Wastes

In terms of quantity by weight, the greatest quantities are those from categories designated by hazardous waste numbers preceded by F and K, respectively. The former are those from nonspecific sources and include the following examples:

- F001—The spent halogenated solvents used in degreasing: tetrachloroethylene, trichloroethylene, methylene chloride, 1,1,1-trichloroethane, carbon tetrachloride, and the chlorinated fluorocarbons and sludges from the recovery of these solvents in degreasing operations
- F004—The spent nonhalogenated solvents: cresols, cresylic acid, and nitrobenzene and still bottoms from the recovery of these solvents

- F007— Spent cyanide plating-bath solutions from electroplating operations
- F010—Quenching-bath sludge from oil baths from metal heat treating operations

The "K-type" hazardous wastes are those from specific sources produced by industries such as the manufacture of inorganic pigments, organic chemicals, pesticides, explosives, iron and steel, and nonferrous metals and from processes such as petroleum refining or wood preservation; some examples are as follows:

- K001—Bottom sediment sludge from the treatment of wastewaters from wood-preserving processes that use creosote and/or pentachlorophenol
- K002—Wastewater treatment sludge from the production of chrome yellow and orange pigments
- K020—Heavy ends (residue) from the distillation of vinyl chloride in vinyl chloride monomer production
- K043—2,6-Dichlorophenol waste from the production of 2,4-D
- K047—Pink/red water from trinitrotoluene (TNT) operations
- K049—Slop oil emulsion solids from the petroleum refining industry
- K060—Ammonia lime still sludge from coking operations
- K084—Wastewater treatment sludges generated during the production of veterinary pharmaceuticals from arsenic or organoarsenic compounds

The remainder of wastes consists of reactive wastes; corrosive wastes; toxic wastes; ignitable wastes; and "P" wastes (discarded commercial chemical products, off-specification species, containers, and spill residues), "U" wastes, and unspecified types.

The EPA National Waste Minimization Program has designated 31 Priority Chemicals that are encountered in products and wastes.[5] In keeping with the best practice of green chemistry, the minimization program emphasizes finding ways to eliminate or substantially reduce the use of these chemicals in production and in recovering or recycling them in cases where their use cannot be eliminated. The priority list consists of the following 28 organic chemicals and 3 heavy metals: 1,2,4-trichlorobenzene; 1,2,4,5-tetrachlorobenzene; 2,4,5-trichlorophenol; 4-bromophenyl phenyl ether; acenaphthene; acenaphthylene; anthracene; benzo(g,h,i)perylene; dibenzofuran; dioxins/furans; endosulfan, alpha and beta; fluorene (a polycyclic aromatic hydrocarbon); heptachlor and heptachlor epoxide, hexachlorobenzene; hexachlorobutadiene; hexachlorocyclohexane, gamma-(lindane); hexachloroethane; methoxychlor; naphthalene; pendimethalin; pentachlorobenzene; pentachloronitrobenzene (quintozene); pentachlorophenol; phenanthrene; polycyclic aromatic compounds; polychlorinated biphenyls (PCBs); pyrene; trifluralin; cadmium and its compounds; lead and its compounds; and mercury and its compounds.

15.3.2 Hazardous Waste Generators

Several hundred thousand companies generate hazardous wastes in the United States, but most of these generators produce only small quantities. Hazardous waste generators are unevenly distributed geographically across the continental United States, with a relatively large number located in the industrialized upper Midwest, including the states of Illinois, Indiana, Ohio, Michigan, and Wisconsin.

Industry types of hazardous waste generators can be divided among the seven following major categories, each containing some 10–20% of hazardous waste generators: chemicals and allied products manufacture, petroleum-related industries, fabricated metals, metal-related products, electrical equipment manufacture, "all other manufacturing," and nonmanufacturing and nonspecified generators. About 10% of the generators produce more than 95% of all hazardous wastes. Whereas, the number of hazardous waste generators is distributed relatively evenly among several major types of industries, 70–85% of the *quantities* of hazardous wastes are generated by the chemical and petroleum industries. Of the remainder, about three-fourths comes from metal-related industries and about one-fourth from all other industries.

15.4 FLAMMABLE AND COMBUSTIBLE SUBSTANCES

Most chemicals that are likely to burn accidentally are liquids. Liquids form **vapors**, which are usually more dense than air, and thus tend to settle. The tendency of a liquid to ignite is measured by a test in which the liquid is heated and periodically exposed to a flame until the mixture of vapor and air ignites at the liquid's surface (Figure 15.2). The temperature at which ignition occurs momentarily under these conditions is called the **flash point**. The temperature at which combustion is sustained for at least 5 seconds is called the **fire point**.

With these definitions in mind, it is possible to divide ignitable materials into four major classes. A **flammable solid** is one that can ignite from friction or from heat remaining from its manufacture or that might cause a serious hazard if ignited. Explosive materials are not included in this classification. A **flammable liquid** is one having a flash point below 60.5°C (141°F). A **combustible liquid** has a flash point in excess of 60.5°C, but below 93.3°C (200°F). Where gases are substances that exist entirely in the gaseous phase at 0°C and 1 atm pressure, a **flammable compressed gas** meets specified criteria for lower flammability limit (LFL), flammability range, and flame projection.

In considering the ignition of vapors, two important concepts are those of flammability limit and flammability range. Values of the vapor–air ratio below which ignition cannot occur because of insufficient fuel define the **LFL**. Similarly, values of the vapor–air ratio above which ignition cannot occur because of insufficient air define the **upper flammability limit** (**UFL**). The difference between UFL and LFL at a specified temperature is the **flammability range**. Table 15.1 gives some examples of these values for common liquid chemicals. The percentage of flammable substance for best combustion (most explosive mixture) may be called "optimal." In the case of acetone, for example, the optimal flammable mixture is 5.0% acetone.

One of the more disastrous problems that can occur with flammable liquids is a boiling liquid expanding vapor explosion (BLEVE). These are caused by rapid pressure buildup in closed containers of flammable liquids heated by an external source. The explosion occurs when the pressure buildup is sufficient to break the container walls.

15.4.1 COMBUSTION OF FINELY DIVIDED PARTICLES

Finely divided particles of combustible materials are somewhat analogous to vapors with respect to flammability. One such example is a spray or mist of hydrocarbon liquid in which oxygen has the opportunity for intimate contact with the liquid particles causing the liquid to ignite at a temperature below its flash point.

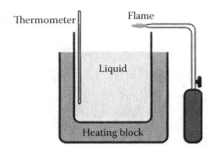

FIGURE 15.2 Schematic of the Cleveland open-cup method for the determination of flash point and fire point.

TABLE 15.1

Flammabilities of Some Common Organic Liquids

| Liquid | Flash Point (°C)[a] | Volume Percent in Air | |
		LFL[b]	UFL[b]
Diethyl ether	−43	1.9	36
Pentane	−40	1.5	7.8
Acetone	−20	2.6	13
Toluene	4	1.27	7.1
Methanol	12	6.0	37
Gasoline (2,2,4-tri-methylpentane)	—	1.4	7.6
Naphthalene	157	0.9	5.9

[a] Closed-cup flash point test.
[b] LFL, lower flammability limit; UFL, upper flammability limit at 25°C.

Dust explosions can occur with a large variety of solids that have been ground to a finely divided state. Many metal dusts, particularly those of magnesium and its alloys, zirconium, titanium, and aluminum, can burn explosively in air. In the case of aluminum, for example, the reaction is as follows:

$$4Al(powder) + 3O_2 (from\ air) \rightarrow 2Al_2O_3 \qquad (15.1)$$

Coal dust and grain dusts have caused many fatal fires and explosions in coal mines and grain elevators respectively. Dusts of polymers such as cellulose acetate, polyethylene, and polystyrene can also be explosive.

15.4.2 OXIDIZERS

Combustible substances are reducing agents that react with **oxidizers** (oxidizing agents or oxidants) to produce heat. Diatomic oxygen, O_2, from air is the most common oxidizer. Many oxidizers are chemical compounds that contain oxygen in their formulas. The halogens (periodic table group 7A) and many of their compounds are oxidizers. Some examples of oxidizers are given in Table 15.2.

An example of a reaction of an oxidizer is that of concentrated HNO_3 with copper metal, which gives toxic NO_2 gas as a product:

$$4HNO_3 + Cu \rightarrow Cu(NO_3)_2 + 2H_2O + 2NO_2 \qquad (15.2)$$

15.4.3 SPONTANEOUS IGNITION

Substances that catch fire spontaneously in air without an ignition source are called **pyrophoric**. These include several elements—white phosphorus, the alkali metals (group 1A), and powdered forms of magnesium, calcium, cobalt, manganese, iron, zirconium, and aluminum. Also included are some organometallic compounds, such as ethyllithium (LiC_2H_5) and phenyllithium (LiC_6H_5), and some metal carbonyl compounds such as iron pentacarbonyl, $Fe(CO)_5$. Another major class of pyrophoric compounds consists of metal and metalloid hydrides, including lithium hydride, LiH;

TABLE 15.2

Examples of Some Oxidizers

Name	Formula	State of Matter
Ammonium nitrate	NH_4NO_3	Solid
Ammonium perchlorate	NH_4ClO_4	Solid
Bromine	Br_2	Liquid
Chlorine	Cl_2	Gas (stored as liquid)
Fluorine	F_2	Gas
Hydrogen peroxide	H_2O_2	Solution in water
Nitric acid	HNO_3	Concentrated solution
Nitrous oxide	N_2O	Gas (stored as liquid)
Ozone	O_3	Gas
Perchloric acid	$HClO_4$	Concentrated solution
Potassium permanganate	$KMnO_4$	Solid
Sodium dichromate	$Na_2Cr_2O_7$	Solid

pentaborane, B_5H_9; and arsine, AsH_3. Moisture in air is often a factor in spontaneous ignition. For example, lithium hydride undergoes the following reaction with water from moist air:

$$LiH + H_2O \rightarrow LiOH + H_2 + heat \qquad (15.3)$$

The heat generated from this reaction can be sufficient to ignite the hydride so that it burns in air:

$$2LiH + O_2 \rightarrow Li_2O + H_2O \qquad (15.4)$$

Some compounds with organometallic character are also pyrophoric. An example of such a compound is diethylethoxyaluminum:

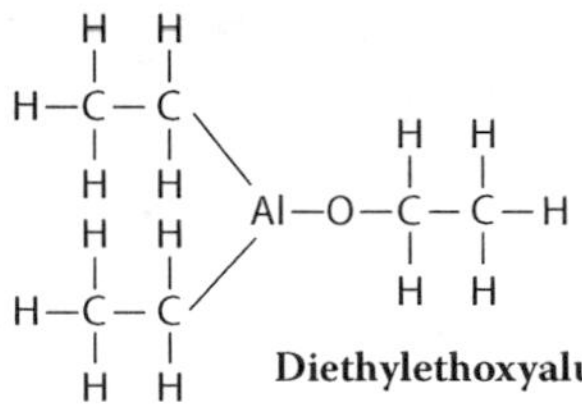

Diethylethoxyaluminum

Many mixtures of oxidizers and oxidizable chemicals catch fire spontaneously and are called **hypergolic mixtures**. Nitric acid and phenol form such a mixture.

15.4.4 TOXIC PRODUCTS OF COMBUSTION

Some of the greater dangers of fires are from toxic products and by-products of combustion. The most obvious of these is carbon monoxide, CO, which can cause serious illness or death because it forms carboxyhemoglobin with hemoglobin in the blood so that the blood no longer carries oxygen to body tissues. Toxic SO_2, P_4O_{10}, and HCl are formed by the combustion of sulfur, phosphorus, and organochlorine compounds respectively. A large number of noxious organic compounds such as aldehydes are generated as by-products of combustion. In addition to forming carbon monoxide, combustion under oxygen-deficient conditions produces polycyclic aromatic hydrocarbons consisting of fused ring structures. Some of these compounds, such as benzo[a]pyrene, in the following figure, are precarcinogens that are acted upon by enzymes in the body to yield cancer-producing metabolites.

Benzo(a)pyrene

15.5   REACTIVE SUBSTANCES

Reactive substances are those that tend to undergo rapid or violent reactions under certain conditions. Such substances include those that react violently or form potentially explosive mixtures with water. An example is sodium metal, which reacts strongly with water as follows:

$$2Na + 2H_2O \rightarrow 2NaOH + H_2 + heat \tag{15.5}$$

This reaction usually generates enough heat to ignite the sodium and hydrogen. Explosives constitute another class of reactive substances. For regulatory purposes, substances that produce toxic gases or vapors when they react with water, acids, or bases are also classified as reactive. Hydrogen sulfide and hydrogen cyanide are the most common toxic substances released in this manner.

Heat and temperature are usually very important factors in reactivity. Many reactions require energy of activation to get them started. The rates of most reactions tend to increase sharply with increasing temperature, and most chemical reactions give off heat. Therefore, once a reaction is started in a reactive mixture lacking an effective means of heat dissipation, the rate may increase exponentially with time, leading to an uncontrollable event. Other factors that may affect reaction rate include physical form of reactants (e.g., a finely divided metal powder that reacts explosively with oxygen, whereas a single mass of metal barely reacts), rate and degree of mixing of reactants, degree of dilution with nonreactive media (solvent), presence of a catalyst, and pressure.

Some chemical compounds are self-reactive in that they contain an oxidant and a reductant in the same compound. Nitroglycerin, a strong explosive with the formula $C_3H_5(ONO_2)_3$, decomposes spontaneously to CO_2, H_2O, O_2, and N_2 with a rapid release of a very large amount of energy. Pure nitroglycerin has such a high inherent instability that the slightest blow might be sufficient to detonate it. TNT is also an explosive with a high degree of reactivity. However, it is inherently relatively stable because some sort of detonating device is required to cause it to explode.

15.5.1   CHEMICAL STRUCTURE AND REACTIVITY

As shown in Table 15.3, some chemical structures are associated with high reactivity. High reactivity in some organic compounds results from unsaturated bonds in the carbon skeleton, particularly where multiple bonds are adjacent (allenes, C=C=C) or separated by only one carbon–carbon single bond (dienes, C=C–C=C). Some organic structures involving oxygen are very reactive. Examples are oxiranes such as ethylene oxide, hydroperoxides (ROOH), and peroxides (ROOR′), where R and R′ stand for hydrocarbon moieties such as the methyl group –CH$_3$. Many organic compounds containing nitrogen along with carbon and hydrogen are very reactive. Included are triazenes containing a functionality with three nitrogen atoms (R–N=N–N), some azo compounds (R–N=N–R′), and some nitriles in which a nitrogen atom is triply bonded to a carbon atom:

$$R–C \equiv N \quad \textbf{Nitrile}$$

TABLE 15.3

Examples of Reactive Compounds and Structures

Name	Structure or Formula
Organic	
Allenes	C=C=C
Dienes	C=C–C=C
Azo compounds	C–N=N–C
Triazenes	C–N=N–N
Hydroperoxides	R–OOH[a]
Peroxides	R–OO–R′
Alkyl nitrates	R–O–NO$_2$
Nitro compounds	R–NO$_2$
Inorganic	
Nitrous oxide	N$_2$O
Nitrogen halides	NCl$_3$, NI$_3$
Interhalogen compounds	BrCl
Halogen oxides	ClO$_2$
Halogen azides	ClN$_3$
Hypohalites	NaClO

[a] R and R′ denote hydrocarbon groups such as –CH$_3$.

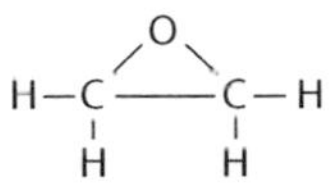

$$H-\underset{\underset{H}{|}}{C}\underbrace{\qquad}_{}\overset{O}{\diagup\diagdown}\underset{\underset{H}{|}}{C}-H \quad \textbf{Ethylene oxide}$$

Functional groups containing both oxygen and nitrogen tend to impart reactivity to an organic compound. Examples of such functional groups are alkyl nitrates (R–NO$_2$), alkyl nitrites (R–O–N=O), nitroso compounds (R–N=O), and nitro compounds (R–NO$_2$).

Many different classes of inorganic compounds are reactive. These include some of the halogen compounds of nitrogen (shock-sensitive nitrogen triiodide, NI$_3$, is a spectacular example), compounds with metal-nitrogen bonds (NaN$_3$), halogen oxides (ClO$_2$), and compounds with oxyanions of the halogens. An example of the last group of compounds is ammonium perchlorate, NH$_4$ClO$_4$, a solid rocket propellant that was involved in a series of massive explosions that destroyed 8 million lb of the compound and demolished a 40 million lb/year U.S. rocket fuel plant near Henderson, Nevada, in 1988. (By late 1989, a new \$92 million plant for the manufacture of ammonium perchlorate had been constructed near Cedar City in a remote region of southwest Utah. Prudently, the buildings at the new plant have been placed at large distances from each other!)

Explosives such as nitroglycerin or TNT are single compounds containing both oxidizing and reducing functions in the same molecule. Such substances are commonly called **redox compounds**. Some redox compounds have even more oxygen than is needed for a complete reaction and are said to have a positive balance of oxygen, some have exactly the stoichiometric quantity of oxygen required (zero balance and maximum energy release), and others have a negative balance and require oxygen from outside sources to completely oxidize all components. TNT has a substantial negative balance of oxygen; ammonium dichromate ((NH$_4$)$_2$Cr$_2$O$_7$) has a zero balance, reacting with exact stoichiometry to H$_2$O, N$_2$, and Cr$_2$O$_3$; and treacherously explosive nitroglycerin has a positive balance, as shown by the following reaction:

$$4C_3H_5N_3O_9 \rightarrow 12CO_2 + 10H_2O + 6N_2 + O_2 \tag{15.6}$$

15.6 CORROSIVE SUBSTANCES

Corrosive substances are regarded as those that dissolve metals or cause oxidized material to form on the surface of metals—rusted iron is a prime example—and, more broadly, cause deterioration of materials, including living tissue, that they contact. Most corrosives belong to at least one of the four following chemical classes: (1) strong acids, (2) strong bases, (3) oxidants, and (4) dehydrating agents. Table 15.4 lists some of the major corrosive substances and their effects.

15.6.1 SULFURIC ACID

Sulfuric acid is a prime example of a corrosive substance. As well as being a strong acid, concentrated sulfuric acid is also a dehydrating agent and oxidant. The tremendous affinity of H_2SO_4 for water is illustrated by the heat generated when water and concentrated sulfuric acid are mixed. If this is done incorrectly by adding water to the acid, localized boiling and spattering can occur, which could result in personal injury. The major destructive effect of sulfuric acid on skin tissue is removal of water with accompanying release of heat. Sulfuric acid decomposes carbohydrates by removal of water. In contact with sugar, for example, concentrated sulfuric acid reacts to leave a charred mass. The reaction is

$$C_{12}H_{22}O_{11} \xrightarrow{(Conc.\,H_2SO_4)} 11H_2O\,(H_2SO_4) + 12C + heat \tag{15.7}$$

Some dehydration reactions of sulfuric acid can be very vigorous. For example, the reaction with perchloric acid produces unstable Cl_2O_7, and a violent explosion can result. Concentrated sulfuric acid produces dangerous or toxic products with a number of other substances such as toxic carbon monoxide (CO) from reaction with oxalic acid, $H_2C_2O_4$; toxic bromine and sulfur dioxide (Br_2 and SO_2) from reaction with sodium bromide, NaBr; and toxic, unstable chlorine dioxide (ClO_2) from reaction with sodium chlorate, $NaClO_3$.

Contact with sulfuric acid causes severe tissue destruction resulting in a severe burn that may be difficult to heal. Inhalation of sulfuric acid fumes or mists damages tissues in the upper respiratory tract and eyes. Long-term exposure to sulfuric acid fumes or mists has caused erosion of teeth.

TABLE 15.4
Examples of Some Corrosive Substances

Name and Formula	Properties and Effects
Nitric acid, HNO_3	Strong acid and strong oxidizer, corrodes metal, and reacts with protein in tissue to form yellow xanthoproteic acid; lesions are slow to heal
Hydrochloric acid, HCl	Strong acid, corrodes metals, and gives off HCl gas vapor, which can damage respiratory tract tissue
Hydrofluoric acid, HF	Corrodes metals, dissolves glass, and causes particularly bad burns to flesh
Alkali metal hydroxides, NaOH and KOH	Strong bases, corrode zinc, lead, and aluminum, substances that dissolve tissue and cause severe burns
Hydrogen peroxide, H_2O_2	Oxidizer, all but very dilute solutions cause severe burns
Interhalogen compounds such as ClF and BrF_3	Powerful corrosive irritants that acidify, oxidize, and dehydrate tissue
Halogen oxides such as OF_2, Cl_2O, and Cl_2O_7	Powerful corrosive irritants that acidify, oxidize, and dehydrate tissue
Elemental fluorine, chlorine, and bromine (F_2, Cl_2, Br_2)	Very corrosive to mucous membranes and moist tissue, strong irritants

15.7 TOXIC SUBSTANCES

Toxicity is of the utmost concern in dealing with hazardous substances. This includes both long-term chronic effects from continual or periodic exposures to low levels of toxicants and acute effects from a single large exposure. Some toxic substances and their toxicological chemistry are covered in Chapter 2.

For regulatory and remediation purposes, a standard test is needed to measure the likelihood of toxic substances getting into the environment and causing harm to organisms. The U.S. EPA specifies a test called the **toxicity characteristic leaching procedure** (**TCLP**) designed to determine the toxicity hazard of wastes.[6] The test was designed to estimate the availability to organisms of both inorganic and organic species in hazardous materials present as liquids, solids, or multiple-phase mixtures and does not test for the direct toxic effects of wastes. Basically, the procedure consists of leaching a material with a solvent designed to mimic leachate generated in a municipal waste-disposal site, followed by chemical analysis of the leachate.

15.8 PHYSICAL FORMS AND SEGREGATION OF WASTES

Three major categories of wastes based on their physical forms are **organic materials**, **aqueous wastes**, and **sludges**. These forms largely determine the course of action taken in treating and disposing of the wastes. The **level of segregation**, a concept illustrated in Figure 15.3, is very important in treating, storing, and disposing of different kinds of wastes. It is relatively easy to deal with wastes that are not mixed with other kinds of wastes, that is, those that are highly segregated. For example, spent hydrocarbon solvents can be used as fuel in boilers. However, if these solvents are mixed with spent organochlorine solvents, the production of contaminant hydrogen chloride during combustion may prevent fuel use and require disposal in special hazardous waste incinerators. Further mixing with inorganic sludges adds mineral matter and water. These impurities complicate the treatment processes required by producing mineral ash in incineration or lowering the heating value of the material incinerated because of the presence of water. Among the most difficult types

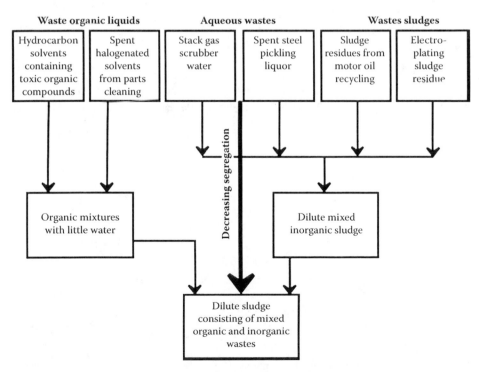

FIGURE 15.3 Illustration of waste segregation.

of wastes to handle and treat are those with the least segregation, of which a "worst case scenario" would be "dilute sludge consisting of mixed organic and inorganic wastes," as shown in Figure 15.3.

Concentration of wastes is an important factor in their management. A waste that has been concentrated or, preferably, never diluted is generally much easier and more economical to handle than one that is dispersed in a large quantity of water or soil. Dealing with hazardous wastes is greatly facilitated when the original quantities of wastes are minimized and the wastes remain separated and concentrated as much as possible.

15.9 ENVIRONMENTAL CHEMISTRY OF HAZARDOUS WASTES

The properties of hazardous materials, their production, and what makes a hazardous substance a hazardous waste were discussed in Sections 15.1–15.8 of this chapter. Hazardous materials normally cause problems when they enter the environment and have detrimental effects on organisms or other parts of the environment. Therefore, this chapter deals with the environmental chemistry of hazardous materials. In discussing the environmental chemistry of hazardous materials, it is convenient to consider the following five aspects based on the definition of environmental chemistry:

1. Origins
2. Transport
3. Reactions
4. Effects
5. Fates

It is also useful to consider the five environmental spheres as defined and discussed in other chapters of this book:

1. Anthrosphere
2. Geosphere
3. Hydrosphere
4. Atmosphere
5. Biosphere

Hazardous materials almost always originate in the anthrosphere, are often discarded into the geosphere, and are frequently transported through the hydrosphere or the atmosphere. The greatest concern for their effects is usually on the biosphere, particularly human beings. Figure 15.4 summarizes these relationships.

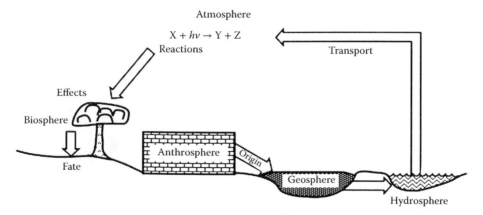

FIGURE 15.4   Scheme of interactions of hazardous wastes in the environment.

There are several ways in which hazardous wastes get into the environment. Although now much more controlled by pollution prevention laws, hazardous substances have been deliberately added to the environment by humans. Wastewater containing a variety of toxic substances has been discharged in large quantities into waterways. A particularly egregious example is the estimated 590,000 kg of PCBs (see Chapter 4, Section 4.12) that were discharged into the Hudson River in New York by a General Electric electrical equipment manufacturing operation before PCBs were banned in 1977. Although the dumping was entirely legal at the time, the U.S. EPA declared a 317-km section of the Hudson River from Hudson Falls, New York, to the southern tip of Manhattan to be a Superfund hazardous waste site with an ongoing cleanup involving dredging of sediment and shipping of it to a disposal site in Texas that may cost several hundred million dollars before it is completed.

Hazardous gases and particulate matter have been discharged into the atmosphere through stacks from power plants, incinerators, and a variety of industrial operations. Hazardous wastes have been deliberately spread on soil or placed in landfills in the geosphere. Evaporation and wind erosion may move hazardous materials from waste dumps into the atmosphere, or they may be leached from waste dumps into groundwater or surface waters. Underground storage tanks or pipelines have leaked a variety of materials into soil. Accidents, fires, and explosions may distribute dangerous materials into the environment. Another source of such materials consists of improperly operated waste treatment or storage facilities.

15.10 TRANSPORT, EFFECTS, AND FATES OF HAZARDOUS WASTES

The transport of hazardous wastes is largely a function of their physical properties, the physical properties of their surrounding matrix, the physical conditions to which they are subjected, and chemical factors. Highly volatile wastes are obviously more likely to be transported through the atmosphere and more soluble ones to be carried by water. Wastes will move farther and faster in porous, sandy formations than in denser soils. Volatile wastes are more mobile under hot, windy conditions and soluble ones during periods of heavy rainfall. Wastes that are more chemically and biochemically reactive will not move so far as less reactive wastes before breaking down.

The major properties of hazardous substances and their surroundings that determine the environmental transport of such substances are as follows:

- Physical properties of the substances, including vapor pressure and solubility.
- Physical properties of the surrounding matrix.
- Physical conditions to which wastes are subjected. Higher temperatures and erosive wind conditions enable volatile substances to move more readily.
- Chemical and biochemical properties of wastes. Substances that are less chemically reactive and less biodegradable will tend to move farther before breaking down.

15.10.1 PHYSICAL PROPERTIES OF WASTES

The major physical properties of wastes that determine their amenability to transport are volatility, solubility, and the degree to which they are sorbed to solids, including soil and sediments.

The distribution of hazardous waste compounds between the atmosphere and the geosphere or hydrosphere is largely a function of compound volatility. Usually, in the hydrosphere, and often in soil, hazardous waste compounds are dissolved in water; therefore, the tendency of water to hold the compound is a factor in its mobility. For example, although ethyl alcohol has a higher vapor pressure and lower boiling temperature (77.8°C) than toluene (110.6°C), vapor of the latter compound is more readily evolved from soil because of its limited solubility in water compared with ethanol, which is totally miscible with water.

15.10.2 CHEMICAL FACTORS

The environmental movement, effects, and fates of hazardous waste compounds are strongly related to their chemical properties. For example, a toxic heavy metal cationic species, such as Pb^{2+} ion, may be strongly held by negatively charged soil solids. If the lead is chelated by the chelating ethylenediaminetetraacetic acid (EDTA) anion, represented as Y^{4-}, it becomes much more mobile as PbY^{2-}, an anionic form. Oxidation state can be very important in the movement of hazardous substances. The reduced states of iron and manganese, Fe^{2+} and Mn^{2+}, respectively, are water soluble and relatively mobile in the hydrosphere and the geosphere. However, in their common oxidized states, Fe(III) and Mn(IV), these elements are present as insoluble $Fe_2O_3 \cdot xH_2O$ and MnO_2, which have virtually no tendency to move. Furthermore, these iron and manganese oxides will sequester heavy metal ions, such as Pb^{2+} and Cd^{2+}, preventing their movement in the soluble form.

15.10.3 ENVIRONMENTAL EFFECTS OF HAZARDOUS WASTES

The effects of hazardous wastes in the environment can be divided among effects on organisms, effects on materials, and effects on the environment. These are addressed briefly here and in greater detail through Section 15.15.

The ultimate concern with wastes has to do with their toxic effects on animals, plants, and microbes. Virtually all hazardous waste substances are poisonous to a degree, some extremely so. The toxicity of a waste is a function of many factors, including the chemical nature of the waste, the matrix in which it is contained, circumstances of exposure, the species exposed, manner of exposure, degree of exposure, and time of exposure. The toxicities of some of the substances found in hazardous wastes are discussed in more detail in Chapter 2.

As defined in Section 15.6, many hazardous wastes are *corrosive* to materials, usually because of extremes of pH or because of dissolved salt content. Oxidant wastes can cause combustible substances to burn uncontrollably. Highly reactive wastes can explode, causing damage to materials and structures. Contamination by wastes, such as by toxic pesticides in grain, can result in substances becoming unfit for use.

In addition to their toxic effects in the biosphere, hazardous wastes can damage air, water, and soil. Wastes that get into air can cause deterioration of air quality, either directly or by the formation of secondary pollutants. Hazardous waste compounds dissolved in, suspended in, or floating as surface films on the surface of water can render it unfit for use and for sustenance of aquatic organisms.

Soil exposed to hazardous wastes can be severely damaged by alteration of its physical and chemical properties and ability to support plants. For example, soil exposed to concentrated brines from petroleum production may become unable to support plant growth so that the soil becomes extremely susceptible to erosion.

15.10.4 FATES OF HAZARDOUS WASTES

The fate of a hazardous waste substance in water is a function of the substance's solubility, density, biodegradability, and chemical reactivity. Dense, water-immiscible liquids may simply sink to the bottoms of bodies of water or aquifers and accumulate there as "blobs" of liquid. This has happened, for example, with hundreds of tons of PCB wastes that have accumulated in sediments in the Hudson River in New York State noted in Section 15.9. Biodegradable substances are broken down by bacteria, a process for which the availability of oxygen is an important variable. Substances that readily undergo bioaccumulation are taken up by organisms, exchangeable cationic materials become bound to sediments, and organophilic materials may be sorbed by organic matter in sediments.

The fates of hazardous waste substances in the atmosphere are often determined by photochemical reactions. Ultimately, such substances may be converted to nonvolatile, insoluble matter and precipitate from the atmosphere onto soil or plants.

15.11 HAZARDOUS WASTES AND THE ANTHROSPHERE

As the part of the environment where humans process substances, the anthrosphere is the source of most hazardous wastes. Releases of hazardous wastes from the anthrosphere commonly occur through incidents such as spills of liquids, accidental discharge of gases or vapors, fires, and explosions. Under regulations enforced as part of the RCRA act, hazardous waste generators in the United States are required to have specified equipment, trained personnel, and procedures that protect human health in the event of a release and that facilitate remediation if a release occurs. An effective means of communication for summoning help and giving emergency instruction must be available. Also required are firefighting capabilities including fire extinguishers and adequate water. To deal with spills, a facility is required to have on hand absorbents, such as granular vermiculite clay, or absorbents in the form of pillows or pads. Neutralizing agents for corrosive substances that may be used should be available as well.

In addition to originating in the anthrosphere, to a large extent, hazardous wastes move, have effects, and end up in the anthrosphere as well. Large quantities of hazardous substances are moved through the anthrosphere by truck, rail, ship, and pipeline. Spills and releases from such movement, ranging from minor leaks from small containers to catastrophic releases of petroleum from wrecked tanker ships, are a common occurrence. Much effort in the area of environmental protection can be profitably devoted to minimizing and increasing the safety of the transport of hazardous substances through the anthrosphere.

In the United States, the transportation of hazardous substances is regulated through the U.S. Department of Transportation (DOT). One of the ways in which this is done is through a **manifest** system of documentation designed to track shipments of wastes; provide information regarding proper actions in the event of emergencies such as collisions, spills, fires, or explosions; and provide documentation for record keeping and reporting.

Many of the adverse effects of hazardous substances occur in the anthrosphere. One of the main examples of such effects occurs as corrosion of materials that are strongly acidic or basic or that otherwise attack materials. Fire and explosion of hazardous materials can cause severe damage to anthrospheric infrastructure.

The fate of hazardous materials is often in the anthrosphere. One of the main examples of a material dispersed in the anthrosphere consists of lead-based anticorrosive paints used to be widely coated on steel structural members.

15.12 HAZARDOUS WASTES IN THE GEOSPHERE

The sources, transport, interactions, and fates of contaminant hazardous wastes in the geosphere involve a complex scheme, some aspects of which are illustrated in Figure 15.5. As illustrated in the figure, there are numerous vectors by which hazardous wastes can get into groundwater. Leachate from a landfill can move as a waste plume carried along by groundwater, in severe cases draining into a stream or into an aquifer where it may contaminate well water. Sewers and pipelines may leak hazardous substances into the geosphere. Such substances seep from waste lagoons into geological strata, eventually contaminating groundwater. Wastes leaching from sites where they have

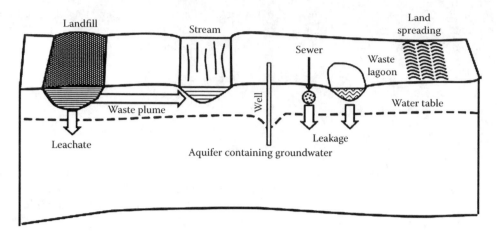

FIGURE 15.5   Sources and movement of hazardous wastes in the geosphere.

been spread on land for disposal or as a means of treatment can contaminate the geosphere and groundwater. In some cases, wastes are pumped into deep wells as a means of disposal.

The movement of hazardous waste constituents in the geosphere is largely by the action of flowing water in a waste plume, as shown in Figure 15.5. The speed and degree of waste flow depend on numerous factors. Hydrologic factors such as water gradient and permeability of the solid formations through which the waste plume moves are important. The rate of flow is usually rather slow, typically several centimeters per day. An important aspect of the movement of wastes through the geosphere is **attenuation** by the mineral strata. This occurs because waste compounds are sorbed to solids by various mechanisms. A measure of the attenuation can be expressed by a **distribution coefficient**, K_d,

$$K_d = \frac{C_s}{C_w} \tag{15.8}$$

where C_s and C_w are the equilibrium concentrations of the constituent on solids and in solution respectively. This relationship assumes relatively ideal behavior of the hazardous substance that is partitioned between water and solids (the sorbate). A more empirical expression is based on the Freundlich equation

$$C_s = K_F C_{eq}^{1/n} \tag{15.9}$$

where and K_F and $1/n$ are empirical constants.

Several important properties of the solid determine the degree of sorption. One obvious factor is surface area. The chemical nature of the surface is also important. Among the important chemical factors are presence of sorptive clays, hydrous metal oxides, and humus (particularly important for the sorption of organic substances).

In general, sorption of hazardous waste solutes is higher above the water table in the unsaturated zone of soil. This region tends to have a higher surface area and to favor oxic biodegradation processes.

The chemical nature of the leachate is important in sorptive processes of hazardous substances in the geosphere. Organic solvents or detergents in leachates will solubilize organic materials, preventing their retention by solids. Acidic leachates tend to dissolve metal oxides

$$M(OH)_2(s) + 2H^+ \rightarrow M^{2+} + 2H_2O \tag{15.10}$$

thus preventing sorption of metals in insoluble forms. This is a reason that leachates from municipal landfills, which contain weak organic acids, are particularly prone to transport metals. Solubilization by acids is particularly important in the movement of heavy metal ions.

Heavy metals are among the most dangerous hazardous waste constituents that are transported through the geosphere. Many factors affect their movement and attenuation. The temperature, pH, and reducing nature (as expressed by the negative log of the electron activity, pE) of the solvent medium are important. The nature of the geospheric solids, especially the inorganic and organic chemical functional groups on the surface, the cation-exchange capacity, and the surface area of the solids largely determine the attenuation of heavy metal ions. In addition to being sorbed and undergoing ion exchange with geospheric solids, heavy metals may undergo oxidation-reduction processes, precipitate as slightly soluble solids (especially sulfides), and in some cases, such as occurs with mercury, undergo microbial methylation reactions that produce mobile organometallic species.

The importance of chelating agents interacting with metals and increasing their mobilities has been illustrated by the effects of chelating EDTA on the mobility of radioactive heavy metals, especially ^{60}Co. The EDTA and other chelating agents, such as diethylenetriaminepentaacetic acid (DTPA) and nitrilotriacetic acid (NTA), were used to dissolve metals in the decontamination of radioactive facilities and were codisposed with radioactive materials at Oak Ridge National Laboratory (Tennessee) during the period 1951–1965. Unexpectedly, high rates of radioactive metal mobility were observed, which was attributed to the formation of anionic species such as ^{60}CoT$^-$ (where T^{3-} is the chelating NTA anion). Whereas unchelated cationic metal species are strongly retained in soil by precipitation reactions and cation-exchange processes

$$Co^{2+} + 2OH^- \rightarrow Co(OH)_2(s) \tag{15.11}$$

$$2Soil\}^-H^+ + Co^{2+} \rightarrow (Soil\}^-)_2 Co^{2+} + 2H^+ \tag{15.12}$$

anion bonding processes are very weak, so that the chelated anionic metal species are not strongly bound. Naturally occurring humic acid chelating agents may also be involved in the subsurface movement of radioactive metals and the less soluble humic substances may serve to immobilize radioactive metals.

Soil can be severely damaged by hazardous waste substances. Such materials may alter the physical and chemical properties of soil and thus its ability to support plants. Some of the more catastrophic incidents in which soil has been damaged by exposure to hazardous materials have arisen from soil contamination from SO_2 emitted from copper or lead smelters or from brines from petroleum production. Both of these contaminants stop the growth of plants and, without the binding effects of viable plant root systems, topsoil is rapidly lost by erosion.

Unfortunate cases of the improper disposal of hazardous wastes into the geosphere have occurred throughout the world. This often occurs in poorer countries where wastes are dumped from concerns in more developed countries.

15.13 HAZARDOUS WASTES IN THE HYDROSPHERE

Hazardous waste substances can enter the hydrosphere as leachate from waste landfills, drainage from waste ponds, seepage from sewer lines, or runoff from soil. Deliberate release into waterways also occurs and is a particular problem in countries with lax environmental enforcement. There are, therefore, numerous ways by which hazardous materials may enter the hydrosphere.

For the most part, the hydrosphere is a dynamic, moving system, so that it provides, perhaps, the most important variety of pathways for moving hazardous waste species in the environment. Once in the hydrosphere, hazardous waste species can undergo a number of processes by which they are degraded, retained, and transformed. These include the common chemical processes of precipitation–dissolution, acid-base reactions, hydrolysis, and oxidation-reduction reactions. Also included is a wide variety of biochemical processes that, in most cases, reduce hazards, but in some cases, such as the biomethylation of mercury, they greatly increase the risks posed by hazardous wastes.

The unique properties of water have a strong influence on the environmental chemistry of hazardous wastes in the hydrosphere. Aquatic systems are subject to constant change. Water moves with groundwater flow, stream flow, and convection currents. Bodies of water become stratified so that low-oxygen reducing conditions may prevail in the bottom regions of a body of water, and there is a constant interaction of the hydrosphere with the other environmental spheres. There is a continuing exchange of materials between water and the other environmental spheres. Organisms in water may have a strong influence on even poorly biodegradable hazardous waste species through bioaccumulation mechanisms.

Figure 15.6 shows some of the pertinent aspects of hazardous waste materials in bodies of water, with emphasis on the strong role played by sediments. An interesting kind of hazardous waste material that may accumulate in sediments consists of dense, water-immiscible liquids that can sink to the bottoms of bodies of water or aquifers and remain there as "blobs" of liquid as was the case with the PCB wastes dumped into the Hudson River, noted in Section 15.9.

Hazardous waste species undergo a number of physical, chemical, and biochemical processes in the hydrosphere that strongly influence their effects and fates. The major ones are listed as follows:

- **Hydrolysis reactions** are those in which a molecule is cleaved with the addition of a molecule of H_2O. An example of a hydrolysis reaction is the hydrolysis of dibutyl phthalate, Hazardous Waste Number U069:

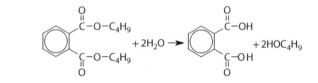

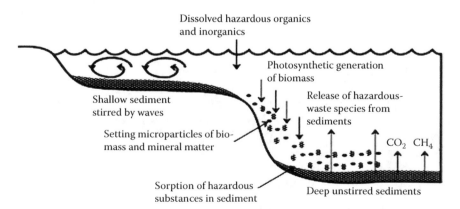

FIGURE 15.6 Aspects of hazardous wastes in surface water in the hydrosphere. The deep unstirred sediments are anoxic, and the site of hydrolysis reactions and reductive processes that may act on hazardous waste constituents is sorbed to the sediment.

Another example is the hydrolysis of bis(chloromethyl)ether to produce HCl and formaldehyde:

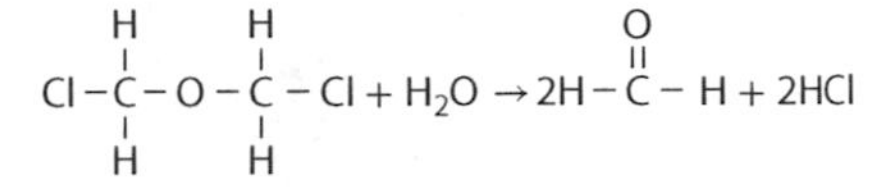

$$Cl-\overset{\overset{\text{H}}{|}}{\underset{\underset{\text{H}}{|}}{C}}-O-\overset{\overset{\text{H}}{|}}{\underset{\underset{\text{H}}{|}}{C}}-Cl + H_2O \rightarrow 2H-\overset{\overset{\text{O}}{||}}{C}-H + 2HCl$$

Compounds that hydrolyze are normally those, such as esters and acid anhydrides, that are originally formed by joining two other molecules with the loss of H_2O.

- **Precipitation reactions**, such as the formation of insoluble lead sulfide from soluble lead(II) ion in the anoxic regions of a body of water:

$$Pb^{2+} + HS^- \rightarrow PbS(s) + H^+$$

An important part of the precipitation process is normally **aggregation** of the colloidal particles first formed to produce a cohesive mass. Precipitates are often relatively complicated species such as the basic salt of lead carbonate, $2PbCO_3 \cdot Pb(OH)_2$. Heavy metals, a common ingredient of hazardous waste species precipitated in the hydrosphere, tend to form hydroxides, carbonates, and sulfates with the OH^-, HCO_3^-, and SO_4^{2-} ions that commonly are present in water, and sulfides are likely to be formed in bottom regions of bodies of water where sulfide is generated by anoxic bacteria. Heavy metals are often coprecipitated as a minor constituent of some other compound or are sorbed by the surface of another solid.

- **Oxidation-reduction reactions** commonly occur with hazardous waste materials in the hydrosphere, generally mediated by microorganisms. An example of such a process is the oxidation of ammonia to toxic nitrite ion mediated by *Nitrosomonas* bacteria:

$$NH_3 + 3/2O_2 \rightarrow H^+ + NO_2^-(s) + H_2O$$

- **Biochemical processes**, which often involve hydrolysis and oxidation-reduction reactions. Organic acids and chelating agents, such as citrate, produced by bacterial action may solubilize heavy metal ions. Bacteria also produce methylated forms of metals, particularly mercury and arsenic.
- **Photolysis reactions** and miscellaneous chemical phenomena. Photolysis of hazardous waste compounds in the hydrosphere commonly occurs on surface films exposed to sunlight on the top of water.

Hazardous waste compounds have a number of effects on the hydrosphere. Perhaps the most serious of these is the contamination of groundwater, which in some cases can be almost irreversible. Waste compounds accumulate in sediments such as river or estuary sediments. Hazardous waste compounds dissolved in, suspended in, or floating as surface films on the surface of water can render it unfit for use and for sustenance of aquatic organisms.

Many factors determine the fate of a hazardous waste substance in water. Among these are the substance's solubility, density, biodegradability, and chemical reactivity. As noted in Section 15.13 and in Chapter 3, Section 3.14, biodegradation largely determines the fate of hazardous waste substances in the hydrosphere. In addition to biodegradation, some substances are concentrated in organisms by bioaccumulation processes and may become deposited in sediments as a result. Organophilic materials may be sorbed by organic matter in sediments. Cation-exchanging sediments have the ability to bind cationic species, including cationic metal ions and organics that form cations.

15.14 HAZARDOUS WASTES IN THE ATMOSPHERE

Hazardous waste chemicals can enter the atmosphere by evaporation from hazardous waste sites, by wind erosion, or by direct release. Hazardous waste chemicals usually are not evolved in large enough quantities to produce secondary air pollutants. (As noted in Chapter 7, Section 7.1, secondary air pollutants are formed by chemical processes in the atmosphere. Examples are sulfuric acid formed from emissions of sulfur oxides and oxidizing photochemical smog formed under sunny conditions from nitrogen oxides and hydrocarbons.) Therefore, species from hazardous waste sources are usually of most concern in the atmosphere as primary pollutants emitted in localized areas at a hazardous waste site. Plausible examples of primary air pollutant hazardous waste chemicals include corrosive acid gases, particularly HCl; toxic organic vapors such as vinyl chloride (U043); and toxic inorganic gases such as HCN potentially released by the accidental mixing of waste cyanides with acid:

$$H_2SO_4 + 2NaCN \rightarrow Na_2SO_4 + 2HCN(g) \tag{15.13}$$

Primary air pollutants such as these are almost always of concern to people only adjacent to the site or to workers involved in site remediation. One such substance that has been responsible for fatal poisonings at hazardous waste sites, usually tanks that are undergoing cleanup or demolition, is the highly toxic hydrogen sulfide gas, H_2S.

An important characteristic of a hazardous waste material that enters the atmosphere is its **pollution potential**. This refers to the degree of environmental threat posed by the substance acting as a primary pollutant or to its potential to cause harm from secondary pollutants.

Another characteristic of a hazardous waste material that determines its threat to the atmosphere is its **residence time**, which can be expressed by an estimated atmospheric half-life, $t_{1/2}$. Among the factors that go into estimating atmospheric half-lives are water solubilities, rainfall levels, and atmospheric mixing rates.

Hazardous waste compounds in the atmosphere that have significant water solubilities are commonly removed from the atmosphere by **dissolution** in water. The water may be in the form of very small cloud or fog particles or it may be present as rain droplets.

Some hazardous waste species in the atmosphere are removed by **adsorption onto aerosol particles**. Typically, the adsorption process is rapid so that the lifetime of the species is that of the aerosol particles (typically a few days). Adsorption onto solid particles is the most common removal mechanism for highly nonvolatile constituents such as benzo(a)pyrene.

Dry deposition is the name given to the process by which hazardous waste species are removed from the atmosphere by impingement onto soil, water, or plants on the Earth's surface. These rates are dependent on the type of substance, the nature of the surface that they contact, and weather conditions.

A significant number of hazardous waste substances leave the atmosphere much more rapidly than predicted by dissolution, adsorption onto particles, and dry deposition, meaning that chemical processes must be involved. The most important of these are photochemical reactions, commonly involving hydroxyl radical, HO·. Other reactive atmospheric species that may act to remove hazardous waste compounds are ozone (O_3), atomic oxygen (O), peroxyl radicals (HOO·), alkylperoxyl radicals (ROO·), and NO_3. Although its concentration in the troposphere is relatively low, HO· is so reactive that it tends to predominate in the chemical processes that remove hazardous waste species from air. Hydroxyl radical undergoes **abstraction reactions** that remove H atoms from organic compounds

$$R - H + HO· \rightarrow R· + H_2O \tag{15.14}$$

and may react with those containing unsaturated bonds by addition as illustrated by the following reaction:

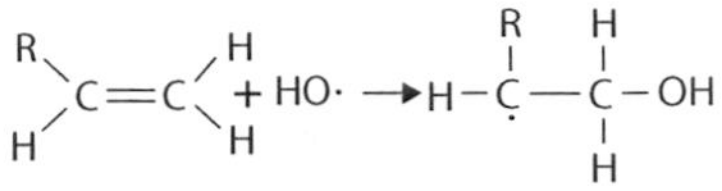

$$\underset{H}{\overset{R}{\diagdown}}C=C\underset{H}{\overset{H}{\diagup}}+\;HO\cdot\;\longrightarrow H-\underset{H}{\overset{R}{\underset{|}{C}}}-\underset{H}{\overset{H}{\underset{|}{C}}}-OH \tag{15.15}$$

The free radical products are very reactive. They react further to form oxygenated species, such as aldehydes, ketones, and dehalogenated organics, eventually leading to the formation of particles or water-soluble materials that are readily scavenged from the atmosphere.

Direct photodissociation of hazardous waste compounds in the atmosphere may occur by the action of the shorter wavelength light that reaches to the troposphere and is absorbed by a molecule with a light-absorbing group called a **chromophore**:

$$R-X+h\nu \rightarrow R\cdot +X\cdot$$

Among the factors involved in assessing the effectiveness of direct absorption of light to remove species from the atmosphere are light intensity, quantum yields (chemical reactions per quantum absorbed), and atmospheric mixing. The requirement of a suitable chromophore limits direct photolysis as a removal mechanism for most compounds other than conjugated alkenes, carbonyl compounds, some halides, and some nitrogen compounds, particularly nitro compounds, all of which commonly occur in hazardous wastes.

15.15 HAZARDOUS WASTES IN THE BIOSPHERE

Microorganisms, bacteria, fungi, and, to a certain extent, protozoa may act metabolically on hazardous waste substances in the environment. Most of these substances are **anthropogenic** (made by human activities), and most are classified as **xenobiotic** molecules that are foreign to living systems. Although by their nature xenobiotic compounds are degradation-resistant, almost all classes of them—nonhalogenated alkanes, halogenated alkanes (trichloroethane and dichloromethane), non-halogenated aryl compounds (benzene, naphthalene, and benzo[a]pyrene), halogenated aryl compounds (hexachlorobenzene and pentachlorophenol), phenols (phenol and cresols), PCBs, phthalate esters, and pesticides (chlordane and parathion)—can be at least partially degraded by various microorganisms.

Bioaccumulation occurs in which wastes are concentrated in the tissues of organisms. It is an important mechanism by which wastes enter food chains. **Biodegradation** occurs when wastes are converted by biological processes to generally simpler molecules; the complete conversion to simple inorganic species, such as CO_2, NH_4^+, $H_2PO_4^-/HPO_4^-$, and SO_4^{2-}, is called **mineralization.** The production of a less toxic product by biochemical processes is called **detoxification**. An example is the bioconversion of highly toxic organophosphate paraoxon to p-nitrophenol, which is only about 1/200 as toxic:

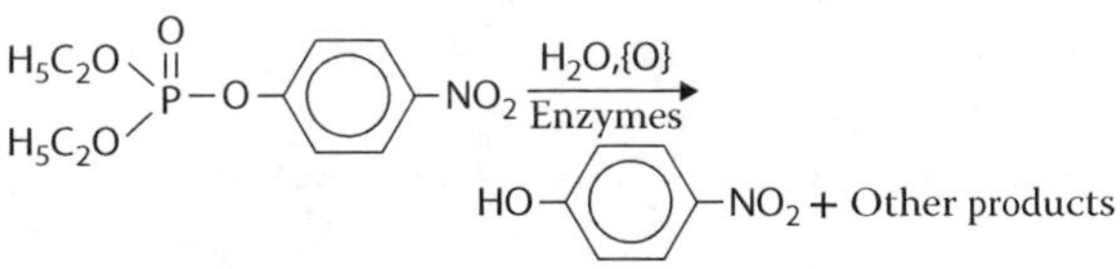

$$\underset{H_5C_2O}{\overset{H_5C_2O}{\diagdown}}\overset{\overset{O}{\parallel}}{P}-O-\!\!\!\bigcirc\!\!\!-NO_2 \xrightarrow[\text{Enzymes}]{H_2O,\{O\}}$$
$$HO-\!\!\!\bigcirc\!\!\!-NO_2 + \text{Other products} \tag{15.16}$$

15.15.1 Microbial Metabolism in Waste Degradation

The following terms and concepts apply to the metabolic processes by which microorganisms biodegrade hazardous waste substances:

- **Biotransformation** is the enzymatic alteration of a substance by microorganisms.
- **Metabolism** is the biochemical process by which biotransformation is carried out.
- **Catabolism** is an enzymatic process by which more complex molecules are broken down into less complex ones.
- **Anabolism** is an enzymatic process by which simple molecules are assembled into more complex biomolecules.

Two major divisions of biochemical metabolism that operate on hazardous waste species are **oxic processes**, which use molecular O_2 as an oxygen source, and **anoxic processes**, which make use of another oxidant. For example, when sulfate ion acts as an oxidant (electron receptor), the transformation $SO_4^{2-} \rightarrow H_2S$ occurs. (This has the benefit of providing sulfide, which precipitates insoluble metal sulfides in the presence of hazardous waste heavy metals.) Because molecular oxygen does not penetrate to such depths, anoxic processes predominate in the deep sediments, as shown in Figure 15.6.

For the most part, anthropogenic compounds resist biodegradation much more strongly than do naturally occurring compounds. Given the nature of xenobiotic substances, there are very few enzyme systems in microorganisms that act specifically on these substances, especially in making an initial attack on the molecule. Therefore, most xenobiotic compounds are acted upon by a process called **cometabolism**, which occurs concurrently with normal metabolic processes. An interesting example of cometabolism is provided by the white rot fungus, *Phanerochaete chrysosporium*, which has been promoted for the treatment of hazardous organochlorides such as PCBs, DDT, and chlorodioxins. This fungus uses dead wood as a carbon source and has an enzyme system that breaks down wood lignin, a degradation-resistant biopolymer that binds the cellulose in wood. Under appropriate conditions, this enzyme system attacks organochloride compounds and enables their mineralization.

The susceptibility of a xenobiotic hazardous waste compound to biodegradation depends on its physical and chemical characteristics. Important physical characteristics include water solubility, hydrophobicity (aversion to water), volatility, and lipophilicity (affinity for lipids). In organic compounds, certain structural groups—branched carbon chains, ether linkages, meta-substituted benzene rings, chlorine, amines, methoxy groups, sulfonates, and nitro groups—impart particular resistance to biodegradation.

Microorganisms vary in their ability to degrade hazardous waste compounds; virtually never does a single microorganism have the ability to completely mineralize a waste compound. Abundant oxic bacteria of the **Pseudomonas** family are particularly adept at degrading synthetic compounds such as biphenyl, naphthalene, DDT, and many other compounds. **Actinomycetes**, microorganisms that are morphologically similar to both bacteria and fungi, degrade a variety of organic compounds, including degradation-resistant alkanes and lignocellulose, as well as pyridines, phenols, nonchlorinated aryls, and chlorinated aryls.

Because of their requirement for oxygen-free (anoxic) conditions, anoxic bacteria are fastidious and difficult to study. However, they can play an important role in degrading biomass, particularly through hydrolytic processes in which molecules are cleaved with addition of H_2O. Anoxic bacteria reduce oxygenated organic functional groups. As examples, they convert nitro compounds to amines, degrade nitrosamines, promote reductive dechlorination, reduce epoxide groups to alkenes, and break down aryl structures. Partial dechlorination of PCBs by bacteria growing anoxically in PCB-contaminated river sediments such as those in New York's Hudson River has been reported. PCB waste remediation schemes have been proposed that make use of anoxic dechlorination of the more highly chlorinated PCBs and oxic degradation of the less highly chlorinated products.

Fungi are particularly noted for their ability to attack long-chain and complex hydrocarbons and are more successful than bacteria in the initial attack on PCB compounds. The potential of the white rot fungus, *P. chrysosporium*, to degrade biodegradation-resistant compounds, especially organo-chloride species, was previously noted.

Phototrophic microorganisms, algae, photosynthetic bacteria, and cyanobacteria that perform photosynthesis have lipid bodies that accumulate lipophilic compounds. There is some evidence to suggest that these organisms can induce photochemical degradation of the stored compounds.

Biologically, the greatest concern with wastes has to do with their toxic effects on animals, plants, and microbes. Virtually all hazardous waste substances are poisonous to a degree, some extremely so. Toxicities vary markedly with the physical and chemical nature of the waste; the matrix in which it is contained; the type and condition of the species exposed; and the manner, degree, and time of exposure.

15.16 HAZARDOUS SUBSTANCES AND ENVIRONMENTAL HEALTH AND SAFETY

Health effects are a major concern with hazardous substances and hazardous wastes generated in the anthrosphere. These include worker exposure during the generation, use, treatment, and disposal of hazardous materials; worker exposure resulting from remediation of disposal sites; exposure of people to emissions from disposal sites; and exposure of the public to hazardous substances as a result of transportation.

Children are especially vulnerable to the health effects of hazardous substances. As noted in the discussion of the Love Canal waste disposal site in Section 15.1, a particular concern with this notorious site was the exposure of children in residential areas and even in a school that was constructed essentially on top of some of the buried wastes. Another concern involving children is the possibility of increased birth defects due to exposure to toxic substances in hazardous wastes.

Contaminated soil is a particular concern with respect to exposure of children to hazardous substances. The most common such contaminant is lead, especially in areas around lead smelters. Soil contaminants from hazardous waste sources also include arsenic, carcinogenic benzo(a)pyrene, and PCBs.[7]

QUESTIONS AND PROBLEMS

1. Match the following kinds of hazardous substances on the left with a specific example of each from the right, as follows:

1. Explosives	(a) Oleum, sulfuric acid, and caustic soda
2. Compressed gases	(b) White phosphorus
3. Radioactive materials	(c) NH_4ClO_4
4. Flammable solids	(d) Hydrogen and sulfur dioxide
5. Oxidizing materials	(e) Nitroglycerin
6. Corrosive materials	(f) Plutonium and cobalt-60

2. Of the following, the property that is **not** a member of the same group as the other properties listed is (a) substances that are liquids whose vapors are likely to ignite in the presence of ignition sources, (b) nonliquids that may catch fire from friction or contact with water and that may burn vigorously or persistently, (c) ignitable compressed gases, (d) oxidizers, and (e) substances that exhibit extremes of acidity or basicity.

3. In what respects can it be said that measures taken to alleviate air and water pollution tend to aggravate hazardous waste problems?

4. Why is attenuation of metals likely to be very poor in acidic leachate? Why is attenuation of anionic species in soil less than that of cationic species?

5. Discuss the significance of LFL, UFL, and flammability range in determining the flammability hazards of organic liquids.

6. Concentrated HNO_3 and its reaction products pose several kinds of hazards. What are these?

7. What are substances that catch fire spontaneously in air without an ignition source called?

8. Name four or five hazardous products of combustion and specify the hazards posed by these materials.

9. What kind of property tends to be imparted to a functional group of an organic compound containing both oxygen and nitrogen?

10. Match the corrosive substance from the column on the left, in the following, with one of its major properties from the right column:

1. Alkali metal hydroxides	(a) Reacts with protein in tissue to form yellow xanthoproteic acid
2. Hydrogen peroxide	(b) Dissolves glass
3. Hydrofluoric acid, HF	(c) Strong bases
4. Nitric acid, HNO_3	(d) Oxidizer

11. Rank the following wastes in increasing order of segregation (a) mixed halogenated and hydrocarbon solvents containing little water, (b) spent steel pickling liquor, (c) dilute sludge consisting of mixed organic and inorganic wastes, (d) spent hydrocarbon solvents free of halogenated materials, and (e) dilute mixed inorganic sludge.

12. List and discuss some of the important processes determining the transformations and ultimate fates of hazardous chemical species in the hydrosphere.

13. In what form would a large quantity of hazardous waste PCB likely be found in the hydrosphere?

14. The TCLP was originally devised to mimic a "mismanagement scenario" in which hazardous wastes were disposed of along with biodegradable organic municipal refuse. Discuss how this procedure reflects the conditions that might arise from circumstances in which hazardous wastes and actively decaying municipal refuse were disposed of together.

15. What are three major properties of wastes that determine their amenability to transport?

16. List and discuss the significance of major sources for the origin of hazardous wastes, that is, their main modes of entry into the environment. What are the relative dangers posed by each of these? Which part of the environment would each be most likely to contaminate?

17. What is the influence of organic solvents in leachates upon attenuation of organic hazardous waste constituents?

18. What features or characteristics should a compound possess for direct photolysis to be a significant factor in its removal from the atmosphere?

19. Describe the particular danger posed by codisposal of strong chelating agents with radionuclide wastes. What can be said about the chemical nature of the latter with regard to this danger?

20. Describe a beneficial effect that might result from the precipitation of either $Fe_2O_3 \cdot xH_2O$ or $MnO_2 \cdot xH_2O$ from hazardous wastes in water.

21. Why are secondary air pollutants from hazardous waste sites usually of only limited concern as compared with primary air pollutants? What is the distinction between the two?

22. Match the following physical, chemical, and biochemical processes dealing with the transformations and ultimate fates of hazardous chemical species in the hydrosphere on the left with the description of the process on the right, as follows:

1. Precipitation reactions	(a) Molecule is cleaved with the addition of H_2O
2. Biochemical processes	(b) Generally accompanied by aggregation of colloidal particles suspended in water
3. Oxidation-reduction	(c) Generally mediated by microorganisms
4. Hydrolysis reactions	(d) By sediments and by suspended matter
5. Sorption	(e) Often involve hydrolysis and oxidation-reduction

23. As applied to hazardous wastes in the biosphere, distinguish among biodegradation, biotransformation, detoxification, and mineralization.

24. What is the potential role of *P. chrysosporium* in treatment of hazardous waste compounds? For which kinds of compounds might it be most useful?

25. Which part of the hydrosphere is most subject to long-term, largely irreversible contamination from the improper disposal of hazardous wastes in the environment?

26. Several physical and chemical characteristics are involved in determining the amenability of a hazardous waste compound to biodegradation. These include hydrophobicity, solubility, volatility, and affinity for lipids. Suggest and discuss ways in which each one of these factors might affect biodegradability.

27. Look up the U.S. EPA website that gives the National Priorities List for waste cleanup: (http://www.epa.gov/superfund/sites/query/queryhtm/nplfin2.htm). Select some of the Site Listing Narratives for specific descriptions of the various sites. Can you find any sites where heavy metals, chromium(VI), organochlorine solvents, or alkaline wastes are listed as contaminants? What other kinds of wastes can you find?

LITERATURE CITED

1. Brown, Michael Harold, *Laying Waste: The Poisoning of America by Toxic Chemicals*, Random House, Westminster, MD, 1983.
2. Clapp, Jennifer, *Toxic Exports: The Transfer of Hazardous Wastes from Rich to Poor Countries*, Cornell University Press, Ithaca, NY, 2010.
3. Galuszka, Agnieszka, Zdzislaw Migaszewski, and Piotr Manecki, Pesticide Burial Grounds in Poland: A Review, *Environment International* **37**, 1265–1272 (2011).
4. Applegate, John S., and Jan G. Laitos, *Environmental Law: RCRA, CERCLA, and the Management of Hazardous Waste*, Foundation Press, New York, 2006.
5. Priority Chemicals, http://www.epa.gov/osw/hazard/wastemin/, 2012.
6. Toxicity Characteristic Leaching Procedure, U.S. Environmental Protection Agency, http://www.epa.gov/osw/hazard/testmethods/sw846/pdfs/1311.pdf, 2008.
7. Mielke, Howard W., Jan Alexander, Marianne Langedal, and Rolf Tore Ottesen, Children, Soils, and Health: How Do Polluted Soils Influence Children's Health?, in *Mapping the Chemical Environment of Urban Areas*, Christopher C. Johnson, Alecos Demetriades, Juan Locutura, and Rolf Tore Ottesen, Eds., Wiley, Hoboken, NJ, 2011, pp. 134–150.

SUPPLEMENTARY REFERENCES

Adeola, Francis O., *Hazardous Wastes, Industrial Disasters, and Environmental Health Risks: Local and Global Environmental Struggles*, Palgrave Macmillan, Hampshire, UK, 2011.

Applegate, John S., Jan G. Laitos, Jeffrey M. Gaba, and Noah M. Sachs, *The Regulation of Toxic Substances and Hazardous Wastes*, 2nd ed., Foundation Press, New York, 2011.

Bevelacqua, Armando S., *Hazardous Materials Chemistry*, 2nd ed., Thomson Delmar Learning, Clifton Park, NY, 2006.

Booking, Edward C., Ed., *Trends in Hazardous Materials Research*, Nova Science Publishers, New York, 2007.

Cabaniss, Amy D., *Handbook on Household Hazardous Waste*, Government Institutes, Lanham, MD, 2008.

Hudson, Robert C., Ed., *Hazardous Materials in the Soil and Atmosphere: Treatment, Removal, and Analysis*, Nova Science Publishers, New York, 2006.

Lavelle, James R., Ed., *Waste Management: Research, Technology, and Developments*, Nova Science, New York, 2008.

Lewinsky, Allison A., Ed., *Hazardous Materials and Wastewater: Treatment, Removal and Analysis*, Nova Science Publishers, Inc., New York, 2006.

Maczulak, Anne, *Cleaning Up the Environment: Hazardous Waste Technology*, Facts on File, New York, 2009.

Nemerow, Nelson Leonard, Ed., *Industrial Waste Treatment*, Elsevier/Butterworth-Heinemann, Amsterdam, 2007.

Occupational Safety and Health Administration, *Hazardous Waste Operations and Emergency Response*, Kindle edition, U.S. Department of Labor, Washington, DC, 2011.

Patnaik, Pradyot, *A Comprehensive Guide to the Hazardous Properties of Chemical Substances*, 3rd ed., John Wiley, Hoboken, NJ, 2007.

Pichtel, John, *Waste Management Practices: Municipal, Hazardous, and Industrial*, CRC Press, Boca Raton, FL, 2005.

Pichtel, John, *Fundamentals of Site Remediation: For Metal and Hydrocarbon-Contaminated Soils*, Government Institutes, Lanham, MD, 2007.

Pohanish, Richard P., Ed., *Sittig's Handbook of Toxic and Hazardous Chemicals and Carcinogens*, 6th ed., Elsevier/William Andrew, Amsterdam, 2012.

Reisch, Mark, and David M. Bearden, *Superfund and the Brownfields Issue*, Novinka Books, New York, 2003.

Rosenfeld, Paul E., and Lydia Feng, *Risks of Hazardous Wastes*, Elsevier/William Andrew, Amsterdam, 2011.

Shafer, Donald A., *Hazardous Materials Characterization: Evaluation Methods, Procedures, and Considerations*, Wiley-Interscience, Hoboken, NJ, 2006.

VanGuilder, Cliff, *Hazardous Waste Management*, Mercury Learning and Information, Dulles, VA, 2011.

Wang, Lawrence K., Ed., *Waste Treatment in the Process Industries*, CRC/Taylor & Francis, Boca Raton, FL, 2006.

Wang, Lawrence K., Ed., *Hazardous Industrial Waste Treatment*, CRC/Taylor & Francis, Boca Raton, FL, 2007.

Wareym Phillip B., Ed., *New Research on Hazardous Materials*, Nova Science Publishers, New York, 2006.

Woodard & Currant, Inc., *Industrial Waste Treatment Handbook*, 2nd ed., Elsevier/Butterworth-Heinemann, Amsterdam, 2006.

16 Industrial Ecology and Green Chemistry for Sustainable Management of the Anthrosphere

16.1 MANAGING THE ANTHROSPHERE FOR SUSTAINABILITY

This chapter addresses management of the anthrosphere for sustainability. It is important to manage the anthrosphere sustainably for two major reasons. One of these is that the anthrosphere tends to be a voracious consumer of materials and Earth's natural capital. Therefore, it is very important to manage the anthrosphere in ways that control its appetite for consumption and conserve and preserve Earth's limited resources in the most sustainable way possible. A second reason is that the anthrosphere has a strong tendency to produce pollutants and wastes (Figure 16.1). For this reason, a major goal of sustainability must be to manage the anthrosphere in a way that minimizes or completely eliminates wastes, especially hazardous wastes.

The first part of this chapter addresses utilization of materials by the anthrosphere. Emphasis is placed on renewable materials, especially those from the biosphere, and on conservation and recycling of metals and other nonrenewable materials. The second part of this chapter discusses minimization of the impact of anthrospheric wastes with emphasis on elimination of hazardous wastes and treatment of wastes to make them nonhazardous.

16.2 FEEDING THE ANTHROSPHERE

As noted in Chapter 14, Section 14.14, **feedstocks** are defined as the main ingredients that go into the production of chemical products and **reagents** are defined as substances that act upon feedstocks in the manufacture of materials; often the two are not readily distinguished. This chapter addresses the sources, processing, and uses of feedstocks in more detail. Closely related to feedstocks are **fuels**, which are burned to produce energy; their sources, processing, and utilization are addressed in Chapter 17 in the discussion on energy. The anthrosphere is in fact "fed" by feedstocks and fuels.

Since the Industrial Revolution gained impetus around 1800, and especially around 1900 with the development of the chemical industry, the anthrosphere has developed a voracious appetite for materials. This is especially true of the vast petrochemicals industry fed by materials from petroleum and producing huge quantities of polymers, plastics, synthetic detergents, agricultural chemicals, and many other products. The era of petrochemicals must eventually come to an end because sources of petroleum cannot indefinitely sustain the enormous appetite of the anthrosphere for petrochemicals. The demand for the kinds of products now produced from petroleum will not go away, so alternate means of providing the materials desired by humans will have to be found. The only real alternative is biomaterials, which in fact provided most of the stuff that humans used until very recently in the history of humankind. Although challenging, this shift in raw materials sources promises to be a very exciting one for chemistry. And it provides an opportunity for chemists and

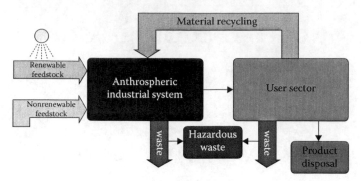

FIGURE 16.1 Industrial systems operating in the anthrosphere take in materials (feedstocks) and produce wastes, including hazardous wastes. Both these activities have important implications for sustainability.

engineers to "get it right" by applying the principles of green chemistry, green engineering, and industrial ecology in ways that will ensure a sustainable future.

Feedstock selection largely dictates the reactions and conditions that are employed in a chemical synthesis and is, therefore, of utmost importance in the practice of green chemistry. A feedstock should be as safe as possible. The source of a feedstock can largely determine its environmental impact, and the acquisition of the feedstock should not strain Earth's resources. The process of isolating and concentrating a feedstock can add to the potential harm of otherwise safe materials. This is true of some metal ores in which corrosive and toxic reagents (in the case of gold, e.g., cyanide; see Section 16.4) are used to isolate the desired material.

As a general rule, it is best if feedstocks come from renewable sources rather than depletable resources. A biomass feedstock, for example, can be obtained as a renewable resource grown by plants on land, whereas a petroleum-based feedstock is obtained from depletable crude oil resources. However, the environmental trade-offs between these two sources may be more complex than they first appear in that the petroleum feedstock may be simply pumped from a few wells in Saudi Arabia, whereas the biomass may require large areas of land, significant quantities of fertilizer, and large volumes of irrigation water for its production. Another important decision is whether or not the feedstock should be made entirely from virgin materials or at least in part from recycled material.

In the United States, petroleum amounts to all but about 2% of the raw material used for the manufacture of organic chemicals and the many products made from them, such as textiles, plastics, and rubber. To a degree, petroleum is an ideal feedstock for this purpose; during the last 100 years, it has been readily available and relatively inexpensive except during times of temporary supply disruption. There are of course disadvantages to the use of petroleum as a feedstock, not the least of which is the fact that eventually available supplies will become exhausted. The transportation and refining of petroleum consumes large amounts of energy, amounting to more than 15% of total energy use in the United States. Chemically, a consideration in the use of petroleum as a raw material is that the hydrocarbon molecules that compose petroleum are in a highly reduced chemical state. In order to be utilized as feedstocks, petroleum hydrocarbons often must be oxidized. The oxidation process entails a net consumption of energy and often requires the use of severe and hazardous reagents.

Much of the challenge and potential environmental harm in obtaining feedstocks is in separating the feedstocks from other materials. This is certainly true with petroleum, which consists of many different hydrocarbons only one of which may be needed as the raw material for a particular kind of product. Some metals occur at levels of less than 1% in their ores, requiring energy-intensive means of separating out the metals from huge quantities of rock. The smelting of copper and lead ores releases significant quantities of impurity arsenic with the flue dust, which must be collected from the smelting operation. Indeed, this by-product arsenic provides more than enough of the arsenic needed in commerce. Biobased materials are also generally mixtures that require separation. Cellulose from wood, which can be converted to paper and a variety of chemicals, is mixed intimately with lignin, from which it is separated only with difficulty.

In evaluating the suitability of a feedstock, it is not sufficient to consider just the hazards attributable to the feedstock itself and its acquisition. This is because different feedstocks require different processing and synthetic operations downstream that may add to their hazards. If feedstock A requires the use of a particularly hazardous material to convert it to product, whereas feedstock B can be processed by relatively benign processes, feedstock B should be chosen. This kind of consideration points to the importance of considering the whole life cycle of materials rather than just one aspect of them.

An example of the use of a hazardous material in further processing of feedstock is the synthesis of 2,2,4-trimethylpentane, commonly called isooctane, from isobutane and butene (Reaction 16.1). As shown by its structural formula, 2,2,4-trimethylpentane is a highly branched hydrocarbon and an important ingredient of gasoline because it burns smoothly and does not cause engine "knocking." The favored synthesis of this hydrocarbon uses a catalyst of hydrogen fluoride, HF, a volatile, highly reactive, and corrosive toxic substance that can be fatal if inhaled and that causes severe chemical burns that are difficult to heal. The use of concentrated HF mixed with a small amount of water has been a subject of concern with respect to possible terrorist attacks on petroleum refineries.

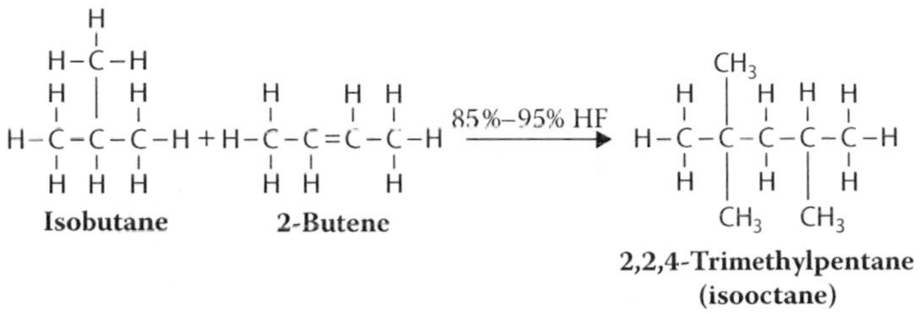

$$\text{Isobutane} \quad \text{2-Butene} \xrightarrow{85\%-95\% \text{ HF}} \text{2,2,4-Trimethylpentane (isooctane)} \tag{16.1}$$

16.2.1 Utilization of Feedstocks

Before considering alternative sources of feedstocks, it is useful to consider how those feedstocks can be used in the least polluting, most sustainable way possible. Feedstocks are modified by chemical processes to produce new chemical materials with commercial uses: The ideal feedstock is renewable and poses no hazards. And it can be converted to the desired product using a few steps with 100% yield and 100% atom economy. This should be done with minimum quantities of reagent using only safe media in which the reaction occurs.

There are three major categories of reactions that are involved in the chemical processing of feedstocks, as shown in a general sense in Figure 16.2. In an **addition reaction**, all feedstock material becomes part of the product and there are no by-products. These are the best kinds of reactions from the viewpoint of green chemistry because, when they work ideally, there are no wastes. A **substitution reaction** uses a reagent to replace a functional group on the feedstock molecule. As its name implies, an **elimination reaction** removes a functional group from a feedstock molecule. Both these latter kinds of reactions produce by-product materials from the feedstock and from the spent reagent. Their impacts can be reduced by reclaiming by-products, if some use can be found for them, and by regenerating the reagent, when this is possible. In some cases, elimination reactions can be carried out without the use of a reagent, reducing the impact of this kind of reaction.

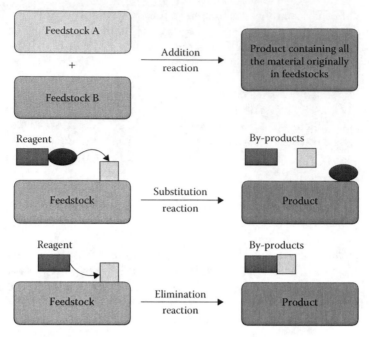

FIGURE 16.2 Illustration of three major categories of reaction processes by which feedstocks are acted upon by reagents to produce desired products.

16.3 KEY FEEDSTOCK: ABUNDANT ELEMENTAL HYDROGEN FROM SUSTAINABLE SOURCES

Of all possible feedstocks, none is more important than elemental hydrogen, H_2. This is because elemental hydrogen has so many uses in chemical synthesis and in making synthetic fuels. From an environmental viewpoint, hydrogen's big advantage, whether it is used in a fuel cell or in a turbine or piston engine, is that its only exhaust product is water:

$$2H_2 + O_2 \rightarrow 2H_2O \tag{16.2}$$

This characteristic has led to much interest in some quarters on the possibility of a "hydrogen economy" in which hydrogen would be the fuel of choice for virtually all energy applications.[1] Although fuel-cell-powered buses are in operation in some locations around the world and hydrogen-powered automobiles are common in Iceland, hydrogen has not become the fuel of choice for two important reasons. The first of these is the limited storage capacity for hydrogen, which normally must be contained as a highly pressurized gas. This is not a problem in Iceland, an island where one cannot drive very far from a fueling station and elemental hydrogen is generated cheaply from electricity produced with geothermal energy, but it is very limiting in most other parts of the world. The second limitation is that to date most hydrogen is generated from steam reforming of natural gas

$$CH_4 + 2H_2O \rightarrow 4H_2 + CO_2 \tag{16.3}$$

or from the reaction of steam with hot carbon from coal for which the overall reaction is as follows:

$$C + 2H_2O \rightarrow 2H_2 + CO_2 \tag{16.4}$$

Although the aforementioned processes, especially steam reformation of methane, are the current methods of choice for the production of industrial hydrogen, they both use fossil fuels and produce greenhouse gas carbon dioxide. Therefore, a sustainable, low-cost means of H_2 production is needed. The best means for producing hydrogen gas sustainably is through electrolysis (see Chapter 14, Section 14.14.4) by passing a direct current through water made conducting by the addition of a nonreactive salt such as Na_2SO_4:

$$2H_2O + \text{electrical energy} \rightarrow 2H_2(g) + O_2(g) \tag{16.5}$$

Current commercial electrolysis cells are about 70% efficient in utilizing electricity to produce H_2. In most regions, the most likely electrical energy source for this application is "free" wind power driving generators. Wind is entirely renewable and not subject to uncertainties in the market, including international markets subject to political pressures. In this application, the intermittent characteristic of wind power is not an issue in that hydrogen produced in abundance during times of strong wind can be pumped underground under pressure to be withdrawn for subsequent use during periods of lesser wind speed (see Chapter 14, Figure 14.16). Another advantage of the electrolysis reaction is the production of by-product O_2, which can be used to burn fossil fuels from which the exhaust CO_2 product can be captured without dilution with N_2 from air, thus enabling sequestration of the carbon dioxide or even its use as a concentrated carbon source (see Chapter 17, Section 17.11).

Sustainably generated hydrogen has many uses in chemical synthesis and in synthetic fuels manufacture. In addition to its applications in fuel cells, it can be burned directly in gas turbines and internal combustion engines. It is the fuel of choice for fuel cells and, combined chemically with N_2 from air, it is used to make ammonia for fertilizer and synthetic nitrogen compounds. It is a key feedstock for processes to convert biomass materials, such as sugars, to useful synthetic chemicals. And elemental H_2 is a key ingredient in making synthetic fuels from coal, biomass, and even CO_2.

16.4   FEEDSTOCKS FROM THE GEOSPHERE

The geosphere is a source of a large number of feedstocks that are taken from it in the form of minerals. These include all the metals that are refined from metal-bearing ores including sources of iron; aluminum; copper; lead; zinc; and, as discussed in Chapter 9, Section 9.3, the rare earths that are very important in modern technology-based societies. Also taken from the geosphere are fuels including fossil fuels such as coal and uranium used in nuclear reactors, which are discussed in Chapter 17.

The environmental and toxicological effects of mineral extraction from the geosphere are substantial. A major portion of these effects results from physical disturbances of the geosphere. In strip mining, soil and rock may be removed that cover a mineral deposit; the mineral dug up; and, with the best practice of modern mining technology, the overburden replaced along with a layer of topsoil followed by revegetation of the whole area. In some countries, these restoration practices are not followed as they should be and the geosphere is scarred by many improperly operated surface mines from the past.

Underground mining has also caused numerous problems. Surface subsidence can be a problem with many relatively shallow underground mines. Aquifers have been disturbed and contaminated by improperly designed and operated underground mines. Water contamination can come from several sources. In the case of metal mines, water may become polluted with metals such as lead, copper, and zinc. A substantial water pollution problem has resulted from the exposure of iron pyrite, FeS_2, to air. As discussed in some detail in Chapter 3, Section 3.5.1 microbial and chemical action produces sulfuric acid and soluble iron from this mineral, causing substantial water pollution problems.

Desired minerals are always mixed with other rocks that must be separated. In many cases, particularly for more valuable metals, the mineral that is sought is less than 1% of the ore. This requires processing of the ore, a procedure called **beneficiation**, which produces relatively large amounts of finely divided by-product rock. For economic reasons, beneficiation is usually carried out at the mine site with the residues returned to the mine or left in piles at the site. As a result, water pollution problems may develop from the leaching of mine spoils. Some communities in the vicinity of lead mines have been contaminated by lead from mine residues, with particular concern over the health of exposed children. Spoils from iron mining in Minnesota have contaminated water with an asbestos-like mineral associated with the iron ore, requiring remedial action that has cost millions of dollars. Enormous piles of tar-contaminated sand are left over from the extraction by hot water of heavy crude oil from tar sands in the Canadian province of Alberta.

16.4.1 Occupational and Public Health Aspects of Mining

The occupational and public health aspects of mining are very important. In some countries, there is still a high toll of injuries and deaths from mining. Coal mining in China takes a shocking number of lives each year. One of the lowest recent annual tolls from mining in China was in 2010 when "only" 2433 deaths were reported, down from 6000 deaths in both 2004 and 2005. U.S. coal mining deaths typically range from 20 to 50 each year. The year 2010 was a bad one with 48 deaths being reported including 29 from the April 5, 2010, blast at Massey Energy Company's Upper Big Branch Mine in West Virginia. The major hazard in coal mining results from explosions of methane gas released from coal seams.

The occupational health aspects of mining can be substantial. In coal mining, the main problem is "black lung disease" from the inhalation of coal dust. Miners of several mineral commodities may suffer lung silicosis from the inhalation of silica dust released in the mining process. Uranium miners have developed lung cancer from the inhalation of α-particle-emitting noble gas radon and its daughter product polonium. Inhalation of radium from the dust released in uranium mining can also cause lung cancer.

16.4.2 Toxic Hazards of Cyanide in Gold Recovery

Metal recovery and processing can pose some significant toxicological hazards. One such potential hazard is from cyanide (CN^-), which is used to extract gold (chemical symbol Au) from low-grade ores. In such ores, the gold is present in the elemental form, but it is so finely divided and dispersed that it cannot be separated mechanically from the ore. To extract the gold, the ore is leached with a solution of sodium cyanide in the presence of atmospheric air, for which the reaction is as follows:

$$4Au + 8CN^- + O_2 + 2H_2O \rightarrow 4Au(CN)_2^- + 4OH^- \qquad (16.6)$$

The gold is then chemically precipitated from the dissolved $Au(CN)_2^-$ in the leachate solution.

An obvious concern with the cyanide extraction of gold, as well as with metal electroplating processes that use cyanide in the plating baths, is the toxicity of cyanide, both as soluble cyanide salts and as hydrogen cyanide gas, HCN (see Chapter 7, Section 7.3). Given the rather frightening picture of thousands of liters of deadly cyanide solution seeping through beds of gold ore, surprisingly few fatal cases of cyanide poisoning have been reported as a result of gold ore processing, although there have been numerous reports of massive fish kills in streams into which the cyanide leachate has spilled.

16.5   BIOLOGICAL FEEDSTOCKS

As discussed in Chapter 13, Section 13.2, there are several major categories of **biomass** that can be used as raw material to replace petroleum as a feedstock for the organic chemicals industry:

1. Carbohydrate, which has the general formula of approximately CH_2O. Carbohydrate is the biomass that is produced initially as glucose sugar from water and carbon dioxide during photosynthesis. It is contained in the structural parts of plants as cellulose, a biopolymer.
2. Lignin, a biological polymer with a complex structure, which occurs with carbohydrate cellulose in woody parts of plants where it binds fibers of cellulose together. Relatively few uses have been found for lignin, and it poses impurity problems in extracting cellulose for feedstock use.
3. Lipid oils extracted from seeds, including soybeans, sunflowers, and corn.
4. Hydrocarbon terpenes produced by rubber trees, pine trees, and some other kinds of plants.
5. Proteins, produced in relatively small quantities but potentially valuable as nutrients and in other uses.

Biological materials used as sources of feedstocks are usually complex mixtures, which makes separation of desired materials difficult. However, in some biological starting materials, nature has done much of the synthesis of products from biological feedstocks. Most biomass materials are partially oxidized as is the case with carbohydrates, which contain approximately one oxygen atom per carbon atom (compared to petroleum hydrocarbons, which have no oxygen). This can avoid expensive, sometimes difficult, oxidation steps, which may involve potentially hazardous reagents and conditions.

There are several main pathways by which feedstocks can be obtained from biomass. The most straightforward of these is a simple physical separation of biological materials, such as squeezing oil from oil-bearing biomass or tapping latex from rubber trees. Only a slightly more drastic treatment consists of extraction of oils by organic solvents. Physical and chemical processes can be employed to remove useful biomass from the structural materials of plants, which consist of lignocellulose, a mixture of cellulose and the related carbohydrate polymer hemicellulose bound together by lignin "glue."

The most abundant biomass feedstocks are carbohydrates, which are discussed in more detail in Sections 16.6 and 16.8. It follows that one of the most promising pathways to obtain useful raw materials and fuels from biomass is their synthesis directly from carbohydrates. Carbohydrates come in several forms. Sucrose sugar, $C_{12}H_{22}O_{11}$, can be squeezed from sugar cane as sap and can be extracted from sugar beets and sugar cane with water. The exceptional photochemical productivity of sugar cane makes sucrose from this source an attractive option. Larger amounts of carbohydrates are available in starch, a polymer of glucose readily isolated from grains, such as corn, or from potatoes. An even more abundant source is cellulose, which occurs in woody parts of plants.[2] It is relatively easy to break down starch molecules with the addition of water (hydrolysis) to give simple sugar glucose. Breaking down cellulose is more difficult but can be accomplished by the action of cellulase enzymes from microorganisms.

Lipid oils are extracted from the seeds of some plants (see Chapter 13, Figure 13.3, for the structural formula of a typical lipid oil). Volatile solvents, most commonly the six-carbon straight-chain alkane n-hexane, C_6H_{14}, are used to extract oils. In this process, the solvents are distilled off from the extract and recirculated through the process. Care has to be exercised to prevent worker exposure to excessive levels of n-hexane because it can cause **polyneuropathy**, that is, multiple disorders of the peripheral nervous system (see Chapter 2, Section 2.12). This occurs as the result of loss of myelin (a fatty substance constituting a sheath around certain nerve fibers) and degeneration of axons (part of a nerve cell through which nerve impulses are transferred out of the cell). This malady afflicted workers in Japan shortly after World War II who were involved in home production of sandals from used tires using glue with n-hexane solvent.

The hydrocarbon terpenes that occur in rubber trees can be tapped from the trees as a latex suspension in tree sap. Steam treatment and distillation can be employed to extract terpenes from sources such as pine or citrus tree biomass. Terpenes from plants are discussed as potential air pollutants and several structural formulas of terpenes are given in Chapter 7, Section 7.7.

Grain seeds are rich sources of protein, almost always used for food, and are potentially useful as chemical feedstocks for specialty applications. An exciting possibility just now coming to fruition in a practical sense is to transplant genes into plants so that they will make specialty proteins, such as medicinal agents (see Chapter 13, Section 13.3).

Genetic engineering can be very useful in producing biomass for feedstocks. One area in which there is much room for improvement is in enhancing the efficiency of photosynthesis. Grain crops can be bred to increase the amount of by-product biomass along with the grain they produce. Dedicated crops can be developed for the production of large quantities of biomass alone. This has already been done using conventional plant breeding techniques to develop rapidly growing hybrid poplar trees that produce large quantities of lignocellulosic wood.

Single-cell algae may well turn out to be the best option for the production of large quantities of biomass, renewable chemical feedstocks, and biomass fuels.[3] Algae are significantly more productive than rooted plants and can be grown in saltwater or brackish water in desert regions that are not suitable for other plants. Algae can serve as sources of lipids including fatty acid esters of glycerol, phospholipids, and glycolipids; carbohydrates including starch and complex polysaccharides; and proteins including nutrient amino acids and secondary metabolites that may be sources of antibiotics and other natural products.

16.6 MONOSACCHARIDE FEEDSTOCKS: GLUCOSE AND FRUCTOSE

The most abundant biomass feedstocks are carbohydrates. It follows that one of the most promising pathways to obtain useful raw materials and fuels from biomass is their synthesis directly from carbohydrates. The monosaccharides such as glucose and fructose, as well as xylose, a five-carbon sugar that is the monomer of hemicellulose, which makes up almost one-third of typical plant biomass

are produced in abundance by plants. These compounds are excellent platforms for a number of different organic syntheses. With their relatively high oxygen contents, they are particularly advantageous where a partially oxidized product is made, as is often the case in organic synthesis. Monosaccharides contain hydroxyl groups, $-OH$, around the molecule, which act as convenient sites for the attachment of various functionalities. Glucose is metabolized by essentially all organisms, so it serves as an excellent starting material for biosynthesis reactions using enzymes, and it and many of its products are biodegradable, adding to their environmental acceptability.

Glucose can be obtained by enzyme-catalyzed processes from other sugars, including fructose, and from disaccharide sucrose and from starch, a macromolecular polymer of glucose produced by plants for the storage of food and energy. A large fraction of the glucose that is now used is obtained from the enzymatic hydrolysis of cornstarch, and approximately one-third of U.S. corn production is now consumed to produce glucose for fermentation to ethanol fuel. This diversion has led to distortions in the corn market and higher food prices. Alarmed by the disruptive price increases for corn caused by production of biofuels from cornstarch, in 2007 China stopped this practice and switched to cassava, a root crop that is a prolific producer of starch.[4] The percentage of Chinese cassava going into biofuels jumped from 10% in 2008 to 52% in 2010. Cassava is a root crop that serves as a food energy source for approximately 500 million people in Africa and in some areas of Asia including China and Thailand. An interesting toxicological aspect regarding cassava as a food source is that it can be toxic because the starch is mixed with two cyanide-producing glycosides, linamarin and lotaustralin. Cyanide-forming linamarin in cassava leaves protects the plants from insects and animals. Food made from cassava, such as the traditional African dishes gari and fufu, is subjected to soaking, boiling, fermentation, and other processes that break down the cyanide-producing glycosides and make the cassava safe to eat. Some people have been poisoned by HCN fumes released in the processing of cassava.

In addition to starch as a source of glucose, it is also possible to obtain glucose by the enzymatic hydrolysis of cellulose, although it has proved to be economically and technically challenging on an industrial scale because of the refractory nature of the cellulose polymer (see Section 16.8). Nevertheless, the enormous quantities of cellulose available in wood and other biomass sources make glucose from cellulose an attractive prospect.

The greatest use of glucose and fructose (which is readily converted to glucose by enzymes) for chemical synthesis is by fermentation with yeasts to produce ethanol

$$
\begin{array}{ccc}
 & H & H \\
 & | & | \\
H- & C- & C-OH \\
 & | & | \\
 & H & H
\end{array} \quad \textbf{Ethanol}
$$

an alcohol that is widely used as a gasoline additive, solvent, and chemical feedstock. A by-product of this fermentation process is carbon dioxide, the potential of which in green chemical applications as a supercritical fluid solvent is discussed in Chapter 14, Section 14.13.

Glucose is widely used as a starting material for the biological synthesis of a number of different biochemical compounds. These include ascorbic acid, citric acid, and lactic acid. Several amino acids used as nutritional supplements, including lysine, phenylalanine, threonine, and tryptophan, are biochemically synthesized starting with glucose. The vitamins folic acid, ubiquinone, and enterochelin are also made biochemically from glucose.

In addition to the aforementioned predominantly biochemical applications of glucose, monosaccharides can be used to make feedstocks for chemical manufacture. The possibilities for doing so are now greatly increased by the availability of genetically engineered microorganisms that can be made to express genes for the biosynthesis of a number of products from glucose. Sophisticated genetic engineering is required to make chemical feedstocks because these are materials not ordinarily produced biologically.

A study by the U.S. Department of Energy Pacific Northwest Regional Laboratory has identified "top twelve value-added chemicals" that can be made enzymatically from monosaccharides, especially glucose and fructose.[5] Listed in Table 16.1, these chemicals could form the main feedstocks for future biorefineries that would generate an abundance of products that are currently made largely from petrochemicals. Several possible syntheses of commercially valuable chemicals starting with glucose are discussed here.

In 2011, San Diego-based Genomatica demonstrated industrial-scale microbial production from glucose of 1,4-butanediol, an organic alcohol used around the world in quantities of about 1 million metric tons per year as a solvent and in the manufacture of plastics, polyurethane, polyesters, tetrahydrofuran, and other materials. The conventional chemical synthesis of

TABLE 16.1

Top 12 Chemical Feedstocks That Can Be Made Enzymatically from Monosaccharides

Name and Structural Formula	Examples of Products
Four carbon 1,4-diacids	
Succinic acid	Tetrahydrofuran
Fumaric acid	2,5-Bis(aminomethyl)-tetrahydrofuran
Malic acid	Methyl acrylate
2,5-Furandicarboxylic acid	Aspartame (artificial sweetener)
3-Hydroxypropionic acid	
Aspartic acid	5-Hydroxymethyl-furfural
Glucaric acid	Proline
Glutamic acid	3-Methylpyrrolidone
Itaconic acid	Acrylic acid
Levulinic acid	
3-Hydroxybutyro-lactone	Epoxylactone
Glycerol	Propylene glycol
Sorbitol	Sorbitol itself has numerous uses in foods (as a low-calorie sweetener) and in cosmetics
Xylitol	Ethylene glycol

1,4-butanediol involves the reaction of acetylene with formaldehyde followed by hydrogenation of the product:

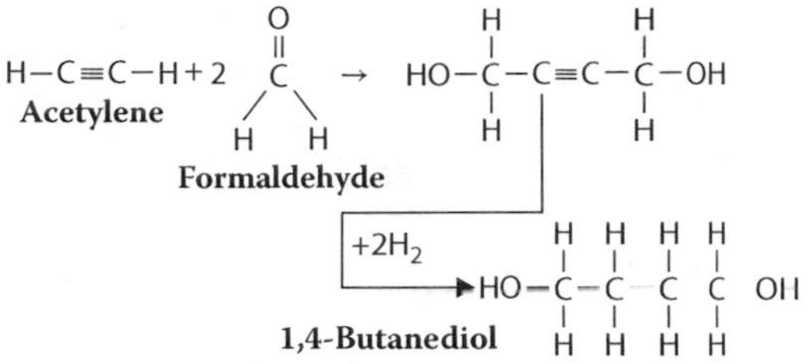

$$
\text{(16.7)}
$$

Although this synthesis is "green" in that it is carried out by addition reactions without by-products, it involves explosive acetylene and hydrogen gas and toxic formaldehyde. The sugar-based biosynthesis of 1,4-butanediol is carried out by genetically engineered *Escherichia coli* (a type of bacteria that inhabits the lower intestines of warm-blooded animals) acting upon glucose:[6]

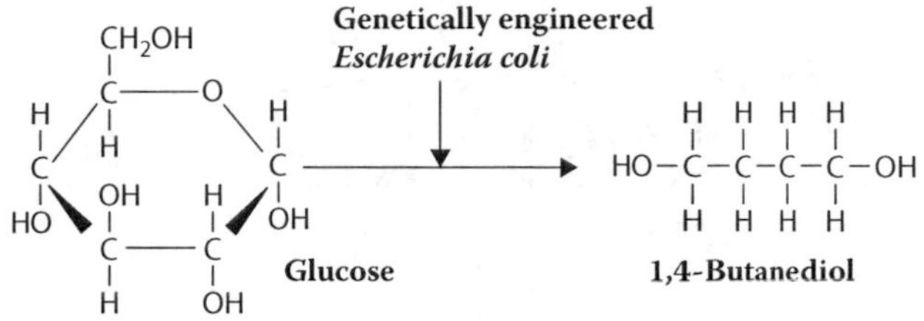

$$
\text{(16.8)}
$$

As another example of the potential of glucose for making important feedstocks, consider the synthesis from glucose of adipic acid

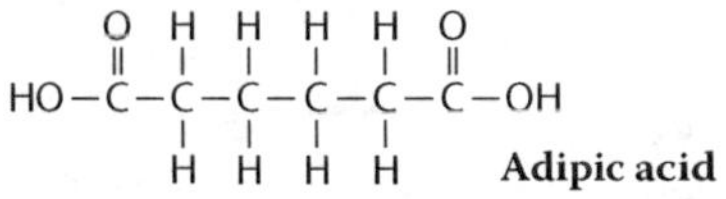

a feedstock consumed in large quantities to make nylon. The conventional chemical synthesis of this compound starts with benzene, a volatile, flammable hydrocarbon that is believed to cause leukemia in humans. The synthesis involves several steps using catalysts at high pressure and corrosive oxidant nitric acid, which releases air pollutant nitrous oxide, N_2O. Throughout the synthesis process, elevated temperatures of approximately 250°C are employed. The N_2O released by the synthesis of adipic acid in the manufacture of nylon accounts for a significant fraction of worldwide N_2O releases. The potential dangers and environmental problems with this synthesis are obvious.

As an alternative to the aforementioned chemical synthesis of adipic acid, a biological synthesis using genetically modified *E. coli* bacteria and a simple hydrogenation reaction has been described.[7] The bacteria convert glucose to *cis, cis*-muconic acid:

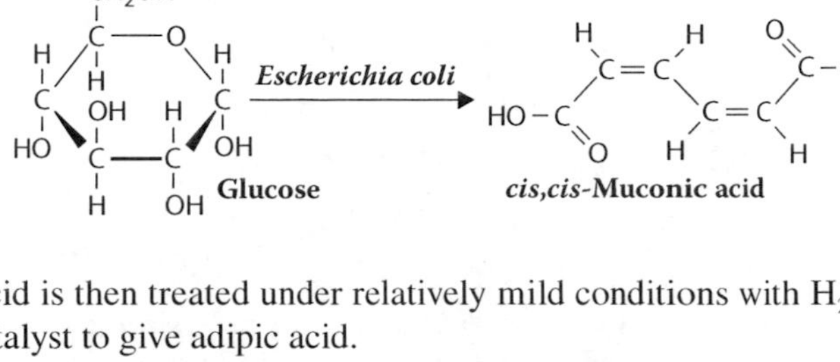

$$
\text{(16.9)}
$$

The muconic acid is then treated under relatively mild conditions with H_2 under 3-atm pressure over a platinum catalyst to give adipic acid.

16.7 HYDROCARBONS AND SIMILAR MATERIALS FROM SUGARS

Hydrocarbons and materials that have properties similar to hydrocarbons are among the most valuable feedstocks and fuels, and efforts are under way to produce these materials from abundant simple sugars. One of the more promising end products from the chemical modification of carbohydrates is dimethylfuran, an oxygen-containing cyclic organic compound that has most of the desirable properties of hydrocarbons as a raw material and fuel. Compared to ethanol, dimethylfuran has a relatively low boiling temperature, has high energy content per unit mass, does not absorb water, and exhibits combustion characteristics comparable to those of commonly used hydrocarbon fuels. As shown in Figure 16.3, structurally, dimethylfuran with its five-membered ring resembles the abundant monosaccharide fructose, which also has a five-membered ring, and the conversion of fructose to dimethylfuran is basically a matter of removing oxygen from the monosaccharide by dehydration and hydrogenation. This has been done by dehydration in a biphasic catalytic system employing mineral acids (HCl, H_2SO_4, H_3PO_4) to first produce 5-hydroxymethylfurfural, which is extracted into the organic phase of the biphasic system, followed by hydrogenation over a CuRu catalyst. The reaction scheme is illustrated in Figure 16.3.

As with many syntheses of feedstocks from sugars, the aforementioned synthesis of dimethylfuran involves chemical processes, especially expensive chemical hydrogenation, usually with metal catalysts. It is known that some organisms produce hydrocarbons, so the biochemical conversion of simple sugars to hydrocarbons has been an area of active research. Hydrocarbons and organics with hydrocarbon-like properties are produced by a variety of organisms and include constituents of plant cuticular waxes, insect pheromones (sex attractants), and other compounds with as yet unknown functions. Many of these are generated by plants, but microorganisms as well can synthesize hydrocarbons. One such microorganism is cyanobacterium, a highly productive photosynthetic bacterium of which many strains are commonly found in water. A study of the hydrocarbons biosynthesized by selected cyanobacteria has shown the production of heptadecane, a straight-chain $C_{17}H_{36}$ alkane hydrocarbon that apparently derives from C_{18} fatty acids. Some shorter-chain alkanes were found as well.[8]

The 2010 Presidential Green Chemistry Award for Small Business was given to LS9, a South San Francisco, California, company that published the aforementioned study of alkanes from cyanobacteria and has developed a one-step fermentation process to convert sugars to alkanes, alkenes, long-chain alcohols, and esters useful as feedstocks and fuels. As shown in Figure 16.4, the process uses a single-stage fermenter fed by a simple sugar feedstock and containing suspended biochemical whole-cell catalysts to produce alkanes, alkenes, long-chain alcohols, and fatty esters. The products are preferentially soluble in hydrocarbons and are extracted into a light water-immiscible hydrocarbon phase for removal to the product stream.

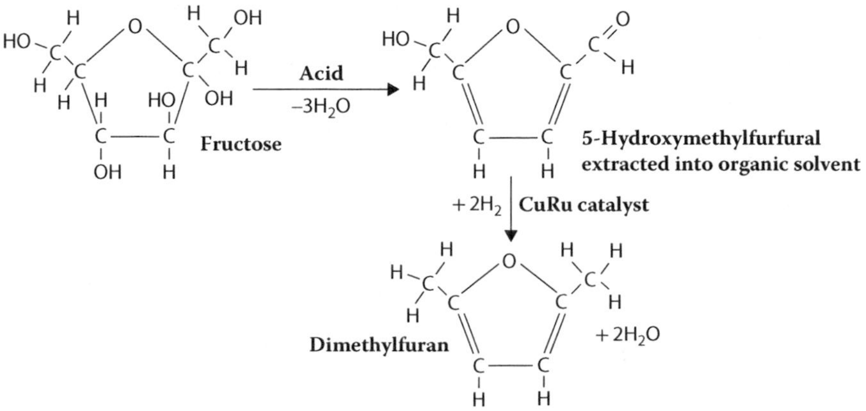

FIGURE 16.3 Synthesis from fructose of dimethylfuran, an organooxygen compound with many of the desirable properties of petroleum hydrocarbons: fructose is a carbohydrate produced in abundance by plants.

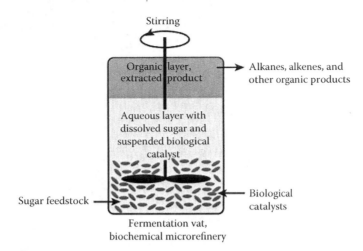

FIGURE 16.4 A fermentation process for the production of hydrocarbons and other organics by biochemically catalyzed reaction of simple sugars.

Assuming that it works well on a commercial scale, the aforementioned process for direct biosynthesis of hydrocarbons represents the best of green chemical production. It uses a renewable feedstock that can be produced in large quantities photosynthetically. It takes place under ambient conditions of temperature and pressure with little or no input of external energy. It does not require metal catalysts and does not produce waste by-products. The diesel fuel produced is claimed to be "ultraclean" and does not contain potentially pollutant or toxic benzene, sulfur, and the heavy metals that can cause problems in petroleum-based fuels.

16.8 CELLULOSE

The most abundant natural material produced by organisms is **cellulose**, which is synthesized biologically by the joining of glucose molecules with the loss of one H_2O molecule for each bond formed (see Figure 16.5). This makes the chemical formula of cellulose $(C_6H_{10}O_5)_n$, where n ranges from about 1500 to 6000 or more. Most cellulose is made by plants, with total amounts exceeding 500 billion metric tons per year worldwide. Cellulose makes up the sturdy cell walls of plants. Wood is about 40% cellulose; leaf fibers about 70%; and cotton, one of the purest sources of cellulose, about 95%. Cellulose occurs in different forms and is generally associated with **hemicellulose**, a material also composed of carbohydrate polymers, and lignin, a biopolymer of varied composition and bonding composed largely of aromatic units (see the structural formula of lignin in Chapter 13, Figure 13.2).

The first major step in cellulose utilization, such as extraction of cellulose fibers for making paper, consists of separating the cellulose from its matrix of lignocellulose (hemicellulose and lignin). This step has been the cause of many problems in utilizing cellulose because of the harsh chemical processing that is employed. Lignin residues impart color to the cellulose, so wood pulp used in making paper has to be bleached with oxidants that alter the structure of the coloring agents. Bleaching used to be done almost entirely with elemental chlorine, Cl_2, and with salts of hypochlorite ion, ClO^-. However, the bleaching of biomass with these chlorine-based materials produces chlorinated organic impurities and pollutants. Therefore, ozone and hydrogen peroxide are preferred as bleaching agents.

A finely divided form of cellulose called **microcrystalline cellulose** is produced by appropriate physical and chemical processing of cellulose. This material has many uses in foods in which it imparts smoothness, stability, and a quality of thickness. Microcrystalline cellulose is also used in pharmaceutical preparations and cosmetics. When added to food, indigestible cellulose contributes bulk and retains moisture.

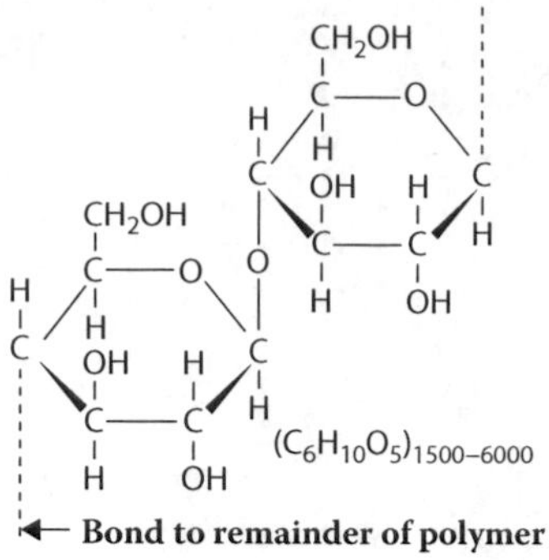

Bond to remainder of polymer

FIGURE 16.5 Segment of the cellulose molecule in which nearly 1500 to several thousand anhydroglucose units (glucose molecules less H_2O) are bonded together.

Chemically modified cellulose is used to make a wide variety of materials. Like the glucose that comprises it, cellulose has an abundance of −OH groups to which various other groups can be bonded to impart a variety of properties. One of the oldest synthetic fabrics, rayon, is made by treating cellulose with a base and carbon disulfide, CS_2, and then extruding the product through fine holes to make thread. In a similar process, chemically treated cellulose is extruded through a long, narrow slot to form a sheet of transparent film called cellophane.

As seen by the structural formula in Figure 16.5, each unit of the cellulose polymer has three −OH groups that are readily attached to other functional groups, leading to chemically modified cellulose. One of the most common such products is cellulose acetate, an ester (see Chapter 20) used primarily for apparel and home furnishings fabrics in which most of the −OH groups on cellulose are replaced by acetate groups by reaction with acetic anhydride (see the following structural formulas):

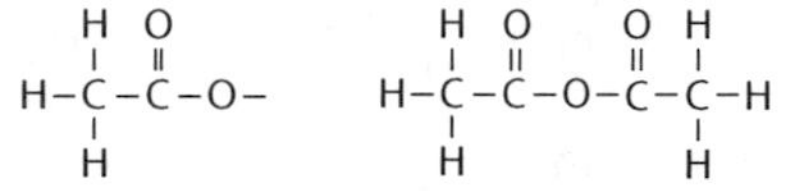

Acetate group   Acetic anhydride reagent

Although the cellulose feedstock for cellulose acetate synthesis is certainly a "green" material, acetic anhydride used to make the acetate is a corrosive, toxic chemical that produces poorly healing wounds on exposed flesh. Furthermore, potentially hazardous solvents, such as dichloromethane, are used in some processes for making cellulose acetate.

Another cellulose ester that is widely manufactured is cellulose nitrate, in which the −OH groups on cellulose are replaced by −ONO_2 groups by treating cellulose with a mixture of nitric acid, HNO_3, and sulfuric acid, H_2SO_4. Cellulose nitrate makes transparent film and was used in the early days of moving pictures for movie film. However, one of the other major uses of this material is as an explosive, so cellulose nitrate can burn violently giving off highly toxic fumes of NO_2 gas. In the early years of movie film, this characteristic led to several tragic fires involving human fatalities. The use of cellulose nitrate is now largely restricted to lacquer coatings, explosives, and propellants. Although the cellulose raw material is green, neither the process for making cellulose nitrate involving strong acids nor the flammable product would qualify as green.

From this discussion, it is apparent that cellulose is an important raw material for the preparation of a number of materials. The reagents and conditions used to convert cellulose to other products are in some cases rather severe. It is hoped that advances in the science of transgenic organisms will result in alternative biological technologies that will enable conversion of cellulose to a variety of products under relatively mild conditions.

16.8.1 FEEDSTOCKS FROM CELLULOSE WASTES

Large quantities of cellulose-rich waste biomass are generated as by-products of crop production in the form of straw remaining from grain harvest, bagasse residue from the extraction of sucrose from sugar cane, and other plant residues. This biomass represents an abundant source of raw material that potentially can be converted to chemical feedstocks. Cellulose can be hydrolyzed to produce simple sugars, a process called **saccharification**. Glucose and other simple sugars obtained in this way can be fermented to produce ethanol or used directly as chemical or biochemical feedstocks. Saccharification has long been carried out with strong mineral acids, a rather expensive process that requires handling of potentially hazardous reagents. Saccharification can also be performed with enzymes, although this has proved to be difficult and normally must be carried out in dilute media, which increases the energy required for product separation. The bacteria and protozoa in the digestive systems of ruminant animals such as cattle effectively convert cellulose to organic acids, including acetic and propionic acids; microbial protein biomass; and other materials with potential uses as feedstocks. Research is under way to duplicate this process artificially with rumen microorganisms in digesters and extraction of the substances produced as feedstock materials.

Another product of rumen fermentation is methane, CH_4. As much as 6% or 7% of the food energy ingested by ruminant animals is converted to methane gas, which is expelled from the animals to the atmosphere, a significant source of greenhouse warming gases. Methane is a premium fuel and, although it is not practical to harvest it from the digestive systems of cows, presumably it could be taken from artificial digesters.

16.9 LIGNIN

Lignin, a chemically complex biopolymer that is associated with cellulose in plants and that serves to bind cellulose in the plant structure, ranks second in abundance only to cellulose and hemicellulose as a biomass material produced by plants. Lignin is normally regarded as a troublesome waste in the processing and utilization of cellulose, although efforts are being made to utilize this material in biotechnology and bioenergy applications.[9] The characteristic that makes lignin so difficult to handle in chemical processing is its inconsistent, widely variable molecular structure as shown by the segment of lignin polymer in Chapter 13, Figure 13.2. This structure shows that much of the carbon is present in aromatic rings that are bonded to oxygen-containing groups, and lignin is the only major plant biopolymer that is largely aromatic. Because of this characteristic, lignin is of considerable interest as a source of aromatic compounds including phenolic compounds, which have the −OH group bonded to aromatic rings or even aromatic hydrocarbons. The abundance of hydroxyl, −OH; methoxyl, −OCH$_3$; and carbonyl, C=O, groups in lignin also suggests potential chemical uses for the substance. A significant characteristic of lignin is its resistance to biological attack. This property, combined with lignin's highly heterogeneous nature, makes it a difficult substrate to use for the enzyme-catalyzed reactions favored in the practice of green chemistry to give single pure products useful as chemical feedstocks.

Since lignin is a major component of most plant biomass, significant fractions of this biopolymer must be dealt with in biorefineries. Lignin generated as a by-product in the extraction of cellulose from wood is now largely burned for fuel, the lowest level of use for this material. By retaining much of the lignin molecule intact, use may be made of larger-molecular-mass segments of the molecule; such has been done for some uses for binders to hold materials together in coherent masses, fillers, resin extenders, and dispersants. There is also some potential to use lignin as a degradation-resistant structural material, such as in circuit boards. Potentially, the most profitable use for lignin is to make small aromatic molecules useful for chemical synthesis. For this to be practical, special techniques need to be developed to partially break down the lignin molecule without destroying the relatively valuable aromatic molecule segments in it.

16.10 BIOSYNTHESIS OF CHEMICALS

In the provision of specialty and commodity chemicals and feedstocks, there are two main biological sources of materials. One of these consists of plants and algae, which make huge quantities of cellulose and lesser quantities of other materials by photosynthesis. The other source is microorganisms, especially bacteria and yeasts.

The status of biotechnology in chemical synthesis has been the subject of an extensive review,[10] and several examples of chemical syntheses involving biotechnology are given here.

Enzyme-catalyzed reactions (biocatalytic conversions) can be carried out in a number of configurations, including suspensions of microorganisms in a liquid medium, microorganisms held on a solid support, and isolated enzymes. A biocatalytic conversion may be carried out as a one-step process using a single enzyme or as a multistep process using several enzymes in sequence.[11] Where applicable, biosynthesis using enzymes offers a number of potential advantages over conventional chemical synthesis. The most important of these are the following:

1. Unlike some chemical catalysts, especially those based on precious metals, enzymes come from renewable biological sources.
2. Coming from biological systems, enzymes have to be compatible with biological systems and are therefore generally not toxic or chemically hazardous.
3. Because they have evolved in organisms, enzymes function under moderate conditions of temperature, pressure, and pH, which contributes substantially to the safety of enzymatically catalyzed processes.
4. Enzymes are highly selective in the reactions they catalyze and often result in shorter syntheses with fewer steps.
5. Enzyme-catalyzed reactions generally produce pure products with minimal by-product impurities and generate relatively little waste.
6. Enzymes function in aqueous media and minimize the need for organic solvents.

16.10.1 FERMENTATION AND INDUSTRIAL MICROBIOLOGY

Fermentation or **industrial microbiology** refers to the action of microorganisms on nutrients under controlled conditions to produce desired products. A general outline of the fermentation process used for chemical synthesis is shown in Figure 16.6. Fermentation has a number of applications in green chemical synthesis.[12]

Fermentation for some products is anoxic (absence of O_2); for other products, oxic fermentation may be required. Fermentation processes have been used for thousands of years to produce alcoholic beverages, sauerkraut, vinegar, pickles, cheese, yogurt, and other foods. More recently, fermentation has been applied to the production of a wide variety of organic acids, antibiotics, antibodies, complex plant metabolites, enzymes, and vitamins. Ethanol, the alcohol in alcoholic beverages, is the most widely produced chemical made by fermentation.

Another very important chemical that has been made from biological feedstocks by fermentation for many years is lactic acid. Lactic acid is the most important hydrocarboxylic acid and is used to make a number of chemicals including pyruvic acid, acrylic acid, 1,2-propanediol, lactate esters, and polylactic acid polymer. Fermentation of glucose with the intermediate production of pyruvic acid is the leading process for making lactic acid, as shown in Reaction 16.10:[13]

$$(16.10)$$

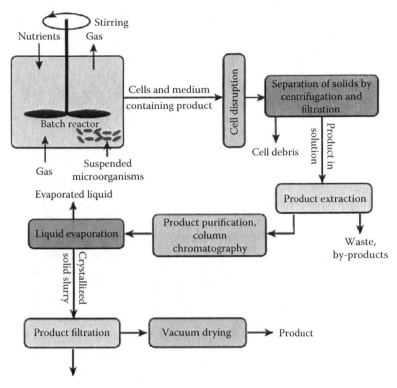

FIGURE 16.6 General outline of an industrial microbiology process for the synthesis of chemicals such as antibiotics: details of this overall scheme may vary depending on the product made.

This reaction summarizes a rather complex set of metabolic processes that, depending on the kinds of bacteria involved, may also produce ethanol and carbon dioxide.

Great impetus was given to the technology of industrial microbiology in the 1940s with the establishment of fermentation processes to produce antibiotic penicillin, a bactericidal pharmaceutical that has saved millions of lives. Following penicillin, fermentation processes were developed for the production of numerous other significant antibiotics and other pharmaceuticals.

Selection of the appropriate microorganism is the most important consideration of a successful fermentation production process. Conditions such as those of temperature, exposure to oxygen, supply of nutrients, and sterility demand exacting control.

Fermentation is undergoing tremendous development with the use of transgenic microorganisms to which genes have been transferred to make specific kinds of substances. The most common and valuable substances made by transgenic microorganisms consist of a variety of proteins. These include proteins and smaller-molecule polypeptides that are used as pharmaceuticals. The best example of such a substance is human insulin, which is now produced in large quantities by transgenic microorganisms.

There is increasing interest in using fermentation and biotechnology to make commodity chemicals used on a large scale. The large-scale production of ethanol as a gasoline additive from the fermentation of glucose sugar by yeasts and the production of lactic acid by fermentation are important examples of the production of commodity chemicals by fermentation. Advances in transgenic microbiology have increased the possibilities for using fermentation to produce a variety of chemicals and chemical feedstocks, several examples of which are discussed in this chapter.

The first, and a very important, example of the enzymatic manufacture of a bulk synthetic chemical is the biosynthesis of acrylamide from acrylonitrile (Reaction 16.11). The chemical pathway that has been used for this transformation is definitely not a green process in that it involves the hydration of acrylonitrile over a Raney copper catalyst (porous copper made by dissolving aluminum from a copper/aluminum alloy with concentrated NaOH). This leaves a residue of potentially toxic wastes and may release highly toxic HCN, and makes a product that requires extensive purification. As an alternative, an industrial process has now been developed that uses immobilized *Rhodococcus rhodochrous* bacteria with 100% product yield to convert acrylonitrile to acrylamide.

$$
\underset{\text{Acrylonitrile}}{\text{H}_2\text{C=CH–C}\equiv\text{N}} + \text{H}_2\text{O} \xrightarrow[\text{rhodochrous}]{\textit{Rhodococcus}} \underset{\text{Acrylamide}}{\text{H}_2\text{C=CH–C(=O)–NH}_2}
\tag{16.11}
$$

Acrylonitrile is highly flammable and toxic, burning to release deadly hydrogen cyanide gas and nitrogen oxides. By doing biosynthesis under conditions that enable enzyme action, the transformation to acrylamide avoids the severe conditions that might result in the release of acrylonitrile or its combustion products. Unlike acrylonitrile, acrylamide does not release hydrogen cyanide in an organism so that it is relatively less toxic than acrylonitrile. However, acrylamide does form adducts with blood hemoglobin by binding with sulfhydryl, –SH, groups on hemoglobin. The presence of these adducts in blood has been used as a biomarker of occupational exposure to acrylamide. An interesting footnote is that acrylamide was discovered at levels up to 3.5 mg/kg in potato chips and french fries by scientists in Sweden in 2002. It is believed that the acrylamide forms by the reaction of amino acid asparagine with glucose and other carbohydrates at elevated temperatures.

16.10.2 Metabolic Engineering and Chemical Biosynthesis

Metabolic engineering is the name given to the optimization of cellular genetic and regulatory processes to produce specific substances (Figure 16.7).[14] Metabolic engineering has the potential to enable the production of both large quantities of commodity chemicals (e.g., butanol that can substitute for gasoline) and high-value "un-natural" chemicals, such as some kinds of pharmaceuticals. In principle, this may be done by transferring genetic material from different hosts into a single microorganism in which the enzymes and metabolic pathways function to produce desired products.[15] The potential exists to transfer genetic material from rare or genetically intractable organisms into microorganisms that are commonly used for industrial fermentation processes.

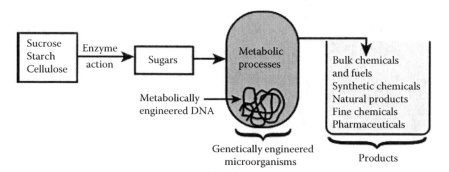

FIGURE 16.7 Production of chemical products by metabolically engineered microorganisms acting on simple sugars produced from sucrose, starch, or cellulose: synthetic chemicals are chemicals that are not normally synthesized biologically but that organisms potentially may be engineered to make. Natural products are substances that are biosynthesized by a variety of organisms, some of which have potential pharmaceutical applications.

16.10.3 PRODUCTION OF MATERIALS BY PLANTS

The uses of microorganisms operating in fermentation processes to generate commodity chemicals were discussed in Section 16.10.2. Plants are the other kind of organisms that can be used for large-scale production of chemicals. Indeed, the nutrients used for fermentation processes originally come from plants. Fermentation is in a sense not a very efficient means of producing chemicals because of the consumption of nutrients to support the microorganisms and their reproduction and because of the generation of large quantities of by-products. Plants, which generate their own biomass from atmospheric carbon dioxide and water are very efficient producers of materials. Wood and the cellulose extracted from it are prime examples of such materials.

In addition to their efficient production of biomass, plants offer distinct advantages in their production and harvesting. Genetics determine the materials that a plant makes and, once a crop is growing in a field, the products it is programmed for will be produced without fear of contamination by other organisms, which is always a concern in fermentation. Plants can be grown by relatively untrained personnel using well-known agricultural practices. Plant matter is generally easy to harvest in the form of grains, stalks, and leaves, which can be taken to a biorefinery (see Section 16.12) to extract needed materials.

The production of feedstocks and other chemical commodities from plants has been limited by the genetic restrictions inherent to plants. Now, however, transgenic plants can be bred to produce a variety of materials directed by genes transplanted from other kinds of organisms. For example, as discussed in Section 16.12, plants have even been developed to synthesize plastics. Another limitation of the production of materials by plants has been the mixture of these materials with other matter generated by plants. The intimate mixture of useful wood cellulose with lignin, for which uses are still being sought, is a prime example of this problem. Again, transgenic technology can be expected to be helpful in developing plants that produce relatively pure products (such as the almost pure cellulose in cotton).

The potential of plants to produce useful products has been greatly increased by the development of hybrid plants with spectacular capacities to generate biomass photosynthetically. Corn is one of the more productive field crops, and hybrid varieties produce large quantities of grain and plant biomass (leaves, stalks, husks, and cobs commonly called corn stover). Sugar cane is noted for its ability to produce biomass, some in the form of sucrose sugar and much more in cane stalk biomass. The sugar cane stalk residues left after extracting sugar from them (bagasse) have had relatively few uses, other than for fuel, but they can potentially produce large quantities of chemical feedstocks in biorefineries. One of the more prolific producers of biomass is the hybrid poplar tree, which, nourished by minimal amounts of fertilizer and watered by economical trickle irrigation systems, grows within a few years to a harvestable size for the production of wood pulp, wood for plywood, and cellulose that potentially can be converted to glucose. The ability of these trees and other even more productive hybrid plants to generate cellulose that can be converted to glucose means that they may serve as the basis of an entire plant-based chemicals industry.

16.11 DIRECT BIOSYNTHESIS OF POLYMERS

Cellulose in wood and cotton is only one example of the numerous significant polymers that are made biologically by organisms. Other important examples are wool and silk, which are protein polymers. (An interesting case is spiderweb polymer, which is stronger than steel of a comparable cross section. Unfortunately, spiders are finicky and impossible to train for making significant amounts of spiderwebs. Genes involved in making spiderwebs have been transferred to goats that then produced the material in their milk, but not in useful quantities.) A big advantage of biopolymers from an environmental viewpoint is that polymers made biologically are also the ones that are most likely to be biodegradable. Attempts have been made to synthesize synthetic polymers that are biodegradable. These efforts have centered on those prepared from biodegradable monomers, such as lactic acid.

Much of the green chemistry enterprise is devoted to the sustainable synthesis of polymers, especially those from biobased materials using biocatalytic syntheses. From the standpoint of green chemistry, it is ideal to have polymers that are made by organisms in a form that is essentially ready to use. Recently, interest has focused on poly(hydroxyalkanoate) compounds,[16] of which the most common are polymers of 3-hydroxybutyric acid:

3-Hydroxybutyric acid

This compound and some related ones have both a carboxylic acid, $-CO_2H$, group and an alcohol, $-OH$, group. A carboxylic acid can bond with an alcohol with the elimination of a molecule of H_2O, forming an **ester linkage** (see Chapter 20, Section 20.4). Since hydroxyalkanoates have both functional groups, the molecules can bond with each other to form polymer chains:

Ester group amenable to biological attack

Segment of poly(3-hydroxybutyrate) polymer

Ester groups are among the most common in a variety of biological compounds, such as fats and oils, and organisms possess enzyme systems that readily attack ester linkages. Therefore, the poly(hydroxyalkanoate) compounds are amenable to biological attack. Aside from their biodegradability, polymers of 3-hydroxybutyric acid and related organic acids that have $-OH$ groups on their hydrocarbon chains (alkanoates) can be engineered to have a variety of properties ranging from rubberlike to hard solid materials.

It was first shown in 1923 that some kinds of bacteria make and store poly(hydroxyalkanoate) ester polymers as a reserve of food and energy. In the early 1980s, it was shown that these materials have thermoplastic properties, which means that they melt when heated and resolidify when cooled. This kind of plastic can be very useful, and the thermoplastic property is rare in biological materials. The Metabolix Corporation, Cambridge, Massachusetts, received the 2005 Presidential Green Challenge Award for its work in synthesizing these plastics from biological sources. One commercial operation was set up for the biological synthesis of a polymer in which 3-hydroxybutyrate groups alternate with 3-hydroxyvalerate groups, where valeric acid has a five-carbon atom chain. This process uses a bacterium called *Ralstonia eutropia* that is fed glucose and the sodium salt of propionic acid, a three-carbon carboxylic acid, to make the polymer in fermentation vats. Although the process works, costs are high because of problems common to most microbial fermentation synthesis processes: the bacteria have to be provided with a source of food, yields are relatively low, and it is difficult to isolate the product from the fermentation mixture.

Developments in genetic engineering have raised the possibility of producing poly(hydroxyalkanoate) polymers in plants. The plant *Arabidopsis thaliana* has accepted genes from the bacterial species *Alcaligenes eutrophus*, which has resulted in plant leaves containing as much as 14% poly(hydroxybutyric acid) on a dry mass basis. Transgenic *Arabidopsis thaliana* and *Brassica napus* (canola) have shown production of the copolymer of 3-hydroxybutyrate and 3-hydroxyvalerate. If yields can be raised to acceptable levels, plant-synthesized poly(hydroxyalkanoate) materials would represent a tremendous advance in the biosynthesis of polymers because of the ability of photosynthesis to provide the raw materials used to make the polymers.

16.12 BIOREFINERIES AND BIOMASS UTILIZATION

Just as crude oil is a complex mixture of hydrocarbons and other organic compounds that must be run through a petroleum refinery to separate and chemically modify the materials in it to produce fuel and petrochemical feedstocks, chemicals from biological sources are usually complex mixtures that require refining and chemical processing to provide needed feedstocks. The separation and processing of biomaterials from plants or from bacteria cultured in digesters is accomplished in a **biorefinery**, the output of which consists of organic chemicals and fuels. In a biorefinery, the various products that can be obtained from biomass are separated and subjected to chemical treatment, generally with the objective of reducing the chemically bound oxygen contents of organic liquids. Although the large number of compounds generated from biomass poses a challenge for separations, it also provides the opportunity to produce smaller quantities of high-value chemicals.

Chapter 13, Figure 13.4, illustrates a biorefinery showing the main ways in which feedstocks may be obtained from biomass. The simplest and least energy consumptive of these is **extraction** in which an organic solvent or, in more sophisticated operations, supercritical carbon dioxide is used to dissolve materials from biomass. This approach is widely used to extract oils from some kinds of oilseeds. Terpene hydrocarbons can be extracted from pinewood and other terpene-producing plants. A very attractive possibility is to extract hydrocarbons and other oils from algae that produce these materials. Single-cell algae are particularly efficient photosynthesizers and the potential is high to genetically engineer strains that can produce specific classes of extractable organics.

Other than by extraction, the pathway to useful chemicals from biomass involves breaking down complex biomass polymers. One way in which this is done is with **enzymatic processing**, employing microorganisms or the action of isolated enzymes, for example, to produce glucose sugar from starch or cellulose. This step may require the addition of nutrients including nitrogen, phosphorus, and potassium to enable microorganisms to grow. The glucose and other monomers isolated by enzymatic action can be subjected to additional processing, the most common example of which is fermentation of glucose to alcohol.

Hydrogenation of biomass involves reaction with elemental H_2 under high pressure and at elevated temperatures. This approach can be used with hydrogen generated relatively inexpensively by electrolysis of water employing renewable sources of electricity, especially from wind power (see Section 16.3). Direct hydrogenation of biomass produces a wide variety of organics including oxygenated compounds, some of which have direct uses and others that may be chemically modified to give desired products.

Pyrolysis involves heating biomass externally or with a hot gas stream to evolve liquid and gas products. An external heat source including even solar energy focused and concentrated on a reactor may be employed. As with hydrogenation, pyrolysis generates a variety of products including oxygenated compounds. It also produces large amounts of residual carbon, which can be used directly as fuel or gasified with steam and oxygen to produce synthesis gas.

Gasification, which is discussed in more detail as a source of clean fuel from coal and biomass in Chapter 17, involves the reaction of hot carbon with steam

$$H_2O + C \rightarrow H_2 + CO \qquad (16.12)$$

yielding a synthesis gas mixture of H_2 and CO, from which additional H_2 may be generated by the reaction of CO with H_2O. The carbon is usually heated by partial combustion with O_2, and the CO in the synthesis gas is reacted with steam to increase the ratio of H_2 to CO. Biomass can be gasified directly by reaction with a minimal amount of O_2. Since biomass has the approximate empirical formula of $\{CH_2O\}$, the water required for gasification is largely in the biomass itself. Such direct gasification of biomass also produces large quantities of organics that are processed downstream in the biorefinery.

An important consideration in biorefineries is the use of catalysts. Insofar as possible, biorefineries should use heterogeneous catalysts that do not get into the product and enzyme catalysts that operate at moderate temperatures.

16.13 GREEN CHEMISTRY AND INDUSTRIAL ECOLOGY IN WASTE MANAGEMENT

Having discussed in Sections 16.1–16.12 the raw materials and feedstocks that go into the chemical industry enterprise, attention in the rest of the chapter is dedicated to the reducing and processing of the wastes and by-products generated in chemical manufacture. With its tendency to produce large quantities of wastes, including those that are potentially hazardous, an important aspect of sustaining the anthrosphere is to avoid producing wastes and to deal effectively with the waste materials that are produced. Chapter 15 addresses the nature and sources of hazardous wastes and their environmental chemistry and points out some of the major problems associated with such wastes. The remainder of this chapter deals with means for minimizing wastes, utilizing materials that might go into wastes, and treating and disposing of wastes, the generation of which cannot be avoided. The practice of industrial ecology combined with green chemistry is all about not producing wastes and, instead, utilizing wastes for beneficial purposes. Therefore, in dealing with wastes, it is essential in the modern age to consider the contributions of industrial ecology and green chemistry.

In sustaining the anthrosphere, it is important to apply environmental chemistry, industrial ecology, and green chemistry to hazardous waste management in order to develop measures by which chemical wastes can be minimized, recycled, treated, and disposed of (Figure 16.8). In descending order of desirability, hazardous waste management attempts to accomplish the following:

- Do not produce it.
- If making it cannot be avoided, produce only minimum quantities.
- Recycle it.
- If it is produced and cannot be recycled, treat it, preferably in a way that makes it nonhazardous.

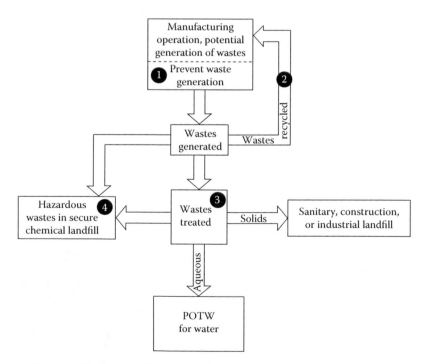

FIGURE 16.8 Order of effectiveness of waste treatment management options: the darkened circles indicate the degree of effectiveness from the most desirable (1) to the least (4). In the figure, POTW refers to a publicly owned treatment works.

- If it cannot be rendered nonhazardous, dispose of it in a safe manner.
- Once it is disposed of, monitor it for leaching and other adverse effects.

Industrial ecology is all about the efficient use of materials (see Chapter 14, Section 14.2). Therefore, by its nature, a system of industrial ecology is also a system of waste reduction and minimization. In reducing the quantities of wastes, it is important to take the broadest possible view. This is because dealing with one waste problem in isolation may simply create another. Early efforts to control air and water pollution resulted in problems from hazardous wastes isolated from industrial operations.

Many hazardous waste problems can be avoided at early stages by **waste reduction** (cutting down quantities of wastes from their sources) and **waste minimization** (utilization of treatment processes that reduce the quantities of wastes requiring ultimate disposal). This section outlines some basic approaches to waste minimization and reduction.

There are several ways in which quantities of wastes can be reduced, including source reduction, waste separation and concentration, resource recovery, and waste recycling. The most effective approaches to minimizing wastes center around careful control of manufacturing processes, taking into consideration discharges and the potential for waste minimization at every step of manufacturing. Viewing the process as a whole (as outlined for a generalized chemical manufacturing process in Figure 16.9) often enables crucial identification of the source of a waste, such as a raw material impurity, catalyst, or process solvent, so that measures may be taken to reduce wastes at the source.

Modifications of the manufacturing process can yield substantial waste reduction. Some such modifications are of a chemical nature. Changes in chemical reaction conditions can minimize the production of by-product hazardous substances. In some cases, potentially hazardous catalysts, such as those formulated from toxic substances, can be replaced by catalysts that are nonhazardous or that can be recycled rather than discarded. Wastes can be minimized by volume reduction, for example, through dewatering and drying sludge.

A crucial part of the process for reducing and minimizing wastes is the development of a material balance, which is an integral part of the practice of industrial ecology. Such a balance addresses various aspects of waste streams, including sources, identification, and quantities of wastes, and methods and costs of handling, treatment, recycling, and disposal. Priority waste streams can then be subjected to detailed process investigations to obtain the information needed to reduce wastes.

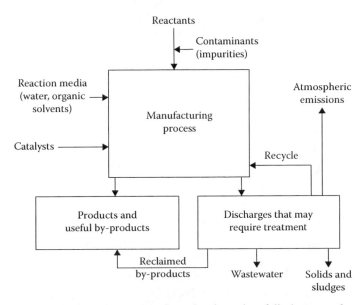

FIGURE 16.9   Chemical manufacturing process from the viewpoint of discharges and waste minimization.

16.14 RECYCLING

Wherever possible, recycling and reuse should be accomplished on site because it avoids having to move wastes and because a process that produces recyclable materials is often the most likely to have use for them. Wastes can be recycled by direct recycling of unconsumed leftover raw feedstock to a process; (2) transfer of waste material to another process where it may serve as feedstock; (3) utilization for waste treatment, such as waste alkali used to neutralize acid; and (4) energy recovery, such as incineration of combustible wastes to generate electrical power.

Some specific examples of recycling include **ferrous metals** (scrap steel) as feedstock for electric arc furnaces; **nonferrous metals**, particularly aluminum, copper, and lead; metal compounds (salts); **inorganic substances** such as steel pickling liquor acid; glass, which makes up about 10% of municipal refuse; **plastics** making up a substantial fraction of municipal refuse; rubber; **catalysts** such as those used in petroleum processing; substances with **agricultural** uses such as phosphate-containing sludges used for fertilizers; and **organic substances** including solvents and hydraulic and lubricating oils.

16.14.1 WASTE OIL UTILIZATION AND RECOVERY

Waste oil generated from lubricants and hydraulic fluids is one of the more commonly recycled materials. A significant fraction of the approximately 4 billion liters of waste oil produced annually in the United States is burned as fuel, much is recycled, and lesser quantities are disposed of as waste. The collection, recycling, treatment, and disposal of waste oil are all complicated by the fact that it comes from diverse, widely dispersed sources and contains several classes of potentially hazardous contaminants. These are divided between organic constituents (polycyclic aromatic hydrocarbons, chlorinated hydrocarbons) and inorganic constituents (aluminum, chromium, and iron from wear of metal parts; barium and zinc from oil additives; and formerly lead from leaded gasoline).

As a minimum, waste oil should be burned for its fuel value, a practice in some automobile servicing shops. Waste cooking oil can serve as a feedstock for diesel engine fuel. Generally, this is done by chemical treatment and esterification of the fatty acids from cooking oils and fats (see Chapter 17). Some enterprising individuals fuel their diesel-powered vehicles directly with waste oil scavenged from restaurants.

Waste lubricating oils from vehicles can be processed to produce an acceptable recycled lubricating oil. Normally, this involves distillation and vacuum distillation to remove water and lighter hydrocarbon impurities, leaving some heavy residue that can be burned, purification by treatment with solvents such as methylethyl ketone, and hydrotreatment with H_2 to eliminate unsaturated hydrocarbons and compounds containing O, N, or S.

16.14.2 WASTE SOLVENT RECOVERY AND RECYCLING

Among the many solvents listed as hazardous wastes and potentially recoverable from wastes are dichloromethane, tetrachloroethylene, trichloroethylene, 1,1,1-trichloroethane, benzene, liquid alkanes, 2-nitropropane, methylisobutyl ketone, and cyclohexanone. For reasons of both economics and pollution control, many industrial processes that use solvents are equipped for solvent recycling. The basic scheme for solvent reclamation and reuse is shown in Figure 16.10.

16.14.3 RECOVERY OF WATER FROM WASTEWATER

It is usually desirable to reclaim water from wastewater generated in manufacturing and other anthrospheric processes, including water in hazardous wastes, to reduce the quantity of wastes requiring disposal and to conserve water. Many hazardous waste materials are predominantly water, and the purification and reclamation of water from these sources can greatly reduce the quantity of wastes requiring further treatment or disposal.

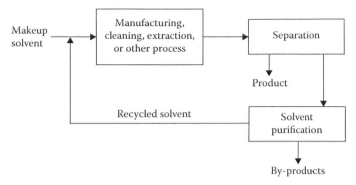

FIGURE 16.10   Overall process for recycling solvents.

The degree and type of treatment applied to wastewater depends on the water's intended use. Water used for industrial quenching and washing usually requires the least treatment, and wastewater from some other processes may be suitable for these purposes without additional treatment. At the other end of the scale, boiler makeup water, potable (drinking) water, water used to directly recharge aquifers, and water that people will directly contact (in boating, water skiing, and similar activities) must be of very high quality.

The treatment processes applied to anthrospheric wastewater are the same as those discussed in Chapter 5. For water associated with hazardous wastes, special attention must be given to the removal of potentially toxic constituents including heavy metals and carcinogenic substances.

The ultimate water quality is achieved by processes that remove essentially all solutes from water, leaving pure H_2O. A combination of activated carbon treatment to remove organics, cation exchange to remove dissolved cations, and anion exchange to remove dissolved anions can provide very high–quality water from wastewater. Reverse osmosis (see Chapter 5, Section 5.10) can accomplish the same objective. However, these processes generate spent activated carbon, ion exchange resins that require regeneration, and concentrated brines (from reverse osmosis) that require disposal, all of which have the potential to end up as hazardous wastes.

16.15   HAZARDOUS WASTE TREATMENT PROCESSES

As shown in Figure 16.11, waste treatment may occur at three major levels—**primary**, **secondary**, and **polishing**—somewhat analogous to the treatment of wastewater (see Chapter 5).[17] Primary treatment is generally regarded as preparation for further treatment, although it can result in the removal of by-products and the reduction of the quantity and hazard of wastes. Secondary treatment detoxifies, destroys, and removes hazardous constituents. Polishing usually refers to treatment of water that is removed from wastes so that it can be safely discharged, but may refer to other hazardous waste materials as well.

16.16   METHODS OF PHYSICAL TREATMENT

For the most part, methods of physical treatment of wastes involve transfers of materials between phases, sometimes with a phase change involved and, in the case of membrane separations, movement across a physical barrier.

The most straightforward means of physical treatment involves **phase separation** of the components of a mixture that are already in two different phases. **Sedimentation** and **decanting** are easily accomplished with simple equipment. In many cases, the separation must be aided by mechanical means, particularly **filtration** or **centrifugation**. **Flotation** is used to bring suspended organic matter or finely divided particles to the surface of a suspension. In the process of **dissolved air flotation** (DAF), air is dissolved in the suspending medium under pressure and comes out of solution when

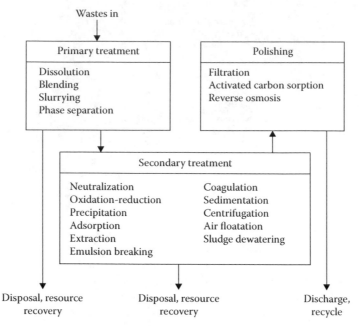

FIGURE 16.11   Major phases of waste treatment.

the pressure is released as minute air bubbles attached to suspended particles, which causes the particles to float to the surface (see Chapter 5, Figure 5.2).

An important and often difficult waste treatment step is **emulsion breaking**, which causes colloidal-sized **emulsions** to aggregate and settle from suspension. Agitation, heat, acid, and the addition of **coagulants** consisting of organic polyelectrolytes, or inorganic substances such as an aluminum salt, can be used for this purpose. The chemical additive acts as a flocculating agent to cause the particles to stick together and settle out.

A second major class of physical separation is that of **phase transition**, in which a material changes from one physical phase to another. It is best exemplified by **distillation**, which is used in treating and recycling solvents, waste oil, aqueous phenolic wastes, xylene contaminated with paraffin from histological laboratories, and mixtures of ethylbenzene and styrene. Distillation produces **distillation bottoms** (still bottoms), which are often hazardous and polluting. These consist of unevaporated solids, semisolid tars, and sludges from distillation. Specific examples with their hazardous waste numbers (see Chapter 15) are distillation bottoms from the production of acetaldehyde from ethylene (hazardous waste number K009) and still bottoms from toluene reclamation distillation in the production of disulfoton (K036). The landfill disposal of these and other hazardous distillation bottoms used to be widely practiced but is now severely limited, and it is much better to treat the waste to make them nonhazardous.

Evaporation is usually employed to remove water from an aqueous waste to concentrate it. A special case of this technique is **thin-film evaporation** in which volatile constituents are removed by heating a thin layer of liquid or sludge waste spread on a heated surface. **Drying** is employed to remove solvent or water from a liquid, suspension, or solid or semisolid (sludge). In **freeze drying**, the solvent, usually water, is sublimed from a frozen material. **Stripping** is a means of separating volatile components from less volatile ones in a liquid mixture by the partitioning of the more volatile materials to a gas phase of air or steam (steam stripping). Examples include air stripping of benzene and dichloromethane from water and volatile ammonia gas from water treated with base.

Physical precipitation is used here as a term to describe processes in which a solid forms from a solute in solution as a result of a physical change in the solution such as evaporation of solvent, or alteration of solvent composition. **Phase transfer** consists of the transfer of a solute in a mixture from one phase to another. An important type of phase transfer process is **solvent extraction**, a

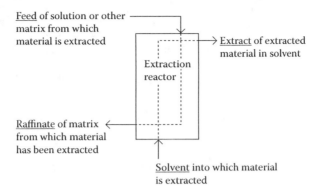

FIGURE 16.12   Outline of solvent extraction/leaching process with important terms underlined.

process in which a substance is transferred from solution in one solvent (usually water) to another (usually an organic solvent) without any chemical change taking place (see Figure 16.12). A common example is extraction into an organic solvent of phenol from water by-product of coal coking. **Supercritical carbon dioxide** (see Chapter 14, Section 14.13) can be used for solvent extraction and leaching of materials from waste solids. After a substance has been extracted from a waste into supercritical CO_2, the pressure can be released, resulting in the separation of the substance extracted. The fluid can then be compressed again and recirculated through the extraction system.

The transfer of a substance from a solution to a solid phase is called **sorption**. The most important solid sorbent is **activated carbon** used for several purposes in waste treatment, commonly to remove organics from wastewater. In some cases, it is adequate for complete treatment and can also be applied to pretreatment of waste streams going into processes such as reverse osmosis to improve treatment efficiency and reduce fouling.

A third major class of physical separation is **molecular separation**, which is often based on **membrane processes** in which dissolved contaminants or solvents pass through a size-selective membrane under pressure.[18] The products are a relatively pure solvent phase (usually water) and a concentrate rich in the solute impurities. Membrane processes including the special case of **reverse osmosis** to remove salts from water are discussed for the treatment of water in Chapter 5, Section 5.10. **Electrodialysis**, employing membranes alternately permeable to cations and to anions and driven by the passage of an electrical current (see Chapter 5, Section 5.10), is sometimes used to concentrate metal plating wastes and to reclaim dissolved metals.

16.17   CHEMICAL TREATMENT

The applicability of chemical treatment to wastes depends on the chemical properties of the waste constituents, particularly acid-base, oxidation-reduction, precipitation, and complexation behaviors; reactivity; flammability/combustibility; corrosivity; and compatibility with other wastes. The chemical behavior of wastes translates to various unit operations for waste treatment that are based on chemical properties and reactions.

Waste acids and bases are treated by **neutralization**:

$$H^+ + OH^- \rightarrow H_2O \tag{16.13}$$

Lime, $Ca(OH)_2$, is widely used as a base for treating acidic wastes, and sulfuric acid, H_2SO_4, is a relatively inexpensive acid for treating alkaline wastes. Acetic acid, CH_3COOH, is sometimes preferred to treat alkaline wastes. It has the advantage of being a weak acid so that an excess of it does little harm, and it is also a natural product and is biodegradable. Neutralization, or pH adjustment, is often required prior to the application of other waste treatment processes such as biochemical treatment.

Chemical precipitation is used in hazardous waste treatment primarily for the removal of heavy metal ions from water. The most widely used means of precipitating metal ions is by the addition of base ($Ca(OH)_2$, NaOH, or Na_2CO_3), leading to the formation of hydroxides such as chromium(III) hydroxide

$$Cr^{3+} + 3OH^- \rightarrow Cr(OH)_3(s) \qquad (16.14)$$

or basic salt precipitates such as basic copper(II) sulfate, $CuSO_4 \cdot 3Cu(OH)_2$, which is formed as a solid when hydroxide is added to a solution containing Cu^{2+} and SO_4^{2-} ions. The solubilities of some heavy metal sulfides are extremely low, so precipitation of heavy metal sulfides such as PbS by H_2S or other sulfides can be a very effective means of treatment. Hydrogen sulfide is a toxic gas that is itself considered to be a hazardous waste (U135). Iron(II) sulfide (ferrous sulfide) can be used as a safe source of sulfide ion to produce sulfide precipitates with other metals that are less soluble than FeS. Some metals can be precipitated from solution in the elemental metal form by the action of a reducing agent such as sodium borohydride or through the reaction of active elemental metals in a process called **cementation**:

$$Cd^{2+} + Zn \rightarrow Cd + Zn^{2+} \qquad (16.15)$$

Coprecipitation, most commonly with iron(III) hydroxide, may be used to remove heavy metals from wastes. An example is removal of lead from lead battery industry wastewater with $Fe(OH)_3$ to which a soluble Fe^{3+} salt has been added.

As shown by the reactions in Table 16.2, **oxidation** and **reduction** can be used for the treatment and removal of a variety of inorganic and organic wastes. Some waste oxidants can be used to treat oxidizable wastes in water and cyanides. Ozone, O_3, is a strong "green" oxidant that can be generated on site by an electrical discharge through dry air or oxygen (see Chapter 5, Section 5.11 and Figure 5.7) and is employed to treat a variety of oxidizable contaminants, effluents, and wastes, including wastewater and sludges containing oxidizable constituents.

16.17.1 Electrolysis

As shown in Figure 16.13, **electrolysis** is a process in which one species in solution (usually a metal ion) is reduced by electrons at the cathode and another gives up electrons to the anode and is oxidized there. In hazardous waste applications, electrolysis is most widely used in the recovery of cadmium, copper, gold, lead, silver, and zinc from sources such as spent metal plating solutions. An electrolysis procedure has been described for the recovery of copper from cyanide-containing

---

TABLE 16.2

Oxidation-Reduction Reactions Used to Treat Wastes

Waste Substance Reaction with Oxidant or Reductant

Oxidation of organics
Organic matter: $\{CH_2O\} + 2\{O\} \rightarrow CO_2 + H_2O$
Aldehyde: $CH_3CHO + \{O\} \rightarrow CH_3COOH$ (acid)
Oxidation of inorganics
Cyanide: $2CN^- + 5OCl^- + H_2O \rightarrow N_2 + 2HCO_3^- + 5Cl^-$
Iron(II): $4Fe^{2+} + O_2 + 10H_2O \rightarrow 4Fe(OH)_3 + 8H^+$
Sulfur dioxide: $2SO_2 + O_2 + 2H_2O \rightarrow 2H_2SO_4$
Reduction of inorganics
Chromate: $2CrO_4^{2-} + 3SO_2 + 4H^+ \rightarrow Cr_2(SO_4)_3 + 2H_2O$
Permanganate: $MnO_4^- + 3Fe^{2+} + 7H_2O \rightarrow MnO_2(s) + 3Fe(OH)_3(s) + 5H^+$

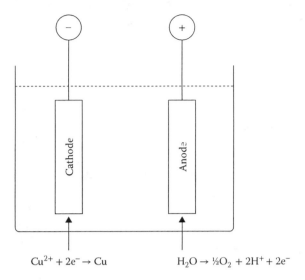

$$Cu^{2+} + 2e^- \rightarrow Cu \qquad\qquad H_2O \rightarrow \tfrac{1}{2}O_2 + 2H^+ + 2e^-$$

Net reaction: $Cu^{2+} + H_2O \rightarrow Cu + \tfrac{1}{2}O_2 + 2H^+$

FIGURE 16.13 Electrolysis of copper solution.

spent plating bath water at the Brazilian mint.[19] Destruction of cyanide also was observed, from the electrolytic oxidation of CN^- at the anode.

16.17.2 HYDROLYSIS

One of the ways to dispose of chemicals that are reactive with water is **hydrolysis**, that is, the reaction with water under controlled conditions. Inorganic chemicals that can be treated by hydrolysis include metals that react with water; metal carbides, hydrides, amides, alkoxides, and halides; and nonmetal oxyhalides and sulfides. An example of a waste chemical treated by hydrolysis is the reaction with water of sodium aluminum hydride (used as a reducing agent in organic chemical reactions):

$$NaAlH_4 + 4H_2O \rightarrow 4H_2 + NaOH + Al(OH)_3 \qquad (16.16)$$

Care must be exercised in the handling of the explosive elemental hydrogen product from this reaction.

Organic chemicals can also be treated by hydrolysis. For example, toxic acetic anhydride is hydrolyzed to relatively safe acetic acid:

$$\begin{array}{c} \text{H O} \quad\quad \text{O H} \qquad\qquad\qquad \text{H O} \\ | \; \| \quad\quad \| \; | \qquad\qquad\qquad | \; \| \\ \text{H}-\text{C}-\text{C}-\text{O}-\text{C}-\text{C}-\text{H}+\text{H}_2\text{O} \longrightarrow 2\text{H}-\text{C}-\text{C}-\text{OH} \\ | \qquad\qquad | \qquad\qquad\qquad | \\ \text{H} \qquad\qquad \text{H} \qquad\qquad\qquad \text{H} \end{array} \qquad (16.17)$$

Acetic anhydride
(an acid anhydride)

16.17.3 CHEMICAL EXTRACTION AND LEACHING

Chemical extraction or **leaching** in hazardous waste treatment is the removal of a hazardous constituent by chemical reaction with an extractant in solution. Poorly soluble heavy metal salts can be extracted by reaction of the salt anions with H^+, as illustrated by the following reaction:

$$PbCO_3 + H^+ \rightarrow Pb^{2+} + HCO_3^- \qquad (16.18)$$

Acids also dissolve basic organic compounds such as amines and aniline. Chelating agents, such as dissolved ethylenediaminetetraacetate (EDTA, HY^{3-}), dissolve insoluble metal salts by forming soluble species with metal ions:

$$FeS + HY^{3-} \rightarrow FeY^{2-} + HS^- \tag{16.19}$$

The EDTA and its metal chelates are poorly biodegradable. As an alternative to EDTA, ethylenediaminesuccinic acid may be used for metal extraction.[20] This chelating agent is about as effective for heavy metal extraction as EDTA, but succinic acid is an intermediate in metabolic processes and ethylenediaminesuccinic acid and its metal chelates are readily biodegradable, making ethylenediaminesuccinic acid a greener alternative to EDTA for metal extraction.

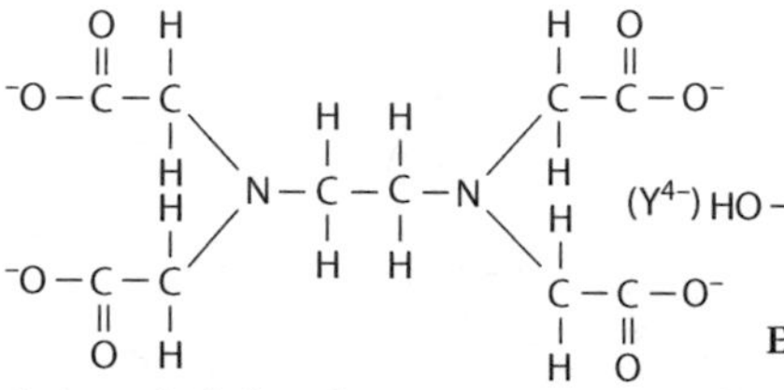

Anion of ethylenediaminetetraacetic acid 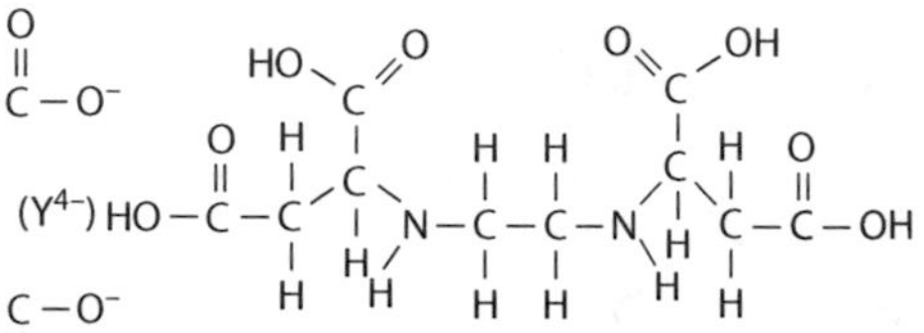   **Ethylenediaminesuccinic acid**

Heavy metal ions in soil contaminated by hazardous wastes may be present in a coprecipitated form with insoluble iron(III) and manganese(IV) oxides, Fe_2O_3 and MnO_2, respectively. These oxides can be dissolved by reducing agents, such as solutions of sodium dithionate/citrate or hydroxylamine. This results in the production of soluble Fe^{2+} and Mn^{2+} and the release of heavy metal ions, such as Cd^{2+} or Ni^{2+}, which are removed with water.

16.17.4 ION EXCHANGE

Ion exchange is a means of removing cations or anions from solution onto a solid resin, which can be regenerated by treatment with acids, bases, or salts. The greatest use of ion exchange in hazardous waste treatment is for the removal of low levels of heavy metal ions from wastewater:

$$2H^{+-}\{CatExchr\} + Cd^{2+} \rightarrow Cd^{2+-}\{CatExchr\}_2 + 2H^+ \tag{16.20}$$

Ion exchange is used in the metal plating industry to purify rinse water and spent plating bath solutions. Cation exchangers remove cationic metal species, such as Cu^{2+}, from such solutions. Anion exchangers remove anionic cyanide metal complexes, such as $Ni(CN)_4^{2-}$, and chromium(VI) species, such as CrO_4^{2-}. Radionuclides can be removed from radioactive and mixed radioactive/hazardous chemical waste by ion exchange resins.

16.18 PHOTOLYTIC REACTIONS

Photolytic (photochemical) **reactions** are discussed as a means of destroying refractory organic water pollutants in Chapter 5, Section 5.8.3 and as ideal green chemistry reagents in Chapter 14, Section 14.12. **Photolysis** can be used to destroy a number of kinds of hazardous wastes. In such applications, it is most useful in breaking chemical bonds in refractory organic compounds. As discussed in Chapter 5, Section 5.15, and shown in Figure 5.12, photolysis by ultraviolet radiation is used to break down chemical compounds such as pharmaceuticals and their metabolites in the treatment of wastewater for recycling. Ultraviolet photolysis is also used along with oxidants to break down organic compounds in water in the analytical determination of total organic carbon in water (see Chapter 18, Section 18.17 and Figure 18.15).

An initial photolysis reaction can result in the generation of reactive intermediates that participate in **chain reactions** that lead to the destruction of a compound. One of the most important reactive intermediates is the hydroxyl radical, HO·. In some cases, **sensitizers** are added to the reaction mixture to absorb radiation and generate reactive species that destroy wastes. Hazardous waste substances other than 2,3,7,8-tetrachlorodibenzo-*p*-dioxin (TCDD) that have been destroyed by photolysis are herbicides (atrazine); 2,4,6-trinitrotoluene (TNT); and polychlorinated biphenyls (PCBs). The addition of a chemical oxidant, such as potassium peroxydisulfate, $K_2S_2O_8$, enhances destruction by oxidizing active photolytic products.

16.19 THERMAL TREATMENT METHODS

Thermal treatment of hazardous wastes can be used to accomplish most of the common objectives of waste treatment—volume reduction; removal of volatile, combustible, mobile organic matter; and destruction of toxic and pathogenic materials. The most widely applied means of thermal treatment of hazardous wastes is **incineration**. Incineration utilizes high temperatures, an oxidizing atmosphere, and often turbulent combustion conditions to destroy wastes.

16.19.1 INCINERATION

Hazardous waste incineration is a process that involves the exposure of waste materials to oxidizing conditions at a high temperature, usually in excess of 900°C. Normally, the heat required for incineration comes from the oxidation of organically bound carbon and hydrogen contained in the waste material or in supplemental fuel. These reactions destroy organic matter and generate heat required for endothermic reactions, such as the breaking of C–Cl bonds in organochlorine compounds.

The four major components of hazardous waste incineration systems are shown in Figure 16.14. Hazardous waste incinerators can be divided as follows, based on the type of combustion chamber: about 40% of U.S. incinerators are of the **rotary kiln** type consisting of a rotating cylinder lined

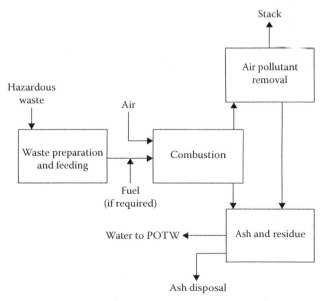

FIGURE 16.14 Major components of a hazardous waste incinerator system. In the figure, POTW refers to a publicly owned treatment works, usually a municipal wastewater treatment facility.

with refractory materials and equipped with a downstream afterburner to complete the destruction of the wastes. Another approximately 40% of U.S. hazardous waste incinerators are of the **liquid injection** type that burn pumpable liquid wastes dispersed as small droplets. **Fixed-hearth** incinerators burn solid wastes on single or multiple hearths. **Fluidized-bed** incinerators in which combustion of wastes is carried out on a bed of granular solid (such as limestone) maintained in a suspended (fluid-like) state by injection of air retain pollutant acid gas and ash products in the bed material and are especially effective for air pollution control. A number of kinds of **advanced design incinerators** have been designed. These include **plasma incinerators** that make use of an extremely hot plasma of ionized air injected through an electrical arc; **electric reactors** that use resistance-heated incinerator walls at around 2200°C to heat and pyrolyze wastes by radiative heat transfer; **infrared systems**, which generate intense infrared radiation by passing electricity through silicon carbide resistance heating elements; **molten salt combustion** that uses a bed of molten sodium carbonate at about 900°C to destroy the wastes and retain gaseous pollutants; and **molten glass processes** that use a pool of molten glass to transfer heat to the wastes and to retain products in a poorly leachable glass form.

16.19.2 EFFECTIVENESS OF INCINERATION

EPA standards for hazardous waste incineration are based on the effectiveness of destruction of **principal organic hazardous constituents** (POHCs). Measurement of these compounds before and after incineration gives the **destruction removal efficiency** (DRE) according to the formula

$$\text{DRE} = \frac{W_{in} - W_{out}}{W_{in}} \times 100 \tag{16.21}$$

where W_{in} and W_{out} are the mass flow rates of the POHC input and output (at the stack downstream from emission controls), respectively. U.S. EPA regulations call for destruction of 99.99% of POHCs and 99.9999% ("six nines") destruction of 2,3,7,8-tetrachlorodibenzo-*p*-dioxin, commonly called TCDD or "dioxin."

16.19.3 HAZARDOUS WASTE FUEL

Many industrial wastes, including hazardous wastes, are burned as **hazardous waste fuel** for energy recovery in industrial furnaces and boilers and in incinerators for nonhazardous wastes, such as sewage sludge incinerators. This process is called **coincineration**, and more combustible wastes are utilized by it than are burned solely for the purpose of waste destruction. In addition to heat recovery from combustible wastes, it is a major advantage to use an existing on-site facility for waste disposal rather than a separate hazardous waste incinerator.

16.20 BIODEGRADATION OF HAZARDOUS WASTES

Through **biodegradation**, wastes can be converted enzymatically to simple inorganic molecules (mineralization) and, to a certain extent, to biological materials. Usually, the products of biodegradation are molecular forms that tend to occur in nature and that are in greater thermodynamic equilibrium with their surroundings than are the starting materials. **Detoxification** refers to the biological conversion of a toxic substance to a less toxic species. Microbial bacteria and fungi possessing enzyme systems that are required for the biodegradation of wastes are usually best obtained from populations of indigenous microorganisms at a hazardous waste site where the microorganisms have developed the ability to degrade particular kinds of molecules.

The **biodegradability** of a compound is influenced by its physical characteristics, such as solubility in water and vapor pressure, and by its chemical properties, including molar mass, molecular structure, and presence of various kinds of functional groups, some of which provide a "biochemical handle" for the initiation of biodegradation. With the appropriate organisms and under the right conditions, even substances such as phenol that are considered to be biocidal to most microorganisms can undergo biodegradation.

Recalcitrant or **biorefractory** substances are those that resist biodegradation and tend to persist and accumulate in the environment. Such materials are not necessarily toxic to organisms but simply resist their metabolic attack. However, even some compounds regarded as biorefractory can be degraded by microorganisms that have had the opportunity to adapt to their biodegradation, for example, dichlorodiphenyltrichloroethane (DDT) is degraded by properly acclimated *Pseudomonas*. Chemical pretreatment, especially by partial oxidation, can make some kinds of recalcitrant wastes much more biodegradable.

Properties of hazardous wastes and their media can be changed to increase biodegradability. This can be accomplished by adjustment of conditions to optimum temperature, pH (usually in the range of 6–9), stirring, oxygen level, and material load. Biodegradation can be aided by the removal of toxic organic and inorganic substances, such as heavy metal ions.

16.20.1 Oxic and Anoxic Waste Biodegradation

Oxic (aerobic) **waste treatment** processes utilize bacteria and fungi that require molecular oxygen, O_2. These processes are often favored by microorganisms, in part because of the high energy yield obtained when molecular oxygen reacts with organic matter. Oxic waste treatment is well adapted to the use of an activated sludge process. It can be applied to hazardous wastes such as chemical process wastes and landfill leachates. Some systems use powdered activated carbon as an additive to absorb organic wastes that are not biodegraded by microorganisms in the system.

Contaminated soils can be mixed with water and treated in a bioreactor to eliminate biodegradable contaminants in the soil. It is possible, in principle, to treat contaminated soils biologically in place by pumping oxygenated, nutrient-enriched water through the soil in a recirculating system.

Anoxic waste treatment in which microorganisms degrade wastes in the absence of oxygen can be practiced on a variety of organic hazardous wastes. Compared with the aerated activated sludge process, anoxic digestion requires less energy; yields less sludge by-product; generates hydrogen sulfide, H_2S, which precipitates toxic heavy metal ions; and produces methane gas, CH_4, which can be used as an energy source.

The overall process for anoxic digestion is a generally metabolically complicated fermentation process in which organic matter is both oxidized and reduced; it is especially suited to the destruction of oxygenated compounds, such as acetaldehyde or methylethyl ketone. The simplified reaction for the anoxic fermentation of a hypothetical organic substance, $\{CH_2O\}$, is as follows:

$$2\{CH_2O\} \rightarrow CO_2 + CH_4 \qquad (16.22)$$

Reductive dehalogenation is a mechanism by which halogen atoms are removed from organohalide compounds by anoxic bacteria converting the organically bound chlorine to innocuous Cl^- ion. It is an important means of detoxifying alkyl halides (particularly solvents), aryl halides, and organochlorine pesticides, all of which are important hazardous waste compounds and which were discarded in large quantities in some of the older waste disposal dumps. Reductive dehalogenation is the only means by which some of the more highly halogenated waste compounds are biodegraded; such compounds include tetrachloroethene, hexachlorobenzene, pentachlorophenol, and the more highly chlorinated PCB congeners.

The two general processes by which reductive dehalogenation occurs are **hydrogenolysis**, as shown by the example in Reaction 16.23:

$$\text{DRE} = \frac{W_{\text{in}} - W_{\text{out}}}{W_{\text{in}}} \times 100 \qquad (16.23)$$

and **vicinal reduction**, which removes two adjacent halogen atoms and works only on alkyl halides, not aryl halides:

$$(16.24)$$

16.20.2 LAND TREATMENT AND COMPOSTING

Soil has physical, chemical, and biological characteristics that can enable waste detoxification, biodegradation, chemical decomposition, and physical and chemical fixation. Therefore, **land treatment** of wastes can be accomplished by mixing the wastes with soil under appropriate conditions.

The microorganisms in soil that biodegrade wastes are bacteria, including those from the genera *Agrobacterium*, *Arthrobacteri*, *Bacillus*, *Flavobacterium*, *Pseudomonas*, fungus-like *Actinomycetes*, and fungi, usually present naturally in sufficient numbers to provide the inoculum required for their growth. The growth of these indigenous microorganisms can be stimulated by adding nutrients and an electron acceptor to act as an oxidant (for anoxic degradation) accompanied by mixing. The most commonly added nutrients are nitrogen and phosphorus. For oxic biodegradation, oxygen can be added by pumping air underground or by treatment with hydrogen peroxide, H_2O_2. In some cases, such as for treatment of hydrocarbons on or near the soil surface, simple tillage provides both oxygen and the mixing required for optimum microbial growth.

Wastes that are amenable to land treatment are biodegradable organic substances. However, in soil contaminated with hazardous wastes, bacterial cultures may develop that are effective in degrading normally recalcitrant compounds through acclimation over a long period of time. Land treatment is most used for petroleum-derived wastes and biodegradable organic chemical wastes, including some organohalide compounds.

Composting of hazardous wastes is the biodegradation of solid or solidified materials in a medium other than soil. Bulking material, such as plant residue, paper, municipal refuse, or sawdust, can be added to retain water and enable air to penetrate to the waste material. Successful composting of hazardous wastes depends on a number of factors, including those discussed earlier under land treatment. Once a successful composting operation is under way, a good inoculum is maintained by recirculating spent compost to each new batch. The composting process generates heat; so, if the mass of the compost pile is sufficiently high, it can be self-heating under most conditions. Some wastes are deficient in nutrients such as nitrogen, which must be supplied from commercial sources or from other wastes.

16.21   PREPARATION OF WASTES FOR DISPOSAL

Although it is far better to find uses for hazardous waste constituents and put them into a nonhazardous form, in the past, and even today, wastes were disposed of to the geosphere, usually in landfill. This requires treatment of wastes to get them into a form that is suitable for long-term disposal, especially to avoid leaking or leaching into groundwater. Several techniques for doing this are discussed here.

Immobilization includes physical and chemical processes that reduce surface areas, solubility, and volatility of wastes to minimize leaching. It isolates the wastes from their environment, especially groundwater, so that they have the least possible tendency to migrate. **Stabilization** or **fixation** means the conversion of a waste from its original form to a physically and chemically more stable material that is less likely to cause problems during handling and disposal, and is less likely to be mobile after disposal. **Solidification** may involve chemical reaction of the waste with a solidification agent, mechanical isolation in a protective binding matrix, or a combination of chemical and physical processes. It can be accomplished by evaporation of water from aqueous wastes or sludges, sorption onto solid material, reaction with portland cement, reaction with silicates, encapsulation, or embedding in polymers or thermoplastic materials. Hazardous waste liquids, emulsions, sludges, and free liquids in contact with sludges can be solidified and stabilized by fixing onto solid **sorbents**, including activated carbon (for organics), fly ash, kiln dust, clays, vermiculite, and various proprietary materials.

Some wastes can be immobilized in **thermoplastics**, which are solids or semisolids that become liquefied at elevated temperatures and that cool to rigid but deformable forms. The thermoplastic material most used for this purpose is asphalt bitumen.

Vitrification, or **glassification**, consists of embedding wastes in a glass material. In this application, glass can be regarded as a high-melting-temperature inorganic thermoplastic. Molten glass can be used, or glass can be synthesized in contact with the waste by mixing and heating with glass constituents—silicon dioxide, SiO_2; sodium carbonate, Na_2CO_3; and calcium oxide, CaO. Vitrification is relatively complicated and expensive, the latter because of the energy consumed in fusing glass. Despite these disadvantages, it is the favored immobilization technique for some special wastes and has been promoted for the solidification of radionuclear wastes because glass is chemically inert and resistant to leaching.

Water-insoluble **pozzolanic substances** containing oxyanionic silicon such as SiO_3^{2-} and usually reacted with calcium hydroxide are used for waste solidification. The most abundant and commonly used pozzolanic substance is fly ash from coal combustion. Other pozzolanic substances, which are usually waste products from various processes include flue dust and ground slag from blast furnaces.

Similar to pozzolanic substances, **portland cement** is widely used for the solidification of hazardous wastes and is most applicable to inorganic sludges. In this application, portland cement provides a solid matrix for the isolation of wastes, chemically binds water from sludge wastes, and may react chemically with wastes (e.g., the calcium and the base in portland cement react chemically with inorganic arsenic sulfide wastes to reduce their solubilities). Some kilns in which the sand, limestone, and clay ingredients are heated to make portland cement are partially fueled by combustible hazardous wastes. The very high temperatures required to make the cement ensure destruction of even the most refractory organic wastes, and the mineral ingredients of portland cement effectively sequester acid gas products and heavy metals.

16.22 ULTIMATE DISPOSAL OF WASTES

Regardless of the destruction, treatment, and immobilization techniques used, there will always remain from hazardous wastes some material that has to be put somewhere. This section briefly addresses the ultimate disposal of ash, salts, liquids, solidified liquids, and other residues that must be placed where their potential to do harm is minimized.

In some cases, hazardous wastes are disposed of above the ground, essentially in a mound resting on a layer of compacted clay covered with impermeable membrane liners laid somewhat above the original soil surface and shaped to allow leachate flow and collection. In a properly designed aboveground disposal facility, any leachate that is produced drains quickly by gravity to the leachate collection system, where it can be detected and treated.

Historically, **landfill** has been the most common way of disposing of solid hazardous wastes and some liquids, although it is being severely limited today in many nations by new regulations and high land costs. Landfill involves disposal that is at least partially underground in excavated cells, quarries, or natural depressions lined with impermeable clay and synthetic membranes to stop the permeation of leachate. Usually, fill is continued above the ground to utilize space most efficiently and provide a grade for drainage of precipitation.

Many liquid hazardous wastes, slurries, and sludges are placed in **surface impoundments**, which usually serve for treatment and often are designed to be filled in eventually as landfill disposal sites. Most liquid hazardous wastes and a significant fraction of solids are placed in surface impoundments in some stage of treatment, storage, or disposal. The construction of a surface impoundment is similar to that discussed earlier for landfills in that its bottom and walls should be impermeable to liquids and provision must be made for leachate collection.

Deep-well disposal of liquids consists of their injection under pressure to underground strata isolated by impermeable rock strata from aquifers. Early experience with this method was gained in the petroleum industry where disposal is required of large quantities of saline wastewater coproduced with crude oil. The method was later extended to the chemical industry for the disposal of brines, acids, heavy metal solutions, organic liquids, and other liquids. As of 2012, a controversial deep-well disposal practice was that of the disposal of large quantities of contaminated water left over from hydraulic fracturing ("fracking") (see Chapter 17, Section 17.9) of tight shale formations to recover natural gas.

16.23   LEACHATE AND GAS EMISSIONS

The production of contaminated **leachate** is a possibility with most disposal sites. Leachate consists of water that has become contaminated by wastes as it passes through a waste disposal site. It contains waste constituents that are soluble, not retained by soil, and not readily degraded chemically or biochemically. Therefore, new hazardous waste landfills require leachate collection/treatment systems, and many older sites are required to have such systems retrofitted to them. Leachate is collected in perforated pipes that are embedded in granular drain material.

The best approach to leachate management is to prevent its production by limiting the infiltration of water into the site. Rates of leachate production may be very low when sites are selected, designed, and constructed with minimal production of leachate as a major objective. A well-maintained, low-permeability cap over a landfill is very important for leachate minimization.

If leachate is released from a hazardous waste landfill, it may be necessary to install a leachate treatment system. Such a facility is a sophisticated wastewater treatment system using biological, physical, and chemical treatment processes as described in Chapter 5.

Gas emissions may be produced from hazardous waste disposal sites. In some cases, gases come directly from materials disposed of in the site. If biodegradable wastes are present, the anoxic biodegradation of biomass, represented by $\{CH_2O\}$, may produce significant quantities of CO_2, explosive CH_4, and toxic H_2S:

$$2\{CH_2O\} \rightarrow CH_4(g) + CO_2(g) \tag{16.25}$$

$$SO_4^{2-} + 2\{CH_2O\} + 2H^+ \rightarrow H_2S + 2CO_2 + 2H_2O \tag{16.26}$$

These gases may carry volatile wastes, such as benzene and chlorinated hydrocarbons, from the disposal site into the atmosphere and cause air pollution problems or pose health hazards near the site. In some cases, it may be necessary to collect the gases and treat them as discussed in Section 16.24. Filtration over activated carbon can be employed, and in some cases the gases have been pumped with intake air into internal combustion engines where they are destroyed by combustion.

16.24   IN SITU TREATMENT OF DISPOSED HAZARDOUS WASTES

In a number of cases, the treatment of improperly disposed hazardous wastes may be required. An expensive option is to dig up the wastes and treat them or place them in modern, properly designed landfills. Soil contaminated by waste leachate may require treatment such as digging it up, washing with water, and putting it back into place. In some cases, contaminated soil has even been incinerated.

Another approach to treating disposed hazardous wastes is in situ treatment in which the wastes are treated in place. Heavy metal contaminants including lead, cadmium, zinc, and mercury can be immobilized by chemical precipitation as their sulfides by treatment with gaseous H_2S or alkaline Na_2S solution. Heavy metal ions can also be immobilized by the oxidation of soluble Fe^{2+} and Mn^{2+} to their insoluble hydrous oxides, $Fe_2O_3 \cdot xH_2O$ and $MnO_2 \cdot xH_2O$, respectively, which can coprecipitate other heavy metal ions. Chelation can convert metal ions to less mobile forms, although with most agents chelation has the opposite effect. The insoluble humin fraction of soil humic substances immobilizes metal ions.

In situ solidification can be used as a remedial measure at hazardous waste sites. One approach is to inject soluble silicates followed by reagents that cause them to solidify. For example, injection of soluble sodium silicate followed by calcium chloride or lime forms solid calcium silicate.

Many important wastes have relatively high vapor pressures and can be removed by **vapor extraction**. Vapor extraction involves pumping air into injection wells in soil and withdrawing it, along with the volatile components that it has picked up, through extraction wells. The substances vaporized from the soil are removed by activated carbon or by other means. In some cases, the air is pumped through an engine (which can be used to run the air pumps) and vapors are destroyed by conditions in the engine's combustion chambers. Vapor extraction is most applicable to the removal of volatile organic compounds (VOCs) such as chloromethanes, chloroethanes, chloroethylenes (such as trichloroethylene), benzene, toluene, and xylene.

16.24.1   TREATMENT IN SITU

Some groundwater plumes contaminated by dissolved wastes can be treated by a permeable bed of material placed in a trench through which the groundwater must flow. Limestone in a permeable bed neutralizes acid and precipitates some heavy metal hydroxides or carbonates. Synthetic ion exchange resins can be used in a permeable bed to retain heavy metals and even some anionic species, although competition with ionic species present naturally in the groundwater can reduce the effectiveness of this treatment method.

Soil washing may be performed in situ to decontaminate soil that has absorbed hazardous waste materials. The washing medium may consist of pure water or it may contain acids (to leach out metals or neutralize alkaline soil contaminants), bases (to neutralize contaminant acids), chelating agents (to solubilize heavy metals), surfactants (to enhance the removal of organic contaminants from the soil and improve the ability of the water to emulsify insoluble organic species), or reducing agents (to reduce oxidized species). Soil contaminants may dissolve, form emulsions, or react chemically. Heavy metal salts; lighter aromatic hydrocarbons, such as toluene and xylenes; lighter organohalides, such as trichloro- or tetrachloroethylene; and light- to medium-molar-mass aldehydes and ketones can be removed from soil by washing.

QUESTIONS AND PROBLEMS

Access to and use of the Internet is assumed in answering all questions, including general information, statistics, constants, and mathematical formulas required to solve problems. These questions are designed to promote inquiry and thought rather than just finding material in the chapter. So in some cases there may be several "right" answers. Therefore, if your answer reflects intellectual effort and a search for information from available sources, it may be considered to be right.

1. What is a fundamental chemical difference between petroleum and biological feedstocks?
2. Potassium permanganate, sodium dichromate, and hydrogen peroxide may all three be used as reagents to react with organic molecules. What kind of reaction do they cause to occur? Which of these would be regarded as the greenest reagent, and why is that so?
3. Name three kinds of reactions used in processing feedstocks. Which is best from the viewpoint of green chemistry?
4. Name several fractions of biomass that can be used for feedstocks. Which of these is the least useful?
5. How are oils extracted from plant sources?
6. Use chemical formulas to show that carbohydrates are a more oxidized chemical feedstock than hydrocarbons.
7. Before the petrochemicals industry developed, the use of coal in making steel gave rise to a major organic chemicals industry. Explain why this was so, and list some of the products that were made.
8. With respect to Problem 7, see if you can find a toxicological concern that developed out of the organic chemicals industry based on coal coking by-products.
9. Name some categories of chemicals routinely produced by fermentation.
10. Which pharmaceutical material has been produced by fermentation for many years?
11. What is the first, most important, consideration in developing a fermentation process for the production of a chemical?
12. What is the significance of temperature in fermentation processes? What happens if the temperature is too high?
13. Of all the commodity chemicals made by fermentation, which is made in largest quantities? Explain.
14. In which fundamental respect are plants more efficient producers of material than fermentation?
15. Which relatively recent advance in biotechnology has greatly increased the scope of materials potentially produced by plants?
16. Why are hybrid poplar trees particularly important in the production of raw materials? Are there other more productive alternatives?
17. Describe the structural characteristics of glucose and other carbohydrates that make them good platforms for chemical synthesis.
18. Give a disadvantage and an advantage of the use of cellulose as a source of glucose.
19. Using Internet resources, try to find an example of an operational biorefinery that is either in commercial operation or at a pilot plant level. What materials does it process, and how does it process them? What are its main products?

20. Give a major concern with the use of benzene as a feedstock.

21. Although the chemical formula of glucose is $C_6H_{12}O_6$, the formula of the cellulose polymer made from glucose is $(C_6H_{10}O_5)_n$, where n is a large number. Since cellulose is made from glucose, why is the cellulose formula not $(C_6H_{12}O_6)_n$?

22. Give some examples of useful chemically modified cellulose. Which of these has proved to be rather dangerous?

23. Why is it difficult to deal with lignin as a source of chemical feedstocks? What is the current main use of waste lignin?

24. Give the main advantage of biopolymers from an environmental viewpoint.

25. Which structural feature of hydroxyalkanoates enables them to make polymeric molecules?

26. What was the original source of poly(hydroxyalkanoate) polymers? How is it now proposed to produce them?

27. Although enzymes have not developed specifically to act upon synthetic compounds, they have some specific advantages that make them attractive for carrying out chemical processes on synthetic compounds. What are some of these advantages?

28. Name two chemicals for which it has been shown that enzymatic processes can actually convert synthetic raw materials to chemical products normally made by nonbiological chemical reactions.

29. Look up the chemical synthesis of ammonia, NH_3. Is the final reaction an addition, an elimination, or a substitution reaction? Are there other kinds of reactions that may be involved in the overall process of ammonia synthesis? What are the environmental and resource implications of these kinds of reactions?

30. A hazardous waste incinerator monitored over a 24-hour period took in a total of 1 kg of hexachlorobenzene. Analysis of the exhaust gas from the facility showed levels of 0.25 μg m^3 of hexachlorobenzene at an exhaust gas flow rate of 0.200 m^3/s. What is the DRE of the process?

31. Pozzolanic substances were mentioned in this chapter as materials used to solidify and immobilize wastes. What is the origin of the term pozzolanic? How were pozzolanic substances used in ancient Rome?

32. Look up deep-well disposal of gases. Which gases have been disposed of in deep wells? What are the criteria for the rock formations into which waste gases may be pumped?

33. Place the following in descending order of desirability for dealing with wastes and discuss your rationale for doing so (explain): (A) reducing the volume of remaining wastes by measures such as incineration; (B) placing the residual material in landfills, properly protected from leaching or release by other pathways; (C) treating residual material as much as possible to render it nonleachable and innocuous; (D) reduction of wastes at the source; and (E) recycling as much waste as is practical.

34. Match the waste recycling process or industry from the column on the left with the kind of material that can be recycled from the list on the right in the following table:

A. Recycle as raw material to the generator	1. Waste alkali
B. Utilization for pollution control or waste treatment	2. Hydraulic and lubricating oils
C. Energy production	3. Incinerable materials
D. Materials with agricultural uses	4. Incompletely consumed feedstock material
E. Organic substances	5. Waste lime or phosphate-containing sludge

35. What is the most important operation in solvent purification and recycle that is used to separate solvents from impurities, water, and other solvents?

36. The DAF is used in the secondary treatment of wastes. What is the principle of this technique? For what kinds of hazardous waste substances is it most applicable?

37. Match the process or industry from the column on the left with its phase of waste treatment from the list on the right in the following table:

A. Activated carbon sorption	1. Primary treatment
B. Precipitation	2. Secondary treatment
C. Reverse osmosis	3. Polishing
D. Emulsion breaking	
E. Slurrying	

38. Distillation is used in treating and recycling a variety of wastes, including solvents, waste oil, aqueous phenolic wastes, and mixtures of ethylbenzene and styrene. What is the major hazardous waste problem that arises from the use of distillation for waste treatment?

39. Supercritical fluid technology has a great deal of potential for the treatment of hazardous wastes. What are the principles involved in the use of supercritical fluids for waste treatment? Why is this technique especially advantageous? Which substance is most likely to be used as a supercritical fluid in this application? For which kinds of wastes are supercritical fluids most useful?

40. What are some advantages of using acetic acid, compared, for example, to sulfuric acid, as a neutralizing agent for treating waste alkaline materials?

41. Designate which of the following would be **least likely** to be produced by, or used as, a reagent for the removal of heavy metals by their precipitation from solution (explain): (A) Na_2CO_3, (B) CdS, (C) $Cr(OH)_3$, (D) KNO_3, and (E) $Ca(OH)_2$.

42. Both $NaBH_4$ and Zn are used to remove metals from solution. How do these substances remove metals? What are the forms of the metal products?

43. Of the following, thermal treatment of wastes is **not** useful for (explain) (A) volume reduction; (B) destruction of heavy metals; (C) removal of volatile, combustible, mobile organic matter; (D) destruction of pathogenic materials; and (E) destruction of toxic substances.

44. From the following, choose the waste liquid that is the least amenable to incineration and explain why it is not readily incinerated: (A) methanol, (B) tetrachloroethylene, (C) acetonitrile, (D) toluene, (E) ethanol, and (F) acetone.

45. Name and give the advantages of the process that is used to destroy more hazardous wastes by thermal means than are burned solely for the purpose of waste destruction.

46. What is the major advantage of fluidized-bed incinerators from the standpoint of controlling pollutant by-products?

47. Explain the best way to obtain microorganisms to be used in the treatment of hazardous wastes by biodegradation.

48. What are the principles of composting? How is it used to treat hazardous wastes?

49. How is portland cement used in the treatment of hazardous wastes for disposal? What might be some disadvantages of such a use?

50. What are the advantages of aboveground disposal of hazardous wastes as opposed to burying wastes in landfills?

51. Describe and explain the best approach to managing leachate from hazardous waste disposal sites.

52. An incinerator is primarily operated to destroy chlorophenols as the POHCs fed along with other less hazardous constituents at an average rate of 10 kg/h. The exhaust gas coming from the incinerator stack at a rate of 10 m^3/min contains 1 $\mu g/m^3$ of chlorophenols. What is the DRE of the incinerator for the POHC?

53. Can phytoremediation be described as a bioremediation process? Is it a biodegradation process? If not, how can it be used to treat hazardous wastes?

LITERATURE CITED

1. Ball, Michael, and Martin Wietschel, Eds., *The Hydrogen Economy: Opportunities and Challenges*, Cambridge University Press, Cambridge, UK, 2010.
2. Wertz, Jean-Luc, Jean P. Mercier, and Olvier Bédué, *Cellulose Science and Technology*, EFPL Press, Lausanne, Switzerland, 2010.
3. Foley, Patrick M., Evan S. Beach, and Julie Zimmerman, Algae as a Source of Renewable Chemicals: Opportunities and Challenges, *Green Chemistry* **13**, 1399–1405 (2011).
4. Rosenthal, Elizabeth, Rush to Use Crops as Fuel Raises Food Prices and Hunger Fears, *New York Times*, April 6, 2011, p, A1.
5. Werpy, T., and G. Petersen, *Top Value Added Chemicals from Biomass*, U.S. Department of Energy Pacific Northwest Regional Laboratory, Richland, WA, 2004, Vol. 1, available from http://www.osti.gov/bridge (Vol. 2 on lignin 2007).
6. Genomatica a Step Closer to Biobased Butanediol for Use in Polyesters and Polyurethanes, *Plastics Today*, June 15, 2011.
7. Picataggio, Stephen, and Tom Beardslee, Verdezyne, Inc., Genetic Engineering of Microorganisms for Fermentative Preparation of Adipic Acid, International Patent Number 2011003034, January 6, 2011.
8. Schirmer, Andreas, Mathew A. Rude, Xuezhi Li, Emanuela Popova, and Stephen B. del Cardayre, Microbial Biosynthesis of Alkanes, *Science* **329**, 559–562 (2010).
9. Paterson, Ryan, *Lignin: Properties and Applications in Biotechnology and Bioenergy*, Nova Science Publishers, Hauppauge, NY, 2011.
10. Wenda, Stefanie, Sabine Illner, Annett Mell, and Udo Kragl, Industrial Biotechnology–The Future of Green Chemistry?, *Green Chemistry* **13**, 3007–3047 (2011).
11. Lopez-Gallego, Fernando, and Claudia Schmidt-Dannert, Multi-Enzymatic Synthesis, *Current Opinion in Chemical Biology* **14**, 174–183 (2010).
12. Baltz, Richard H., Julian E. Davies, and Arnold L. Demain, Eds., *Manual of Industrial Microbiology and Biotechnology*, 3rd ed., ASM Press, Washington, DC, 2010.
13. Gao, Chao, Cuiqing Ma, and Ping Xu, Biotechnological Routes Based on Lactic Acid Production from Biomass, *Biotechnology Advances* **29**, 930–939 (2011).
14. Villadsen, John, Jens Nielsen, and Gunnar Lidén, *Bioreaction Engineering Principles*, Elsevier, Berlin, 2011.
15. Keasling, Jay D., Manufacturing Molecules Through Genetic Engineering, *Science* **330**, 1355–1358 (2010).
16. Pollet, Eric, and Luc Averous, Production, Chemistry and Properties of Polyhydroxyalkanoates, in *Biopolymers: New Materials for Sustainable Films and Coatings*, David Plackett, Ed., Wiley, Hoboken, NJ, pp. 65–86.
17. Tang, Walter Z., *Physicochemical Treatment of Hazardous Wastes*, CRC Press, Boca Raton, FL, 2007.
18. Schrotter, Jean-Christophe, and Bengu Bozkaya-Schrotter, Current and Emerging Membrane Processes for Water Treatment, in *Membrane Technology: Membranes for Water Treatment*, Vol. 4, Klaus-Viktor Peinemann and Suzana Pereira Nunes, Eds., Wiley-VCH, Weinheim, Germany, 2010, pp. 53–91.
19. Dutra, A. J. P., G. P. Rocha, and F. R. Pombo, Copper Recovery and Cyanide Oxidation by Electrowinning from a Spent Copper-Cyanide Electroplating Electrolyte, *Journal of Hazardous Materials* **152**, 648–655 (2008).
20. Kolodynska, Dorotoa, The Effects of the Treatment Conditions on Metal Ions Removal in the Presence of Complexing Agents of a New Generation, *Desalination* **263** 159–169 (2010).

SUPPLEMENTARY REFERENCES

Basu, Prabir, *Biomass Gasification and Pyrolysis: Practical Design and Theory*, Academic Press, Burlington, MA, 2010.

Belgacem, Mohamed Naceur, and Alessandro Gandini, Eds., *Monomers, Polymers and Composites from Renewable Resources*, Elsevier, Amsterdam, 2008.

Blackman, William C., *Basic Hazardous Waste Management*, CRC Press/Lewis Publishers, Boca Raton, FL, 2001.

Brown, Robert C., Ed., *Thermochemical Processing of Biomass Conversion into Fuels, Chemicals and Power*, Wiley, Hoboken, NJ (2011).

Cheng, H. N., and Richard A. Gross, Eds., *Green Polymer Chemistry: Biocatalysis and Biomaterials*, American Chemical Society, Washington, DC, 2010.

Gallant, Brian, *Hazardous Waste Operations and Emergency Response Manual*, Wiley-Interscience, Hoboken, NJ, 2006.

Garrett, Theodore L., Ed., *The RCRA Practice Manual*, 2nd ed., American Bar Association, Section of Environment, Energy, and Resources, Chicago, 2004.

Haggerty, Alfred P., Ed., *Biomass Crops: Production, Energy, and the Environment*, Nova Science Publisher's, Hauppage, NY, 2011.

Hawley, Chris, *Hazardous Materials Air Monitoring and Detection Devices*, 2nd ed., Thomson/Delmar Learning, Clifton Park, NY, 2007.

Hawley, Chris, *Hazardous Materials Handbook: Awareness and Operations Levels*, Delmar Cengage, Clifton Park, NY, 2008.

Hawley, Chris, *Hazardous Materials Incidents*, 3rd ed., Thomson/Delmar Learning, Clifton Park, NY, 2008.

Henderson, Oscar P., Ed., *Biomass for Energy*, Nova Science Publishers, Hauppage, NY, 2011.

Hood, Elizabeth E., Peter Nelson, and Randy Powell, Eds., *Plant Biomass Conversion*, Wiley-Blackwell, Ames, IA, 2011.

LaGrega, Michael D., Phillip L. Buckingham, and Jeffrey C. Evans, *Hazardous Waste Management*, Waveland Press Inc, Long Grove, IL, 2010.

Meyer, Eugene, *Chemistry of Hazardous Materials*, 5th ed., Prentice-Hall, Upper Saddle River, NJ, 2009.

Nag, Ahinda, Ed., *Biofuels Refining and Performance*, McGraw-Hill, New York, 2008.

Nakaya, Andreas, *Energy Alternatives*, Reference Point Press, San Diego, CA, 2008.

Nemerow, Nelson Leonard, Ed., *Industrial Waste Treatment: Contemporary Practice and Vision for the Future*, Elsevier/Butterworth-Heinemann, Amsterdam, 2007.

Nemerow, Nelson Leonard, and Franklin J. Agardy, *Strategies of Industrial and Hazardous Waste Management*, Van Nostrand Reinhold, New York, 1998.

Network Environmental Systems Staff, *Hazardous Waste Compliance Manual*, 2nd ed., Network Environmental Systems, Folsom, CA, 2004.

Office of Technology Assessment, *Technologies and Management Strategies For Hazardous Waste Control*, University Press of the Pacific, Honolulu, 2005.

Pohanish, Richard P., *Sittig's Handbook of Toxic and Hazardous Chemicals and Carcinogens*, 5th ed., Noyes Publications, Westwood, NJ, 2007.

Romm, Joseph J., *The Hype About Hydrogen: Fact and Fiction in the Race to Save the Climate*, Island Press, Washington, DC, 2004.

Sara, Martin N., *Site Assessment and Remediation Handbook*, 2nd ed., CRC Press/Lewis Publishers, Boca Raton, FL, 2003.

Sheha, Reda R., and Hanan H. Someda, Eds., *Hazardous Waste: Classifications and Treatment Technologies*, Nova Science Publishers, New York, 2008.

Spencer, Amy Beasley, and Guy R. Colonna, *NFPA Pocket Guide to Hazardous Materials*, National Fire Protection Association, Quincy, MA, 2003.

Soetaert, Wim, and Erick Vandamme, *Biofuels*, Wiley, Hoboken, NJ, 2008.

Tang, Walter Z., *Physicochemical Treatment of Hazardous Wastes*, CRC Press, Boca Raton, FL, 2003.

Urben, P. G., Ed., *Bretherick's Handbook of Reactive Chemical Hazards*, 7th ed., Elsevier, Amsterdam, 2007.

VanGuilder, Cliff, *Hazardous Waste Management*, Mercury Learning and Information, Dulles, VA, 2011.

Vertes, Alain, Nasib Qureshi, Hideaki Yukawa, and Hans Blashek, Eds., *Biomass to Biofuels: Strategies for Global Industries*, Wiley, Hoboken, NJ, 2010.

Weiss, Charles, and William B. Bonvillian, *Structuring an Energy Technology Revolution*, MIT Press, Cambridge, MA, 2009.

Wang, Lawrence K., Nazih K. Shammas, and Yung-Tse Hung, Eds., *Advances in Hazardous Industrial Waste Treatment*, CRC/Taylor & Francis, Boca Raton, FL, 2008.

Wang, Lawrence K., Yung-Tse Hung, and Nazih K. Shammas, Eds., *Handbook of Advanced Industrial and Hazardous Wastes Treatment*, CRC Press, Boca Raton, FL, 2009.

Woodside, Gayle, *Hazardous Materials and Hazardous Waste Management*, John Wiley & Sons, New York, 1999.

Worldwatch Institute, *Biofuels for Transport: Global Potential and Implications for Energy and Agriculture*, Earthscan, Sterling, VA, 2007.

17 Sustainable Energy
The Key to Everything

17.1 ENERGY PROBLEM

It is a rather bold claim that sustainable energy is "the key to everything." But a convincing argument can be made that most problems in the physical environment can be solved to at least a large degree if enough energy is available, if it is inexpensive enough, and if it can be used without doing irreparable environmental harm. Consider the following environmental and sustainability problems that can be solved at least in part with sufficient sustainable energy:

- *Water*: With enough energy, wastewater can be reclaimed to drinking water standards by distillation, reverse osmosis, and other energy-consuming technologies, and seawater can be desalinated.
- *Food*: With enough energy, marginal land can be reclaimed by measures such as leveling, terracing, and rock removal and irrigation water can be pumped from long distances or desalinated to grow food. Greenhouses can be heated even during winter months to grow high-value specialty foods.
- *Wastes*: The disposal of hazardous organic wastes in landfills, though widely practiced, is not a good idea. With sufficient energy, such wastes can be converted to forms that cannot do harm.
- *Transportation*: With sufficient sustainable energy, transportation problems can be solved by technologies such as electrified railways.
- *Fuels*: Biomass sources of fixed carbon can be converted to hydrocarbon fuels for applications for which there are no viable alternatives (such as aircraft) without adding any net amounts of greenhouse gas carbon dioxide to the atmosphere.

The preceding list can be extended to many other areas and to a large number of sustainability and environmental problems. The great challenge is, of course, that systems of energy utilization have developed that are unsustainable. One of the most obvious sustainability challenges is that humankind will run out of the energy sources on which its economic systems are based. High petroleum prices and price fluctuations show the economic uncertainties of reliance on petroleum as an energy source, especially by nations that do not have domestic sources. There are still abundant resources of coal, but their utilization to provide for energy with current technology will certainly cause unacceptable global warming. Therefore, the great challenge facing humankind during the next several decades is to develop sources of energy that will meet energy needs and that will not ruin Earth and its climate.

There are energy alternatives to fossil fuels that can be developed, that are environmentally safe (or can be made so), and that, taken in total, can be adequate to supply energy needs.[1] These include wind, solar, biomass, geothermal, and nuclear energy sources. Some other miscellaneous sources, such as tidal energy, may contribute as well. Fossil fuels will continue to be used and may contribute sustainably for decades with sequestration of greenhouse gas carbon dioxide to reduce global warming. And, of course, energy conservation and greatly enhanced efficiency of energy use will make substantial contributions. This chapter discusses the energy alternatives listed above with emphasis on energy sustainability.

17.2 NATURE OF ENERGY

Energy is the capacity to do work (basically, to move matter around) or **heat** in the form of the movement of atoms and molecules. **Kinetic energy** is contained in moving objects. One such is the energy contained in a rapidly spinning flywheel, a device with significant potential for energy storage to even out the energy flow from intermittent solar and wind sources. **Potential energy** is stored energy, such as in an elevated reservoir of water used as a means of storing hydroelectric energy for later use that can be run through a hydroelectric turbine to generate electricity as needed.

A very important form of potential energy is **chemical energy** stored in the bonds of molecules and released, usually as heat, when chemical reactions occur. For example, in the case of methane, CH_4, in natural gas, when the methane burns

$$CH_4 + 2O_2 \rightarrow CO_2 + 2H_2O \tag{17.1}$$

the difference between the bond energies in the CO_2 and H_2O products and the CH_4 and O_2 reactants is released, primarily in the form of heat. If the heat is released by combustion of methane in a gas turbine, part of the heat energy can be converted to **mechanical energy** in the form of the rapidly spinning turbine and an electrical generator to which it is attached. The generator, in turn, converts the mechanical energy to **electrical energy**.

The standard unit of energy is the **joule**, abbreviated as **J**. A total of 4.184 J of heat energy will raise the temperature of 1 g of liquid water by 1°C. This amount of heat is equal to 1 **calorie** of energy (1 cal = 4.184 J), the unit of energy formerly used in scientific work. [The "calorie" commonly used to express the energy value of food (and its potential to produce fat) is actually a kilocalorie, kcal, equal to 1000 cal.] A joule is a small unit, and the kilojoule, kJ, equal to 1000 J, is widely used in describing chemical processes. The exajoule (EJ), equal to 1×10^{18} J, is used to express large quantities of energy; Earth's total primary energy consumption is about 450 EJ/year.

Power refers to energy generated, transmitted, or used per unit time. The unit of power is the **watt**, equal to an energy flux of 1 J/s. A compact fluorescent light bulb adequate to illuminate a desk area might have a rating of 21 watts. A large power plant may put out electricity at a power level of 1000 **megawatts** (MW, where 1 MW is equal to 1 million watts). Power on a national or global scale is often expressed in **gigawatts** (GW), each one of which is equal to a billion watts or even **terawatts** (TW), where a TW is equal to a trillion watts.

The science that deals with energy in its various forms and with work is **thermodynamics**. There are some important laws of thermodynamics. The **first law of thermodynamics** states that energy is neither created nor destroyed. This law is also known as the **law of conservation of energy**. The first law of thermodynamics must always be kept in mind in the practice of green technology, the best practice of which requires the most efficient use of energy. Thermodynamics enables calculation of the amount of usable energy. As described by the laws of thermodynamics, only a relatively small amount of the potential energy in fuel can be converted to mechanical energy or electrical energy with the remaining energy from the combustion of the fuel dissipated as heat. With the application of green technology, much of this heat is salvaged for applications such as district heating of homes.

Although energy is neither created nor destroyed, the amount of useful energy that can be obtained from a system tends to become dissipated. Useful mechanical energy can be produced, for example, in a heat engine by harnessing part of the energy flow from a hot part to a cooler part of the system (see Section 17.5 and Equation 17.2). The amount of this useful energy is called **exergy** and it can go to zero in a system that has reached equilibrium.

17.3   SUSTAINABLE ENERGY: AWAY FROM THE SUN AND BACK AGAIN

The key to sustainability is abundant, environmentally safe energy. The evolution of humankind's utilization of energy is illustrated in Figure 17.1. Until very recently in the history of humankind, we have depended on the sun to meet our energy needs. The sun has kept most of the landmass of Earth at a temperature that enables human life to exist. Solar radiation has provided the energy for photosynthesis that converts atmospheric carbon dioxide to plant biomass providing humans with food, fiber, and wood employed for dwelling construction and fuel. Animals feeding on this biomass provided meat for food and hides and wool that humans used for clothing. Eventually, humans developed means of indirectly using solar energy. This was especially true of wind driven by solar heating of air masses and used to propel sailing vessels and eventually to power windmills employed for power. The solar-powered hydrologic cycle provided flowing water, the energy of which was harnessed by water wheels. Virtually all the necessities of life came from utilization of solar energy.

17.3.1   THE BRIEF ERA OF FOSSIL FUELS

Dating from around 1800, humans began to exploit fossil fuels for their energy needs. Initially, coal was burned for heating and to power newly developed steam engines for mechanical energy used in manufacturing, steamships, and steam locomotives. After about 1900, petroleum developed rapidly as a source of fuel and, with the development of the internal combustion engine, became the energy source of choice for transportation needs. Somewhat later, natural gas developed as an energy source. The result was a massive shift from solar and biomass energy sources to fossil fuels.

Utilization of fossil carbon-based materials resulted in a revolution that went far beyond just energy utilization. One important example was the invention by Carl Bosch and Fritz Haber in Germany in the early 1900s of a process for converting atmospheric elemental nitrogen from air to ammonia, NH_3, by the reaction of N_2 with H_2. This high-pressure, high-temperature process required large amounts of fossil fuel to provide energy and to react with steam to produce elemental

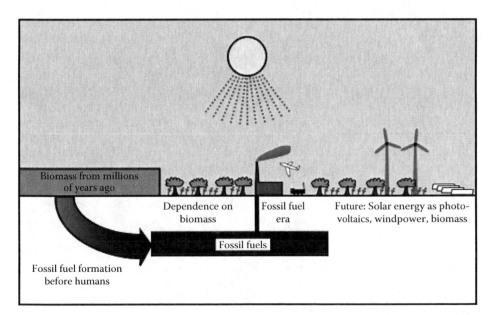

FIGURE 17.1   Evolution of the use of energy from solar and biomass sources through the brief, but spectacular, era of fossil fuels and on to renewable, largely solar-based sources.

hydrogen. The discovery of synthetic nitrogen fixation enabled the production of huge quantities of relatively inexpensive nitrogen fertilizer, and the resulting increase in agricultural production may well have saved Europe, with a rapidly developing population at the time, from widespread starvation. (It also enabled the facile synthesis of great quantities of nitrogen-based explosives that killed millions of people in World War I and subsequent conflicts.) Fossil fuel, in a sense "fossilized sunshine," resulted in an era of unprecedented material prosperity and an increase in human population from around 1 billion to over 6 billion.

By the year 2000, it had become obvious that the era of fossil fuels was not sustainable. One reason is that fossil fuel is a depleting resource that cannot last indefinitely as the major source of energy for the industrial society to which it has led. Although new discoveries of petroleum and new techniques for utilizing previously inaccessible petroleum sources in recent years have somewhat eased shortages of oil, a large fraction of the world's total petroleum resource has already been consumed so that petroleum will continue to become more scarce and expensive and can last for only a limited time as the dominant fuel and organic chemicals' raw material. Coal is much more abundant, but its utilization leads to the second reason that the era of fossil fuels must end because it is the major source of anthropogenic atmospheric carbon dioxide, greatly increased levels of which will almost certainly lead to global warming and massive climate change. Natural gas (methane, CH_4) is an ideal, clean-burning fossil fuel that produces the least amount of carbon dioxide per unit energy generated. Rapidly expanding new discoveries of natural gas largely from previously inaccessible tight shale formations means that it can serve as a "bridging fuel" for several decades until other sources can be developed. Nuclear energy, properly used with nuclear fuel reprocessing, can take on a greater share of energy production, especially for base load electricity generation. But it is clear that drastic shifts must occur in the ways in which energy is obtained and used.

17.3.2 BACK TO THE SUN

With the closing of the brief but spectacular era of fossil hydrocarbons, the story of humankind and its relationship to Planet Earth is becoming one of "from the sun to fossil fuels and back again" as humankind returns to the sun as the dominant source of energy and photosynthetic energy to convert atmospheric carbon dioxide to biomass raw materials. In addition to direct uses for solar heating and for photovoltaic power generation, there is enormous potential to use the sun for the production of energy and material. Arguably, the fastest growing energy source in the world is wind-generated electricity. The wind is produced when the sun heats masses of air causing the air to expand. Once the dominant source of energy and materials, biomass produced by solar-powered photosynthesis is beginning to live up to its potential as a source of feedstocks to replace petroleum in petrochemicals manufacture and of energy in synthetic fuels. Biomass is still evolving as a practical source of liquid fuels. The main two of these are fermentation to produce ethanol and synthesis of biodiesel fuel made from plant lipid oils. Unfortunately, although ethanol made from sugar derived from sugarcane that grows prolifically in some areas such as Brazil is an economical gasoline substitute, the net energy gain from ethanol derived from cornstarch relies on the grain, the most valuable part of the plant that is otherwise used for food and animal feed, and the net energy gain from corn-based ethanol is marginal at best. The economics of producing synthetic biodiesel fuel from sources such as soybeans may be somewhat better. However, production of this fuel from oil palm trees in countries such as Malaysia is resulting in destruction of rain forests and diversion of palm oil from the food supply. However, as discussed in Section 17.16, practical means do exist to utilize biomass for energy and materials without seriously disrupting the food supply

Future scientific discoveries and technological advances will play key roles in the achievement of energy sustainability. Three areas in which Nobel-level breakthroughs are needed in the achievement of energy sustainability were expressed in a February 2009 interview by Dr. Steven Chu, a

Nobel Prize-winning physicist who had just been appointed Secretary of Energy in U.S. President Barack Obama's new administration. The first of these is in solar power in which the efficiency of solar energy capture and conversion to electricity needed to improve several-fold. A second area of need is for improved electric batteries to store electrical energy generated by renewable means and to enable practical driving ranges in electric vehicles. A third area in need of a quantum leap is for improved crops capable of converting solar energy to chemical energy in biomass by photosynthesis at much higher efficiencies. In this case, the potential for improvement is enormous because most plants convert less than 1% of the solar energy falling on them to chemical energy through photosynthesis. Through genetic engineering, it is likely that this efficiency could be improved several-fold leading to vastly increased generation of biomass. Clearly, the achievement of sustainability employing high-level scientific developments will be an exciting development in decades to come.

17.4   SOURCES OF ENERGY USED IN THE ANTHROSPHERE: PRESENT AND FUTURE

As of 2011, world power consumption stood at 17.8 trillion watts or 17.8 TW. Figure 17.2 shows U.S. and world energy sources used annually as of the year 2000. The predominance of **fossil fuel** petroleum, natural gas, and coal is obvious. The potential of various energy alternatives is presented here. An outstanding and comprehensive survey of world energy resources is given in a detailed publication on that topic.[2]

World energy resources may be divided between **finite resources**, of which only a limited amount will ever be available, and **perpetual resources**, which are continuously being renewed (Table 17.1). Finite resources are predominantly fossil fuels including coal and petroleum as well as uranium and thorium, which can be used as nuclear reactor fuel. Perpetual resources are mostly based on solar energy, which provides energy directly for solar heating or solar voltaic cells or indirectly such as from wind or ocean waves. Other perpetual sources include geothermal or tidal energy. Intermediate sources include peat, which can be harvested from deposits that are thousands of years old, but is also being renewed by fresh plant material. If it could ever be developed as a practical energy source, nuclear fusion power from the hydrogen isotope deuterium would be finite in that a fixed quantity of deuterium is present in the Earth System, but in a sense is a perpetual resource in that the amount is so vast that there would never be a chance of running out of it.

Since about 2000, two sources of energy have emerged as growth areas in the energy area, good news with respect to sustainability. The first of these is natural gas, new discoveries of which from formerly inaccessible sources have led to spectacular increases in supply. The second

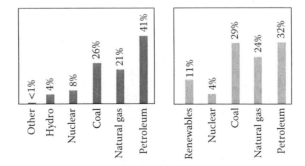

FIGURE 17.2   U.S. (left) and world (right) sources of energy. Percentages of total rounded to the nearest 1%. Figures for the United States are as of 2000 and those from the world are those projected for 2015. Since 2000, the fraction of U.S. energy provided by "other" sources has become greater because of the development of renewable sources, especially wind power.

TABLE 17.1

Earth's Major Energy Resources[a]

Resource	Availability and Potential
Finite resources	
Coal	World reserves of about 860 billion metric tons, about half bituminous coal, remainder subbituminous, lignite brown coal
Crude oil and natural gas hydrocarbon liquids	Proven world reserves of about 1.2 billion barrels, about a 40-year supply at current production rate of 85 million barrels per day
Shale oil	From heating oil shale containing organic kerogen, world resources around 5 trillion barrels, but not now produced on a large scale
Extra-heavy oil and bitumen	Viscous hydrocarbon containing heavy metals and chemically bound nitrogen, oxygen, and sulfur, as much as 5.5 trillion barrels may be available globally
Peat	Plant remains accumulated under cold, wet conditions or in some tropical conditions, a global peat resource of around 500 billion tons (of carbon), somewhat renewable
Uranium	Identified world resources of around 5.4 million metric tons
Perpetual resources	
Hydropower	875 gigawatts (875,000 megawatts) world capacity in 2008
Bioenergy	About 50 exajoules (EJ, 1 EJ = 10^{18} joules) worldwide annually, or about 10% world's total primary energy consumption of about 450 EJ
Solar energy	Potentially vast energy resource, total solar energy falling on Earth's surface about 7500 times primary energy consumption
Geothermal energy	Realistic estimate of up to 210 gigawatts of electricity
Wind energy	A total of 239 gigawatts capacity in 2011
Tidal energy	A 240-MW tidal power plant at La Rance in Brittany, France has operated since 1968 and a 254-MW tidal plant began operation in Sihwa, South Korea, in 2010; potential for many more
Wave power	Potential for power equal to total world electricity production
Ocean thermal power	Generate mechanical energy based on temperature differences of at least 20°C between warm surface water and deep ocean water, no operational units but immense potential

[a] Joule (J) is the basic unit of energy; an exajoule (EJ) is 1×10^{18} joules; watt (W) is the basic unit of power (energy per unit time); a megawatt (MW) is 1000 W and a gigawatt is 1000 MW; 1 gigawatt year is equivalent to 0.03156 EJ.

is wind power, which has developed very rapidly and has become economically competitive with other sources of energy. Fortunately, both of these energy sources are favorable from the sustainability viewpoint. Natural gas is the best fossil fuel from the standpoint of emissions including those of greenhouse warming carbon dioxide. Wind energy is completely renewable and arguably has the least environmental impact of all the perpetual resources. The footprint of wind turbines is quite small and they can be installed on agricultural land with virtually no impact on its food-producing capacity. Unlike water power, which is the most commonly used perpetual resource, wind power does not involve putting dams on waterways and the damage to aquatic ecosystems that may cause.

17.5   ENERGY DEVICES AND CONVERSIONS

Energy occurs in many forms and its utilization requires conversion to other forms. Many devices exist for the utilization of energy and its conversion to other forms. The most common of these are shown in Figure 17.3. The types of energy available, the forms in which it is utilized, and the processes by which it is converted to other forms have a number of implications for green technology and sustainability. For example, the wind turbine shown in Figure 17.3a, once in place, continues to pump electricity into the power grid with virtually no harm to the environment (though some people regard them as unsightly while others think they are picturesque), whereas the steam power plant shown in Figure 17.3b requires mining depletable coal, combustion of the fossil fuel with its potential for air pollution, control of air pollutants, and means for cooling the steam exiting the turbine, with its potential for thermal pollution of waterways.

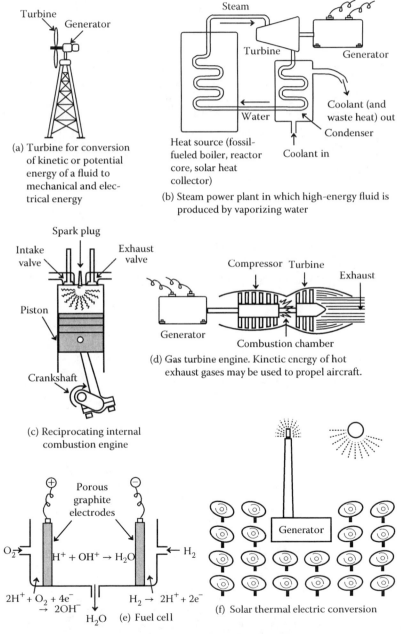

FIGURE 17.3 Examples of many devices for the collection of energy and its conversion to other forms.

An important aspect of energy utilization is its conversion to usable forms. For example, gasoline burned in automobile engines comes originally from petroleum pumped from underground, the petroleum constituents are separated, molecules with properties suitable for engine fuel are produced chemically, the gasoline product is burned in an internal combustion engine converting chemical energy to mechanical energy, and the mechanical energy is transmitted to the wheels of the automobile in the form of kinetic energy that moves the automobile. Significantly less than half of the energy in the gasoline is actually converted to mechanical energy of the automobile's motion; the rest is dissipated as waste heat through the engine's cooling system.

Figure 17.4 illustrates major forms of energy and conversions between them. A significant point of this illustration is the very large ranges of energy conversion efficiencies from just a few percent or less to almost 100%. These differences suggest areas in which improvements may be sought. One of the most striking inefficiencies is the less than 0.5% conversion of light energy to chemical energy by photosynthesis. Despite such a low conversion efficiency, photosynthesis has generated the fossil fuels from which industrialized societies now get their energy and provides a significant fraction of energy in areas where wood and agricultural wastes are used. Doubling photosynthesis efficiency with genetically engineered plants could be a major factor in making biomass a more desirable energy source. Replacement of woefully inefficient incandescent light bulbs with fluorescent bulbs that are five to six times more efficient in converting electrical energy to light can save large amounts of energy.

A particularly important energy conversion carried out in the anthrosphere is that of heat, such as from chemical combustion of fuel, to mechanical energy that is used to propel a vehicle or run an electrical generator. This occurs, for example, when gasoline in a gasoline engine burns, generating hot gases that move pistons in the engine connected to a crankshaft, converting the up-and-down movement of the piston to rotary motion that drives a vehicle's wheels. It also occurs when hot steam generated at high pressure in a boiler flows through a turbine connected directly to an electrical generator. A device, such as a steam turbine, in which heat energy is converted to mechanical energy is called a heat engine. Unfortunately, the laws of thermodynamics dictate

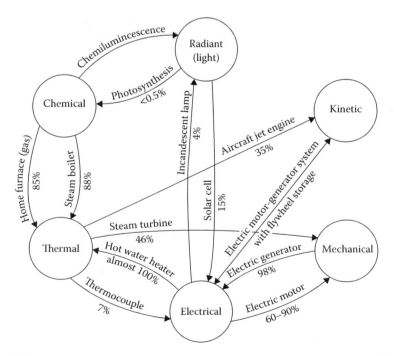

FIGURE 17.4 Major types of energy and conversions between them showing conversion efficiencies.

that the conversion of heat to mechanical energy is always much less than 100% efficient. The efficiency of this conversion is given by the Carnot equation

$$\text{Percent efficiency} = \frac{T_1 - T_2}{T_1} \times 100 \tag{17.2}$$

in which T_1 is the inlet temperature (e.g., of steam into a steam turbine) and T_2 is the outlet temperature, both expressed in Kelvin (°C + 273). Consider a steam turbine as shown in Figure 17.5. Substitution into the Carnot equation of 875 K for T_1 and 335 K for T_2 gives a maximum theoretical efficiency of 62%. However, it is not possible to introduce all the steam at the highest temperature and friction losses occur so that the energy conversion efficiency of most modern steam turbines is just below 50%. About 80% of the chemical energy released by combustion of fossil fuel in a boiler is actually transferred to water to produce steam so that the net efficiency for conversion of chemical energy in fossil fuels to mechanical energy to produce electricity is about 40%. The overall conversion of chemical energy to electricity is essentially the same because an electrical generator converts virtually all of the energy of a rotating turbine to electricity. Because nuclear reactor peak temperatures are limited for safety reasons, their conversion of nuclear energy to electricity is only about 30%.

A particularly important machine for converting chemical energy to mechanical energy is the internal combustion piston engine shown in Figure 17.6. Most internal combustion engines operate

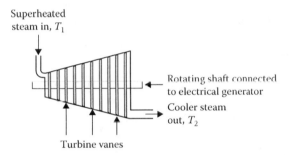

FIGURE 17.5 In a steam turbine, superheated steam impinges on vanes attached to a shaft to produce mechanical energy. For generation of electricity, the shaft is coupled to an electrical generator.

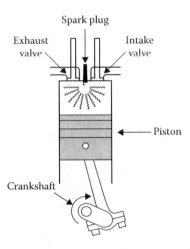

FIGURE 17.6 An internal combustion piston engine in which a very rapidly burning mixture of air and fuel drives a piston downward during the power stroke and this motion is converted to rotary mechanical motion by the crankshaft.

on a cycle of four strokes. In the first of these, the piston moves downward drawing air or an air–fuel mixture into the cylinder. Next, with both valves closed, the air or air–fuel mixture is compressed as the piston moves upward. With the piston near the top of the cylinder (a point at which fuel may be injected if only air is compressed), ignition occurs and the burning fuel creates a mass of highly pressurized combustion gas in the cylinder, which drives the piston down in the third stroke. The exhaust valve then opens and the exhaust gas is expelled as the piston moves upward during the exhaust stroke.

The efficiency of the internal combustion engine increases with the peak temperature reached by the burning fuel, which increases with the degree of compression during the compression stroke. This temperature is highest for the diesel engine in which the compression is so high (around 20:1) that fuel injected into the combustion chamber ignites without a spark plug ignition source. Whereas a standard gasoline engine is typically about 25% efficient in converting chemical energy in fuel to mechanical energy, a diesel engine is typically 37% efficient, with some reaching higher values.

Although highly superior from the standpoint of efficiency, diesel engines do have some disadvantages with respect to emissions. The first of these is that the combustion zone is not homogeneous because the fuel is injected into the highly compressed air at the top of the compression stroke resulting in incomplete combustion and production of carbon particles; improperly adjusted diesel engines are a major source of particle air pollution in urban areas. In addition, because of their very high combustion temperatures and high ignition pressures, diesel engines tend to produce elevated levels of air pollutant nitrogen oxides. However, recent advances in diesel engine design, computerized control, and exhaust pollutant control devices have greatly reduced diesel engine emissions.

17.5.1 Fuel Cells

Fuel cells are devices that convert the energy released by electrochemical reactions directly to electricity without going through a combustion process and electricity generator. Fuel cells are the primary means for utilizing hydrogen fuel (see Section 17.18) and are becoming more common as electrical generators. A fuel cell has an anode at which elemental hydrogen is oxidized, releasing electrons to an external circuit, and a cathode at which elemental oxygen is reduced by electrons introduced from the external circuit, as shown by the half-reactions in Figure 17.7. The H^+ ions generated at the anode migrate to the cathode through a solid membrane permeable to protons. The net reaction is

$$2H_2 + O_2 \rightarrow 2H_2O + \text{electrical energy} \tag{17.3}$$

and the only product of the fuel cell reactions is water.

Although elemental hydrogen is the ultimate fuel for fuel cells, it may be produced by the chemical breakdown of hydrogen-rich fuels, such as methane, methanol, or even gasoline, a process that also generates carbon dioxide. Tubular-style solid-oxide fuel cells, such as those manufactured by Siemens Westinghouse, operate at an elevated temperature of about 1000°C and produce an exhaust that is hot enough to drive a turbine or even to cogenerate steam. Such systems may be able to develop overall efficiencies of up to 80%.

17.6 GREEN TECHNOLOGY AND ENERGY CONVERSION EFFICIENCY

One of the best ways to conserve fuel resources is through increasing the efficiency of energy conversion including that of chemical energy to mechanical energy with the intermediate step of production of heat energy. Many advances have been made in this area since the late 1800s.

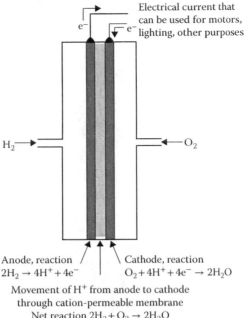

Electrical current that
can be used for motors,
lighting, other purposes

e^- e^-

$H_2 \rightarrow$ $\leftarrow O_2$

Anode, reaction Cathode, reaction
$2H_2 \rightarrow 4H^+ + 4e^-$ $O_2 + 4H^+ + 4e^- \rightarrow 2H_2O$

Movement of H^+ from anode to cathode
through cation-permeable membrane
Net reaction $2H_2 + O_2 \rightarrow 2H_2O$

FIGURE 17.7 Cross-sectional diagram of a fuel cell in which elemental hydrogen can be reacted with elemental oxygen to produce electricity directly with water as the only chemical product.

Part of the increase in conversion of fuel energy to electricity going from around 4% conversion in 1900 to more than 40% in the most modern engines resulted from increasing the input temperature (T_1 in the Carnot equation) in the heat engines driving electrical generators. Energy use efficiency increased by more than fourfold when picturesque steam engines were replaced by diesel/electric locomotives during the 1940s and 1950s. Substitution of diesel engines for gasoline engines in trucks and farm and construction equipment has resulted in gains in energy efficiency.

Much of the increased efficiency in fuel utilization has come from improved materials that allow higher operating temperatures. In addition to high-temperature-tolerant metals in engines, a contribution has been made by lubricating oils that do not break down at high temperatures. Much of the progress has been achieved with better engineering, now greatly aided by computerized design, evaluation, and manufacturing of engines. Engineers of a century ago had never heard of green technology and probably would not have cared had they known about it. But they did understand costs of fuel (which on the basis of constant value currency were often higher then than they are now) and they welcomed the greater efficiencies they achieved on the basis of costs.

A key aspect of the most efficient conversion of chemical energy to mechanical energy in engines is the precise control of operational aspects such as ignition timing, valve timing, and fuel injection. In modern engines, key operating parameters are controlled by computer leading to optimum efficiency in engine operation.

As an inevitable consequence of the thermodynamics described by the Carnot equation, engines that convert heat to mechanical energy cannot utilize much of the heat, and the waste heat is carried away by an engine cooling system. Typically, a small portion of this heat is used in automotive heaters on cold days. On a broader scale, such as municipal electrical systems, this heat can be used for heating buildings. Such efficiencies are discussed in Section 17.19.

17.7 ENERGY CONSERVATION AND RENEWABLE ENERGY SOURCES

Any consideration of energy needs and production must take energy conservation into consideration. This does not have to mean cold classrooms with thermostats set at 60°F in mid-winter, nor swelteringly hot homes with no air-conditioning, nor total reliance on the bicycle for transportation, although these and even more severe conditions are routine in many countries. The fact remains that the United States and several other industrialized nations have wasted energy at a deplorable rate. For example, U.S. energy consumption is higher per capita than that of some other countries that have equal, or significantly better, living standards. Obviously, a great deal of potential exists for energy conservation that will ease the energy problem.

Efficient use of energy can in fact correlate positively with higher economic standards. Figure 17.8 shows a plot of the ratio of energy use per unit of gross domestic product (GDP) in developed industrialized nations and illustrates a steady and favorable decrease of energy required relative to economic output. Whereas in 2000, 1.7 barrels of oil equivalents were required per $1000 GDP in developed nations, the corresponding figure for developing nations lacking advanced means of using energy efficiently was 5.2 barrels or three times as much. These figures indicate the substantial potential for decreased energy consumption by energy-conscious development of the less industrially advanced nations as well as the additional conservation that can be achieved if citizens of industrialized nations can be persuaded to forego wasteful energy practices such as excessively large and inefficient vehicles and overly large dwellings.

Transportation is the economic sector with the greatest potential for increased efficiencies. The private auto and airplane are only about one-third as efficient as buses or trains for transportation. Transportation of freight by truck requires about 3800 Btu/ton-mile compared to only 670 Btu/ton-mile for a train. Truck transport is terribly inefficient compared to rail transport (as well as dangerous, labor-intensive, and environmentally disruptive). Major shifts in current modes of transportation in the United States will not come without anguish, but energy conservation dictates that they be made.

Figure 17.9 shows the trend in U.S. automobile fuel economy during recent decades. The gains through about 1990 were very impressive, but then dropped off as less fuel-efficient vehicles became more popular. If the same trends from this period would have been maintained, the U.S. automobile fleet would by 2012 would have been close to 40 miles per gallon (MPG). Such a figure is readily achievable without seriously compromising safety or comfort and, as is obvious from Figure 17.9, with much lower emissions from pollutants compared to 1970. In 2007, the U.S. Congress passed legislation mandating higher fuel economy standards for vehicles sold in the United States.

Household and commercial uses of energy are relatively efficient. Here again, appreciable savings can be made. The all-electric home requires much more energy (considering the percentage

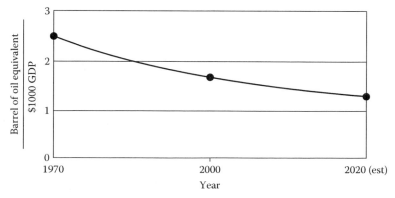

FIGURE 17.8 Plot of barrel of oil equivalent required per $1000 gross domestic product (GDP) as a function of year in industrialized nations.

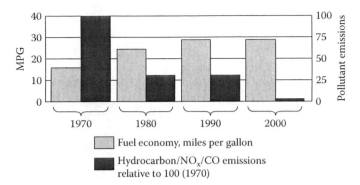

FIGURE 17.9   U.S. auto fleet fuel economy and emissions over 3 decades. Fuel economy has improved markedly while emissions have been greatly reduced.

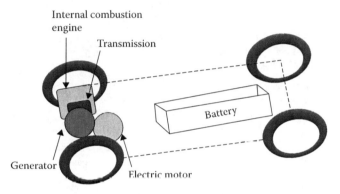

FIGURE 17.10   Major components of a hybrid automobile in which an internal combustion engine drives a generator to produce electricity to propel a vehicle. Electrical energy is stored in a battery. Electrical energy is also salvaged for storage from braking the vehicle.

wasted in generating electricity) than a home heated with fossil fuels. The sprawling ranch house–style home uses much more energy per person than does an apartment unit, row house, or even a home of comparable floor area built in a compact format (more like a square box). Improved insulation, sealing around the windows, and other measures can conserve a great deal of energy. Electric-generating plants centrally located in cities can provide waste heat for commercial and residential heating and cooling and, with proper pollution control, can use municipal refuse for part of their fuel, thus reducing quantities of solid wastes requiring disposal.

One of the greatest contributions to energy conservation and energy use efficiency in very recent years has been the **hybrid vehicle** that uses an internal combustion engine to produce electricity that is stored for propulsion of the vehicle in a nickel-metal-hydride battery and, in latest models, lithium-ion batteries (Figure 17.10). Although not greatly more efficient for prolonged driving at highway speeds, these vehicles have achieved improvements up to 50% in stop-and-go driving in traffic. For routine operation, the internal combustion engine supplies all the power needed plus additional power, if required, to run the generator to recharge a battery (larger than the battery in a conventional automobile but significantly smaller than the battery in an all-electric vehicle). When a surge of power is required, electricity from the storage battery drives the electric motor to produce the additional power. The braking system also generates electricity that is stored in the battery. When the vehicle is stopped, the internal combustion engine does not run, which also saves fuel.

Vastly reduced fuel consumption can be achieved by hybrid vehicles now becoming more popular that have batteries that can be charged by an external electrical source as well as the onboard engine. Such vehicles give around 40–50 km driving range from an overnight charge, alone, sufficient for the majority of routine driving. As the battery runs down, the internal combustion engine recharges it.

Although gasoline engines are now employed in hybrid vehicles, even greater fuel economy could be achieved with an inherently more efficient diesel engine as the internal combustion engine component. By allowing the diesel engine to run at a generally steady rate, the output of exhaust pollutants, which are produced at higher levels by diesel engines as the engine speed is changed, could be greatly reduced. Furthermore, diesel engines idle with remarkably little fuel consumption, so that the diesel engine would not need to be turned off when the vehicle is stopped, thus staying hot and further reducing emissions when it is brought up to speed.

Still greater fuel economy could be achieved employing a stratified charge engine.[3] It might be feasible to design a "three-way hybrid" vehicle with a battery and a dual-fuel system of methane and diesel fuel. With the proposed engine system, a methane–air mixture, too lean in methane to allow for spark ignition, is taken into the engine cylinder and highly compressed as in a conventional diesel engine. At the top of the compression stroke, a minimum amount of diesel fuel is injected that causes ignition of the mixture for the power stroke. Although having two fuel sources would be a complication, such an engine could be designed to run temporarily on a relatively richer spark-fired mixture of air and methane if it ran out of diesel fuel or in the conventional diesel mode if it were out of methane. The engine could be a three-cylinder air-cooled unit, which would run hotter and thus more efficiently than a fluid-cooled unit, and three cylinders would be adequate, but relatively less costly and complicated than a larger engine.

As scientists and engineers undertake the crucial task of developing alternative energy sources to replace dwindling petroleum and natural gas supplies, energy conservation must receive proper emphasis. In fact, zero energy-use growth, at least on a per capita basis, is a worthwhile and achievable goal. Such a policy would go a long way toward solving many environmental problems. With ingenuity, planning, and proper management, it could be achieved while increasing the standard of living and quality of life.

Closely related to energy conservation is the concept of **renewable energy** from sources that do not run out. Essentially, all of these depend on energy from the sun including direct solar energy, wind driven by the solar heating of air masses, falling water from the solar-powered hydrologic cycle, and biomass formed from photosynthesis. For most of its lifetime on Earth, humankind has depended entirely on renewable sources of energy, and most countries are again emphasizing these sources.

Enlightened sustainable energy policies are being implemented in many developing countries. China implemented a new law on renewable energy at the beginning of 2006 and updated in 2009. This policy encourages renewable energy alternatives including wind power, biomass energy, and biomethane generation.[4] Long known for its utilization of wastes (including even use of human wastes as fertilizer for growing vegetables), China has constructed many waste-to-methane generators in rural areas with 40 million rural households served by such facilities by 2011. Experimental biopower projects burning crop biomass by-products have been undertaken and China's biomass power capacity reached 3.2 GW in 2009. As of 2006, China had 80 million square meters of solar collectors to heat water, equivalent to the energy from 10 million tons per year of coal. China's total wind power capacity reached 45 GW in 2011, about 5% of its total electricity-generating capacity. Combined with at least 200 GW of hydropower and much smaller amounts of biomass and grid-connected photovoltaic systems, China now gets about one-fourth of its electrical power from other than fossil fuel sources (including nuclear power).

17.8 PETROLEUM HYDROCARBONS AND NATURAL GAS LIQUIDS

Currently, the largest source of energy that is used in the anthrosphere consists of hydrocarbon liquids, the majority of which are composed of hydrocarbons pumped from the ground as liquid petroleum. An increasing share of these liquids now comes from heavy oil and bitumen, which, in the case of tar sand oil from Canada, have to be mined and processed above ground. The various sources of hydrocarbon liquids are discussed in this section.

Liquid **petroleum** occurs in rock formations ranging in porosity from 10% to 30%. Up to half of the pore space is occupied by water. The oil in these formations must flow over long distances to an approximately 15-cm-diameter well from which it is pumped. The rate of flow depends on the permeability of the rock formation, the viscosity of the oil, the driving pressure behind the oil, and other factors. Because of limitations in these factors, **primary recovery** of oil yields an average of about 30% of the oil in the formation, although it is sometimes as little as 15%. More oil can be obtained using **secondary recovery** techniques, which involve forcing water under pressure into the oil-bearing formation to drive the oil out. Primary and secondary recovery together typically extract somewhat less than 50% of the oil from a formation. Finally, **tertiary recovery** can be used to extract even more oil, normally through the injection of pressurized carbon dioxide, which forms a mobile solution with the oil and allows it to flow more easily to the well. Other chemicals, such as detergents, may be used to aid in tertiary recovery. Currently, about 300 billion barrels of U.S. oil are not available through primary recovery alone. A recovery efficiency of 60% through secondary or tertiary techniques could double the amount of available petroleum. Much of this would come from fields that have already been abandoned or essentially exhausted using primary recovery techniques.

Proven world reserves of crude oil and natural gas liquids are about 1.2 trillion barrels. An annual production rate of 85 million barrels per day provides an estimated 40-year supply of these hydrocarbons. However, these hydrocarbons will be used much longer than that because more will be found, substitutes will be developed, and use will decrease as supply limits are approached.

Estimates of crude oil resources around the world are in a state of flux and have generally increased since 2000 with substantial new discoveries. The high Arctic regions across the northern parts of Alaska, Canada, Norway, and Russia are believed to contain very large petroleum resources. The largest oil find in 2011 was that of an estimated 250 million barrels of premium sweet crude oil beneath the Barents Sea located north of Norway and Russia. The major challenge to petroleum development in the Arctic is to produce petroleum in the challenging, often brutally cold, surroundings and to do so without harming the fragile Arctic environment. With horizontal drilling and hydraulic fracturing of oil-bearing formations, the large tight shale Bakken formation in North Dakota was undergoing explosive growth in crude oil production by 2011 reaching levels of 400,000 barrels per day at the beginning of 2012, the third ranking state. (The top-ranking state for oil production in the United States is Texas at around 1.1 million barrels per day and California is third at around 567,000 barrels per day. Somewhat surprisingly, the United States is third in the world for petroleum production at 9.1 million barrels per day, just behind Russia at 9.9 million barrels per day and Saudi Arabia at 9.7 million barrels per day.)

17.8.1 Heavy Oil

Natural bitumen is a tarry material that has been mined for centuries and used as a sealant and road paving material. Natural bitumen and **extra-heavy oil** are the remains of biodegradation of the lighter fractions of crude oil and are relatively dense and viscous hydrocarbon materials containing heavy metals and chemically bound nitrogen, oxygen, and sulfur. Extra-heavy oil presents challenges to petroleum refiners largely because it requires coking (pyrolysis that leaves a carbon solid residue) to obtain hydrocarbon liquids, but is being used increasingly because of its availability and competitive price. A total of about 5.5 trillion barrels of bitumen and heavy oil may be available worldwide.

As of 2012, the largest single source of oil available to the United States was heavy oil from tar sands mined and processed in Alberta, Canada. This hydrocarbon source is environmentally troubling because of the large amount of solid material that is dug up and moved for processing to remove the hydrocarbons and the huge residues of sand left from the hot water extraction of the heavy oil. A more environmentally acceptable alternative that is being developed is **steam-assisted gravity drainage** as illustrated in Figure 17.11 for extraction of heated bitumen from a sandstone formation. This process takes advantage of developments in the area of horizontal drilling that have been refined in recent

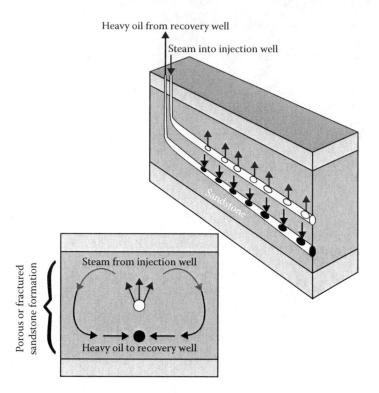

FIGURE 17.11   Steam-assisted gravity drainage for in situ extraction of steam-heated bitumen from a sandstone formation.

years. For extraction of heavy oil, two horizontal wellbores are drilled with the upper one approximately 5 m directly above the lower in the bitumen-bearing formation. Steam injected into the upper borehole heats the bitumen, which liquefies it and enables it to flow by gravity into the lower bore hole. The flow may be assisted by additives that have a detergent and mobilizing effect on the bitumen.

17.8.2   SHALE OIL

Shale oil is a mixture consisting primarily of hydrocarbons, but containing hetero elements including nitrogen compounds as well, that is produced by heating fine-grained sedimentary oil shales to about 500°C. The carbonaceous material in oil shale is a substance called **kerogen**. Shale oil is more costly and environmentally damaging than liquid petroleum because the rock is mined and processed in retorts releasing enormous quantities of greenhouse gas carbon dioxide and leaving a residue with a tendency to leach salts into water. Optimistic predictions of large-scale shale oil production in the United States from around the early 1970s have fallen flat because of unfavorable economics and environmental impact. There has been some limited production in Brazil, China, Estonia, Germany, and Israel. Nevertheless, estimates of shale oil potentially available in the world are enormous, around 5 trillion barrels.

17.8.3   NATURAL GAS LIQUIDS

The **natural gas liquids** that can be condensed from natural gas (methane) include ethane, propane, butanes, pentanes, and natural gasoline. All of these hydrocarbons are clean-burning fuels. Ethane can be converted to ethylene, C_2H_4, an alkene with a double bond that is a valuable raw material for the production of polyethylene plastic and several other important petrochemicals. With the rapid development of natural gas resources since about 2000, increased production of natural gas liquids will occur.

17.9 NATURAL GAS

Natural gas, consisting largely of methane, CH_4, is a very attractive fuel that produces few pollutants and less carbon dioxide per unit energy than any other fossil fuel. In addition to its use as a fuel, natural gas can be converted to many other hydrocarbon materials. It can be used as a raw material for the Fischer–Tropsch synthesis of gasoline. As of the early 2000s, increased demand for natural gas had led to tight supplies in the United States. However, by 2012, exploitation of natural gas in "tight" shale formations opened by horizontal drilling into the formations and hydraulic fracturing with highly pressurized water, sand, and additives had resulted in increased supply of natural gas in the United States and some other countries. This has led to a dramatic increase in natural gas production in the United States including parts of Arkansas, Louisiana, Pennsylvania, and Texas with the result that from 2000 to 2011 the United States surpassed Russia as the world's largest producer of natural gas, leading to a glut of gas in U.S. markets and falling prices for that commodity. Many other areas of the world have a large potential for production of shale gas. Proposals have been made to repurpose harbor terminals designed for importing natural gas into the United States to facilities for natural gas export.

Estimates of world reserves of natural gas are in a constant state of flux, largely because of rapid exploitation of new natural gas sources, especially tight shale deposits. A reasonable estimate of world resources as of 2012 was around 200 trillion standard cubic meters (tcm) with annual world consumption of about 3.1 tcm as of 2012. In the 3-year period of 2005–2008, almost 70% more natural gas was added to reserves in the United States than were consumed. In 2008, the United States proven reserves of shale gas increased by more than 50%, reaching a total of 13.4% of proved U.S. reserves by the year's end.

Environmental considerations are important in exploiting natural gas resources. Some of the most abundant new sources are in fragile or challenging environments. For example, Statoil of Norway operates a gas field called Snow White in the Barents Sea 340 miles north of the Arctic Circle. The melting of Arctic ice due to global warming is leading to increased possibilities for the discovery and development of natural gas deposits in far northern regions. Production of natural gas from coal seams in Wyoming has required pumping saline, alkaline water from the seams, which has caused water pollution problems. Hydraulic fracturing of tight shale formations to extract their natural gas content has led to accusations of water pollution and even a few spectacular demonstrations of flames issuing from ignited methane gas from running-water faucets. Some of the recently developing tight shale formations from which natural gas is being extracted or proposed for development in the United States are in relatively highly populated regions of Arkansas, Louisiana, Pennsylvania, Texas (including substantial sources beneath the Fort Worth metropolitan area), and upper New York State.

As efforts are under way to reduce greenhouse gas emissions of carbon dioxide and to develop energy sources that do not release carbon dioxide, natural gas can serve as a "bridging fuel" that emits less carbon dioxide than petroleum and much less than coal per unit of heat energy generated. Methane can be used as a fuel in existing electrical power–generating facilities that are now fueled by petroleum or coal. Very high efficiency in methane utilization can be achieved by combined power cycles in which methane is used as a fuel in gas turbines coupled to generators and the hot exhaust gas used to raise steam for steam turbines. As discussed in Section 17.6, methane can be used as automotive fuel with especially high efficiencies achievable in hybrid vehicles. Natural gas will be a very significant factor in world energy for many decades to come.

17.10 COAL

The general term **coal** describes a large range of solid fossil fuels derived from partial degradation of plants. Coal is differentiated largely by **coal rank** based on percentage of fixed carbon, percentage of volatile matter, and heating value, dividing coal into several categories including bituminous coal, anthracite, subbituminous coal, lignite, and brown coal. The approximate average empirical formula of coal is $CH_{0.8}$ and coal typically contains from 1% to several percent sulfur, nitrogen, and oxygen. Of these elements, sulfur bound to the organic coal molecule and mixed with coal as mineral pyrite, FeS_2, presents major environmental problems because of production of air pollutant sulfur dioxide during combustion. Much of the FeS_2 can be removed physically from coal before combustion and sulfur dioxide can be removed from stack gas by various scrubbing processes (see Chapter 8, Section 8.7).

From Civil War times until World War II, coal was the dominant energy source behind industrial expansion in the United States and most developed nations. However, after World War II, the greater convenience of lower-cost petroleum resulted in a decrease in the use of coal for energy in the United States and in a number of other countries. Annual coal production in the United States fell by about one-third, reaching a low of approximately 400 million tons in 1958, but since then, domestic U.S. coal production for electricity generation has reached about 1 billion tons per year. (As concerns over the global warming potential from coal use grow and alternate energy sources, especially natural gas, grow in importance, coal use in the United States may have peaked in 2011.) About one-third of the world's energy and around 50% of electrical energy is provided by coal. Coal makes up from one-fourth to one-third of world energy consumption. World coal reserves are about 860 billion metric tons of which about half is bituminous coal and anthracite with the remainder divided among subbituminous coal, lignite, and brown coal, adequate to meet world needs for many decades.

17.10.1   COAL CONVERSION

As shown in Figure 17.12, coal can be converted to gaseous, liquid, or low-sulfur, low-ash solid fuels such as coal char (coke) or solvent-refined coal (SRC). Coal conversion is an old idea; a house belonging to William Murdock at Redruth, Cornwall, England, was illuminated with coal gas in 1792. The first municipal coal gas system was employed to light Pall Mall in London in 1807. The coal gas industry began in the United States in 1816. The early coal gas plants used coal pyrolysis (heating in the absence of air) to produce a hydrocarbon-rich product particularly useful for illumination. Later in the 1800s, the water gas process was developed, in which steam was impinged upon hot coal coke to produce a mixture consisting primarily of a (deadly) mixture of H_2 and CO. It was necessary to add volatile hydrocarbons to this "carbureted" water gas to bring its illuminating power up to that of gas prepared by coal pyrolysis. The United States had 11,000 coal gasifiers operating in the 1920s. At the peak of its use in 1947, the water gas method accounted for 57% of U.S. manufactured gas. The gas was made in low-pressure, low-capacity gasifiers that by today's standards would be inefficient and environmentally unacceptable (many sites of these old plants have been designated as hazardous waste sites because of residues of coal tar and other wastes). This was definitely not a green technology because of the high toxicity of carbon monoxide in the gas product, and many people were killed by release of CO in their houses. During World War II, Germany developed a major synthetic petroleum industry based on coal, which reached a peak capacity of 100,000 barrels per day in 1944. A synthetic petroleum plant operating in Sasol, South Africa, reached a capacity of several tens of thousands of tons of coal per day in the 1970s and currently produces hydrocarbons and feedstocks equivalent to about 150,000 barrels of petroleum per day.

A number of environmental implications are involved in the widespread use of coal conversion. These include strip mining, water consumption in arid regions, lower overall energy conversion compared to direct coal combustion, and increased output of atmospheric carbon dioxide. These plus economic factors have prevented coal conversion from being practiced on a very large scale. However, coal conversion does enable relatively facile carbon sequestration (see Section 17.11), which could enable much more sustainable coal utilization.

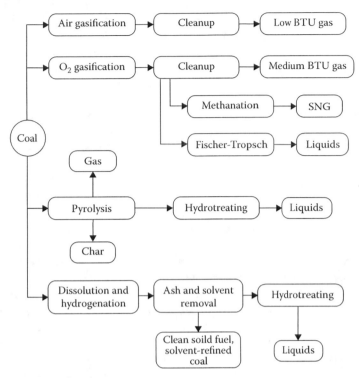

FIGURE 17.12 Routes to coal conversion. BTU refers to British thermal units, a measure of the heat energy that can be obtained from a fuel. Methanation means synthesis of CH_4 gas. Hydrogenation and hydrotreating refer to reaction with elemental H_2 gas.

17.11 CARBON SEQUESTRATION FOR FOSSIL FUEL UTILIZATION

Carbon sequestration, which prevents carbon dioxide generated by fossil fuels from entering the atmosphere, holds the promise of enabling utilization of fossil fuels, especially coal, without contributing to greenhouse warming. Basically, the various schemes that have been proposed entail capturing carbon dioxide from a product or waste stream and sequestering it in a place where it cannot enter the atmosphere. Several approaches have been suggested or tried for capturing carbon dioxide and there are several possibilities for sequestration. The term "carbon sequestration" is also used in a more general sense to apply to removal of carbon dioxide from the atmosphere, especially by photosynthesis.

There are several possible sinks in which carbon dioxide can be sequestered. The largest of these, a natural sink for the gas, is the ocean. Earth's oceans have an almost inexhaustible capacity for carbon dioxide. However, lowering the average pH of the oceans by as little as 0.1 pH unit by adding acidic carbon dioxide could have a serious adverse effect on ocean life and productivity; there is some evidence to suggest that shells of some small sea creatures have become thinner because of higher levels of dissolved carbon dioxide in ocean water. Deep saline formations also have a very high capacity for carbon dioxide sequestration. Depleted oil and gas reservoirs and unmineable coal seams have much lower, but still significant, carbon dioxide capacities.

Geological carbon dioxide sequestration can be accomplished by injecting the gas into porous sedimentary formations at depths exceeding approximately 1000 m. Experience in the petroleum industry with underground disposal of carbon dioxide and injection of the gas into oil-bearing formations for petroleum recovery has provided the technology required for geological carbon dioxide sequestration. The carbon dioxide injected into sedimentary formations rises and is confined by poorly permeable caprock. Breaches in caprock, such as those from abandoned oil wells, can result in carbon dioxide release. Eventually, the carbon dioxide dissolves and is stabilized in the generally saline pore waters in the sedimentary formation into which it is injected. Chemical reactions in the water and with the surrounding geological strata can result in long-term stability of the carbon dioxide. It is much more effective and less expensive to force carbon dioxide into unsaturated saline groundwater where it is dissolved than to compress the gas and store it underground as a highly pressurized gas.

The first commercial application of carbon dioxide sequestration has operated since 1996 in the North Sea about 240 km from the Norwegian coast in a region known as the Sleipner oil and gas field. The natural gas product from this field is about 9% carbon dioxide, a value that must be reduced to 2.5% for commercial distribution of the gas. Whereas all other gas-producing operations simply discharge the carbon dioxide removed into the atmosphere, at Sleipner, it is pumped under pressure into a 200-m-thick layer of sandstone, the Utsira formation, that is about 1000 m below the seabed. A mixture of carbon dioxide and hydrogen sulfide removed from sour natural gas that is abundant in Alberta, Canada, is now being sequestered underground.

The easiest sources of carbon dioxide to capture are those from industrial processes that generate the gas in high concentrations. An example of such a process is the fermentation of carbohydrates to produce ethanol for fuel or other uses. This source provides much of the carbon dioxide that is used commercially at present. The largest anthropogenic source of carbon dioxide discharged into the atmosphere is generated in power plants fueled with fossil fuels. These sources present a substantial challenge for carbon dioxide removal because they are so dilute. A power plant fueled with carbon-rich coal produces an exhaust stream that is 13–15% carbon dioxide, whereas a boiler burning hydrogen-rich methane produces only 3–5% carbon dioxide. A third possibility is to capture carbon dioxide released from the gasification of fossil fuels, particularly coal (see Section 17.10.1). Normally, gasification is performed using oxygen as an oxidant, and the initial product consists of carbon dioxide and combustible H_2 and CO gases. Carbon monoxide in the synthesis gas product can be reacted with steam

$$CO + H_2O \rightarrow H_2 + CO_2 \tag{17.4}$$

to produce nonpolluting elemental hydrogen fuel and carbon dioxide.

In late 2007, the U.S. Department of Energy announced funding for the construction of the first U.S. power plant with carbon dioxide sequestration, the FutureGen coal-powered power plant to be located at a site in Mattoon, Illinois. The project was abruptly cancelled in early 2008 allegedly because of greatly increased costs, but a Government Accountability Office report released in March 2009 showed that the costs had been overestimated by $500 million because of an accounting cost mathematics error! The 275-MW power plant was to be large enough to supply electricity to 275,000 households and to sequester 1–2 million tons of carbon dioxide per year. As with all carbon sequestration projects, spreading and leakage of the disposed carbon dioxide is of some concern.[5]

Figure 17.13 is a schematic of a coal-fired power plant with carbon sequestration such as the FutureGen power plant. A gas turbine powered by hydrogen fuel produced by coal gasification generates electricity. Hot exhaust gas from the gas turbine is used to raise steam in a boiler and this steam powers a steam turbine that is also coupled to a generator. This combination results in very efficient electrical power generation. The production of intermediate hydrogen adds options for powering fuel cells or for using hydrogen in chemical synthesis or manufacture of synthetic hydrocarbon fuels.

By-product hydrogen sulfide must be removed from the hydrogen product and can be used for making sulfuric acid, an important industrial chemical, or can be sequestered with the carbon dioxide.

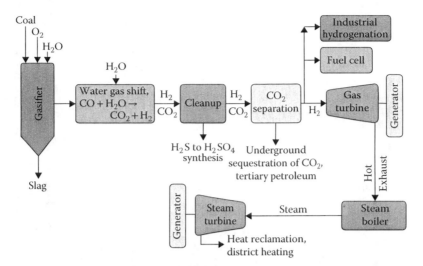

FIGURE 17.13 A coal-based carbon sequestration operation. Electricity is generated at generators connected to both the gas turbine and the steam turbine. Reclaimed hydrogen sulfide is used to synthesize sulfuric acid. Hydrogen product can be employed for fuel cells and industrial hydrogenation as well as to power the gas turbine. Sequestered carbon dioxide can be used for petroleum recovery or simply disposed underground. Waste heat is reclaimed for district heating and industrial heating applications.

The carbon dioxide by-product is sequestered in mineral formations at depths up to around 2000 m. For power plants located near petroleum-bearing formations, carbon dioxide can be pumped into oil-bearing mineral formations for tertiary petroleum recovery.

17.12 GREAT PLAINS SYNFUELS PLANT: INDUSTRIAL ECOLOGY IN PRACTICE TO PRODUCE ENERGY AND CHEMICALS

An excellent example of a system of industrial ecology for sustainable utilization of fossil fuels is provided by the Great Plains Synfuel Plant operated by Dakota Gasification Company near Beulah, North Dakota, the only U.S. commercial plant making methane synthetically.[6] This complex is designed to take advantage of abundant deposits of lignite, a relatively low-heating-value, high-water-content form of coal that occurs in abundance in North Dakota. This plant was constructed in the early 1980s for an investment of $2.1 billion and has been in operation since 1984. Since 1988, $477 million has been invested in the plant to improve efficiency, for new product development, and to ensure environmental compliance. A schematic diagram of the system is shown in Figure 17.14. The heart of the plant consists of 14 Lurgi-type gasifiers that are 13 m high and 4 m in diameter processing 16,000 metric tons of lignite per day at temperatures up to 1200°C to produce a synthesis gas mixture of combustible H_2 and CO along with by-product CO_2, water, and smaller quantities of tars, oils, phenolic compounds, ammonia, and H_2S. Useful fuel hydrocarbons, phenol, cresols, ammonia, and sulfur compounds are extracted from the water, which is used for cooling water. The gas mixture is subjected to a shift reaction that increases the ratio of H_2 to CO, and this mixture is reacted to produce methane (CH_4) that is fed into a natural gas pipeline. Part of the synthesis gas stream is diverted to an ammonia synthesis plant to make this valuable fertilizer product.

A key feature of the Great Plains Synfuel Plant is the extraction of CO_2 from the synthesis gas. Since 2000, the Great Plains plant has been sending about half of the carbon dioxide that it generates by pipeline to oil fields in Saskatchewan, Canada, for enhanced oil production (see Section 17.8) at a current rate of 4.3 million cubic meters per day (2.5–3 million metric tons per year) and by

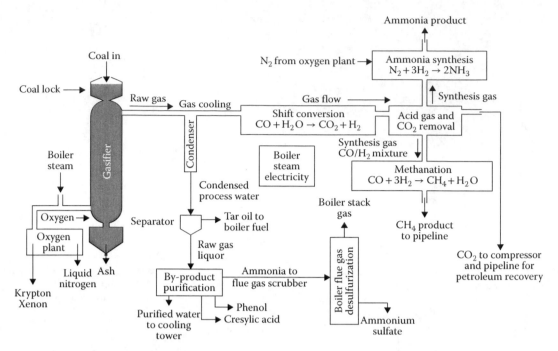

FIGURE 17.14   Schematic of the Great Plains lignite coal gasification complex.

2011 had captured more than 20 million metric tons of CO_2. Unlike carbon dioxide from combustion sources, the CO_2 from the Great Plains plant is not diluted with nitrogen and does not contain a significant amount of water vapor, which makes it very usable and economical to transport by pipeline.

The Great Plains Synfuels Plant provides an excellent example of a diversified industrial ecology complex. It utilizes an abundant resource (lignite coal) in a way that significantly reduces the greenhouse gas emissions and other potential environmental impacts from this resource. Rather than having to transport large quantities of lignite by rail to distant power plants, the energy content of the lignite is converted to methane, the most environmentally friendly fossil fuel, which is moved with minimum environmental disruption through pipelines. An even higher value product, fertilizer ammonia, is synthesized at the site using elemental hydrogen made from gasifying the lignite and elemental nitrogen isolated from air in the air liquefaction operation required to prepare elemental oxygen used for gasification. The carbon dioxide by-product is not released into the atmosphere where it could aggravate the global warming problem but is transported by pipeline to oil fields where it is pumped underground to increase crude oil recovery. Commercially valuable ammonium sulfate by-product, a useful source of nitrogen and sulfur soil nutrients, is recovered from the water released during lignite gasification. In an area where water shortages can occur, the relatively large amount of water released by gasification of lignite (which typically contains 35% water) is used for cooling and as boiler feedwater.

17.13   NUCLEAR ENERGY

The awesome power of the atom nucleus revealed at the end of World War II held out enormous promise for the production of abundant, cheap energy. This promise has never really come to full fruition, although nuclear energy currently provides a significant percentage of electric energy in many countries, and it may be the only source of electrical power that can meet world demand without unacceptable environmental degradation, particularly through the generation of greenhouse gases.

Nuclear fission for power production is carried out in nuclear power reactors in which the fission (splitting) of uranium-235 or plutonium nuclei occurs. Each such event generates two radioactive fission product atoms of roughly half the mass of the nucleus fissioned, an average of 2.5 neutrons, plus an enormous amount of energy compared to normal chemical reactions. The neutrons, initially released as fast-moving, highly energetic particles, are slowed to thermal energies in a moderator medium. For a reactor operating at a steady state, exactly one of the neutron products from each fission is used to induce another fission reaction in a chain reaction (Figure 17.15).

The energy from these nuclear reactions is used to heat water in the reactor core and produce steam to drive a steam turbine, as shown in Figure 17.16. As noted in Section 17.5, temperature limitations make nuclear power less efficient in converting heat to mechanical energy and, therefore, to electricity, than fossil energy conversion processes.

A limitation of fission reactors is the fact that only 0.71% of natural uranium is fissionable uranium-235. This situation could be improved by the development of **breeder reactors**, which convert uranium-238 (natural abundance 99.28%) to fissionable plutonium-239.

A major consideration in the widespread use of nuclear fission power is the production of large quantities of highly radioactive waste products. These remain lethal for thousands of

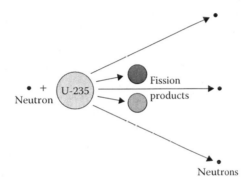

FIGURE 17.15 Fission of a uranium-235 nucleus.

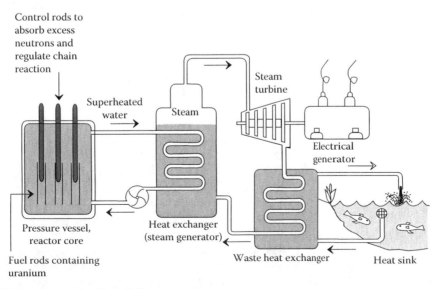

FIGURE 17.16 A typical nuclear fission power plant.

years. They must either be stored in a safe place or disposed of permanently in a safe manner. At the present time, spent fuel elements are being stored under water at the reactor sites. Under current regulations in most countries, the wastes from this fuel will eventually have to be buried. An alternative favored by many investigators is to process the material in the spent fuel elements to remove radioactive products from uranium fuel, isolate the relatively short-lived fission products that decay spontaneously within several hundred years, and bombard the longer-lived nuclear wastes with neutrons in nuclear reactors. The absorption of neutrons by the nuclei of the nuclear waste elements causes **transmutation** in which the elements are converted to other elements or fission products with shorter half-lives resulting in relatively rapid production of stable isotopes. Radioactive waste elements for which transmutation is feasible include plutonium, americium, neptunium, curium, technetium-99, and iodine-129. Plutonium, americium, neptunium, and curium are heavy actinide elements that are fissionable and add fuel value in a nuclear reactor.

Another problem to be faced with nuclear fission reactors is their eventual decommissioning. There are three possible solutions. One is dismantling soon after shutdown, in which the fuel elements are removed, various components are flushed with cleaning fluids, and the reactor is cut up by remote control and buried. "Safe storage" involves letting the reactor stand 30–100 years to allow for radioactive decay, followed by dismantling. The third alternative is entombment, encasing the reactor in a concrete structure, which is essentially what has been done with the remains of the reactor destroyed in the 1986 Chernobyl catastrophe (discussed in the following paragraph).

The course of nuclear power development was altered significantly by three accidents. The first of these occurred on March 28, 1979, with a partial loss of coolant water from the Metropolitan Edison Company's nuclear reactor located on Three Mile Island (TMI) in the Susquehanna River, 28 miles outside of Harrisburg, Pennsylvania. The result was a loss of control, overheating, and partial disintegration of the reactor core. Some radioactive xenon and krypton gases were released and some radioactive water was dumped into the Susquehanna River. Eventually, the reactor building was sealed. A much worse accident occurred at Chernobyl in the Soviet Union in April of 1986 when a reactor blew up and burned, spreading radioactive debris over a wide area and killing a number of people from acute radiation poisoning (officially 30, but possibly significantly more). Thousands of people were evacuated and the entire reactor structure had to be entombed in concrete and steel plate. Food was seriously contaminated as far away as northern Scandinavia. More recently, the disastrous magnitude 9.0 earthquake that occurred on March 11, 2011, off the coast of Honshu, Japan, about 373 km (231 miles) northeast of Tokyo and the devastating tsunami that followed destroyed the Fukushima Daiichi Nuclear Power Station in Japan. The three reactors that were running at the plant shut down automatically when the quake hit, as they were designed to do. However, the quake destroyed the electrical power infrastructure leading to the plant so that water could not be pumped to cool the reactor core and the spent fuel rods normally stored under water close to the core. Zirconium metal alloy cladding the hot fuel rods reacted with water to generate hydrogen gas, which exploded, destroying structures that sheltered the reactors. Significant amounts of radioactivity were released leading to the mandatory evacuation of land in a 20 km radius around the power plant and a ban on sale of food crops raised in the area.

As of 2006, 33 years had passed since a new nuclear electric power plant had been ordered in the United States, in large part because of the projected high costs of new nuclear plants. Although this tends to indicate hard times for the nuclear industry, pronouncements of its demise may be premature and by 2008 several U.S. utilities had begun the process of ordering new nuclear power plants and a significant number of such plants were under construction around the world. Properly designed nuclear fission reactors can generate large quantities of electricity reliably and safely and have done so for decades in U.S. naval submarines and carrier ships and in France, which gets about 80% of its electricity from nuclear sources. The single most important factor that may lead to renaissance of nuclear energy is the threat to the atmosphere from greenhouse gases produced in large quantities by fossil fuels. It can be argued that nuclear energy is the only proven alternative that can provide the amounts of energy required within acceptable limits of cost, reliability, and environmental effects.

New designs for nuclear power plants can enable power reactors that are much safer and environmentally acceptable than those built with older technologies. The latest designs incorporate built-in passive safety features that work automatically in the event of problems that could lead to incidents such as TMI or Chernobyl with older reactors. These devices—which depend on passive phenomena such as gravity feeding of coolant, evaporation of water, or convection flow of fluids—give the reactor the desirable characteristics of **passive stability**. They have also enabled significant simplification of hardware, with only about half as many pumps, pipes, and heat exchangers as are contained in older power reactors.

According to the European Nuclear Society, as of the year 2012, there were 435 nuclear power plants operating around the world.[7] With an installed electric net capacity of about 368 GW, these reactors were running in 31 countries. The country with the largest nuclear capacity is the United States with 104 power reactors followed by France (58), Japan (50, though by May 2012, all of these had been temporarily taken out of commission because of concern over the Fukushima Daiichi nuclear plant accident), and the Russian Federation (33). As of 2012, there were 63 nuclear power plants with a total installed capacity of 61 GW under construction in 15 countries.

After a generation had passed during which no new nuclear power plants had been approved in the United States, in February 2012, the U.S. Nuclear Regulatory Commission approved licensing of two new 1100-MW nuclear reactors to be constructed by the Southern Company at its existing Alvin W. Vogtle Nuclear Power Plant near Augusta, Georgia.[8] These are to be new state-of-the-art Westinghouse AP1000 reactors designed to withstand plane crashes and earthquakes. They will have passive safety features that can function even in the event of the kind of power failure that doomed the Fukushima Daiichi reactors. A major safety feature with the AP1000 reactor is a 3.2-million-liter water tank above the reactor, the drain valves of which are held shut electrically. When electrical power fails, the valves open spontaneously cooling the reactor core with water. This passive cooling system can operate for up to 3 days before power is required to pump additional cooling water. The reactor system also has an air space between the heavy concrete shield and the steel liner to allow cooling by passive convection circulation of air.

17.13.1 THORIUM-FUELED REACTORS

Thorium, Th, atomic number 90, exists primarily as the ^{232}Th isotope, which is slightly radioactive and undergoes alpha decay with a half-life of 14 billion years. This form of thorium is not itself fissionable, but a ^{232}Th nucleus can absorb a neutron and the product undergoes two successive beta decays to produce fissionable ^{233}U, which can serve as a fissionable fuel in nuclear power reactors. Proponents of thorium-fueled reactors, who contend that thorium is "the nuclear fuel of the future," point out that, compared to uranium, thorium is safer and more abundant (about three times the abundance in Earth's crust than uranium).[9] Also, there is a much lower potential to produce dangerous plutonium and actinides in a thorium fuel cycle than in a uranium fuel cycle, and hence a lower potential to produce fissionable material for weapons. (The fact that thorium cannot be used readily to produce nuclear weapons was a major factor in the thorium cycle **not** being chosen for nuclear fuel in the early days of atomic energy development following 1945 in which the ability to make weapons-grade fissionable material was a very important goal.) Also, the fission products that result from a thorium fuel cycle are much less long-lived than those based on ^{235}U or ^{239}Pu and would decay to safe levels after only a few hundred years. A potentially highly sustainable application of thorium to fuel power reactors uses circulating hot liquid thorium fluoride salts as fuel.

17.13.2 NUCLEAR FUSION

The fusion of a deuterium nucleus and a tritium nucleus releases a lot of energy as shown in Equation 17.5, where Mev stands for million electron volts, a unit of energy:

$$_1^2H + {}_1^3H \rightarrow {}_2^4He + {}_0^1n$$

$$+17.6\,\text{Mev (energy released per fusion)} \tag{17.5}$$

This reaction is responsible for the enormous explosive power of the "hydrogen bomb." So far, it has eluded efforts at containment for a practical continuous source of energy. And since physicists have been trying to make it work on a practical basis for the past approximately 50 years, it will probably never be done. (Within about 15 years after the discovery of the phenomenon of nuclear fission, it was being used in a power reactor to power a nuclear submarine.) However, the tantalizing possibility of using the essentially limitless supply of deuterium, an isotope of hydrogen, from Earth's oceans for nuclear fusion still give some investigators hope of a practical nuclear fusion reactor and some research continues toward that end.

Nuclear fusion was the subject of one of the greatest scientific embarrassments of modern times that occurred in 1989. This incident came about when investigators at the University of Utah announced that they had accomplished so-called cold fusion of deuterium during the electrolysis of deuterium oxide (heavy water). This resulted in an astonishing flurry of activity as scientists throughout the world sought to repeat the results, whereas others ridiculed the idea. Unfortunately, for the dream of a cheap and abundant source of energy, the skeptics were right, and the whole story of cold fusion stands as a lesson in the (temporary) triumph of wishful technological thinking over scientific good sense (although there are still scientists with reputable educations who contend that cold fusion is real).

17.14 GEOTHERMAL ENERGY

Underground heat in the form of steam, hot water, or hot rock used to produce steam has been employed as an energy resource for about a century and can be regarded as largely renewable. This energy was first harnessed for the generation of electricity at Larderello, Italy, in 1904, and has since been developed in Japan, Russia, New Zealand, the Philippines, and at the Geysers in northern California.

Underground dry steam is relatively rare, but it is the most desirable from the standpoint of power generation. More commonly, energy reaches the surface as superheated water and steam. In some cases, the water is so pure that it can be used for irrigation and livestock; in others, it is loaded with corrosive, scale-forming salts. Utilization of the heat from contaminated geothermal water generally requires that the water be reinjected into the hot formation after heat removal to prevent contamination of surface water.

The utilization of hot dry rocks for energy requires fracturing of the hot formation, followed by injection of water and withdrawal of steam. This technology is still in the experimental state but promises approximately 10 times as much energy production as steam and hot water sources.

Land subsidence and seismic effects, such as the mini-earthquakes that occur when water is pumped under extreme pressure into hot rock formations that fracture as a consequence, are environmental factors that may hinder the development of geothermal power. However, this energy source holds considerable promise, and its development continues.

An interesting possibility that has yet to be demonstrated is the use of supercritical carbon dioxide as a working fluid for the extraction of energy from hot rocks (Figure 17.17).

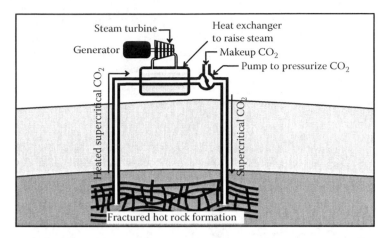

FIGURE 17.17 A system for extracting heat from hot dry rock formations using supercritical or dense phase carbon dioxide. The carbon dioxide is pumped into a fractured hot rock formation through an injection well and is removed through a production well. The hot carbon dioxide is run through a heat exchanger to raise steam to run a turbine that is coupled to an electrical generator and through which water and steam are recycled. Makeup carbon dioxide is added to the carbon dioxide line downstream from the heat exchanger and the carbon dioxide is repressurized and recycled through the hot rock. Because of loss of carbon dioxide to the underground rock formations, this system is best coupled with a carbon sequestration system (see Section 17.11).

17.15   SUN: AN IDEAL, RENEWABLE ENERGY SOURCE

Solar power is an ideal source of energy that is unlimited in supply, widely available, and inexpensive. It does not add to the Earth's total heat burden or produce chemical air and water pollutants. On a global basis, utilization of only a small fraction of solar energy reaching the Earth could provide for all energy needs. In the United States, for example, with conversion efficiencies ranging from 10% to 30%, it would only require collectors ranging in area from one-tenth down to one-thirtieth that of the state of Arizona to satisfy present U.S. energy needs. (This is still an enormous amount of land, and there are economic and environmental problems related to the use of even a fraction of this amount of land for solar energy collection. Certainly, many residents of Arizona would not be pleased at having so much of the state covered by solar collectors, and some environmental groups would protest the resultant shading of rattlesnake habitat.)

There are vast land areas available in the United States and throughout the world that receive exceptional levels of solar energy suitable for power generation. Factors involved in evaluating areas for solar energy generation include proximity to the Equator, relatively high altitude, and consistent absence of cloud cover. The terrestrial area that has optimum conditions is in the Sahara desert of southeast Niger in Africa, which receives an average of 6.78 kWh/m^2 of solar energy input each day, which is close to the amount of energy required to heat water for a typical U.S. household.

A common means of generating electricity from solar energy depends on the collection of solar thermal energy followed by conversion to electrical energy. The simplest such approach involves focusing sunlight on a steam-generating boiler (see Illustration 6 in Figure 17.3). One of the world's largest solar power stations is the 150-MW Andasol complex in Granada, southern Spain. It collects and concentrates solar heat with rows of parabolic mirrors covering several square kilometers of area focused on pipes through which a heat-stable synthetic oil flows to collect solar heat energy. Part of the hot oil goes through a heat exchanger where it heats water to produce steam to generate electrical power. Another fraction of the hot oil flows through another heat exchanger where it heats a molten mixture of 60% sodium nitrate and 40% potassium nitrate from 290°C to 390°C for storage in insulated tanks to provide heat energy for generating electricity when the sun is not shining. The relatively inexpensive liquid salt mixture used for heat storage is stable up to 600°C, has a high density and high heat capacity, and is readily pumped through pipes, making it an excellent medium for collecting and storing heat energy in a relatively small volume. The system generates electricity at a cost of about $0.17/kWh, which is about the same cost as photovoltaic systems (see Section 17.15.1) and about three times the cost of systems powered by natural gas (around $0.06/kWh in 2012).

Another solar-powered heat engine used to generate electricity is the **Stirling engine**, which has achieved close to 30% overall efficiency in the conversion of solar energy to electricity. The Stirling engine uses hydrogen gas as a working fluid in a sealed system in which pistons connected by rods in cylinders pump the gas back and forth in the system. The gas is heated at a heater head kept at a relatively high temperature by sunlight focused from an array of reflective mirrors that concentrate sunlight on the heater head. The resultant pressure forces one of the pistons down in a power stroke. The gas is then transferred to a piston on the cooled side of the engine through a regenerator consisting of a material that captures some of the heat to be used in the next cycle, thereby significantly increasing efficiency. As of 2012, plans to install a commercial-scale solar-driven energy system in the United States had been put on hold, in large part because of competition from photovoltaic systems that had continued to decrease in cost.

17.15.1 Solar Photovoltaic Energy Systems

Solar power cells (photovoltaic cells) for the direct conversion of sunlight to electricity have been developed and are widely used for energy in space vehicles. With present technology, however, they are relatively expensive in most places for large-scale generation of electricity, although the economic gap is narrowing. The direct conversion of energy in sunlight to electricity is accomplished by special solar voltaic cells. Most types of photovoltaic cells depend on the electronic properties of silicon atoms containing low levels of other elements. In a typical photovoltaic cell, the cell consists of two layers of silicon, a donor layer that is doped with about 1 part per million of arsenic atoms and an acceptor layer doped with about 1 part per million of boron. Lewis symbols use dots to represent the outermost valence electrons of atoms, those that can be lost, gained, or shared in chemical bonds (see the discussion of atomic structure in Chapter 19, Section 19.2). Examination of the Lewis symbols of silicon, arsenic, and boron

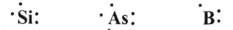

shows that substitution of an arsenic atom with its five valence electrons for a silicon atom with its four valence electrons in the donor layer gives a site with an excess of 1 electron, whereas substitution of a boron atom with only three electrons for a silicon atom in the acceptor layer gives a site "hole" that is deficient in one electron. The surface of a donor layer in contact with an acceptor layer contains electrons that are attracted to the acceptor layer. When light shines on this area, the energy of the photons of light can push these electrons back onto the donor layer, from which they can go through an external circuit back to the acceptor layer as shown in Figure 17.18. This flow of electrons constitutes an electrical current that can be used for energy.

Solar voltaic cells based on crystalline silicon have operated with a 30% efficiency for experimental cells and 15–20% for commercial units available in 2008, at a cost of around 15 cents/kWh, compared to 4–7 cents/kWh for fossil fuel–fired power plants and 6–9 cents for those fired by biomass. Costs of photovoltaic electricity have shown a continuous downward trend.[10] Part of the high cost in the past has resulted from the fact that the silicon used in the cells must be cut as small wafers from silicon crystals for mounting on the cell surfaces. Significant advances in costs and technology are being made with thin-film photovoltaics, which use an amorphous silicon alloy. These cells are only about half as efficient as those made with crystalline silicon, but cost only about 25% as much. A newer approach to the design and construction of amorphous silicon film photovoltaic devices

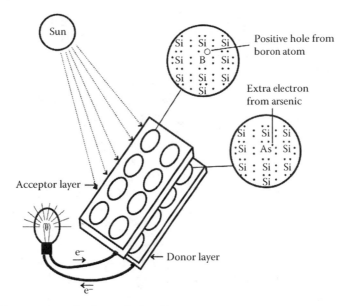

FIGURE 17.18 The operation of a photovoltaic cell.

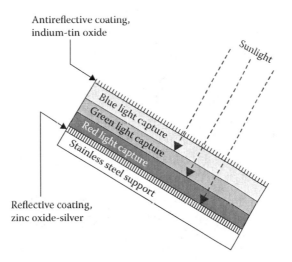

FIGURE 17.19 High-efficiency thin-film solar photovoltaic cell using amorphous silicon.

uses three layers of amorphous silicon to absorb, successively, short wavelength ("blue"), intermediate wavelength ("green"), and long wavelength ("red") light, as shown in Figure 17.19. Thin-film solar panels constructed with this approach have achieved solar-to-electricity energy conversion efficiencies just over 10%, lower than those using crystalline silicon, but higher than other amorphous film devices. The low cost and relatively high conversion efficiencies of these solar panels should enable production of electricity at only about twice the cost of conventional electrical power, which would be competitive in some situations.

   Systems have been developed in which solar cells lining glass tubes are used to generate electricity. Such a configuration is well adapted to rooftops because it captures light from all angles including even light from below reflected from a white roof. It has been estimated that covering all flat rooftops in the United States with such collectors could provide about 150 GW of electricity, 15% of U.S. consumption.

A major disadvantage of solar energy is its intermittent nature. However, flexibility inherent in an electric power grid would enable it to accept up to 15% of its total power input from solar energy units without special provision for energy storage. Existing hydroelectric facilities may be used for pumped-water energy storage in conjunction with solar electricity generation. Heat or cold can be stored in water, in a latent form in water (ice) or eutectic salts, or in beds of rock. Enormous amounts of heat can be stored in water as a supercritical fluid contained at high temperatures and very high pressures deep underground. Mechanical energy can be stored with compressed air or flywheels. Utilization of solar energy to produce elemental hydrogen as a means to store, transfer, and utilize energy as discussed in Section 17.18 will probably come into widespread use.

No real insurmountable barriers exist to block the development of solar energy, such as might be the case with fusion power. In fact, the installation of solar space and water heaters became widespread in the late 1970s, and research on solar energy was well supported in the United States until after 1980, when it became fashionable to believe that free-market forces had solved the "energy crisis." At present, shortages of energy and concern over the effects of global warming due to the use of fossil fuels is leading to much increased interest in solar energy. With the installation of more heating devices and the probable development of some cheap, direct solar electrical generating capacity, it is likely that during the coming century, solar energy will be providing an appreciable percentage of energy needs in areas receiving abundant sunlight.

17.15.2 ARTIFICIAL PHOTOSYNTHESIS FOR CAPTURING SOLAR ENERGY

The production of biomass fuel is one of the main ways that solar energy is utilized indirectly. As discussed in Chapter 12, Section 12.3, virtually all the biomass produced by organisms is generated by photosynthesis using solar light energy ($h\nu$) by the fixation of carbon from carbon dioxide. The initial product of most photosynthesis reactions is energy-rich, carbon-rich glucose sugar, $C_6H_{12}O_6$, from which starch, cellulose, proteins, and other biomolecules are biosynthesized. The overall chemical reaction for photosynthesis is the following:

$$6CO_2 + 6H_2O + h\nu \rightarrow C_6H_{12}O_6 + 6O_2 \qquad (17.6)$$

Described by a simple chemical equation, photosynthesis is in reality an extremely complicated biochemical process carried out under ambient conditions without expensive precious metal catalysts. It has been a "holy grail" of energy research to perform photosynthesis artificially, that is, **artificial photosynthesis**, a feat that so far has defied the efforts of large numbers of scientists throughout the world.[11]

There is not much point in performing artificial photosynthesis to make glucose and other biochemicals; plants do that very well. However, it would be very useful to artificially photosynthesize elemental hydrogen, H_2, and perhaps even carbon monoxide, CO, along with elemental O_2, from H_2O and CO_2. Elemental hydrogen can be used as a nonpolluting fuel to generate electricity in fuel cells, and CO and H_2 can be used as feedstocks in methanation and Fischer–Tropsch syntheses to synthetically produce methane, methyl and ethyl alcohols, and higher hydrocarbons that can be employed as fuels. Indeed, under some specific conditions, some cyanobacteria do in fact generate H_2 photosynthetically.

The initial step in biological photosynthesis is harvesting of visible light energy by an array of pigments (the antenna system) to break up water molecules and produce O_2, H^+, and electrons. The electrons produced by this process can act as chemical-reducing agents and are transferred by an electron transport chain to a second light-harvesting process where they are further energized to provide energy for CO_2 fixation, H_2 generation, or, potentially in an artificial photosynthetic system, for reduction of CO_2 to CO. Ideally, for synthetic fuels production, artificial photosynthetic systems could be made to work efficiently for generation of elemental H_2 from water, for which the overall reaction would be the following:

$$2H_2O + h\nu \rightarrow 2H_2 + O_2 \tag{17.7}$$

A much less likely possibility, but potentially very useful to provide feedstock for synthetic fuels manufacture if it could be accomplished, is production of CO from CO_2 and H_2O:

$$2CO_2 + h\nu \rightarrow 2CO + O_2 \tag{17.8}$$

17.16   ENERGY FROM EARTH'S TWO GREAT FLUIDS IN MOTION

In Chapter 1, Section 1.6, it was mentioned that the Earth System has two "great fluids" that are in constant motion, atmospheric air and surface water. The movement of wind in the atmosphere, seawater in waves and tides, and flowing surface water involves huge amounts of potentially capturable energy. The harvesting of these renewable energy sources for use in the anthrosphere is discussed here.

17.16.1   SURPRISING SUCCESS OF WIND POWER

Wind power using huge turbines mounted on high towers and coupled to electrical generators is emerging at a somewhat surprising rate as a source of renewable energy (Figure 17.20). Although used for centuries with windmills that drove grain grinding and water-pumping operations and during the early 1900s for small-scale electricity generation, especially in remote locations, modern large-scale wind-powered electrical generators emerged during the 1990s as economical means of generating electrical power. Wind power is completely renewable and nonpolluting. It is an indirect means of utilizing solar energy because winds are caused by the movement of air masses heated by the sun.

FIGURE 17.20   Wind-powered electrical generators mounted on towers are becoming increasingly common sights in the world in areas where consistent wind makes this nonpolluting source of renewable energy practical.

The potential of wind power to provide energy is enormous. A conservative estimate of the wind resource is that wind power generators installed on just 1% of Earth's land would provide all of the anthrosphere's electrical energy. (The footprint of wind generators is small and essentially all the land upon which they are placed can still be used for agriculture and other purposes.) At the beginning of 2012, total installed world wind energy capacity was 239 GW, sufficient to supply energy to 11 million average U.S. homes. China ranked as the world leader in installed wind capacity at 62 GW with the United States second at 44-GW capacity.

Wind power is especially attractive for some agricultural regions. One reason is that such regions often have low population densities so that there are fewer objections to wind power installations. A commercial wind turbine generating 3 MW is a formidable machine, typically mounted on towers around 100 m high to take advantage of higher, more consistent wind speeds at such heights, and may have blades 50 m long. However, the footprints of these structures are relatively small and do not occupy much farmland. The electricity is conveyed from the turbines in underground lines eliminating surface power lines. Adding to the potential attractiveness of wind power in agricultural regions is the fact that electricity generated by wind energy can be used to electrolyze water to produce elemental H_2 and O_2, an application not handicapped by wind's intermittent nature. The H_2 currently required to make NH_3 fertilizer is usually produced from steam reacting with methane (natural gas), a relatively expensive process, and its production from water by inexpensive wind energy could help keep the price of ammonia fertilizer at reasonable levels. Furthermore, both elemental H_2 and O_2 can be used to convert crop by-product biomass to hydrocarbon fuels as discussed in Section 17.16.

Northern regions, including parts of Alaska, Canada, the Scandinavian countries, and Russia, often have consistently strong wind conditions conducive to the generation of wind power. Isolation from other sources of energy makes wind power attractive for many of these regions. Severe climate conditions in these regions pose special challenges for wind generators. One problem can be the buildup of rime consisting of ice condensed directly on structures from supercooled fog in air. (In warmer regions, the remains of insects zapped by the rotating turbine blades have built up to the point of reducing the aerodynamic efficiency of the blades.)

Although wind turbines have been growing in size leading to much greater economy in power production, there is also a market for small wind turbines that can be used for an individual home or commercial building. Small wind turbines have been installed in a number of locations including Logan International Airport in Boston and the Brooklyn Naval Yard. Although the cost of these turbines is much less than that of large commercial turbines, the cost of electricity produced by the small turbines is much higher.

17.16.2 ENERGY FROM MOVING WATER

Water flowing in contact with a device called a waterwheel is one of the oldest sources of power other than humans or animals. Grain mills driven by waterpower existed in ancient Greece and Rome, and large water wheels developing up to 50 horsepower were constructed in the Middle Ages. In colonial North America, water wheels drove gristmills and sawmills and were further applied to leather, textile, and machine shop operations. Because of problems with low water flow in the summer and ice formation in the winter, these operations were rapidly displaced when steam engines became available in the early 1800s.

With the development of electric power in the late 1800s, waterpower underwent a spectacular renaissance to drive electrical generators. The first practical hydroelectric plant went into operation on the Fox River near Appleton, Wisconsin, in 1882. Hydroelectric power grew rapidly as an energy source from that time and by 1980 accounted for about 25% of world electrical production and 5% of total world energy use. The potential to construct hydroelectric plants is favored by mountainous terrain with large river valleys and is distributed relatively evenly around the world. China has about one-tenth of the world's potential for hydroelectric power. About 99% of Norway's electric power is hydroelectric accounting for about 50% of that country's total energy use.

The largest hydroelectric project in the world is the Three Gorges installation on the huge Yangtze River in China. Located at the end of a number of steep canyons, the dam spans 2.3 km across the river valley and reaches a height of 185 m. At full capacity, the reservoir formed by the dam extends for 630 km with an average width of 1.3 km. When the dam was finished in 2009, it had 26 generating units each capable of generating 700 MW of hydroelectric power, a total capacity of 18.2 GW. After 2009, 6 more units were constructed in a subterranean powerhouse bringing the total generating capacity to 22.4 GW in 2012.

The sustainability and environmental acceptability of hydroelectric power present a mixed picture. In the modern age, construction of water impoundments tends to displace significant numbers of people, more than 1 million for the Chinese Three Gorges project. Altering the flow of rivers can change their aquatic ecology. Esthetics can be harmed by filling scenic river valleys with impounded water. In some cases in the United States, dams are being dismantled to restore river valleys to their former state. However, hydroelectric power prevents release of greenhouse gases from equivalent fossil energy–powered plants. Reservoirs can provide recreational facilities and serve as sources of fish protein and water used by municipalities, industries, and agriculture.

17.16.3 Energy from Moving Water without Dams

A promising approach to the utilization of energy from moving water without building dams is provided by **hydrokinetic** and **wave energy conversion** devices that tap the kinetic energy of moving water in river flows, tides, or ocean waves. These devices can be put in natural streams, tidal estuaries, ocean currents, and constructed waterways. One promising device consists of a turbine with large blades spaced relatively far apart and connected directly to a generator that can be placed in a water current, such as in a river. Such turbines can be attached directly to structures constructed on waterway beds or can be attached to bridge supports. Bidirectional turbines have been designed for use in tidal currents that flow back and forth.

An interesting way of harnessing water energy is **pressure-retarded osmosis** in which saline ocean water and freshwater are separated by a water-permeable membrane, and the flow of water through the membrane from the freshwater to the saline water side builds pressure in the latter that can be harnessed to produce electricity. Pressure-retarded osmosis is illustrated in Figure 17.21. Although the process operates on a continuous basis, it is shown as a stepwise process in Figure 17.21 to more clearly illustrate the operating principle. The world's first osmotic plant, a demonstration unit with a minuscule capacity, went into operation in Tofte, Norway, in November 2009. Pressure-retarded osmosis plants can be located in almost any of the huge number of locations worldwide where freshwater flows into the sea.

17.17 BIOMASS ENERGY: AN OVERVIEW OF BIOFUELS AND THEIR RESOURCES

Until about 1800, humans existed in what may be termed a **premodern biomass economy** in which most energy came from burning wood and from the muscle power of humans and animals, to a degree augmented by wind and water power. Considering all of its uses, including as a primary source of cooking fuel in many societies, biomass is the leading renewable energy source and has the potential to supply much more energy.[12] The largest single use of biofuel has been primarily that of wood burned in relatively inefficient cooking and heating stoves in developing countries, where biofuel can make up as much as 20% of primary energy production, compared to only 3% in industrialized nations. This application of wood fuel is not an entirely "green" technology because large tracts of land have been stripped of trees and become eroded to provide fuel for cooking in

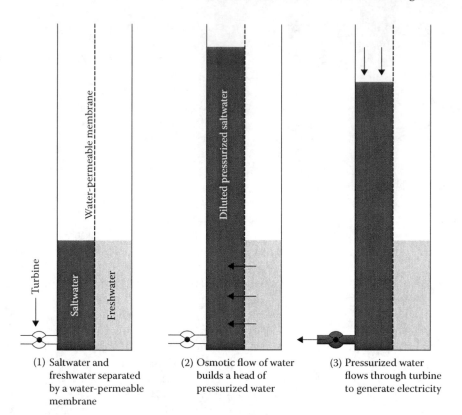

FIGURE 17.21   Illustration of the principle of pressure-retarded osmosis power generation that is based on the difference in osmotic pressure between saltwater and freshwater. Although it is shown here as a stepwise process, this mode of power generation operates on a continuous basis.

sub-Saharan Africa, Haiti, and other areas where wood is the predominant fuel used for cooking. Furthermore, the inhalation of smoke particles, carcinogenic polycyclic aromatic hydrocarbons, and carbon monoxide is suspected of causing lung maladies and other adverse health effects in people who use poorly vented wood-fueled cooking fires.

Starting with a mix of biomass raw materials, including abundant lignocellulose, and algae biomass, integrated biorefineries can produce biofuels including liquid transport fuels, chemical feedstocks, electrical power, and heat. In countries with substantial agricultural and forestry potential, such as the United States, biomass fuel offers the following:

- Significantly increased energy supplies with improved energy security and balance of payments
- Increased diversity, versatility, and redundancy in energy sources
- Environmental benefits including significantly reduced greenhouse gas emissions
- Enhanced economic opportunities and improved social conditions in rural societies
- Reductions in wastes such as those from livestock feeding operations

Figure 17.22 shows the distribution of bioenergy sources worldwide. A vast variety of kinds of biomass can be used for fuel, including wood, straw, sawdust, peanut husks, bagasse (sugarcane stalk residue), and water hyacinth. Wood that is used primarily for heating and cooking in generally more rural and less industrialized societies has long been the predominant source of bioenergy. As dedicated energy crops and food crop by-products are used increasingly for bioenergy production, it is likely that the segment from agricultural crops and by-products will increase as a fraction of bioenergy supply.

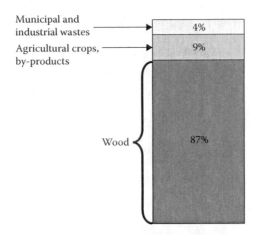

FIGURE 17.22 Distribution of bioenergy sources worldwide. The predominance of wood energy, primarily for cooking and heating in less developed countries, is obvious.

Fossil fuels originally came from photosynthetic processes. Photosynthesis is a promising source of combustible chemicals to be used for energy production, including transportation fuel, and could certainly produce all needed organic raw materials to substitute for petroleum in the current petrochemicals industry. It suffers from the disadvantage of being a very inefficient means of solar energy collection (a collection efficiency of only a fraction of a percent by photosynthesis is typical of most common plants). However, the overall energy conversion efficiency of several plants, such as sugarcane, is around 0.6%. Furthermore, some plants, such as *Euphorbia lathyris* (gopher plant), a small bush growing wild in California, produce hydrocarbon emulsions directly. The fruit of the Philippine plant, *Pittsosporum resiniferum*, can be burned for illumination due to its high content of hydrocarbon terpenes primarily α-pinene and myrcene. Conversion of agricultural plant residues to energy could be employed to provide much of the energy required for agricultural production. Indeed, until about 1900, virtually all of the energy required in agriculture—hay and oats for horses, home-grown food for laborers, and wood for home heating—originated from plant materials produced on the land. (An interesting exercise is to calculate the number of horses required to provide the energy used for transportation at the present time in the Los Angeles basin. It can be shown that such a large number of horses would fill the entire basin with manure at a rate of several feet per day.)

Annual world production of biomass is estimated at 146 billion metric tons, mostly from uncontrolled plant growth. Many farm crops and trees can produce around 2 metric tons per acre per year of dry biomass, and some algae and grasses can produce significantly more. The heating value of this biomass is 5000–8000 Btu/lb (11,600–18,600 kJ/kg) about half of typical values for coal. However, most biomass contains virtually no ash or sulfur, both problems with coal. Another sustainability advantage of biomass is that all of the carbon in it is taken from carbon dioxide in the atmosphere so that biomass combustion does not add any net quantities of carbon dioxide to the atmosphere. Indeed, use of biomass to produce hydrogen-rich methane or elemental hydrogen along with sequestration of by-product carbon dioxide would result in an overall loss of carbon dioxide from the atmosphere.

As it has been throughout history, biomass is significant as heating fuel, and in some parts of the world is the fuel most widely used for cooking. About 15% of Finland's energy needs are provided by wood and wood products (including black liquor by-product from pulp and paper manufacture). Despite the charm of a wood fire and the somewhat pleasant odor of wood smoke, air pollution from wood-burning stoves and furnaces is a significant problem in some areas. Currently, wood provides about 8% of world energy needs. This percentage could increase through the development of energy plantations consisting of trees grown solely for their energy content.

The potential for biomass fuel is huge with credible estimates of as much as 1500 EJ/year. More realistic estimates of terrestrial sources taking into account sustainability considerations are 200–500 EJ/year. Between 50 and 150 EJ/year would come from organic wastes including municipal solid waste, agricultural by-products (straw and cornstalks), and forestry residues with the remainder provided by surplus forest production and dedicated energy crops. These estimates do not account for the potentially very large production by algae. Among the sources of biomass that could be used for energy and chemical production are grains and sugar crops (for ethanol manufacture), oilseeds, animal by-products, manure, and sewage (the last two for methane generation by anoxic fermentation). The biggest potential source of chemicals is the lignocellulose making up the bulk of most plant material (see Section 17.17.7). Biomass could certainly be used to replace much of the petroleum and natural gas currently consumed in the manufacture of nonfuel primary chemicals in the world each year, of which the largest volume feedstock is ethylene with annual world consumption of slightly more than 100 million metric tons per year.

Biomass such as wood offers several important advantages as a fuel and feedstock including relatively high abundance, competitive cost, and carbon neutrality (no net production of atmospheric-warming carbon dioxide). Disadvantages of biomass fuel include low heating value and energy density, high moisture content, and tendency to produce soot during combustion.

17.17.1 Processing of Biofuel to More Compact Forms

As it comes from the field, biomass is a bulky material with a relatively low energy content, high water content, and tendencies to undergo microbial decay and absorption of moisture. All of these factors lower the value of biomass as a fuel and for processing to gaseous and liquid fuels and feedstocks. Furthermore, bulky biomass with a high water content is relatively costly to transport. Therefore, it is desirable to dry and convert biomass to more compact forms before transporting it or using it as a fuel or feedstock. The volume of loose biomass, such as straw or cornstalks, can be reduced approximately twofold by baling it and another twofold reduction in bulk can be achieved by compressing the material into pellets. Both of these processes add cost to the material.

Traditionally, the problems with biomass fuel described in the preceding paragraph have been largely overcome by converting wood to charcoal consisting mostly of elemental carbon and ash. Charcoal has been the cooking fuel of choice in many societies for centuries. Although charcoal is prepared in a number of ways under a variety of conditions, it is typically made by heating biomass in the absence of air or in an oxygen-deficient atmosphere to about 500°C, which converts most of the cellulose, hemicellulose, and lignin in the wood to combustible gas (CO, H_2, some CH_4, and combustible tar constituents that are burned to pyrolyze the biomass) and leaves a solid consisting mostly of elemental carbon and mineral ash that makes a fuel that burns cleanly at high temperatures. A disadvantage of charcoal is that at best, only slightly over 50% of the original heat content of the biomass is retained in the charcoal.

Some of the disadvantages of using biomass directly for fuel or conversion to charcoal can be overcome by **torrefaction**, a process in which biomass is heated in the absence of oxygen for 2–3 hours in a temperature range of 250–300°C, generally lower than the temperatures used to make charcoal.[13] Such treatment largely destroys the carbohydrate fraction of biomass and leaves much of the lignin, which serves as a binding agent for the solid product. Torrefaction tends to remove the biomass −OH groups, which are sites to which water binds in biomass, thus greatly reducing the tendency of biomass to absorb moisture. In addition to being a more dense, higher heat content material, torrefied biomass burns more readily and with less soot production than charcoal, thus offering the advantages of charcoal, but more chemically reactive than charcoal and thus more amenable to further processing, such as gasification and liquefaction by reaction with H_2.

Figure 17.23 illustrates the conversion of solid biofuel to progressively more compact forms. Starting with loose biofuel such as wheat straw or cornstalks, the yield of energy per unit volume

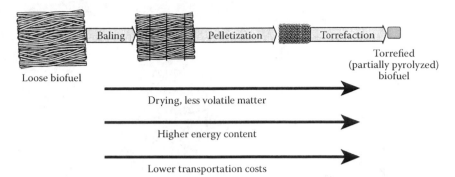

FIGURE 17.23 Processing of biofuel to more compact forms with higher energy content per unit volume and mass.

and mass is increased progressively by (1) drying and compaction, which also decreases susceptibility to physical deterioration and decay; (2) grinding and compression into pellets, which produces a form readily injected into furnaces or co-fired with coal; and (3) torrefaction, partial pyrolysis to give a material with excellent qualities for combustion or gasification that burns cleanly with minimal soot production.

17.17.2 DECARBONIZATION WITH BIOMASS UTILIZATION

Decarbonization of the atmosphere is one of the primary objectives of the harvesting and utilization of biomass, such as in fuel production. For the most part, the utilization of biomass is carbon neutral in that the carbon dioxide released in burning biomass was originally taken from the atmosphere by photosynthesis. The utilization of biomass can result in a net removal of carbon from the atmosphere in two ways. In the first such application, biomass is reacted with O_2 and steam to produce H_2 and the carbon dioxide by-product is sequestered by pumping underground. In the second such application, biomass is pyrolyzed yielding some combustible gaseous fuel and leaving a nondegradable carbon residue commonly called **biochar** that can be added to soil to enhance soil quality and fertility.[14] Biochar can be enriched with fertilizer to further add to its value for soil fertility enhancement. Figure 17.24 shows a biochar-based system in which biochar is used to absorb nutrients from animal and human wastes and added to soil to increase fertility.

Proper management can significantly increase the ability of forests to remove atmospheric carbon dioxide and produce biofuel. Removal of forest residue with its collection for fuel can improve site conditions for planting additional trees. Thinning of trees from excessively dense stands improves the productivity of the remaining stands and reduces the risks of destructive wildfires. Old-growth trees are no longer very effective for removal of carbon dioxide from the atmosphere and production of additional biomass. Therefore, their removal enables planting new trees and vegetation that are much more photosynthetically active. However, the harvesting of old growth forests by clear-cutting can result in increased carbon dioxide release in the near term from decay of exposed biomass and uncontrolled burning of forest wastes.

17.17.3 CONVERSION OF BIOMASS TO OTHER FUELS

It has become standard practice to label as first-generation biofuels those that are made from the food fractions of biomass, especially ethanol produced by fermenting sugars and biodiesel fuel prepared from chemical processing of plant oils and to call second-generation biofuels those that are made from lignocellulose, for example, ethanol made by fermenting sugars that come from enzymatic breakdown of lignocellulose that composes the structural materials in wood or straw. It is

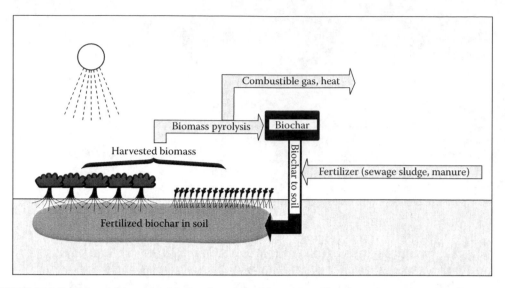

FIGURE 17.24 A system in which biochar is prepared from plant biomass, mixed with sewage sludge and animal manure, and then added to soil to increase fertility. Combustible gas and heat that can be used for district heating can be recovered as part of the charring process.

more accurate to call these two classes of biofuels second- and third-generation biofuels, respectively, and to label as first generation those that consist of the biomass itself or solid biomass that has been dried or partially pyrolyzed. As noted above, these first-generation biofuels are largely consumed in simple cooking and heating facilities and in boilers to raise steam for electrical power generation.

17.17.4 ETHANOL FUEL

A major option for converting photosynthetically produced biochemical energy to forms suitable for internal combustion engines is the production of ethanol, C_2H_6O, by fermentation of sugars from biomass. Using the terminology defined in Section 17.17.3, ethanol would be classified as a second-generation biofuel. Suitably designed internal combustion engines can burn pure ethanol or a mixture of 85% ethanol and 15% gasoline called E85. More commonly, ethanol is blended in proportions of around 10% with gasoline to give **gasohol**, a fuel that can be used in existing internal combustion engines with little or no adjustment.

Gasohol boosts octane rating and reduces emissions of carbon monoxide. From a resource viewpoint, because of its photosynthetic origin, alcohol may be considered a renewable resource rather than a depletable fossil fuel. Ethanol is most commonly produced biochemically by fermentation of carbohydrates. Brazil, a country that produces copious amounts of fermentable sugar from sugarcane, has been a leader in the manufacture of ethanol for fuel uses, with an annual production rate of about 24 billion liters. However, due to sugarcane crop shortfalls in 2009–2010, Brazil actually had to import some ethanol from the United States in early 2011 to make up for a deficiency in this fuel.

All motor fuels in Brazil contain at least 24% ethanol and some fuel is essentially pure ethanol. Most of the gasoline consumed in the United States is supplemented with ethanol as mandated by law.

Although most of the ethanol that has been produced for fuel has been made from the fermentation of grain or sugar, there is legitimate concern that, considering the energy that goes into producing grain ethanol, there is no net energy gain. A potentially much more abundant and cheaper source of ethanol consists of biomass generated as a by-product of crop production including straw from wheat or rice production or cornstalks from growing corn, making it a third-generation biofuel. In the past, much rice straw from commercial production in the United States was simply burned to save the cost of cultivating it back into the soil. Straw cannot be fermented directly, but must be broken down to hexose and pentose sugars for fermentation. This has traditionally been done with acid treatment, which is expensive, although technologies exist for recycling the acid. The major alternative is enzymatic hydrolysis with cellulase enzyme to produce the required sugars. The Canadian Iogen Corporation has a means for obtaining fermentable sugars from wheat straw and other plant materials and has attempted to develop a cost-effective commercial process.

17.17.5 BIODIESEL FUEL

Biodiesel fuel is a growing source of renewable liquid hydrocarbon fuels. Unlike fuel ethanol, which must be transported in corrosion-resistant truck tanker trailers or railroad tankers, biodiesel fuel is readily moved through existing pipelines. Rudolf Diesel developed the high-compression, compression-ignited diesel engine in the late 1800s and first operated the engine in Augsburg, Germany, in 1893. Demonstrated at the World Fair in Paris in 1900, the engine got the "Grand Prix" (highest prize) for the invention. Interestingly, peanut oil fuel was used in Diesel's demonstrations, and vegetable oils were the main source of fuel for diesel engines during the first 2 decades of their use.

Vegetable oils were eventually replaced by petroleum-based hydrocarbons. More recently, diesel fuel has been developed from vegetable oil sources that are derivatives of the fatty acids in the oils. Vegetable oils from soybeans and other biological sources are used to make biodiesel fuel as discussed in the following paragraph.

Vegetable oils are fatty acid esters of glycerol, a 3-carbon alcohol with 3-OH groups attached. To produce biodiesel fuel, the glycerol esters are hydrolyzed by strong base (NaOH) in the presence of methanol alcohol ($HOCH_3$) and the fatty acids are converted to their methyl esters, the molecules composing biodiesel fuel:

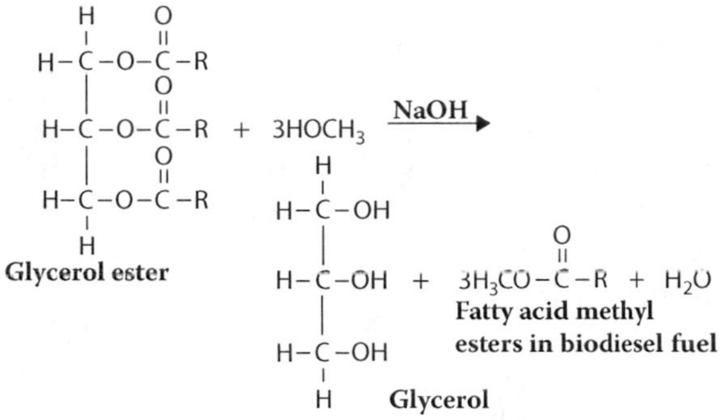

$$(17.9)$$

In this reaction, R stands for a long hydrocarbon chain in one of a number of fatty acids including stearic acid, linoleic acid, oleic acid, lauric acid, and behenic acid. For example, in stearic acid, R is a straight chain with 17 carbon atoms, $C_{17}H_{35}$. Many useful **lipid oils** are extracted from a variety of plant seeds including rapeseed, soybeans, sunflowers, and corn. In addition to their food uses, these oils are used in a large variety of applications including raw materials for making other chemical products, lubricants, as well as for the synthesis of biodiesel fuels. The extraction from plants and the utilization of lipid oils are discussed in some detail in Chapter 13, Section 13.2.

17.17.6 FUEL FROM ALGAE

An interesting possibility for biodiesel fuel production is algae, especially rapidly growing single-cell microalgae (Figure 17.25) that may have an oil content exceeding 50%.[15] Such algae can grow profusely in ponds fed with nutrient-rich effluent from wastewater treatment plants. Oil-producing

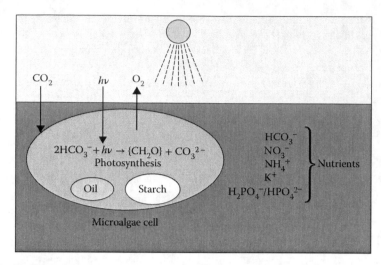

FIGURE 17.25 Rapidly reproducing single-cell microalgae use solar energy and simple inorganic nutrients to generate large quantities of biomass, oils, and starch by photosynthesis.

algae have also been grown in the carbon-dioxide-rich atmosphere provided by power plant stack gases. Unlike terrestrial plants that require freshwater for irrigation, algae can thrive in oceanic saltwater, brackish saline water, and wastewater. By growing algae in treated wastewater within a carbon-dioxide-rich atmosphere from power plants, nutrients can be removed from wastewater, thus reducing eutrophication in receiving waters, and some of the carbon dioxide can be removed from stack gas emissions.

The potential productivity of algae for biomass fuel production is spectacular. Whereas soybean and palm oil sources typically produce, respectively, up to 200 and 2500 L/acre annually of biodiesel fuel, algae have the potential to yield as much as 40,000 L/acre yearly of the fuel. Realistically, a reasonable goal for algae productivity is 50 metric tons of dry biomass per hectare per year with a 25% triglyceride oil content, which translates to 14,000 L of oil per hectare annually.[16] Furthermore, with adequate supplies of water, algae can be grown on desert lands that are unsuitable for growing soil-based crops. One experimental system grows algae in water that circulates through transparent plastic tubes. Some algae thrive in saltwater, which can come from the sea or underground formations bearing brackish water.

Currently, algae including species of *Spirulina*, *Chlorella*, *Dunaliella*, and *Haematococcus* are cultivated for nutritional products. World production is around 10,000 metric tons per year, around half of which is in China. This is a significant amount, though trivial compared to quantities required for meaningful energy production. Microalgae consisting of suspensions of single-cell algae grow profusely in many wastewater treatment ponds where they function to treat wastewater, especially in removing phosphate and fixed nitrogen that contribute to the eutrophication potential of wastewater effluent. Virtually all commercial algae production is done in open ponds, which are much less expensive and simpler than closed systems and offer maximum access to sunlight and atmospheric carbon dioxide. Open ponds suffer from the disadvantage of susceptibility to contamination by wild strains of algae that produce relatively little oil and are otherwise not suitable for energy production.

There are several pathways by which algae may potentially produce biofuels. The most direct way is through the production of biomass that potentially could be dried and used directly as a first-generation biofuel. A second major pathway for making algal biofuels is through direct production of oils, especially lipid oils that are produced profusely by some algae and that can be converted to biodiesel fuel, the focus of most of the current research in algal biofuel production. Some investigations have examined the possibility of algae producing fermentable carbohydrate that could be excreted directly into growth media and fermented to produce the alcohol. Ideally, algae could produce ethanol directly. A "far-out" possibility is to genetically engineer a symbiotic combination of photosynthetic algae and ethanol-producing fungal yeasts (lichens that act to weather rocks are a symbiotic combination of algae and fungi). An intriguing possibility is the direct production of hydrogen, H_2, by algae, although so far it has eluded efforts of investigators in any practical sense.

The strains of algae used are crucial in the production of biofuels. Major criteria must include productivity of biomass, oil content, harvestability, robust character and resistance to contamination, and tolerance to high levels of oxygen produced in photosynthesis. To be utilized to produce fuels, microalgae must be flocculated and concentrated by at least 30-fold; therefore, harvestability is a daunting challenge.

A major advantage in the development of microalgae with respect to genetically engineering improved strains is their very short generation time measured in hours. This enables the growth of literally hundreds of generations of algae per year compared to, at most, three or four generations per year of terrestrial plants under even the most favorable greenhouse and artificial lighting conditions.

17.17.7 UNREALIZED POTENTIAL OF LIGNOCELLULOSE FUELS

Both grain- and sugar-based ethanol and vegetable oils used to make biodiesel fuel are not the best candidates for biomaterial fuels because they use only relatively small fractions of the plants consisting of the parts that have the most value for food, animal feed, and raw materials. A much more abundant source of fuel, a third-generation biofuel using the terminology employed in this chapter, consists of the **lignocellulose** parts of plants. Lignocellulose composed of large polymeric molecules with an approximate empirical formula of CH_2O that makes up the structural members of plants including tree wood, stalks, straw, corncobs, and leaves. The three major components of lignocellulose and their approximate empirical formulas are cellulose/starch, $[C_6(H_2O)_5)]_n$; hemicellulose, $[C_5(H_2O)_4)]_n$; and lignin, $[C_{10}(H_{12}O)_3)]_n$.

Large amounts of crop by-product biomass are generated annually. Assuming conservatively that the amount of this material available in the United States is equal to the mass of corn grain produced, about 230 million tons of crop by-product biomass could be made available for fuel each year, a figure that underestimates the amount of by-product potentially available from crops. An advantage of using crop biomass by-product lignocellulose for biofuel lies in its collection. The harvesting of any crop, such as grain from corn or wheat, requires processing the crop, usually by a combine harvesting machine that traverses the field and through which the crop is run and the grain separated out. Rather than returning all the crop biomass separated from the grain back to the land surface as is the current practice, the material can be collected as part of the harvesting operation and processed for biofuel generation.

The amount of biomass that could be generated from dedicated trees and grass is very high, an estimated 2240 million tons per year in the United States alone. A major advantage of this source is that it comes from perennial plants that can be grown on erodible land, much of which has been taken from agricultural production as the result of government programs. One of the plants that is remarkably productive of biomass consists of hybrid poplars, from the genus *Populus* that includes cottonwoods and aspens. These trees may grow more than 2 m/year and will establish new growth from the stumps of harvested trees, retaining their root systems that prevent soil erosion.

The grass most commonly considered for its biomass productivity is switchgrass. Native to North America, switchgrass is disease- and pest-resistant and requires little fertilizer. It tolerates both drought and flooding very well. Upland varieties of switchgrass grow up to 2 m tall in one growing season on well-drained soils. Lowland varieties can reach heights of 4 m and grow best on heavy soils in bottomlands. Improved varieties of switchgrass have been developed for animal forage and yield around 8 tons per acre of biomass each year, about twice the output of forest biomass.

Another high-yielding grass native to swampy regions, such as the Florida Everglades, is sawgrass (*Cladium jamaicense*), which gets its name from the sawlike serrations on its leaves. It is

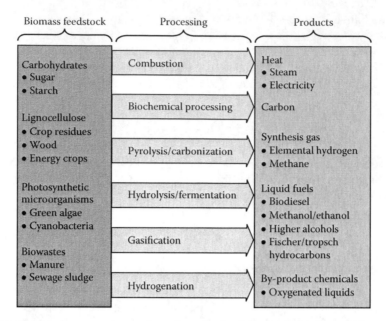

FIGURE 17.26 Feedstocks, processes, and products involved in the conversion of biomass to fuels and energy.

well adapted to cultivation in wet areas where other crops cannot be grown and has the additional advantage of providing good wildlife cover.

There are numerous pathways by which biomass and biowastes (including sewage and animal wastes) can be converted to energy, fuels, and feedstocks. These are the following as summarized in Figure 17.26:

- Direct combustion to produce heat and to raise steam to generate electricity
- Fermentation to produce ethanol (from sugar) or methane
- Pyrolysis to yield gaseous fuels (particularly methane), liquids (including hydrocarbons and oxygenated species), and solid carbon
- Thermochemical gasification to produce CO, H_2, CH_4, by-product liquids, and solid carbon
- Fischer–Tropsch synthesis of hydrocarbons from CO and H_2 derived from biomass
- Hydrogenation of oxygenated liquids from biomass to produce hydrocarbons
- Methyl esterification of oils to produce methyl ester biodiesel fuels

Sustainable preparation of biofuels from biomass makes it desirable to recover nutrients such as phosphorus, potassium, and micronutrient elements from the biomass during processing and return these materials to soil. Chemically fixed nitrogen, predominantly in the form of proteins in the biofuels, is lost during biofuel production, but nitrogen supply from the vast atmospheric reservoir of N_2 makes nitrogen loss a nonproblem.

17.17.8   Chemical Conversion of Biomass to Synthetic Fuels

The most direct way to use biomass for fuel is direct combustion as a first-generation biofuel with a heating value on a dry mass basis about half that of coal. Compared to coal, biomass is low in sulfur and ash (although rice straw has a high ash content, primarily as silica) and the biomass ash does not contain toxic elements such as the arsenic that occur in some coals. Elemental carbon obtained by pyrolysis of biomass can serve as a concentrated form of fuel or as a raw material for gasification. Biomass can be converted to other high-value fuels including hydrocarbons by gasification.

One approach to biomass gasification begins with combustion of part of the biomass (represented by the formula $\{CH_2O\}$) with pure molecular oxygen oxidant (to avoid diluting the product gas with N_2 from air)

$$\{CH_2O\} + O_2 \rightarrow CO_2 + H_2O + \text{heat} \tag{17.10}$$

yielding heat required for the rest of the gasification process. Under the oxygen-deficient conditions through which gasification is carried out, part of the biomass is partially oxidized to combustible carbon monoxide, CO:

$$\{CH_2O\} + 1/2O_2 \rightarrow CO + H_2O + \text{heat} \tag{17.11}$$

Part of the biomass is pyrolyzed by the heat produced by Reaction 17.10

$$\{CH_2O\} + \text{heat} \rightarrow C + H_2O \tag{17.12}$$

yielding hot carbon and combustible gaseous and liquid by-products. The hot carbon reacts with steam

$$C + H_2O + \text{heat} \rightarrow CO + H_2 \tag{17.13}$$

yielding a **synthesis gas** mixture of CO and H_2. Biomass may also react as it is heated

$$\{CH_2O\} + \text{heat} \rightarrow CO + H_2 \tag{17.14}$$

yielding synthesis gas. The carbon monoxide in synthesis gas can be subjected to the water gas shift reaction

$$CO + H_2O \rightarrow CO_2 + H_2 \tag{17.15}$$

giving elemental H_2 as the only gaseous product. The CO_2 generated in Reactions 17.10 and 17.15 can be separated and sequestered, such as by pumping into underground petroleum formations to enable recovery of petroleum, thereby preventing the release of this greenhouse warming gas to the atmosphere.

Elemental hydrogen can be used as an end product of biomass gasification directly as a fuel in gas turbines and other heat engines or to generate electricity in fuel cells. And H_2 can be used to synthesize ammonia, NH_3, an important industrial chemical and fertilizer. A mixture of CO and H_2 can be reacted over a catalyst in a methanation reaction

$$CO + 3H_2 \rightarrow CH_4 + H_2O \tag{17.16}$$

to produce methane, which, made by this method, is called **synthetic natural gas** (SNG). With different proportions of hydrogen and CO reactants and a different catalyst, a mixture of CO and H_2 can react according to the Fischer–Tropsch reaction to yield a variety of hydrocarbons including gasoline, jet fuel, and diesel fuel as shown by Reaction 17.17 for the synthesis of octane, one of the hydrocarbons in gasoline:

$$8CO + 17H_2 \rightarrow C_8H_{18} + 8H_2O \tag{17.17}$$

A similar reaction can also be used to make ethanol and methanol, CH_3OH, which can be used as a fuel, gasoline additive, and to produce H_2 for fuel cells in vehicles.

An interesting possibility for increasing the amount of hydrocarbon fuel that can be obtained from biomass is to use H_2 and O_2 produced by the electrolysis of water with electricity generated by wind power

$$2H_2O + \text{electrical energy} \rightarrow 2H_2(g) + O_2(g) \qquad (17.18)$$

to react with biomass for gasification. The pure elemental oxygen can be used to produce energy from biomass as shown by Reaction 17.10 without diluting the gas product with elemental nitrogen, N_2, as would be the case with air oxidant, thus enabling sequestration of the CO_2 product. The elemental hydrogen generated by electrolysis of water can be reacted directly with biomass. Ideally, direct hydrogenation of biomass would produce predominantly hydrocarbons such as methane generated by the following reaction:

$$\{CH_2O\} + 2H_2 \rightarrow CH_4 + H_2O \qquad (17.19)$$

More commonly, hydrogenation produces a mixture of many liquids, most of which contain oxygen in compounds such as ethers, alcohols, and ketones. For fuel use, these compounds must be further treated with hydrogen to remove oxygen and leave hydrocarbon liquids. Such a process based upon elemental hydrogen from water (Reaction 17.18) provides a means for using the energy originally generated by wind power through the intermediate of elemental H_2 to make a high-energy hydrocarbon fuel from biomass that can be used for transportation or other purposes.

17.17.9 BIOGAS

A significant source of clean-burning methane can be obtained from the anoxic (oxygen-free) bacterial fermentation of biomass of a variety of kinds. Representing biomass as $\{CH_2O\}$, the biochemical reaction is the following:

$$2\{CH_2O\} \rightarrow CH_4 + CO_2 \qquad (17.20)$$

This reaction has long been used in the anoxic digesters of sewage treatment plants to reduce the amount of degradable organic matter in excess sewage sludge. A well-balanced such plant makes enough methane to provide for all its energy needs. Large livestock feeding operations may have digesters to produce methane from livestock manure and other biological wastes associated with the feeding operation. Another source of methane generated by anoxic fermentation is obtained by burying collector pipes in mounds of municipal solid wastes and collecting the off-gas.

An outstanding example of "grassroots" energy sustainability is provided by the widespread use of biogas generators in rural China, an area chronically short of fuel. The Chinese have built millions of anoxic digesters located just below the ground surface in which all kinds of biomass including animal wastes, human excrement, and vegetable and crop wastes are degraded in the absence of air to produce combustible methane gas. The digesters are usually located in proximity to dwellings and to sources of waste, including latrines and pigstys. Covered by a thin layer of soil, these facilities take advantage of solar heating to accelerate fermentation. In addition to producing fuel, anoxic digestion alleviates waste disposal and health problems through the destruction of pathogens in the waste by the fermentation process, which often takes place at a somewhat elevated temperature that kills most pathogenic organisms. An additional advantage is that the fertilizer value of the wastes is conserved so that residues from the digesters can be placed on soil to serve as fertilizer for crops without much concern over spreading pathogens. The contents of both nitrogen and phosphorus in the digester residues are much higher than in biomass that has been treated by mulching in an air environment.

The gas from the Chinese biogas facilities is used largely for cooking and lamps. In larger installations, internal combustion engines have been adapted to run in part on biogas. In the case of gasoline engines, biogas is fed into carburetors through an adjustable feed valve, and in the case of diesel engines, it is fed into the manifold to supplement the minimal amount of diesel fuel injected into the combustion chamber to initiate combustion during engine operation.

The world's largest livestock-based biogas installation is located at the Huishan Dairy in Northeast China where part of the wastes from the dairy's 250,000 cows is subjected to anoxic biodegradation to produce methane used to run large internal combustion engines connected to generators that produce 5.66 MW of power.[17] Hydrogen sulfide is removed from the gas product over iron oxide. The system reportedly reduces carbon emissions to the atmosphere by about 180,000 metric tons per year and generates about 160,000 metric tons of fertilizer annually.

17.17.10 BIOREFINERIES AND SYSTEMS OF INDUSTRIAL ECOLOGY FOR UTILIZING BIOMASS

As discussed in Chapter 13, Section 13.2, biorefineries are facilities in which materials taken from the biosphere are extracted, isolated, refined, and processed to produce useful products.[18] Figure 13.4 is an outline of a biorefinery. It shows several pathways by which biomass can be utilized ranging all the way from simply squeezing sugar-rich sap from sugarcane stalks or oils from oilseeds to sophisticated biochemical processes for processing biomass to products including alcohols and other synthetic fuels. In addition to producing biomaterials, biorefineries can be used to make a variety of fuels from biomass.

17.17.11 SYSTEM OF INDUSTRIAL ECOLOGY FOR METHANE PRODUCTION FROM RENEWABLE SOURCES

Figure 17.27 illustrates an industrial ecosystem based on the production of methane fuel from renewable biomass and wind energy sources. Methane is the ideal clean-burning fuel for virtually all applications that currently use fossil fuels except for aircraft. With ultra-efficient hybrid automobile technology, methane can be used as motor fuel giving an acceptable range between refueling. With

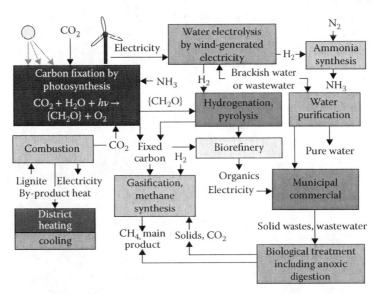

FIGURE 17.27 An industrial ecosystem based on the synthesis of methane and other fuels from biomass using wind-generated electrical power as the major energy source driving the system.

previously untapped sources coming on line in the United States and a number of other countries, methane is relatively available and inexpensive. Pipeline systems used for natural gas are in place for its distribution. And, when methane sources are eventually exhausted, the gas can be synthesized from biomass and elemental hydrogen produced from the electrolysis of water.

As shown in Figure 17.27, biomass is used as a source of fixed carbon that is reacted with elemental hydrogen made by electrolyzing water using renewable wind energy. Another source of methane in the system is anoxic fermentation of wastes including sewage. Instead of municipal refuse in the community being disposed in landfill, it is first subjected to anoxic fermentation to generate methane after which the residues are thermochemically gasified to provide feedstock for chemical synthesis. Other aspects of the system are largely self-explanatory from the figure.

17.18 HYDROGEN AS A MEANS TO STORE AND UTILIZE ENERGY

Discussed as a key feedstock in Chapter 16, Section 16.3, hydrogen gas, H_2, is an ideal chemical fuel in some respects that may serve as a storage medium for energy generated by intermittent sources including solar energy and wind energy. There are several ways to generate elemental hydrogen.[19] As noted in the Section 17.17, solar- or wind-generated electricity can be used to electrolyze water:

$$2H_2O + \text{electrical energy} \rightarrow 2H_2(g) + O_2(g) \qquad (17.21)$$

The hydrogen fuel product, and even elemental oxygen produced as an electrolysis by-product, can be piped some distance and the hydrogen burned without pollution, or it may be used in a fuel cell (Figure 17.7). This may, in fact, make possible a "hydrogen economy." Disadvantages of using hydrogen as a fuel include its low heating value per unit volume and the wide range of explosive mixtures it forms with air. Although not yet economical, photochemical processes can be used to split water to H_2 and O_2 that can be used to power fuel cells.

Fuel cell–powered vehicles are now practical in some applications and Honda began production of the first commercial fuel cell automobile in 2008. One of the greatest barriers to their widespread adoption has been their inability to carry sufficient hydrogen for an acceptable range. Several solutions to this problem are now being investigated. One is the potentially problematic use of very cold liquid hydrogen as a fuel source. Another is the use of very high pressure containers composed of multilayer cylinders wrapped with carbon composite and filled with hydrogen at pressures up to 10,000 psi (about 670 × atmospheric pressure!) reputed to contain sufficient hydrogen to propel an automobile 300 miles. Other systems use catalysts to break down liquid fuels, such as methanol or gasoline, to generate hydrogen for fuel cells.

Although there has been much enthusiasm in some quarters for hydrogen fuel and predictions of a new hydrogen economy, some of the more enthusiastic arguments for hydrogen fuel may be too optimistic. The most important point is that, unlike fossil fuels, hydrogen is not a primary source of energy and has to be made by processes such as the electrolysis of water (Reaction 17.18) that use other sources of energy. Most of the 6 million metric tons of elemental hydrogen used in the United States each year is made from steam reforming of methane from natural gas:

$$CH_4 + H_2O \rightarrow 3H_2 + CO \qquad (17.22)$$

The carbon monoxide product can be reacted with steam as noted in Section 17.17.8, to produce additional H_2, and the CO_2 product of this reaction can be sequestered as discussed in Section 17.11:

$$CO + H_2O \rightarrow CO_2 + H_2 \qquad (17.23)$$

In principle, the preceding described process and the utilization of elemental hydrogen in fuel cells can provide a transport fuel that is pollution-free. However, methane gas is easier to store than elemental hydrogen, and the modern internal combustion engine with associated emissions control equipment is virtually pollution-free. So the intermediate production of elemental hydrogen is unlikely to be the most practical approach. Production of elemental hydrogen by electrolysis of water using electricity from renewable sources, such as photovoltaics and wind power, is essentially nonpolluting, but it requires balancing the relatively inefficient electrolysis process against the value of the electricity that it consumes.

17.19   COMBINED POWER CYCLES

Combined power cycles enable much more efficient utilization of combustible fuels by first using the heat of combustion in a turbine coupled to an electrical generator, raising steam in a boiler with the hot exhaust gas from this turbine, using the steam to power a second turbine linked to a generator, and finally using the steam and hot water from the steam turbine for applications such as processing in the chemical industry, heating commercial buildings, or heating homes. A schematic diagram of a combined power cycle system is illustrated in Figure 17.28. The water condensed from the steam used for heating is pure and is recycled to the boiler, thus minimizing the amount of makeup boiler feedwater, which requires expensive treatment to make it suitable for use in boilers. The use of steam leaving a steam turbine for heating, a concept known as **district heating**, is commonly practiced in Europe (and many university campuses in the United States) and can save large amounts of fuel otherwise required for heating. Such a system as the one described is in keeping with the best practice of industrial ecology and should be employed whenever it is practical to do so.

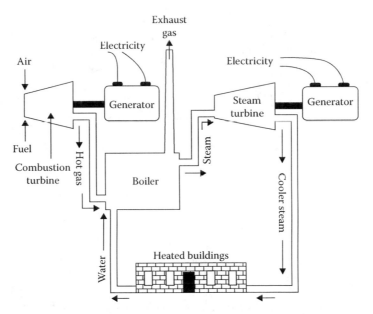

FIGURE 17.28   A combined power cycle in which combustible gas or oil is first used to fire a gas turbine connected to an electrical generator. The hot gases from this turbine are fed to a boiler to raise steam, which drives a steam turbine, also connected to a generator. The still hot exhaust steam from the steam turbine is used for process heat or conveyed to commercial or residential buildings for heating. The water condensed from the steam in this final application is returned to the power plant to generate more steam, thus conserving water and avoiding the necessity to treat more water to the high purity standards required by the boiler.

17.20   ENVIRONMENTAL HEALTH ASPECTS OF ENERGY PRODUCTION AND UTILIZATION

All means of energy production, whether from depleting or renewable sources, have important implications for the environmental health of workers and the public exposed to energy production and utilization activities. The energy enterprise is huge, involving millions of workers around the world. Their safety and well-being is an important part of the energy picture.

17.20.1   COAL

The greatest impact of all the energy production and utilization activities has to do with coal. All aspects of coal have important environmental health implications.

Subsurface coal mining is the most deadly energy industry worldwide with thousands of miners killed each year from blasts, rockfalls, floods, and other causes. Many coal seams release explosive levels of methane gas, explosions of which probably cause the majority of coal-mining deaths. Of all countries, China has the largest coal mining death toll accounting for 80% of the world's total while only producing about 35% of the world's total coal. According to the Chinese State Administration of Work Safety, 2433 coal miners were killed in China in 2010, a typical figure for the annual toll in that country. In 2006, an astounding number of 4749 coal miners were killed in China's coal mines. In comparison, the lowest number of coal miners killed in one year in the United States during the past century was in 2009 with a total of 18 miners lost. The most deadly recent year in the United States was 2010 when 48 miners lost their lives, including 29 at Massey Energy's Upper Big Branch Mine.

A significant toll of coal miners is caused by **black lung disease**, a potentially debilitating occupational disease caused by prolonged breathing of coal dust. This malady is a form of pneumoconiosis, a type of interstitial lung disease caused by long-term inhalation of generally mineral irritants in which the lung air sacs become inflamed and scarring of the tissue between the sacs stiffens the lungs, severely reducing lung function and making the victim more susceptible to bacterial and viral lung disease. Two main stages of black lung disease are recognized: (1) simple coal miner's pneumoconiosis, which produces few symptoms and (2) complicated coal miner's pneumoconiosis, a debilitating massive pulmonary fibrosis. Fatalities resulting from black lung disease are uncertain. On a website, the United Mine Workers of America[20] states that "Today it is estimated that 1500 former coal miners each year die an agonizing death in often isolated rural communities, away from the spotlight of publicity."

The utilization of coal has some important environmental health implications for the general public. Vast improvements have been made in controlling the emissions from coal-fired power plants during the past century. The most deadly example of the damage that coal combustion without proper pollution controls can cause is provided by the "killer smog" that afflicted London for several days beginning on December 5, 1952.[21] Thousands of tons of black soot, tar particles, and sulfur dioxide raised the levels of PM_{10} (particulate matter less than 10 μm in size) in the London atmosphere up to 14,000 mg/m^3, at least 50 times normal levels, and sulfur dioxide levels reached 700 parts per billion. Death rates from pneumonia, bronchitis, tuberculosis, and heart failure soared, reaching levels of 900 per day at their peak. A total of at least 4000 excess deaths occurred during the entire incident while thousands more probably died in the months following due to exposure to the deadly air. Those at most risk were people already afflicted with respiratory and cardiovascular conditions, suffering lung inflammation and damage, impaired breathing, increased numbers of asthma attacks, asphyxiation, and cardiac distress. The incident caused the British government to enact air pollution control legislation that largely eliminated air pollution problems from coal combustion.

Mercury, a potent neurotoxin, is released during the combustion of coal. This mercury poses some direct threat to human health. Various species of wildlife are subject to harm from this mercury. Some long-lived fish species tend to accumulate levels of mercury that are potentially harmful to humans who eat them in large quantities. Estimates of mercury emissions from coal-fired power plants vary. Releases in the United States from coal combustion have been estimated by the U.S. Department of Energy at around 50 tons per year as elemental mercury vapor and mercury compounds. Wet scrubbers typically remove about half of the mercury from flue gas, and baghouse collection of particles can also remove significant amounts of mercury. One of the most effective technologies for mercury removal is injection of powdered activated carbon followed by collection in a baghouse.

17.20.2 PETROLEUM AND NATURAL GAS

There are significant health and safety implications associated with the drilling, transport, and refining of petroleum. The 2010 blowout and fire of BP's Deepwater Horizon well in the Gulf of Mexico killed 11 workers outright. Only 14 of 67 workers were rescued from the December 2011 sinking of a Russian deep sea oil rig when the rig sank in heavy seas near the island of Sakhalin off Russia's east coast. It is possible that some refinery workers suffer adverse health effects from exposure to toxic organics from petroleum and its refined products. Of these, arguably one of the more hazardous is benzene, prolonged exposure to which can cause blood abnormalities and possibly leukemia (it is now required that benzene be removed from the mix of hydrocarbons going into gasoline).

The greatest hazard to the biosphere from petroleum production and transport results from oil spills that kill wildlife. Most such incidents occur in coastal areas, affecting birds, fish, and shellfish. Two of the more notorious incidents of oil spills that caused harm to wildlife were the 1989 Exxon Valdez tanker oil spill in Prince William Sound, Alaska, which may have spilled up to 750,000 barrels of crude oil, and the 2010 blowout and fire of BP's Deepwater Horizon well in the Gulf of Mexico, which some credible sources have estimated released as much as 5 million barrels of crude oil into the fragile Gulf waters.

A toxicity hazard associated with petroleum and natural gas production involves highly toxic hydrogen sulfide gas (H_2S) associated with these fuels. Fatalities have occurred from releases of this gas from oil wells in which pressurized carbon dioxide was being used for petroleum recovery. As noted in Chapter 7, Section 7.3, more than 240 people died from hydrogen sulfide poisoning when H_2S was released from a natural gas well in Gaoqiao, China, in 2003. About 9,000 people required hospital treatment and 64,000 people were evacuated from the area. The hydrogen sulfide problem was alleviated by igniting the gas, converting the hydrogen sulfide to sulfur dioxide, which, though an irritant to the respiratory system, did not cause fatalities.

Another toxic gas associated with the use of petroleum and natural gas as fuels is carbon monoxide, CO. Poisoning by this colorless, odorless gas results in around 20,000 hospitalizations and about 450 deaths in the United States each year. As noted in the discussion of coal conversion in Section 17.10, inhalation of the mixture of CO and H_2 produced by the manufacture of synthesis gas from coal and piped into homes in the 1800s and early 1900s was once a leading cause of accidental or intentional (suicide) deaths. Now most cases of carbon monoxide poisoning are due to improperly functioning stoves, furnaces, and hot water heaters and from automobile exhausts, the latter often the result of suicide. Cases of carbon monoxide poisoning have occurred from the use of portable electric generator powered by gasoline engines that are commonly used in the event of power failures. High-level exposure to CO has a number of effects including disorientation, collapse, cardiorespiratory failure, and death. A significant fraction of survivors of carbon monoxide poisoning later suffer neurocognitive disorders including memory loss.

17.20.3 NUCLEAR ENERGY

Significant health hazards have resulted from exposure to radioisotopes in the mining of uranium used as nuclear reactor fuel. The main hazards are from inhalation of radioactive uranium decay products. The most significant of these are radium-226 and radon-226. The radium is carried by dust particles and radon is a noble gas. Both of these radioisotopes are alpha particle emitters, the most dangerous form of radioactivity if it gets inside the human body. The main adverse health effects from exposure to these radioisotopes in mining and uranium mineral processing are lung cancer (strongly synergistic with smoking) and bone cancer.

Because of radiation hazards, accidents at nuclear power reactors have the potential to cause deaths and illness. The worst nuclear accident of all time, the 1986 explosion and fire of Power Reactor Number 4 at Chernobyl in the former Soviet Union (now part of Ukraine) caused 30 deaths outright in personnel exposed at the site, two from the force of the explosion and 28 from acute radiation poisoning soon afterward. A total of 134 cases of acute radiation poisoning were diagnosed among workers at the site. About 265,000 people, including all the residents of the plant operators' town of Pripyat 3 km from the reactor were evacuated in a sparsely populated 4300-km^2 area around the power plant. The new town of Slavutich located 30 km from the reactor has been constructed to house workers at the power plant and those involved in cleanup.

The main contaminant radioisotopes of health concern in the areas that received fallout from the Chernobyl incident are cesium-137 with a half-life of 30 years and strontium-90 a biological "bone-seeker" with a half-life of 29 years (the other fission product of most health concern, iodine-131, has a half-life of only 8 days and was essentially all gone from the affected area within a few weeks after the fire was extinguished). In 2010, the government of Belarus, a small country adjacent to Chernobyl that received much of the fallout from the reactor explosion and fire, announced a program to resettle residents in the contaminated area. The plan calls for conversion of much of the affected area to forests, something that has taken place to a large extent spontaneously following abandonment of the area. After studies to establish areas where crops can be grown without contamination by cesium-137 and strontium-90, agriculture is to be reestablished. Use of local firewood has been banned (officially) and natural gas lines are being installed to provide fuel for the new residents.

As discussed in Section 17.13, the explosion and meltdown of reactors at the Fukushima Daiichi Nuclear Power Station in Japan that resulted from the March 11, 2011, earthquake and tidal wave off the coast of Honshu, Japan, resulted in the release of large quantities of radioactive fission products. To protect human health, evacuation of land in a 20-km radius around the power plant was mandated and sale of food crops raised in the area was banned. For several days following the incident, levels of iodine-131 in some of Tokyo's municipal water supplies were elevated leading to advisories against use of the water for infant formula. Unlike Chernobyl, nobody was killed by acute exposure to radiation at Fukushima. However, a number of workers were exposed to significant, but non-life-threatening, levels of radiation at the reactor site as they worked to bring the damaged reactors under control, and two workers were briefly hospitalized for radiation exposure to their feet and lower legs after they stepped into water contaminated by fission products. As many as 3000 workers have subsequently been exposed to significant amounts of radiation during cleanup efforts at the site (Figure 17.29).

QUESTIONS AND PROBLEMS

Access to and use of the Internet is assumed in answering all questions including general information, statistics, constants, and mathematical formulas required to solve problems. These questions are designed to promote inquiry and thought rather than just finding material in the text. So in some cases, there may be several "right" answers. Therefore, if your answer reflects intellectual effort and a search for information from available sources, it may be considered to be "right."

1. The tail of a firefly glows, although it is not hot. Explain the kind of energy transformation that is most likely involved in the firefly's producing light.
2. What is the standard unit of energy? What unit did it replace? What is the relationship between these two units?
3. Which law states that energy is neither created nor destroyed?

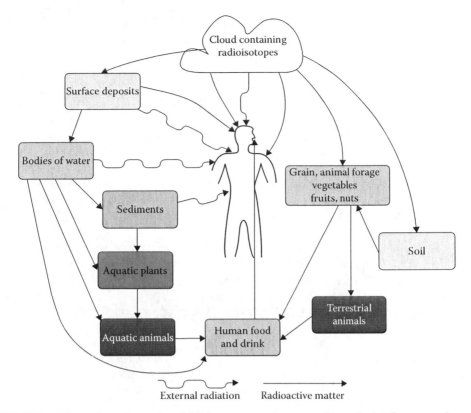

FIGURE 17.29 The main pathways by which humans may be exposed to radiation from a nuclear reactor incident. The wavy lines indicate direct exposure to radiation and the solid lines show transfer of radioactive matter. The two major ways in which humans are exposed are inhalation of radioactive particles from the cloud of radioactive materials or from windblown dust and from ingestion from food and drink.

4. What is the special significance of 1340 watts with respect to potential energy sources?
5. What is the reaction in nature by which solar energy is converted to chemical energy?
6. In what respects is wind both one of the oldest and one of the newest sources of energy?
7. What are two major problems with reliance on coal and petroleum for energy?
8. Why does natural gas contribute less to greenhouse warming than petroleum and much less than coal?
9. How might coal be utilized for energy without producing greenhouse gas carbon?
10. What is a large limiting factor in growing biomass for fuel, and in what respect does this limit hold hope for the eventual use of biomass fuel?
11. What relationship describes the limit to which heat energy can be converted to mechanical energy?
12. Why does a diesel-powered vehicle have significantly better fuel economy than a gasoline-powered vehicle of similar size?
13. Why is a nuclear power plant less efficient in converting heat energy to electricity than is a fossil-fueled power plant?
14. Instead of having a spark plug that ignites the fuel, a diesel engine has a glow plug that operates only during engine start-up. Explain the operation of the glow plug.
15. Cite two examples of vastly increased efficiency of energy utilization that took place during the 1900s.
16. Describe a combined power cycle. How may it be tied with district heating?
17. What are three reactions used in biomass gasification?

18. What is a major proposed use of liquid methanol as a fuel for the future?
19. Describe a direct and an indirect way to produce electricity from solar energy.
20. What is the distinction between donor and acceptor layers in photovoltaic cells?
21. Using Internet resources for information, list some possible means for storing energy generated from solar radiation?
22. What are the advantages of *P. resiniferum* and *E. lathyris* for the production of biomass energy?
23. Corn produces biomass in large quantities during its growing season. What are two potential sources of biomass fuel from corn, one that depends on the corn grain and the other that does not?
24. Does the use of biomass for fuel contribute to greenhouse gas carbon dioxide? Explain.
25. What fermentation process is used to generate a fuel from wastes such as animal wastes?
26. What are two potential pollution problems that accompany the use of geothermal energy to generate electricity?
27. What basic phenomenon is responsible for nuclear energy? What keeps the process going?
28. What is the biggest problem with nuclear energy? Why is it not such a bad idea to store spent nuclear fuel at a reactor site for a number of years before moving it?
29. What is meant by passive stability in nuclear reactor design?
30. What is the status of thermonuclear fusion for power production?
31. Arrange the following energy conversion processes in order from the least to the most efficient: (a) electric hot water heater, (b) photosynthesis, (c) solar cell, (d) electric generator, (e) aircraft jet engine.
32. Considering the Carnot equation and common means for energy conversion, what might be the role of improved materials (metal alloys and ceramics) in increasing energy conversion efficiency?
33. As it is now used, what is the principle or basis for the production of energy from uranium by nuclear fission? Is this process actually used for energy production? What are some of its environmental disadvantages? What is one major advantage?
34. What would be at least two highly desirable features of nuclear fusion power if it could ever be achieved in a controllable fashion on a large scale?
35. Justify describing the sun as "an ideal energy source." What are two big disadvantages of solar energy?
36. What are some of the greater implications of the use of biomass for energy? How might such widespread use affect greenhouse warming? How might it affect agricultural production of food?
37. On the Internet look up the Bakken tight shale formation in North Dakota as an emerging source of petroleum. Was it producing any oil in 2000? What is its current rate of production? Why was it not developed earlier, such as during the first "energy crisis" in the 1970s?
38. From a search on the Internet, look up the largest coal-based synthetic fuels plant in the United States. Where is it located? What does it make? How long has it been operating? Get the same information for the largest coal-based synthetic fuel plant in the world. Is there at least one other synthetic fuels plant in the world that uses natural gas as a feedstock? What is the future of such plants given recent increased development of previously inaccessible natural gas resources?

LITERATURE CITED

1. Yergin, Daniel, *The Quest: Energy, Security, and the Remaking of the Modern World*, Penguin Press, London, 2011.
2. World Energy Council, *2010 Survey of Energy Resources*, World Energy Council, London, 2010, http://www.worldenergy.org/documents/ser_2010_report_1.pdf.

3. Davy, Michael, Robert L. Evans, and Andrew Mezo, The Ultra Lean Burn Partially Stratified Charge Natural Gas Engine, *Proceedings of the Ninth International Conference on Engines and Vehicles*, SAE International, Warrendale, PA, 2009.

4. Martinot, Eric, Renewable Power for China: Past, Present, and Future, *Frontiers of Energy and Power Engineering in China*, **4**, 287–294 (2010).

5. Tollefson, Jeff, Is FutureGen Betting on the Wrong Rock? *Nature*, **472**, 398–399 (2011).

6. Great Plains Synfuels Plant, Dakota Gasification Company, http://www.dakotagas.com/, 2012.

7. European Nuclear Society, http://www.euronuclear.org/, 2012.

8. Wald, Matthew, Federal Regulators Approve Two Nuclear Reactors in Georgia, *New York Times*, February 10, 2012, p. B3.

9. Martin, Richard, *SuperFuel: Thorium, the Green Energy Source for the Future*, Palgrave Macmillan, New York, 2012.

10. Kadra Branker, Michael J. M. Pathak, and Joshua M. Pearce, A Review of Solar Photovoltaic Levelized Cost of Electricity, *Renewable and Sustainable Energy Reviews*, **15**, 4470–4482 (2011).

11. Andreiadis, Eugen S., Murielle Chavarot-Kerlidou, Marc Fontecave, and Vincent Artero, Artificial Photosynthesis: From Molecular Catalysts for Light-Driven Water Splitting to Photoelectrochemical Cells, *Photochemistry and Photobiology*, **87**, 946–964 (2011).

12. Rosillo-Calle, Frank, and Francis X. Johnson, Eds., *Food versus Fuel: An Informed Introduction to Biofuels*, Zed Books, London, 2011.

13. Ben, Haoxi, and Arthur J. Ragauskas, Torrefaction of Loblolly Pine, *Green Chemistry*, **14**, 72–76 (2012).

14. Lehmann, Johannes, and Stephen Joseph, Eds., *Biochar for Environmental Management: Science and Technology*, Routledge, London, 2009.

15. Williams, Peter, Ian Joint, and Carole A. Llewellyn, *Algae for Biofuels, Biomass, and Fine Chemicals*, Earthscan/Routledge, London, 2012.

16. Benemann, John R., Opportunities and Challenges in Algae Biofuels Production, A Position Paper by Dr. John R. Benemann in line with Algae World, 2008, http://www.fao.org/uploads/media/algae_positionpaper.pdf.

17. Boyle, Rebecca, Chinese Cow Manure Generates Electricity in Largest-Ever Methane Capture System, http://www.popsci.com/technology/article/2010-11/chinese-cow-manure-generates-electricity-largest-ever-methane-capture-system, 2010.

18. Aresta, Michele, Angela Dibenedetto, and Frank Dumeignil, Eds., *Biorefinery: From Biomass to Chemicals and Fuels*, Walter de Gruyter, Berlin, 2012.

19. Holladay, Jamie D., Jinxin Hu, David L. King, and Yuan Wang, An Overview of Hydrogen Production Technologies, *Catalysis Today*, **139**, 244–260 (2009).

20. United Mine Workers of America website "Black Lung," http://www.umwa.org/?q=content/black-lung

21. De Angelo, Laura, London Smog Disaster, England, *Encyclopedia of Earth*, 2008, http://www.eoearth.org/article/London_smog_disaster,_England.

SUPPLEMENTARY REFERENCES

Botkin, Daniel B., *Powering the Future: A Scientist's Guide to Energy Independence*, F. T. Press, Upper Saddle River, NJ, 2010.

Brenes, Michael D., Ed., *Biomass and Bioenergy: New Research*, Nova Science Publishers, New York, 2006.

Capehart, Barney L., *Guide to Energy Management*, 7th ed., Fairmont Press, Lilburn, GA, 2011.

Coley, David A., *Energy and Climate Change: Creating a Sustainable Future*, Wiley, Hoboken, NJ, 2008.

Ginley, David S., and David Cahen, Eds., *Fundamentals of Materials for Energy and Environmental Sustainability*, Cambridge University Press, Cambridge, UK, 2011.

Goetzberger, Adolf, and Volker U. Hoffman, *Photovoltaic Solar Energy Generation*, Springer, Berlin, 2005.

Harvey, L. D. Danny, Ed., *A Handbook on Low-Energy Buildings and District-Energy System: Fundamentals, Techniques and Examples*, Earthscan, Sterling, VA, 2006.

Hau, Erich, *Windturbines: Fundamentals, Technologies, Application, and Economics*, 2nd ed., Springer, Berlin, 2006.

Herbst, Alan M., and George W. Hopley, *Nuclear Energy Now: Why the Time Has Come for the World's Most Misunderstood Energy Source*, Wiley, Hoboken, NJ, 2007.

Hinrichs, Roger, *Energy: Its Use and the Environment*, 5th ed., Brooks Cole, Belmont, CA 2012.

Hoffman, Peter, *Tomorrow's Energy: Hydrogen, Fuel Cells, and the Prospects for a Cleaner Planet*, MIT Press, Cambridge, MA, 2012.

Hofman, Konrad A., Ed., *Energy Efficiency, Recovery and Storage*, Nova Science Publishers, New York, 2007.

Infield, David, and Leon Freris, *Renewable Energy in Power Systems*, Wiley, Hoboken, NJ, 2008.

Krauter, Stefan C. W., *Solar Electric Power Generation—Photovoltaic Energy Systems*, Springer, New York, 2006.

Kruger, Paul, *Alternative Energy Resources: The Quest for Sustainable Energy*, Wiley, Hoboken, NJ, 2006.

Laughlin, Robert B., *Powering the Future: How We Will (Eventually) Solve the Energy Crisis and Fuel the Civilization of Tomorrow*, Basic Books, New York, 2011.

Lester, Richard K., and David M. Hart, *Unlocking Energy Innovation: How America Can Build a Low-Cost, Low-Carbon, Energy System*, The MIT Press, Cambridge, MA, 2011.

Manwell, James F., Jon G. McGowan, and Anthony L. Rogers, *Wind Energy Explained: Theory, Design, and Application*, 2nd ed., Wiley, Hoboken, NJ, 2010.

McGowan, Thomas, *Biomass and Alternate Fuel Systems: An Engineering and Economic Guide*, Hoboken, NJ, 2009.

Nag, Ahinda, Ed., *Biofuels Refining and Performance*, McGraw-Hill, New York, 2008.

O'Hayre, Ryan, *Fuel Cell Fundamentals*, Wiley, Hoboken, NJ, 2006.

Patel, Mukund R., *Wind and Solar Power Systems: Design, Analysis, and Operation*, 2nd ed., Taylor & Francis, London, 2006.

Peinke, Joachim, *Wind Energy*, Springer, New York, 2006.

Savage, Lorraine, Ed., *Geothermal Power*, Greenhaven Press, Detroit, MI, 2007.

Soetaert, Wim, and Erick Vandamme, *Biofuels*, Wiley, Hoboken, NJ, 2008.

Sørensen, Bent, *Renewable Energy Conversion, Transmission, and Storage*, Elsevier/Academic Press, Amsterdam, 2007.

Stolten, Detlef, Ed., *Hydrogen and Fuel Cells*, Wiley-VCH, Weinheim, Germany, 2010.

Suppes, Galen J., and Truman S. Storvick, *Sustainable Nuclear Power*, Elsevier/Academic Press, Amsterdam, 2007.

Tabak, John, *Solar and Geothermal Energy*, Facts On File, New York, 2009.

U. S. Energy Information Agency, Energy Explained, Your Guide to Understanding Energy, http://205.254.135.7/energyexplained/, 2012.

Wachtel, Alan, *Geothermal Energy (Energy Today)*, Chelsea House Publications, New York, 2010.

Weiss, Charles, and William B. Bonvillian, *Structuring an Energy Technology Revolution*, MIT Press, Cambridge, MA, 2009.

Wolfson, Richard, *Energy, Environment, and Climate*, 2nd ed, W. W. Norton, New York, 2011.

Worldwatch Institute, *Biofuels for Transport: Global Potential and Implications for Energy and Agriculture*, Earthscan, Sterling, VA, 2007.

Yergin, Daniel, *The Quest: Energy, Security, and the Remaking of the Modern World*, Penguin Press, London, 2011.

Zini, Gabriele, and Paolo Tartarini, *Solar Hydrogen Energy Systems: Science and Technology for the Hydrogen Economy*, Springer, Berlin, 2012.

18 Analytical Chemistry and Industrial Hygiene

18.1 ANALYTICAL CHEMISTRY

Analytical chemistry is that branch of the chemical sciences employed to determine the composition of a sample of material. A **qualitative analysis** is performed to determine *what* is in a sample. The amount, concentration, composition, or percentage of a substance present in a sample is determined by **quantitative analysis**. Often, both qualitative and quantitative analyses are performed as part of the same process.

Analytical chemistry is important in practically all areas of human endeavor and in all spheres of the environment. Industrial raw materials and products processed in the anthrosphere are assayed by chemical analysis, and analytical monitoring is employed to monitor and control industrial processes. Hardness, alkalinity, and trace-level pollutants (see Chapters 3–5) are measured in water by chemical analysis. Nitrogen oxides, sulfur oxides, oxidants, and organic pollutants (see Chapters 6–8) are determined in air by chemical analysis. In the geosphere (see Chapters 9–11), fertilizer constituents in soil and commercially valuable minerals in ores are measured by chemical analysis. In the biosphere, xenobiotic materials and their metabolites (see Chapters 2 and 12) are monitored by chemical analysis. As discussed further in this chapter, analytical chemistry is very important in the area of occupational health and the practice of industrial hygiene.

Analytical chemistry is a dynamic discipline. New chemicals and increasingly sophisticated instruments and computational capabilities are constantly coming into use to improve the ways in which chemical analyses are done. Some of these improvements involve the determination of ever smaller quantities of substances, others simplify the procedures for analysis and greatly shorten the time required for analysis, and some enable analysts to tell with much greater specificity the identities of a large number of compounds in a complex sample.

This chapter is designed to provide an overview of analytical chemistry particularly as it applies to environmental/toxicological chemistry and to the practice of industrial hygiene. Whereas classical analytical chemistry involves measurements of masses (gravimetry) and volumes (volumetric analysis), the current practice of analytical chemistry is largely based on the use of instruments (instrumental analysis). Modern analytical instruments are computer controlled, and analytical data are processed by computers. Given the enormous number of pollutants and other substances that need to be determined in the hydrosphere, atmosphere, geosphere, biosphere, and anthrosphere, the literature of analytical chemistry fills many volumes. For readers who may become directly involved in doing chemical analyses, a more detailed coverage of the topic and specific analytical procedures are discussed in the reference works listed at the end of this chapter.

18.2 INDUSTRIAL HYGIENE AND ANALYTICAL CHEMISTRY

In Chapter 14, Section 14.15, **industrial hygiene** is defined as the discipline devoted to recognizing and preventing conditions in the anthrospheric workplace that may have detrimental effects on workers and on people in the immediate surroundings. A crucial aspect of industrial hygiene and occupational health is to recognize the hazards posed by anthrospheric activities. This often requires the application of analytical chemistry. Because of the strong role that analytical chemistry plays in the practice of industrial hygiene, the two areas are considered together in this chapter.

18.2.1 What Is Industrial Hygiene?

The workplace is potentially a source of substances, conditions, and stresses that may adversely affect workers and the public through impaired health, a lowered quality of well-being, trauma, increased rate of aging, or discomfort.[1] Industrial hygiene is devoted to anticipating, recognizing, evaluating, and controlling physical, chemical, biological, or ergonomic stressors from the anthrosphere that may adversely affect human well-being. Industrial hygiene employs corrective measures to reduce or eliminate exposure to hazards. This may be accomplished by the implementation of protective measures, of which wearing safety glasses; hard hats; ear protection; respirators; or, in extreme cases, fully protective "moon suits" are examples. Preferably, hazards are reduced by such measures as installation of ventilating equipment, substitution of safer processes for more hazardous ones, and using less hazardous materials. Often, hazards are reduced by good housekeeping measures and prompt, safe disposal of potentially hazardous wastes.

18.2.2 Laws and Regulations Pertaining to Occupational Safety and Health

Prior to 1970, regulation of workplace health and safety in the United States was primarily the responsibility of the states with the result that enforcement was often poor and not uniform. In 1970, the U.S. Congress enacted the Occupational Safety and Health Act (OSHA) with the stated goal to "assure so far as possible every working man and woman in the nation safe and healthful working conditions and to preserve our human resources."[2] The OSHA legislation obligated each employer to "furnish to each employee a place of employment, which is free from recognized hazards that are causing or are likely to cause death or serious harm to their employees," and "comply with occupational safety and health standards under the Act." For employees, it was stipulated, "Each employee shall comply with occupational and health standards and all rules, regulations, and orders issued pursuant to the Act which are applicable to his own actions and conduct."[2] Concurrent with the effective date of the Act on April 28, 1971, the Occupational Safety and Health Administration (OSHA) came into existence as an agency of the U.S. Department of Labor. As part of the OSHA, the National Institute for Occupational Safety and Health (NIOSH) was established within the Centers for Disease Control and Prevention (CDCs) of the U.S. Public Health Service.

18.3 CATEGORIES OF WORKPLACE HAZARDS

There are four broad categories of hazards that can pose risks to workers. These categories are listed as follows:

- **Physical hazards**, including nonionizing radiation (extremely low frequency radiation, such as the 60-Hz radiation from power lines, radio-frequency radiation, microwave radiation, infrared radiation, visible light, ultraviolet radiation, and laser hazards), ionizing radiation (particulate radioactivity including α and β particles and electromagnetic radiation including higher frequency ultraviolet rays, x-rays, γ-rays), noise, vibration, extreme temperatures, and excessively high or low pressures
- **Ergonomic hazards**, including procedures, work areas, and tools that may cause injury or fatigue in performing routine operations
- **Biological hazards**, consisting of living organisms, especially pathogenic microorganisms, that may cause illness in exposed humans
- **Chemical hazards**, especially airborne gases, vapors, dusts, or fumes that pose inhalation hazards as well as skin irritants and toxicants that may be absorbed through the skin

Of the aforementioned hazards, the ones to which analytical chemistry is most pertinent are chemical hazards.

18.4 CHEMICAL HAZARDS

The majority of occupational maladies come from inhaling toxic substances present in the microenvironment, and a significant fraction of maladies arise from skin contact that may damage the skin directly or that may serve as a pathway for the entry of systemic toxicants into the body system. Depending on qualities such as their solubility or particle size, inhalation of some substances may irritate the upper respiratory tract, damage the terminal passages or air sacs in the lungs, or act as systemic poisons after inhalation and absorption through the lungs.

Some substances pose a direct hazard in the workplace, an example of which is vinyl chloride, the inhalation of which may cause a characteristic form of liver cancer. Other substances are those liberated in the processing of otherwise benign materials, for example, toxic by-products liberated during the machining or heating of otherwise harmless polymers. An important tool to warn against the harmful effects of substances consists of the material safety data sheet (MSDS), which, under OSHA regulations, must be provided to vendors of materials.

A common class of workplace toxicants consists of **asphyxiants** that act to deprive the body and its tissues of oxygen. Simple asphyxiants act by simply diluting the oxygen in the air to the point that blood pumped from the lungs lacks enough oxygen to support normal respiratory processes. Chemical asphyxiants act by depriving blood of its normal ability to carry oxygen or interfere with the metabolism (respiration) of oxygen in tissues. An example of the former is carbon monoxide, which binds strongly to blood hemoglobin so that it no longer carries oxygen, and an example of the latter is hydrogen cyanide, which, after getting into the blood stream by inhalation, acts systemically by binding with an enzyme required for the utilization of oxygen in body tissues.

Solvents of various kinds pose particular hazards in the workplace. Solvents come from a variety of organic chemical families including hydrocarbons, chlorinated hydrocarbons, alcohols, aldehydes, and ketones as well as some inorganics including carbon disulfide, liquid carbon dioxide, and liquid ammonia. These solvents are used in a variety of applications including the polymers industry, extraction of fats and oils from biomaterials, dry cleaning, and painting. By their nature, solvents are volatile and pose a variety of health hazards when inhaled or when absorbed through the skin. Solvents are distributed systemically in the body and tend to accumulate in lipid-rich tissues including the central nervous system, liver, and bone marrow. In addition to toxicity, solvents may be highly flammable or form explosive mixtures with air.

Materials may be hazardous to personnel for a variety of reasons including reactivity, instability and spontaneous decomposition, flammability, or volatility, as discussed in Chapter 15, Sections 15.4–15.7. **Flammable substances** are generally liquids, such as gasoline, with a strong tendency to burn; many solvents are flammable liquids. **Oxidizing substances** tend to release oxygen and can form hazardous mixtures with combustible organic matter. **Explosives** are substances that react on their own (generally containing both chemically reducing and oxidizing functionality in the same compound or mixture) with a sudden release of energy, gases, and a shock wave characteristic of an explosion. **Corrosives** react destructively with materials or tissues.

18.4.1 EXPOSURE LIMITS

An important aspect of hazardous materials, especially toxicants, in the workplace microatmosphere consists of **threshold limit values**® **(TLVs)**, a term copyrighted by the American Conference of Governmental Industrial Hygienists (ACGIH) to express the limits to which workers may be (presumably) safely exposed without causing irritation, tissue damage, or significant narcosis (reduced level of consciousness). A **short-term exposure limit** (STEL) is generally regarded as a concentration limit to which workers may be exposed for a short period of time, generally 15 minutes (up to four times daily with 60-minute intervals in between), without adverse effects.

For longer term exposures throughout the day and workweek, the **time-weighted average** (**TWA**) is used, which for a standard 8-hour day is calculated as

$$\text{TWA} = \frac{C_a T_a + C_b T_b + \ldots + C_n T_n}{8} \tag{18.1}$$

in which C_x is the concentration of the contaminant in the atmosphere over a time interval T_x.

18.5 WORKPLACE SAMPLING AND PERSONAL MONITORING

An important interface between analytical chemistry and industrial hygiene is the monitoring of workplaces for toxic substances and hazards. This may be done with devices that are mounted in place and that warn directly of hazards. Perhaps the most common example is the carbon monoxide detector widely used in homes to warn of dangerous levels of CO. Such a device gives off a very loud alarm when dangerous levels of carbon monoxide are detected. A similar monitor is used to warn of potentially explosive levels of methane in coal mines. For less immediate detection of hazards, sampling devices that pump ambient air through tubes packed with adsorbent materials may be used. Periodically, the tubes are removed and air contaminants collected in them are thermally desorbed into a gas chromatograph (Section 18.13) for analysis. Such devices give time-weighted average concentrations but do not sound alarms of short-term hazardous high levels of contaminants.

A wide variety of monitors have been developed to measure radiation and radioactivity to which personnel may be exposed. Radiation dosimeters are typically carried in pockets by workers and measure cumulative exposure to β radiation, γ radiation, and x-rays. A similar device is used to measure exposure to radio-frequency (RF) radiation, especially among workers exposed to communications equipment. A related device can be used to measure exposure to microwaves.

Figure 18.1 shows a battery-powered personal air sampling device that can be worn by a worker and that allows for the collection and concentration of substances from ambient air in a tube packed with adsorbent. Typically, the tube is removed and heated with a carrier gas flowing through it to carry the sample into a gas chromatograph or mass spectrometer to enable a qualitative and quantitative analysis of potentially hazardous substances in workplace air. By running the device at a steady flow for a long period of time, a time-weighted average value may be obtained for the contaminant.

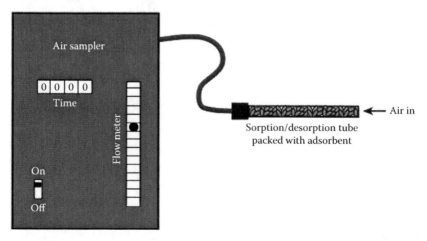

FIGURE 18.1 Battery-operated air sampler for collecting air samples in a tube packed with adsorbent from which analyte can be thermally desorbed into a gas chromatograph.

18.6 CHEMICAL ANALYSIS PROCESS

It is important to regard chemical analysis within the framework of an overall **chemical analysis process** rather than an isolated "laboratory experiment." Each step in the analysis process is crucial to getting accurate and meaningful results. Figure 18.2 outlines the process.

The first step in the analytical process is to obtain the right **sample** or samples, that is, the portion of matter on which the analysis is performed. The sample should be a **representative sample**, the composition of which is as close as possible to the whole mass of whatever is being analyzed. Obtaining a good sample is a crucial step in the chemical analysis process. Failure to obtain and properly preserve a good sample means that the whole analysis may be incorrect, rendering all the other steps involved worthless.

Sample processing is performed to get the sample into a form that can be analyzed. For a few kinds of analysis, the sample is analyzed without further processing or after minimal steps, such as grinding and mixing. Often, sample processing requires putting the sample into solution. Sample dissolution can be as simple as stirring a soil sample with hydrochloric acid to dissolve potassium commonly added to soil as fertilizer. Or it can be complicated and severe, such as oxidizing and dissolving fish tissue for metals analysis with a hot concentrated mixture of HNO_3 and $HClO_4$.

The preceding discussion on sample processing leads to the definition of two kinds of analyses, depending on what is done with the sample. When a sample is oxidized, dissolved in acid, or otherwise greatly altered as part of the analytical process, the chemical analysis is termed **destructive**. In some cases, such as those where evidence of a crime is involved, it is important to preserve the sample in an unaltered form. This requires **nondestructive** methods of analysis, such as can be performed by making the sample radioactive by irradiation with neutrons in a nuclear reactor and measuring the energies and intensities of γ radiation given off by the activated elements (neutron activation analysis).

After sample processing, it is often necessary to eliminate **interferences** from substances in the sample that can cause erroneous results. This can be done by removing interfering substances or by treating the sample with substances that react with interferences to render them noninterfering.

After all the aforementioned steps have been performed, the actual measurement of whatever is being determined is performed. The substance that is measured, such as calcium in a water sample or *trans,trans*-muconic acid measured in blood as evidence of occupational exposure to toxic benzene, is called the **analyte**. The specific measurement of an analyte is referred to as a **determination**, whereas the total process to which the sample is subjected to is called an **analysis**.

The final step in a chemical analysis is **calculation** of results. This step may consist of a few simple calculations, or it may involve a complicated data processing operation that calculates analyte levels and compensates for interferences in the method. In addition to providing a number for the quantity or percentage of analyte in a sample, the calculation of results usually involves an evaluation of the reliability of the data (precision and accuracy) of the analytical values. In modern analytical laboratories, results are calculated and stored by computers, frequently as part of the process by which analyte levels are measured with an appropriate instrument.

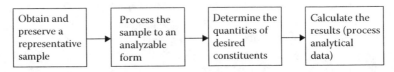

FIGURE 18.2   Schematic representation of the major steps involved in the chemical analysis process.

18.7　MAJOR CATEGORIES OF CHEMICAL ANALYSIS

Both qualitative and quantitative analyses are divided between **classical** methods involving primarily chemical reactions and simple measurements of mass and volume and **instrumental methods** that use instruments of varying degrees of complexity and sophistication to measure quantities of analytes. Classical methods are often **wet chemical** procedures using reagents in solution and reactions of dissolved analytes. Instrumental methods use various devices to measure physical manifestations of chemical species and chemical reactions, such as absorption of light, electrical potentials, or small changes in temperature.

Analytical chemistry can also be divided between **chemical** and **physical** methods of analysis. Chemical methods almost always involve the measurement of a mass of a chemical species or a volume of a reagent solution produced or consumed by a chemical reaction. For example, the acid in an acid mine water sample can be determined by adding exactly enough of a solution of base (NaOH) of accurately known concentration to exactly neutralize the strong acid in the sample

$$H^+ + Na^+ + OH^- \rightarrow H_2O + Na^+ \tag{18.1}$$

exactly measuring the amount of NaOH required, and calculating the quantity of acid neutralized; such a procedure is an acid–base **titrimetric** procedure.

Physical methods of analysis normally involve a measurement of a physical parameter other than mass or volume. For example, a water sample suspected of being polluted with hexavalent chromium can be injected into an inductively coupled plasma atomic emission spectrometer (ICP/AES) and the intensity of light given off by the very hot chromium atoms emitted by the sample measured to give the chromium concentration. Or fluoride in a water sample can be determined by measuring the potential versus a reference electrode of a fluoride ion–selective electrode immersed in the sample and comparing that value with the potential measured in a standard F^- solution to give the value of $[F^-]$.

18.8　ERROR AND TREATMENT OF DATA

A chemical analysis is only as good as the numbers that go into calculating and expressing the result. Therefore, data analysis is a crucial aspect of chemical analysis. All analytical measurements are subject to greater or lesser degrees of error. So every reasonable effort is made to reduce the amount of error in an analytical measurement. Since some error is inevitable, it is important to know the degree of error and express it correctly in the final result.

One of the major objectives of analytical measurements is to obtain *reproducible results*. For example, if three determinations of the percentage of iron in the same iron ore sample gave values of 18.76%, 18.71%, and 18.73%, the analyst would have a relatively high degree of confidence in the validity of the results because the three values are so close together. The degree to which numbers in a group of analytical results are in agreement with each other is the **precision** of the group of numbers. A lack of precision may indicate the presence of **indeterminate**, or **random**, **errors**. Such errors vary randomly in direction and magnitude and are from sources that cannot be determined.

However, just because the results of a set of analyses are in close agreement does not necessarily mean that the values are correct. This is because of the possibility of **determinate errors**. Such errors have a definite cause (although it may be unknown to the analyst), and each type of determinate error is always in the same direction. For example, if an analytical chemist is using a pipette rated to deliver 25.00 mL of solution that through a manufacturing mistake actually delivered 25.35 mL, a determinate error would be introduced into the analysis; in this case, it could readily be detected by calibrating the pipette.

The extent to which the data or the average value of a set of data agrees with the true value being determined is the **accuracy** of the data or the set of data. The relationship between accuracy and precision is shown graphically in Figure 18.3. Although the average of a set of randomly scattered, imprecise results may be close to the true value, normally the average of a set of imprecise results is inaccurate as well.

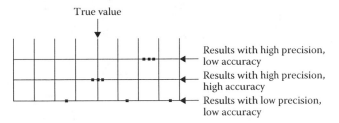

FIGURE 18.3 Illustration of precision and accuracy in chemical analysis.

In doing analytical calculations and expressing analytical results, it is important to know and correctly handle and express the **uncertainties** of the numbers used. For example, a skilled analyst can read the volume delivered by a laboratory burette to the nearest 0.01 mL. Therefore, a volume expressed as 36.27 mL implies that the volume is within ±0.01 mL of 36.27 mL and has an uncertainty of 0.01 mL. It would be incorrect to express the volume as 36.270 mL because it is not known to the nearest 0.001 mL as that number would imply. It would also be incorrect to express the volume as 36.3 mL because the value is known more accurately than ±0.1 mL.

18.9 GRAVIMETRIC ANALYSIS

Conceptually, the most straightforward kind of quantitative analysis, **gravimetric analysis**, consists of isolating in a pure form a species produced stoichiometrically by an analyte, weighing it, and calculating the percentage of the analyte in the sample. Obtaining a pure, weighable product is often a complicated process. A number of ways of doing this have been developed. The most common of these is formation of a precipitate by a reaction of the analyte in solution. As an example, the chloride content of a weighed, water-soluble sample can be determined by precipitating the chloride in the dissolved sample with excess silver nitrate solution:

$$\underbrace{Ag^+(aq) + NO_3^-(aq)}_{\substack{\text{Silver nitrate}\\\text{reagent}}} + \underset{\substack{\text{Chloride}\\\text{analyte}}}{Cl^-(aq)} \rightarrow \underset{\substack{\text{Silver chloride}\\\text{precipitate}}}{AgCl(s)} + NO_3^-(aq) \qquad (18.2)$$

The silver chloride precipitate, which can be produced in a very pure form, is collected on a weighed filter crucible (Figure 18.4) and washed to remove extraneous residual salts. After drying to remove excess water, the crucible and the precipitate are weighed to get the mass of the precipitate and the percentage of chloride is calculated by stoichiometric calculation. Where the atomic mass of chloride is 35.45 g/mol and the molar mass of AgCl is 143.32 g/mol, the calculation is

$$\text{Mass chloride} = \text{mass AgCl} \times \frac{1\,\text{mol Ag Cl}}{143.32\,\text{g Ag Cl}} \times \frac{1\,\text{mol Cl}}{1\,\text{mol AgCl}} \times \frac{35.45\,\text{g}}{1\,\text{mol Cl}} \qquad (18.3)$$

$$\text{Percent chloride} = \frac{\text{mass chloride}}{\text{mass sample}} \times 100 \qquad (18.4)$$

Over the decades before modern instrumental methods of analysis were developed, gravimetric techniques were developed for a wide range of analytes. Examples include sulfate determined by precipitating $BaSO_4$ with $BaCl_2$ reagent; calcium precipitated as the oxalate CaC_2O_4, which, in turn, could be heated to produce weighable $CaCO_3$ or CaO; and magnesium precipitated as an ammonium phosphate salt, which is then heated to produce $Mg_2P_2O_7$.

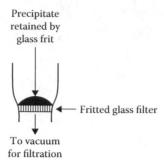

FIGURE 18.4 Filter crucible used to collect and weigh precipitates for gravimetric analysis.

The versatile nature of organic chemistry led to a number of organic reagents used to form precipitates with analytes, especially metals. One widely used reagent was 8-hydroxyquinoline, which is shown forming a precipitate with zinc ion in the following reaction:

$$\text{(18.5)}$$

8-Hydroxyquinoline

Precipitate of the 8-hydroxyquinolate chelate of zinc(II)

Two 8-hydroxyquinoline anions (formed by the loss of H^+) bind with Zn^{2+} ion as shown in the preceding reaction to produce a chelate species that can be weighed to calculate the amount of zinc in the sample. Organic precipitants with high molar masses, such as 8-hydroxyquinoline, offer an advantage for gravimetric analysis in that they produce a relatively high mass of precipitate from comparatively little analyte. Higher masses of precipitates translate to less relative error in weighing, thus increasing the accuracy of the determination.

18.10 VOLUMETRIC ANALYSIS: TITRATION

Other than gravimetric analysis, the other major type of classical wet chemical analysis techniques consists of measuring the volume of a reagent required to react with an analyte. Such a procedure is called **titration**, which involves the following steps:

- A measured quantity of sample that may consist of a weighed quantity of a solid, or a measured volume of a solution of the unknown, is placed in solution.
- A **standard solution** of known concentration of a reagent that reacts quantitatively with the analyte is added to the unknown with a burette so that the volume of the standard solution can be measured accurately.
- An **indicator** consisting of a dye that changes color, or some other means, is used to detect an **end point**, the experimental representation of the **equivalence point** at which exactly the stoichiometric amount of reagent required to react with the analyte occurs. The volume at which the end point occurs is recorded from the burette.
- The quantity or concentration of the analyte is calculated from the stoichiometry of the titration reaction.

The most common type of titration reaction consists of acid-base titration in which an unknown quantity or concentration of acid is titrated with standard base or vice versa. For example, a solution of hydrochloric acid of unknown concentration can be titrated with a standard solution of sodium hydroxide base using phenolphthalein indicator, which changes from colorless to pink at the end point. The physical steps involved in the titration are shown in Figure 18.5. The reaction between HCl and NaOH is as follows:

$$H^+ + Cl^- + Na^+ + OH^- \rightarrow H_2O + Na^+ + Cl^- \tag{18.6}$$

Before the end point, there is excess H^+, so the pH is less than 7. Beyond the end point, there is excess base and the pH is greater than 7. Since HCl is a strong (completely ionized) acid and NaOH

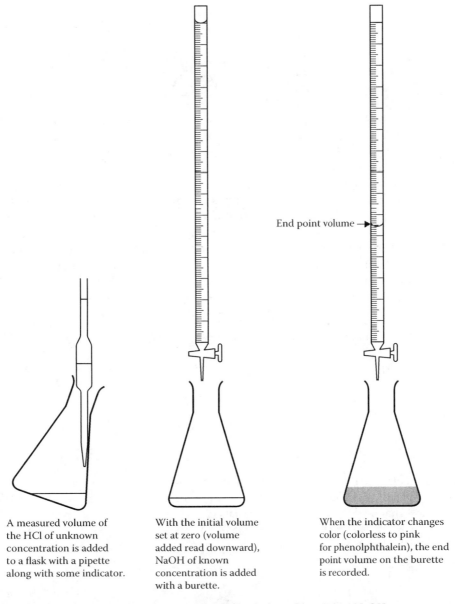

End point volume →

A measured volume of the HCl of unknown concentration is added to a flask with a pipette along with some indicator.

With the initial volume set at zero (volume added read downward), NaOH of known concentration is added with a burette.

When the indicator changes color (colorless to pink for phenolphthalein), the end point volume on the burette is recorded.

FIGURE 18.5   Steps in the titration of an unknown HCl solution with standard NaOH.

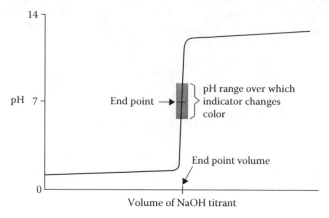

FIGURE 18.6 Titration curve as a plot of pH versus volume of titrant for the titration of HCl with NaOH titrant: an indicator changes color over a pH range, which indicates the end point.

is a strong (completely ionized) base, the pH at the end point is exactly 7. Furthermore, the pH changes very markedly by several pH units with the addition of just a few drops of sodium hydroxide titrant at a volume that is in the immediate vicinity of the end point. This change is reflected by the abrupt change in color of phenolphthalein indicator from colorless to pink at the end point, where the addition of titrant is stopped and the end point volume recorded. From this volume, the concentration of the standard solution, and the amount of the unknown analyzed, the amount of analyte and the composition of the sample may be calculated.

As noted in the preceding discussion, the course of acid-base titrations is reflected by the pH of the solution in the reaction flask as titrant is added. By using a device called a pH meter, the pH can be recorded and plotted as a function of added titrant. The result is a **titration curve** (Figure 18.6).

Several different kinds of reactions other than acid-base reactions can be used for titrations. Oxidation reactions can be used, for example, in determining dissolved oxygen levels in water; the precipitation of Cl^- ion with a standard solution of Ag^+ is one of the oldest titration procedures, and chelation with the anion of the strong chelating agent ethylenediaminetetraacetic acid (EDTA) can be used for determining Ca^{2+} ion concentration in water (water hardness) (Chapter 3, Section 3.10).

18.11 SPECTROPHOTOMETRIC METHODS OF ANALYSIS

The remaining methods of analysis addressed in this chapter are instrumental methods of analysis. The first group of these to be addressed are spectrophotometric methods based on the absorption and emission of photons of electromagnetic radiation including visible light, ultraviolet radiation, and infrared radiation.

18.11.1 ABSORPTION SPECTROPHOTOMETRY

Absorption spectrophotometry of light-absorbing species in solution, historically called colorimetry when visible light is absorbed, is still used for the analysis of many water, and some air, pollutants. Basically, absorption spectrophotometry consists of measuring the percent transmittance (%T) of monochromatic light passing through a light-absorbing solution compared with the amount passing through a blank solution containing everything in the medium but the sought-for constituent (100%). The absorbance (A) is defined as follows:

$$A = \log \frac{100}{\%T} \tag{18.7}$$

The relationship between A and the concentration (C) of the absorbing substance is given by Beer's law:

$$A = abC \qquad (18.8)$$

where a is the absorptivity, a wavelength-dependent parameter characteristic of the absorbing substance; b is the path length of the light through the absorbing solution; and C is the concentration of the absorbing substance. A linear relationship between A and C at constant path length indicates adherence to Beer's law. In many cases, analyses can be performed even when Beer's law is not obeyed, if a suitable calibration curve is prepared. A color-developing step usually is required in which the sought-for substance reacts to form a colored species, and in some cases a colored species is extracted into a nonaqueous solvent to provide a more intense color and a more concentrated solution.

Absorption spectrophotometric analysis procedures have been developed for a number of environmental species. For water contaminants alone, these include procedures for the determination of arsenic, boron, bromide ion, cyanide, fluoride, nitrate, phenols, phosphate, selenium, silica, sulfide, surfactants, and tannin and lignin.[3] A typical such procedure is the spectrophotometric determination of phenol in water by the reaction with 4-aminoantipyrene

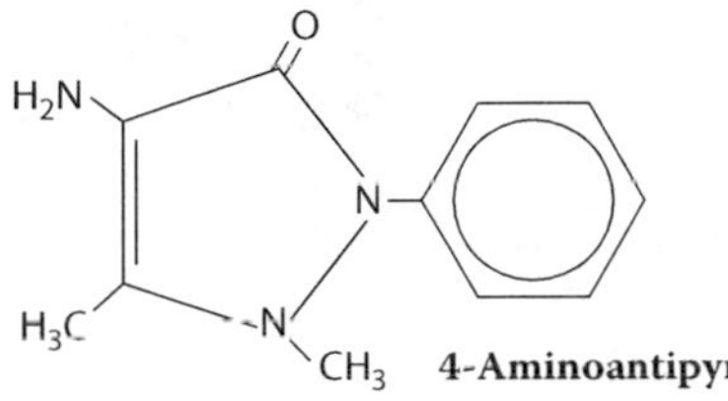

at pH 10 in the presence of oxidant potassium ferricyanide, $K_3Fe(CN)_6$, to form an antipyrine dye that is extracted into pyridine and measured spectrophotometrically at 460 nm.

18.11.2 ATOMIC ABSORPTION AND EMISSION ANALYSES

Atomic absorption analysis is commonly used for the determination of metals in environmental samples. This technique is based on the absorption of monochromatic light by a cloud of atoms of the analyte metal. The monochromatic light can be produced by a source composed of the same atoms as those being analyzed. The source produces intense electromagnetic radiation with a wavelength that is exactly the same as that absorbed by the atoms, resulting in extremely high selectivity. The basic components of an atomic absorption instrument are shown in Figure 18.7. The key element is the hollow cathode lamp in which atoms of the analyte metal are energized such that they become electronically excited and emit radiation with a very narrow wavelength band characteristic of the metal. This radiation is guided by appropriate optics through a flame into which the sample is aspirated. In the flame, most metallic compounds are decomposed, and the metal is reduced to the elemental state, forming a cloud of atoms. These atoms absorb a fraction of radiation in the flame. The fraction of radiation absorbed increases with the concentration of the sought-for element in the sample according to the Beer's law relationship (Equation 18.8). The attenuated light beam next goes to a monochromator to eliminate extraneous light resulting from the flame, and then to a detector.

Atomizers other than a flame can be used. The most common of these is the graphite furnace, an electrothermal atomization device that consists of a hollow graphite cylinder placed so that the light

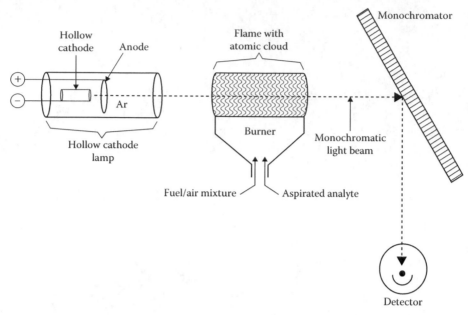

FIGURE 18.7   The basic components of a flame atomic absorption spectrophotometer.

beam passes through it. A small sample of up to 100 μL is inserted in the tube through a hole in the top. An electric current is passed through the tube to heat it—gently at first to dry the sample and then rapidly to vaporize and excite the metal analyte. The absorption of metal atoms in the hollow portion of the tube is measured and recorded as a spike-shaped signal. A diagram of a graphite furnace with a typical output signal is shown in Figure 18.8. The major advantage of the graphite furnace is that it gives detection limits up to 1000 times lower (better) than those of conventional flame devices.

A special technique for the flameless atomic absorption analysis of mercury involves room-temperature reduction of mercury to its elemental state by tin(II) chloride in solution, followed by sweeping the mercury with air into an absorption cell having end windows made of ultraviolet-transparent silica (SiO_2). Nanogram (10^{-9} g) quantities of mercury can be determined by measuring mercury absorption at 253.7 nm.

18.11.3 ATOMIC EMISSION TECHNIQUES

Metals can be determined in water, atmospheric particulate matter, and biological samples very well by observing the spectral lines emitted when they are heated to a very high temperature. An especially useful atomic emission technique is ICP/AES. The "flame" in which analyte atoms are excited in plasma emission consists of an incandescent plasma (ionized gas) of argon heated inductively by RF energy at 4–50 MHz and 2–5 kW (Figure 18.9). A stream of ionized argon absorbs the energy from an induction coil, producing temperatures up to 10,000 K. The sample atoms are subjected to temperatures around 7000 K, twice those of the hottest conventional flames (e.g., nitrous oxide–acetylene operates at 3200 K). Since emission of light increases exponentially with temperature, lower detection limits are obtained. Furthermore, the technique enables emission analysis of some of the environmentally important metalloids such as arsenic, boron, and selenium. Interfering chemical reactions and interactions in the plasma are minimized compared with flames. Of greatest significance, however, is the capability of analyzing as many as 30 elements simultaneously, enabling a true multielement analysis technique. Plasma atomization combined with mass

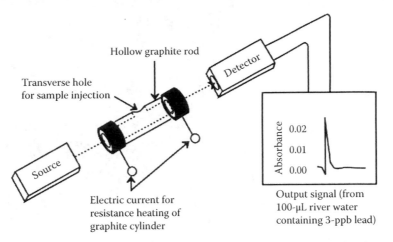

FIGURE 18.8 Graphite furnace for atomic absorption analysis, and typical output signal.

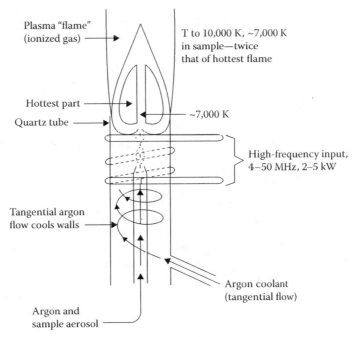

FIGURE 18.9 Schematic diagram showing inductively coupled plasma used for optical emission spectroscopy.

spectrometric measurement of analyte elements (ICP/MS) is a relatively new technique that is an especially powerful means for multielement analysis.

18.12 ELECTROCHEMICAL METHODS OF ANALYSIS

Several useful techniques for water analysis utilize electrochemical sensors. These techniques can be potentiometric, voltammetric, or amperometric. Potentiometry is based on the general principle that the relationship between the electrical potential of a measuring electrode and that of a reference electrode is a function of the logarithm of the activity of an ion in solution. For a measuring electrode responding selectively to a particular ion, this relationship is given by the Nernst equation

$$E = E^0 + \frac{2.303RT}{zF} \log(a_z) \qquad (18.9)$$

where E is the measured potential, E^0 is the standard electrode potential, R is the gas constant, T is the absolute temperature, z is the signed charge (+ for cations, − for anions), F is the Faraday constant, and a is the activity of the ion being measured. At a given temperature, the quantity $2.303RT/F$ has a constant value; at 25°C, it is 0.0592 V (59.2 mV). At constant ionic strength, the activity, a, is directly proportional to concentration, and the Nernst equation can be written as follows for electrodes responding to Cd^{2+} and F^-, respectively:

$$E\,(\text{in mV}) = E^0 + \frac{59.2}{2} \log[Cd^{2+}] \qquad (18.10)$$

$$E = E^0 - 59.2 \log \left[F^- \right] \qquad (18.11)$$

Electrodes that respond more or less selectively to various ions are called **ion-selective electrodes**. Generally, the potential-developing component is a membrane of some kind that allows for selective exchange of the sought-for ion. The glass electrode used for the measurement of hydrogen-ion activity and pH is the oldest and most widely used ion-selective electrode. The potential is developed at a glass membrane that selectively exchanges hydrogen ion in preference to other cations, giving a Nernstian response to hydrogen ion activity, a_{H^+}:

$$E = E^0 + 59.2 \log \left(a_{H^+} \right) \qquad (18.12)$$

Of the ion-selective electrodes other than glass electrodes, the fluoride electrode, in which the sensing membrane is a crystal of LaF_3, is the most successful. It is well behaved, relatively free of interferences, and has an adequately low detection limit and a long range of linear response. Like all ion-selective electrodes, its electrical output is in the form of a potential signal that is proportional to the logarithm of concentration. A small error in E leads to a variation in the logarithm of concentration, which translates to relatively high concentration errors.

Voltammetric techniques, the measurement of current resulting from the potential applied to a microelectrode, have found some applications in water analysis. One such technique is differential pulse polarography, in which the potential is applied to the microelectrode in the form of small pulses superimposed on a linearly increasing potential. The current is read near the end of the voltage pulse and compared with the current just before the pulse was applied. It has the advantage of minimizing the capacitive current from charging the microelectrode surface, which sometimes obscures the current due to the reduction or oxidation of the species being analyzed. Anodic stripping voltammetry involves deposition of metals on an electrode surface over a period of several minutes, followed by stripping them off very rapidly using a linear anodic sweep. The electrodeposition concentrates the metals on the electrode surface, and increased sensitivity results. An even better technique is to strip the metals off using a differential pulse signal. A differential-pulse anodic-stripping voltammogram of copper, lead, cadmium, and zinc at subparts-per-billion levels in tap water is shown in Figure 18.10.

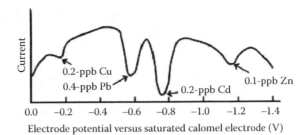

Electrode potential versus saturated calomel electrode (V)

FIGURE 18.10 Differential-pulse anodic-stripping voltammogram of tap water at a mercury-plated, wax-impregnated graphite electrode showing detection of metals at subparts-per-billion levels.

18.13 CHROMATOGRAPHY

First described in the literature in the early 1950s, gas chromatography has played an essential role in the analysis of organic materials. Gas chromatography is both a qualitative and a quantitative technique; for some analytical applications of environmental importance, it is remarkably sensitive and selective. Gas chromatography is based on the principle that when a mixture of volatile materials transported by a carrier gas is passed through a column containing an adsorbent solid phase or, more commonly, an absorbing liquid phase coated on a solid material, each volatile component will be partitioned between the carrier gas and the solid or liquid. The length of time required for the volatile component to traverse the column is proportional to the degree to which it is retained by the nongaseous phase. Since different components may be retained to different degrees, they will emerge from the end of the column at different times. If a suitable detector is available, the time at which the component emerges from the column and the quantity of the component are both measured. A recorder trace of the detector response appears as peaks of different sizes, depending on the quantity of material producing the detector response. Both quantitative and (within limits) qualitative analyses of the sought-for substances are obtained.

The essential features of a gas chromatograph are shown schematically in Figure 18.11. The carrier gas generally is argon, helium, hydrogen, or nitrogen. The sample is injected as a single compact plug into the carrier gas stream immediately ahead of the column entrance. If the sample is liquid, the injection chamber is heated to vaporize the liquid rapidly. The separation column may consist of a metal or glass tube packed with an inert solid of high surface area covered with a liquid

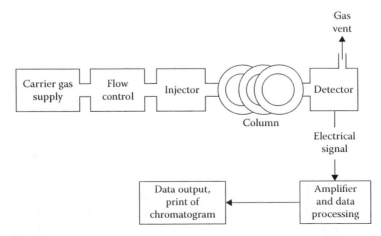

FIGURE 18.11 Schematic diagram of the essential features of a gas chromatograph.

phase, or it may consist of an active solid, which enables the separation to occur. More commonly, capillary columns are now employed that consist of very long coiled tubes of very small diameter in which the liquid phase is coated on the inside of the column.

The component that primarily determines the sensitivity of gas chromatographic analysis and, for some classes of compounds, the selectivity as well, is the detector. One such device is the thermal conductivity detector, which responds to changes in the thermal conductivity of gases passing over it. The electron-capture detector, which is especially useful for halogenated hydrocarbons and phosphorus compounds, operates through the capture of electrons emitted by a β-particle source. The flame-ionization gas chromatographic detector is very sensitive for the detection of organic compounds. It is based on the phenomenon by which organic compounds form highly conducting fragments, such as C^+, in a flame. Application of a potential gradient across the flame results in a small current that can be readily measured. The mass spectrometer, described in Section 18.14, can be used as a detector for a gas chromatograph. A combined gas chromatograph/mass spectrometer (GC/MS) instrument is an especially powerful analytical tool for organic compounds.

Gas chromatographic analysis requires that a compound exhibit at least a slight vapor pressure at the highest temperature at which it is stable. In many cases, organic compounds that cannot be passed through a chromatographic column directly can be converted to derivatives that are amenable to gas chromatographic analysis. It is seldom possible to analyze organic compounds in water by direct injection of the water into the gas chromatograph; a higher concentration of the compounds is usually required. Two techniques commonly employed to remove volatile compounds from water and concentrate them are (1) extraction with solvents and (2) purging volatile compounds with a gas, such as helium; concentrating the purged gases on a short column; and driving them off by heat into the chromatograph.

18.13.1 High-Performance Liquid Chromatography

A liquid mobile phase used with very small column-packing particles enables high-resolution chromatographic separation of materials in the liquid phase. Very high pressures up to several thousand pounds per square inch are required to get a reasonable flow rate in such systems. Analysis using such devices is called **high-performance liquid chromatography** (HPLC), and it offers an enormous advantage in that the materials analyzed need not be changed to the vapor phase, a step that often requires preparation of a volatile derivative or results in decomposition of the sample. The basic features of a high-performance liquid chromatograph are the same as those of a gas chromatograph, shown in Figure 18.11, except that a solvent reservoir and a high-pressure pump are substituted for the carrier gas source and regulator. A hypothetical HPLC chromatogram is shown in Figure 18.12. Refractive index and ultraviolet detectors are both used for the detection of peaks coming from the liquid chromatograph column. Fluorescence detection can be especially sensitive for some classes of compounds. Mass spectrometric detection of HPLC effluents has led to the development of LC/MS analysis. Somewhat difficult in practice, this technique can be a powerful

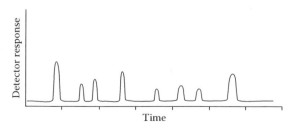

FIGURE 18.12   Hypothetical high-performance liquid chromatography chromatogram.

tool for the determination of analytes that cannot be subjected to gas chromatography. HPLC has emerged as a very useful technique for the analysis of a number of water pollutants.

18.13.2 Ion Chromatography

The liquid chromatographic determination of ions, particularly anions, has enabled the measurement of species that, despite their widespread occurrence in water, used to pose special challenges for water chemists to determine. This technique is **ion chromatography**, and its development has been facilitated by special detection techniques using so-called suppressors to enable the detection of analyte ions in the chromatographic effluent. Ion chromatography has been developed for the determination of most of the common anions, including arsenate, arsenite, borate, carbonate, chlorate, chlorite, cyanide, the halides, hypochlorite, hypophosphite, nitrate, nitrite, phosphate, phosphite, pyrophosphate, selenate, selenite, sulfate, sulfite, sulfide, trimetaphosphate, and tripolyphosphate. Cations, including the common metal ions, can also be determined by ion chromatography, although they are relatively easy to determine by other means.

18.13.3 Chromatography-Based Methods of Analysis for Water Pollutants

The U.S. Environmental Protection Agency has developed and published a variety of largely chromatography-based methods for the analysis of water, commonly known as EPA Series 500, 600, and 1600 Methods.[4] As examples, these include EPA Method 601 for purgeable halocarbons by gas chromatography, EPA Method 625 for base/neutrals and acids (more than 70 organic compounds) by gas chromatography with mass spectrometric detection, EPA Method 1624 for volatile organic compounds by gas chromatography with mass spectrometric detection, EPA Method 610 for polycyclic aromatic hydrocarbons by HPLC, and EPA Method 501.1 for trihalomethanes in drinking water by a purge-and-trap procedure followed by gas chromatographic separation and halogen-specific detection.

18.14 MASS SPECTROMETRY

Mass spectrometry is particularly useful for the identification of specific organic pollutants. It depends on the production of ions by an electrical discharge or a chemical process, followed by separation based on the charge-to-mass ratio and measurement of the ions produced. The output of a mass spectrometer is a mass spectrum, such as the one shown in Figure 18.13. A mass spectrum

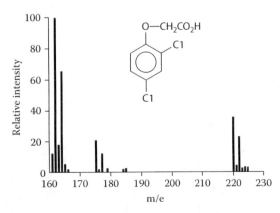

FIGURE 18.13 Partial mass spectrum of the herbicide 2,4-dichlorophenoxyacetic acid (2,4-D), which may be found as a water pollutant.

is characteristic of a compound and serves to identify it. Computerized data banks for mass spectra have been established and are stored in computers interfaced with mass spectrometers. Identification of a mass spectrum depends on the purity of the compound from which the spectrum is taken. Prior separation by gas chromatography with continual sampling of the column effluent by a mass spectrometer, commonly called GC/MS, is particularly effective in the analysis of organic pollutants.

18.15   AUTOMATED ANALYSES

Huge numbers of analyses must often be performed to get meaningful results and for reasons of economics. This has resulted in the development of a number of automated procedures in which traditional wet chemical methods of analysis have been adopted to automate procedures. With such procedures, the samples are introduced through a sampler and the analyses performed and results posted without manual manipulation of reagents and apparatus. Such procedures have been developed and instruments marketed for the determination of a number of analytes. In water, automated analyses have been developed from wet chemical procedures for alkalinity, sulfate, ammonia, nitrate/nitrite, and metals. The somewhat cumbersome West-Gaeke determination of sulfur dioxide in air has been adapted to automated analyzers. Colorimetric (absorption spectrophotometric) procedures are popular for such automated analytical instruments, using simple, rugged colorimeters for absorbance measurements.

Figure 18.14 shows an automated analytical system for the determination of water alkalinity (Chapter 3, Section 3.7). The reagents and sample liquids are transported through the analyzer by a peristaltic pump. This relatively simple device consists basically of rollers moving over flexible tubing, which "squeezes" solutions through the tubing. By using different sizes of tubing and varying the speed of the rollers, the flow rates of the reagents are proportioned. Air bubbles are introduced into the liquid stream to aid mixing and to separate one sample from the other. Mixing of the sample and various reagents is accomplished in mixing coils. Since many color-developing reactions are not rapid, a delay coil is provided that allows the color to develop before reaching the colorimeter.

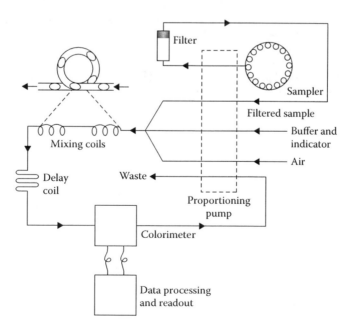

FIGURE 18.14   Automated analyzer system for the determination of total alkalinity in water: addition of a water sample to a methyl orange solution buffered to pH 3.1 causes a loss of color in proportion to the alkalinity in the sample.

Bubbles are removed from the liquid stream by a debubbler prior to its introduction into the flow cell for colorimetric analysis.

18.16   IMMUNOASSAY SCREENING

Immunoassay has emerged as a useful technique for screening samples, such as hazardous waste residues, for specific kinds of pollutants. Commercial immunoassay techniques have been developed that permit very rapid analyses of large numbers of samples. A variety of immunoassay techniques have been developed. These techniques all use biologically produced antibodies that bind specifically to analytes or classes of analytes. This binding is combined with chemical processes that enable detection through a signal-producing species (reporter reagent) such as enzymes, chromophores, fluorophores, and luminescent compounds. The reporter reagent binds with the antibody. When an analyte is added to the antibody to displace the reagent, the concentration of displaced reagent is proportional to the level of the analyte displacing it from the antibody. Detection of the displaced reporter reagent enables quantification of the analyte.

Immunoassay techniques are divided into the two major categories of heterogeneous and homogeneous; the former requires a separation (washing) step, whereas the latter does not require such a step. Typically, when heterogeneous procedures are used, the antibody is immobilized on a solid support on the inner surface of a disposable test tube. The sample is contacted with the antibody displacing the reporter reagent, which is removed by washing. The amount of reagent displaced, commonly measured spectrophotometrically, is proportional to the amount of analyte added. Very widely used enzyme immunoassays make use of reporter reagent molecules bound with enzymes, and kits are available for enzyme-linked immunosorbent assays (ELISAs) of a number of organic species likely to be found in hazardous wastes.

Immunoassay techniques have been approved for the determination of numerous analytes commonly found in hazardous wastes. Where the EPA method numbers are given in parentheses in the following list, these include pentachlorophenol (4010); 2,4-dichlorophenoxyacetic acid (4015); polychlorinated biphenyls (4020); petroleum hydrocarbons (4030); polycyclic aromatic hydrocarbons (4035); toxaphene (4040); chlordane (4041); dichlorodiphenyltrichloroethane (DDT) (4042); trinitrotoluene (TNT) explosives in soil (4050); and hexahydro-1,3,5-trinitro-1,3,5-triazine (RDX) in soil (4051). ELISAs have been reported for monitoring pentachlorophenol and BTEX (benzene; toluene; ethylbenzene; and *o*-, *m*-, and *p*-xylene) in industrial effluents.

18.17   TOTAL ORGANIC CARBON IN WATER

Dissolved organic carbon exerts an oxygen demand in water, often is in the form of toxic substances, and is a general indicator of water pollution. Therefore, its measurement is quite important, and the measurement of total organic carbon (TOC) is now recognized as the best means of assessing the organic content of a water sample. The measurement of TOC is accomplished by oxidizing the organic matter in water to CO_2 and measuring the quantity of this gas produced from a measured volume of the water sample (Figure 18.15). The dissolved carbon is oxidized by potassium peroxydisulfate, $K_2S_2O_8$, and energized by intense ultraviolet radiation impinging on the water. Before the organic matter is oxidized, phosphoric acid is added to the sample, which is sparged with air or nitrogen to drive off CO_2 formed from HCO_3^- and CO_3^{2-} in solution. After sparging, the sample is pumped to a chamber containing a lamp emitting ultraviolet radiation of 184.9 nm. This radiation produces reactive free radical species such as hydroxyl radical, $HO\cdot$ (see Chapter 5, Section 5.8). These active species bring about the rapid oxidation of dissolved organic compounds as shown in the following general reaction:

$$Organics + HO\cdot \rightarrow CO_2 + H_2O \tag{18.13}$$

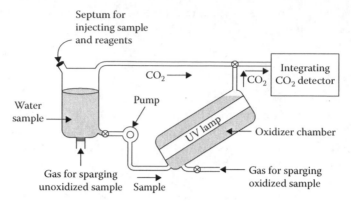

FIGURE 18.15   Total organic carbon analyzer employing ultraviolet-promoted sample oxidation.

After oxidation is complete, the CO_2 is sparged from the system and measured with a gas chromatographic detector or by absorption in ultrapure water followed by a conductivity measurement.

18.18   MEASUREMENT OF RADIOACTIVITY IN WATER

There are several potential sources of radioactive materials that can contaminate water (see Chapter 4, Section 4.14). Radioactive contamination of water is normally detected by measurements of gross β activity and gross α activity, a procedure that is simpler than detecting individual isotopes. The measurement is made from a sample formed by evaporating water to a very thin layer on a small pan, which is then inserted inside an internal proportional counter. This setup is necessary because β particles can penetrate only very thin detector windows, and α particles have essentially no penetrating power. More detailed information can be obtained for radionuclides that emit γ-rays by the use of gamma spectrum analysis. This technique employs solid-state detectors to resolve rather closely spaced γ peaks characteristic of specific isotopes in the sample's spectra. In conjunction with multichannel spectrometric data analysis, it is possible to determine a number of radionuclides in the same sample without chemical separation. This method requires minimal sample preparation.

18.19   ANALYSIS OF WASTES AND SOLIDS

The analysis of hazardous wastes of various kinds for a variety of potentially dangerous substances is one of the most important aspects of hazardous waste management. These analyses are performed for a number of reasons including tracing the sources of wastes, assessing the hazards posed by the wastes to surroundings and to waste remediation personnel, and determining the best means of waste treatment. Here, wastes are broadly defined to include all kinds of solids, semisolids, sludges, liquids, contaminated soils, sediments, and other kinds of materials that are either wastes themselves or contaminated by wastes.

For the most part, the substances determined as part of waste analysis, the *analytes*, are measured by techniques that are used for the determination of the same analytes in water and, to a lesser extent, in air. However, the preparation techniques that must be employed for waste analysis are usually more complex than those used for the same analytes in water. This is because the matrices in which the waste analytes are contained are usually relatively complicated, which makes it difficult to recover all the analytes from the waste and which introduces interfering substances. As a result, the lower limits at which substances can be measured in wastes (a parameter known as the practical quantitation limit) are usually significantly higher than in water.

There are several distinct steps in the analysis of a waste. Compared with water, wastes are often highly heterogeneous, making the collection of representative samples difficult. Whereas water samples can often be introduced into an analytical instrument with minimal preparation, the processing of hazardous wastes to get a sample that can be introduced into an instrument is often relatively complicated. Such processing can consist of dilution of oily samples with an organic solvent, extraction of organic analytes into an organic solvent, evolution and collection of volatile organic carbon analytes, or digestion of solids with strong acids and oxidants to extract metals for analysis. The products of these processes must often be subjected to relatively complicated sample cleanup procedures to remove the contaminants that might interfere with the analysis or damage the analytical instrument.

Over a number of years, the U.S. Environmental Protection Agency has developed specialized methods for the characterization of wastes. These methods are given in the publication entitled *Test Methods for Evaluating Solid Waste, Physical/Chemical Methods* (SW-846), which is periodically updated.[5,6] Because of the difficult and exacting nature of many of the procedures in this work and because of the hazards associated with the use of reagents such as strong acids and oxidants employed for sample digestion and solvents used to extract organic analytes, anyone attempting analyses of hazardous waste materials should use this resource and follow the procedures carefully with special attention to precautions.

18.19.1 Toxicity Characteristic Leaching Procedure

The **toxicity characteristic leaching procedure** (TCLP) is specified to determine the potential toxicity hazard of various kinds of wastes.[5] It was designed to estimate the availability to organisms of both inorganic and organic species in hazardous materials present as liquids, solids, or multiple-phase mixtures by producing a leachate, the TCLP extract, which is analyzed for a number of specified toxicants in the categories of heavy metals, such as lead; organics including benzene; and several specified pesticides, such as chlordane.

The procedure for conducting the TCLP test is rather involved. The procedure need not be run at all if a total analysis of the sample reveals that none of the pollutants specified in the procedure could exceed regulatory levels. At the opposite end of the scale, analysis of any of the liquid fractions of the sample showing that any regulated species would exceed regulatory levels even after the dilutions involved in the TCLP measurement have been carried out designates the sample as hazardous, and the TCLP measurement is not required.

The procedures used in the TCLP test vary with the nature of the sample. If free liquids are present, they may be determined separately. An extract is prepared from solid or semisolid materials using an extractant consisting of dilute acetic acid or an acetic acid/sodium acetate mixture. The extraction procedure is designed to mimic conditions that may be obtained with respect to a "waste mismanagement scenario" in which hazardous wastes are codisposed with actively decomposing municipal refuse.

18.20 ATMOSPHERIC MONITORING

The maintenance of atmospheric quality requires monitoring air pollutants, especially the **criteria air pollutants** known to injure health, harm the environment, and cause property damage: carbon monoxide, lead, nitrogen dioxide, ozone, particulate matter, and sulfur dioxide. In addition, it is important to determine the approximately 60 hydrocarbons and aldehydes that are precursors to photochemical smog formation.

The atmosphere is a particularly difficult analytical system because of the very low levels of substances to be analyzed; sharp variations in pollutant levels with time and location (see Figure 18.16); differences in temperature and humidity; and difficulties encountered in reaching desired sampling points, particularly those substantially above ground level. These conditions make the acquisition of representative atmospheric samples particularly challenging. The ideal atmospheric analysis techniques are those that work successfully without sampling, such as some

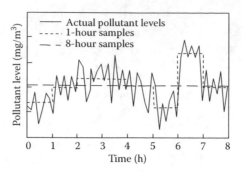

FIGURE 18.16 The effect of duration of sampling on observed values of air pollutant levels.

kinds of direct spectrophotometric measurements. In some cases, air samples may be collected and analyzed automatically and the results transmitted electronically to a central receiving station. Often, however, a batch sample is collected for later chemical analysis.

18.20.1 METHODS FOR SAMPLING AND ANALYZING ATMOSPHERIC POLLUTANTS

The U.S. Environmental Protection Agency specifies reference methods of analysis for selected air pollutants to determine compliance with the primary and secondary national ambient air quality standards for those pollutants. These methods are published annually in the *Code of Federal Regulations.*[7] These methods are not necessarily state of the art and are, in some cases, outdated and cumbersome. However, they provide proven reliable measurements for regulatory and legal purposes.

18.20.2 DETERMINATION OF ATMOSPHERIC SULFUR DIOXIDE BY THE WEST–GAEKE METHOD

The reference method for the analysis of sulfur dioxide is the spectrophotometric West-Gaeke pararosaniline method. This somewhat involved method uses a collecting solution of 0.04-M potassium tetrachloromercurate to collect sulfur dioxide according to the following reaction:

$$HgCl_4^{2-} + SO_2 + H_2O \rightarrow HgCl_2SO_3^{2-} + 2H^+ + 2Cl^- \qquad (18.14)$$

The $HgCl_2SO_3^{2-}$ complex stabilizes the readily oxidized sulfur dioxide from the reaction with oxidants such as ozone and nitrogen oxides. For analysis, sulfur dioxide in the scrubbing medium is reacted with formaldehyde:

$$HCHO + SO_2 + H_2O \rightarrow HOCH_2SO_3H \qquad (18.15)$$

The adduct formed is then reacted with uncolored organic pararosaniline hydrochloride to produce a red-violet dye that is measured spectrophotometrically.

When performed manually, the West-Gaeke method for sulfur dioxide analysis is cumbersome and complicated. However, the method has been refined to the point that it can be done automatically with continuous-monitoring equipment. A block diagram of such an analyzer is shown in Figure 18.17.

18.20.3 ATMOSPHERIC PARTICULATE MATTER

The method commonly used for determining the quantity of total suspended particulate matter in the atmosphere draws air over filters that remove the particles. This device, called a **Hi-Vol sampler** (Figure 18.18), is essentially a glorified vacuum cleaner that draws air through a filter. The samplers are usually placed under a shelter that excludes precipitation and particles larger than about 0.1 mm

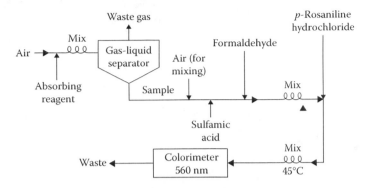

FIGURE 18.17   Block diagram of an automated system for the determination of sulfur dioxide by the para-rosaniline method.

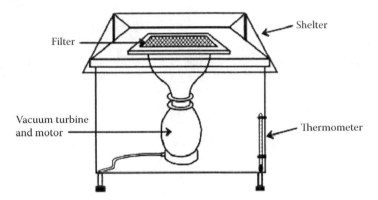

FIGURE 18.18   Hi-Vol sampler for the collection of particulate matter from the atmosphere for analysis.

in diameter, favoring the collection of particles up to 25–50 μm in diameter. These devices efficiently collect particles from a large volume of air, typically 2000 m³ and typically over a 24-hour period. Particles collected by a Hi-Vol sampler or some other means of sample collection may be subjected to additional chemical analysis such as determination of heavy metals, such as lead, or organics, such as polycyclic aromatic compounds.

18.20.4   NITROGEN OXIDES IN THE ATMOSPHERE

Gas-phase chemiluminescence is the favored method of NO_x analysis. The general phenomenon of chemiluminescence is one in which two chemical species react to produce an energized excited species that loses energy by emitting electromagnetic radiation, such as visible light, the intensity of which may be measured as a means of quantitating the analyte causing the chemiluminescence. In the chemiluminescent measurement of atmospheric NO, ozone is used to bring about the reaction

$$NO + O_3 \rightarrow NO_2^* + O_2 \tag{18.16}$$

producing electronically excited nitrogen dioxide, which loses energy and returns to the ground state through the emission of light in the 600–3000-nm range. The emitted light is measured by a photomultiplier; its intensity is proportional to the concentration of NO. A schematic diagram of the device used is shown in Figure 18.19.

Since the chemiluminescence detector system depends on the reaction of O_3 with NO, it is necessary to convert NO_2 to NO in the sample prior to analysis. This is accomplished by passing

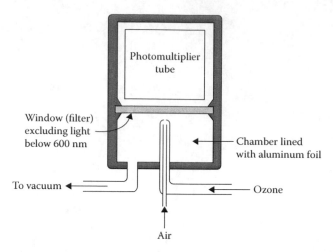

FIGURE 18.19   Chemiluminescence detector for NO$_x$.

the air sample over a thermal converter, which brings about the desired conversion. Analysis of such a sample gives NO$_x$, the sum of NO and NO$_2$. Chemiluminescence analysis of a sample that has not been passed over the thermal converter gives NO. The difference between these two results is NO$_2$.

This analysis technique is illustrative of chemiluminescence analysis in general. Chemiluminescence is an inherently desirable technique for the analysis of atmospheric pollutants because it avoids wet chemistry, is basically simple, and lends itself well to continuous monitoring and instrumental methods. Another chemiluminescence method, which is employed for the analysis of ozone, is described Section 18.20.5.

18.20.5   DETERMINATION OF ATMOSPHERIC OXIDANTS

The atmospheric oxidants that are commonly determined include ozone, hydrogen peroxide, organic peroxides, and chlorine. The classic manual method for the analysis of oxidants is based on their oxidation of I$^-$ ion followed by spectrophotometric measurement of the product. The sample is collected in 1% KI buffered at pH 6.8. Oxidants react with I$^-$ ion as shown by the following reaction of ozone:

$$O_3 + 2H^+ + 3I^- \rightarrow I_3^- + O_2 + H_2O \tag{18.17}$$

The absorbance of the colored I$_3^-$ product is measured spectrophotometrically at 352 nm. Generally, the level of oxidant is expressed in terms of ozone, although it should be noted that not all oxidants—peroxyacetyl nitrate (PAN), for example—react with the same efficiency as O$_3$.

The currently favored method for oxidant analysis uses the chemiluminescent reaction between ozone and ethylene. This reaction emits light at a maximum intensity of 435 nm. The intensity of emitted light is directly proportional to the level of ozone.

18.20.6   ATMOSPHERIC CARBON MONOXIDE BY INFRARED ABSORPTION

Carbon monoxide is analyzed in the atmosphere by nondispersive infrared spectrometry. This technique depends on the fact that carbon monoxide absorbs infrared radiation strongly at certain wavelengths. Therefore, when such radiation is passed through a long (typically 100 cm) cell containing trace levels of carbon monoxide, more of the infrared radiant energy is absorbed.

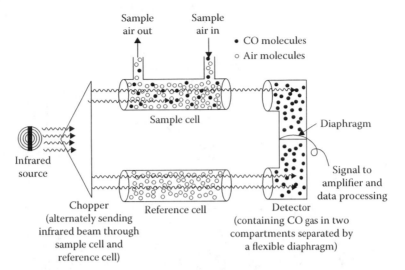

FIGURE 18.20 Nondispersive infrared spectrometer for the determination of carbon monoxide in the atmosphere.

A nondispersive infrared spectrometer differs from standard infrared spectrometers in that the infrared radiation from the source is not dispersed according to wavelength by a prism or grating. The nondispersive infrared spectrometer is made very specific for a given compound, or type of compound, by using the sought-for material as part of the detector, or by placing it in a filter cell in the optical path. A diagram of a nondispersive infrared spectrometer selective for CO is shown in Figure 18.20. Radiation from an infrared source is "chopped" by a rotating device so that it alternately passes through a sample cell and a reference cell. In this particular instrument, both beams of light fall on a detector filled with CO gas and are separated into two compartments by a flexible diaphragm. The relative amounts of infrared radiation absorbed by the CO in the two sections of the detector depend on the level of CO in the sample. The difference in the amount of infrared radiation absorbed in the two compartments causes slight differences in heating so that the diaphragm bulges slightly toward one side. Even very slight movements of the diaphragm can be detected and recorded. By means of this device, carbon monoxide can be measured from 0 to 150 ppm, with a relative accuracy of ±5% in the optimum concentration range.

18.20.7 DETERMINATION OF HYDROCARBONS AND ORGANICS IN THE ATMOSPHERE

Monitoring of hydrocarbons in atmospheric samples takes advantage of the very high sensitivity of the hydrogen-flame-ionization detector to measure this class of compounds. Known quantities of air are run through the flame-ionization detector 4–12 times per hour to provide a measure of total hydrocarbon content.

In some cases, it is important to have a method to determine individual organics because of their toxicities and abilities to form photochemical smog, as indicators of photochemical smog, and as a means of tracing sources of pollution. Numerous techniques have been published for the determination of organic compounds in the atmosphere. For example, whole-air samples can be collected in Tedlar bags, the organic analytes can be concentrated cryogenically at −180°C, and then they can be thermally desorbed and measured with high-resolution capillary column gas chromatography.

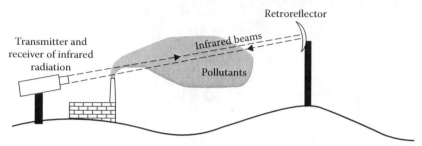

FIGURE 18.21   Fourier transform infrared system for remote sensing of air pollutants.

18.20.8   Direct Spectrophotometric Analysis of Gaseous Air Pollutants

Direct spectrophotometric techniques, such as nondispersive infrared spectroscopy, are highly desirable when they are available and when they are capable of accurate analysis at the low levels required.[8] Three such spectrophotometric methods are Fourier transform infrared spectroscopy, tunable diode laser spectroscopy, and differential optical absorption spectroscopy. These techniques can be used for point air monitoring, in which a sample is monitored at a given point, generally by measurement in a long absorption cell. In-stack monitoring can be performed to measure effluents. A final possibility is the collection of long-line data (sometimes using sunlight as a radiation source), yielding concentrations in units of concentration-length (parts per million–meters). If the path length is known, the concentration can be calculated. This approach is particularly useful for measuring concentrations in stack plumes.

The low levels of typical air constituents require long path lengths, sometimes up to several kilometers, for spectroscopic measurements. These can be achieved by locating the radiation source some distance from the detector, by the use of a distant retroreflector to reflect the radiation back to the vicinity of the source or by the use of cells in which a beam is reflected multiple times to achieve a long path length.

A typical open-path Fourier transform infrared system for remote monitoring of air pollutants uses a single unit (telescope) that functions as both a transmitter and a receiver of infrared radiation (Figure 18.21). The radiation is generated by a silicon carbide glower; modulated by a Michelson interferometer; and transmitted to a retroreflector, which reflects it back to the telescope, where its intensity is measured. The modulated infrared signal, called an inteferogram, is processed by a mathematical algorithm, the Fourier transform, to give a spectrum of the absorbing substances. This spectrum is fitted mathematically to spectra of the absorbing species to give their concentrations.

18.21   ANALYSIS OF BIOLOGICAL MATERIALS AND XENOBIOTICS

As defined in Chapter 2, Section 2.9, a xenobiotic species is one that is foreign to living systems. The determination of xenobiotic substances in biological materials is very important in environmental/toxicological chemistry. The greatest concern with such substances is their presence in human tissues and other samples of human origin. They may also be determined in plant or microbiological samples. The measurement of xenobiotic substances and their metabolites in blood, urine, breath, and other samples of biological origin to determine exposure to toxic substances is called **biological monitoring**. For more detailed current information regarding this topic, the reader is referred to several reviews of the topic[9,10] and several books on biological monitoring such as those by Angerer, Baselt, Conti, Emanuel, Kneip, and Ott and coauthors, which are listed in the back of this chapter under "Supplementary References" and are available as well.

18.21.1 INDICATORS OF EXPOSURE TO XENOBIOTICS

Both the type of sample and the type of analyte are influenced by what happens to a xenobiotic material when it gets into the body. For some exposures, the entry site composes the sample. This is the case, for example, in exposure to asbestos fibers in the air, which is manifested by lesions to the lung. More commonly, the analyte may appear at some distance from the site of exposure, such as lead that was originally taken in by the respiratory route in bone. In other cases, the original xenobiotic is not even present in the analyte. An example of this is methemoglobin in blood, the result of exposure to aniline absorbed through the skin.

The two major kinds of samples analyzed for xenobiotics exposure are blood and urine. Both these kinds of samples are analyzed for systemic xenobiotics, which are those that are transported in the body and metabolized in various tissues. Blood is of unique importance as a sample for biological monitoring.

Phase I and Phase II reaction products of xenobiotics (Chapter 2, Section 2.10) tend to be converted to more polar and water-soluble metabolites. These are eliminated with urine, making it a good sample to analyze as evidence of exposure to xenobiotic substances. Urine has the advantage of being a simpler matrix than blood and one that subjects are usually more willing to give for analysis. Other kinds of samples that can be analyzed include breath (for volatile xenobiotics and volatile metabolites), air or nails (for trace elements, such as selenium), adipose tissue (fat), and milk (obviously limited to lactating females). Various kinds of organ tissue can be analyzed in cadavers to determine the toxic agent that has caused death by poisoning.

The choice of the analyte actually measured varies with the xenobiotic substance to which the subject has been exposed. Therefore, it is convenient to divide xenobiotic analysis on the basis of the type of chemical species determined. The most straightforward analyte is, of course, the xenobiotic itself. This applies to elemental xenobiotics, especially metals, which are almost always determined in the elemental form. In a few cases, organic xenobiotics can also be determined as the parent compound. However, organic xenobiotics are commonly determined as Phase I and Phase II reaction products. Often, the Phase I reaction product is measured, frequently after it is hydrolyzed from the Phase II conjugate, using enzymes or acid hydrolysis procedures. Thus, for example, *trans,trans*-muconic acid

trans,trans-Muconic acid

can be measured as evidence of exposure to the parent compound benzene. In other cases, a Phase II reaction product is measured, for example, hippuric acid,

Toluene → Metabolic formation of Phase I and Phase II reaction products → Hippuric acid

determined as evidence of exposure to toluene. Some xenobiotics or their metabolites form adducts with endogenous materials in the body, which are then measured as evidence of exposure. A simple example is the adduct formed between carbon monoxide and hemoglobin, that is, carboxyhemoglobin. More complicated examples are the adducts formed by the carcinogenic Phase I reaction products of polycyclic aromatic hydrocarbons with deoxyribonucleic acid (DNA) or hemoglobin. Another class of analytes consists of endogenous substances that are produced on exposure to a xenobiotic material. Methemoglobin, a derivative of hemoglobin that is useless for transporting oxygen in blood, which is formed as a result of exposure to nitrobenzene, aniline, and related compounds, is an example of such a substance that does not contain any of the original xenobiotic material. Another class of substances causes measurable alterations in enzyme activity. The most common example of this is the inhibition of acetylcholinesterase enzyme by organophosphates or carbamate insecticides.

18.21.2 Immunological Methods of Xenobiotics Analysis

As discussed in Section 18.16, immunoassay methods offer distinct advantages in specificity, selectivity, simplicity, and costs. Although they are used in simple test kits for blood-glucose and pregnancy testing, immunoassay methods have been limited in biological monitoring of xenobiotics, in part because of interferences in complex biological systems. Because of their inherent advantages, however, it can be anticipated that immunoassays will grow in importance for biological monitoring of xenobiotics in the future.[11]

QUESTIONS AND PROBLEMS

1. Doing some research on the Internet, determine the status of industrial hygiene and occupational safety and health prior to about 1900. What were some of the earlier efforts to address these issues?
2. What are two specific examples of cancer caused by workplace exposure to carcinogens by inhalation?
3. Choose a specific material used industrially and look up the MSDS pertaining to it. What are some of the specific issues addressed by the MSDS?
4. Both an atmosphere composed almost entirely of elemental nitrogen gas and one containing several parts per thousand of CO can be fatal to workers breathing it. In what ways are the mechanisms of toxicity similar? How do they differ?
5. A common expression is "canary in a coal mine." What does this expression mean? What is the history of the canary in the coal mine?
6. Monitored hourly over an 8-hour day, the level of a solvent vapor in milligrams per liter showed values of 0.80, 1.22, 1.06, 0.75, 1.56, 1.88, 0.91, and 1.09 mg/L. What is the TWA concentration of the vapor?
7. The sulfur content of a unit trainload of coal is to be determined to see if the fuel meets pollution standards. Suggest how a representative sample of the coal might be taken.
8. What is the function of a hollow cathode in the atomic absorption determination of a metal? Of what kind of material is a hollow cathode composed?
9. Lead may be determined in water by using atomic absorption with both a flame atomizer and a graphite tube atomizer. Suggest how the output signal, absorbance as a function of time, would appear differently in these two modes of analysis.
10. What is a titrant? How are titrants used in chemical analysis?
11. Describe the general characteristics of a titration curve for the titration of HCl with NaOH.
12. How is the end point found in the titration of HCl with NaOH using either a titration curve or an indicator?
13. A 3.471-g sample of a compound containing C, H, and O was ignited in a stream of O_2 and CO_2 and H_2O were collected. Masses of 8.758 g of CO_2 and 1.537 g of H_2O were collected. Calculate the percentages of C and H in the compound.
14. Based on material covered in Chapter 19 and on information from the Internet regarding the nature of electrons in atoms, quantum chemistry, and photochemistry, attempt to explain the phenomena of atomic absorption and atomic emission discussed as analytical techniques in this chapter.

15. Considering the Nernst equation as it applies to the measurement of pH, calculate the voltage change at a glass electrode used to measure pH for each unit change in pH.

16. Distinguish among electron-capture detector, flame-ionization detector, and mass spectrometer as detectors for gas chromatographic separations.

17. What is required to get a reasonably fast flow rate in a high-performance liquid chromatographic separation?

18. What is the basis of separations made in mass spectrometry? Why is mass spectrometry one of the most specific means of detecting organic compounds?

19. What are the main components of an automated analyzer system? What are the functions of each?

20. What is the basis of immunoassay analysis? What is meant by it being classified as a good screening technique?

21. A sample of a colored analyte at a concentration of 3.60×10^{-3} mol/L shows 34.2% transmittance in a 2.00-cm cell. What is the value of a in the Beer's law equation for this substance at the wavelength measured? If a sample of the colored analyte of unknown concentration gives an absorbance, A, of 0.520 in the same cell at the same wavelength, what is the concentration of the analyte in this solution?

22. What are the similarities between the measurement of TOC and that of biochemical oxygen demand (BOD) in water? What are the differences between them? Which is likely to be indicative of livestock feedlot pollution, and which would suggest water pollution from industrial wastes?

23. In a sense, TCLP is a somewhat strange test based on a "mismanagement scenario." What is the origin of the test, and what is the mismanagement scenario?

24. Suggest how the standard method for the determination of atmospheric sulfur dioxide is not in keeping with the best practice of green chemistry.

25. Suggest a conceptual similarity between the determination of lead by atomic absorption spectrophotometry and the determination of carbon dioxide in air by nondispersive infrared analysis.

26. Match each kind of substance or phenomenon actually measured that is listed in the right column with the potentially toxic substance or pollutant responsible for it from the left column:

A. Carbon monoxide	1. Methemoglobin
B. Toluene	2. Acetylcholinesterase activity
C. Aniline	3. Hemoglobin adduct
D. Nerve gas	4. Hippuric acid

LITERATURE CITED

1. Plog, Barbara A., and Patricia J. Quinlan, *Fundamentals of Industrial Hygiene*, 5th ed, National Safety Council Press, Itasca, IL, 2002.

2. United States Department of Labor Occupational Safety and Health Administration, http://www.osha.gov/, 2012.

3. Clesceri, Lenore, S., Arnold E. Greenberg, Andrew D. Eaton, and Mary Ann H. Franson, Eds., *Standard Methods for the Examination of Water and Wastewater*, 21st ed, American Public Health Association, Washington, DC, 2005.

4. U.S. Environmental Protection Agency, *Methods and Guidance for the Analysis of Water (including EPA Series 500, 600, and 1600 Methods), on CD-ROM*, National Technical Information Service, Alexandria, VA, 2012.

5. *Test Methods for Evaluating Solid Waste, Physical/Chemical Methods*, EPA Publication SW-846, 3rd ed, (1986), as amended by Updates I, II, IIA, IIB, III, IIIA and IIIB, IVA, and IVB updated to 2012, National Technical Information Service, Springfield, VA, available on CD/ROM.
6. *The Guide to Environmental Analytical Methods*, 5th ed, The Genium Group, Inc., Amsterdam, NY, 2012.
7. 40 *Code of Federal Regulations*, Part 50, Office of the Federal Register, National Archives and Records Administration, Washington, DC, July 1, annually.
8. Jones, Charlotte, Chemistry in the Atmosphere, *Chemistry Review* **19**, 16–18 (2010).
9. Angerer, Juergen, Ulrich Ewers, and Michael Wilhelm, Human Biomonitoring: State of the Art, *International Journal of Hygiene and Environmental Health* **210**, 201–228 (2007).
10. Atio, A., Special Issue: Biological Monitoring in Occupational and Environmental Health, *Science of the Total Environment* **199**, 1–226 (1997).
11. Van Emon, Jeanette M., Ed., *Immunoassay and Other Bioanalytical Techniques*, CRC Press/Taylor & Francis, Boca Raton, FL, 2007.

SUPPLEMENTARY REFERENCES

Angerer, J. K., and K. H. Schaller, *Analyses of Hazardous Substances in Biological Materials*, Vol. 1, VCH, Weinheim, Germany, 1985.
Angerer, J. K., and K. H. Schaller, *Analyses of Hazardous Substances in Biological Materials*, Vol. 2, VCH, Weinheim, Germany, 1988.
Angerer, J. K., and K. H. Schaller, *Analyses of Hazardous Substances in Biological Materials*, Vol. 3, VCH, Weinheim, Germany, 1991.
Angerer, J. K., and K. H. Schaller, *Analyses of Hazardous Substances in Biological Materials*, Vol. 4, VCH, Weinheim, Germany, 1994.
Angerer, J. K., and K. H. Schaller, *Analyses of Hazardous Substances in Biological Materials*, Vol. 5, John Wiley & Sons, New York, 1996.
Angerer, J. K., and K. H. Schaller, *Analyses of Hazardous Substances in Biological Materials*, Vol. 6, John Wiley & Sons, New York, 1999.
Angerer, Jurgen K., and Karl-Heinz Schaller, *Analyses of Hazardous Substances in Biological Materials*, Vol. 7, John Wiley & Sons, New York, 2001.
Angerer, Jurgen K., and Karl-Heinz Schaller, *Analyses of Hazardous Substances in Biological Materials*, Vol. 8, John Wiley & Sons, New York, 2003.
Angerer, Jurgen K., and Michael Muller, Eds., *Analyses of Hazardous Substances in Biological Materials*, Vol. 9, *Markers of Susceptibility*, John Wiley & Sons, New York, 2004.
Angerer, Jurgen K., *Essential Biomonitoring Methods from the MAK Collection for Occupational Health and Safety*, Wiley-VCH, Weinheim, Germany, 2006.
Angerer, Jürgen, and Helmut Greim, Eds., *The MAK-Collection for Occupational Health and Safety: Part IV: Biomonitoring Methods*, Wiley-VCH, Weinheim, Germany, 2010.
AWWA Staff, *Simplified Procedures for Water Examination*, 5th ed., American Water Works Association, Denver, CO, 2002.
Baselt, Randall C., *Disposition of Toxic Drugs and Chemicals in Man*, 6th ed., Biomedical Publications, Foster City, CA, 2002.
Baselt, Randall C., *Biological Monitoring Methods for Industrial Chemicals*, 2nd ed., PSG Publishing Company, Inc., Littleton, MA, 1988.
Cazes, Jack, Ed., *Ewing's Analytical Instrumentation Handbook*, 3rd ed., Marcel Dekker, New York, 2005.
Christian, Gary D., *Analytical Chemistry*, 6th ed., Wiley, Hoboken, NJ, 2004.
Conti, M. E., Ed., *Biological Monitoring: Theory and Applications*, WIT Press, Billerica, MA, 2008.
Crompton, T. R., *Analysis of Seawater: A Guide for the Analytical and Environmental Chemist*, Springer, Berlin, 2006.
Crompton, T. R., *Chromatography of Natural and Treated Waters*, Taylor & Francis/CRC Press, Boca Raton, FL, 2003.
Crompton, T. R., *Determination of Anions in Natural and Treated Waters*, Taylor & Francis/CRC Press, Boca Raton, FL, 2002.
Crompton, T. R., *Determination of Metals in Natural and Treated Waters*, Taylor & Francis/CRC Press, Boca Raton, FL, 2001.
Emanuel, Peter, Jason W. Roos, and Kakoli Niyogi, Eds., *Sampling for Biological Agents in the Environment*, ASM Press, Washington, DC, 2008.

Guzzi, R., Ed., *Exploring the Atmosphere by Remote Sensing Techniques*, Springer-Verlag, Berlin, 2003.

Day, R. A. and Arthur L. Underwood, *Quantitative Analysis*, 6th ed., Prentice Hall, Upper Saddle River, NJ, 1991.

Hage, David S., and James D. Carr, *Analytical Chemistry and Quantitative Analysis*, Prentice Hall, Upper Saddle River, NJ, 2010.

Harris, Daniel C., *Quantitative Chemical Analysis*, 8th ed., W. H. Freeman & Co., New York, 2010.

Khan, JaVed I., Thomas J. Kennedy, and Donnell R. Christian, *Basic Principles of Forensic Chemistry*, Humana Press, New York, 2012.

Li, Yungcong, and Kati Migliaccio, Eds., *Water Quality Concepts, Sampling, and Analysis*, Taylor & Francis/CRC Press, Boca Raton, FL, 2010.

Manahan, Stanley E., *Quantitative Chemical Analysis*, Brooks Cole Publishing Co., Monterey, CA, 1986.

Na, Li, John J. Hefferen, and Li Ke'an, *Quantitative Chemical Analysis*, Peking University Press, Bejing, 2009.

Nollet, Leo M. L., *Handbook of Water Analysis*, 2nd ed., CRC Press, Boca Raton, FL, 2007.

Popek, E. P., *Sampling and Analysis of Environmental Chemical Pollutants: A Complete Guide*, Academic Press, Boston, MA, 2003.

Quevauviller, Philippe, and K. Clive Thompson, Eds., *Analytical Methods for Drinking Water: Advances in Sampling and Analysis*, Wiley, Hoboken, NJ, 2005

Rubinson, Kenneth A. and Judith Faye Rubinson, *Contemporary Instrumental Analysis*, Prentice Hall, Upper Saddle River, NJ, 2000.

Settle, Frank A., Ed., *Handbook of Instrumental Techniques for Analytical Chemistry*, Prentice Hall, Upper Saddle River, NJ, 1997.

Settle, Frank, A., Brian D. Lamp, David L. McCurdy, Mark K. Vitha, Brian W. Gregory, and Yinfa Ma, *Instrumental Methods of Analysis*, 8th ed., Wiley-Interscience, Hoboken, NJ, 2011.

Skoog, Douglas A., James F. Holler, and Stanley R. Crouch, *Fundamentals of Analytical Chemistry*, 8th ed., Thomson-Brooks/Cole, Belmont, CA, 2004.

Skoog, Douglas A., James F. Holler, and Stanley R. Crouch, *Principles of Instrumental Analysis*, 6th ed., Thomson Brooks/Cole, Belmont, CA, 2007.

19 Fundamentals of Chemistry

19.1 SCIENCE OF MATTER

This chapter is designed to give those readers who have had little exposure to chemistry the basic knowledge needed to understand the material in the rest of the book. For the reader who has had no exposure to chemistry, this chapter provides the basic concepts and terms in general chemistry. A larger category of readers consist of those who have had at least one chemistry course, but whose chemistry background, for various reasons, is inadequate. By learning the material in this chapter and that in Chapter 9, these readers can comprehend the chemistry in this book. For a more complete coverage of chemistry fundamentals, readers should consult one of a number of basic chemistry books such as *Fundamentals of Sustainable Chemical Science*[1] and other supplementary references listed at the end of the chapter.

Chemistry is the science of matter. Therefore, it deals with all of the things that surround humankind and with all aspects of the environment. Chemical properties and processes are central to environmental science. A vast variety of chemical reactions occur in water, for example, including acid-base reactions, precipitation reactions, and oxidation–reduction reactions largely mediated by microorganisms. Atmospheric chemical phenomena are largely determined by photochemical processes and chain reactions. A large number of organic chemical processes occur in the atmosphere. The geosphere, including soil, is the site of many chemical processes, particularly those that involve solids. The biosphere obviously is where the many biochemical processes crucial to the environment and to the toxic effects of chemicals occur. Much of the anthrosphere is based on things that are made chemically and chemical processes.

This chapter emphasizes several aspects of chemistry. It begins with a brief discussion of the nature of matter and the states of matter. Next follows a discussion of the fundamental subatomic particles that make up all matter and explains how these are assembled to produce atoms. In turn, atoms join together to make compounds. Chemical reactions and chemical equations that represent them are discussed. Solution chemistry is especially important to aquatic chemistry and is addressed in a separate section. The important, vast discipline of organic chemistry is crucial to all parts of the environment and is addressed in Chapter 20.

19.1.1 STATES OF MATTER

The forms of matter, whether solid, liquid, or gas, are called the **states of matter**. A **solid** has a definite shape and volume, a **liquid** has a definite volume (is essentially noncompressible), and a **gas** takes on the shape and volume of its container. These states of matter are very much related to the chapters in this book. The hydrosphere with its vast oceans (Chapter 4) is normally considered as liquid, although significant portions of it are in the solid state as ice in glaciers and polar ice caps and part of it is in the gas form as water vapor in the atmosphere. The atmosphere (Chapter 6) is composed of gas with a small fraction of liquid water droplets in clouds and some solids suspended as particulate matter in it. The geosphere (Chapter 9) is regarded as solid rock and soil, although much of it at greater depths is liquid rock and it contains liquids as groundwater and liquid petroleum as well as gases in natural gas.

Liquids tend to dissolve solids, gases, and other liquids to produce **solutions**. Solutions, especially those in liquid water, are very important in environmental and toxicological chemistry and are discussed in more detail in Section 19.5. The molecules and ions dissolved in solutions are mobile, which enable them to come into contact and react with one another. For this reason, liquids are widely used as media in which chemical processes occur.

19.1.2 GASES AND THE GAS LAWS

The behavior of gases in the atmosphere is governed by several fundamental **gas laws**, which are covered briefly here. In using these laws, it should be kept in mind that the quantity of gas is most usefully expressed in numbers of moles. There are many units of pressure, but one that is very meaningful conceptually is the **atmosphere** (atm) where 1 atmosphere is the average pressure of air in the atmosphere at sea level. (Air has pressure because of the mass of all the molecules of air pressing down from the atmosphere above; at higher altitudes, the pressure becomes progressively lower.) For calculations involving temperature, the **absolute** temperature scale is used in which each degree is the same size as a degree Celsius (or centigrade, the temperature scale used for scientific measurements and for temperature readings in most of the world), but zero is 273° below the freezing point of water, which is taken as zero on the Celsius scale. Three important gas laws are as follows:

1. **Avogadro's law:** At constant temperature and pressure, the volume of a gas is directly proportional to the number of moles; doubling the number of moles at a constant temperature and pressure doubles the volume.
2. **Charles's law:** At constant pressure, the volume of a fixed number of moles of gas is directly proportional to the absolute temperature (degrees Celsius + 273) of the gas; doubling the absolute temperature at constant pressure doubles the volume.
3. **Boyle's law:** At constant temperature, the volume of a fixed number of moles of gas is inversely proportional to the pressure; doubling the pressure halves the volume.

These three laws are summarized in the **general gas law** relating volume (V), pressure (P), number of moles (n), and absolute temperature (T) expressed as follows:

$$PV = nRT \tag{19.1}$$

where R is a constant.

Mathematical calculations involving the gas laws are simple. One of the most common such calculation is that of changes in volume resulting from changes in pressure, temperature, or moles of gas. The parameter that does not change is the constant R. Using subscripts to represent conditions before and after a change yields the following relationship:

$$R = \frac{P_1 V_1}{n_1 T_1} = \frac{P_2 V_2}{n_2 T_2} \tag{19.2}$$

This equation can be arranged in a form that can be solved for a new volume resulting from changes in P, n, or T:

$$V_2 = V_1 \times \frac{n_2 T_2 P_1}{n_1 T_1 P_2} \tag{19.3}$$

As an example, calculate the volume of a fixed number of moles of gas initially occupying 12.0 L when the temperature is changed from 10°C to 90°C at constant pressure. To use these temperatures, they must be changed to absolute temperature by adding 273°. Therefore, $T_1 = 10° + 273° = 283°$, and $T_2 = 90° + 273° = 363°$. Since n and P remain constant, they cancel out of the equation yielding the following relationship:

$$V_2 = V_1 \times \frac{T_2}{T_1} = 12.0 \text{ L} \times \frac{363°}{283°} = 15.4 \text{ L} \tag{19.4}$$

As another example, consider the effects of a change of pressure, holding both the temperature and the number of moles constant. Calculate the new volume of a quantity of gas occupying initially 16.0 L at a pressure of 0.900 atm when the pressure is changed to 1.20 atm. In this case, both n and T remain the same and cancel out of the equation giving the following relationship:

$$V_2 = V_1 \times \frac{P_1}{P_2} = 16.0 \text{ L} \times \frac{0.900 \text{ atm}}{1.20 \text{ atm}} = 12.0 \text{ L} \tag{19.5}$$

Note that an increase in temperature *increases* the volume and an increase in pressure *decreases* the volume.

19.2 ELEMENTS

All substances are composed of only about a hundred fundamental kinds of matter called **elements**. Elements, themselves, may be of environmental concern. The "heavy metals," including lead, cadmium, and mercury, are well recognized as toxic substances in the environment. Elemental forms of otherwise essential elements may be very toxic or cause environmental damage. Oxygen in the form of ozone, O_3, is the agent most commonly associated with atmospheric smog pollution and is very toxic to plants and animals. Elemental white phosphorus is highly flammable and toxic.

Each element is made up of very small entities called **atoms**; all atoms of the same element have identical chemical behavior. The study of chemistry, therefore, can logically begin with elements and the atoms of which they are composed. Each element is designated by an **atomic number**, a **name**, and a **chemical symbol**, such as carbon, C; potassium, K (for its Latin name kalium); or cadmium, Cd. Each element has a characteristic **atomic mass** (atomic weight), which is the average mass of all atoms of the element. Atomic numbers of the elements are integrals ranging from 1 for hydrogen, H, to somewhat more than 100 for some of the transuranic elements (those beyond uranium). Atomic number is a unique, important way of designating each element, and it is equal to the number of protons in the nuclei of each atom of the element (see Section 19.2.1).

19.2.1 SUBATOMIC PARTICLES AND ATOMS

Figure 19.1 represents an atom of deuterium, a form of hydrogen. It is seen that such an atom is made up of even smaller subatomic particles—positively charged protons, negatively charged electrons, and uncharged (neutral) neutrons.

The subatomic particles differ in mass and charge. Their masses are expressed by the atomic mass unit, u (also called the dalton), which is also used to express the masses of individual atoms and molecules (aggregates of atoms). The atomic mass unit is defined as a mass equal to exactly one-twelfth that of an atom of carbon-12, the isotope of carbon that contains six protons and six neutrons in its nucleus.

The proton, p, has a mass of 1.007277 u and a unit charge of +1. This charge is 1.6022×10^{-19} coulombs, where a coulomb is the amount of electrical charge involved in a flow of electrical current of 1 ampere for 1 second. The neutron, n, has no electrical charge and a mass of 1.009665 u. The proton and the neutron each have a mass of essentially 1 u and are said to have a **mass number** of 1.

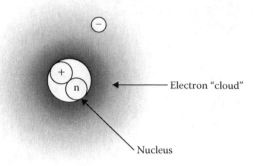

FIGURE 19.1   Representation of a deuterium atom. The nucleus contains one proton (+) and one neutron (n). The electron (−) is in constant, rapid motion around the nucleus, forming a cloud of negative electrical charge, the density of which drops off with increasing distance from the nucleus.

TABLE 19.1

Properties of Protons, Neutrons, and Electrons

Subatomic Particle	Symbol	Unit Charge	Mass Number	Mass (u)	Mass (g)
Proton[a]	p	+1	1	1.007277	1.6726×10^{-24}
Neutron[a]	n	0	1	1.008665	1.6749×10^{-24}
Electron[a]	e	−1	0	0.000549	9.1096×10^{-28}

[a] The mass number and the charge of each of these kinds of particles may be indicated by a superscipt and a subscript, respectively, as in the symbols $_1^1\text{p}$, $_0^1\text{n}$, and $_{-1}^0\text{e}$.

(Mass number is a useful concept expressing the total number of protons and neutrons, as well as the approximate mass, of a nucleus or subatomic particle.) The electron, *e*, has a unit electrical charge of −1. It is very light, however, with a mass of only 0.00054859 u, about 1/1840 that of the proton or the neutron. Its mass number is 0. The properties of protons, neutrons, and electrons are summarized in Table 19.1.

Although it is convenient to think of the proton and the neutron as having the same mass, and each is assigned a mass number of 1, it is seen in Table 19.1 that their exact masses differ slightly from each other. Furthermore, the mass of an atom is not exactly equal to the sum of the masses of subatomic particles composing the atom. This is because of the energy relationships involved in holding the subatomic particles together in atoms so that the masses of the atom's constituent subatomic particles do not add up to exactly the mass of the atom.

19.2.2   ATOM NUCLEUS AND ELECTRON CLOUD

Protons and neutrons, which have relatively high masses compared to electrons, are contained in the positively charged **nucleus** of the atom. The nucleus has essentially all of the mass but occupies virtually none of the volume of the atom. An uncharged atom has the same number of electrons as protons. The electrons in an atom are contained in a cloud of negative charge around the nucleus that occupies most of the volume of the atom. These concepts are emphasized in Figure 19.2.

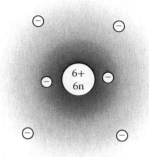 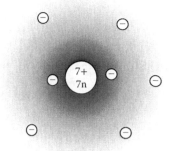

An atom of carbon, symbol C.
Each C atom has 6 protons (+) in its
nucleus, so the atomic number is 6.
The atomic mass of C is 12.

An atom of nitrogen, symbol N.
Each N atom has 7 protons (+) in its
nucleus, so the atomic number is 7.
The atomic mass of C is 14.

FIGURE 19.2   Atoms of carbon and nitrogen.

19.2.3   ISOTOPES

Atoms with the *same* number of protons but *different* numbers of neutrons in their nuclei are called **isotopes**. They are chemically identical atoms of the same element but have different masses and may differ in their nuclear properties. Some isotopes are **radioactive isotopes** or **radionuclides**, which have unstable nuclei that give off charged particles and gamma rays in the form of **radioactivity**. Radioactivity may have detrimental, or even fatal, health effects; a number of hazardous substances are radioactive and they can cause major environmental problems. The most striking example of such contamination resulted from a massive explosion and fire at a power reactor in the Ukrainian city of Chernobyl in 1986. A more recent episode of radionuclide contamination took place as the result of the catastrophic March 11, 2011, earthquake and tsunami followed by hydrogen gas explosions that destroyed four reactors and exposed spent fuel rods at the Fukushima Daiichi Nuclear Power Plant in Japan, requiring the evacuation of thousands of nearby residents.

19.2.4   IMPORTANT ELEMENTS

An abbreviated list of a few of the most important elements that could be useful for the reader to learn at this point is given in Table 19.2. A complete list of elements is given in Table 19.3 at the end of this chapter.

19.2.5   PERIODIC TABLE

When elements are considered in order of increasing atomic number, it is observed that their properties are repeated in a periodic manner. For example, elements with atomic numbers 2, 10, and 18 are gases that do not undergo chemical reactions and consist of individual atoms, whereas those with atomic numbers larger by 1—elements with atomic numbers 3, 11, and 19—are unstable, highly reactive metals. An arrangement of the elements in a manner that reflects this recurring behavior is known as the **periodic table** (Figure 19.3). The periodic table is extremely useful in understanding chemistry and predicting chemical behavior. As shown in Figure 19.3, the entry for each element in the periodic table gives the element's atomic number, name, symbol, and atomic mass. More detailed versions of the table include other information as well.

TABLE 19.2

List of Some of the More Important Common Elements

Element	Symbol	Atomic Number	Atomic Mass	Significance
Aluminum	Al	13	26.9815	Abundant in Earth's crust
Argon	Ar	18	39.948	Noble gas
Arsenic	As	33	74.9216	Toxic metalloid
Bromine	Br	35	79.904	Toxic halogen
Cadmium	Cd	48	112.40	Toxic heavy metal
Calcium	Ca	20	40.08	Abundant essential element
Carbon	C	6	12.011	"Life element"
Chlorine	Cl	17	35.453	Halogen
Copper	Cu	29	63.54	Useful metal
Fluorine	F	9	18.998	Halogen
Helium	He	2	4.00260	Lightest noble gas
Hydrogen	H	1	1.008	Lightest element
Iodine	I	53	126.904	Halogen
Iron	Fe	26	55.847	Important metal
Lead	Pb	82	207.19	Toxic heavy metal
Magnesium	Mg	12	24.305	Light metal
Mercury	Hg	80	200.59	Toxic heavy metal
Neon	Ne	10	20.179	Noble gas
Nitrogen	N	7	14.0067	Important nonmetal
Oxygen	O	8	15.9994	Abundant, essential nonmetal
Phosphorus	P	15	30.9738	Essential nonmetal
Potassium	K	19	39.0983	Alkali metal
Silicon	Si	14	28.0855	Abundant metalloid
Silver	Ag	47	107.87	Valuable nonreactive metal
Sodium	Na	11	22.9898	Essential, abundant alkali metal
Sulfur	S	16	32.064	Essential element, occurs in air pollutant SO_2
Tin	Sn	50	118.69	Useful metal
Uranium	U	92	238.03	Fissionable metal, nuclear fuel
Zinc	Zn	30	65.37	Useful metal

19.2.6 ELECTRONS IN ATOMS

The periodic table has several important features including the following: **Groups** of elements having similar chemical behavior are contained in vertical columns in the periodic table. **Main group** elements may be designated as A groups (1A and 2A on the left, 3A through 8A on the right). **Transition elements** are those between main groups 2A and 3A. **Noble gases** (group 8A), a group of gaseous elements that are virtually chemically unreactive, are in the far right column. The chemical similarities of elements in the same group are especially pronounced for groups 1A, 2A, 7A, and 8A. Horizontal rows of elements in the periodic table are called **periods**, the first of which consists of only hydrogen (H) and helium (He). The second period begins with atomic number 3 (lithium) and terminates with atomic number 10 (neon), whereas the third goes from atomic number 11 (sodium) through 18 (argon). The fourth period includes the first row of transition elements, whereas lanthanides and actinides are listed separately at the bottom of the table.

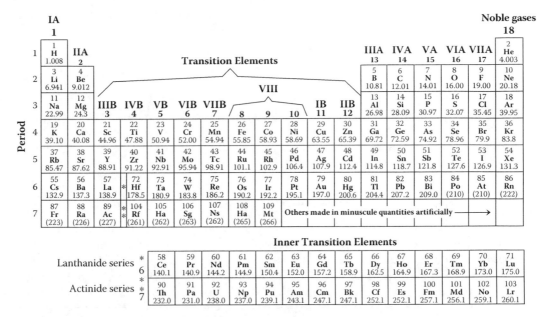

FIGURE 19.3 The periodic table of the elements.

Although a detailed discussion of the placement of electrons in atoms determines how the atoms behave chemically and, therefore, the chemical properties of each element, it is beyond the scope of this chapter to discuss electronic structure in detail. Several key points pertaining to this subject are mentioned here.

Electrons in atoms are present in orbitals in which the electrons have different energies, orientations in space, and average distances from the nucleus. Each orbital may contain a maximum of two electrons. The placement of electrons in their orbitals determines the chemical behavior of an atom; in this respect, the outermost orbitals and the electrons contained in them are the most important. These outer electrons are the ones beyond those of the immediately preceding noble gas in the periodic table. They are of particular importance because they become involved in the sharing and transfer of electrons through which chemical bonding occurs that results in the formation of huge numbers of different substances from only a few elements.

Much of environmental chemistry and water chemistry are concerned with electrons in atoms. In the atmosphere, the absorption of electromagnetic radiation, primarily ultraviolet radiation, promotes electrons to higher energy levels, forming reactive excited species and reactive free radicals with unpaired electrons. These phenomena can result in photochemical reactions such as the formation of stratospheric ozone, which is an essential filter for solar ultraviolet radiation. Atomic absorption and emission methods of elemental analysis, important in the study of pollutants, involve transitions of electrons between energy levels.

19.2.7 Lewis Structures and Symbols of Atoms

Outer electrons are called **valence electrons** and are represented by dots in **Lewis symbols**, as shown for carbon and argon in Figure 19.4.

The four electrons shown for the carbon atom are those added beyond the electrons possessed by the noble gas that immediately precedes carbon in the periodic table (helium, atomic number 2). Eight electrons are shown around the symbol of argon. This is an especially stable electron configuration for noble gases known as an **octet**. (Helium is the exception among noble gases in that it has a stable shell of only two electrons.) When atoms interact through the sharing, loss, or gain of

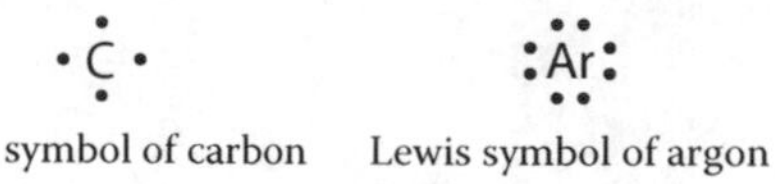

$$\cdot \overset{\cdot}{\underset{\cdot}{C}} \cdot \qquad\qquad\qquad :\overset{\cdot\cdot}{\underset{\cdot\cdot}{Ar}}:$$

Lewis symbol of carbon   Lewis symbol of argon

FIGURE 19.4   Lewis symbols of carbon and argon.

electrons to form molecules and chemical compounds (see Section 19.3), many attain an octet of outer shell electrons. This tendency is the basis of the **octet rule** of chemical bonding. (Two or three of the lightest elements, most notably hydrogen, attain stable helium-like electron configurations containing two electrons when they become chemically bonded.)

19.2.8 METALS, NONMETALS, AND METALLOIDS

Elements are divided between metals and nonmetals; a few elements with intermediate character are called metalloids. **Metals** are elements that are generally solid, shiny in appearance, electrically conducting, and malleable—that is, they can be pounded into flat sheets without disintegrating. They tend to have only one to three outer electrons, which they may lose in forming chemical compounds. Examples of metals are iron, copper, and silver. Most metallic objects that are commonly encountered are not composed of just one kind of elemental metal but are alloys consisting of homogeneous mixtures of two or more metals. **Nonmetals** often have a dull appearance, are not at all malleable, and frequently occur as gases or liquids. Colorless oxygen gas, green chlorine gas (transported and stored as a liquid under pressure), and brown bromine liquid are common nonmetals. Nonmetals tend to have close to a full octet of outer shell electrons, and in forming chemical compounds, they gain or share electrons. **Metalloids**, such as silicon or arsenic, are elements with properties intermediate between those of metals and nonmetals. Under some conditions, a metalloid may exhibit properties of metals, and under other conditions, properties of nonmetals.

19.3   CHEMICAL BONDING

Only a few elements, particularly the noble gases, exist as individual atoms; most atoms are joined by chemical bonds to other atoms. For example, elemental hydrogen exists as **molecules**, each consisting of two H atoms linked by a **chemical bond** as shown in Figure 19.5. Because hydrogen molecules contain two H atoms, they are said to be diatomic and are denoted by the **chemical formula**, H_2. The H atoms in the H_2 molecule are held together by a **covalent bond** made up of two electrons, each contributed by one of the H atoms, and shared between the atoms. (Bonds formed by transferring electrons between atoms are described later in this section.) The shared electrons in the covalent bonds holding the H_2 molecule together are represented by two dots between the H atoms in Figure 19.5. By analogy with Lewis symbols defined in Section 19.2.7, such a representation of molecules showing outer shell and bonding electrons as dots is called a **Lewis formula**.

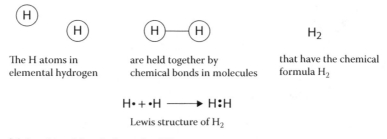

| The H atoms in elemental hydrogen | are held together by chemical bonds in molecules | that have the chemical formula H_2 |

H• + •H ⟶ H:H

Lewis structure of H_2

FIGURE 19.5   Molecule and Lewis formula of H_2.

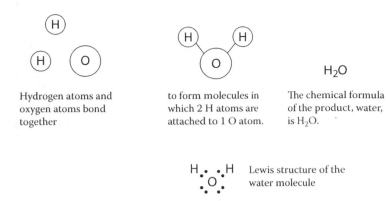

Hydrogen atoms and oxygen atoms bond together

to form molecules in which 2 H atoms are attached to 1 O atom.

The chemical formula of the product, water, is H_2O.

H_2O

H .. .. H   Lewis structure of the
   .. O ..   water molecule

FIGURE 19.6   Formation and Lewis formula of a chemical compound, water.

19.3.1 CHEMICAL COMPOUNDS

Most substances consist of two or more elements joined by chemical bonds. As an example, consider the chemical combination of hydrogen and oxygen shown in Figure 19.6. Oxygen, chemical symbol O, has an atomic number of 8 and an atomic mass of 16.00; it exists in the elemental form as diatomic molecules of O_2. Hydrogen atoms combine with oxygen atoms to form molecules in which two H atoms are bonded to one O atom in a substance with a chemical formula of H_2O (water). A substance such as H_2O that consists of a chemically bonded combination of two or more elements is called a **chemical compound**. In the chemical formula for water, the letters H and O are the chemical symbols of the two elements in the compound and the subscript 2 indicates that there are two H atoms per O atom. (The absence of a subscript after the O denotes the presence of just one O atom in the molecule.)

As shown in Figure 19.6, each of the hydrogen atoms in the water molecule is connected to the oxygen atom by a chemical bond composed of two electrons shared between the hydrogen and the oxygen atoms. For each bond, one electron is contributed by the hydrogen and one by the oxygen. The two dots located between each H and O in the Lewis formula of H_2O represent the two electrons in the covalent bond joining these atoms. Four of the electrons in the octet of electrons surrounding O are involved in H–O bonds and are called bonding electrons. The other four electrons shown around the oxygen that are not shared with H are nonbonding outer electrons.

19.3.2 MOLECULAR STRUCTURE

As implied by the representations of the water molecule in Figure 19.6, the atoms and bonds in H_2O form an angle somewhat greater than 90°. The shapes of molecules are referred to as their **molecular geometry**, which is crucial in determining the chemical and toxicological activity of a compound and structure–activity relationships.

As shown in Figure 19.7, the transfer of electrons from one atom to another produces charged species called **ions**. Positively charged ions are called **cations** and negatively charged ions are called **anions**. Ions that make up a solid compound are held together by **ionic bonds** in a crystalline lattice consisting of an ordered arrangement of the ions in which each cation is largely surrounded by anions and each anion by cations. The attracting forces of the oppositely charged ions in the crystalline lattice constitute ionic bonds in the compound. The formation of the ionic compound magnesium oxide is shown in Figure 19.7. In naming this compound, the cation is simply given the name of the element from which it was formed, magnesium. However, the ending of the name of the anion, oxide, is different from that of the element from which it was formed, oxygen.

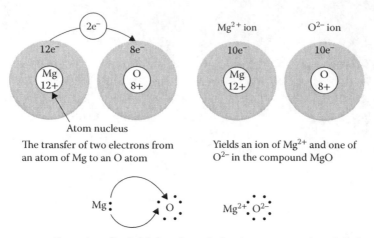

Formation of ionic MgO as shown by Lewis structures and symbols. In
MgO, Mg has lost 2 electrons and is in the +2 oxidation state, Mg(II),
and O has gained 2 electrons and is in the −2 oxidation state.

FIGURE 19.7   Ionic bonds are formed by the transfer of electrons and the mutual attraction of oppositely
charged ions in a crystalline lattice.

Rather than individual atoms that have lost or gained electrons, many ions are groups of atoms
bonded together covalently and having a net charge. A common example of such an ion is the
ammonium ion, NH_4^+,

$$
\begin{array}{c}
\text{H+} \\
| \\
\text{H--N--H} \\
| \\
\text{H} \qquad \text{Ammonium ion, } NH_4^+
\end{array}
$$

which consists of four hydrogen atoms covalently bonded to a single nitrogen (N) atom and having
a net electrical charge of +1 for the whole cation, as shown by its structural formula above.

19.3.3   Summary of Chemical Compounds and the Ionic Bond

The beginning of this chapter covered some material on chemical compounds and bonds that are
essential to understanding chemistry. To summarize, these are as follows:

- Atoms of two or more different elements can form *chemical bonds* with each other to yield
  a product that is entirely different from the elements.
- Such a substance is called a *chemical compound*.
- The *formula* of a chemical compound gives the symbols of the elements and uses subscripts to show the relative numbers of atoms of each element in the compound.
- *Molecules* of some compounds are held together by *covalent bonds* consisting of shared
  electrons.
- Another kind of compound consists of *ions* composed of electrically charged atoms or
  groups of atoms held together by *ionic bonds* that exist because of the mutual attraction of
  oppositely charged ions.

19.3.4   Molecular Mass

The average mass of all molecules of a compound is its **molecular mass** (formerly called molecular
weight). The molecular mass of a compound is calculated by multiplying the atomic mass of each
element by the relative number of atoms of the element and then adding all the values obtained for

each element in the compound. For example, the molecular mass of NH_3 is $14.0 + 3 \times 1.0 = 17.0$. As another example, consider the following calculation of the molecular mass of ethylene, C_2H_4:

- The chemical formula of the compound is C_2H_4.
- Each molecule of C_2H_4 consists of 2 C atoms and 4 H atoms.
- From the periodic table or Table 19.2, the atomic mass of C is 12.0 and that of H is 1.0.
- Therefore, the molecular mass of C_2H_4 is

$$\underbrace{12.0 + 12.0}_{\text{From 2 C atoms}} + \underbrace{1.0 + 1.0 + 1.0 + 1.0}_{\text{From 4 H atoms}} = 28.0$$

19.3.5 MOLE AND MOLAR MASS

It is convenient to consider quantities of chemicals in compounds in units proportional to the number of molecules. Such a unit is the **mole**, which is expressed as the same number of grams of a compound as the molecular mass of the compound. Using this definition, a mole of C_2H_4, calculated in the preceding discussion to have a molecular mass of 28.0, is 28.0 g and the **molar mass** of C_2H_4 is 28.0 g/mol. The molecular mass of H_2O is 18.0 and its molar mass is 18.0 g/mol. The definition applies equally well to ionic compounds that do not consist of distinct molecules. Sodium chloride, NaCl, consists of aggregates of equal numbers of Na^+ and Cl^- ions. The molecular mass of NaCl is 23.0 (from Na) + 35.5 (from Cl) = 58.5. Therefore, the molar mass of NaCl is 58.5 g/mol. In the case of elements that are composed of individual atoms, the molar mass is given by the atomic mass. The molar mass of helium, atomic mass 4.0, is 4.0 g/mol. Elemental hydrogen, atomic mass 1.0 consists of molecules each composed of two atoms (H_2) and has a molecular mass of 2.0. The molar mass of H_2 is 2.0 g/mol.

19.3.6 OXIDATION STATE

The loss of two electrons from the magnesium atom as shown in Figure 19.7 is an example of **oxidation**, and the Mg^{2+} ion product is said to be in the +2 **oxidation state**. (A positive oxidation state or oxidation number is conventionally denoted by a Roman numeral in parentheses following the name or symbol of an element as in magnesium(II) and Mg(II)). In gaining two negatively charged electrons in the reaction that produces magnesium oxide, the oxygen atom is **reduced** and is in the −2 oxidation state. (Unlike positive oxidation numbers, negative ones are not conventionally shown by Roman numerals in parentheses.) In chemical terms, an **oxidizer** is a species that takes electrons from a reducing agent in a chemical reaction. Many hazardous waste substances are oxidizers or strong reducers, and oxidation–reduction is the driving force behind many dangerous chemical reactions. For example, the reducing tendencies of the carbon and hydrogen atoms in propane cause it to burn violently or explode in the presence of oxidizing oxygen in air. The oxidizing ability of concentrated nitric acid, HNO_3, enables it to react destructively with organic matter such as cellulose or skin. As discussed in Chapter 3, oxidation–reduction phenomena are very important in aquatic chemical processes.

Covalently bonded atoms that have not actually lost or gained electrons to produce ions are also assigned oxidation states. This can be done because in covalent compounds electrons are not shared equally. Therefore, an atom of an element with a greater tendency to attract electrons is assigned a negative oxidation number compared to the positive oxidation number assigned to an element with a lesser tendency to attract electrons. For example, Cl atoms attract electrons more strongly than do H atoms so that in hydrogen chloride gas, HCl, the Cl atom is in the −1 oxidation state and the H atoms are in the +1 oxidation state. **Electronegativity** values are assigned to elements on the basis of their tendencies to attract electrons.

The oxidation state (oxidation number) of an element in a compound may have a strong influence on the hazards posed by the compound. For example, chromium from which each atom has lost three electrons to form a chemical compound, designated as chromium(III) or Cr(III), is not toxic, whereas Cr in the +6 oxidation state (CrO_4^{2-}, chromate) is regarded as a cancer-causing chemical when inhaled.

19.4 CHEMICAL REACTIONS AND EQUATIONS

Chemical reactions occur when substances are changed to other substances through the breaking and formation of chemical bonds. For example, water is produced by the chemical reaction of hydrogen and oxygen:

Hydrogen plus oxygen yields water

Chemical reactions are written as **chemical equations**. The chemical reaction between hydrogen and water is written as the **balanced chemical equation**

$$2H_2 + O_2 \rightarrow 2H_2O \tag{19.6}$$

in which the arrow is read as "yields" and separates the hydrogen and oxygen **reactants** from the water **product**. Note that because elemental hydrogen and elemental oxygen occur as *diatomic molecules* of H_2 and O_2, respectively, it is necessary to write the equation in a way that reflects these correct chemical formulas of the elemental form. All correctly written chemical equations are **balanced** in that *the same number of each kind of atom must be shown on both sides of the equation*. Equation 19.6 is balanced because of the following:

On the left
- There are two H_2 *molecules* each containing two H *atoms*, a total of four H atoms on the left.
- There is one O_2 *molecule* containing two O *atoms* for a total of two O atoms on the left.

On the right
- There are two H_2O *molecules* each containing two H *atoms* and one O *atom* for a total of four H atoms and two O atoms on the right.

19.4.1 REACTION RATES

Most chemical reactions give off heat and are classified as exothermic reactions. The rate of a reaction may be calculated by the Arrhenius equation, which contains absolute temperature (K = °C + 273) in an exponential term. Typically, the speed of a reaction doubles for each 10°C increase in temperature. Reaction rate factors are important factors in fires or explosions involving hazardous chemicals. Rates of reactions are also very important in atmospheric chemical processes such as those involved in the formation of photochemical smog (see Chapter 7, Section 7.8).

Catalysts are materials that speed up chemical reactions or even enable them to proceed without themselves being consumed. An example of a catalyst is the one located in the catalytic converter of an automobile exhaust system that consists of small quantities of precious metals coated onto the solid honeycomb-like surface that acts as a catalyst support. Such a catalyst speeds the reaction with oxygen of toxic carbon monoxide from the engine exhaust to produce nontoxic carbon dioxide:

$$2CO + O_2 \rightarrow 2CO_2 \tag{19.7}$$

As discussed in Section 2.7, the most sophisticated catalysts consist of special proteins called enzymes that occur in living organisms. A common enzyme-catalyzed process is the reaction of glucose (blood sugar, $C_6H_{12}O_6$) with molecular oxygen to produce energy:

$$C_6H_{12}O_6 + 6O_2 \rightarrow 6CO_2 + 6H_2O + \text{energy} \tag{19.8}$$

This is the important process of oxic respiration carried out by all organisms, including humans, that live in contact with air and utilize oxygen from air to react with food materials. Although the overall reaction for oxic respiration can be written very simply, the actual process requires many steps, and several catalytic enzymes are used.

Catalysts are very important in green chemistry. One reason that this is so is because catalysts enable reactions to be carried out very specifically. Also, the right catalyst can enable reactions to occur with relatively less energy consumption and at relatively lower temperatures.

19.5 SOLUTIONS

A liquid **solution** is formed when a substance in contact with a liquid becomes dispersed homogeneously throughout the liquid at the molecular level. The substance, called a **solute**, is said to **dissolve**. The liquid is called a **solvent**. There may be no readily visible evidence that a solute is present in the solvent; for example, a deadly poisonous solution of sodium cyanide in water looks like pure water. Some solutions have a strong color, as is the case for intensely purple solutions of potassium permanganate, $KMnO_4$. A solution may have a strong odor such as that of ammonia, NH_3, dissolved in water. Solutions may consist of solids, liquids, or gases dissolved in a solvent. Technically, it is even possible to have solutions in which a solid is a solvent.

19.5.1 SOLUTION CONCENTRATION

The quantity of a solute relative to that of a solvent or solution is called the **solution concentration**. Concentrations are expressed in numerous ways. Very high concentrations are often given as percent by mass. For example, commercial concentrated hydrochloric acid is 36% HCl, meaning that 36% of the mass has come from dissolved HCl and 64% from water solvent. Concentrations of very dilute solutions, such as those of hazardous waste leachate containing low levels of contaminants, are expressed as mass of solute per unit volume of solution. Common units are milligrams per liter (mg/L) or micrograms per liter (μg/L). Since a liter of water weighs essentially 1000 g, a concentration of 1 mg/L is equal to 1 part per million (ppm) and a concentration of 1 μg/L is equal to 1 part per billion (ppb).

Chemists often express concentrations in moles per liter, or **molarity**, M. Molarity is given by the relationship

$$M = \frac{\text{Number of moles of solute}}{\text{Number of liters of solution}} \tag{19.9}$$

The number of moles of a substance is its mass in grams divided by its molar mass. For example, the molecular mass of ammonia, NH_3, is 14 + 1 + 1 + 1, so a mole of ammonia has a mass of 17 g. Therefore, 17 g of NH_3 in 1 L of solution has a value of M equal to 1 mole/L.

19.5.2   WATER AS A SOLVENT

Water is a uniquely important solvent in the environment and in living systems. Water has some unique properties as a solvent that arise from its molecular structure as represented by the following Lewis structure of the water molecule:

$$(+) \quad H \quad \ddot{O} \quad (-)$$

Polar water molecule

The H atoms are not on opposite sides of the O atom and the two H–O bonds form an angle of 105°. Furthermore, the O atom (−2 oxidation state) is able to attract electrons more strongly than the two H atoms (each in the +1 oxidation state) so that the molecule is **polar**, with the O atom having a partial negative charge and the end of the molecule with the two H atoms having a partial positive charge. This means that water molecules can cluster around ions with the positive ends of the water molecules attracted to negatively charged anions and the negative end to positively charged cations. This kind of interaction is part of the general phenomenon of **solvation**. It is specifically called **hydration** when water is the solvent and is partially responsible for water's excellent ability to dissolve ionic compounds including acids, bases, and salts.

Water molecules form a special kind of bond called a **hydrogen bond** with each other and with solute molecules that contain O, N, or S atoms. As its name implies, a hydrogen bond involves a hydrogen atom held between two other atoms of O, N, or S. Hydrogen bonding is partly responsible for water's ability to solvate and dissolve chemical compounds capable of hydrogen bonding.

As shown by its structural formula, the water molecule is a polar species, which affects its ability to act as a solvent. Solutes may likewise have polar character. In general, solutes with polar molecules are more soluble in water than nonpolar ones. The polarity of an impurity solute in wastewater is a factor in determining how it may be removed from water. Nonpolar organic solutes are easier to take out of water by an adsorbent species such as activated carbon than are more polar solutes.

19.5.3   SOLUTIONS OF ACIDS, BASES, AND SALTS

Acids are substances that produce H^+ ion in water solution and **bases** are substances that produce OH^- ion. **Salts** are substances that are composed of ions and that contain a cation other than H^+ and an anion other than OH^-. Salts can be produced by the reaction between H^+ ion from an acid and OH^- ion from a base, a **neutralization** reaction. As a specific example, consider the reaction of H^+ from a solution of sulfuric acid, H_2SO_4, with OH^- from a solution of calcium hydroxide:

$$\underset{\text{Base, source of } OH^- \text{ ion}}{\underset{\uparrow}{H_2SO_4}} + Ca(OH)_2 \longrightarrow \underset{\text{Salt}}{\underset{\uparrow}{2H_2O + CaSO_4}} \tag{19.10}$$

Acid, source of H^+ ion   Water

In addition to water, which is always the product of a neutralization reaction, the other product is calcium sulfate, $CaSO_4$. This compound is a **salt** composed of Ca^{2+} ions and SO_4^{2-} ions held together by ionic bonds. A salt, consisting of a cation other than H^+ and an anion other than the OH^- ion, is the other product in addition to water produced when an acid and a base react. Some salts are hazardous substances and environmental pollutants because of their dangerous or harmful properties. Some examples are the following:

- Ammonium perchlorate, NH_4ClO_4, (reactive oxidant)
- Barium cyanide, $Ba(CN)_2$ (toxic)
- Lead acetate, $Pb(C_2H_3O_2)_2$ (toxic)
- Thallium(I) carbonate, Tl_2CO_3 (toxic)

19.5.4 CONCENTRATION OF H^+ ION AND pH

Acids such as HCl and sulfuric acid (H_2SO_4) produce H^+ ion, whereas bases such as sodium hydroxide and calcium hydroxide [NaOH and $Ca(OH)_2$ respectively] produce hydroxide ion, OH^-. Molar concentrations of water solutions of hydrogen ion, [H^+], range over many orders of magnitude and are conveniently expressed by pH defined as follows:

$$pH = -\log\left[H^+\right] \tag{19.11}$$

In absolutely pure water at 25°C, the value of [H^+] is exactly 1×10^{-7} mole/L, the pH is 7.00, and the solution is **neutral** (neither acidic nor basic). **Acidic** solutions have pH values of less than 7, and **basic** solutions have pH values of greater than 7.

Strong acids and strong bases are **corrosive** substances that exhibit extremes of pH. They are destructive to materials and flesh. Strong acids can react with cyanide and sulfide compounds to release highly toxic hydrogen cyanide (HCN) or hydrogen sulfide (H_2S) gases respectively. Strong bases such as NaOH liberate noxious ammonia gas (NH_3) from solid ammonium compounds.

19.5.5 METAL IONS DISSOLVED IN WATER

Metal ions dissolved in water have some unique characteristics that influence their properties as natural water constituents and heavy metal pollutants and in biological systems. The formulas of metal ions are usually represented by the symbol for the ion followed by its charge. For example, iron(II) ion [from a compound such as iron(II) sulfate, $FeSO_4$] dissolved in water is represented as Fe^{2+}. Actually, in water solution, each iron(II) ion is strongly solvated by bonding to water molecules, so that the formula is more correctly shown as $Fe(H_2O)_6^{2+}$. Many metal ions have a tendency to lose hydrogen ions from the solvating water molecules as shown by the following reaction:

$$Fe\left(H_2O\right)_6^{2+} \rightarrow Fe\left(OH\right)\left(H_2O\right)_5 + H^+ \tag{19.12}$$

Ions of the next higher oxidation state, iron(III), have such a tendency to lose H^+ ion in aqueous solution that, except in rather highly acidic solutions, they precipitate out as solid hydroxides, such as iron(III) hydroxide, $Fe(OH)_3$:

$$Fe\left(H_2O\right)_6^{3+} \rightarrow Fe\left(OH\right)\left(s\right)_3 + 3H_2O + 3H^+ \tag{19.13}$$

19.5.6 COMPLEX IONS DISSOLVED IN WATER

It was noted in Section 19.5.5 that metal ions are solvated (hydrated) by binding to water molecules in aqueous solution. Some species in solution have a stronger tendency than water to bond to metal ions. An example of such a species is cyanide ion, CN^-, which displaces water molecules from some metal ions in solution as shown in the following reaction (where the double arrow denotes that the reaction is reversible):

$$Ni(H_2O)_4^{2+} + 4CN^- \leftrightarrow Ni(CN)_4^{2-} + 4H_2O \tag{19.14}$$

The species that bonds to the metal ion, cyanide in this case, is called a **ligand**, and the product of the reaction is a **complex ion** or metal complex. The overall process is called **complexation**. (See also the discussion of complexation and metal chelates in Chapter 3, Section 3.11.)

19.5.7 COLLOIDAL SUSPENSIONS

Very small particles on the order of 1 μm or less in size, called **colloidal particles**, may stay suspended in a liquid for an indefinite period of time. Such a mixture is a **colloidal suspension** and it behaves in many respects like a solution. As discussed in Chapter 3, Section 3.8, colloids play an important role in the environmental chemistry of the hydrosphere. The three major kinds of colloidal suspensions in water are shown in Figure 3.11. Colloidal suspensions are used in many industrial applications. Many waste materials are colloidal and are often emulsions consisting of colloidal liquid droplets suspended in another liquid, usually wastewater. One of the challenges in dealing with colloidal wastes or in purifying water that contains colloidal solids is to remove a relatively small quantity of colloidal material from a large quantity of water by precipitating the colloid. This process is called **coagulation** or **flocculation** and is often brought about by the addition of chemical agents.

19.5.8 SOLUTION EQUILIBRIA

Many of the phenomena in aquatic chemistry and geochemistry (chemical phenomena in the geosphere) involve solution equilibrium. In a general sense, solution equilibrium deals with the extent to which **reversible** acid-base, solubilization–precipitation, complexation, or oxidation–reduction reactions proceed in a forward or backward direction. This is expressed for a generalized equilibrium reaction

$$aA + bB \leftrightarrow cC + dD \tag{19.15}$$

by solution concentrations in the following **equilibrium constant expression**:

$$\frac{[C]^c[D]^d}{[A]^a[B]^b} = K \tag{19.16}$$

where K is the equilibrium constant.

A reversible reaction may approach equilibrium from either direction. In the preceding example, if A were mixed with B, or C were mixed with D, the reaction would proceed in a forward or reverse direction such that the concentrations of species—[A], [B], [C], and [D]—substituted into the equilibrium expression gave a value equal to K.

As expressed by **Le Châtelier's principle**, a stress placed upon a system in equilibrium will shift the equilibrium to relieve the stress. For example, adding product "D" to a system in equilibrium will cause Reaction 19.15 to shift to the left, consuming "C" and producing "A" and "B," until the equilibrium constant expression is again satisfied. This **mass action effect** is the driving force behind many environmental chemical phenomena.

When concentrations and pressures are used in equilibrium constant expression calculations, K is not exactly constant with varying concentrations and pressures; it is an **approximate equilibrium constant** that applies only to limited conditions. **Thermodynamic equilibrium constants** are more exact forms derived from thermodynamic data that make use of **activities** in place of concentrations. At a specified temperature, the value of a thermodynamic equilibrium constant is applicable over a wide concentration range. The activity of a species, commonly denoted as a_X for species "X," expresses how effectively it interacts with its surroundings, such as other solutes or electrodes in solution. Activities approach concentrations at low values of concentration. The thermodynamic equilibrium constant expression for Reaction 19.15 is expressed as the following in terms of activities:

$$\frac{a_C^c a_D^d}{a_A^a a_B^b} = K \tag{19.17}$$

There are several major kinds of equilibria in aqueous solution. One of these is acid-base equilibrium as exemplified by the ionization of acetic acid, HAc,

$$HAc \leftrightarrow H^+ + Ac^- \tag{19.18}$$

for which the acid dissociation constant is the following where the subscript "a" denotes an acid dissociation constant:

$$\frac{[H^+][Ac^-]}{[HAc]} = K_a \tag{19.19}$$

Very similar expressions are obtained for the formation and dissociation of metal complexes or complex ions formed by the reaction of a metal ion in solution with a complexing agent or ligand, both of which are capable of independent existence in solution (see Section 19.5.6). This can be shown by the reaction of iron(III) ion and thiocyanate ligand

$$Fe^{3+} + SCN^- \leftrightarrow FeSCN^{2+} \tag{19.20}$$

for which the **formation constant expression** is

$$\frac{[FeSCN^{2+}]}{[Fe^{3+}][SCN^-]} = K_f = 1.07 \times 10^{-3} \quad \text{(at 25°C)} \tag{19.21}$$

The bright red color of the $FeSCN^{2+}$ complex formed is used to test for the presence of iron(III) in polluted acid mine water.

An example of an **oxidation–reduction reaction**, which involves the transfer of electrons between species, is

$$MnO_4^- + 5Fe^{2+} + 8H^+ \leftrightarrow Mn^{2+} + 5Fe^{3+} + 4H_2O \tag{19.22}$$

for which the equilibrium expression is

$$\frac{[Mn^{2+}][Fe^{3+}]^5}{[MnO_4^-][Fe^{2+}]^5[H^+]^8} = K = 3 \times 10^{62} \quad \text{(at 25°C)} \tag{19.23}$$

19.5.9   DISTRIBUTION BETWEEN PHASES

Many important environmental chemical phenomena involve distribution of species between phases. This most commonly involves the equilibria between species in solution and in a solid phase. **Solubility equilibria** deal with reactions such as

$$AgCl(s) \leftrightarrow Ag^+ + Cl^- \tag{19.24}$$

in which one of the participants is a solid that is slightly soluble (virtually insoluble). For the preceding example, the equilibrium constant is

$$\left[Ag^+\right]\left[Cl^-\right] = K_{sp} = 1.82 \times 10^{-10} \quad (\text{at } 25^\circ C) \tag{19.25}$$

a **solubility product**. Note that in the equilibrium constant expression, there is no value given for the solid AgCl. This is because the activity of a solid is constant at a specific temperature and is contained in the value of K_{sp}.

An important example of distribution between phases is that of a hazardous waste species partitioned between water and a body of immiscible organic liquid in a hazardous waste site. The equilibrium for such a reaction

$$X(aq) \leftrightarrow X(org) \tag{19.26}$$

is described by the **distribution law** expressed by a **distribution coefficient** or **partition coefficient** in the following form:

$$\frac{[X(org)]}{[X(aq)]} = K_d \tag{19.27}$$

Another important example of distribution between phases is that between a gas and the gas species dissolved in water. Gas solubilities are described by Henry's law as discussed for oxygen solubility in water in Chapter 3, Section 3.8.1.

QUESTIONS AND PROBLEMS

Access to and use of the Internet is assumed in answering all questions including general information, statistics, constants, and mathematical formulas required to solve problems.

1. What distinguishes a radioactive isotope from a "normal" stable isotope?
2. Why is the periodic table so named?
3. After examining Figure 19.7, consider what might happen when an atom of sodium (Na), atomic number 11, loses an electron to an atom of fluorine, (F), atomic number 9. What kinds of particles are formed by this transfer of a negatively charged electron? Is a chemical compound formed? If so, what is it called?
4. Match the following:

1. O_2	(a) Element consisting of individual atoms
2. NH_3	(b) Element consisting of chemically bonded atoms
3. Ar	(c) Ionic compound
4. NaCl	(d) Covalently bound compound

5. Consider the following atom:

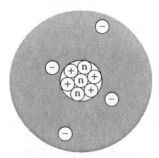

How many electrons, protons, and neutrons does it have? What is its atomic number? Give the name and chemical symbol of the element of which it is composed.

6. Give the chemical formula and molecular mass of the molecule represented in the following:

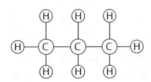

7. Calculate the molecular masses of (a) C_2H_2, (b) N_2H_4, (c) Na_2O, (d) O_3 (ozone), (e) PH_3, and (f) CO_2.

8. Calculate the new volume of a quantity of gas occupying 15.6 L of volume at 0°C and 1.10 atm pressure when conditions are changed to (a) 80°C and 1.10 atm pressure and then to (b) 160°C and 0.90 atm pressure.

9. Is the equation $H_2 + O_2 \rightarrow H_2O$ a balanced chemical equation? Explain. Point out the reactants and products in the equation. If the equation is not balanced, what is the balanced form?

10. An uncharged atom has the same number of _____ as protons. The electrons in an atom are contained in a cloud of _____ around the nucleus that occupies most of the _____ of the atom.

11. Match:

1. Argon	(a) A halogen
2. Hydrogen	(b) Fissionable element
3. Uranium	(c) Noble gas
4. Chlorine	(d) Has an isotope with a mass number of 2
5. Mercury	(e) Toxic heavy metal

12. The entry for each element in the periodic table gives the element's _____, _____, and _____ and the periodic table is arranged horizontally in _____ and vertically in _____.

13. Electrons in atoms occupy _____ in which electrons have different _____ _____. Each orbital may contain a maximum of _____ electrons.

14. The Lewis symbol of carbon is _____ in which each dot represents _____.

15. Elements that are generally solid, shiny in appearance, electrically conducting, and malleable are _____, whereas elements that tend to have a dull appearance, are not at all malleable, and frequently occur as gases or liquids are _____. Elements with intermediate properties are called _____.

16. In the Lewis formula for the elemental hydrogen molecule in the following, what do the two dots represent?

$$H:H$$

17. Explain why, even though it consists of molecules containing chemical bonds, H_2 is not a chemical compound whereas H_2O is a chemical compound.
18. Using examples, distinguish between covalent and ionic bonds.
19. In terms of c, h, o and the appropriate atomic masses, write a mathematical formula for the molecular mass of a compound with a general formula of $C_cH_hO_o$.
20. Considering oxidation–reduction phenomena, when Al reacts with O_2 to produce Al_2O_3, which contains the Al^{3+} ion, the Al is said to have been _____ and is in the _____ oxidation state.
21. Calculate the concentration in moles per liter of (a) a solution that is 27.0% H_2SO_4 by mass and that has a density of 1.198 g/L and (b) of a solution that is 1 mg/L H_2SO_4 having a density of 1000 g/L.
22. Calculate the pH of the second solution described in the preceding problem, keeping in mind that each H_2SO_4 molecule yields two H^+ ions.
23. Write a balanced neutralization reaction between NaOH and H_2SO_4.
24. Distinguish between solutions and colloidal suspensions.
25. What is the nature of the Fe^{3+} ion? Why are solutions containing this ion acidic?
26. What kind of species is $Ni(CN)_4^{2-}$?
27. What is the solubility product expression for lead sulfate, $PbSO_4$, in terms of concentrations of Pb^{2+} ion and SO_4^{2-} ion?

TABLE 19.3
List of the Elements[a]

Atomic Number	Name	Symbol	Atomic Mass	Atomic Number	Name	Symbol	Atomic Mass
1	Hydrogen	H	1.00794	56	Barium	Ba	137.327
2	Helium	He	4.0026	57	Lanthanum	La	138.9055
3	Lithium	Li	6.941	58	Cerium	Ce	140.115
4	Beryllium	Be	9.01218	59	Praseodymium	Pr	140.9077
5	Boron	B	10.811	60	Neodymium	Nd	144.24
6	Carbon	C	12.011	61	Promethium	Pm	145
7	Nitrogen	N	14.0067	62	Samarium	Sm	150.36
8	Oxygen	O	15.9994	63	Europium	Eu	151.965
9	Fluorine	F	18.9984	64	Gadolinium	Gd	157.25
10	Neon	Ne	20.1797	65	Terbium	Tb	158.925
11	Sodium	Na	22.9898	66	Dysprosium	Dy	162.50
12	Magnesium	Mg	24.305	67	Holmium	Ho	164.9303
13	Aluminum	Al	26.98154	68	Erbium	Er	167.26
14	Silicon	Si	28.0855	69	Thulium	Tm	168.9342
15	Phosphorus	P	30.9738	70	Ytterbium	Yb	173.04
16	Sulfur	S	32.066	71	Lutetium	Lu	174.967
17	Chlorine	Cl	35.4527	72	Hafnium	Hf	178.49
18	Argon	Ar	39.948	73	Tantalum	Ta	180.9497
19	Potassium	K	39.0983	74	Tungsten	W	183.85
20	Calcium	Ca	40.078	75	Rhenium	Re	186.207
21	Scandium	Sc	44.9559	76	Osmium	Os	190.2

TABLE 19.3 (*Continued*)
List of the Elements[a]

Atomic Number	Name	Symbol	Atomic Mass	Atomic Number	Name	Symbol	Atomic Mass
22	Titanium	Ti	47.88	77	Iridium	Ir	192.22
23	Vanadium	V	50.9415	78	Platinum	Pt	195.08
24	Chromium	Cr	51.9961	79	Gold	Au	196.9665
25	Manganese	Mn	54.938	80	Mercury	Hg	200.59
26	Iron	Fe	55.847	81	Thallium	Tl	204.383
27	Cobalt	Co	58.9332	82	Lead	Pb	207.2
28	Nickel	Ni	58.6934	83	Bismuth	Bi	208.98
29	Copper	Cu	63.546	84	Polonium	Po	209
30	Zinc	Zn	65.39	85	Astatine	At	210
31	Gallium	Ga	69.723	86	Radon	Rn	222
32	Germanium	Ge	72.61	87	Francium	Fr	223
33	Arsenic	As	74.9216	88	Radium	Ra	226.0254
34	Selenium	Se	78.96	89	Actinium	Ac	227.0278
35	Bromine	Br	79.904	90	Thorium	Th	232.038
36	Krypton	Kr	83.8	91	Protactinium	Pa	231.0359
37	Rubidium	Rb	85.4678	92	Uranium	U	238.0289
38	Strontium	Sr	87.62	93	Neptunium	Np	237.048
39	Yttrium	Y	88.9056	94	Plutonium	Pu	244
40	Zirconium	Zr	91.224	95	Americium	Am	243
41	Niobium	Nb	92.9064	96	Curium	Cm	247
42	Molybdenum	Mo	95.94	97	Berkelium	Bk	247
43	Technetium	Tc	98	98	Californium	Cf	251
44	Ruthenium	Ru	101.07	99	Einsteinium	Es	252
45	Rhodium	Rh	102.9055	100	Fermium	Fm	257.1
46	Palladium	Pd	106.42	101	Mendelevium	Md	258.1
47	Silver	Ag	107.8682	102	Nobelium	No	255
48	Cadmium	Cd	112.411	103	Lawrencium	Lr	260
49	Indium	In	114.82	104	Rutherfordium	Rf	261.11
50	Tin	Sn	118.710	105	Dubnium	Db	262.11
51	Antimony	Sb	121.757	106	Seaborgium	Sg	263.12
52	Tellurium	Te	127.60	107	Bohrium	Bh	262.12
53	Iodine	I	126.9045	108	Hassium	Hs	265
54	Xenon	Xe	131.29	109	Meitnerium	Mt	266
55	Cesium	Cs	132.9054				

[a] Elements above atomic number 92 have been made artificially.

LITERATURE CITED

1. Manahan, Stanley E., *Fundamentals of Sustainable Chemical Science*, Taylor & Francis/CRC Press, Boca Raton, FL, 2009.

SUPPLEMENTARY REFERENCES

Brown, Lawrence S., and Thomas A. Holme, *Chemistry for Engineering Students*, 2nd ed., Brooks/Cole Cengage Learning, Belmont, CA, 2011.
Brown, Theodore L., H. Eugene Lemay, Bruce E. Bursten, Catherine J. Murphy, and Patrick Woodward, *Chemistry: The Central Science*, 12th ed., Prentice Hall, Upper Saddle River, NJ, 2012.

Denniston, Katherine J., Joseph J. Topping, and Robert L. Caret, *General, Organic, and Biochemistry*, 7th ed., McGraw-Hill, New York, 2011.

Frost, Laura D., Todd S. Deal, and Karen C. Timberlake, *General, Organic, and Biological Chemistry: An Integrated Approach*, Pearson Education, Upper Saddle River, NJ, 2011.

Guinn, Denise, and Rebecca Brewer, *Essentials of General, Organic, and Biochemistry: An Integrated Approach*, W. H. Freeman and Co., New York, 2009.

Hill, John W., Terry W. McCreary, and Doris Kolb, *Chemistry for Changing Times*, 13th ed., Prentice Hall, Upper Saddle River, NJ, 2013.

McMurry, John, David S. Ballantine, Carl A. Hoeger, Virginia E. Peterson, and Mary E. Castellion, *Fundamentals of General, Organic, and Biological Chemistry*, 6th ed., Prentice-Hall, Upper Saddle River, NJ, 2009.

Mickulecky, Peter J., Katherine Brutlag, Michelle Gilman, and Brian Peterson, *Chemistry Workbook for Dummies*, Wiley, Hoboken, NJ, 2008.

Middlecamp, Catherine, Steve Keller, Karen Anderson, Anne Bentley, and Michael Cann, *Chemistry in Context*, 7th ed., McGraw-Hill, Dubuque, IA, 2011.

Moore, John, *Chemistry for Dummies*, 2nd ed., Wiley, Hoboken, NJ, 2011.

Moore, John W., Conrad L. Stanitski, and Peter C. Jurs, *Chemistry: The Molecular Science*, 4th ed., Brooks/Cole Cengage Learning, Belmont, CA, 2011.

Raymond, Kenneth, *General Organic and Biological Chemistry*, Wiley, Hoboken, NJ, 2010.

Smith, Janice G., *Principles of General, Organic, and Biochemistry*, McGraw-Hill, New York, 2011.

20 Organic Chemistry

20.1 ORGANIC CHEMISTRY

Most carbon-containing compounds are **organic chemicals** and are addressed by the subject of **organic chemistry**. Organic chemistry is a vast, diverse discipline because of the enormous number of organic compounds that exist as a consequence of the versatile bonding capabilities of carbon. Such diversity is due to the ability of carbon atoms to bond to each other through single (two shared electrons) bonds, double (four shared electrons) bonds, and triple (six shared electrons) bonds in a limitless variety of straight chains, branched chains, and rings.

Among organic chemicals are included the majority of important industrial compounds, synthetic polymers, agricultural chemicals, biological materials, and most substances that are of concern because of their toxicities and other hazards. Pollution of the water, air, and soil environments by organic chemicals is an area of significant concern.

Chemically, most organic compounds can be divided among hydrocarbons, oxygen-containing compounds, nitrogen-containing compounds, sulfur-containing compounds, organohalides, phosphorus-containing compounds, or combinations of these kinds of compounds. Each of these classes of organic compounds is discussed briefly in this chapter.

All organic compounds of course contain carbon. Virtually all also contain hydrogen and have at least one C–H bond. The simplest organic compounds, and those easiest to understand, are those that contain only hydrogen and carbon. These compounds are called **hydrocarbons** and are addressed first among the organic compounds discussed in this chapter. Hydrocarbons are used here to illustrate some of the most fundamental points of organic chemistry, including organic formulas, structures, and names.

20.1.1 MOLECULAR GEOMETRY IN ORGANIC CHEMISTRY

The three-dimensional shape of a molecule, that is, its molecular geometry, is particularly important in organic chemistry. This is because its molecular geometry determines in part the properties of an organic molecule, particularly its interactions with biological systems and how it is metabolized by organisms. Shapes of molecules are represented in drawings by lines of normal, uniform thickness for bonds in the plane of the paper; broken lines for bonds extending away from the viewer; and heavy lines for bonds extending toward the viewer. These conventions are shown by the example of dichloromethane, CH_2Cl_2, an important organochloride solvent and extractant, illustrated in Figure 20.1.

20.1.2 CHIRALITY AND THE SHAPES OF ORGANIC MOLECULES

An important aspect of molecular geometry is **chirality**; **chiral molecules** are otherwise identical but have three-dimensional structures such that a molecule cannot be directly imposed on its mirror image. A simple analogy is the human hand in which the left hand is a mirror image of the right. Placed palm-to-palm (mirror images), they match, but placing one hand palm down over the other results in a mismatch with the thumb of one over the little finger of the other. Chiral molecules have different groups arranged around an atom, usually of carbon, that constitutes a chiral center. Two chiral molecules of the same compound are called **enantiomers**, commonly designated as R and S. The physical and chemical properties of enantiomers are generally identical, having the same melting points, boiling points, and solubilities. However, the biochemical properties of the two enantiomers

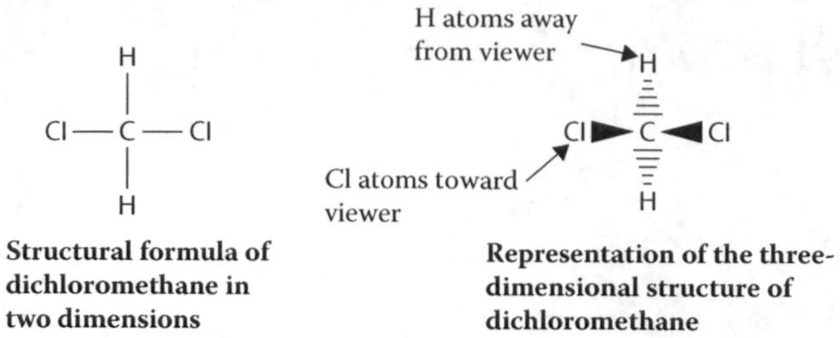

H atoms away
from viewer

Cl atoms toward
viewer

**Structural formula of
dichloromethane in
two dimensions**

**Representation of the three-
dimensional structure of
dichloromethane**

FIGURE 20.1 Structural formulas of dichloromethane, CH_2Cl_2; the formula on the right provides a three-dimensional representation.

may be quite different in that one may fit exactly onto the active site of an enzyme and be acted upon by it, whereas the other enantiomer does not fit and is not acted upon by the enzyme. This may be especially important with respect to toxic effects, pharmaceutical action, and pesticide action. For example, one enantiomer of a chiral pharmaceutical may function very well, whereas the other does not work at all or may even be toxic. Chirality as related to herbicidal mecoprop is discussed in Chapter 14, Section 14.11, and illustrated in Figure 14.13.

20.2 HYDROCARBONS

The tremendous variety and diversity of organic chemistry is due to the ability of carbon atoms to bond with each other in a variety of straight chains, branched chains, and rings and of adjacent carbon atoms to be joined by single, double, or triple bonds. This bonding ability can be illustrated with the simplest class of organic chemicals, the **hydrocarbons** consisting only of hydrogen and carbon. Figure 20.2 shows some hydrocarbons in various configurations. Hydrocarbons are the major ingredients of petroleum and are pumped from the ground as crude oil or extracted as natural gas. They have two major uses. The first of these is combustion as a source of fuel. The most abundant hydrocarbon in natural gas, methane, CH_4, is burned in home furnaces, electrical power plants, and even in vehicle engines

$$CH_4 + 2O_2 \rightarrow CO_2 + 2H_2O + \text{heat energy} \tag{20.1}$$

to provide energy. The second major use of hydrocarbons is as a raw material for making rubber, plastics, polymers, and many other kinds of materials. Given the value of hydrocarbons as a material, it is unfortunate that so much of hydrocarbon production is simply burned to provide energy, which could be generated by other means.

There are several major classes of hydrocarbons, all consisting of only hydrogen and carbon. **Alkanes** have only single bonds between carbon atoms. Cyclohexane, *n*-heptane, and 3-ethyl-2, 5-dimethylhexane, shown in Figure 20.2, are alkanes; cyclohexane is a cyclic hydrocarbon. **Alkenes**, such as propene shown in Figure 20.2, have at least one double bond consisting of four shared electrons between two of the carbon atoms in the molecule. **Alkynes** have at least one triple bond between the carbon atoms in the molecule as shown for acetylene in Figure 20.2. Acetylene is an important fuel for welding and cutting torches. A fourth class of hydrocarbons consists of **aromatic** compounds, which have rings of carbon atoms with special bonding properties as discussed in Section 20.2.3.

20.2.1 ALKANES

The molecular formulas of noncyclic alkanes are C_nH_{2n+2}. By counting the numbers of carbon and hydrogen atoms in the molecules of alkanes shown in Figure 20.2, it is seen that the molecular

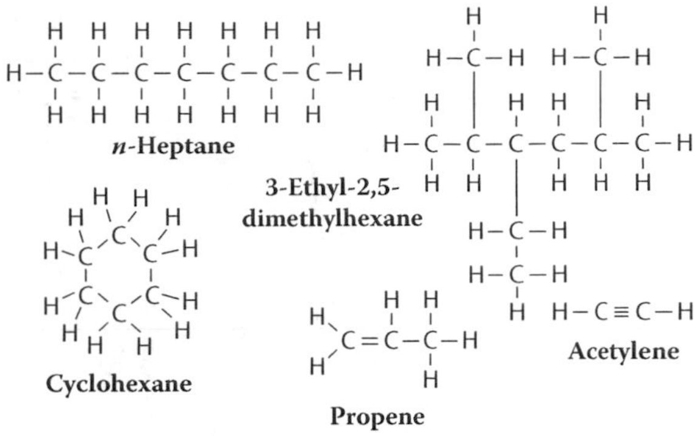

FIGURE 20.2 Some typical hydrocarbons: these formulas illustrate the bonding diversity of carbon, which gives rise to an enormous variety of hydrocarbons and other organic compounds.

formula of n-heptane is C_6H_{16} and that of 3-ethyl-2,5-dimethylhexane is $C_{10}H_{22}$, both of which fit the aforementioned general formula. The general formula of cyclic alkanes is C_nH_{2n}; the molecular formula of cyclohexane, the most common cyclic alkane, is C_6H_{12}. These formulas are **molecular formulas**, which give the number of carbon and hydrogen atoms in each molecule, but do not tell anything about the structure of the molecule. The formulas given in Figure 20.2 are **structural formulas**, which show how a molecule is assembled. The structure of n-heptane is that of a straight chain of carbon atoms; each carbon atom in the middle of the chain is bound to two H atoms, and the two carbon atoms at the ends of the chain are each bound to three H atoms. The prefix *hep* in the name denotes seven carbon atoms and the n- indicates that the compound consists of a single straight chain. This compound can be represented by a **condensed structural formula** as $CH_3(CH_2)_5CH_3$ representing seven carbon atoms in a straight chain. In addition to methane, mentioned in the beginning of Section 20.2, the lower alkanes include the following:

Ethane: CH_3CH_3 Propane: $CH_3CH_2CH_3$

Butane: $CH_3(CH_2)_2CH_3$ n-Pentane: $CH_3(CH_2)_3CH_3$

For alkanes with five or more carbon atoms, the prefix (*pen* for five, *hex* for six, *hept* for seven, *oct* for eight, and *non* for nine) shows the total number of carbon atoms in the compound, and n- may be used to denote a straight-chain alkane. Condensed structural formulas may be used to represent branched-chain alkanes as well. The condensed structural formula of 3-ethyl-2,5-dimethylhexane is

$$CH_3CH(CH_3)CH(C_2H_5)CH_2CH(CH_3)CH_3$$

In this formula, the C atoms and their attached H atoms that are not in parentheses show carbons that are part of the main hydrocarbon chain. The (CH_3) after the second C in the chain shows a methyl group attached to it, the (C_2H_5) after the third carbon atom in the chain shows an ethyl group attached to it, and the (CH_3) after the fifth carbon atom in the chain shows a methyl group attached to it.

Compounds that have the same molecular formula but different structural formulas are **structural isomers**. For example, the straight-chain alkane with the molecular formula $C_{10}H_{22}$ is n-decane

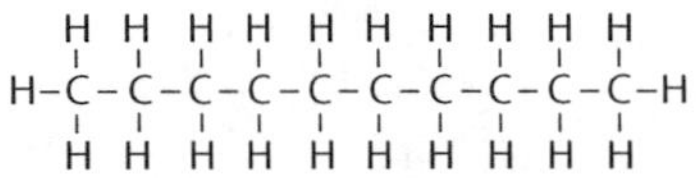

n-Decane

which is a structural isomer of 3-ethyl-2,5-dimethylhexane.

The names of organic compounds are commonly based on the structure of the hydrocarbon from which they are derived, using the longest continuous chain of carbon atoms in the compound as the basis for the name. For example, the longest continuous chain of carbon atoms in 3-ethyl-2,5-dimethylhexane shown in Figure 20.2 is six carbon atoms, so the name is based on *hexane*. The names of the chain branches are also based on the alkanes from which they are derived. As shown here

$$\begin{array}{cccc} H & H & H\ H & H\ H \\ | & | & |\ \ | & |\ \ | \\ H\!-\!C\!-\!H & -\!C\!-\!H & H\!-\!C\!-\!C\!-\!H & -\!C\!-\!C\!-\!H \\ | & | & |\ \ | & |\ \ | \\ H & H & H\ H & H\ H \end{array}$$

Methane (CH_4)   **Methyl group (CH_3)**   **Ethane (C_2H_6)**   **Ethyl group (C_2H_5)**

the two shortest-chain alkanes are methane with one carbon atom and ethane with two carbon atoms. Removal of one of the H atoms from methane gives the **methyl** group, and removal of one of the H atoms from ethane gives the **ethyl** group. These terms are used in the name 3-ethyl-2,5-dimethylhexane to show the groups attached to the basic hexane chain. The carbon atoms in this chain are numbered sequentially from left to right. An ethyl group is attached to the third carbon atom, yielding the "3-ethyl" part of the name, and methyl groups are attached to the second and fifth carbon atoms, which gives the "2,5-dimethyl" part of the name.

The aforementioned names are **systematic names**, which are based on actual structural formulas of molecules. In addition, there are **common names** of organic compounds that do not indicate their structural formulas. Naming organic compounds is a complex topic, and no attempt is made in this chapter to teach it to the reader. However, from the names of compounds given in this chapter and later chapters, some appreciation of the rationale for organic compound names should be obtained.

Other than burning them for energy, the major kind of reaction with alkanes consists of **substitution reactions** such as

$$C_2H_6 + 2Cl_2 \rightarrow C_2H_4Cl_2 + 2HCl \qquad (20.2)$$

in which one or more H atoms are displaced by another kind of atom. This is normally the first step in converting alkanes to compounds containing elements other than carbon or hydrogen for use in synthesizing a wide variety of organic compounds.

20.2.2 ALKENES

Four common alkenes are shown in Figure 20.3. Alkenes have at least one C=C double bond per molecule and may have more. The first of the alkenes in Figure 20.3, ethylene, is a very widely produced hydrocarbon used to synthesize polyethylene plastic and other organic compounds. About 25 billion kilograms of ethylene are processed in the United States each year. About 14.5 billion kilograms of propylene are used in the United States each year to produce polypropylene plastic and other chemicals. The two 2-butene compounds illustrate an important aspect of alkenes: the possibility of *cis-trans* isomerism. Whereas carbon atoms and the groups substituted onto them joined by single bonds can freely rotate relative to each other as though they are joined by a single shaft, carbon atoms connected by a double bond behave as though they are attached by two parallel

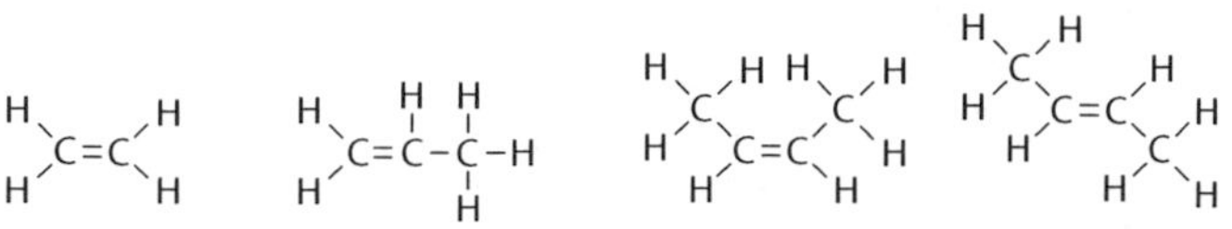

Ethylene   **Propylene (propene)**   ***Cis*-2-Butene**   ***trans*-2-Butene**

FIGURE 20.3   Examples of alkene hydrocarbons.

shafts and are not free to rotate. So, *cis*-2-butene in which the two end methyl, $-CH_3$, groups are on the same side of the molecule is a different compound from *trans*-2-butene in which they are on opposite sides. These two compounds are *cis-trans* isomers.

Alkenes are chemically much more active than alkanes. This is because the double bond is **unsaturated** and has electrons available to form additional bonds with other atoms. This leads to **addition reactions** in which a molecule is added across a double bond. For example, the addition of H_2O to ethylene

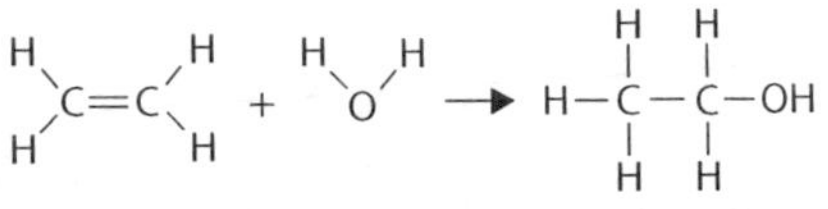

$$(20.3)$$

Ethylene   Water   Ethanol

yields ethanol, the same kind of alcohol that is in alcoholic beverages. The tendency to undergo addition reactions adds greatly to the chemical versatility and reactivity of alkenes and other organic molecules that have the C=C double bond. The presence of C=C double bonds adds to the biochemical and toxicological activity of compounds in organisms. Because of the ability to undergo rapid addition reactions, alkenes are quite reactive in the atmosphere during the formation of photochemical smog and are important atmospheric pollutants. As shown by Reaction 7.27 in the discussion of photochemical smog in Chapter 7, Section 7.8, alkenes very readily add hydroxyl radical (HO·, where the dot denotes an unpaired electron), a highly reactive species characteristic of smog-forming conditions, producing reactive organic free radicals that undergo smog-forming chain reactions. Ozone, O_3, which is a species characteristic of photochemical smog, also adds across C=C bonds in alkenes present in the atmosphere.

Because of their double bonds, alkenes can undergo **polymerization** reactions in which large numbers of individual molecules add to each other to produce big molecules called **polymers** (see Section 20.5). For example, three ethylene molecules can add together as follows:

$$(20.4)$$

This is a process that can continue, forming longer and longer chains and resulting in the formation of huge molecules of polyethylene.

20.2.3 AROMATIC HYDROCARBONS

A special class of hydrocarbons consists of rings of carbon atoms, almost always containing six C atoms, which can be viewed as having alternating single and double bonds as shown here:

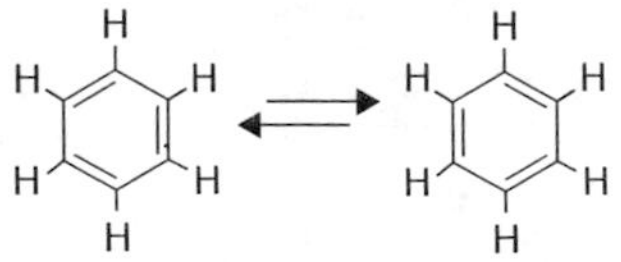

These structures show the simplest aromatic hydrocarbon, benzene, C_6H_6. Although the benzene molecule is represented with three double bonds, chemically it differs greatly from alkenes, for example, it undergoes substitution reactions rather than addition reactions. **Aromaticity** is the term given to the special properties of aromatic compounds. The two structures shown earlier are equivalent **resonance** structures, which can be viewed as having atoms that stay in the same places, but in which the bonds joining the atoms can shift positions with the movement of electrons composing the bonds. Since benzene has different chemical properties from those implied by either of the afore-mentioned structures, it is commonly represented as a hexagon with a circle in the middle:

Many aromatic hydrocarbons have two or more rings. The simplest of these is **naphthalene**

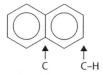

Naphthalene, $C_{10}H_8$

a two-ringed compound in which two benzene rings share the carbon atoms at which they are joined; these two carbon atoms do not have any H attached, each of the other eight C atoms in the compound has one H attached. Aromatic hydrocarbons with multiple rings, called **polycyclic aromatic hydrocarbons** (PAHs), are common and are often produced as by-products of combustion. One of the most studied of these is **benzo(a)pyrene**

Benzo(a)pyrene, $C_{20}H_{12}$

found in tobacco smoke, diesel exhaust, and charbroiled meat. This compound is toxicologically significant because it is partially oxidized by enzymes in the body to produce a cancer-causing metabolite.

The presence of hydrocarbon groups and of elements other than carbon and hydrogen bonded to an aromatic hydrocarbon ring gives a variety of **aromatic compounds**. Three examples of common aromatic compounds are given here. Toluene is widely used for chemical synthesis and as a solvent. The practice of green chemistry now calls for substituting toluene for benzene wherever possible because benzene can cause leukemia, whereas the body is capable of metabolizing toluene to harmless metabolites. Several hundred million kilograms of aniline are made in the United States each year as an intermediate in the synthesis of dyes, pesticides, and other organic chemicals. Phenol is a relatively toxic oxygen-containing aromatic compound, which, despite its toxicity to humans, was the first antiseptic used in the 1800s.

In naming aromatic compounds, numbers are used to denote atoms around the aromatic rings where substituent groups may be attached as shown for phenol here. In addition, positions 2, 3, and 4 around the benzene ring may be denoted by *ortho*, *meta*, and *para*, respectively.

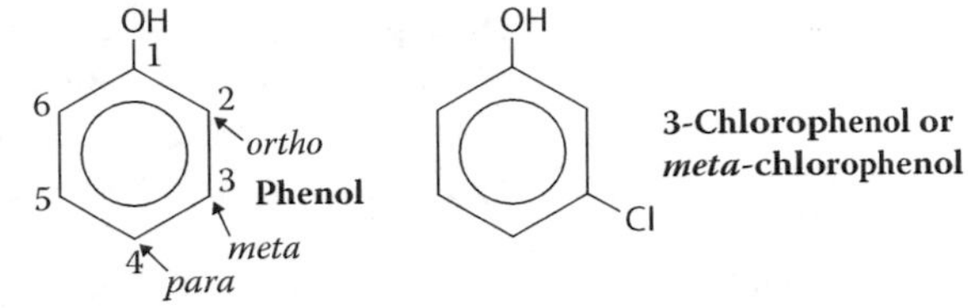

3-Chlorophenol or
meta-chlorophenol

20.3   USING LINES TO SHOW STRUCTURAL FORMULAS

The aromatic structures shown in Section 20.2.3 use a hexagon with a circle in it to denote an aromatic benzene ring. Organic chemistry uses lines to show other kinds of structural formulas as well. The reader who may have occasion to look up organic formulas will probably run into this kind of notation, so it is important to be able to interpret these kinds of formulas. Some line formulas are shown in Figure 20.4.

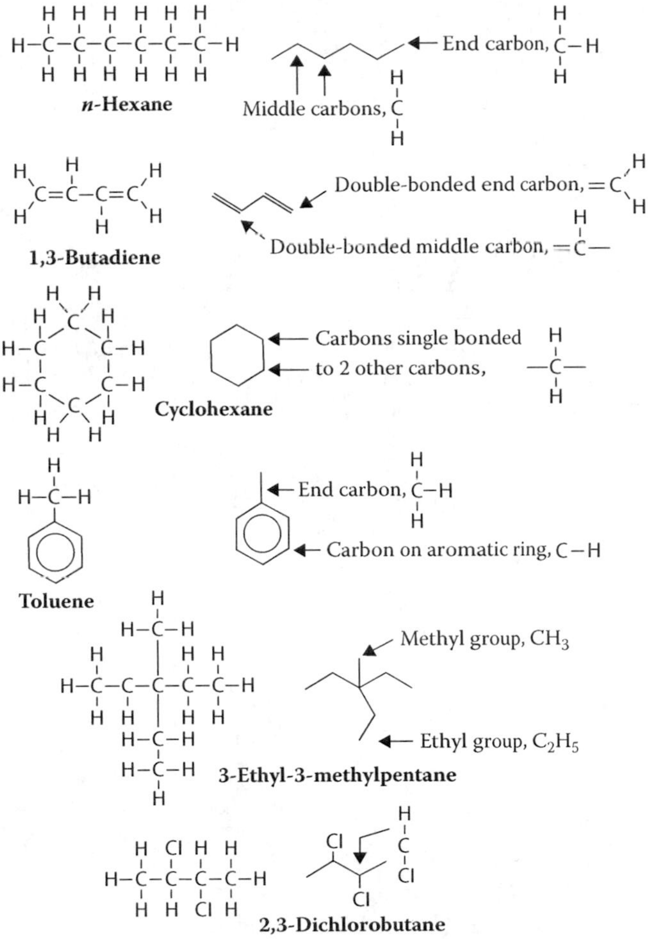

FIGURE 20.4   Representation of organic structural formulas with lines: the structural formulas showing atoms are on the left and the corresponding line formulas on the right. Each line corner and end represents a carbon atom unless otherwise specified. Each C atom has four covalent bonds or equivalent attached, and the presence of H atoms is implied to provide the required bonds.

In using lines to represent organic structural formulas, the corners where lines intersect and the ends of lines represent C atoms, and each line stands for a covalent bond (two shared electrons). It is understood that each C atom at the end of a single line has three H atoms attached, each C atom at the intersection of two lines has two C atoms attached, each C at the intersection of three lines has one H atom attached, and the intersection of four lines denotes a C atom with no H atoms attached. Multiple lines represent multiple bonds as shown for the double bonds in 1,3-butadiene. Substituent groups are shown by their symbols (for individual atoms), or formulas of functional groups consisting of groups of atoms; it is understood that each such group substitutes for a hydrogen atom as shown in the formula of 2,3-dichlorobutane in Figure 20.4. The six-carbon-atom aromatic ring is denoted by a hexagon with a circle in it.

Exercise

What is the structural formula of the compound represented on the left, here?

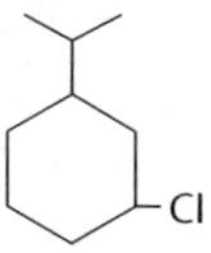

Answer:

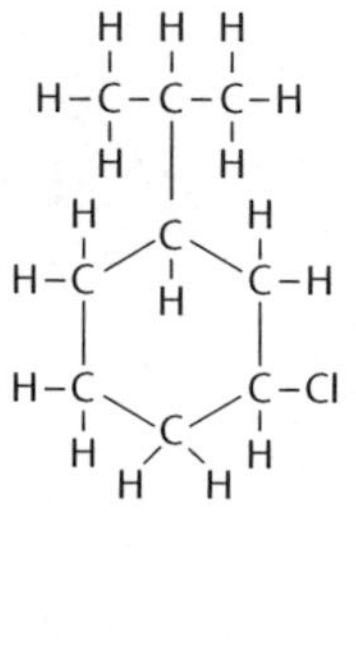

20.4   FUNCTIONAL GROUPS

Numerous elements in addition to carbon and hydrogen occur in organic compounds. These are contained in **functional groups**, which define various classes of organic compounds. The $-NH_2$ group in aniline and the $-OH$ groups in phenol mentioned in Section 20.2.3 are examples of functional groups. The same organic compound may contain two or more functional groups. Among the elements common in functional groups are O, N, Cl, S, and P. There is not space here to discuss all the possible functional groups and the classes of organic compounds that they define. Some important examples are given to provide an idea of the variety of organic compounds with various functional groups. Other examples are encountered in Sections 20.4.1–20.4.4.

20.4.1   ORGANOOXYGEN COMPOUNDS

Figure 20.5 shows several important classes of organic compounds that contain oxygen. **Ethylene oxide** is a sweet-smelling, colorless, flammable, explosive gas. It is an **epoxide** characterized by an

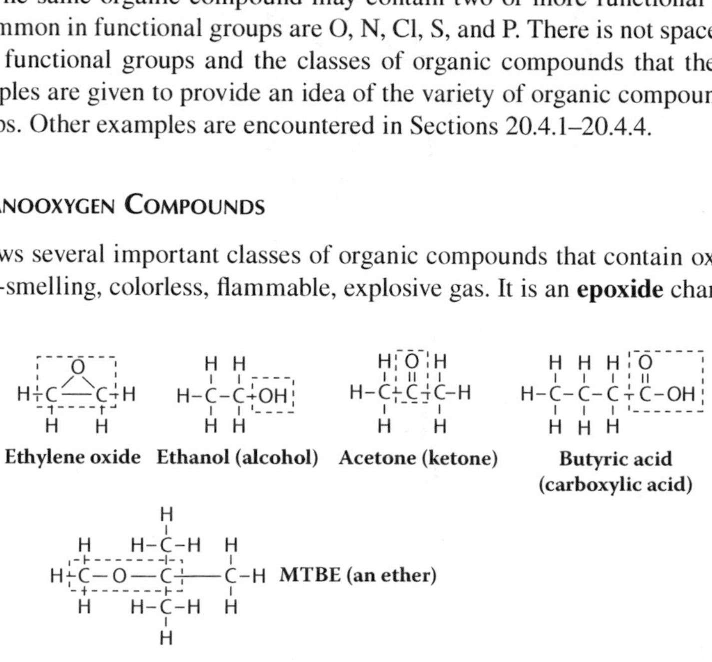

Ethylene oxide   Ethanol (alcohol)   Acetone (ketone)   Butyric acid (carboxylic acid)

MTBE (an ether)

FIGURE 20.5   Examples of important classes of organic compounds with oxygen-containing functional groups: the functional groups characteristic of various classes of compounds are outlined by dashed lines.

oxygen atom bridging two carbon atoms that are also bonded with each other. Ethylene oxide is toxic and is used as a sterilant and fumigant, as well as a chemical intermediate. Because of the toxicity and flammability of this compound, the practice of green chemistry tries to avoid its generation and use. **Ethanol**, which occurs in alcoholic beverages, is an **alcohol**, a class of compounds in which the −OH group is bonded to an alkane or an alkene (attachment of the −OH group to an aromatic hydrocarbon molecule gives a **phenolic** compound). **Acetone** is a ketone, a class of compounds that has the C=O functional group in the middle of a hydrocarbon chain. Acetone is an excellent organic solvent and is relatively safe. **Butyric** acid, which occurs in butter, is an organic carboxylic acid, which, like all organic carboxylic acids, contains the functional group

$$
\begin{array}{c} O \\ \parallel \\ -C-OH \end{array} \quad \begin{array}{l} \text{Carboxylic acid} \\ \text{functional group} \end{array}
$$

which can release the H^+ ion characteristic of acids. **Methyltertiarybutyl ether** (MTBE) is an example of an ether in which an O atom connects two C atoms. When highly toxic tetraethyllead was phased out of gasoline as an octane booster, MTBE was chosen as a substitute. It was subsequently found to be a particularly noxious water pollutant, and its use has been largely banned.

The C=O group in the middle of an organic molecule is characteristic of ketones. When this group is located at the end of a molecule and the carbon is also bonded to H, the compound is an **aldehyde**. The two lowest aldehydes are formaldehyde and acetaldehyde

$$
\begin{array}{cc}
\begin{array}{c} O \\ \parallel \\ H-C-H \end{array} &
\begin{array}{c} H \ \ O \\ | \ \ \parallel \\ H-C-C-H \\ | \\ H \end{array} \\
\textbf{Formaldehyde} & \textbf{Acetaldehyde}
\end{array}
$$

of which formaldehyde is the most widely produced. Despite its many uses, formaldehyde lacks characteristics of green chemicals because it is a volatile, toxic, noxious substance. Formaldehyde tends to induce hypersensitivity (allergies) in people who inhale the vapor or whose skin is exposed to it.

The reaction of an alcohol and an organic acid

$$
\begin{array}{c}
\begin{array}{c} H \ \ H \ \ H \\ | \ \ \ | \ \ \ | \\ H-C-C-C-OH \\ | \ \ \ | \ \ \ | \\ H \ \ H \ \ H \end{array} +
\begin{array}{c} O \ \ H \\ \parallel \ \ | \\ HO-C-C-H \\ | \\ H \end{array} \rightarrow
\begin{array}{c} H \ \ H \ \ H \ \ \ \ O \ \ H \\ | \ \ \ | \ \ \ | \ \ \ \parallel \ \ | \\ H-C-C-C-O-C-C-H \\ | \ \ \ | \ \ \ | \ \ \ \ \ \ \ \ \ H \\ H \ \ H \ \ H \end{array} + H_2O
\end{array} \quad (20.5)
$$

$$
\textbf{Propyl alcohol} \qquad \textbf{Acetic acid} \qquad \textbf{Propyl acetate ester}
$$

produces an important kind of organic compound called **esters**. The linkage characteristic of esters is outlined by the dashed box in the aforementioned structure of propyl acetate. A large number of the naturally occurring esters made by plants are noted for their pleasant odors. Propyl acetate, for example, gives pears their pleasant odor. Other fruit odors due to esters include methyl butyrate, apple; ethyl butyrate, pineapple; and methyl benzoate, ripe kiwi fruit.

20.4.2   ORGANONITROGEN COMPOUNDS

Methylamine

$$
\begin{array}{c}
H \ \ \ \ \ H \\ | \ \ \ \ \diagup \\ H-C-N \\ | \ \ \ \ \diagdown \\ H \ \ \ \ \ H
\end{array} \quad \textbf{Methylamine}
$$

is the simplest of the amines, compounds in which an N atom is bonded to a hydrocarbon group. In an amine, the N atom may be bonded to two H atoms, or one or both of these H atoms may be substituted by hydrocarbon groups as well. Although it is widely used in chemical synthesis because no suitable substitutes are available, methylamine is definitely not compatible with the practice of green chemistry. This is because it is highly flammable and toxic. It is a severe irritant to skin, eyes, and mucous membranes of the respiratory tract. It has a noxious odor and is a significant contributor to the odor of rotten fish. In keeping with the reputation of amines as generally unpleasant compounds, another amine, putrescine, gives decayed flesh its characteristic odor.

Many organonitrogen compounds contain oxygen as well. One such compound is nitromethane

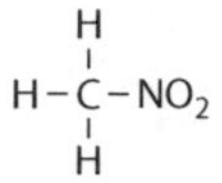

$$H-\overset{\overset{\displaystyle H}{|}}{\underset{\underset{\displaystyle H}{|}}{C}}-NO_2 \quad \textbf{Nitromethane}$$

used in chemical synthesis and as a fuel in some race cars. As seen in the aforementioned structural formula, the nitro group, $-NO_2$, is the functional group in this compound and related nitro compounds. Another class of organonitrogen compounds also containing oxygen consists of the **nitrosamines**, or N-nitroso compounds, which have figured prominently in the history of green chemistry before it was defined as such. These are compounds that have the $N-N=O$ functional group, which are of concern because several are known carcinogens (cancer-causing agents). The most well-known of these is **dimethylnitrosamine**, which is shown here:

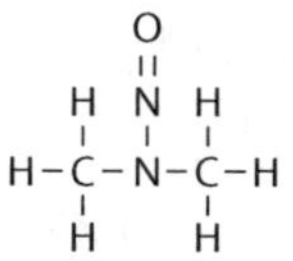

$$H-\overset{\overset{\displaystyle H}{|}}{\underset{\underset{\displaystyle H}{|}}{C}}-\overset{\overset{\displaystyle \overset{\displaystyle O}{\|}}{N}}{N}-\overset{\overset{\displaystyle H}{|}}{\underset{\underset{\displaystyle H}{|}}{C}}-H \quad \textbf{Dimethylnitrosamine}$$

This compound used to be employed as an industrial solvent and was used in cutting oils. However, workers exposed to it suffered liver damage and developed jaundice, and the compound, as well as other nitrosamines, was found to be a carcinogen. A number of other nitrosamines were later found in industrial materials and as by-products of food processing and preservation. Because of their potential as carcinogens, nitrosamines are avoided in the practice of green chemistry.

20.4.3 ORGANOHALIDE COMPOUNDS

Organohalides, exemplified by those shown in Figure 20.6, are organic compounds that contain halogens, F, Cl, Br, or I, but usually chlorine, on alkane, alkene, or aromatic molecules. Organohalides have been widely produced and distributed for a variety of applications, including industrial solvents, chemical intermediates, coolant fluids, pesticides, and other applications. They are for the most part environmentally persistent and, because of their tendency to accumulate in adipose (fat) tissue, they tend to undergo bioaccumulation and biomagnification in organisms.

Carbon tetrachloride, CCl_4, is produced when all four H atoms on methane, CH_4, are substituted by Cl. This compound was once widely used and was even sold to the public as a solvent to remove stains and in fire extinguishers, where the heavy CCl_4 vapor smothers fires. It was subsequently found to be very toxic, causing severe liver damage, and its uses are severely restricted. **Dichlorodifluoromethane** is a prominent member of the **chlorofluorocarbon** class of compounds, popularly known as Freons. Developed as refrigerant fluids, these compounds are notably unreactive and nontoxic. However, as discussed in Chapter 7, Section 7.9, they are found to be indestructible in the lower atmosphere, persisting to very high altitudes in the stratosphere where chlorine split from them by ultraviolet radiation destroys stratospheric ozone. So the manufacture of chlorofluorocarbons is now prohibited. **Vinyl chloride**, an alkene-based organohalide compound, is widely used to make polyvinylchloride polymers and pipe. Unfortunately, it is a known human carcinogen, so human exposure to it is severely limited. **Trichloroethylene** is an excellent organic solvent that is nonflammable. It is used as a dry-cleaning solvent and for degreasing manufactured parts and

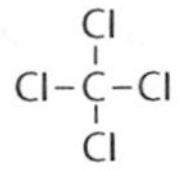

Carbon tetrachloride

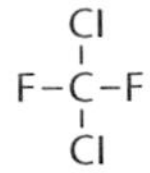

Dichlorodifluoromethane

(Both of these compounds are alkyl halides.)

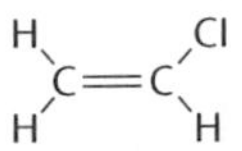

Vinyl chloride

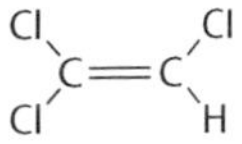

Trichloroethylene

(These compounds are alkenyl halides.)

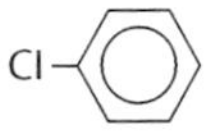

Chlorobenzene

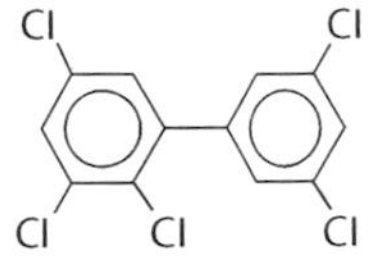

A PCB

(These compounds are aromatic halides.)

FIGURE 20.6 Examples of important organohalide compounds including alkyl halides based on alkanes, alkenyl halides based on alkenes, and aromatic halides.

was formerly used for food extraction, particularly to decaffeinate coffee. **Chlorobenzene** is the simplest aromatic organochloride. In addition to its uses in making other chemicals, it serves as a solvent and as a fluid for heat transfer. It is extremely stable, and its destruction is a common test for the effectiveness of hazardous waste incinerators. The **polychlorinated biphenyl (PCB)** compound shown in Figure 20.6 is one of 209 PCB compounds that can be formed by substituting from 1 to 10 Cl atoms onto the basic biphenyl (two-benzene-ring) carbon skeleton. These compounds are notably stable and persistent, leading to their uses in electrical equipment, particularly as coolants in transformers and in industrial capacitors, as hydraulic fluids, and in other applications. Their extreme environmental persistence has led to their being banned. Sediments in New York's Hudson River are badly contaminated with PCBs that were (at the time, legally) dumped or leaked into the river from electrical equipment manufacture from the 1950s to the 1970s.

From this discussion, it is obvious that many organohalide compounds are definitely not green because of their persistence and biological effects. A lot of the effort in the development of green chemistry has been devoted to finding substitutes for organohalide compounds. A 2001 United Nations treaty formulated by approximately 90 nations in Stockholm, Sweden, designated a "dirty dozen" of 12 organohalide compounds of special concern as persistent organic pollutants (POPs); other compounds have subsequently been added to this list.

20.4.4 Organosulfur and Organophosphorus Compounds

A number of organosulfur and organophosphorus compounds have been synthesized for various purposes including pesticidal applications. A common class of organosulfur compounds consists of thiols, the simplest of which is methanethiol:

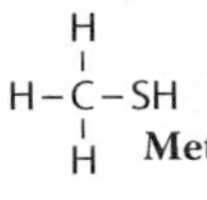

Methanethiol

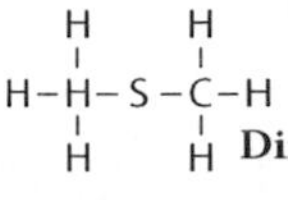

Dimethylsulfide

As with other thiols, which contain the –SH group, this compound is noted for its foul odor. Thiols are added to natural gas so that their odor can warn of gas leaks. Dimethylsulfide, also shown here, is a volatile compound released by ocean-dwelling microorganisms to the atmosphere in such quantities that it constitutes the largest flux of sulfur-containing vapors from Earth to the atmosphere.

Among the most prominent organophosphorus compounds are the organophosphates as shown by methyl parathion and malathion (shown later in this section). Both these compounds are insecticides and contain sulfur as well as phosphorus. Parathion was developed during the 1940s and was once widely used as an insecticide in place of DDT because parathion is very biodegradable, whereas DDT is not and undergoes bioaccumulation and biomagnification in ecosystems. Unfortunately, parathion has a high toxicity to humans and other animals, and some human fatalities have resulted from exposure to it. Like other organophosphates, it inhibits acetylcholinesterase, an enzyme essential for nerve function (the same mode of action as its deadly cousins, the nerve gas military poisons, such as sarin). Because of its toxicity, parathion is now banned from general use. Malathion is used in its place and is only about 1/100 as toxic as parathion to mammals because they—although not insects—have enzyme systems that can break it down.

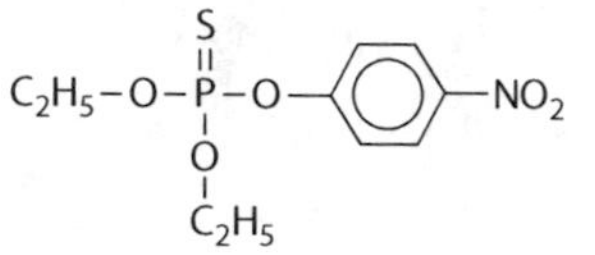

Parathion

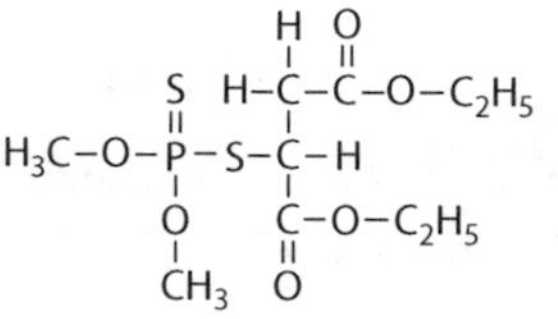

Malathion

20.5   GIANT MOLECULES FROM SMALL ORGANIC MOLECULES

Reaction 20.4 shows the bonding together of molecules of ethylene to form larger molecules. This process, widely practiced in the chemical and petrochemical industries, is called *polymerization*, and the products are *polymers*. Many other unsaturated molecules, usually based on alkenes, undergo polymerization to produce synthetic polymers used as plastics, rubber, and fabrics. As an example, tetrafluoroethylene polymerizes as shown in Figure 20.7 to produce a polymer (Teflon) that is exceptionally resistant to heat and chemicals and that can be used to form coatings to which other materials will not stick (e.g., frying pan surfaces).

Polyethylene and polytetrafluoroethylene are both **addition polymers** in that they are formed by the chemical addition of the monomers making up the large polymer molecules. Other polymers are **condensation polymers** that join together with the elimination of a molecule of water for each monomer unit joined. A common condensation polymer is **nylon**, which is formed by the bonding together of two different kinds of molecules. There are several forms of nylon, the original form of which is nylon 66 discovered by Wallace Carothers, a DuPont chemist, in 1937 and made by the polymerization of adipic acid (mentioned as a feedstock that can be made from glucose is Chapter 16, Section 16.6) and 1,6-hexanediamine:

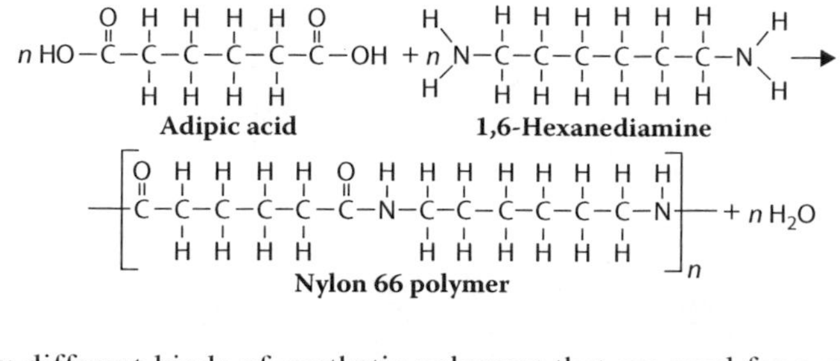

Adipic acid   **1,6-Hexanediamine**

Nylon 66 polymer

(20.6)

There are many different kinds of synthetic polymers that are used for a variety of purposes. Some examples in addition to the ones already discussed in this chapter are given in Table 20.1.

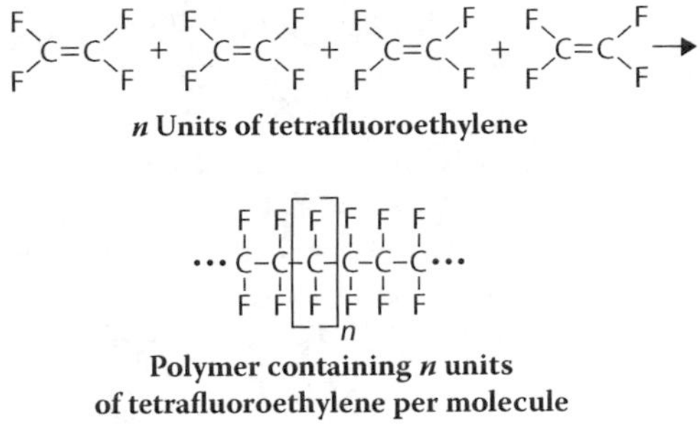

n Units of tetrafluoroethylene

Polymer containing *n* units
of tetrafluoroethylene per molecule

FIGURE 20.7 Polymerization of tetrafluoroethylene to produce large molecules of a polymer commonly known as Teflon.

TABLE 20.1

Some Typical Polymers and the Monomers from Which They Are Formed

Monomer	Monomer Formula	Polymer	Applications
Propylene (polypropylene)	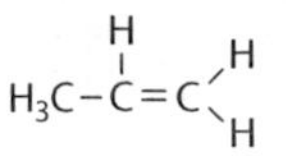	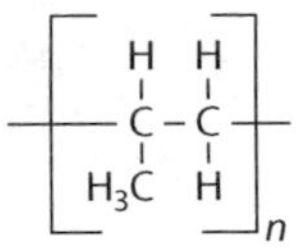	Applications requiring harder plastic, luggage, bottles, outdoor carpet
Vinyl chloride (polyvinylchloride)	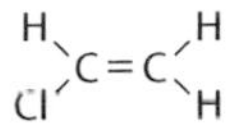	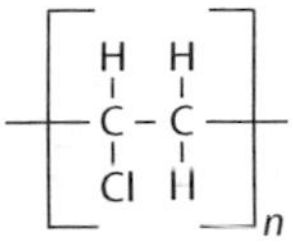	Thin plastic wrap, hose, flooring, PVC pipe
Styrene (polystyrene)	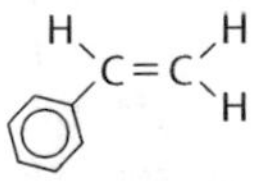	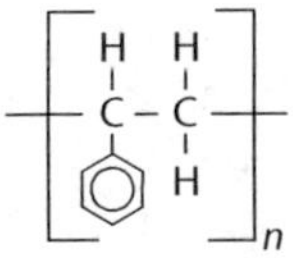	Plastic furniture, plastic cups and dishes, blown to produce Styrofoam plastic products
Acrylonitrile (polyacrylonitrile)	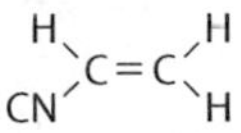	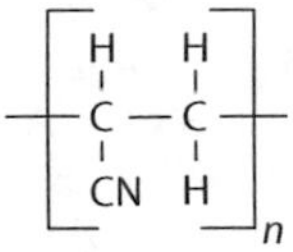	Synthetic fabrics (Orlon, Acrilan, Creslan), acrylic paints
Isoprene (polyisoprene)		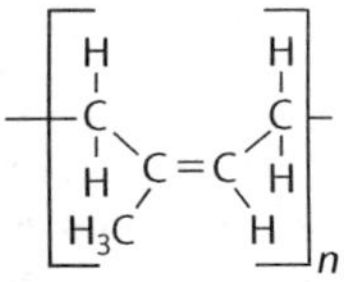	Natural rubber

---

Polymers and the industries on which they are based are of particular concern in the practice of green chemistry for a number of reasons. The foremost of these is because of the huge quantities of materials consumed in the manufacture of polymers. In addition to the enormous quantities of ethylene and propylene previously cited in this chapter, the United States processes about 1.5 billion kilograms of acrylonitrile, 5.4 billion kilograms of styrene, 2.0 billion kilograms of butadiene, and 1.9 billion kg of adipic acid (for nylon 66) each year to make polymers containing these monomers. These and similarly large quantities of monomers used to make other polymers place significant demands on petroleum resources and the energy, materials, and facilities required to make the monomers.

There is a significant potential for the production of pollutants and wastes from monomer processing and polymer manufacture. Some of the materials contained in documented hazardous waste sites are by-products of polymer manufacture. Monomers are generally volatile organic compounds with a tendency to evaporate into the atmosphere, and this characteristic, combined with the presence of reactive C=C bonds, tends to make monomer emissions active in the formation of photochemical smog (see Chapter 7, Section 7.8). Polymers, including plastics and rubber, pose problems for waste disposal, as well as opportunities and challenges for recycling. On the positive side, improved polymers can provide long-lasting materials that reduce material use and have special applications, such as liners in waste disposal sites that prevent waste leachate migration and liners in lagoons and ditches that prevent water loss. Strong, lightweight polymers are key components of the blades and other structural components of huge wind generators that are making an increased contribution to renewable energy supplies around the world.

As shown by the example of di(2-ethylhexyl)phthalate

Di(2-ethylhexyl) phthalate

polymers typically contain additives to improve their performance and durability. The most notable of these are **plasticizers**, normally blended with plastics to improve flexibility, such as to give polyvinylchloride the flexible characteristics of leather. The plasticizers are not chemically bound as part of the polymer and they leak from the polymer over a period of time, which can result in human exposure and environmental contamination. The most widely used plasticizers are phthalates, esters of phthalic acid such as di(2-ethylhexyl) phthalate. Although not particularly toxic, these compounds are environmentally persistent, resistant to treatment processes, and prone to undergo bioaccumulation. They are found throughout the environment and have been implicated by some toxicologists as possible estrogenic agents that mimic the action of female sex hormone and cause premature sexual development in young female children and feminization of male aquatic organisms including frogs and alligators.

Recent concern about plasticizers in the environment and their toxicological effects have centered around bisphenol-A, a potential estrogenic agent that was widely used in consumer plastics including even baby bottles. This compound is discussed in Chapter 2, Section 2.15, and its structural formula is shown in Figure 2.21.

QUESTIONS AND PROBLEMS

Access to and use of the Internet is assumed in answering all questions, including general information, statistics, constants, and mathematical formulas required to solve problems. These questions are designed to promote inquiry and thought rather than just finding material in the chapter. So in some cases there may be several "right" answers. Therefore, if your answer reflects intellectual effort and a search for information from available sources, it may be considered to be right.

1. What are two major reactions of alkanes?
2. What is the difference between molecular formulas and structural formulas of organic compounds?
3. What is the difference between ethane and the ethyl group?
4. What is the structural formula of 3-ethyl-2,3-dimethylpentane?
5. What is a type of reaction that is possible with alkenes but not with alkanes?
6. What is represented by the following structural formula?

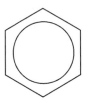

7. Suggest a name for the following compound, which is derived from hydrocarbon toluene:

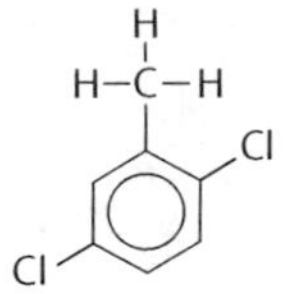

8. What is a health concern with the following aromatic compound?

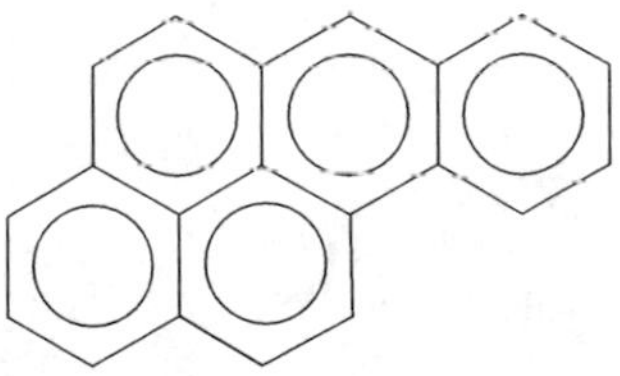

9. What do the groups of atoms outlined by dashed lines represent in the following structure?

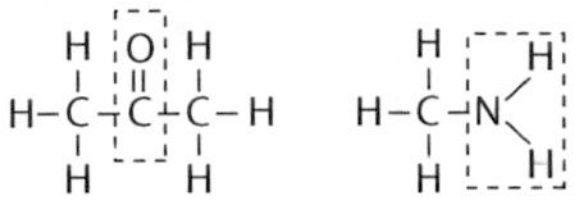

10. Based on the structures shown in Figure 20.5, what are the similarities and differences between organic oxides and ethers.
11. What are three separate kinds of groups that are characteristic of organonitrogen compounds?
12. What is a class of organochlorine compounds consisting of many different kinds of molecules that is noted for environmental persistence?
13. What is a notable characteristic of organosulfur thiols?
14. What is a particularly toxic organophosphorus compound? What is a biochemical molecule containing phosphorus?
15. What are polymers, and why are they important?
16. The examination of the formulas of many of the monomers used to make polymers reveals a common characteristic. What is this characteristic, and how does it enable polymer formation? Does nylon illustrate a different pathway to monomer formation? Explain.

17. Write complete structural formulas corresponding to each of the following line structures:

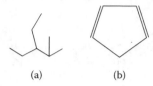

(a)   (b)

18. Some of the most troublesome organic pollutant compounds are organochlorine compounds including the dirty dozen persistent organic pollutants mentioned in this chapter. Organohalides involving a halogen other than chlorine have emerged as significant pollutants. Doing some research on the Internet, find which class of compounds these are and why they are significant pollutants.

SUPPLEMENTARY REFERENCES

Armstrong, James, *General, Organic, and Biochemistry*, Brooks/Cole, Cengage Learning, Belmont, CA, 2010.
Bettelheim, Frederick A., *Introduction to General, Organic, and Biochemistry*, 9th ed., Brooks/Cole, Cengage Learning, Belmont, CA, 2010.
Brown, William H., and Thomas Poon, *Introduction to Organic Chemistry*, Wiley, Hoboken, NJ, 2011.
Denniston, Katherine J., Joseph J. Topping, and Robert L. Caret, *General, Organic, and Biochemistry*, 7th ed., McGraw-Hill, New York, 2011.
Guinn, Denise, and Rebecca Brewer, *Essentials of General, Organic, and Biochemistry: An Integrated Approach*, W. H. Freeman and Co., New York, 2009.
McMurry, John, David S. Ballantine, Carl A. Hoeger, Virginia E. Peterson, and Mary E. Castellion, *Fundamentals of General, Organic, and Biological Chemistry*, 6th ed., Prentice-Hall, Upper Saddle River, NJ, 2009.
Seager, Spencer L., and Michael R. Slabaugh, *Organic and Biochemistry for Today*, 7th ed., Brooks/Cole, Cengage Learning, Belmont, CA, 2010.
Smith, Janice G., *Principles of General, Organic, and Biochemistry*, McGraw-Hill, New York, 2011.
Solomons, T. W. Graham, and Craig Fryhle, *Organic Chemistry*, 10th ed., Wiley, Hoboken, NJ, 2011.
Stoker, H. Stephen, *General, Organic, and Biological Chemistry*, 5th ed., Brooks/Cole, Cengage Learning, Belmont, CA, 2010.
Winter, Arthur, *Organic Chemistry 1 for Dummies*, Wiley, Hoboken, NJ, 2005.

Index